Current and Future Developments in Food Science

(*Volume 2*)

Maria Luisa Marina (Ed.)

Capillary Electrophoresis in Food Analysis

Edited by

María Castro-Puyana

Department of Analytical Chemistry,
Physical Chemistry and Chemical Engineering
University of Alcalá, Alcalá de Henares (Madrid), Spain

Miguel Herrero

Laboratory of Foodomics
Institute of Food Science Research (CIAL)
Spanish National Research Council (CSIC)
Madrid, Spain

María Luisa Marina

Department of Analytical Chemistry,
Physical Chemistry and Chemical Engineering
University of Alcalá, Alcalá de Henares (Madrid), Spain

Current and Future Developments in Food Science

Volume # 2

Capillary Electrophoresis in Food Analysis

Editors: María Castro-Puyana, Miguel Herrero and María Luisa Marina

ISBN (Online): 978-981-5036-15-2

ISBN (Print): 978-981-5036-16-9

ISBN (Paperback): 978-981-5036-17-6

Bentham Science Publishers Pte. Ltd.
80 Robinson Road #02-00
Singapore 068898
Singapore
Email: subscriptions@benthamscience.net

BENTHAM
SCIENCE

CONTENTS

FOREWORD

The present book on Capillary Electrophoresis in Food Analysis edited by Dr. Maria Castro-Puyana, Dr. Miguel Herrero and Prof. Maria Luisa Marina represents very comprehensive and updated source of information on the topic.

The book opens with the chapter on the basic principles of Capillary Electrophoresis written by Kalaycıoğlu and Erim. This chapter is a good reading for every researcher interested in this technique and not just limited to food scientists. The aspects such as separation mechanism, basic instrument, separation parameters, modes of Capillary Electrophoresis, on-line sensitivity enhancement, detection methods and separation strategies for some groups of analytes are discussed in a very concise and clear way.

The second chapter by Li and co-authors deals with sample preparation techniques used in food analysis. In particular, the updated overview on the techniques based on phase separation, field-assisted extraction, membrane separation, chemical conversion and online sample preparation is provided.

Following five chapters deal with the analysis of specific groups of compounds such as lipids, carbohydrates and proteins (Chapter 3 by Oliveira and coauthors), peptides (Chapter 4 by Stepanova and Kasicka), amino acids (Chapter 5 by Castro-Puyana and Marina), vitamins (Chapter 6 by Van Schepdael, Wang and co-authors) and polyphenols (Chapter 7 by Marina and co-authors). Together, these chapters provide the most recent and important advances in the analysis of such relevant analytes as powerful tools for food characterization.

The book also comprises of interesting chapters devoted to very important aspects in the field of food analysis such as the application of Capillary Electrophoresis to the quantification of food additives (Chapter 8 by Rios and co-authors), food authenticity (Chapter 11 by Kvasnicka), controlling of chemical food safety (Chapter 12 by Hernández-Mesa and co-authors) or quality and safety of dietary supplements (Chapter 13 by Donati and Aturki).

Three chapters dealing with advanced methodological aspects in Capillary Electrophoresis complete the book, including Chapter 9 by Marina and co-authors on the application of chiral Capillary Electrophoresis to food analysis, Chapter 10 by Crevillén, Escarpa and co-authors on the application of Microchip Electrophoresis to food analysis and the concluding Chapter 14 by Recber and Celebier on the potential of Capillary Electrophoresis-Mass Spectrometry as analytical methodology to perform metabolomic approaches to food analysis.

I would like to congratulate the editors, all the contributing authors, as well as the readers of this book for publishing this timely and well-balanced book on application of Capillary Electrophoresis in food analysis. The strength of this book is based on the long time professional activity of all three editors in the field covered by the book. This allowed them to select a very strong team of authors for covering almost all the most important aspects of food analysis based on Capillary Electrophoresis.

This book will definitely become a reference source for everyone using Capillary Electrophoresis for food analysis and will promote further use of this technique, as well as its advanced modalities such as Capillary Electrophoresis-Mass Spectrometry, Microchip Electrophoresis, multichannel- and multidimensional Capillary Electrophoresis for sophisticated problem solving in food analysis.

Bezhan Chankvetadze
Professor of Physical Chemistry and Director
Institute of Physical and Analytical Chemistry
School of Exact and Natural Sciences
Tbilisi State University
Tbilisi
Georgia

PREFACE

Modern food analysis comprises different important fields within Food Science, including food quality, food safety and traceability assessment, detection of frauds, authentication of origin, determination of the nutritional value of food products, or even the assessment of the potential bioactivities that food ingredients may exert. Moreover, food analysis has a significant social impact due to the consumers' concern about everything related to food. Among these fields, the researches aimed to observe the effects that bioactive compounds present in food may confer are probably those that have increased their importance the most. This is not only related to an increase of interest in this particular subfield but also in the improvement and evolution of modern analytical tools that have allowed to tackle analytical challenges not reachable some decades ago. In this regard, the development of faster, more sensitive, accurate, and powerful analytical methods capable to providing information about all these aspects is crucial.

In this context, capillary electrophoresis (CE) is positioned among other more extended separative tools, mainly those based on gas and liquid chromatography. However, despite not being so broadly used, capillary electromigration methods, with CE leading the way, possess exceptional properties related to their high analysis speed and separation efficiencies combined with the small samples and reagent requirements and a great variety of potential applications.

The present volume is aimed at providing an updated overview of the current state-of-the-art related to the use of CE in the field of food analysis. The book is structured, including some general chapters dealing with the basic principles of capillary electrophoresis, providing an overview of the most relevant and important characteristics and potential of this analytical tool. The theoretical background of capillary electromigration is included as well as the description of the basic instrumentation needed and the different separation modes that make CE such a versatile tool. Moreover, sample preparation aspects, specifically directed towards the subsequent application of CE are also presented, which are relevant considering the great influence of sample preparation steps and procedures on the analytical results that are attainable. A second group of chapters is focused on the required approaches for the analysis of important groups of food components, such as lipids, carbohydrates, proteins, peptides, amino acids, vitamins, polyphenols, and food additives. Each chapter includes information showing the advantages of the use of CE as well as its different separation modes for those applications compared to other separation and analytical techniques. Two additional chapters deal with more specific developments within the general use of CE, namely, the development of approaches for chiral separations and the use of miniaturized devices. Indeed, CE has demonstrated very good capabilities for the efficient separation of chiral compounds. The different approaches and compounds that can be used as chiral selectors are presented and discussed. Besides, CE has been demonstrated as one of the most suitable analytical tools to be miniaturized and to construct lab-on-a-chip devices based on electromigration even allowing to perform multiple simultaneous analyses. Instrumentation, detection, and microchip design aspects are included in the chapter devoted to microchip CE. Finally, several chapters are focused on specific fields of the study showing the latest developments and applications presented in each topic. Those chapters include information related to the use of CE for food authentication, food safety, the analysis of dietary supplements as well as the use of CE coupled to mass spectrometry for its use in food metabolomics-related researches.

Together, the chapters included in this volume, written by renowned experts in their respective fields, provide a wide perspective of the use of capillary electrophoresis-based approaches in the field of food analysis. Readers with a background in Food Science, from Ph.D. students to experts and researchers working in food and analytical chemistry, as well as chemists working in food control laboratories, might find this information interesting.

María Castro-Puyana
Department of Analytical Chemistry
Physical Chemistry and Chemical Engineering
University of Alcalá
Alcalá de Henares (Madrid)
Spain

Miguel Herrero
Laboratory of Foodomics
Institute of Food Science Research (CIAL)
Spanish National Research Council (CSIC)
Madrid
Spain

&

María Luisa Marina
Department of Analytical Chemistry
Physical Chemistry and Chemical Engineering
University of Alcalá
Alcalá de Henares (Madrid)
Spain

List of Contributors

Alberto Escarpa	Department of Analytical Chemistry, Physical Chemistry and Chemical Engineering, University of Alcalá, Alcala de Henares, Madrid E-28871, Spain Chemical Research Institute "Andrés M. del Río" (IQAR), University of Alcalá, Alcalá de Henares, Madrid E-28805, Spain
Agustín G. Crevillén	Department of Analytical Sciences, Faculty of Sciences, Universidad Nacional de Educación a Distancia (UNED), Madrid E-28040, Spain
Ana M. García-Campaña	Department of Analytical Chemistry, Faculty of Sciences, University of Granada, Granada, Spain
Andrea Martin-Ortiz	Departamento de Química Analítica, Química Física e Ingeniería Química, Universidad de Alcalá, Ctra., Madrid-Barcelona Km. 33.600, 28871 Alcalá de Henares (Madrid), Spain
Ángel Ríos	Analytical Chemistry and Food Technology Department, University of Castilla-La Mancha, Camilo José Cela Avenue, E-13005, Ciudad Real, Spain Regional Institute for Applied Scientific Research, IRICA, Camilo José Cela Avenue, E-13005, Ciudad Real, Spain
Ann Van Schepdael	Department of Pharmaceutical and Pharmacological Sciences, Pharmaceutical Analysis, KU Leuven - University of Leuven, Leuven, Belgium
David Moreno-González	ISAS-Leibniz-Institut für analytische Wissenschaften, Bunsen-Kirchhoff-Str. 11, 44139 Dortmund, Germany
Elena Sánchez-López	Department of Human Genetics, Leiden University Medical Center, Leiden, The Netherlands
Enrica Donati	Istituto per I Sistemi Biologici, Consiglio Nazionale delle Ricerche, Italy
F. Bedia Erim	Department of Chemistr, Istanbul Technical University, Maslak, Turkey
Fei Zhao	State Key Laboratory of Food Nutrition and Safety, College of Food Science and Engineering, Tianjin University of Science and Technology, Tianjin-300457, China
Francisco J. Lara	Department of Analytical Chemistry, Faculty of Sciences, University of Granada, Granada, Spain
František Kvasnička	Department of food preservation, University of Chemistry and Technology, Technicka 3, 16628 Prague 6, Czech Republic
Gaud Dervilly	LABERCA, Oniris, INRAE, Nantes, France
Gongke Li	School of Chemistry, Sun Yat-sen University, Guangzhou-510275, China
Guilherme de Paula Campos	Federal University of Juiz de Fora, Juiz de Fora, MG, Brazil
Jéssica Cordeiro Queiroz de Souza	Federal University of Juiz de Fora, Juiz de Fora, MG, Brazil

Jinhui Li	State Key Laboratory of Food Nutrition and Safety, College of Food Science and Engineering, Tianjin University of Science and Technology, Tianjin-300457, China
Ling Xia	School of Chemistry, Sun Yat-sen University, Guangzhou-510275, China
María Castro-Puyana	Departamento de Química Analítica, Química Física e Ingeniería Química, Universidad de Alcalá, Ctra, Madrid-Barcelona Km. 33.600. 28871 Alcalá de Henares (Madrid), Spain Instituto de Investigación Química Andrés M. del Río (IQAR), Universidad de Alcalá, Ctra, Madrid-Barcelona Km. 33.600. 28871 Alcalá de Henares (Madrid), Spain
María Luisa Marina	Departamento de Química Analítica, Química Física e Ingeniería Química, Universidad de Alcalá, Ctra, Madrid-Barcelona Km. 33.600. 28871 Alcalá de Henares (Madrid), Spain Instituto de Investigación Química Andrés M. del Río (IQAR), Universidad de Alcalá, Ctra, Madrid-Barcelona Km. 33.600. 28871 Alcalá de Henares (Madrid), Spain
Marcone Augusto Leal de Oliveira	Federal University of Juiz de Fora, Juiz de Fora, MG, Brazil
María Castro-Puyana	Departamento de Química Analítica, Química Física e Ingeniería Química, Universidad de Alcalá, Ctra, Madrid-Barcelona Km. 33.600,28871 Alcalá de Henares (Madrid), Spain Instituto de Investigación Química Andrés M. del Río (IQAR), Universidad de Alcalá, Ctra, Madrid-Barcelona Km. 33.600, 28871 Alcalá de Henares (Madrid), Spain
María Luisa Marina	Departamento de Química Analítica, Química Física e Ingeniería Química, Universidad de Alcalá, Ctra, Madrid-Barcelona Km. 33.600,28871 Alcalá de Henares (Madrid), Spain Instituto de Investigación Química Andrés M. del Río (IQAR), Universidad de Alcalá, Ctra, Madrid-Barcelona Km. 33.600, 28871 Alcalá de Henares (Madrid), Spain
Maria Patrícia do Nascimento	Federal University of Juiz de Fora, Juiz de Fora, MG, Brazil
Maykel Hernández-Mesa	Department of Analytical Chemistry, Faculty of Sciences, University of Granada, Granada, Spain LABERCA, Oniris, INRAE, Nantes, France
Merichel Plaza	Departamento de Química Analítica, Química Física e Ingeniería Química, Universidad de Alcalá, Ctra., Madrid-Barcelona Km. 33.600, 28871 Alcalá de Henares (Madrid), Spain Instituto de Investigación Química Andrés M. del Río (IQAR), Universidad de Alcalá, Ctra, Madrid-Barcelona Km. 33.600. 28871 Alcalá de Henares (Madrid), Spain
Mohammed Zougagh	Regional Institute for Applied Scientific Research, IRICA, Ciudad Real, Spain Analytical Chemistry and Food Technology Department, Faculty of Pharmacy, University of Castilla-La Mancha, Albacete, Spain

Mustafa Çelebier	Hacettepe University, Faculty of Pharmacy, Department of Analytical Chemistry, Ankara, Turkey
Nerilson Marques Lima	Federal University of Juiz de Fora, Juiz de Fora, MG, Brazil
Olívia Brito de Oliveira Moreira	Federal University of Juiz de Fora, Juiz de Fora, MG, Brazil
Paula Rocha Chellini	Federal University of Juiz de Fora, Juiz de Fora, MG, Brazil
Samah Lahouidak	Analytical Chemistry and Food Technology Department, University of Castilla-La Mancha, Camilo José Cela Avenue, E-13005, Ciudad Real, Spain Regional Institute for Applied Scientific Research, IRICA, Camilo José Cela Avenue, E-13005, Ciudad Real, Spain
Samuel Bernardo-Bermejo	Departamento de Química Analítica, Química Física e Ingeniería Química, Universidad de Alcalá, Ctra, Madrid-Barcelona Km. 33.600,28871 Alcalá de Henares (Madrid), Spain
Sille Štěpánová	Institute of Organic Chemistry and Biochemistry of the Czech Academy of Sciences, Flemingovo n. 2, 166 10 Prague 6, Czechia
Simin Huang	School of Chemistry, Sun Yat-sen University, Guangzhou-510275, China
Silvia Dortez	Department of Analytical Chemistry, Physical Chemistry and Chemical Engineering, University of Alcalá, Alcala de Henares, Madrid E-28871, Spain
Tania Sierra	Department of Analytical Chemistry, Physical Chemistry and Chemical Engineering, University of Alcalá, Alcala de Henares, Madrid E-28871, Spain
Tatiane Lima Amorim	Federal University of Juiz de Fora, Juiz de Fora, MG, Brazil
Tuba Reçber	Department of Analytical Chemistry, Faculty of Pharmacy, Hacettepe University, Ankara, Turkey
Václav Kašička	Institute of Organic Chemistry and Biochemistry of the Czech Academy of Sciences, Flemingovo n. 2, 166 10 Prague 6, Czechia
Xu Wang	State Key Laboratory of Food Nutrition and Safety, College of Food Science and Engineering, Tianjin University of Science and Technology, Tianjin-300457, China
Zeineb Aturki	Istituto per I Sistemi Biologici, Consiglio Nazionale delle Ricerche, Italy
Zeynep Kalaycıoğlu	Department of Chemistry, Istanbul Technical University, Maslak, Turkey
Zhuoting Liu	State Key Laboratory of Food Nutrition and Safety, College of Food Science and Engineering, Tianjin University of Science and Technology, Tianjin-300457, China

CHAPTER 1

Capillary Electrophoresis: Basic Principles

Zeynep Kalaycıoğlu[1] and **F. Bedia Erim**[1,*]

[1] *Department of Chemistry, Istanbul Technical University, Maslak, Istanbul, Turkey*

Abstract: Capillary Electrophoresis (CE) is a powerful separation and analysis technique that has been rapidly progressing since it was first introduced. The application range of CE is so diverse that it ranges from small analytes to large and complex macromolecules. This chapter aims to provide a deep understanding of the basic principles of CE. The first part of the chapter involves the theoretical basis, instrumentation, and separation mechanism of CE. The second part focuses on capillary electrophoretic separation modes and the third part describes the detection methods in CE. The fourth and final part covers capillary electrophoretic strategies for specific analyte groups.

Keywords: Capillary electrophoresis, CEC, CGE, Chiral separation, Contactless conductivity detector, CZE, Indirect detection, LIF, MEKC, Proteins, Sample stacking, Small anions, Small organic acids.

INTRODUCTION

Capillary Electrophoresis (CE) is a separation method that takes the basis of separation principle from classical electrophoresis and has the device design of modern chromatographic techniques. CE was first described as a new separation technique between 1980-1990, then a relatively new separation technique by comparison with High-Performance Liquid Chromatography (HPLC) and Gas Chromatography (GC). Today, CE has been widely applied to real samples and can be defined as a modern separation technique.

Classical electrophoresis was first introduced by the Swedish chemist Arne Tiselius in 1930 [1]. This invention earned Tiselius a Nobel Prize in 1948. Electrophoresis, in its simplest definition, means the migration of charged particles under the electric field.

* **Corresponding author F. Bedia Erim:** Department of Chemistry, Istanbul Technical University, Maslak, Istanbul, Turkey; E-mail: erim@itu.edu.tr

María Castro-Puyana, Miguel Herrero and María Luisa Marina (Eds.)
Capillary Electrophoresis in Food Analysis

Classical electrophoresis experiments began in U-shaped tubes with electrodes attached to both ends, containing an electrolyte inside. Later, the separation media preferred were gels, as the reproducibility of migration velocities in open solution was low. This method, is known today as slab-gel electrophoresis

and it is an essential tool of biochemistry laboratories, especially in protein separations. The technology has developed over the years and made it possible to produce silica columns with a diameter of micrometers. Thus, reproducible migration velocities have been obtained in open solutions in capillary columns. J.

W. Jorgensen and K. D. Lukacs introduced the CE technique in 1981 with an article published in Analytical Chemistry [2]. The method first evolved with hand-made devices in research labs, and later, modern CE devices were introduced to the market.

The present chapter describes the basic principles and mechanisms of capillary electrophoretic techniques. It covers instrumentation, different separation modes, and detection methods in CE, and separation strategies for specific analyte groups, such as small inorganic anions and organic acids, proteins, metal ions, and chiral molecules.

SEPARATION MECHANISM

In this section, general and some unique principles, strengths, and weaknesses of the CE technique will be given. Although the CE separation principle is based on classical electrophoresis, there are essential differences between the two methods. The high electrical resistance of the capillary column makes it possible to apply a very high voltage (HV) between the two electrodes. In commercial CE devices, up to 30 kV voltage can be achieved. The high voltage applied gives the CE technique a tremendous separation speed. This high speed also increases the separation efficiency of the method. An advantage of CE over other chromatographic techniques is that separation can be done in an aqueous medium. If necessary, hydrophilic organic solvents can also be added to the aqueous separation medium. Another advantage of CE is that different types of analytes can be separated in the same silica column using the same instrument design. One designation that has been used since the early years of CE is that the species can be separated in a broad spectrum from Li-ion to DNA by capillary electrophoretic methods. Indeed, many species in inorganic, organic, and macromolecular structures have been shown to be separated by CE. The different applications of CE can be seen in the most recent reviews [3 - 5]. Another advantage of CE compared to liquid chromatography methods is that silica capillary columns are less costly. Since the injection volume in CE is in nL size, it is an economical method in terms of capillary columns and the small amounts of sample and chemical consumption. The capillary column can be regenerated by washing with separation electrolytes and suitable solvents

between each injection. Due to this advantage, many sample solutions can be injected directly into the capillary column without the need for pre-purification.

The weakness of CE is the detection limit. In CE, the separation column is also the detection cell. Since the light path of the detection cell is in micrometers, if the molar absorptivity of the analyte is not high enough in UV detection, detection limits will be high. There are unique methods to overcome this obstacle in CE. These methods will also be mentioned in this section. The schematic device design of CE is shown in Fig. (**1**).

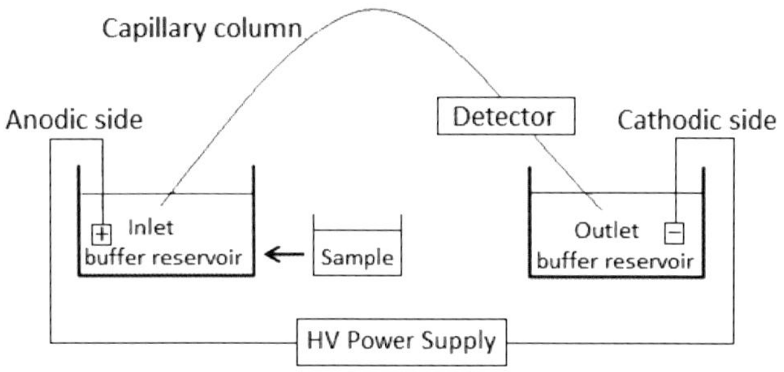

Fig. (1). Basic components of capillary electrophoresis instrumentation.

Fig. (**1**) shows two small buffer vials and a sample vial. The ends of a fused-silica column are immersed in buffer vials during electrophoresis. For electrophoretic separation, the separation medium must be an electrolyte solution with electrical conductivity. In CE separations, the use of a buffer solution for the background electrolyte medium is preferred due to the formation of H^+ and OH^- ions in the electrode regions as a result of electrolysis. H^+ and OH^- concentrations due to electrolysis are at the micromolar levels. However, the migration mobilities of pH- sensitive analytes and the surface charge of silica capillary column are affected by the pH change of the medium. Therefore, it is common practice to use a buffer as the separation medium in CE applications, with some exceptions.

The two electrodes are connected to a high voltage (HV) source. Voltage up to 30 kV can be provided in CE devices. Capillary columns are generally fused-silica columns. The outer layer of a fused-silica capillary is coated with a thin layer of polymer to prevent breakage and provide flexibility while using it. The detection window is opened by burning the polymer layer about 0.5 cm in length. As seen in Fig. (**1**), in general practice, the detector is placed close to the cathode side. Injections are performed from the anodic side. All the injected species, *i.e.*, positively charged, negatively charged, and neutral species, migrate inside the capillary from the anodic side to the cathodic side, regardless of their charges.

During this migration, the analytes separate from each other, reach the detector, and the data is evaluated with the help of a software.

Upon the application of high voltage between electrodes, ionic species experience an electric force given by eq. (1):

$$F_E = qE \tag{1}$$

where q is ion charge and E is applied electric field which is given by eq. (2):

$$E = \frac{V}{L} \tag{2}$$

where V is the voltage applied across the capillary, and L is the capillary length.

Ionic species in the capillary accelerate towards electrically opposite electrodes. While the velocity of the charged species increase, the friction force caused by the effect of the environment slows down the speed of the species. According to Stokes' law, the friction force for a spherical molecule can be expressed by eq. (3):

$$F_F = 6\pi\eta r v \tag{3}$$

where η is the viscosity of solution inside the capillary, r is the radius of the ion, and *v* is the velocity of the ion.

When a steady state is reached, the forces are equal but opposite, and the ions move with a constant velocity, which is proportional to the electrical field.

$$v = \mu_e E \tag{4}$$

where the proportionality constant μ_e here is called electrophoretic mobility ($cm_, V^{-1}s^{-1}$). After steady-state, *i.e.* $F_E = F_F$, by the combination of equations (1-4), the electrophoretic mobility μ_e of a charged particle is expressed as

$$\mu_e = \frac{q}{6\pi\eta r} \quad \textbf{(5)}$$

The separation of the species from each other is due to the mobility differences during migration in the capillary column. As seen from eq. (5), in a medium where viscosity is constant, the electrophoretic separation of the species is based on differences in their charge/size ratios.

We have mentioned that the injection is generally done from the anodic side, and all species migrate towards the cathode. It is necessary to look inside the capillary to understand the reason for this, which is contrary to Coulomb's law at first sight. The column consists of fused-silica capillaries. When an aqueous solution of electrolytes at around pH $\geq$ 1.5 is in contact with the capillary wall, the silanol groups of the silica on the surface of the capillary inner wall are deprotonated, leaving negative groups on the wall. Hence the inner wall of fused-silica capillary is now negatively charged. The excess of positive ions is located near the capillary wall due to the electrostatic attraction between the negatively charged wall and positively charged ions inside the capillary.

Capillary wall-SiOH + H_2O $\leftrightarrow$ Capillary wall-SiO$_{.}$ + $H_3O_{.}$

When a high potential is applied between the ends of the capillary, excess positive charge inside the capillary migrates to the cathodic side and drags their hydrate solvent molecules. Thus, a solution flow occurs in the capillary. This flow is called electroosmotic flow (EOF). In CE separation techniques, the driving force is EOF. The magnitude of the EOF can be expressed by electroosmotic mobility (EOM). Generally, the mobility of the electroosmotic flow is greater than the electrophoretic mobilities of the injected species. While positively charged species injected from the anodic side move towards the cathode with their electrophoretic mobility, they are also dragged by the electroosmotic flow in the same direction. So the apparent mobility for these species is given with the following equation:

$$\mu_a = \mu_e + \mu_{eo} \quad \textbf{(6)}$$

where

μ_a: apparent mobility

μ_e: electrophoretic mobility μ_{eo}: electroosmotic mobility

In contrast, while the negatively charged species try to move towards the injection direction, that is, the anodic side, with their electrophoretic mobility, the strong electrosmotic flow in the capillary drags them to the cathode direction. The apparent mobility of negatively charged species within the capillary is given by the following equation:

$$\mu_a = \mu_{eo} - \mu_e \quad (7)$$

This process is illustrated in Fig. (**2**). The charged particles will separate from each other during their migration in the cathode direction within the capillaries, if there is a difference in their charge/size ratios. However, all of the injected uncharged species will move to the detector together with apparent mobility equal to electroosmotic mobility, *i.e.* $\mu_a = \mu_{eo}$.

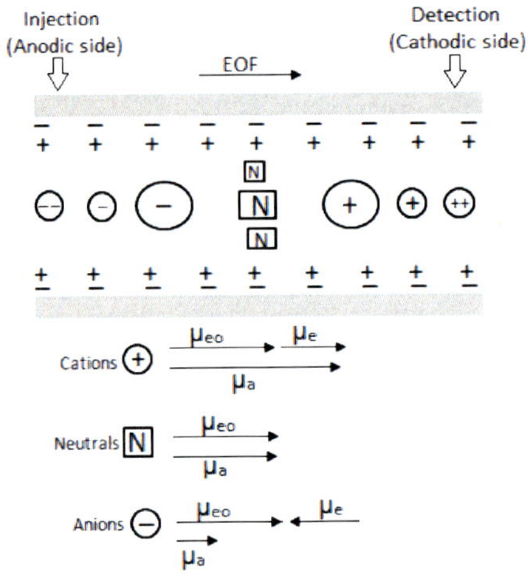

Fig. (2). Migration order of ionic and neutral species with the effect of EOF.

EOF can be easily adjusted with some experimental variables. However, while changing EOF, positive and negative effects of variables on separation should be taken into account. Experimental parameters affecting the speed of EOF are buffer pH, buffer concentration, applied electric field, temperature, buffer additions, and coating the inner wall of the capillary column. The change of buffer pH most easily and effectively controls the speed of the electroosmotic flow in the capillary. As the pH of the separation electrolyte increases, the ionization in the capillary inner wall increases hence the negative charge of the wall increases and EOF accelerates.

Increasing the electric field causes an increase in EOF. However, an increase in the electrical field may increase the current and, consequently, the Joule heating in the capillary. Reducing the electric field reduces separation efficiency and resolution.

EOF decreases with increasing buffer concentration. Increased buffer concentration causes an increased current in the capillary and possible Joule heating. On the other hand, a very low concentration of buffer increases the risk of wall adsorption of some analytes. Increased temperature lowers the viscosity of the separation electrolyte and EOF increases.

Organic solvents, additives, especially surfactants which are added into the buffer, significantly affect the speed of the EOF. Finally, the temporary or permanent coating of the capillary interior wall is an effective method for EOF control.

The most basic CE type, in which charged particles are separated according to mobility differences in free solution, is called Capillary Zone Electrophoresis (CZE). The uncharged species can be separated by other CE types, as will be explained later.

SEPARATION PARAMETERS

Migration Time

The migration time of species in the capillary column is calculated according to the eq. (8):

$$t = \frac{LL'}{V\mu_a} \qquad \textbf{(8)}$$

where

t: migration time (s)

L: capillary length (cm)

L': capillary effective length (the length of capillary from injection point to the detection point) (cm)

V: separation voltage (V).

Separation Efficiency and Peak Broadening

The expression of high separation efficiency is used in defining the CE methods. One of the reasons for the high separation efficiency of CE is the flat profile of EOF. Fig. (**3A**) schematically shows the flat profile of EOF.

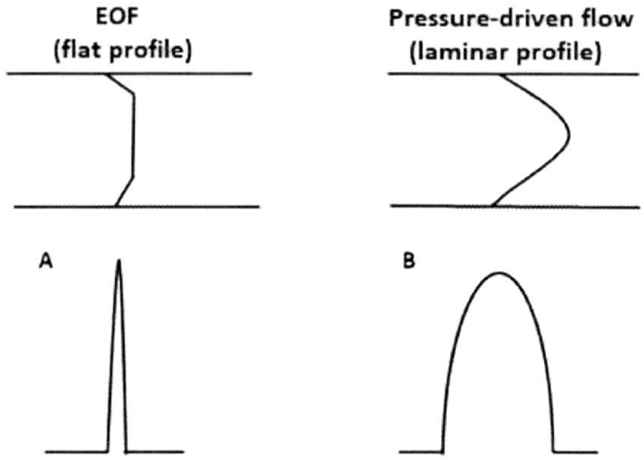

Fig. (3). Flow profiles and corresponding peak shapes.

The pressure-driven flow is a laminar flow resulting in broader peaks. More than one analyte peak may be hidden in the peak in Fig. (**3B**). However, the peaks formed in a flat flow are sharp peaks, and the resolution of the two peaks within a few seconds can be achieved. Peak broadening is an undesirable phenomenon that reduces the separation efficiency in all chromatographic methods. One of the reasons for peak broadening is the diffusion of the injected zone during migration. However, since the capillary has a very small diameter, the diffusion of the injection zone in the direction of the capillary walls can be neglected. The diffusion can be in the longitudinal direction. The smaller the standard deviation of the data of a peak in the form of a Gaussian curve results in a narrow peak. The effect of longitudinal diffusion to the peak broadening is given as the variance of the peak in eq. (9):

$$\sigma_L^2 = 2\,D\,t \qquad \textbf{(9)}$$

where

D: diffusion coefficient of the species.

t: migration time.

The very high electric field applied in the CE method causes short migration times. As t decreases, the effect of longitudinal peak broadening will decrease. The peaks of macromolecules such as protein and DNA with low D values will be very narrow, and high separation efficiencies are achieved for these molecules.

Apart from its longitudinal diffusion, another factor that causes peak broadening is the formation of Joule heating in the capillary. Although the electrical current passing through the capillary is very low, there may be a heat generation in the capillary called Joule heating. The Joule heating formed in the capillaries causes the electrolyte viscosity in the capillary center to be lower than the edges, and the flat profile flow within the capillaries turns into a laminar flow, which causes zone broadening. The way to reduce Joule heating is to lower the operating voltage and buffer concentration. However, a low electric field will reduce separation efficiency and resolution.

On the other hand, lowering the buffer concentration or ionic strength will both decrease the buffer capacity and increase the risk of possible sample adsorption to the wall. Another method for controlling Joule heating is to reduce the internal

capillary diameter. However, with the use of narrow capillaries, sensitivity will decrease for analytes with low molar absorptivity, especially in UV detection. Joule heating is dissipated through the capillary wall if the device has active temperature control.

Sample adsorption to the capillary wall may cause peak tailing. This effect usually results from the electrostatic interaction between positively charged analytes and the negatively charged capillary wall. Increasing the buffer concentration prevents adsorption by shielding the wall charge. However, since CE separation is very fast, there is no wall adsorption problem for small particles. Adsorption is a problem for proteins carrying a large number of charges on them. In protein separations, coating the capillary inner wall with an uncharged or positively charged polymer according to the type of protein is a method applied to prevent wall adsorption.

If the conductivity of the sample zone and the separation buffer are very different from each other, this may cause distortion and tail in the peak shapes. Although this effect is smaller than the other effects that cause peak broadening, it can be effective in low-resolution separations. Additionally, it can be turned into an advantage to increase the sensitivity of some analytes in the form of sample stacking, which will be explained soon.

Precision in Migration Times and Peak Areas

Capillary columns provide reproducible results by rinsing the column with a washing solution between each run. It is sufficient to wash only with the working buffer for some analytes, while special washing strategies should be developed for some others. Sodium hydroxide, hydrochloric acid, and some organic solvents such as methanol are usual washing solutions. Since the EOM is very sensitive to pH, buffer capacity is a critical factor for obtaining reproducible migration times. In the case of initial concentrations of conjugate acid-base pair of the buffer are equal to each other, the buffer capacity is maximum. The buffer works effectively when the pH of the buffer is around ± 1 unit of the pKa value of the conjugated acid. A high concentration of buffers is also required for a good buffer capacity. However, buffer concentration in CE is limited due to possible Joule heating. The CE device automatically cuts off or lowers the voltage when the current limit value is exceeded.

Zwitterionic buffer solutions can be used in relatively high concentrations without increasing the CE current. Tris, MES, CAPS, CAPSO are the most used zwitterionic buffers in CE. A buffer with a low absorbance at the wavelength studied is a desired feature. Phosphate and borate buffers are the most used buffers in capillary electrophoretic separations with their low absorbances and high buffer capacities. Even if the buffer has a high capacity, the buffer containers need to be refreshed occasionally. When the buffer contains an organic solvent, the polymer layer at the capillary end can creep and consequently partially close the capillary entrance, which affects injection volume reproducibility. This problem can be solved by burning the polymer layer at the capillary end approximately half a cm.

pH difference between the washing solution and the buffer solution can cause hysteresis in the capillary column. In such cases, it is preferred to apply an electric field for a while without injecting the new solution or to ignore the first injection results.

The internal standard increases peak area reproducibility. In this case, the calibration curve is drawn as peak areas/internal standard area ratios *vs*. analyte concentrations. However, it is not always possible to find a stable and suitable internal standard that migrates near analyte peaks. In this case, the calibration curve can be created between the corrected peak areas and concentrations. Corrected peak areas are obtained by dividing the area of each peak by migration time.

The most common solutions for reproducibility of migration time and peak area are summarized in Tables **1** and **2**, respectively.

Table 1. Common solutions to increase migration time reproducibility.

Cause	Solution
Viscosity and consequently EOM change due to temperature change in the capillary	• Effective temperature control
Buffer composition change due to pH change, evaporation, and mixing of washing solution into the buffer	• Increase buffer capacity • Replenish buffer solution occasionally • Use caps for buffer vials • Use a separate outlet vial for wash waste • Dip the inlet end of capillary into water vial after washing step
Hysteresis of wall charge due to pH difference between buffer and washing solution	• Give sufficient equilibrium time to the running buffer
EOM change due to wall adsorption	• Find an effective wash solution • Increase the concentration of the buffer • Coat the capillary wall

Table 2. Common solutions to increase peak area reproducibility.

Cause	Solution
Injected volume change due to temperature change	• Effective temperature control
Sample concentration change due to evaporation	• Effective temperature control • Use caps for sample vials • Refresh sample solutions occasionally
When the buffer solution contains an organic solvent, the outer polymer layer softens and partially covers the capillary inlets	• Remove polyimide layer from the capillary ends
Change of sample composition in electrokinetic injection	• Use hydrodynamic injection • Use a new sample solution for each injection
Extra sample inlet from the outer capillary wall at the injection tip	• Dip capillary inlet end in water or buffer in a separate vial after sample injection
Low signal to noise ratio	• Increase the sample concentration • Apply sample stacking methods

MODES OF CE

Capillary Zone Electrophoresis (CZE)

The separation of analytes in free solution according to mobility differences, as described above, is CZE. CZE is the basic separation mode of CE. Other capillary modes use the device design of CZE.

Micellar Electrokinetic Chromatography (MEKC)

If the separation is mainly based on the interaction between analytes and some separation medium additives, such separations in CE are commonly referred to as electrokinetic chromatography or electrokinetic capillary chromatography. Micellar Electrokinetic Chromatography (MEKC) is the dominantly used form of electrokinetic chromatography.

After Terabe *et al.* introduced MEKC [6], it has become a CE method primarily used in the separation of uncharged species. In MEKC, a surfactant is added to the separation buffer above its critical micelle concentration (CMC). Aggregates called micelles are formed inside the capillaries, and these aggregates act as a fixed phase in the liquid chromatography. However, since the micelles also have an electrophoretic mobility, they move within the capillary. Micellar aggregates can be called a pseudo-stationary phase. Sodium dodecyl sulfate (SDS) is the most widely used surfactant in MEKC applications. The hydrophilic ends of the micelles formed in the aqueous buffer medium are directed towards the aqueous phase and the hydrophobic tails towards the middle of the micelle.

A schematic separation scheme of MEKC is given in Fig. (**4**). A small neutral hydrophilic molecule injected from the anodic side will be dragged to the detector with the electroosmotic flow (t_0). The peak will appear in the electropherogram at t_0 time. Neutral markers such as acetone, formamide, mesityl oxide can be used in order to determine the mobility of the EOF in both MEKC and CZE. The negatively charged micelles, in that example, tend towards the anodic side with their electrophoretic mobility. However, they will be driven towards the cathodic side, that is, the detector direction with the electroosmotic flow in the capillary.

When a very hydrophobic dyestuff is injected into the medium, it will remain in the hydrophobic core of the micelle. It will appear as a peak in the electropherogram when the micelle reaches the detector (t_m). Sudan dye is commonly used as a marker for this purpose. In MEKC separations, the partition of species between the hydrophobic core of micelles and the solution phase occurs. All uncharged analytes will migrate between t_0 and t_m window. Those with a more hydrophilic structure will be seen in areas closer to t_0, and those that are more hydrophobic will appear in areas close to t_m. If all hydrophobic analytes tend to spend most of their time in the micelle phase, a certain proportion of water-compatible organic solvent can be added to the buffer to increase the tendency of these analytes to the aqueous phase. Alcohols and acetonitrile are the most commonly used organic solvents in the studies. In MEKC, the organic solvent added should not exceed 20-30% concentration because micelles can be dispersed in the organic solvent medium.

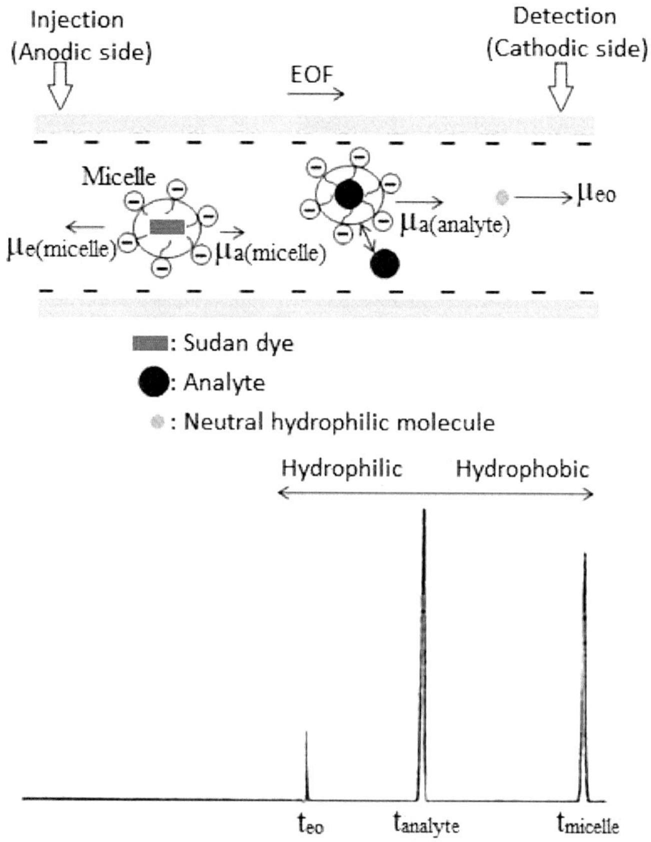

Fig. (4). Separation mechanism of Micellar Electrokinetic Chromatography (MEKC).

Alternatively, by adding cyclodextrins to the buffer together with a surfactant, separation of hydrophobic molecules can be achieved by the partition of molecules between the micellar phase and cyclodextrin cavity. MEKC has been used in the analysis of a large number of compounds in food, plant, biological, and environmental samples.

Nonaqueous Capillary Electrophoresis (NACE)

In NACE, the separation medium is an organic solvent instead of water. Methanol, acetonitrile, formamide, dimethylformamide are the most common organic solvents in NACE studies. Organic solvent affects the solubility of analytes. However, an important factor in NACE separations is the significant change of acid-base properties of analytes relative to the water environment. The organic solvent also affects the ionization of the capillary wall. Since the conductivity of the organic solvents used is lower compared to water, the high electric field in the capillary increases the separation speed. of $HClO_4$. Thus, with the selected NACE electrolyte, acrylamide migrates electrophoretically in the capillary as a positively charged molecule. Fig. (**1**) NACE analysis of acrylamide (A) a French fry extract and (B) a spiked extract with 2 mg/L acrylamide standard. * shows the acrylamide peak. Buffer: 30 mmol/L $HClO_4$, 218 mmol/L CH_3COOH, and 0.129% w/v water. Voltage: 25 kV (injection is from the positive electrode side). Capillary, 75 mm, (52.5 x 44) cm; wavelength: 200 nm; injection: hydrodynamic, 40 mbar, 6 s. Reprinted from [7] with permission.

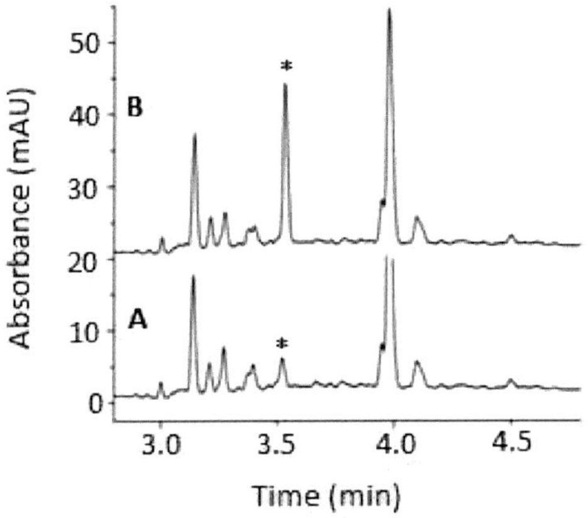

Fig. (5). NACE analysis of acrylamide **(A)** a French fry extract and **(B)** a spiked extract with 2 mg/L acrylamide standard. * shows the acrylamide peak. Buffer: 30 mmol/L $HClO_4$, 218 mmol/L CH_3COOH, and 0.129% w/v water. Voltage: 25 kV (injection is from the positive electrode side). Capillary, 75 mm, (52.5 x 44) cm; wavelength: 200 nm; injection: hydrodynamic, 40 mbar, 6 s. Reprinted from [7] with permission.

Capillary Gel Electrophoresis (CGE)

Capillary Gel Electrophoresis (CGE) is a type of CE mostly used to separate proteins and DNA according to their size differences. DNA is negatively charged due to its phosphate groups, whereas proteins gain negative charge by being treated with SDS. Since the sizes of macromolecules such as DNA are huge, their electrophoretic separation based on charge/size differences is not possible. These molecules are separated according to their size differences only in a gel environment [8].

In CGE, a gel medium exists in the capillary column. The most commonly used gel medium in CGE applications is crosslinked polyacrylamide (PAA). The separation in the CGE method is based on the ability of the molecules to pass through the pores of the gel. Macromolecules having small sizes can easily migrate through the pores of the gel, whereas the bigger molecules look for pores that can pass through, and their migration paths increase. In CGE, the movement of species is either by the combination of electrophoretic mobility and electroosmotic mobility, or by coating the capillary wall with an uncharged polymer, EOF is suppressed, and migration occurs only with electrophoretic mobilities. In capillary gel electrophoresis, the separation medium is not always solid gel. Entangled linear polymer solutions can be used as separation media. The separation mechanism in the non-crosslinked polymer network is identical to that of solid gel.

Capillary Electrochromatography (CEC)

CEC is a relatively newly developed hybrid separation method of CE and liquid chromatography. In this mode, the partition of species is between the fixed stationary phase and solution phase. Here again, the driving force is EOF. CEC can be performed in packed, monolithic, and open-tubular columns. As can be seen in the recent review, the development of new fixed phases in CEC is an evolving field [9].

Capillary Isotachophoresis (CITP)

Two buffer systems are used in CITP. The leader electrolyte (LE) has higher mobility than the terminating electrolyte (TE). With the application of voltage, a non-uniform electric field is formed inside the capillary and the analyte zone stacks between LE and TE. CITP provides concentration and separation of the analyte bands at the same time. The overall progress in CITP has been collected in a recent review article [10].

Capillary Isoelectric Focusing (CIEF)

CIEF is mostly used to separate proteins and peptides based on their isoelectric points (pI) [11]. Ampholyte mixtures establish a pH gradient inside the capillary. The protein migrates until it reaches the pH area corresponding to its pI, where it becomes uncharged and immobilized.

ON-LINE SENSITIVITY ENHANCEMENT: SAMPLE STACKING

In CZE, the sample is injected into the capillary column as a short zone. When voltage is applied, analyte zones are separated from each other along the capillary column. Rapid migration time due to the high voltage applied causes the analyte zone to spend less time in the capillaries and less exposure to peak broadening due to the longitudinal diffusion along the capillary. Analyte zones reach the detector as narrow peaks under the influence of the flat character of the EOF.

However, if the molar absorptivities of analytes are low in UV detection, their concentrations may be below the LOD of the analysis method. Injecting larger sample zones to increase detection sensitivity causes broad and overload peaks.

A practical method to increase detection sensitivity in CE includes sample stacking methods where a long sample zone with low conductivity is injected into the capillary column and a high conductivity buffer is selected as the separation buffer. Since the electric field is high in the low conductivity zone, the charged particles move with high mobilities in this zone under the applied voltage. The electric field drops in the high conductivity buffer zone and the analytes slow down at the border of two zones. The analyte molecules are stacked into a narrow zone. This concentrated zone appears in the detector with increased sensitivity.

Sample stacking is an easily applicable technique and different stacking applications can be developed according to the types of samples. For example, the low conductivity sample region can be discarded through the capillary by changing the polarity for a short period of time, or in electrokinetic injection, a short frontal buffer or acetonitrile zone can be hydrodynamically injected. Different names have been used for stacking procedures in the literature, such as Large-Volume Sample Stacking (LVSS), Field-Amplified Sample Stacking (FASS), Field-Amplified Sample Injection (FASI). However, the method used is generally referred to as sample stacking in recent years. The details of the stacking methods are explained in the articles.

In Fig. (**6A and B**), a sample stacking effect is seen for nitrite and nitrate ions for a method that was developed for the simultaneous determination of these ions in canned fish samples [12]. In the method, keeping the pH of the separation electrolyte low, the anions injected from the cathodic side are enabled to move against EOF. The addition of sodium sulfate increases buffer conductivity. The anions moving rapidly in the direction of the anode in the low conductivity large sample zone are slowed down when they reach the high conductivity buffer border. The schematic view of the stacking is given in Fig. (**7**). In the study, sensitivities are increased by 30 times for both ions due to the sample stacking. Since the EOF is to the cathode direction in the method, the sample zone is discarded from the cathode side during electrophoresis.

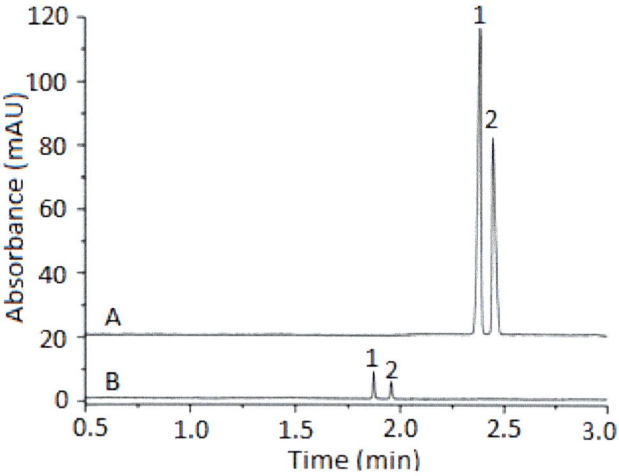

Fig. (6). Comparison of the sample stacking experiment and small-volume injection of a standard sample containing 100 μmol/L nitrate and nitrite. **A)** Injection: 50 mbar, 160 s, Buffer: 30 mmol/L formic acid, 30 mmol/L Na_2SO_4, pH:4. **B)** Injection: 50 mbar, 6 s, Buffer: 30 mmol/L formic acid pH:4. Voltage: −25 kV. Peaks: 1:nitrate, 2:nitrite. Reprinted from [12] with permission.

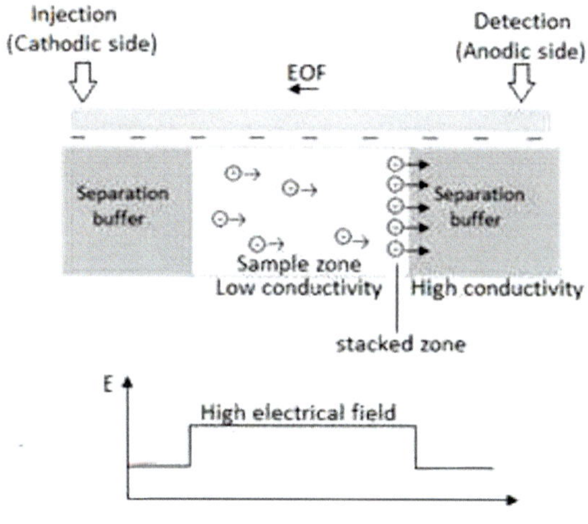

Fig. (7). Schematic illustration of sample stacking method for nitrate and nitrite ions.

Although the easiest and most applied method of stacking is based on the concentration difference of the sample and buffer zone, pH-mediated stacking techniques can be applied for the pH-dependent analytes, where the pH difference of the sample zone and buffer zone is created. Stacking can be carried out by the sweeping method in EKC, especially in MEKC. In this method, the sample zone does not include the micelle phase. When voltage is applied, micelles in the separation electrolyte enter the sample zone and collect analytes with micelle-analyte interaction. The analytes are concentrated at the front of micelles as a short zone. The sample stacking studies in recent years have been collected in review articles [13, 14].

DETECTION METHODS

UV and Indirect UV Detection

The most widely used CE detectors in CE applications are the variable-wavelength ultraviolet (UV) detectors and diode-array (DAD) detectors. The analytes with UV active functional groups can be detected with these detectors during electrophoretic separation. For species that do not contain chromophore groups, a useful and widely used application in CE is the indirect detection method where a strong chromophore group-containing compound is added into the separation buffer. The detector's wavelength is adjusted to the maximum wavelength of the chromophore. The detector sees the baseline signal of the chromophore. Analyte ions displace chromophore ions, and a decrease in absorbance occurs as the analyte zone passes across the detector. The area of this negative peak depends on the analyte concentration [15]. The formation of indirect detection peaks and their symmetry depend on chromophore concentration and mobility. In indirect detection, analyte mobility and chromophore mobility should be compatible with each other. While negative peaks were given on the electropherogram in the first studies with indirect detection, the display of negative peaks in the form of positive peaks with the adjustment of software has been preferred in the last years. The indirect detection method is widely used for organic acid determinations in food samples. Organic acids are present in many food products in small or large amounts. While it gives a sour taste to the juices, the amounts in the commercial juices are also adulteration markers. Fig. (**8**) shows the separation of malic acid and citric acid in a commercial pomegranate juice by indirect detection. 2,6-pyridine dicarboxylic acid (PDC) was used as the chromophore. In this study, 7 commercial pomegranate juices were analyzed, and the significantly high value of the malic acid amount of one pomegranate juice was interpreted as a sign of its adulteration with apple juice [16]. With an additional analysis, the significant difference in Fructose/Glucose ratio, total phenolic amount, and ferric reducing antioxidant power (FRAP) value of this specific juice from others supported adulteration.

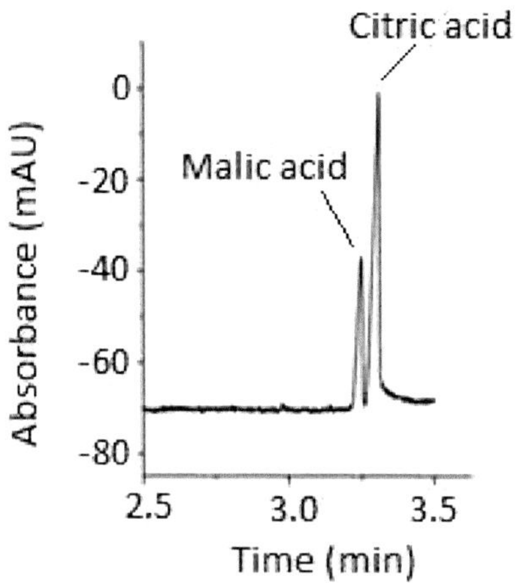

Fig. (8). Separation and indirect detection of organic acids in a commercial pomegranate juice. Separation buffer: 10 mmol/L of PDC + 0.1 mmol/L CTAB. pH: 5.6. Detection wavelength: 350 nm with a reference at 200 nm. Injection: 50 mbar, 6 s. Voltage: -25 kV. Reprinted from [16] with permission.

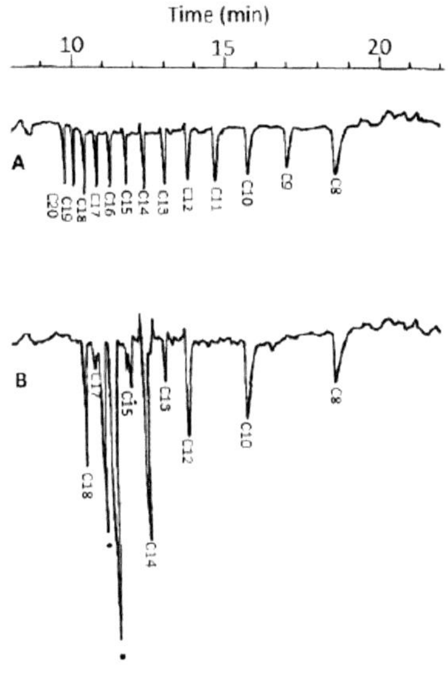

Fig. (9). Separation and indirect detection of fatty acids. **(A)**: the standard solutions of C8-C20 fatty acids, **(B)**: Unsaturated fatty acids in an extract of dairy-fresh butter. Background electrolyte: 10 mM SDBS + 50% acetonitrile + 30 mM Brij. Reprinted from [17] with permission.

Fluorescent and Laser-Induced Fluorescence Detection

Fluorescent and Laser-Induced Fluorescence (LIF) detectors are used for analytes that naturally show fluorescent properties or can be derivatized with a fluorescent dye. A fluorescent detector provides a lower detection limit (LOD) than a UV detector. On the other hand, very low LOD values can be reached with LIF detectors. For example, riboflavin (vitamin B2) which is a fluorescent substance and an essential vitamin, cannot be synthesized in the body so it must be taken from food products. However, the content of riboflavin in many foods is too low to allow detection by a UV detector.

In Fig. (**10**), an electropherogram showing vitamin B2 separation and LIF detection in a saffron extract is given [18].

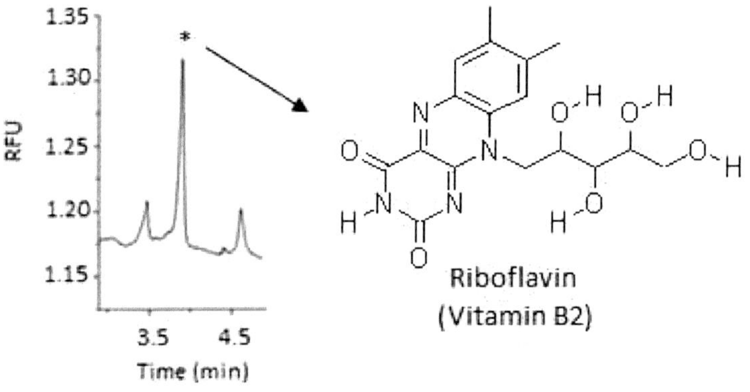

Fig. (10). Separation and LIF detection of vitamin B2 in a saffron sample. Buffer: 20 mM borate at pH 9.5. Injection 50 mbar, 6 s; voltage: 25 kV; Detection: excitation at 488 nm and emission at 520 nm. Riboflavin peak is shown with an asterisk. Reprinted from [18] with permission.

On the other hand, fluorescent detectors are not as easy to handle as UV because some separation medium can result in fluorescent quenching. However, there are cases where fluorescence intensity increases depending on the content of the separation medium. Curcumin is a bioactive substance of turmeric spice that has been very popular in recent years. However, there are two more curcuminoids beside curcumin in turmeric and each has different bioactivities [19].

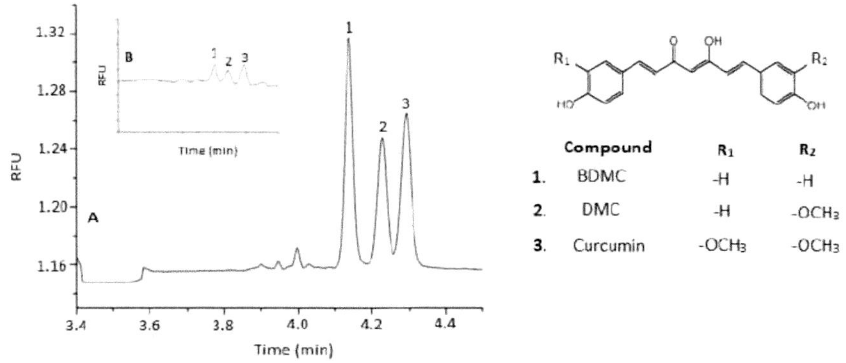

Fig. (11). Separation of curcuminoids and the effect of 2-HP-β-CD in buffer on fluorescence intensities of curcuminoids. (**A**) Buffer: 25 mM borate, 35 mM 2-HP-β-CD; (**B**) Buffer: 25 mM borate. Injection: 50 mbar 6s, Voltage: 25 kV, detection: excitation at 488 nm and emission at 520 nm; 1: BDMC (1.5 µg/mL); 2: DMC (20 µg/mL); 3: Curcumin (20 µg/mL). Reprinted from [20] with permission.

Similarly, the fluorescence intensities of biogenic amines, which are derivatized by fluorescein isothiocyanate (FITC) in MEKC-LIF analysis, increase to between 4 and 21 folds when Brij 35 is substituted for SDS in the same separation medium as seen in Fig. (**12**).

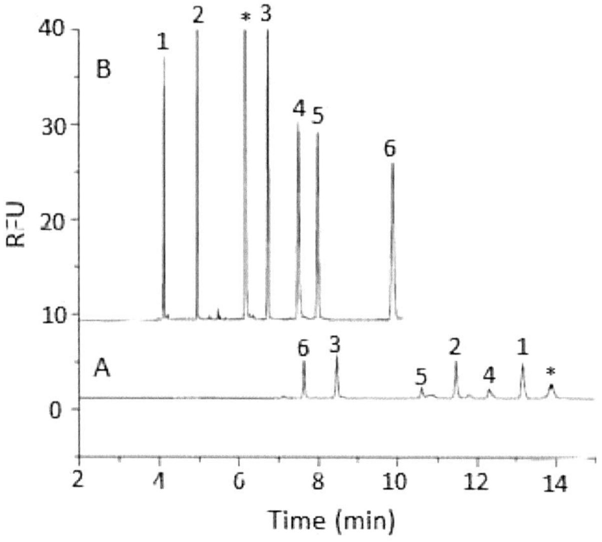

Fig. (12). Effect of Brij 35 in buffer on fluorescence intensities of biogenic amines derivatized with FITC.

(A) Buffer: 75 mM borate, 20 mM SDS, pH 9.7 (B) 75 mM borate, 10 mM Brij, pH 9.7. Voltage: 25 kV, capillary: 50 µm, (63x47) cm; injection: hydrodynamic, 50 mbar, 6s. 1:Trypthophan, 2:Tyrosine, 3:Cadaverine, 4:Spermidine, 5:Histidine, 6:Putrescine. Concentrations: 0.3 µM for each. *:FITC. Reprinted from [21] with permission.

MS Detection

Combining CE with MS allowed many compounds to separate with high separation efficiency and obtain molecular mass information of the separated analytes. Although the most used interface in CE-MS applications until today is Electrospray Ionization Interface (ESI), there are also studies where CE is combined with MALDI-MS and some other ionization techniques. CE-MS combination has been mostly used in proteomics, foodomics, pharmaceutical, and clinical researches. Recent developments and applications in CE-MS have been collected in review articles [22, 23]. Although less common, inductively coupled plasma mass spectrometry coupled with CE (CE-ICP-MS) has also been used for metal- biomolecule relationships [24].

Contactless Conductivity Detection

A Contactless Conductivity Detector (C^4D) for capillary electrophoresis was first introduced in 1998 [25]. Since then, this method has been applied for the detection of inorganic and organic ions in electrophoresis [26]. C^4D allows simple combination with commercial CE instruments and can be performed in an on-capillary mode. The detector contains two electrodes that are placed around the polyimide coating of the silica capillary column. The electrical conductivity changes of the solution between two electrodes are measured.

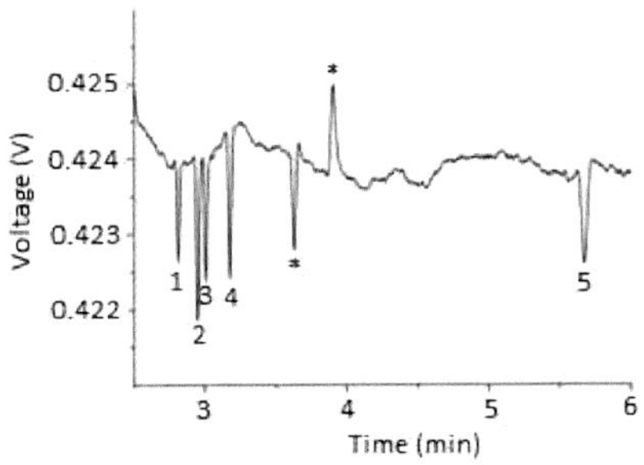

Fig. (13). Separation and CE-C^4D detection of biogenic amines. 1:spermidine, 2: putrescine, 3: histamine, 4: cadaverine, 5: tyramine, and '*' represents system peaks. The concentration of each amine is 2 mg/L, except tyramine. Its concentration is 4 mg/L. Separation electrolyte: 500 mM hydroxyisobutyric acid (HIBA), pH 2.05. Separation conditions: 28 kV, 20 °C. injection: hydrodynamic, 60 mbar for 60 s from the anodic side. Reprinted from [27] with permission.

Other Detection Methods

Although they are not routinely used as the above detection methods, different detection systems have been combined with CE such as amperometric detection, especially in research studies. Amperometric detection is particularly suitable for microchip CE.

2.05. Separation conditions: 28 kV, 20 °C. injection: hydrodynamic, 60 mbar for 60 s from the anodic side. Reprinted from [27] with permission.

SEPARATION STRATEGIES FOR SOME ANALYTE GROUPS

CE is a separation method that can be applied to samples in a broad spectrum with different separation modes. It has been used in sample analysis in many different fields due to its high speed, easy method development, and high separation efficiency. In this section, the separation strategies applicable for some special groups will be briefly mentioned.

Small Inorganic Anions and Organic Acids

As mentioned before, injections are generally performed from the anodic side in CE separations. All the species move towards the cathode direction and reach a detector located near the cathode end of the capillary. The migration behaviors of small inorganic anions and organic acids are exceptions. Due to the high electrophoretic mobility of both groups, EOF can not easily drag them into the cathodic side. Thereby, they cannot reach the detector, or their migration times become very long. Separation options for small negatively charged ions are given schematically in (Fig. **14A-C**). In Fig. (**14A**), a cationic surfactant below its critical micelle concentration (CMC) is added to the separation buffer. The monomers of cationic surfactants are attracted to the negatively charged capillary wall. The hydrophobic tails of surfactant monomers adhering to the wall interact with the tails of the free surfactant molecules in the solution, forming a double layer on the wall. Thus, the capillary inner wall is positively charged. In this case, the direction of the EOF in the capillary is reversed from the cathode to the anode. It is possible to change the polarity of the electrodes in CE devices. Changing the polarity, the injection is made from the cathodic side, and the detector is on the anodic side. When negatively charged small ions are injected from the cathodic side, they migrate to the anodic side due to their electrophoretic mobility. Since the EOF in the capillary is also directed to the anode, a high-speed separation of these ions takes place.

Cetyltrimethylammonium bromide *(*CTAB) is the most commonly used surfactant for dynamically coating the capillary wall. Fig. (**15**) shows the separation of small organic acids in a pollen sample using CTAB in the separation medium [28].

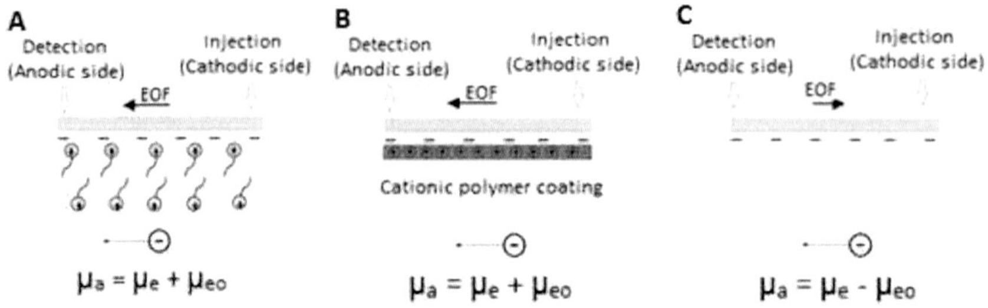

Fig. (14). Electromigration of small negatively charged ion. **A)** Reversed EOF with a cationic surfactant in buffer, **B)** Reversed EOF in a coated capillary with a positively charged polymer. C) Reduced EOF.

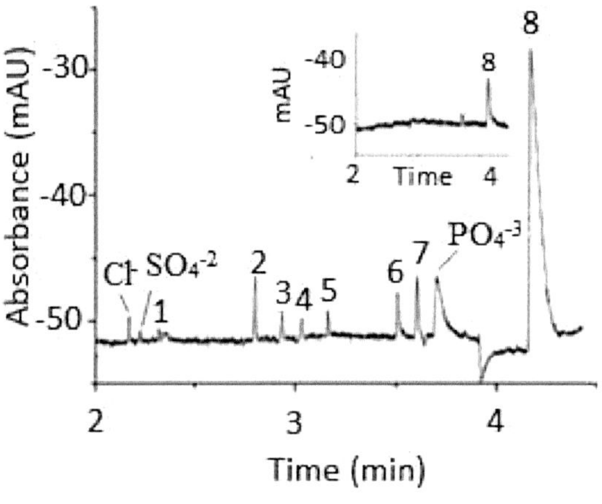

Fig. (15). Separation of organic acids in a pollen extract (Buckwheat pollen). Conditions: capillary 56.5 cm effective length 50 µm I.D; Separation electrolyte: 5 mmol/L of PDC + 0.1 mmol/L CTAB, pH: 5.6. Detection wavelength: 350 nm with a reference at 200 nm. Voltage: -25 kV. Peaks: 1: Oxalic acid, 2: Tartaric acid, 3: Malic acid, 4: Citric acid, 5: Succinic acid, 6: Acetic acid, 7: Lactic acid, 8: Gluconic acid. The on-set electropherogram shows the analysis of the diluted pollen extract for the quantification of gluconic acid. Reprinted from [28] with permission.

As shown schematically in Fig. (**14B**), alternatively, by physically adhering or covalently bonding a positively charged polymer to the wall, the capillary inner wall is positively charged for a long time. (Fig. **16**) shows the separation of nitrate and nitrite ions in a physically coated capillary column with high molecular weight PEI. Thiocyanate ion has been used as the internal standard. As can be seen, separation is extremely fast. This method has been used for the quantification of both ions in processed meat products and some vegetables [29].

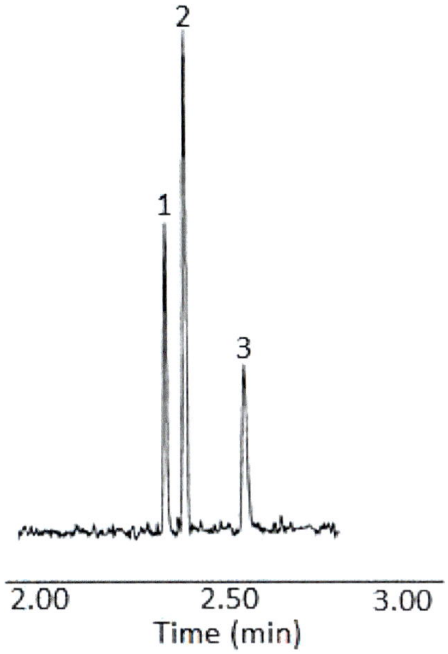

Fig. (16). Simultaneous separation of nitrate and nitrite ions in PEI coated capillary. Buffer: 20 mmol/L Tris, pH: 7.5. Voltage: -28 kV, Injection: 4.10^{-3} MPa. Total length 75 cm, effective length 60 cm. i.d. 75 µm. Direct UV detection at 210 nm. Peaks: 1:nitrite, 2:nitrate, 3:thiocyanate (internal standard). Reprinted from [29] with permission.

Another option is to suppress EOF in the capillaries and allow ions to move against EOF. As the pH of the working buffer decreases, the negative charge density in the capillary wall decreases, as a result of which the speed of the EOF decreases. The pH of the separation buffer is chosen at low values. The sample containing negatively charged small ions is injected from cathodic side. Negatively charged ions move towards the anode side. Meanwhile, although there is an EOF in the direction of the cathode in the capillary, its speed has decreased very much. Small ions with high electrophoretic mobility move rapidly towards the anodic side against EOF. This form of separation is shown schematically in Fig. (**14C**). Fig. (**6**) shows the separation of nitrate and nitrite ions in suppressing EOF mode.

Metal ions

The general approach to separating metal ions, whose charge and mass ratios are close together, is to complex them with a suitable ligand. The separation of metal ions is based on the differences in the charge and size ratios of the complexes as a result of the differences in the complex formation constants. If the ligand used is UV active, direct UV detection is possible. Metal complexes can be determined indirectly using a suitable chromophore together with a UV inactive ligand. Two different methods are applied to the separation of metal ions. If the stability constants of the complexes are high, precolumn complexation is preferred, and the metal ion-ligand mixture is injected into the capillary column. In the case of weak complexes, an on-column complexation is preferred. A ligand is added into the buffer solution, and metal ions are injected into this separation medium. More than one ligand can be used if the charges, ionic diameters of metal ions and the formation constants of complexes formed by a ligand are very close to each other. Separation is accomplished by competing ligands for metal ions. Lanthanide group metal ions are one of the examples for this situation, as seen in Fig. (**17**).

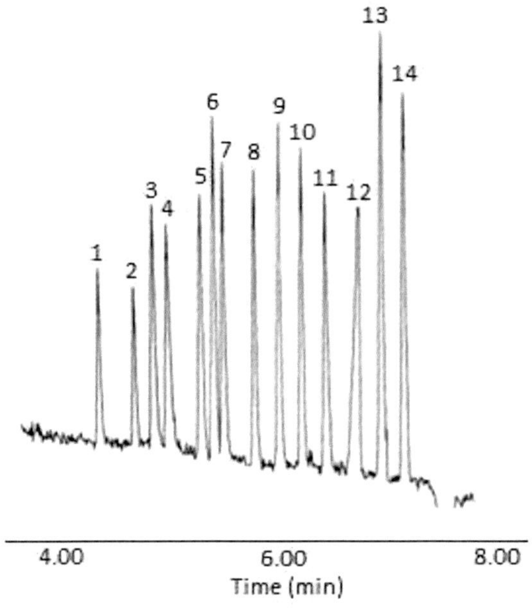

Fig. (17). Electrophoregram of 14 lanthanide group metal ions. Buffer: 15 mmol/L HIBA, 13 mmol/L Tris, 0.1 mmol/L cupferron, pH 4.9. Injection 40 mbar, 0.06 min. Run voltage 28 kV. 1: La, 2: Ce, 3: Pr, 4: Nd, 5: Sm, 6: Eu, 7: Gd, 8: Tb, 9: Dy, 10: Ho, 11: Er, 12: Tm, 13: Yb, 14: Lu. Reprinted from [30] with permission.

The separation of lanthanide group metal ions is a challenge of chromatographic methods considering their similarity. In this study, 14 lanthanide group metal ions were separated by the addition of two ligands, namely HIBA and cupferron, into the separation buffer [30]. Direct UV detection of lanthanides complexed with cupferron was possible since cupferron is a UV-active ligand. Since the weak complexes are formed, both ligands were added into the separation medium, and lanthanide ions were then injected into the capillary.

Metal complexes can be determined with conductivity detectors without requiring a UV active ligand or chromophore. Fourteen lanthanide ions, complexed with HIBA and acetic acid, were separated and determined by the CE-C^4D system [31].

Proteins

There are many applications in which proteins are separated by different electromigration modes such as CZE, CITP, CIEF, CEKC, CEC collected in a recent review article [32]. As stated in the CGE section, SDS-treated proteins can also be separated by size differences. In protein separation with CE, adsorption of proteins to the negatively charged capillary wall should be prevented. One method is to suppress the positive charges of the proteins using a separation medium having a pH well above the isoelectric points of the proteins. The more effective and widely used method is to coat the capillary inner wall with a suitable polymeric layer according to the type of protein. Dynamic and permanent coating methods for protein separation in CE and new coating materials with nanoparticles were collected in the last review article [33].

(Fig. **18**) shows effective separation of basic proteins in a capillary coated with a positively charged polymer (PEI) [34].

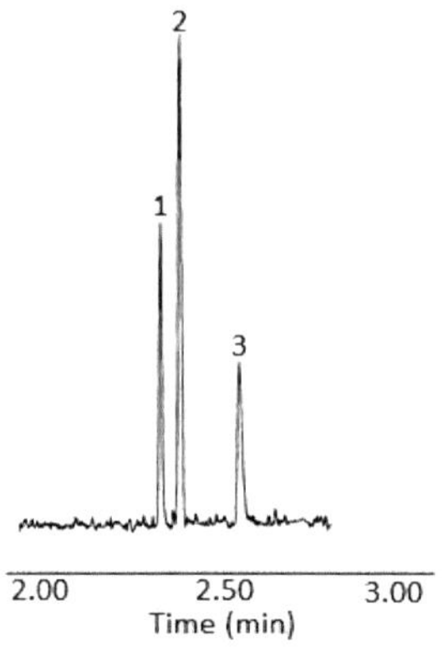

Fig. (18). Separation of five basic proteins in PEI-coated capillary. Total length 63.5 cm, effective length 46.5 cm, I.D. 75 µm. Buffer: 50 mM acetate, pH 5.5. Run voltage -28.8 kV. Injection 10 mbar for 6 s. Sample: 1: trypsinogen 0.32 mg/mL, 2: α-chymotrypsinogen 0.32 mg/mL, 3. ribonuclease A 0.72 mg/mL, 4: cytochrome C 0.32 mg/mL and 5: lysozyme 0.32 mg/mL. UV detection at 214 nm. Reprinted from [34] with permission.

Chiral Molecules

In CE, chiral separation is performed in electrokinetic chromatography mode. A soluble chiral selector in the separation buffer forms a pseudo-stationary phase. During migration, enantiomers are shared between the chiral selector and the solution phase. Interaction differences between the enantiomers and the chiral selector provide the enantioselectivity [35]. Chiral separation with CE is very easy to perform, and high separation efficiencies are achieved. Chiral selective cyclodextrins are the most commonly used in CE separations. Besides numerous commercially available cyclodextrins, research is on the development of new cyclodextrins. However, there are also studies in CE separations where different potential selectors such as ionic liquids are tested [36]. Analytical chiral separation is especially important in the pharmaceutical industry. Therefore, most CE applications in the literature are on chiral drug analysis. In recent years, chiral separations have gained importance also in the field of food [37].

Fig. (**19**). shows the separation of four pairs of chiral aminoacid with the addition of 10 mM betacyclodextrin (β-CD) to the separation medium as the chiral selector.

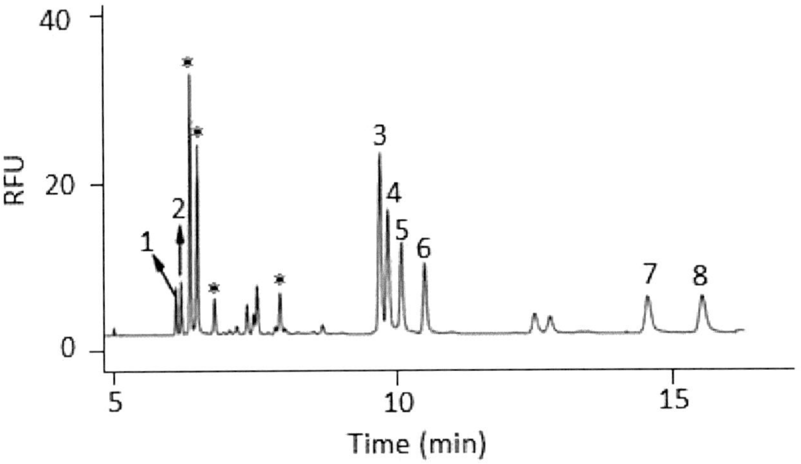

Fig. (19). Electropherogram of amino acids in 50 mM borate buffer at pH 9.6 containing 5 mM SDBS, 10 mM β-CD; Voltage: 25 kV. Temperature: 25 °C. Capillary 50 μm, (60 × 40) cm. Injection: hydrodynamic, 50 mbar, 6 s. 1: D-Arg, 2: L-Arg, 3: L-Pro, 4: D-Pro, 5: D-Ala, 6: L-Ala, 7: D-Asp, 8: L-Asp, □: peaks from FITC. Concentrations: 0.6 μM for each, except Asp as 1.2 μM. Reprinted from [38] with permission.

CONCLUDING REMARKS

CE is a versatile analytical technique because of its different operation modes and plays a great role in the field of separation science based on its multiple applications. Moreover, some of its advantages such as easy operation modes, different detection techniques, enhanced separation efficiency, and separation speed make CE a sophisticated technique. Our approach in this chapter was to give the basic principles of CE and some separation applications for food samples mostly. It appears that CE continues to be developed and be utilized in all branches of chemistry, to expand its application areas.

CONSENT FOR PUBLICATION

Not applicable.

CONFLICT OF INTEREST

The author declares no conflict of interest, financial or otherwise.

ACKNOWLEDGEMENTS

F.B. Erim would like to commemorate the late Prof. J.C. Kraak, who introduced her to CE for the first time in 1993, and to thank Istanbul Technical University, which gave her the opportunity to establish a CE research laboratory.

REFERENCES

[1] Tiselius, A. *The moving boundary method of studying the electrophoresis of proteins,* **1930**.

[2] Jorgenson, J.W.; Lukacs, K.D. Zone electrophoresis in open-tubular glass capillaries. *Anal. Chem.,* **1981**, *53*, 1298-1302.
[http://dx.doi.org/10.1021/ac00231a037]

[3] Harstad, R.K.; Johnson, A.C.; Weisenberger, M.M.; Bowser, M.T. Capillary Electrophoresis. *Anal. Chem.,* **2016**, *88*(1), 299-319.
[http://dx.doi.org/10.1021/acs.analchem.5b04125] [PMID: 26640960]

[4] Voeten, R.L.C.; Ventouri, I.K.; Haselberg, R.; Somsen, G.W. Capillary electrophoresis: trends and recent advances. *Anal. Chem.,* **2018**, *90*(3), 1464-1481.
[http://dx.doi.org/10.1021/acs.analchem.8b00015] [PMID: 29298038]

[5] Álvarez, G.; Montero, L.; Llorens, L.; Castro-Puyana, M.; Cifuentes, A. Recent advances in the application of capillary electromigration methods for food analysis and Foodomics. *Electrophoresis,* **2018**, *39*(1), 136-159.
[http://dx.doi.org/10.1002/elps.201700321] [PMID: 28975648]

[6] Terabe, S.; Otsuka, K.; Ando, T. Electrokinetic chromatography with micellar solution and open-tubular capillary. *Anal. Chem.,* **1985**, *57*, 834-841.
[http://dx.doi.org/10.1021/ac00281a014]

[7] Başkan, S.; Erim, F.B. NACE for the analysis of acrylamide in food. *Electrophoresis,* **2007**, *28*(22), 4108-4113.
[http://dx.doi.org/10.1002/elps.200700421] [PMID: 17960538]

[8] Heller, C. Capillary electrophoresis of proteins and nucleic-acids in gels and entangled polymer-solutions. *J. Chromatogr. A,* **1995**, *698*, 19-31.
[http://dx.doi.org/10.1016/0021-9673(94)01169-F]

[9] Mao, Z.; Chen, Z. Advances in capillary electro-chromatography. *J. Pharm. Anal.,* **2019**, *9*(4), 227-237.

[http://dx.doi.org/10.1016/j.jpha.2019.05.002] [PMID: 31452960]

[10] Malá, Z.; Gebauer, P. Recent progress in analytical capillary isotachophoresis. *Electrophoresis,* **2019**, *40*(1), 55-64.
[http://dx.doi.org/10.1002/elps.201800239] [PMID: 30039625]

[11] Righetti, P.G.; Sebastiano, R.; Citterio, A. Capillary electrophoresis and isoelectric focusing in peptide and protein analysis. *Proteomics,* **2013**, *13*(2), 325-340.
[http://dx.doi.org/10.1002/pmic.201200378] [PMID: 23180512]

[12] Kalaycıoğlu, Z.; Erim, F.B. Simultaneous determination of nitrate and nitrite in fish products with improved sensitivity by sample stacking-capillary electrophoresis. *Food Anal. Methods,* **2016**, *9*, 706-711.
[http://dx.doi.org/10.1007/s12161-015-0241-4]

[13] Šlampová, A.; Malá, Z.; Gebauer, P.; Boček, P. Recent progress of sample stacking in capillary electrophoresis (2014-2016). *Electrophoresis,* **2017**, *38*(1), 20-32.
[http://dx.doi.org/10.1002/elps.201600292] [PMID: 27456212]

[14] Šlampová, A.; Malá, Z.; Gebauer, P. Recent progress of sample stacking in capillary electrophoresis (2016-2018). *Electrophoresis,* **2019**, *40*(1), 40-54.
[http://dx.doi.org/10.1002/elps.201800261] [PMID: 30073675]

[15] Foret, F.; Fanali, S.; Ossicini, L.; Boček, P. Indirect photometric detection in capillary zone electrophoresis. *J. Chromatogr. A,* **1989**, *470*, 299-308.
[http://dx.doi.org/10.1016/S0021-9673(01)83558-5]

[16] Tezcan, F.; Gültekin-Özgüven, M.; Diken, T.; Özçelik, B.; Erim, F.B. Antioxidant activity and total phenolic, organic acid and sugar content in commercial pomegranate juices. *Food Chem.,* **2009**, *115*, 873-877.
[http://dx.doi.org/10.1016/j.foodchem.2008.12.103]

[17] Erim, F.B.; Xu, X.; Kraak, J.C. Application of micellar electrokinetic chromatography and indirect UV detection for the analysis of fatty-acids. *J. Chromatogr. A,* **1995**, *694*, 471-479.
[http://dx.doi.org/10.1016/0021-9673(94)01114-T]

[18] Hashemi, P.; Erim, F.B. Analysis of vitamin B2 in saffron stigmas (*Crocus sativus* L) by capillary electrophoresis coupled with laser-induced fluorescence detector. *Food Anal. Methods,* **2016**, *9*, 2395-2399.
[http://dx.doi.org/10.1007/s12161-016-0430-9]

[19] Kalaycıoğlu, Z.; Gazioğlu, I.; Erim, F.B. Comparison of antioxidant, anticholinesterase, and antidiabetic activities of three curcuminoids isolated from Curcuma longa L. *Nat. Prod. Res.,* **2017**, *31*(24), 2914-2917.
[http://dx.doi.org/10.1080/14786419.2017.1299727] [PMID: 28287280]

[20] Kalaycıoğlu, Z.; Hashemi, P.; Günaydın, K.; Erim, F.B. The sensitive capillary electrophoretic-LIF method for simultaneous determination of curcuminoids in turmeric by enhancing fluorescence intensities of molecules upon inclusion into (2-hydroxypropyl)-β-cyclodextrin. *Electrophoresis,* **2015**, *36*(20), 2516-2521.
[http://dx.doi.org/10.1002/elps.201500253] [PMID: 26178140]

[21] Başkan, S.; Tezcan, F.; Köse, S.; Öztekin, N.; Erim, F.B. Non-ionic micellar electrokinetic chromatography with laser-induced fluorescence: a new method tested with biogenic amines in brined and dry-salted fish. *Electrophoresis,* **2010**, *31*(13), 2174-2179.
[http://dx.doi.org/10.1002/elps.200900683] [PMID: 20533326]

[22] Týčová, A.; Ledvina, V.; Klepárník, K. Recent advances in CE-MS coupling: Instrumentation, methodology, and applications. *Electrophoresis,* **2017**, *38*(1), 115-134.
[http://dx.doi.org/10.1002/elps.201600366] [PMID: 27783411]

[23] Gomes, F.P.; Yates, J.R., III Recent trends of capillary electrophoresis-mass spectrometry in

proteomics research. *Mass Spectrom. Rev.,* **2019**, *38*(6), 445-460. [http://dx.doi.org/10.1002/mas.21599] [PMID: 31407381]

[24] Prange, A.; Pröfrock, D. Application of CE-ICP-MS and CE-ESI-MS in metalloproteomics: challenges, developments, and limitations. *Anal. Bioanal. Chem.,* **2005**, *383*(3), 372-389. [http://dx.doi.org/10.1007/s00216-005-3420-0] [PMID: 16158292]

[25] Zemann, A.J.; Schnell, E.; Volgger, D.; Bonn, G.K. Contactless conductivity detection for capillary electrophoresis. *Anal. Chem.,* **1998**, *70*(3), 563-567. [http://dx.doi.org/10.1021/ac9707592] [PMID: 21644754]

[26] Kubáň, P.; Hauser, P.C. 20th anniversary of axial capacitively coupled contactless conductivity detection in capillary electrophoresis. *TrAC-Trend. Anal. Chem.,* **2018**, *102*, 311-321.

[27] Adımcılar, V.; Öztekin, N.; Erim, F.B. A direct and sensitive analysis method for biogenic amines in dairy products by capillary electrophoresis coupled with contactless conductivity detection. *Food Anal. Methods,* **2018**, *11*, 1374-1379. [http://dx.doi.org/10.1007/s12161-017-1122-9]

[28] Kalaycıoğlu, Z.; Kaygusuz, H.; Döker, S.; Kolaylı, S.; Erim, F.B. Characterization of Turkish honeybee pollens by principal component analysis based on their individual organic acids, sugars, minerals, and antioxidant activities. *Lebensm. Wiss. Technol.,* **2017**, *84*, 402-408. [http://dx.doi.org/10.1016/j.lwt.2017.06.003]

[29] Öztekin, N.; Nutku, M.S.; Erim, F.B. Simultaneous determination of nitrite and nitrate in meat products and vegetables by capillary electrophoresis. *Food Chem.,* **2002**, *76*, 103-106. [http://dx.doi.org/10.1016/S0308-8146(01)00287-4]

[30] Öztekin, N.; Erim, F.B. Separation and direct UV detection of lanthanides complexed with cupferron by capillary electrophoresis. *J. Chromatogr. A,* **2000**, *895*(1-2), 263-268. [http://dx.doi.org/10.1016/S0021-9673(00)00589-6] [PMID: 11105870]

[31] Yelkenci, H.E.; Öztekin, N. Separation and sensitive detection of lanthanides by capillary electrophoresis and contactless conductivity detection. *J. Chromatogr. Sci.,* **2017**, *55*(4), 465-470. [PMID: 27988487]

[32] Štěpánová, S.; Kašička, V. Recent applications of capillary electromigration methods to separation and analysis of proteins. *Anal. Chim. Acta,* **2016**, *933*, 23-42. [http://dx.doi.org/10.1016/j.aca.2016.06.006] [PMID: 27496994]

[33] Hajba, L.; Guttman, A. Recent advances in column coatings for capillary electrophoresis of proteins. *TrAC-Trend. Anal. Chem.,* **2017**, *90*, 38-44.

[34] Erim, F.B.; Cifuentes, A.; Poppe, H.; Kraak, J.C. Performance of a physically adsorbed high-molecular-mass polyethyleneimine layer as coating for the separation of basic-proteins and peptides by capillary electrophoresis. *J. Chromatogr. A,* **1995**, *708*, 356-361. [http://dx.doi.org/10.1016/0021-9673(95)00394-3]

[35] Chankvetadze, B. Contemporary theory of enantioseparations in capillary electrophoresis. *J. Chromatogr. A,* **2018**, *1567*, 2-25. [http://dx.doi.org/10.1016/j.chroma.2018.07.041] [PMID: 30025609]

[36] Yu, R.B.; Quirino, J.P. Chiral selectors in capillary electrophoresis: trends during 2017-2018. *Molecules,* **2019**, *24*, 1-18. [http://dx.doi.org/10.3390/molecules24061135]

[37] Alvarez-Rivera, G.; Bueno, M.; Ballesteros-Vivas, D.; Cifuentes, A. Chiral analysis in food science. *TrAC-Trend. Anal. Chem.,* **2020**, *123*, 115761.

[38] Tezcan, F.; Uzaşçı, S.; Uyar, G.; Öztekin, N.; Erim, F.B. Determination of amino acids in pomegranate juices and fingerprint for adulteration with apple juices. *Food Chem.,* **2013**, *141*(2), 1187-1191. [http://dx.doi.org/10.1016/j.foodchem.2013.04.017] [PMID: 23790902]

CHAPTER 2

Sample Preparation in Capillary Electrophoresis for Food Analysis

Ling Xia[1], **Simin Huang**[1] and **Gongke Li**[1,*]

[1] *School of Chemistry, Sun Yat-sen University, Guangzhou, 510275, China*

Abstract: This chapter introduces sample preparation techniques in Capillary Electrophoresis (CE) for food analysis. Food sample preparation prior to CE analysis aims to transfer target analytes from random statuses in the original food matrix to highly ordered pre-detection statuses, which is an entropy reduction procedure and cannot happen spontaneously. Generally, this is a time-consuming, labor-intensive, and error-prone step in complex sample analysis, especially in food analysis. Nevertheless, to match the fast analysis nature of CE, food samples have to be prepared efficiently in a relatively short time. Therefore, many highly efficient and fast sample preparation techniques were applied in CE for food analysis, including phase separation, field-assisted extraction, membrane separation, chemical conversion, and online coupling of sample preparation/analysis techniques. The principles and operation of each of the above-listed sample preparation techniques and some application examples are shown in different sections.

Keywords: Chemical conversion, Field- assisted extraction, Food analysis, Membrane separation, Sample preparation, capillary electrophoresis, phase partition.

INTRODUCTION

Sample preparation is a critical step in capillary electrophoresis for food analysis, which affects sensitivity, selectivity, accuracy, and speed of analytical results. From an analytical science viewpoint, foodstuff is a type of sample with wide varieties and high complexity, which contain a large number of target analytes and in a very complex matrix [1]. Moreover, these target analytes may be extremely similar to each other but present a trace amount in the food sample matrix [2 - 4]. Before injecting foodstuff into capillary, the preparation process can isolate and/or preconcentrate target analytes from food matrix based on their different physical, chemical, and biological properties, making them suitable for

* **Corresponding author Gongke Li:** School of Chemistry, Sun Yat-sen University, Guangzhou, 510275, China; E-mail: cesgkl@mail.sysu.edu.cn

María Castro-Puyana, Miguel Herrero and María Luisa Marina (Eds.)
Capillary Electrophoresis in Food Analysis

electrophoretic analysis [5]. Sample preparation is a time-consuming process, since transferring analytes from random status in the original sample matrix to highly ordered pre- detection statuses is an entropy reduction procedure which cannot occur spontaneously [6, 7]. Nevertheless, to match the fast analysis nature of capillary electrophoresis (CE), food samples have to be efficiently prepared in a relatively short time [8]. Therefore, many high efficiency and fast sample preparation methods and techniques were applied in CE for food analysis as shown in Fig. (**1**), including phase separation, field-assisted extraction, membrane separation, chemical conversion, and online coupling techniques.

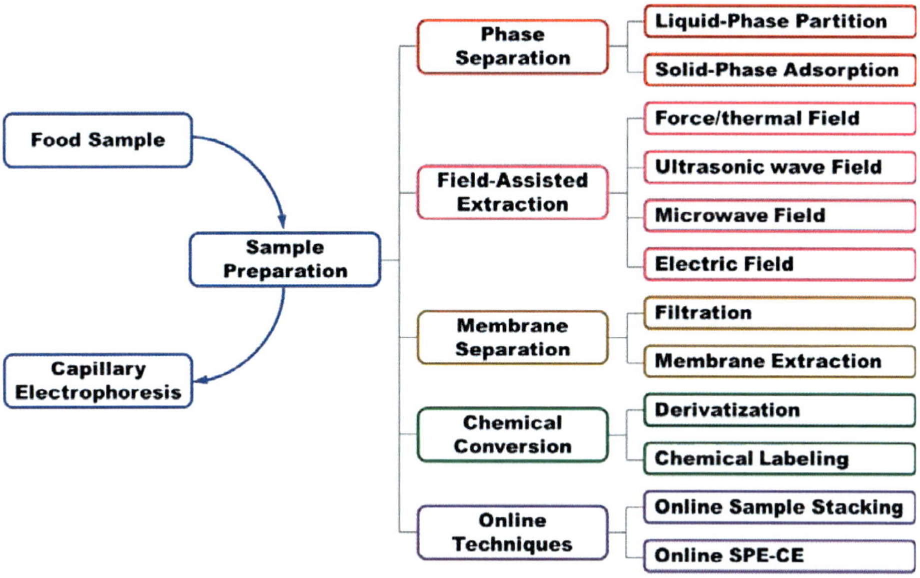

Fig. (1). Sample preparation in capillary electrophoresis for food analysis.

According to the morphology of sample, techniques to pretreat food are divided into solid and liquid sample preparation techniques. Solid sample preparation techniques involve Soxhlet extraction, Supercritical Fluid Extraction (SFE), Accelerated Solvent Extraction (ASE), Ultrasonic-Assisted Extraction (UAE), Microwave-Assisted Extraction (MAE), and so on. Their attributes are shown in Table **1**. Soxhlet extraction is a classic solid sample preparation technique. Due to its shortcomings of time-consuming, large solution consumption, low efficiency and reproducibility, this technique is being replaced by SFE, ASE, UAE, and MAE in food analysis.

Table 1. Sample preparation techniques for solid food samples.

Method	Principle	Advantage	Disadvantage
Soxhlet extraction	Continuous extraction using solvent reflux and siphon	Easy operation, low instrumental cost, deal with a large amount of sample	Time-consuming, large solvent consumption, may cause components decompose
Supercritical Fluid Extraction (SFE)	The supercritical fluid was used as solvent in extraction	Fast, low solvent consumption, high throughput, high selectivity, large concentration multiple, relatively low extraction temperature, suitable for thermo-sensitive analytes	High instrumental cost, complicated operation, may have the risk of sample lost
Accelerated Solvent Extraction (ASE)	Solid-liquid extraction in high-temperature and high-pressure field	Fast, low solvent consumption, easy operation, large concentration multiple	High instrumental cost, low extraction efficiency, not suit for thermally unstable substance
Ultrasonic-Assisted Extraction (UAE)	Ultrasonic field effects are used to accelerate extraction, such as cavitation and thermal effects	Fast, low cost, easy operation, deal with a large amount of sample	Manual operation, require filtration, extraction efficiency is affected by cavitation effect strength, the size and density of solid samples and solvent property
Microwave-Assisted Extraction (MAE)	Microwave field is applied to enhance extraction performance	Fast, low solvent consumption; homogeneous heating; easy operation	Require polar extraction solvent, require filtration and other auxiliary steps

Liquid sample preparation techniques involve Liquid-Liquid Extraction (LLE), Liquid-Phase Microextraction (LPME), Solid-Phase Extraction (SPE), and Solid-Phase Microextraction (SPME). Their attributes are shown in Table **2**. LLE and SPE are traditional liquid sample preparation techniques; they are widely used in food sample analysis. By device miniaturization, their mini versions of LPME and SPME prove to provide better sample preparation performances [9].

PHASE SEPARATION SAMPLE PREPARATION TECHNIQUES

Isolating target analytes from sample matrix through the phase separation process, including liquid-phase partition and solid-phase adsorption, is involved in almost all complicated sample pretreatment before instrumental analysis. Beyond eliminating matrix interference, phase separation sample preparation techniques also enable analytes preconcentration and background solution conversion to enhance the feasibility of electrophoretic analysis. LLE, LPME, SPE, and SPME are phase separation techniques frequently used for food sample preparation.

Table 2. Sample preparation techniques for liquid food samples.

Method	Principle	Advantage	Disadvantage
Liquid-Liquid Extraction (LLE)	Extraction and cleaning of extracts based on the partition of compounds in two immiscible liquid phases because of their solubility	Wide application, high technology maturity, deal with a relatively large amount of sample	Tedious operation, time-consuming, large solvent consumption, may have emulsification or precipitation issue
Liquid-Phase Microextraction (LPME)	Extraction and cleaning of extracts based on the partition of compounds in two immiscible liquid phases because of their solubility	Easy operation, low solvent consumption, high extraction efficiency	Long-term stability is not good enough
Solid-Phase Extraction (SPE)	Target analytes were selectively adsorbed onto solid adsorbents and then were desorbed from adsorbents through thermal desorption or solvent desorption	Easy analyte collection, good concentration effect, deal with low volume sample	Complex sample matrix may cause adsorbents damage, resulting column capacity and penetration volume decreasing, reducing extraction efficiency, recovery rate, and selectivity
Solid-Phase Microextraction (SPME)	Target analytes were selectively adsorbed onto solid adsorbents and then were desorbed from adsorbents through thermal desorption or solvent desorption	Minimum solvent consumption, easy operation, easy online integration with other technologies	Low extraction capacity, expensive, few optional adsorbent coating material, low reproducibility in complex sample matrix application

Liquid-phase partition

Liquid-phase partition sample preparation techniques aim to separate and concentrate target analytes based on the partition of compounds in two immiscible liquid phases. The sample preparation efficiency of these techniques is determined by the partition coefficient of target analytes, volume ratio of two liquid phases, extractant, operation temperature and duration. LLE and its mini version LPME are classical techniques for food sample preparation.

Liquid-liquid extraction

LLE is a mature extraction technique traditionally fulfilled in separatory funnels. Despite being over criticized for time-consuming, labor-intensive, and large solvent consumption, this conventional extraction technique can provide a more suitable prepared sample for CE analysis. For example, Nicolaou *et al.* [10]

compared different sample preparation methods for polyphenols capillary zone electrophoresis (CZE) analysis from wine sample. A simple LLE was carried out by the authors to extract polyphenols from wine using diethyl ether, and followed by solvent evaporation step for analytes preconcentration. The electropherograms of LLE prepared wine sample is shown in Fig. **(2)(1)**, which reveals overwhelming advantages of less matrix interference, higher signal intensity, and better separation resolution than the direct wine sample injection trial Fig. **(2)(2)**.

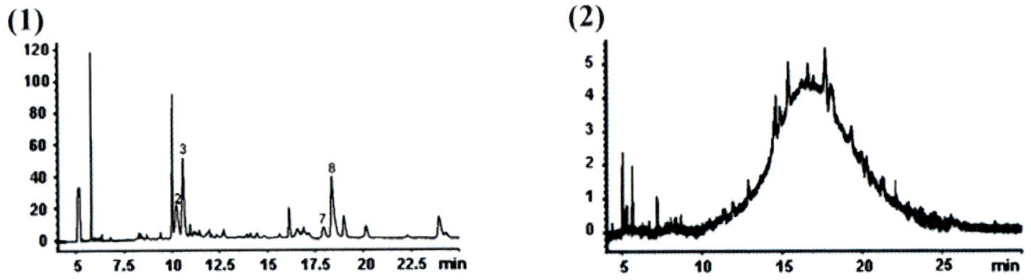

Fig. (2). Electropherograms of the wine samples with (1) and without (2) liquid-liquid extraction in sample preparation prior to capillary zone electrophoresis analysis. Reprinted with permission from [10].

Liquid-phase microextraction

To overcome the shortcoming of large organic solvent consumption in LLE, an approach using microscale extractant to accomplish extraction was carried out by Cantwell and Jeannot in 1996 [11]. This original liquid-phase microextraction was implemented in a hollow Teflon probe with a drop of organic solvent inside. The extraction principle of LPME is similar to LLE, which is based on the partition of compound in two immiscible liquid phases but employed only micro- or nano-liter organic solvent. Benefiting from solvent volume reduction, LPME can combine cleaning, extraction, and concentration all-in-one, as well as perform high efficiency, time-saving, and easy integration with downstream analysis procedures [12]. Various modes of LPME have been developed so far, including Single Droplet Microextraction (SDME), Hollow Fiber Liquid-Phase Microextraction (HF- LPME), and Dispersive Liquid-Liquid Microextraction (DLLME).

SDME is usually accomplished by hanging a small droplet organic solvent on the chromatographic injector needle [13, 14]. Once the extraction is completed, the organic droplet is sucked back into the injector for chromatographic injection. To facilitate CE analysis, headspace-single droplet microextraction (HS-SDME) was carried out by hanging a small droplet organic solvent on the inlet of a capillary. For example, Park *et al.* [15] hanged a drop of acceptor phase at the inlet of capillary 5 mm above the surface of red wine sample for phenolic compounds HS-SDME followed by CE quantitation. A similar sample preparation method was implemented for volatile and semi-volatile organic acids CZE analysis from soy sauce [16] and beverages [17]. The general HS-SDME and CE operation procedures are shown in Fig. (**3**). In HS-SDME step, volatile target analytes evaporated from sample solution and reached the acceptor phase at the inlet of a capillary. The extraction time of these HS-SDME methods can be reduced by applying a temperature controller, *e.g.*, a jacketed beaker. To improve the stability of hanged droplet, Chung's group [18] put the acceptor organic droplet inside the capillary. Before electrophoresis, an additional pre-injection step was carried out to prevent concentrated analytes from sample loss, and a 30 s waiting step was implemented to dissolve the acceptor phase. Overall, this sample preparation approach contributed up to 1100-fold sensitivity improvement to chlorophenols CE analysis from red wine, along with other advantages of time-saving and automation.

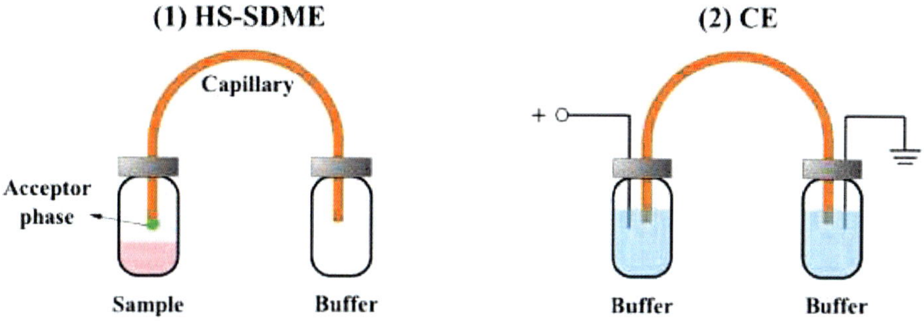

Fig. (3). General headspace-single droplet microextraction capillary electrophoresis procedures.

In 1999, Bjergaard *et al.* [19] completed LPME in a hollow porous fiber on top of the chromatographic injector needle. The introduction of hollow fiber further improves the performance of LPME: hollow fiber can isolate organic phase from the aqueous phase to avoid acceptor loss; porous fiber can help to exclude sample matrix and impurities outside the hollow fiber; the high porosity of porous fiber can increase the contact area between sample and acceptor, resulting in higher extraction efficiency; the disposable hollow porous protects from cross contamination. This HF-LPME sample preparation technique was involved in plasticizers [20] and fungicides [21] electrophoretic analysis in beverages, such as milk, juice, and carbonated drink.

DLLME is implemented by adding dispersant into LLE system. Numerous extractant microdroplets are formed and distributed in sample solution with the attendance of dispersant. Extraction process can be finished in couple of minutes due to the large contact area between sample and extractant and rapid partition equilibrium in microdroplets. Along with other advantages of easy operation and low organic solvent consumption, DLLME turns out to be an alternative approach for food sample preparation. For example, Shu *et al.* [22] applied DLLME for vanillin CE determination from milk powder. In other cases, DLLME were reported to prepare herbicides [23], pesticides [24], and mycotoxin [25] from honey, fruits, and beverages.

Supercritical Fluid Extraction

SFE is a special sample preparation technique using supercritical fluid as medium to extract or clean up target analytes. Supercritical carbon dioxide is the major extractant in SFE. Its advantages (time-saving, high recovery, good concentration effect, and eco-friendly) make SFE an alternate approach for heat-sensitive and easily oxidized compounds isolation and preconcentration from complicated foodstuffs. For example, Andrés *et al.* [26] executed supercritical fluid extraction as the clean-up and extraction for meat sample preparation, and simultaneously detect heterocyclic amines using capillary electrophoresis-fluorimetric detection. In general, SFE is a rapid, automatic, nontoxic, and easy operation liquid-phase partition sample preparation technique.

Solid-Phase Adsorption

Solid-phase adsorption sample preparation techniques perform more complete phase separation *via* solid adsorbents to isolate and concentrate target analytes than liquid-phase partition techniques. Adsorbent material is essential in these techniques and determines selectivity, efficiency, and speed. SPE and SPME are effective solid-phase adsorption sample preparation techniques in CE analysis of foodstuff.

Solid-phase Extraction

SPE is a classical liquid sample preparation technique, and has been applied in food, environmental, and biological analysis [27]. In SPE, target analytes are selectively adsorbed onto solid adsorbents, and then desorbed from adsorbents through thermal desorption or solvent desorption for downstream analysis. The principle of SPE is the distribution of analytes between solid phase (adsorbent) and liquid phase (solvent). SPE is a kind of preparative chromatographic technique, and its operation process is shown in Fig. (**4**).

Typically, SPE includes four steps: preparation of SPE column, loading sample, removing impurities, and elution and collection of analytes.

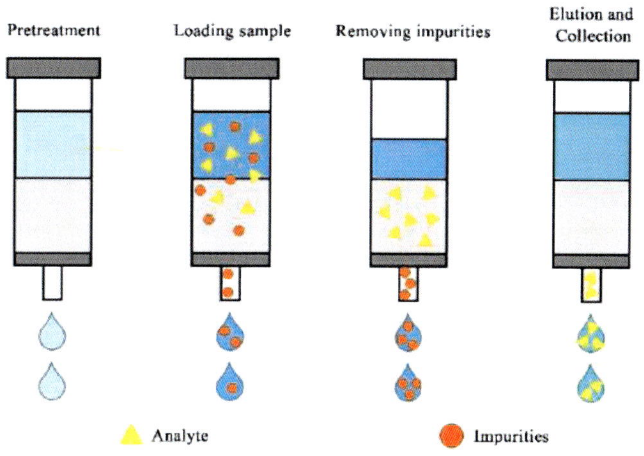

Fig. (4). Schematic of solid-phase extraction operation procedures.

Many SPE modes are working well in food sample preparation, such as Matrix Solid-Phase Dispersion (MSPD), Dispersion Solid-Phase Extraction (DSPE), and Magnetic Solid-Phase Extraction (MSPE). Nevertheless, the adsorbent materials used in SPE is the most significant factor of extraction preference [28 - 31]. The recently reported adsorbent materials and their SPE modes for food sample preparation before CE analysis are summarized in Table **3**. Many commercially available SPE adsorbents were used in food sample preparation before electrophoretic analysis, including poly(benzylpiperidone) polymer based Strata X, poly(divinylbenzene-co-N-vinylpyrrolidone) polymer based Oasis HLB, a hyper cross-linked polystyrene divinylbenzene polymer with a hydroxylated surface based ENV+Isolute, and C_{18}.

Table 3. Adsorbent materials used in SPE for food sample CE analysis.

SPE Method	Adsorbent	Analyte	Sample	Analysis Method	Ref.
SPE	^{C}18	Tetracycline antibiotics	Milk	CE-DAD	32
SPE	^{C}18	Lincosamide antibiotics	Chicken	MEKC-UV	33
SPE	Oasis HLB	Fluoroquinolones	Milk	CE-UV	34
SPE	Oasis HLB	Plant hormones	Honey	CZE-MS	35
SPE	C18, Oasis HLB, Cyanopropylene	Vitamins	Animal feeds	MEKC-UV	36
SPE	Strata X, Oasis HLB, ENV+Isolute	Quinolones antibiotics	Bovine plasma, porcine plasma	CE- DAD	37
SPE	Activated carbon	Plant hormones	Bean sprout, tomato, potato, cucumber, grain	MECK- UV	38
SPE	C18-carbon nanotubes	Sulphonamides	Milk	CE-DAD	39
SPE	Carboxyled single-walled carbon nanotubes	Amino acids	White tea	CE-ELSD	40
SPE	Antibodies immobilized silica gel	Citrinin	Red yeast rice, monascus colour	CZE- UV	41
SPE	Rose bengal imprinted polymer	Rose bengal	Brown sugar	CE-LIF	42
SPE	Trichlorfon-imprinted polymer	Trichlorfon	Cucumber, lettuce, radish	CE-UV	43
MSPD	Florisil, silica gel, C18	Tetracycline antibiotics	Milk	CZE-UV	44
DSPE	Dispersive powder polyamide	Synthetic food colorants	Preserved fruit	CE-C^{4}D	45
QuEChERS	Primary secondary amine, Graphitized carbon blank	Plant growth regulators	Orange, pear, kiwifruit, Watermelon, grape	CE- DAD	46
QuEChERS	^{C}18	Acrylamide	Potato products	CE-C^{4}D	47
QuEChERS	Primary secondary amine, ^{C}18	Isoflavones	Soy biscuits	CE-MS	48
MSPE	Magnetic molecularly, imprinted polymer	Trichlorfon	Cucumber, cauliflower	CE-DAD	49

MSPD was developed for drug residues analysis from animal tissues by Barker' group as early as 1989 [50]. More recently, this easy operation sample preparation technique has been applied in many solid sample pretreatments, including vegetables, fruits, and soil [51]. In MSPD, adsorbents are dispersed into solid, semi- solid, or viscous liquid sample and packed into SPE column for further extraction and elution. By completing sample homogenization, cell lysis, extraction, and purification procedures all-in-one, MSPD avoids unwanted sample loss in traditional sample preparation, and comes with other advantages of no extractant, fast and simplified operation. Mu *et al.* [44] applied MSPD for tetracycline resides CE determination from milk samples. The authors also compared the selectivity of three types of adsorbents in target tetracycline resides isolation, and concluded that $_{C}18$ was the best absorbent in the particular case.

Instead of packing in SPE columns, adsorbents in DSPE are dispersed into sample solution directly. After extraction, adsorbent with target analytes are collected by centrifugation or filtration for further elution. With the combination of LLE and DSPE, an advanced quick, easy, cheap, effective, rugged, and safe (QuEChERS) sample preparation technique was developed and became the standard for pesticide residues pretreatment in vegetables and fruits. In QuEChERS, acetonitrile is added into sample directly for extraction, then magnesium sulfate is introduced to dehydration, followed by DSPE using ethylenediamine-n-propylsilane as adsorbents. QuEChERS has been applied for acrylamide [47], isoflavones [48], and plant growth regulators [46] CE analysis from vegetables and fruits.

By modifying adsorbents onto the surface of magnetic beads, MSPE can accomplish extraction and elution in a short period of time. The good dispersion performance of magnetic adsorbents in sample solution improves the extraction efficiency and speed. The easy magnetic control of adsorbents makes the concentration of analytes convenient. For example, Li and coworkers [49] fabricated magnetic molecularly imprinted polymer adsorbents and implemented MSPE for trace trichlorfon residue CE analysis from vegetables, such as cucumber and cauliflower.

Solid-Phase Microextraction

SPME is a mini version of SPE, using very small amount of solid adsorbent. The very first SPME was carried out by Pawliszyn's group in 1989 on a fused-silica fiber with adsorption coating [52]. This sample preparation technique is integrating sampling, extraction, preconcentration, and injection together. The SPME device is similar to the chromatographic injector, but has an outside layer of adsorbent coating. To perform extraction, SPME fiber is immersed into the sample solution or exposed in the headspace of sample solution. Ferreira *et al.* [53] proposed a SPME-CE technique for oxytetracycline determination in milk products of different brands. Zhang *et al.* [54] applied SPME for residual sulfonamides CE analysis from milk samples. SPME advantages (easy operation, less sample and solvent consumption, and good reproducibility) makes it an optional technique for on-site sample preparation and analysis [55].

FIELD-ASSISTED EXTRACTION

When dealing with complicated food sample, especially in solid or semi-solid matrices, conventional phase separation techniques are not sufficient. The introduction of external assisted field can accelerate mass and heat transfer in sample preparation, resulting in better preparation efficiency and less time consumption. In food sample preparation for CE analysis, force, thermal, ultrasonic wave, microwave, and electrical fields are commonly used energy field to implement field-assisted extraction [56, 57].

Force/thermal-assisted Extraction

Force and thermal fields, including gravity, pressure, centrifugal force, and heat, are the most commonly used assisted fields to accelerate extraction. In many sample pretreatments, however, these techniques are routine. For example, centrifuge is involved in almost all food sample preparation before CE analysis. In CE analysis of antibiotic residue in raw bovine milk, Piñero *et al.* [34] employed centrifugation and SPE for milk sample defatting. The additional centrifuge after extraction also prevent separation capillary of CE from matrix contamination.

ASE or pressurized liquid extraction is a force- and thermal-assisted extraction technique [58]. By applying high pressure, the boiling point of solvent in ASE can be reduced. Therefore, the ASE can be implemented at a higher temperature than the solvent boiling point. Benefiting from both force field and thermal field, ASE can be accomplished fast, effectively, and with low solvent consumption. The extraction efficiency of ASE is related to pressure, temperature, solvent, and cycle number. Chang *et al.* [59] compared extraction performance of ASE and Soxhlet extraction for flavonoids pretreatment from plant. The author concluded that ASE not only reduces the time and solvent consumption, but also is highly automatic when coupling to micellar electrokinetic chromatography (MEKC) analysis. ASE was also reported to accelerate preparation process of seafood [60, 61], tea [62], pharmaceutical and beverages [63] samples.

A thermal field of far infrared irradiation was used by Gan *et al.* [64] to speed up extraction bioactive constituents from herbal drug prior to CE analysis.

The Far infrared-assisted extraction performed better extraction efficiency in 6 min, which is 1/30 of time consumption of conventional solvent extraction.

Ultrasonic-assisted Extraction

Applying ultrasonic wave field to liquid can cause the phenomenon of cavitation. Cavitation increases mass transfer and accelerates dissolution of target analytes. Moreover, combined with the accessory effects of ultrasound, such as mechanical vibration, stirring and thermal effects, the extraction efficiency of target analytes can be significantly improved.

UAE is one of the most frequently used field-assisted sample preparation techniques. The extraction efficiency of UAE is related to the frequency and intensity of applied ultrasonic wave, extraction solvent, sample particle size and extraction time. Compared with conventional phase separation extraction, UAE has the advantages of fast, high efficiency, low temperature, and general applicability. This technique has been reported to accelerate various extractions, including SPE, SPME, LLE, and LLME. By applying ultrasonic field-assisted extraction technique to LLME, Wu *et al.* [65] reported a two-step ultrasonic-assisted liquid-liquid microextraction to accomplish 5-hydroxymethylfurfural CE analysis from high ionic strength foodstuff sample of vinegar and soy sauce. The authors extracted the target analyte from high ionic strength sample matrix using organic accepted phase in the first extraction step, and then back-extracted the target analyte into aqueous solution for CE analysis. With the assistance of ultrasonic wave, this two-step ultrasonic-assisted liquid-liquid microextraction can be finished in 8 min (4 min for each step), which is comparable to CE analysis time. In many other cases, UAE has become routine preparation procedure in solid food sample, including vegetables [66 - 68], fruits [70, 71], plants [71 - 73], fish [74], bread [75], candy [76], milk powder [77, 78], tablets [79, 80], as well as in liquid food samples, such beverages [81, 82], wine [76] and milk [77, 78]. The UAE was usually carried out at a temperature range of 20-30°C for 8-120 min.

Beyond speeding up extraction, ultrasonic wave field-assisted techniques also can be used to accelerate analyte dissolution [83 - 86], speed up hydrolysis [87], liquid degassing [88, 89], and sample clean-up [90] in food sample preparation before injecting into CE apparatus.

Microwave-assisted Extraction

Mass transfer and heat exchange of extraction system can be improved by implementing extraction in microwave field. The MAE was introduced into food sample preparation by Ganzler in 1986 [91]. Compared with conventional solvent extraction, reflux extraction, and UAE, MAE has the advantages of being fast, effective, low cost, and eco-friendly. This advanced sample preparation technique can be coupled with various extraction methods, including SPE, SPME, LLE, and LLME, to enhance extraction efficiency. Moreover, extraction efficiency also relates to microwave radiation strength, solvent, temperature, duration, and sample size.

Zhao and coworkers [92] applied MAE for mercury extraction from dried fish for CE analysis. The mercury extraction was completed in 10 min with a 400 W microwave radiation at 70°C. Li *et al.* [93] compared several extraction methods, such as MAE, UAE, and reflux extraction, for tobacco residues preparation before CE analysis. In this case, MAE was finished at 400 W microwave radiation only in 4 min. The authors concluded that MAE is preferred as sample preparation technique in tobacco residues CE analysis to UAE and reflux extraction due to its advantages of fast and high efficiency. A similar method comparison was carried out by Li *et al.* [94] for tea sample preparation. Employing MAE, extraction of catechin and epicatechin from green tea can be accomplished in 1 min. The extraction time was extended to 60 min in UAE.

Electrical-assisted Extraction

Electric field is not only a separation field in CE analysis, but also an energy field, which can be used to accelerate sample preparation. (Fig. **5**) represents the schematics of two useful electric-assisted sample preparation techniques, electrical- assisted extraction and electromembrane extraction (EME). A preparative electrophoretic technique was reported by Quirino research group for the simultaneous extraction of anionic and cationic analytes before CE analysis [95, 96]. The schematic for this electrical-assisted extraction technique is shown in Fig. (**5**). Adopting electric field as the driving force, target analytes can be quickly extracted and concentrated in conductive hydrogel in micropipettes. This electrical assisted extraction technique presents overwhelming advantages in time and solvent consumption over SPE, and has been used in herbicides residue CE analysis from beer [95] and drinking water [96].

EME is another useful electrical field-assisted extraction technique. Anionic or cationic target analytes can transfer through supporting liquid membrane towards acceptor phased by the electrically driven force. The introduction of supporting liquid membrane in electrical-assisted extraction greatly improves its capability to eliminate matrix interference. Moreover, the method selectivity also close relatives to the supporting liquid membrane.

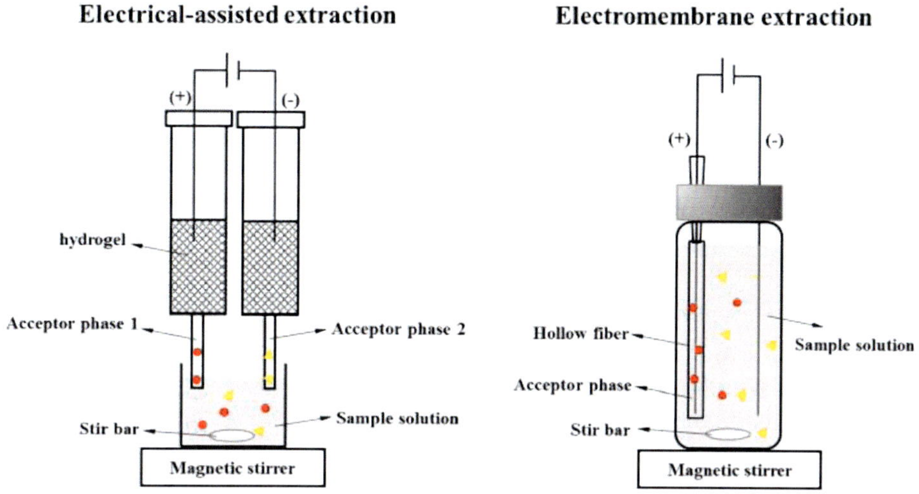

Fig. (5). Schematic of electrical-assisted extraction and electromembrane extraction.

Kubáň *et al.* [97] employed 1-octanol/bis(2- ethylhexyl) phosphonic acid penetrated polypropylene membrane in EME to selective extraction and concentrate heavy metal cations from powdered milk for CE analysis. This EME was completed in 5 min by applying a voltage of 75 V. This electrical-assisted extraction technique is also used in fungicides [98], plastic restricted substances [99], bromate [100], and perchlorate [101] CE analysis from liquid food samples.

MEMBRANE SEPARATION

Membrane separation is a fast, high efficiency, and low-cost separation technique. Filtration, membrane extraction, and dialysis are commonly used membrane separation methods in sample preparation. In food sample preparation, filtration can help eliminate interference from matrix, and membrane extraction can further concentrate target analytes.

By allowing smaller substances pass through porous material but excluding larger species in sample matrix, filtration is included in almost all sample preparation processes before injection into CE apparatus, especially in solid, turbid, and serous food sample pretreatment. For example, pulp can be removed from tomato juice through filtration in sample preparation of amino acid electrophoretic analysis [102]. On the other hand, clean-up running buffer and sample solution by filtration before injection can help protect separation capillary in CE from damage [103]. Various membrane materials can be used in filtration, including nylon [104 - 106], polypropylene [77, 107], acetic acid fiber [108], and polyvinylidene fluoride [109]. The standard pore size range of filtration membrane is 0.2-0.45 μm

[83, 110]. Beyond membrane material and pore size, filtration efficiency is also affected by sample particle shape and size, pressure gradient, and operation temperature.

Membrane extraction is applied to isolate target analytes from sample solution by their different solubility and diffusivity in membrane. Liquid membrane separation and gas membrane separation are two useful membrane extraction techniques in food sample preparation. These sample preparation techniques not only provide a cleaned sample, but also preconcentrated target analytes by adjusting the volume of acceptor solution. A gas membrane separation method, gas-diffusion microextraction, was reported by Lima *et al.* [111] for isolating aldehydes from beverages before CE analysis. In this case, separation, derivatization, and enrichment were accomplished in one-step: volatile aldehydes diffused through the gas membrane (a polytetrafluoroethylene membrane with 0.5 μm pore size) and dissolved in acceptor solution; the acceptor solution of 4-hydrazioinbenzoic acid is the derivatization agent for aldehydes; the derivatization reaction accelerated the aldehydes dissolution resulting in preconcentration of target analytes.

CHEMICAL CONVERSION

Chemical conversion is an important type of sample preparation technique which can transform unstable or nondetectable compounds to more feasible formation for subsequent sample preparation and analysis. In spite of being time-consuming, chemical conversion sample preparation techniques are commonly used in CE analysis. For example, after 12 h derivatization reaction using fluorescein isothiocyanate (FITC) as the derivatization agent, amino acids-FITC derivatives can be sensitively analyzed by CE with laser-induced fluorescence (LIF) detection [112].

Derivatization

By producing derivates, derivatization is the most commonly used chemical conversion sample preparation technique to improve the feasibility to be detected and quantified. In electrophoretic analysis, derivates are more stable and detectable to boost resolving power. The selection of derivatization reagent is the key of this technique, which is closely related to detector in CE. In food sample CE analysis, various derivatization reagents have been reported using different detectors, *e.g.* LIF detector, ultraviolet detector, diode array detector (DAD), and capacitively coupled contactless conductivity detector (C^4D). The derivatization agents used in food sample CE analysis are summarized in Table **4**.

Table 4. Derivatization used in food sample CE analysis.

Derivatization Agent	Derivatization time	Analyte	Sample	Analysis method	Analysis Time	Ref.
FITC	4 h	Biogenic amines	Wines	MEKC-LIF	9 min	112
FITC	4 h	Biogenic amines	Salt-dried-fish	MEKC-LIF	10 min	113
FITC	12 h	Peptide metabolite	Microbial fuel cell	MEKC-LIF	2.5 min	114
FITC	2 h	Ephedra alkaloids	Tablet formulations	MEKC-LIF	40 s	115
8-aminopyrene- 1,3,6-trisulfonic acid	18 h	Oligosaccharides	Human milk	xCGE- LIF	3 h	116
Fluorescein 5-thiosemicarbazide	3 h	Aldehydes	Drinking waters	CZE-LIF	16 min	117
4-hydrazinobenzoic acid	1 h	Aldehydes	Wine	CE-DAD	12 min	111
Cyclohexane-1,3-dione	1 h	Aldehydes	Sugar cane brandy	MEKC-DAD	20 min	118
9-fluorenylmethyl chloroformate	2 h	Azodicarbonamide	Flour	MEKC-DAD	25 min	119
Cysteine	10 min	Acrylamide	Potato	CE-C^4D	2 min	47

To accelerate derivatization involved in sample preparation, several strategies were employed. Yang *et al.* [47] introduce n-butylamine as a catalyst in acrylamide derivatization reaction using cysteine as derivatization agent. The derivatization can be completed in 10 min, and the acrylamide derivate has sensitive C^4D response. With the combination of derivatization and QuEChERS, acrylamide in potato products can be analyzed by CE-C^4D. Beyond acceleration derivatization reaction, integration of a multiple sample preparation process can also reduce time consumption. For example, Lima and coworkers [111] reported a gas-diffusion microextraction method for low-molecular aldehydes one-step derivatization and extraction from alcoholic beverages before CE-DAD analysis. The total sample preparation can be finished in 60 min.

Chemical Labeling

Through chemical labeling, special signal response of target analytes can be obtained. This technique can be used for labeling a certain atom, group of atoms, and functional groups of target analytes. Unlike derivatization, chemical labeling does not produce new compounds. In food safety testing, Jung and coworkers

[120] carried out a useful DNA labeling technique. The DNA fragments were printed onto capsules surface, and can be analyzed using polymerase chain reaction (PCR) coupled with CE for tracking the pharmaceutical supply chain.

Other Chemical Conversion Techniques

Beyond derivatization and chemical labeling, many chemical reactions, *e.g.*, digestion, hydrolysis, saponification, are also useful chemical conversion techniques in food sample preparation. To analyze metal ions in red and rice wines, Zhang and coworkers [121] applied wet digestion in sample preparation prior to CE analysis. Kikonorov and Nikitina [122] used dry ashing and oxidative degradation to pretreat plants for silicon CE determination. The hydrolysis/enzyme hydrolysis was utilized to isolate vitamins from proteins and phosphate groups [123], saccharides and sugar alcohols from fermentation broths [124], four purines from soybean milk [125], as well as amino acids in meat before CE injection. To improve the solubility in CE running buffer, saponification was taken to extract trans fatty acid from cheese [126], cake and biscuit samples [127], as well as omega-3 fatty acids from dietary supplements [128].

The combination technique of PCR and CE is a significant tool in genomics. In foodomics, PCR also reveals good potential. Gomes and coworkers [129] reported a PCR-CE technique for olive oils varietal identification. A similar PCR-CE method was presented by Chen *et al.* [130] for fish meat varietal identification. This technique is also applied for trace-level mycotoxin [131, 132] CE analysis.

ONLINE SAMPLE PREPARATION TECHNIQUES

The performance of food sample preparation can be further improved by the integration of multiple sample preparation steps and instrumental analysis through simultaneous, hyphenated, and online techniques. This strategy can enhance the automaticity of sample preparation, which consequently cuts down operation time, minimizes manual work, promotes assay accuracy and precision, and reduces solvent consumption. In food CE analysis, online sample preparation techniques perform excellent usability and practicality, mainly including online sample stacking and online SPE-CE techniques.

Online Sample Stacking

Online sample stacking is a type of electrophoretic preconcentration technique prior to CE analysis. When passing from high electric field to lower one, velocity of ions will significantly decrease, resulting in the ions stacking at the junction of electric field. This stacking phenomenon can be used to concentrate target analyte and shorten the length of the sample zone in CE analysis, consequently improving the sensitivity and electrophoretic separation efficiency. Generally, four approaches are commonly used to induce the nonuniform electric field and implement online sample stacking, including field amplification, sweeping, pH regulation, and isotachophoresis (ITP).

Field amplification stacking techniques have been reported for the CE analysis of polyphenol [133], insecticide residues [134], vitamin [135], nitrate and nitrite ions [74] from rice [133], fruit juice [134], milk [135], fish [74]. Lin and coworkers [136] applied sweeping to red wine MECK analysis and improve detection sensitivity 1500 times. Dynamic pH junction sample stacking technique was used by Su *et al.* [137] to rapidly preconcentrate the enzyme cofactor in beer sample and reduce the limit of detection. Online transient-ITP preconcentration technique enabled the drugs [138, 139], bromate residual [140], amines [141], and paralytic toxins [142] electrophoretic analysis from beverages [138], drinking water [140, 141], and seafood [142].

Without instrument modification, online sample stacking is a type of simple and useful sample preparation technique for CE analysis, which accomplishes 10- to 1000-fold sensitivity enhancement. Further increase in the sample loading volume or employing multiple stacking techniques together [109], can enable to achieve larger concentration factor by online sample stacking techniques. However, when dealing with complex real samples, *e.g.* food, these techniques may suffer from matrix effect because the matrix is also concentrated.

Online SPE-CE

The online integration of SPE and CE capable of large volume sample loading and target analytes online preconcentrating, not only improve the detection sensitivity but also extend its application in complex real sample analysis. This integration can be realized using one single column or two functional columns (SPE column and CE capillary). As shown in Fig. (**6**), one column SPE-CE or inline SPE-CE have three fabrication modes: (1) coated open tubular capillary inner surface with SPE sorbent at one end of the CE capillary; (2) packed sorbent or monolithic material at one end of the CE capillary; (3) prepared a short SPE column and inserted into CE capillary. In inline SPE-CE, more than 10 μL sample can be loaded and the complete desorbed volume after SPE can be analyzed by CE. For example, Zhang *et al.* [143] coated one end of CE separation capillary with molecularly imprinted material to execute online SPE-CE analysis. Compared to directly CE, the 5 cm SPE coating contributed over 100-fold limit of detection improvement in bisphenol A CE analysis from beverage samples.

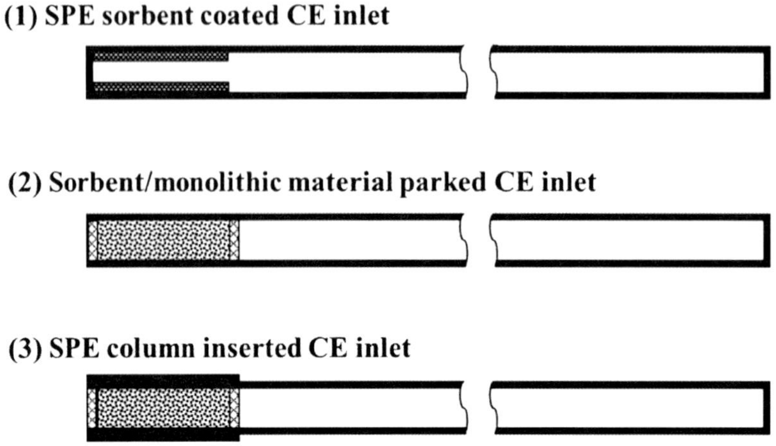

Fig. (6). Fabrication modes of one column SPE-CE.

Despite advantages of inline SPE-CE, using the same column for SPE and CE may induce separation column damage, because sample matrix will pass through column in sample preparation process. Online coupling SPE column and CE capillary through interface by flow injection and valve control, makes online SPE-CE more applicable to real sample analysis. This online technique has been applied to biological and environmental analysis [144, 145], and turns out to be an alternative approach for trace-level target analytes CE analysis from highly complexity food matrix.

Combination of Online and Offline Techniques

To further improve sample preparation performance, online integration strategy can be combined with offline techniques. To accelerate solid food sample preparation prior to CE analysis, Wang and coworkers [60] applied ASE and online sample stacking CE for drug residue analysis from seafood. Benefiting from field-assisted acceleration of ASE and preconcentration effect of online sample stacking, their approach realized rapid and sensitive analysis of residual quinolones and sulfonamides in shrimp and sardine. To minimize sample matrix interference, various SPE modes were offline coupling to online sample stacking CE for plant hormones [38], food additive [45] and drug residue [32] analysis from vegetable [38], grain [38], preserved fruit [45], and milk [32]. For example, Zheng *et al.* [38] adopted activated carbon as SPE adsorbent for indole-3-acetic acid and indole-3- butyric acid extraction from five vegetable and grain samples before online sample stacking CE.

CONCLUDING REMARKS

Foodstuff is a high complexity real sample. The goal of food CE analysis is to obtain more information in a relative short time. Beyond innovation of CE instruments, appropriate sample preparation is an alternative approach to address this goal. For example, phase separation and membrane separation sample preparation techniques can not only eliminate matrix interference, but also selectively isolate and concentrate target analytes; Field-assisted extraction techniques introduce additional energy into system and accelerate the sample preparation procedure; Chemical conversion techniques can transform unstable or nondetectable compounds to more feasible formation for CE analysis; The integration strategy of online coupling techniques allows automation of sample preparation and/or analysis. Various novel sample preparation techniques are continually being developed to improve food CE analysis, which tends to be faster, more efficient, higher throughput, more automated, more eco-friendly, and less costly.

CONSENT FOR PUBLICATION

Not applicable.

CONFLICT OF INTEREST

The author declares no conflict of interest, financial or otherwise.

ACKNOWLEDGEMENTS

This work was supported by the Research and Development Plan for Key Areas of Food Safety in Guangdong Province of China (No. 2019B020211001), National Key Research and Development Program of China (No. 2019YFC1606101), the National Natural Science Foundation of China (No. 21675178, 21804147 and 21976213), and the Guangdong Provincial Natural Science Foundation of China (No.2018A030313241), respectively.

REFERENCES

[1] Nielsen, S.S. *Food analysis,* 5th ed; Springer: Switzerland, **2019**. [chapter 1]

[2] Ntrallou, K.; Gika, H.; Tsochatzis, E. Analytical and sample preparation techniques for the determination of food colorants in food matrices. *Foods,* **2020**, *9*(1), 58. [http://dx.doi.org/10.3390/foods9010058] [PMID: 31936025]

[3] Pérez-Rodríguez, M.; Pellerano, R.G.; Pezza, L.; Pezza, H.R. An overview of the main foodstuff sample preparation technologies for tetracycline residue determination. *Talanta,* **2018**, *182*, 1-21. [http://dx.doi.org/10.1016/j.talanta.2018.01.058] [PMID: 29501128]

[4] Pereira, V.L.; Fernandes, J.O.; Cunha, S.C. Mycotoxins in cereals and related foodstuffs: A review on occurrence and recent methods of analysis. *Trends Food Sci. Technol.,* **2014**, *36*, 96-136.

[http://dx.doi.org/10.1016/j.tifs.2014.01.005]

[5] Liu, Z. Modern separation science. *Beijing: Chemical industry press,* **2017 [chapter 7]**.

[6] Xia, L.; Yang, J.; Su, R.; Zhou, W.; Zhang, Y.; Zhong, Y.; Huang, S.; Chen, Y.; Li, G. Recent progress in fast sample preparation techniques. *Anal. Chem.,* **2020**, *92*(1), 34-48. [http://dx.doi.org/10.1021/acs.analchem.9b04735] [PMID: 31769653]

[7] Xia, L.; Li, Y.; Liu, Y.; Li, G.; Xiao, X. Recent advances in sample preparation techniques in China. *J. Sep. Sci.,* **2020**, *43*(1), 189-201. [http://dx.doi.org/10.1002/jssc.201900768] [PMID: 31587455]

[8] Chen, Y. *Capillary electrophoresis technology and application*; , **2017**. [chapter 4]

[9] Casado, N.; Morante-Zarcero, S.; Pérez-Quintanilla, D.; Câmara, J.; Sierra, I. Two novel strategies in food sample preparation for the analysis of dietary polyphenols: Micro-extraction techniques and new silica-based sorbent materials. *Trends Food Sci. Technol.,* **2020**, *98*, 167-180. [http://dx.doi.org/10.1016/j.tifs.2018.06.020]

[10] Nicolaou, I.N.; Kapnissi-Christodoulou, C.P. Analysis of polyphenols using capillary zone electrophoresis--determination of the most effective wine sample pre-treatment method. *Electrophoresis,* **2010**, *31*(23-24), 3895-3902. [http://dx.doi.org/10.1002/elps.201000316] [PMID: 21064139]

[11] Jeannot, M.A.; Cantwell, F.F. Solvent microextraction into a single drop. *Anal. Chem.,* **1996**, *68*(13), 2236-2240. [http://dx.doi.org/10.1021/ac960042z] [PMID: 21619310]

[12] Andrascíková, M.; Matisová, E.; Hrouzková, S. Liquid phase microextraction techniques as a sample preparation step for analysis of pesticide residues in food. *Separ. Purif. Rev.,* **2015**, *44*, 1-18. [http://dx.doi.org/10.1080/15422119.2013.872125]

[13] AlOthman, Z.A.; Dawod, M.; Kim, J.; Chung, D.S. Single-drop microextraction as a powerful pretreatment tool for capillary electrophoresis: A review. *Anal. Chim. Acta,* **2012**, *739*, 14-24. [http://dx.doi.org/10.1016/j.aca.2012.06.005] [PMID: 22819045]

[14] Tang, S.; Qi, T.; Ansah, P.; Fouemina, J.; Shen, W.; Basheer, C.; Lee, H. Single-drop microextraction. *TrAC. Trends Analyt. Chem.,* **2018**, *108*, 306-313. [http://dx.doi.org/10.1016/j.trac.2018.09.016]

[15] Park, S.T.; Kim, J.; Choi, K.; Lee, H.R.; Chung, D.S. Headspace-single drop microextraction with a commercial capillary electrophoresis instrument. *Electrophoresis,* **2012**, *33*(19-20), 2961-2968. [http://dx.doi.org/10.1002/elps.201200317] [PMID: 23015463]

[16] Ling, D.S.; Xie, H.Y.; He, Y.Z.; Gan, W.E.; Gao, Y. Determination of preservatives by integrative coupling method of headspace liquid-phase microextraction and capillary zone electrophoresis. *J. Chromatogr. A,* **2010**, *1217*(49), 7807-7811. [http://dx.doi.org/10.1016/j.chroma.2010.10.041] [PMID: 21035121]

[17] Yue, M.E.; Lin, Q.; Li, Q.; Xu, J.; Jiang, T.F. Determination of PAEs by integrative coupling method of headspace in-tube microextraction and reverse-flow micellar electrokinetic capillary chromatography. *Food Anal. Methods,* **2017**, *10*, 3565-3571. [http://dx.doi.org/10.1007/s12161-017-0917-z]

[18] Lee, H.R.; Cho, S.M.; Kim, J.; Chung, D.S. Novel and simple headspace in-tube microextraction coupled with capillary electrophoresis. *J. Chromatogr. A,* **2014**, *1346*, 117-122. [http://dx.doi.org/10.1016/j.chroma.2014.04.052] [PMID: 24811149]

[19] Pedersen-Bjergaard, S.; Rasmussen, K.E. Liquid-liquid-liquid microextraction for sample preparation of biological fluids prior to capillary electrophoresis. *Anal. Chem.,* **1999**, *71*(14), 2650-2656. [http://dx.doi.org/10.1021/ac990055n] [PMID: 10424162]

[20] Jiang, T.F.; Liang, T.T.; Yue, M.E.; Wang, Y.H.; Lv, Z.H. Surfactant-free microemulsion reinforced

hollow-fiber liquid-phase microextraction combined with micellar electrokinetic capillary chromatography for detection of phthalic acid esters in beverage and urine. *Food Anal. Methods,* **2016**, *9*, 7-15.
[http://dx.doi.org/10.1007/s12161-015-0145-3]

[21] Barahona, F.; Gjelstad, A.; Pedersen-Bjergaard, S.; Rasmussen, K.E. Hollow fiber-liquid-phase microextraction of fungicides from orange juices. *J. Chromatogr. A,* **2010**, *1217*(13), 1989-1994. [http://dx.doi.org/10.1016/j.chroma.2010.01.077] [PMID: 20181342]

[22] Shu, M.; Man, Y.; Ma, H.; Luan, F.; Liu, H.; Gao, Y. Determination of vanillin in milk powder by capillary electrophoresis combined with dispersive liquid-liquid microextraction. *Food Anal. Methods,* **2016**, *9*, 1706-1712.
[http://dx.doi.org/10.1007/s12161-015-0347-8]

[23] Yue, M.E.; Li, Q.; Xu, J.; Jiang, T.F. Salt de-emulsification dispersive liquid-liquid microextraction and back-extraction combined with sweeping micellar electrokinetic capillary chromatography for detection of triazine herbicides in honey. *Food Anal. Methods,* **2016**, *9*, 699-705.
[http://dx.doi.org/10.1007/s12161-015-0231-6]

[24] Zhang, S.; Li, C.; Song, S.; Feng, T.; Wang, C.; Wang, Z. Application of dispersive liquid-liquid microextraction combined with sweeping micellar electrokinetic chromatography for trace analysis of six carbamate pesticides in apples. *Anal. Methods,* **2010**, *2*, 54-62.
[http://dx.doi.org/10.1039/B9AY00115H]

[25] Víctor-Ortega, M.D.; Lara, F.J.; García-Campaña, A.M.; del Olmo-Iruela, M. Evaluation of dispersive liquid-liquid microextraction for the determination of patulin in apple juices using micellar electrokinetic capillary chromatography. *Food Control,* **2013**, *31*, 353-358.
[http://dx.doi.org/10.1016/j.foodcont.2012.11.003]

[26] De Andrés, F.; Zougagh, M.; Castañeda, G.; Ríos, A. Simultaneous determination of six non-polar heterocyclic amines in meat samples by supercritical fluid extraction-capillary electrophoresis under fluorimetric detection. *Electrophoresis,* **2010**, *31*(13), 2165-2173.
[http://dx.doi.org/10.1002/elps.201000080] [PMID: 20593391]

[27] Chen, J.; Duan, C.; Guan, Y. Sorptive extraction techniques in sample preparation for organophosphorus pesticides in complex matrices. *J. Chromatogr. B Analyt. Technol. Biomed. Life Sci.,* **2010**, *878*(17-18), 1216-1225.
[http://dx.doi.org/10.1016/j.jchromb.2010.02.031] [PMID: 20378426]

[28] Chen, Y.; Xia, L.; Liang, R.; Lu, Z.; Li, L.; Huo, B.; Li, G.; Hu, Y. Advanced materials for sample preparation in recent decade. *TrAC. Trends Analyt. Chem.,* **2019**, *120*, 115652.
[http://dx.doi.org/10.1016/j.trac.2019.115652]

[29] Namera, A.; Nakamoto, A.; Saito, T.; Miyazaki, S. Monolith as a new sample preparation material: recent devices and applications. *J. Sep. Sci.,* **2011**, *34*(8), 901-924.
[http://dx.doi.org/10.1002/jssc.201000795] [PMID: 21394910]

[30] Wang, T.; Ma, J.; Chen, Y.; Li, Y.; Zhang, L.; Zhang, Y. Analysis of melamine and analogs in complex matrices: Advances and trends. *J. Sep. Sci.,* **2017**, *40*(1), 170-182.
[http://dx.doi.org/10.1002/jssc.201600854] [PMID: 27723251]

[31] Ashley, J.; Shahbazi, M.A.; Kant, K.; Chidambara, V.A.; Wolff, A.; Bang, D.D.; Sun, Y. Molecularly imprinted polymers for sample preparation and biosensing in food analysis: Progress and perspectives. *Biosens. Bioelectron.,* **2017**, *91*, 606-615.
[http://dx.doi.org/10.1016/j.bios.2017.01.018] [PMID: 28103516]

[32] Islas, G.; Rodriguez, J.A.; Perez-Silva, I.; Miranda, J.M.; Ibarra, I.S. Solid-phase extraction and large-volume sample stacking-capillary electrophoresis for determination of tetracycline residues in milk. *J. Anal. Methods Chem.,* **2018**, *2018*, 5394527.
[http://dx.doi.org/10.1155/2018/5394527] [PMID: 29675286]

[33] Kowalski, P.; Konieczna, L.; Oledzka, I.; Plenis, A.; Baczek, T. Development and validation of

electromigration technique for the determination of lincomycin and clindamycin residues in poultry tissues. *Food Anal. Methods,* **2014**, *7*, 276-282.
[http://dx.doi.org/10.1007/s12161-013-9624-6]

[34] Piñero, M.Y.; Garrido-Delgado, R.; Bauza, R.; Arce, L.; Valcárcel, M. Easy sample treatment for the determination of enrofloxacin and ciprofloxacin residues in raw bovine milk by capillary electrophoresis. *Electrophoresis,* **2012**, *33*(19-20), 2978-2986.
[http://dx.doi.org/10.1002/elps.201200316] [PMID: 23018278]

[35] Rodríguez-Gonzalo, E.; Domínguez-Alvarez, J.; García-Gómez, D.; García-Jiménez, M.G.; Carabias-Martínez, R. Determination of endocrine disruptors in honey by CZE-MS using restricted access materials for matrix cleanup. *Electrophoresis,* **2010**, *31*(13), 2279-2288.
[http://dx.doi.org/10.1002/elps.200900715] [PMID: 20593405]

[36] Oledzka, I.; Kowalski, P.; Baczek, T.; Muszynska-Furas, B.; Paradziej-Lukowicz, J.; Taciak, M.; Pastuszewska, B. Determination of water soluble vitamins in laboratory animal feeds by micellar electrokinetic chromatography. *Anal. Lett.,* **2012**, *45*, 689-701.
[http://dx.doi.org/10.1080/00032719.2011.653903]

[37] Hermo, M.P.; Nemutlu, E.; Barbosa, J.; Barrón, D. Multiresidue determination of quinolones regulated by the European Union in bovine and porcine plasma. Application of chromatographic and capillary electrophoretic methodologies. *Biomed. Chromatogr.,* **2011**, *25*(5), 555-569.
[http://dx.doi.org/10.1002/bmc.1483] [PMID: 20641008]

[38] Zeng, Y.; Peng, X.; Wu, Y. Field-amplified sample injection and solid-phase extraction for sensitivity improvement of electrophoretic determination of indole-3-acetic acid and indole-3-butyric acid in food samples. *Food Anal. Methods,* **2018**, *11*, 1155-1162.
[http://dx.doi.org/10.1007/s12161-017-1091-z]

[39] Polo-Luque, M.L.; Simonet, B.M.; Valcárcel, M. Solid phase extraction-capillary electrophoresis determination of sulphonamide residues in milk samples by use of C18-carbon nanotubes as hybrid sorbent materials. *Analyst (Lond.),* **2013**, *138*(13), 3786-3791.
[http://dx.doi.org/10.1039/c3an00319a] [PMID: 23671902]

[40] Bouri, M.; Salghi, R.; Zougagh, M.; Ríos, A. Capillary electrophoresis coupled to evaporative light scattering detection for direct determination of underivatized amino acids: application to tea samples using carboxyled single-walled carbon nanotubes for sample preparation. *Electrophoresis,* **2013**, *34*(18), 2623-2631.
[http://dx.doi.org/10.1002/elps.201300145] [PMID: 23857376]

[41] Zhu, D.; Zhang, H.; Bing, X. Preparation of an immunoaffinity column for the clean-up of fermented food samples contaminated with citrinin. *Food Addit. Contam. Part A Chem. Anal. Control Expo. Risk Assess.,* **2013**, *30*(2), 389-394.
[http://dx.doi.org/10.1080/19440049.2012.743042] [PMID: 23157620]

[42] Yuan, K.; Wan, J.; Zhai, H.; Chen, Z.; Huang, L.; Su, Z. Sensitive determination of rose bengal in brown sugar by a molecularly imprinted solid-phase extraction monolithic capillary column coupled with capillary electrophoresis. *Anal. Methods,* **2015**, *7*, 8297-8303.
[http://dx.doi.org/10.1039/C5AY01838B]

[43] Zhao, T.; Gao, H.; Wang, X.; Zhang, L.; Qiao, X.; Xu, Z. Study on a molecularly imprinted solid-phase extraction coupled to capillary electrophoresis method for the determination of trace trichlorfon in vegetables. *Food Anal. Methods,* **2014**, *7*, 1159-1165.
[http://dx.doi.org/10.1007/s12161-013-9729-y]

[44] Mu, G.; Liu, H.; Xu, L.; Tian, L. Matrix Solid-Phase Dispersion Extraction and Capillary Electrophoresis Determination of Tetracycline Residues in Milk. *Food Anal. Methods,* **2012**, *5*, 148-153.
[http://dx.doi.org/10.1007/s12161-011-9225-1]

[45] Yi, J.; Zeng, L.; Wu, Q.; Yang, L.; Xie, T. Sensitive simultaneous determination of synthetic food

colorants in preserved fruit samples by capillary electrophoresis with contactless conductivity detection. *Food Anal. Methods,* **2018**, *11*, 1608-1618. [http://dx.doi.org/10.1007/s12161-017-1141-6]

[46] Chen, Y.; Wu, X.; Li, Y.; Yang, Y.; Yang, D.; Yin, S.; Liu, L.; Sun, C. Simultaneous determination of seven plant growth regulators in melons and fruits by modified QuEChERS coupled with capillary electrophoresis. *Food Anal. Methods,* **2018**, *11*, 2788-2798. [http://dx.doi.org/10.1007/s12161-018-1266-2]

[47] Yang, S.; Li, Y.; Li, F.; Yang, Z.; Quan, F.; Zhou, L.; Pu, Q. Thiol-ene click derivatization for the determination of acrylamide in potato products by capillary electrophoresis with capacitively coupled contactless conductivity detection. *J. Agric. Food Chem.,* **2019**, *67*(28), 8053-8060. [http://dx.doi.org/10.1021/acs.jafc.9b01525] [PMID: 31276393]

[48] Bustamante-Rangel, M.; Delgado-Zamarreño, M.; Pérez-Martín, L.; Carabias-Martínez, L. QuEChERS method for the extraction of isoflavones from soy-based foods before determination by capillary electrophoresis-electrospray ionization-mass spectrometry. *Microchem. J.,* **2013**, *108*, 203-209. [http://dx.doi.org/10.1016/j.microc.2012.10.023]

[49] Li, D.; Qiao, X.; Lu, J.; Xu, Z. Synthesis and evaluation of a magnetic molecularly imprinted polymer sorbent for determination of trace trichlorfon residue in vegetables by capillary electrophoresis. *Adv. Polym. Technol.,* **2018**, *37*, 968-976. [http://dx.doi.org/10.1002/adv.21745]

[50] Barker, S.A.; Long, A.R.; Short, C.R. Isolation of drug residues from tissues by solid phase dispersion. *J. Chromatogr. A,* **1989**, *475*, 353-361. [http://dx.doi.org/10.1016/S0021-9673(01)89689-8] [PMID: 2777960]

[51] Capriotti, A.L.; Cavaliere, C.; Foglia, P.; Samperi, R.; Stampachiacchiere, S.; Ventura, S.; Laganà, A. Recent advances and developments in matrix solid-phase dispersion. *TrAC. Trends Analyt. Chem.,* **2015**, *71*, 186-193. [http://dx.doi.org/10.1016/j.trac.2015.03.012]

[52] Arthur, C.; Pawliszyn, J. Solid phase microextraction with thermal desorption using fused silica optical fibers. *Anal. Chem.,* **1990**, *62*, 2145-2148. [http://dx.doi.org/10.1021/ac00218a019]

[53] Ferreira, T.A.; Flores-Aguilar, J.F.; Santos, E.M.; Rodriguez, J.A.; Ibarra, I.S. New poly(ionic liquid) based fiber for determination of oxytetracycline in milk samples by application of SPME-CE technique. *Molecules,* **2019**, *24*(3), 430. [http://dx.doi.org/10.3390/molecules24030430] [PMID: 30682882]

[54] Zhao, S.; Wang, H.T.; Li, K.; Zhang, J.; Wang, X.Y.; Guo, G.S. Fast determination of residual sulfonamides in milk by in-tube solid-phase microextraction coupled with capillary electrophoresis-laser induced fluorescence. *Chin. J. Anal. Chem.,* **2018**, *46*, e1810-e1816. [http://dx.doi.org/10.1016/S1872-2040(17)61076-4]

[55] Casado, N.; Gañán, J.; Morante-Zarcero, S.; Sierra, I. New advanced materials and sorbent-based microextraction techniques as strategies in sample preparation to improve the determination of natural toxins in food samples. *Molecules,* **2020**, *25*(3), 702. [http://dx.doi.org/10.3390/molecules25030702] [PMID: 32041287]

[56] He, Y.; Xiao, X.; Cheng, Y.; Li, G. Progress in field-assisted extraction and its application to solid sample analysis. *J. Sep. Sci.,* **2016**, *39*(1), 177-187. [http://dx.doi.org/10.1002/jssc.201500938] [PMID: 26390199]

[57] Wen, Y.; Li, J.; Ma, J.; Chen, L. Recent advances in enrichment techniques for trace analysis in capillary electrophoresis. *Electrophoresis,* **2012**, *33*(19-20), 2933-2952. [http://dx.doi.org/10.1002/elps.201200240] [PMID: 23019127]

[58] Mendiola, J.A.; Herrero, M.; Cifuentes, A.; Ibañez, E. Use of compressed fluids for sample

preparation: food applications. *J. Chromatogr. A,* **2007**, *1152*(1-2), 234-246.
[http://dx.doi.org/10.1016/j.chroma.2007.02.046] [PMID: 17353022]

[59] Chang, Y.Q.; Tan, S.N.; Yong, J.W.H.; Ge, L. Surfactant-assisted pressurized liquid extraction for determination of flavonoids from Costus speciosus by micellar electrokinetic chromatography. *J. Sep. Sci.,* **2011**, *34*(4), 462-468.
[http://dx.doi.org/10.1002/jssc.201000766] [PMID: 21254401]

[60] Wang, N.; Su, M.; Liang, S.; Sun, H. Sensitive residue analysis of quinolones and sulfonamides in aquatic product by capillary zone electrophoresis using large-volume sample stacking with polarity switching combined with accelerated solvent extraction. *Food Anal. Methods,* **2016**, *9*, 1020-1028.
[http://dx.doi.org/10.1007/s12161-015-0269-5]

[61] Sun, H.; Qi, H.; Li, H. Development of capillary electrophoretic method combined with accelerated solvent extraction for simultaneous determination of residual sulfonamides and their acetylated metabolites in aquatic products. *Food Anal. Methods,* **2013**, *6*, 1049-1055.
[http://dx.doi.org/10.1007/s12161-012-9509-0]

[62] Wang, N.; Su, M.; Liang, S.; Sun, H. Investigation of six bioactive anthraquinones in slimming tea by accelerated solvent extraction and high performance capillary electrophoresis with diode-array detection. *Food Chem.,* **2016**, *199*, 1-7.
[http://dx.doi.org/10.1016/j.foodchem.2015.11.083] [PMID: 26775937]

[63] Sarazin, C.; Delaunay, N.; Costanza, C.; Eudes, V.; Gareil, P. Application of a new capillary electrophoretic method for the determination of carbohydrates in forensic, pharmaceutical, and beverage samples. *Talanta,* **2012**, *99*, 202-206.
[http://dx.doi.org/10.1016/j.talanta.2012.05.040] [PMID: 22967542]

[64] Guan, Z.; Chen, Q.; Fu, Y.; Chen, G. Determination of bioactive constituents in Flos Sophorae Immaturus and Cortex Fraxini by capillary electrophoresis in combination with far infrared-assisted solvent extraction. *Food Chem.,* **2012**, *130*, 1122-1126.
[http://dx.doi.org/10.1016/j.foodchem.2011.08.018]

[65] Wu, J.H.; Feng, T.T.; Liang, X.; Yu, C.P.; Dong, L.; Du, M.; Tan, M.Q.; Xu, X.B. Determination of HMF in vinegar and soy sauce using two-step ultrasonic assisted liquid-liquid micro-extraction coupled with capillary electrophoresis-ultraviolet detection. *Food Anal. Methods,* **2018**, *11*, 479-485.
[http://dx.doi.org/10.1007/s12161-017-1018-8]

[66] Li, D.; Jiang, M.; Xu, L.; Qiao, X.; Xu, Z. Simultaneous determination of acephate and isocarbophos in vegetables by capillary electrophoresis using ionic liquid and sodium dodecyl sulfate as modifiers. *Food Anal. Methods,* **2017**, *10*, 3368-3374.
[http://dx.doi.org/10.1007/s12161-017-0897-z]

[67] Man, Y.; Shu, M.; Wang, D.; Luan, F.; Liu, H.; Gao, Y. Determination of 6-benzylaminopurine in bean sprouts by capillary electrophoresis compared with HPLC. *Food Anal. Methods,* **2016**, *9*, 3025-3031.
[http://dx.doi.org/10.1007/s12161-016-0496-4]

[68] Anubala, S.; Sekar, R.; Nagaiah, K. Development and validation of an analytical method for the separation and determination of major bioactive curcuminoids in Curcuma longa rhizomes and herbal products using non-aqueous capillary electrophoresis. *Talanta,* **2014**, *123*, 10-17.
[http://dx.doi.org/10.1016/j.talanta.2014.01.017] [PMID: 24725858]

[69] Memon, A.; Solangi, A.; Memon, S.; Mallah, A.; Memon, N.; Memon, A. Simultaneous determination of quercetin, rutin, naringin, and naringenin in different fruits by capillary zone electrophoresis. *Food Anal. Methods,* **2017**, *10*, 83-91.
[http://dx.doi.org/10.1007/s12161-016-0552-0]

[70] Magnuszewska, J.; Krogulec, T. Application of hot platinum microelectrodes for determination of flavonoids in flow injection analysis and capillary electrophoresis. *Anal. Chim. Acta,* **2013**, *786*, 39-46.
[http://dx.doi.org/10.1016/j.aca.2013.05.031] [PMID: 23790290]

[71] Guo, J.; Wang, M.; Guo, H.; Chang, R.; Yu, H.; Zhang, G.; Chen, A. Simultaneous separation and determination of seven isoflavones in Radix Puerariae by capillary electrophoresis with a dual cyclodextrin system. *Biomed. Chromatogr.*, **2019**, *33*(11), e4646. [http://dx.doi.org/10.1002/bmc.4646] [PMID: 31291685]

[72] Xiao, M.W.; Bai, X.L.; Liu, Y.M.; Yang, L.; Hu, Y.D.; Liao, X. Rapid quantification of aloinA and B in aloe plants and aloe-containing beverages, and pharmaceutical preparations by microchip capillary electrophoresis with laser induced fluorescence detection. *J. Sep. Sci.*, **2018**, *41*(19), 3772-3781. [http://dx.doi.org/10.1002/jssc.201800338] [PMID: 30152917]

[73] Rabanes, H.; Guidot, A.; Quirino, J. Micellar electrokinetic chromatography of the constituents in Philippine lagundi (Vitex negundo) herbal products. *Microchem. J.*, **2014**, *112*, 153-158. [http://dx.doi.org/10.1016/j.microc.2013.10.006]

[74] Kalaycıoğlu, Z.; Erim, F.B. Simultaneous determination of nitrate and nitrite in fish products with improved sensitivity by sample stacking-capillary electrophoresis. *Food Anal. Methods*, **2016**, *9*, 706-711. [http://dx.doi.org/10.1007/s12161-015-0241-4]

[75] de Jesus, F.F.S.; Coelho, M.G.; Pallos, L.D.; da Silva, J.A.F.; Daniel, D.; de Jesus, D.P. Simple and fast method for simultaneous determination of propionate and sorbate in bread by capillary electrophoresis with UV spectrophotometric detection. *J. Food Compos. Anal.*, **2018**, *72*, 93-96. [http://dx.doi.org/10.1016/j.jfca.2018.06.010]

[76] Chen, L.; Hu, J.; Zhang, W.; Zhang, J.; Guo, P.; Sun, C. Simultaneous determination of nine banned azo dyes in foodstuffs and beverages by high-performance capillary electrophoresis. *Food Anal. Methods*, **2015**, *8*, 1903-1910. [http://dx.doi.org/10.1007/s12161-014-0074-6]

[77] Lu, Y.; Wang, D.; Kong, C.; Jia, C.; Breadmore, M. Analysis of melamine in milk powder and liquid milk by capillary zone electrophoresis after electrokinetic supercharging. *Food Anal. Methods*, **2015**, *8*, 1356-1362. [http://dx.doi.org/10.1007/s12161-014-9987-3]

[78] Xia, J.; Zhou, N.; Liu, Y.; Chen, B.; Wu, Y.; Yao, S. Simultaneous determination of melamine and related compounds by capillary zone electrophoresis. *Food Control*, **2010**, *10*, 912-918. [http://dx.doi.org/10.1016/j.foodcont.2009.12.009]

[79] Široká, J.; Martincová, A.; Pospíšilová, M.; Polášek, M. Assay of citrus flavonoids, troxerutin, and ascorbic acid in food supplements and pharmaceuticals by capillary zone electrophoresis. *Food Anal. Methods*, **2013**, *6*, 1561-1567. [http://dx.doi.org/10.1007/s12161-013-9573-0]

[80] Akamatsu, S.; Mitsuhashi, T. Development of a simple analytical method using capillary electrophoresis-tandem mass spectrometry for product identification and simultaneous determination of free amino acids in dietary supplements containing royal jelly. *J. Food Compos. Anal.*, **2013**, *30*, 47-51. [http://dx.doi.org/10.1016/j.jfca.2013.01.006]

[81] Wu, Y.; Li, J.; Zhao, S.; Ding, X. Simultaneous determination of 11 food additives by micellar electrokinetic capillary chromatography. *Food Anal. Methods*, **2016**, *9*, 589-595. [http://dx.doi.org/10.1007/s12161-015-0225-4]

[82] Xu, J.J. Ye. L. H., Yi, L.; An, M. R.; Cao, J.; Peng, L. Q., Electrophoretic analysis of natural antioxidants in plant and beverage samples using dynamically coated capillaries with chitosan and multiwall carbon nanotubes. *Food Anal. Methods*, **2017**, *10*, 980-991. [http://dx.doi.org/10.1007/s12161-016-0642-z]

[83] Belal, F. Abd El-Razeq, S.; Fouad. M.; Zayed, S., Fouad, F., Simultaneous determination of five coccidiostats in veterinary powders, feed premixes, and baby food by micellar electrokinetic chromatography: application to chicken tissues and liver. *Food Anal. Methods*, **2018**, *11*, 3531-3541.

[http://dx.doi.org/10.1007/s12161-018-1330-y]

[84] Yaroshenko, I.; Kirsanov, D.; Kartsova, L.; Sidorova, A.; Sun, Q.; Wan, H.; He, Y.; Wang, P.; Legin, A. Exploring bitterness of traditional Chinese medicine samples by potentiometric electronic tongue and by capillary electrophoresis and liquid chromatography coupled to UV detection. *Talanta,* **2016**, *152*, 105-111.
[http://dx.doi.org/10.1016/j.talanta.2016.01.058] [PMID: 26992500]

[85] Cheng, H.; Han, C.; Xu, Z.; Liu, J.; Wang, Y. Sensitivity enhancement by field-amplified sample injection in interfacing microchip electrophoresis with inductively coupled plasma mass spectrometry for bromine speciation in bread. *Food Anal. Methods,* **2014**, *7*, 2153-2162.
[http://dx.doi.org/10.1007/s12161-014-9848-0]

[86] Acosta, G.; Arce, S.; Martínez, L.D.; Llabot, J.; Gomez, M.R. Monitoring of phenolic compounds for the quality control of Melissa officinalis products by capillary electrophoresis. *Phytochem. Anal.,* **2012**, *23*(2), 177-183.
[http://dx.doi.org/10.1002/pca.1340] [PMID: 21898627]

[87] Mu, G.; Luan, F.; Liu, H.; Gao, Y. Use of experimental design and artificial neural network in optimization of capillary electrophoresis for the determination of nicotinic acid and nicotinamide in food compared with high-performance liquid chromatography. *Food Anal. Methods,* **2013**, *6*, 191-200.
[http://dx.doi.org/10.1007/s12161-012-9429-z]

[88] Jiang, T.F.; Chong, L.; Yue, M.E.; Wang, Y.H.; Lv, Z.H. Separation and determination of carbohydrates in food samples by capillary electrophoresis using dynamically coating the capillary with indirect UV detection. *Food Anal. Methods,* **2015**, *8*, 2588-2594.
[http://dx.doi.org/10.1007/s12161-015-0157-z]

[89] Chen, M.L.; Zheng, M.M.; Feng, Y.Q. Preparation of organic-inorganic hybrid silica monolith with octyl and sulfonic acid groups for capillary electrochromatograhpy and application in determination of theophylline and caffeine in beverage. *J. Chromatogr. A,* **2010**, *1217*(21), 3547-3556.
[http://dx.doi.org/10.1016/j.chroma.2010.03.032] [PMID: 20409551]

[90] Sun, H.; Liu, N.; Wang, L.; Wu, Y. Effective separation and simultaneous detection of cyromazine and melamine in food by capillary electrophoresis. *Electrophoresis,* **2010**, *31*(13), 2236-2241.
[http://dx.doi.org/10.1002/elps.201000025] [PMID: 20593399]

[91] Ganzler, K.; Szinai, I.; Salgó, A. Effective sample preparation method for extracting biologically active compounds from different matrices by a microwave technique. *J. Chromatogr. A,* **1990**, *520*, 257-262.
[http://dx.doi.org/10.1016/0021-9673(90)85109-9] [PMID: 2086582]

[92] Zhao, Y.; Zheng, J.; Fang, L.; Lin, Q.; Wu, Y.; Xue, Z.; Fu, F. Speciation analysis of mercury in natural water and fish samples by using capillary electrophoresis-inductively coupled plasma mass spectrometry. *Talanta,* **2012**, *89*, 280-285.
[http://dx.doi.org/10.1016/j.talanta.2011.12.029] [PMID: 22284493]

[93] Li, Z.; Huang, D.; Tang, Z.; Deng, C.; Zhang, X. Fast determination of chlorogenic acid in tobacco residues using microwave-assisted extraction and capillary zone electrophoresis technique. *Talanta,* **2010**, *82*(4), 1181-1185.
[http://dx.doi.org/10.1016/j.talanta.2010.06.037] [PMID: 20801316]

[94] Li, Z.; Huang, D.; Tang, Z.; Deng, C. Microwave-assisted extraction followed by CE for determination of catechin and epicatechin in green tea. *J. Sep. Sci.,* **2010**, *33*(8), 1079-1084.
[http://dx.doi.org/10.1002/jssc.200900647] [PMID: 20175087]

[95] Wuethrich, A.; Haddad, P.R.; Quirino, J.P. Simultaneous electrophoretic concentration and separation of herbicides in beer prior to stacking capillary electrophoresis UV and liquid chromatography-mass spectrometry. *Electrophoresis,* **2016**, *37*(9), 1122-1128.
[http://dx.doi.org/10.1002/elps.201600050] [PMID: 26921124]

[96] Wuethrich, A.; Haddad, P.R.; Quirino, J.P. Green sample preparation for liquid chromatography and

capillary electrophoresis of anionic and cationic analytes. *Anal. Chem.*, **2015**, *87*(8), 4117-4123.
[http://dx.doi.org/10.1021/ac504765h] [PMID: 25831229]

[97] Kubáň, P.; Strieglerová, L.; Gebauer, P.; Boček, P. Electromembrane extraction of heavy metal cations followed by capillary electrophoresis with capacitively coupled contactless conductivity detection. *Electrophoresis,* **2011**, *32*(9), 1025-1032.
[http://dx.doi.org/10.1002/elps.201000462] [PMID: 21449072]

[98] Oliveira, A.M.; Loureiro, H.C.; de Jesus, F.F.S.; de Jesus, D.P. Electromembrane extraction and preconcentration of carbendazim and thiabendazole in water samples before capillary electrophoresis analysis. *J. Sep. Sci.*, **2017**, *40*(7), 1532-1539.
[http://dx.doi.org/10.1002/jssc.201601305] [PMID: 28165201]

[99] Liu, Y.; Guo, L.; Wang, Y.; Huang, F.; Shi, J.; Gao, G.; Wang, X.; Ye, J.; Chu, Q. Electromembrane extraction of diamine plastic restricted substances in soft drinks followed by capillary electrophoresis with contactless conductivity detection. *Food Chem.*, **2017**, *221*, 871-876.
[http://dx.doi.org/10.1016/j.foodchem.2016.11.084] [PMID: 27979287]

[100] Zhang, X.; Guo, L.; Zang, D.; Ge, X.; Ye, J.; Chu, Q. Sensitive determination of bromate in water samples by capillary electrophoresis coupled with electromembrane extraction. *Food Anal. Methods,* **2016**, *9*, 393-400.
[http://dx.doi.org/10.1007/s12161-015-0208-5]

[101] Kiplagat, I.K.; Doan, T.K.; Kubáň, P.; Boček, P. Trace determination of perchlorate using electromembrane extraction and capillary electrophoresis with capacitively coupled contactless conductivity detection. *Electrophoresis,* **2011**, *32*(21), 3008-3015.
[http://dx.doi.org/10.1002/elps.201100279] [PMID: 22002888]

[102] Zgola-Grześkowiak, A.; Grześkowiak, T. Determination of glutamic acid and aspartic acid in tomato juice by capillary isotachophoresis. *Int. J. Food Prop.*, **2012**, *15*, 628-637.
[http://dx.doi.org/10.1080/10942912.2010.494759]

[103] Toutounji, M.R.; Van Leeuwen, M.P.; Oliver, J.D.; Shrestha, A.K.; Castignolles, P.; Gaborieau, M. Quantification of sugars in breakfast cereals using capillary electrophoresis. *Carbohydr. Res.*, **2015**, *408*, 134-141.
[http://dx.doi.org/10.1016/j.carres.2015.03.008] [PMID: 25841355]

[104] Liu, Y.; Sun, G.; Wang, Y.; Yang, L.; Yang, F. Monitoring the quality consistency of Weibizhi tablets by micellar electrokinetic chromatography fingerprints combined with multivariate statistical analyses, the simple quantified ratio fingerprint method, and the fingerprint-efficacy relationship. *J. Sep. Sci.*, **2015**, *38*(12), 2174-2181.
[http://dx.doi.org/10.1002/jssc.201500145] [PMID: 25867134]

[105] Duan, L.P.; Ding, G.S.; Tang, A.N. Preparation of chitosan-modified silica nanoparticles and their applications in the separation of auxins by capillary electrophoresis. *J. Sep. Sci.*, **2015**, *38*(22), 3976-3982.
[http://dx.doi.org/10.1002/jssc.201500810] [PMID: 26377362]

[106] Liu, C.; Feng, X.; Qian, H.; Fang, G.; Wang, S. Determination of norfloxacin in food by capillary electrophoresis immunoassay with laser-induced fluorescence detector. *Food Anal. Methods,* **2015**, *8*, 596-603.
[http://dx.doi.org/10.1007/s12161-014-9936-1]

[107] Yang, L.; Xie, X.; Zhang, J.; Sun, G. Microemulsion electrokinetic chromatography in combination with chemometric methods to evaluate the holistic quality consistency and predict the antioxidant activity of ixeris sonchifolia (bunge) hance injection. *PLoS One,* **2016**, *11*(6), e0157601.
[http://dx.doi.org/10.1371/journal.pone.0157601] [PMID: 27336298]

[108] Keire, D.A.; Trehy, M.L.; Reepmeyer, J.C.; Kolinski, R.E.; Ye, W.; Dunn, J.; Westenberger, B.J.; Buhse, L.F. Analysis of crude heparin by (1)H NMR, capillary electrophoresis, and strong-anio--exchange-HPLC for contamination by over sulfated chondroitin sulfate. *J. Pharm. Biomed. Anal.,*

2010, *51*(4), 921-926.
[http://dx.doi.org/10.1016/j.jpba.2009.10.017] [PMID: 19959313]

[109] Xu, X.; Liu, L.; Jia, Z.; Shu, Y. Determination of enrofloxacin and ciprofloxacin in foods of animal origin by capillary electrophoresis with field amplified sample stacking-sweeping technique. *Food Chem.*, **2015**, *176*, 219-225.
[http://dx.doi.org/10.1016/j.foodchem.2014.12.054] [PMID: 25624227]

[110] Li, Y.Y.; Zhang, Q.F.; Sun, H.; Cheung, N.K.; Cheung, H.Y. Simultaneous determination of flavonoid analogs in Scutellariae Barbatae Herba by β-cyclodextrin and acetonitrile modified capillary zone electrophoresis. *Talanta*, **2013**, *105*, 393-402.
[http://dx.doi.org/10.1016/j.talanta.2012.10.064] [PMID: 23598036]

[111] de Lima, L.F.; Brandão, P.F.; Donegatti, T.A.; Ramos, R.M.; Gonçalves, L.M.; Cardoso, A.A.; Pereira, E.A.; Rodrigues, J.A. 4-hydrazinobenzoic acid as a derivatizing agent for aldehyde analysis by HPLC-UV and CE-DAD. *Talanta*, **2018**, *187*, 113-119.
[http://dx.doi.org/10.1016/j.talanta.2018.04.091] [PMID: 29853022]

[112] Wang, W.; Bai, R.; Cai, X.; Lin, P.; Ma, L. Separation and determination of peptide metabolite of Bacillus licheniformis in a microbial fuel cell by high-speed capillary micellar electrokinetic chromatography. *J. Sep. Sci.*, **2017**, *40*(22), 4446-4452.
[http://dx.doi.org/10.1002/jssc.201700656] [PMID: 28960829]

[113] Uzaşç, S.; Başkan, S.; Erim, F.B. Biogenic amines in wines and pomegranate molasses-a non-Ionic micellar electrokinetic chromatography assay with laser-induced fluorescence detection. *Food Anal. Methods*, **2012**, *5*, 104-108.
[http://dx.doi.org/10.1007/s12161-011-9220-6]

[114] Başkan, S.; Tezcan, F.; Köse, S.; Öztekin, N.; Erim, F.B. Non-ionic micellar electrokinetic chromatography with laser-induced fluorescence: a new method tested with biogenic amines in brined and dry-salted fish. *Electrophoresis*, **2010**, *31*(13), 2174-2179.
[http://dx.doi.org/10.1002/elps.200900683] [PMID: 20533326]

[115] Belder, D.; Tolba, K.; Nagl, S. Rapid quantitative determination of ephedra alkaloids in tablet formulations and human urine by microchip electrophoresis. *Electrophoresis*, **2011**, *32*(3-4), 440-447.
[http://dx.doi.org/10.1002/elps.201000476] [PMID: 21254134]

[116] Kottler, R.; Mank, M.; Hennig, R.; Müller-Werner, B.; Stahl, B.; Reichl, U.; Rapp, E. Development of a high-throughput glycoanalysis method for the characterization of oligosaccharides in human milk utilizing multiplexed capillary gel electrophoresis with laser-induced fluorescence detection. *Electrophoresis*, **2013**, *34*(16), 2323-2336.
[http://dx.doi.org/10.1002/elps.201300016] [PMID: 23716415]

[117] Baños, C.E.; Silva, M. Analysis of low-molecular mass aldehydes in drinking waters through capillary electrophoresis with laser-induced fluorescence detection. *Electrophoresis*, **2010**, *31*(12), 2028-2036.
[http://dx.doi.org/10.1002/elps.200900734] [PMID: 20564695]

[118] Donegatti, T.A.; Lobato, A.; Moreira Gonçalves, L.; Alves Pereira, E. Cyclohexane-1,3-dione as a derivatizing agent for the analysis of aldehydes by micelar electrokinetic chromatography with diode array detection. *Electrophoresis*, **2019**, *40*(22), 2929-2935.
[http://dx.doi.org/10.1002/elps.201900171] [PMID: 31471917]

[119] Chen, L.; Cui, H.; Dong, Y.; Guo, D.; He, Y.; Li, X. Simultaneous detection of azodicarbonamide and the metabolic product semicarbazide in flour by capillary electrophoresis. *Food Anal. Methods*, **2016**, *9*, 1106-1111.

[120] Jung, L.; Hogan, M.E.; Sun, Y.; Liang, B.M.; Hayward, J.A. Rapid authentication of pharmaceuticals *via* DNA tagging and field detection. *PLoS One*, **2019**, *14*(6), e0218314.
[http://dx.doi.org/10.1371/journal.pone.0218314] [PMID: 31194827]

[121] Zhang, C.; Zhang, Y.; Qin, W. Combination of acidfree open-vessel wet digestion and poly(amidoamine) dendrimer-enhanced capillary electrophoresis for determination of metal ions in

wines. *Food Anal. Methods,* **2014**, *7*, 165-171.
[http://dx.doi.org/10.1007/s12161-013-9614-8]

[122] Nikonorov, V.V.; Nikitina, T.G. Capillary electrophoretic determination of silicon in plants. *Microchem. J.,* **2016**, *127*, 7-10.
[http://dx.doi.org/10.1016/j.microc.2016.01.020]

[123] Mazina, J.; Gorbatsov, J. Sample preparation for CE-DAD analysis of the water soluble vitamins in food products. *Procedia Chem.,* **2010**, *2*, 46-53.
[http://dx.doi.org/10.1016/j.proche.2009.12.009]

[124] Oliver, J.D.; Sutton, A.T.; Karu, N.; Phillips, M.; Markham, J.; Peiris, P.; Hilder, E.F.; Castignolles, P. Simple and robust monitoring of ethanol fermentations by capillary electrophoresis. *Biotechnol. Appl. Biochem.,* **2015**, *62*(3), 329-342.
[http://dx.doi.org/10.1002/bab.1269] [PMID: 25040822]

[125] Mu, G.; Luan, F.; Xu, L.; Hu, F.; Liu, H.; Gao, Y. Determination of purines in soybean milk by capillary electrophoresis in comparison with high performance liquid chromatography. *Anal. Methods,* **2012**, *4*, 3386-3391.
[http://dx.doi.org/10.1039/c2ay25488c]

[126] Amorim, T.L.; Duarte, L.M.; Dos Santos, H.F.; de Oliveira, M.A.L. Screening method for simultaneous detection of elaidic and vaccenic trans fatty acid isomers by capillary zone electrophoresis. *Anal. Chim. Acta,* **2019**, *1048*, 212-220.
[http://dx.doi.org/10.1016/j.aca.2018.10.057] [PMID: 30598152]

[127] Prado, T.L.A.; Porto, B.L.S.; Oliveira, M.A.L. Method optimization for trans fatty acid determination by CZE-UV under direct detection with a simple sample preparation. *Anal. Methods,* **2017**, *9*, 958-965.
[http://dx.doi.org/10.1039/C6AY02906J]

[128] Amorim, T.L.; Duarte, L.M.; de la Fuente, M.A.; de Oliveira, M.A.L.; Gómez-Cortés, P. Fast capillary electrophoresis method for determination of docosahexaenoic and eicosapentaenoic acids in marine oils omega-3 supplements. *J. Chromatogr. A,* **2020**, *1613*, 460641.
[http://dx.doi.org/10.1016/j.chroma.2019.460641] [PMID: 31677769]

[129] Gomes, S.; Breia, R.; Carvalho, T.; Carnide, V.; Martins-Lopes, P. Microsatellite high-resolution melting (SSR-HRM) to track olive genotypes: from field to olive oil. *J. Food Sci.,* **2018**, *83*(10), 2415-2423.
[http://dx.doi.org/10.1111/1750-3841.14333] [PMID: 30350554]

[130] Chen, C.H.; Hsieh, C.H.; Hwang, D.F. PCR-RFLP analysis using capillary electrophoresis for species identification of Cyprinidae-related products. *Food Control,* **2013**, *33*, 477-483.
[http://dx.doi.org/10.1016/j.foodcont.2013.03.036]

[131] Luque, M.; Córdoba, J.; Rodríguez, A.; Núñez, F.; Andrade, M. Development of a PCR protocol to detect ochratoxin A producing moulds in food products. *Food Control,* **2013**, *29*, 270-278.
[http://dx.doi.org/10.1016/j.foodcont.2012.06.023]

[132] Rodríguez, A.; Luque, M.I.; Andrade, M.J.; Rodríguez, M.; Asensio, M.A.; Córdoba, J.J. Development of real-time PCR methods to quantify patulin-producing molds in food products. *Food Microbiol.,* **2011**, *28*(6), 1190-1199.
[http://dx.doi.org/10.1016/j.fm.2011.04.004] [PMID: 21645819]

[133] Parveen, S.; Memon, S.; Siyal, A.; Memon, N.; Khuhawar, M. Large-volume sample staking of rice polyphenols prior to their determination by non-aqueous capillary electrophoresis. *Food Anal. Methods,* **2016**, *9*, 2152-2160.
[http://dx.doi.org/10.1007/s12161-015-0394-1]

[134] Santalad, A.; Srijaranai, S.; Burakham, R. Reversed electrode polarity stacking sample preconcentration combined with micellar electrokinetic chromatography for the analysis of carbamate insecticide residues in fruit juices. *Food Anal. Methods,* **2012**, *5*, 96-103.

[http://dx.doi.org/10.1007/s12161-011-9219-z]

[135] Albishri, H.M.; Almalawi, A.M.; Alshitari, W. El-Hady, D. A.; Use of β-cyclodextrin inclusion concurrent with cationic surfactant shielding for the enhancement of ascorbic acid stability followed by ultra-high performance liquid chromatography and online preconcentration capillary electrophoresis. *J. Liq. Chromatogr. Relat. Technol.,* **2018**, *41*, 732-739. [http://dx.doi.org/10.1080/10826076.2018.1511996]

[136] Hsieh, M.C.; Lin, C.H. On-line identification of trans-resveratrol in red wine using a sweeping technique combined with capillary electrophoresis/77 K fluorescence spectroscopy. *Electrophoresis,* **2004**, *25*(4-5), 677-682. [http://dx.doi.org/10.1002/elps.200305587] [PMID: 14981696]

[137] Su, A.K.; Chang, Y.S.; Lin, C.H. Analysis of riboflavin in beer by capillary electrophoresis/blue light emitting diode (LED)-induced fluorescence detection combined with a dynamic pH junction technique. *Talanta,* **2004**, *64*(4), 970-974. [http://dx.doi.org/10.1016/j.talanta.2004.04.011] [PMID: 18969698]

[138] Mikuš, P.; Veizerová, L.; Piešťanský, J.; Maráková, K.; Havránek, E. On-line coupled capillary isotachophoresis-capillary zone electrophoresis in hydrodynamically closed separation system hyphenated with laser induced fluorescence detection. *Electrophoresis,* **2013**, *34*(8), 1223-1231. [http://dx.doi.org/10.1002/elps.201200556] [PMID: 23401242]

[139] Piešťanský, J.; Maráková, K.; Galba, J.; Kováč, A.; Mikuš, P. Comparison of hydrodynamically closed two-dimensional capillary electrophoresis coupled with ultraviolet detection and hydrodynamically open capillary electrophoresis hyphenated with mass spectrometry in the bioanalysis of varenicline. *J. Sep. Sci.,* **2017**, *40*(10), 2292-2303. [http://dx.doi.org/10.1002/jssc.201700098] [PMID: 28322496]

[140] Marák, J.; Staňová, A.; Vaváková, V.; Hrenáková, M.; Kaniansky, D. On-line capillary isotachophoresis-capillary zone electrophoresis analysis of bromate in drinking waters in an automated analyzer with coupled columns and photometric detection. *J. Chromatogr. A,* **2012**, *1267*, 252-258. [http://dx.doi.org/10.1016/j.chroma.2012.07.075] [PMID: 22889601]

[141] Malinina, Y.; Kamentsev, M.Y.; Moskvin, L.N.; Yakimova, N.M.; Kuchumova, I.D. Determination of alkyl- and alkanolamines in drinking and natural waters by capillary electrophoresis with isotachophoretic on-line preconcentration. *J. Anal. Chem.,* **2017**, *72*, 1239-1242. [http://dx.doi.org/10.1134/S1061934817120085]

[142] Abdul Keyon, A.S.; Guijt, R.M.; Bolch, C.J.S.; Breadmore, M.C. Transient isotachophoresis-capillary zone electrophoresis with contactless conductivity and ultraviolet detection for the analysis of paralytic shellfish toxins in mussel samples. *J. Chromatogr. A,* **2014**, *1364*, 295-302. [http://dx.doi.org/10.1016/j.chroma.2014.08.074] [PMID: 25223612]

[143] Zhang, X.F.; Zhu, D.; Huang, C.P.; Sun, Y.H.; Lee, Y.I. Sensitive detection of bisphenol A in complex samples by in-column molecularly imprinted solid-phase extraction coupled with capillary electrophoresis. *Microchem. J.,* **2015**, *121*, 1-5. [http://dx.doi.org/10.1016/j.microc.2015.01.012]

[144] Santaladchaiyakit, Y.; Srijaranai, S.; Burakham, R. Methodological aspects of sample preparation for the determination of carbamate residues: a review. *J. Sep. Sci.,* **2012**, *35*(18), 2373-2389. [http://dx.doi.org/10.1002/jssc.201200431] [PMID: 22997028]

[145] Ramautar, R.; Jong, G.J.; Somsen, G.W. Developments in coupled solid-phase extraction-capillary electrophoresis 2009-2011. *Electrophoresis,* **2012**, *33*(1), 243-250. [http://dx.doi.org/10.1002/elps.201100453] [PMID: 22125291]

CHAPTER 3

Recent Trends in the Analysis of Lipids, Carbohydrates, and Proteins in Food by Capillary Electrophoresis

Marcone Augusto Leal de Oliveira[1,*], Guilherme de Paula Campos[1], Jéssica Cordeiro Queiroz de Souza[1], Maria Patrícia do Nascimento[1], Nerilson Marques Lima[1], Olívia Brito de Oliveira Moreira[1], Paula Rocha Chellini[1] and **Tatiane Lima Amorim[1]**

[1] *Federal University of Juiz de Fora, Juiz de Fora, MG, Brazil*

Abstract: Highly selective and sensitive analytical methods are necessary for food analysis because diverse components can be found in this complex sample matrix, sometimes occurring at only trace levels. Besides, simple and cost-effective methods are needed to meet the requirements of governmental food standards organizations and industries. Capillary Electrophoresis (CE) is a technique that meets these requirements offering high-resolution separations and high-throughput. It only demands small amounts of samples and chemicals for experiments and its versatility due to the different separation modes possible and the combination with different detection systems, has favored its application to determine diverse compounds in food analysis. This chapter summarizes significant issues and challenges involved in the determination of lipids, carbohydrates, and proteins, as well as recent advances in the analysis of these food components by several CE modes and detection systems.

Keywords: Analytical methods, carbohydrates, Electromigration methods, Food quality control, Lipids, Proteins, Sample preparation protocols.

INTRODUCTION

It is essential to study food components in order to ensure the health and safety of consumers. Besides, the determination of certain food compounds is mandatory for compliance with legal standards, quality assurance, nutritional value purpose, and fraud evaluation.

* **Corresponding author Marcone Augusto Leal de Oliveira:** Federal University of Juiz de Fora, Juiz de Fora, MG, Brazil; E-mail: marcone.oliveira@ufjf.edu.br

María Castro-Puyana, Miguel Herrero and María Luisa Marina (Eds.)
Capillary Electrophoresis in Food Analysis

Food nutrition labels generally include information on the provided energy, lipids composition (mainly cholesterol, saturated and *trans* fats), and carbohydrate, sugars, protein and salt contents. Additionally, it is essential to inform in food labels the potentially allergenic nutrients, *e.g.*, gluten and lactose, for celiacs and lactose intolerant people, respectively.

When considering macronutrients which is the focus of this chapter, different approaches can be used to determine a variety of compounds within these classes. Capillary Electrophoresis (CE) is a technique that operates in liquid media employing capillary columns to perform separation and is widely used for macronutrient determination in food. Solvated ions, as well as ionized and neutral species, can be separated using the various separation CE modes, which make CE a very versatile technique.

This chapter describes the application of CE methods for lipids, carbohydrates, and proteins determination in foods over the past five years. The fundamentals of the CE modes and detection systems employed in methods will be discussed, and the main analytical methods conditions will be summarized in tables throughout the chapter. Furthermore, several references will be provided for readers who wish to go deeper into the topic addressed

LIPIDS

Lipids comprise naturally occurring organic compounds that are generally insoluble in water and soluble in organic substances, which present several physicochemical and biological properties [1]. They are essential for energetic, metabolic, and structural activities and play essential roles in nutrition and health [2]. Just like carbohydrates and proteins, lipids contents need to be accurately measured and informed in food packages [3].

Some lipids are considered essential for the organism, such as omega-3 and omega-6 fatty acids. They need to be ingested from the diet because they are not synthesized by the *de novo* synthesis route in the body. These fatty acids are required to maintain cell membranes structure, brain function, the transmission of nerve impulses, and metabolic processes [4]. However, there is considerable awareness that abnormal levels of specific lipids, such as cholesterol, saturated fatty acids, and *trans*-fatty acids, may cause cardiovascular diseases, among other diseases such as obesity and type 2 diabetes [5, 6].

The determination of lipids in food is very challenging due to several varieties of matrices, diverse composition of fatty acids, and a wide range of total fat contents.

However, CE presents some advantages to lipids analysis when compared to chromatographic techniques, *e.g.*, derivatization reactions and specific columns are generally not required, and faster analysis times can be achieved. Besides, the technique follows the green chemistry principles, as small amounts of samples, reagents, and solvents are needed to perform analysis, and less waste is generated [7].

According to Fahy *et al.* [8], lipids can be classified into eight main categories: fatty acids, glycerolipids, glycerophospholipids, sphingolipids, saccharolipids, polyketides, sterols, and prenols. In this section, several methods reported in the literature for the study of the different classes of lipids employing CE will be revised, approaching different CE modes and detection systems, in order to show recent advances in lipids analysis employing the technique. Besides, challenges involving CE analysis for lipids determination in food will be summarized.

Fatty acids are a diverse group of molecules synthesized by chain elongation of an acetyl-CoA primer with malonyl-CoA (or methylmalonyl-CoA) groups [8]. They consist of carbon, hydrogen, and oxygen, arranged as a linear carbon chain skeleton of diverse lengths with a carboxyl group at one end. Fatty acids can be saturated (no double bond), monounsaturated (one double bond), or polyunsaturated (two or more double bonds). In food, they mostly occur as triesters, linked to glycerol forming triacylglycerols [9]. In minor amounts, naturally occurring free fatty acids, such as Hydroxy, Acetylenic, Allenic, and Cumulenic, can be found [10].

Extensive uses of fatty acids have been reported due to their different structures, mainly in food and oleochemicals industries. Nine major commodity oils are tracked worldwide, including coconut, soybean, sunflower, palm, palm kernel, cottonseed, canola, peanut, and olive, that accounts for approximately 97% of total oil production. Additionally, fatty acids are being used to manufacture soaps, detergents, lubricants, coatings, cosmetics, and other products [3]. (Table **1**) shows the methods for fatty acids determination reported to different food matrices, disclosed in chronological order, considering the last five years. Analytes, sample preparation procedure, background electrolyte (BGE), and instrumental conditions are summarized, as well as the CE mode and detection system used to fatty acids separation and identification.

Table 1. Description of analytical methods developed for fatty acid determination employing capillary electrophoresis in the last five years, in chronological order.

Sample	Analytes	Sample preparation	CE Mode / Detection	Background Electrolyte	Instrumental Conditions	Ref
Edible oils	Saturated C6-C18, C18:1 9c, C18:1 9t C18:2 n-6	Dilution in BGE and methanol	CZE-C⁴D	39 mM Tris, 3 mM pelargonic acid, 30 mM Brij 35, 35% acetonitrile, 15% 2-propanol, 2.5% 1-octano, and 300 μM PAMAM G2	Ambient temperature, +15 kV, hydrostatic injection at a height difference of 35 cm for 20 s. Capillary: 55 cm, 50 μm i.d.	11
Vegetable oils	C16:0, C18:1 9c, C18:2 n-6	Dilution with ethanol	CZE-UV	15.0 mM NaH2PO4/Na2HPO4, 4.0 mM SDBS, 8.3 mM Brij 35, 45% acetonitrile, 2.1%	25°C, +19 kV, hydrodynamic injection: 12 mbar for 4 s. Capillary: 48.5 cm, 75 μm i.d.	12
Soybean oil	C18:1 9c	Dilution in BGE	CZE-C⁴D	Methanol/1-propanol (1:6 v/v) containing 4×10^{-2} mM of KOH and 10% (v/v) ethylene glycol	25 °C, -25 kV, hydrostatic injection at a height difference of 10 cm for 30 s. Capillary: 45 cm, 50 μm i. d.	13
Food containing hydrogenated vegetable fat	C18:1 9c, 18:1 9t, C18:2 n-6, C18:3 n-3	Lipid extraction by Hara and Radin's protocol followed by a saponification reaction	CZE-UV	12.0 mM of tetraborate buffer, 12.0 mM of Brij 35, 33% methanol and 17% acetonitrile	27 °C, +27 kV, hydrodynamic injection: 25 mbar for 5 s. Capillary: 48.5 cm, 50 μm i.d.	14
Chocolates	Saturated C14-C16, C18:1 c9 C18:2 n-6	Soxhlet extraction followed by a saponification reaction.	CZE-UV	15.0 mM NaH2PO4/Na2HPO4, 4.0 mM SDBS, 8.3 mM Brij35, 45% acetonitrile, 2.1% octanol	25 °C, +19 kV, hydrodynamic injection: 12 mbar for 4 s. Capillary: 48.5 cm, 75 μm i. d.	15

(Table 1) cont.....

Sample	Analytes	Sample preparation	CE Mode / Detection	Background Electrolyte	Instrumental Conditions	Ref
Milk	Saturated C14-C16, C18:1 9c, C18:2 n-6, C18:3 n-3	Hara and Radin's extraction followed by saponification reaction	CZE-UV	15.0 mM NaH2PO4/Na2HPO4, 4.0 mM SDBS, 8.3 mM Brij35, 45% acetonitrile, 2.1% octanol	25 °C, +19 kV, hydrodynamic injection: 12 mbar for 4 s. Capillary: 48.5 cm, 75 µm i. d.	16
Pequi fruit	C18:1 9c, C16:0, C18:2 n-6	Dilution in ethanol	CZE-UV	15.0 mM of NaH_2PO_4/Na_2HPO_4, 4.0 mM of SDBS, 8.0 mM of Brij 35, 43,5% acetonitrile and 2.1% octanol	25 °C, +19 kV, hydrodynamic injection: 12 mbar for 4 s. Capillary: 48.5 cm, 75 µm i. d.	17
Cheese and coffee	Saturated C2-C22	Dilution in water and acetonitrile, sonication, centrifugation, and filtration	CZE-MS	30 mM ammonium formate in 40% acetonitrile	30 °C, 30 kV, hydrodynamic injection: 50 mbar for 20 s. Capillary: 85 cm, 50 µm i.d.	18
Processed food containing hydrogenated vegetable fat	C18:1 9c, C18:1 9t, C18:2 n-6 C18:3 n-3	Saponification reaction	CZE-UV	12.0 mM of tetraborate buffer, 12.0 mM of Brij 35, 33% methanol and 17% acetonitrile	30 °C, +27 kV, hydrodynamic injection: 25 mbar for 5 s. Capillary: 33.5 cm, 50 µm i.d.	19
Spreadable Cheese	C18:1 9c, C18:1 9t, C18:1 11t	Saponification reaction	CD-CZE-UV	10 mM heptakis (2, 3, 6-tri-o-methyl)-beta cyclodextrin, 24 mM sodium tetraborate buffer solution, 45% methanol and 15% acetonitrile	19 °C, +19 kV, hydrodynamic injection: 25 mbar for 2 s. Capillary: 33.5 cm, 50 µm i.d.	20
Vegetable Oils Supplements	Saturated C14-C16, C18:1 c9, C18:2 n-6, C18:3 n-3	Saponification reaction	CZE-UV	15.0 mM NaH2PO4/Na2HPO4, 4.0 mM SDBS, 8.3 mM Brij35, 45% acetonitrile, 2.1% octanol	25 °C, +19 kV, hydrodynamic injection: 12 mbar for 4 s. Capillary: 48.5 cm, 75 µm i.d.	21

(Table 1) cont.....

Sample	Analytes	Sample preparation	CE Mode / Detection	Background Electrolyte	Instrumental Conditions	Ref
Pumpkin seed oil	Saturated C14-C16, C18:1 c9, C18:2 n-6	Saponification reaction	CZE-UV	15.0 mM NaH2PO4/Na2HPO4, 4.0 mM SDBS, 8.3 mM Brij35, 45% acetonitrile, 2.1% octanol	25 °C, +19 kV, hydrodynamic injection: 12 mbar for 4 s. Capillary: 48.5 cm, 75 µm i.d.	22
Hops	C16:0, C18:0, C18:1 c9, C18:2 n-6	Saponification reaction	CZE-UV	15.0 mM NaH2PO4/Na2HPO4, 4.0 mM SDBS, 8.3 mM Brij35, 45% acetonitrile, 2.1% octanol	25 °C, +19 kV, hydrodynamic injection: 12 mbar for 4 s. Capillary: 48.5 cm, 75 µm i.d.	23
Fish and Krill Oils	C20:5 n-3, C22:6 n-3	Saponification reaction	CZE-UV	30.0 mM of tetraborate buffer, 12.0 mM of Brij 35, 33% methanol and 17% acetonitrile	27 °C, +27 kV, hydrodynamic injection: 25 mbar for 5 s. Capillary: 33.5 cm, 50 µm i.d.	24

SDBS = sodium dodecylbenzene sulphonate, M = mol/L; i.d. = internal diameter

It can be observed that the determination of fatty acids in food samples has been performed mainly by Capillary Zone Electrophoresis (CZE). In this CE mode, the separation of compounds is based on the differentiated mobility in a BGE medium when a particular electric potential is applied, causing each zone to migrate independently [25]. Cyclodextrin Modified Capillary Zone Electrophoresis (CD-CZE) is a modification of the original CZE mode that consists of cyclodextrins' inclusion into the BGE. It is specially employed to achieve challenging isomeric and chiral separations [26]. Some helpful reviews, which provide useful information about the fatty acid determination by CE in detail, can be referred to further studies in this area [7, 27, 28]. Due to the wide variety of fatty acids found in food samples, several CE modes and detection approaches can be employed for fatty acids determination. The present section will place particular emphasis on the CZE and CD-CZE modes employed in the analytical methods described in the last five years for fatty acid determination in food matrices, and the different detection systems used.

Ultraviolet (UV) absorption detection, capacitively coupled contactless conductivity detection (C^4D), and mass spectrometry detection (MS) were the detection systems used to determine fatty acids in the analytical methods approached Table **1**. It is important to remark that BGE optimization depends on the detection system selected. When using UV or C^4D, a wide range of buffers solutions can be selected (pH > 7 are usually used, aiming the dissociation of carboxyl groups and the determination of analytes in the anionic form). However, when using MS detectors, BGE optimization is limited to the use of volatile buffers and other BGE components. The use of organic solvents is generally required in BGE due to the poor solubility of analytes in the aqueous buffer medium. Additionally, cyclodextrins or surfactants have been intensely explored in fatty acids separation, mainly when separating geometrical and/or positional isomers. Furthermore, in the specific case of saturated fatty acid determination employing the UV absorption detection system, the use of a chromophore agent is necessary to promote the indirect detection of analytes, since these compounds lack in chromophore groups [29]. Factorial design can be employed to BGE and instrumental conditions optimization, making it possible to evaluate the interactions of variables in order to obtain optimal results for separation, also employing a reduced number of experiments. Besides, the combination of CE with multivariate analysis is very promising for obtaining fast classificatory models. Some recent analytical methods approached this combination and will be discussed in this chapter. More information about the use of chemometric tools in CE methods optimization can be found in the following references [30, 31].

It can be observed in Table **1** that some methods were reported for the determination of both saturated and unsaturated fatty acids, employing indirect UV absorption detection [12, 16, 21 - 23, 32] or C^4D [11, 13]. On the other hand, when only unsaturated fatty acids were determined, direct UV absorption was employed [14, 19, 20, 24]. In the indirect UV absorption detection, a negative peak is observed when fatty acids cross the detection window. Sodium dodecylbenzene sulphonate (SDBS) has been the most frequent chromophore agent employed, although diethylbarbiturate and p-anisate have also been used in some studies [28]. In C^4D detection, the measurement of the conductance between two or four electrodes is carried out when an alternating current is applied. When the conductivity of the analyte is lower than the conductivity of the BGE, a negative peak is observed.

On the contrary, when the conductivity of analytes is higher than the conductivity of BGE, a positive peak is observed. CE-UV with direct detection of analytes was employed for the determination of *trans* fatty acids and omega-3 fatty acids. Analytical methods were described for the quantification of total C18:1 *trans* fatty acids [14, 19] using CZE, and a screening method employing CD-CZE for the identification of *trans* fatty acids positional isomers from industrial and natural sources was recently published [20].

Moreover, a method for the determination of major omega-3 fatty acids in marine oils (C20:5 and C22:6) was also described using CZE [24].

Fig. (**1A**) depicts an electropherogram where both saturated and unsaturated fatty acids were determined employing C^4D as the detection system. In this Figure, fatty acids were determined in edible oils [11]. Fig. (**1B**) illustrates an electropherogram where direct UV absorption detection was used for the determination of *trans* fatty acids in industrialized food containing hydrogenated vegetable fat [19]. Detailed information about *trans* fatty acids determination by CE and applications is given in the following reference [28].

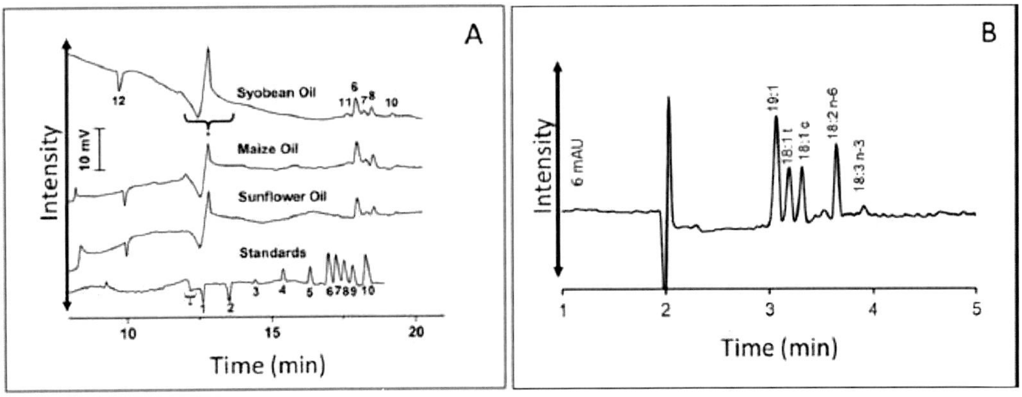

Fig. (1). Electropherograms showing **(A)** saturated and unsaturated fatty acids detection by capacitively coupled contactless conductivity and **(B)** unsaturated fatty acids detection by direct ultraviolet absorption. In A, 1) C6:0; 2) C8:0; 3) C10:0; 4) C12:0; 5) C14:0; 6) C18:2; 7) C16:0; 8) C18:1; 9) C18:1(t); 10) C18:0.

Finally, considering the MS detection system, only one method was developed lately to fatty acids determination in cheese and coffee [18]. In this method, saturated fatty acids, from C2-C22 were determined employing di-cationic ion-pairing reagents forming singly charged complexes with anionic fatty acids. Compounds were analyzed in the positive mode by electrospray ionization-MS. The combination of CE to MS detector has been developed and applied for fatty acids determination. However, CE and MS hyphenation is not straightforward, which justifies the low employability of this detection system in this determination. The review of Stolz *et al.* [33] covers the latest developments on CE and MS hyphenation, instrumental developments, and applications.

Conditions as in Table **1**. A - Adapted with permission from [11]; B - Reproduced with permission from [19].

Lipids possessing a glycerol group can be divided into two categories: Glycerolipids and glycerophospholipids. They are distinguished by the fact that glycerolipids have a glycerol molecule linked to one, two, or three fatty acids, while glycerophospholipids also have a phosphate or a phosphonate group esterified to glycerol [8].

Glycerolipids play structural and functional roles in bacterial, plant, and animal membranes. In foods, they generally occur as triacylglycerols, although small amounts of mono and diacylglycerols can also be found [34]. Triacylglycerols are the main source of energy for the body and take part in important metabolic processes that determine the rate of fatty acid oxidation, plasma levels of free fatty acids, biosynthesis of other lipids, and metabolic fate of lipoproteins [35].

No CE method was found in the literature for the determination of this lipid class recently. Considering that glycerolipids are neutral compounds, their determination by CE is not a simple task and could not be performed by CZE. However, more complex CE modes able to separate neutral compounds by the use of a selector which forms a pseudo stationary phase, such as Electrokinetic Chromatography (EKC) and Micellar Electrokinetic Chromatography (MEKC), or the hybrid. Capillary Electrochromatography (CEC) mode, which combines chromatography and electrophoretic principles, could be possible alternatives.

Glycerophospholipids provide sources of intracellular messengers that participate in the regulation of various key biological and physiological processes in the body [34], involving cell signaling and substrate transport. Besides, they interact with other lipids and proteins, forming bilayers of biological membranes and monolayers of lipoproteins. They are abundant membrane lipids of bacteria and mammalian cells and are used in a variety of foods as emulsifiers, wetting agents, and antioxidants.

High structural diversity is observed among glycerophospholipids. They are amphiphilic compounds which possess polar head groups with a glycerol backbone and nonpolar variable long-chain fatty acids. Glycerophospholipids are also called complex lipids because they yield more than three products on hydrolysis. Only one method was developed to phospholipids determination in food samples recently, applied to olive fruit oil analyses [36]. Non-Aqueous Capillary Electrophoresis (NACE) with electrospray ionization MS detector was used, due to its potential in the separation of water-soluble analytes and detection of compounds without chromophore groups. Table **2** shows method conditions.

Sphingolipids contain a long-chain base as their core structure [8]. They are important components of neuronal membranes, contribute to cellular diversity and functions, and are associated with several neurodegenerative disorders. In foods, they are studied due to their sensorial, rheological, and antioxidant properties. Besides, they have been considered as functional constituents, due to the beneficial effects in inhibiting colon carcinogenesis, reducing serum LDL cholesterol and elevating HDL cholesterol. Sphingolipids concentration in food vary considerably, from a few micromoles per kilogram in fruits to several millimoles per kilogram in some food matrices, such as dairy products, eggs, and soybeans [42]. Some methods were established for the determination of these compounds employing CE, but food samples were poorly studied. New recent developments in food analysis were not found in the literature, and older CE methods by CZE, MEKC, and CEC were revised by Montealegre *et al.* [43].

Table 2. Description of analytical methods developed in the last five years for glycerophospholipids, saccharolipids, and polyketides determination in food matrices.

Sample	Analytes	Sample preparation	CE Mode/Detection	Background Electrolyte	Instrumental Conditions	Ref
Glicerophospholipids						
Olive fruit and oil	Phosphatidylcholine, phosphatidylethanolamine, lysophosphatidylethanolamine, phosphatidylinositol, phosphatidic acid, lysophosphatidic acid, and phosphatidylglycerol	Extraction according to Folch and Stanley's procedure followed by glycerophospholipids extraction.	NACE-MS	100 mM ammonium acetate in 60:40 (v/v) methanol and acetonitrile with 0.5% acetic acid	25 °C, +25 kV, hydrodynamic injection: 50 mbar for 5s. Capillary: 60 cm, 50 µm i.d.	36
Saccharolipids						
Tap water	Lipopolysaccharides	HILIC µElution solid phase extraction with silica-based aminopropyl sorbent, followed by pre-column derivatization	CZE-LIF	50 mM sodium tetraborate buffer (pH 9.30)	25 °C, +30 kV, hydrodynamic injection: 0.5 psi for 5s. Capillary: 39 cm, 50 µm i.d.	37
Polyketides						
Cereals and edible oils	Aflatoxin B1 and Ochratoxin A	Aptamers were diluted in BGE. Samples were diluted with BGE.	MCE-LIF	10.00 mM Tris–HCl, pH 8.2, 10 mM EDTA with 1% Hydroxyethyl cellulose at pH 8.2	+21 kV, separation channel: 35 mm. Injection: 250 V for 30 s	38

(Table 2) cont.....

Sample	Analytes	Sample preparation	CE Mode/Detection	Background Electrolyte	Instrumental Conditions	Ref
Dairy cattle feed samples	Aflatoxins B1, B2, G1, and G2	Methanolic extraction followed by microextraction with 1- octanol and toluene	CZE-LIF	6 mM Na2B4O7, 10 mM Na_2HPO_4, and 30 mM sodium deoxycholate	+20 kV, siphoning injection for 10 s Capillary: 44 cm, 50 µm i.d.	39
Polyketides						
Tap water	Tetracycline, chlorotetracycline, oxytetracycline, and doxycycline	Filtration with 0.22 µm Millipore filter	CZE-UV	30.0 mM Na2HPO4, containing 1.5 mM Na2 EDTA (pH = 11) adjusted by 1.0 mol/L NaOH	+20 kV, injection at 5 psi for 3.8 min Capillary: 60 cm, 75 µm i.d.	40
Milk	Tilmicosin, erythromycin, clarithromycin, roxithromycin, and tylosin	Extraction with acetonitrile and hexane	MEKC-UV	60 mM phosphate, 5 mM sodium cholate, and 1.0 mM cetyltrimethyl ammonium bromide (pH = 7.0)	20 °C, +20 kV, electrokinetic injection: +10 kV for 120 s Capillary: 60 cm, 75 µm i.d.	41

M = mol/L; i.d. = internal diameter

In the saccharolipids class, a monosaccharide substitutes the glycerol backbone present in glycerolipids and glycerophospholipids classes. The most familiar saccharolipids are lipopolysaccharides and lipooligosaccharides, also known as endotoxins. These compounds are components of the outer membrane of Gram-negative bacteria and are released from the bacterial surface *via* outer membrane vesicles, which may be released from the bacterial cell surface, or by lysis and disintegration of the organism [44]. Endotoxins are directly involved in several bacterial diseases in humans. Then, the characterization of these compounds is of vital importance, since their physiological and pathophysiological effects rely on their chemical structure. In food, saccharolipids can be found in supplements and probiotic drugs used as immune-stimulants, as well in water and water-containing foods. The safety of food products containing endotoxins has been intensely questioned. However, when these saccharolipids are orally taken, they do not reach the bloodstream undigested.

Then, the oral uptake of probiotic products containing lipopolysaccharides or Gram-negative bacteria does not seems to pose a health risk, based on the available scientific evidence, as recently reviewed by Wassenaar and Zimmermann [45].

The determination of saccharolipids by CE is facilitated due to their unique molecular conformations and ionic charge distributions [46]. However, food samples were poorly explored by CE, and only one recent method was found for the determination of endotoxins in tap water employing CZE with laser-induced fluorescence (LIF) [37], which conditions are shown in Table **2**. Solid-phase extraction with silica-based aminopropyl sorbent was employed as sample treatment in order to eliminate the influence of matrix components and concentrate endotoxins before analysis. Furthermore, as lipopolysaccharides lack chromophore groups in their chemical structure and the sensitivity of indirect absorption UV detection is limited, an approach of derivatization with an amino-reactive fluorescent dye (fluorescein isothiocyanate) was performed and LIF detection was used since this detector is at least three orders of magnitude more sensitive than UV absorbance detection.

Polyketides are secondary metabolites produced in nature by plants, microbes, insects, and marine organisms [47]. They are important compounds, employed as antibiotics, anticancer agents, immunosuppressants, cholesterol-lowering agents, antifungal agents, and insecticides. In fungi, metabolites derived by the polyketide synthase pathway are commonly produced by the Ascomycetes and by the Fungi Imperfecti such as *Penicillium*, *Aspergillus*, *Fusarium*, and *Alternaria* species [48]. In plants, the most important families of natural secondary metabolites products are chalcone, stilbene, phloroglucinol, chromone, and curcuminoid [49].

Although polyketides are important compounds used as natural products and are medicinally important due to their pharmacological properties, they can also pose a serious health risk to humans and livestock. Aflatoxins are mainly produced by the *Aspergillus flavus* and *A parasiticus* fungi through a polyketide pathway. They are the major group of mycotoxins, a toxic secondary metabolite produced by fungi. Aflatoxins are susceptible to contaminate crops in the regions where high temperature and humidity are optimal for the growth of molds. Thus, the determination of these compounds in foodstuffs is of vital importance. Aflatoxin B1, aflatoxin B2, aflatoxin G1, and aflatoxin G2 are the four major species found, being Aflatoxin B1, the most toxic form. Aflatoxin ingestion can cause serious adverse health effects such as cancer, birth defects, and immunosuppression [50].

Aflatoxins are generally analyzed by MEKC since they are small neutral hydrophobic molecules. Besides, a LIF detector is generally used because these compounds are highly fluorescent substances. Recently, a MEKC-LIF method was proposed for the screening of the four major aflatoxins in dairy cattle feed samples employing the ionic surfactant sodium deoxycholate for micelle formation [39] and vortex assisted low-density solvent–microextraction to preconcentration of analytes. Furthermore, Microchip Capillary Electrophoresis (MCE) with LIF detector was also employed to Aflatoxin B1 and Ochratoxin A determination in foodstuff [38]. The sample preparation and instrumental conditions optimized in each method are shown in (Table **2**).

Methods for the determination of some macrolides and tetracycline antibiotics in foodstuff were also developed recently. These antibiotics exist as cationic, zwitterionic, or anionic species depending on pH. Wu and collaborators [40] developed a CZE-UV method for tetracycline antibiotics in tap water employing large volume sample stacking at pH 11. This procedure is used as a preconcentration method: the electrophoretic mobility of analytes changes suddenly at the interface between two solutions with different conductivity (sample and BGE), resulting in the stacking of analytes. It permitted to obtain preconcentration of analytes in the capillary, and concentrations of 10 ppb were detected, which is close to the level achieved by electro-chemiluminescence or LIF detectors. Moreover, Hong *et al.* [41] developed an analytical method to the determination of five macrolide antibiotic residues in milk employing MEKC-UV, also using large volume sample stacking of analytes. The analytical conditions of both methods are also presented in Table **2**.

Sterols and prenols share a common biosynthetic pathway *via* the polymerization of dimethylallyl pyrophosphate/isopentenyl pyrophosphate and are originated by carbocation-based condensations of isoprene units [8]. Sterols class includes cholesterol and derivatives, steroids, secosteroids, and bile acids and derivatives. These are polycyclic compounds, sharing a similar chemical structure of 17 carbon atoms arranged in a four-ring system. They can have different substituents. Hopanoids are also included in this class. They are natural pentacyclic found in bacteria and primitive organisms, which modify plasma membrane properties influencing its permeability, rigidity, and other characteristics, similarly as sterols do in eukaryotes [51]. Steroids, such as gonane (C17), estrane (C18), androstane (C19), pregnane (C21), cholane (C24), and cholestane (C27) can occur naturally or be processed chemically. They are pollutants observed in the aquatic environment. Then, their determination is essential in water and products containing water because the clean-up processes of drinking water are not precisely established for steroid removal. Besides, they can be detected in animal flesh and fluids. Steroids belong to the endocrine disruptors group and can cause cancer and developmental disorders in humans. They are usually detected employing direct UV absorption detection due to their aromaticity.

All the methods developed recently to sterol determination if food samples employed this detection system, as showed in Table **3**. Besides, MEKC mode was usually used due to analytes neutral characteristics.

El Fellah and collaborators lately proposed a method for steroid determination employing partial-filling micellar electrokinetic chromatography (PF-MEKC) in tap water [53]. In this system, the capillary is filled with the BGE containing only the buffer solution, and then a small plug of a micellar solution is placed before the sample solution. A dual system with polymeric C18 reverse phase columns was used for the extraction of non-ionic steroids. Amino polar phase columns were employed for the extraction of ionic steroids. Microemulsion Electrokinetic Chromatography (MEEKC) was also used to separate some sterols standards [54]. In this approach, the use of different vegetable oils in the preparation of microemulsions was proved, with a comparable performance of traditional emulsion. Additionally, an innovative approach employing molecularly imprinted polymer for the determination of steroids in goat milk by matrix solid-phase dispersion was demonstrated lately [52].

Table 3. Description of analytical methods developed in the last five years for sterols determination in food matrices and standards.

Sample	Analytes	Sample Preparation	CE Mode-Detection	Background Electrolyte	Instrumental Conditions	Ref
Goat milk	Testosterone, estrone, 17β-estradiol, 17α- ethinylestradiol, and progesterone	Molecularly imprinted polymer–matrix solid-phase dispersion	MEKC- UV	Borate buffer (25mM, pH 9.3), SDS (10mM) and acetonitrile (20%)	15 °C, +27 kV, hydrodynamic injection: 0.5 psi for 3s. Capillary: 60.2 cm, 75 µm i.d.	52

(Table 3) cont.....

Sample	Analytes	Sample Preparation	CE Mode-Detection	Background Electrolyte	Instrumental Conditions	Ref
Tap water	Progesterone, testosterone, androstenedione	Solid-phase extraction	PF-MEKC-UV	20 mM ammonium acetate (pH 9.68); Micelle solution: 1000 µL of 20 mM ammonium acetate, 440 µL of 100 mM SDS to 20 mM ammonium acetate 50 µL of 100 mM sodium taurocholate	25 °C, +25 kV, hydrodynamic injection: 34.5 mbar for 6 s. Capillary: 80.0 cm, 50 µm i. d.	53
Standards	Hydrocortisone, androstenedione, 17-α-hydroxyprogesterone, testosterone, 17-α-methyltestosterone and progesterone	Dilution of standards	MEEKC-UV	12 BGE were evaluated: different proportions of oils, 10 mM $Na_2B_4O_7$, SDS, 1-butanol and organic solvents	30 °C, +27 kV, hydrodynamic injection: 0.7 psi for 5 s. Capillary: 60.0 cm, 50 µm i. d.	54

SDS = sodium dodecyl sulphate; M = mol/L; i.d. = internal diameter

Prenol lipids are synthesized from the five carbon precursors isopentenyl diphosphate and dimethylallyl diphosphate that are produced mainly *via* the mevalonic acid pathway [55]. Isoprenoids, quinones, and hydroquinones, and polyprenols are included in this class [8]. Some fat-soluble vitamins, such as vitamins E and K and vitamin A and their carotenoid precursors (C20 isoprenoids), are also classified as prenols lipids. However, Vitamin D is classified inside the sterols class. Wang *et al.* [56] and Lu *et al.* [57] revised the literature about vitamin determination employing CE and these articles can be consulted for more information in this research area. Due to their neutral character and hydrophobicity, MEKC, MEEKC, or CEC are usually employed in the determination of fat-soluble vitamins but NACE has also been used. The use of multiple separation modes was already described for the simultaneous separation of water and fat-soluble vitamins.

Carotenoids, which constitute the predominant class of terpenes, are also highly hydrophobic and neutral molecules, and their separation is as difficult as fat-soluble vitamins. Separation of carotenoids by CE was also revised, and further information can be obtained in the following reference [58]. Recent articles determining prenol lipids in food samples were not found in the literature.

The articles approached in this section are only a demonstration of CE potential and versatility for lipids determination. For further information about lipids analysis employing CE and older applications, the following comprehensive reviews can be consulted [29, 59, 60].

CARBOHYDRATES

Carbohydrates are the most abundant biomolecules on earth, some of them, like sugar and starch, are the main elements of the human diet in many parts of the world, and their oxidation is the main energy production pathway in most non-photosynthetic cells. This group of compounds received this name because they often have the formula $C_x(H_2O)Y$. Carbohydrates are polyhydroxy aldehydes, polyhydroxy ketones, or substances that generate these compounds when hydrolyzed.

There are three main classes of carbohydrates: monosaccharides, disaccharides, and polysaccharides. Monosaccharides, or simple sugars, consist of a single polyhydroxy ketone or polyhydroxy aldehyde unit. The most abundant monosaccharide in nature is a 6-carbon sugar called D-glucose. Oligosaccharides consist of short chains of monosaccharide units, or residues, joined by characteristic bonds called glycosidic bonds. Disaccharides, with two units of monosaccharides such as sucrose (*e.g.,* cane sugar), belong to this category. Lastly, polysaccharides, such as cellulose, are polymers that contain more than 20 monosaccharide units [1, 2].

Table **4** shows all methods developed in the last five years, arranged in chronological order, for carbohydrates determination employing CE in several matrices such as milk, honey, juice, grapevines, and others. CZE, MEKC, and MCE were the CE modes employed in the methods summarized, and several detections systems were used, such as direct and indirect UV absorption detection, C^4D, and amperometric detection (AD).

Table 4. Description of analytical methods developed for carbohydrates determination employing capillary electrophoresis in the last five years.

Sample	Analytes	Sample Preparation	CE Mode/Detection	Background Electrolyte	Instrumental Conditions	Ref
Milk	3-sialyllactose, 6-sialyllactose, and disialyl-lacto-N-tetraose	Solubilization in 30% (v/v) methanol	MEKC-UV	200 mM NaH2PO4 (pH 7.05), 100 mM SDS and 45% methanol	25 °C, +30 kV, injection: 0.7 psi for 6 s Capillary: 50 cm, 50 µm i.d.	62
Pinot Noir grapes	Saccharose, galactose, glucose, rhamnose, fructose, xylose	Dilution in 20 mM NaOH	CZE-UV	130 mM NaOH and 36 mM Na_2HPO_4 (pH 12.6)	15 °C, 10-20 kV, injection: 0.5 psi for 2 s Capillary: 50 cm, 50 µm i.d.	63
Breakfast cereal	Sucrose, maltose, glucose, and fructose	Dilution in water	CZE-UV	130 mM NaOH and 36 mM Na_2HPO_4 (pH 12.6)	15 °C, +16 kV, injection: 34 mbar for 4s Capillary: 61.8 cm	64
Grapevine	Myo-inositol, galactinol, stachyose, raffinose, sucrose, lactose (I-STD), cellobiose, galactose, glucose, mannose, fructose/arabinose, xylose, ribose	Solubilization in water	CZE- UV	130 mM NaOH and 36 mM Na_2HPO_4 (pH 13)	15 °C, 16 kV, samples injection: 0.5 for 5 s, followed by BGE injection at 0.1 psi for 10 s. Capillary: 70 cm, 75 µm i.d.	65
Honey	11 aldoses (maltose, xylose, arabinose, ribose, glucose, rhamnose, fucose, galactose, mannose, glucuronic acid, and galacturonic acid)	Solubilization in water	CZE- UV	200 mM borate buffer with 4% methanol	25 °C, +20 kV, injection: 0.5 psi for 5 s. Fused silica capillary: 58.5 cm, 50 µm i.d.	66
Broccoli-based dietary supplements	Glucoraphanin, glucoiberin, glucoerucin e 4-hydroxyglucobrassicin	Extraction with 50% (v/v) followed by methanol centrifugation and filtration	CZE- UV	100mM Na2B4O7 with 1.0% β-CD	20 °C, + 30 kV, hydrodynamic injection: 0.5 psi for 20 s Capillary: 70 cm, 50 µm i.d.	67

(Table 4) cont.....

Sample	Analytes	Sample Preparation	CE Mode/Detection	Background Electrolyte	Instrumental Conditions	Ref
Dried syrup	D-psicose	Solubilization in water	CZE- UV	130 mM NaOH, 36 mM Na2HPO4 (pH 12.8)	18° C, +29 kV, sample injection: 0.5 psi for 5s followed by BGE injection at 0.1 psi for 10 s Capillary: 60 cm, 50 µm i.d.	68
Honey	Fructose, glucose, and sucrose	Dilution in water and filtration	CZE-UV	20 mM sorbic acid, 0.2 mM CTAB and 40 mM NaOH (pH 12.2)	25 °C, -25kV, Hydrodynamic short- end injection: 50 mbar for 3 s Capillary: 60.0 cm total length, 50 µm i. d.	69
Caprine whey	Lactose, galactose, and glucose	Ultrafiltration	CZE-UV	75 mM borate solution	60 °C, +10 kV, injection: 5 mbar for 3 s Capillary: 34 cm, 50 µm i.d.	70
Commercial juices	Cyclamate, glucose, fructose, sucrose, and sorbitol	Centrifugation followed by dilution in water	CZE-UV	10 mM PCD and 0.5 mM CTAB. (pH 12.6)	15 °C, +25 kV, hydrodynamic injection: 50 mbar for 6 s Capillary: 48,5 cm, 50 µm i.d	71
Milk, juice, honey, candied jujube	32 carbohydrates including mono-, di-, oligosaccharides, and sugar alcohols	Dilution in water and filtration	CZE-UV	20 mM sorbic acid, 0.005% (w/v) HDB, and 40 mM NaOH (pH 12.2)	25 °C, -15 kV, injection: 50 mbar for 5 s Capillary: 60 cm, 50 µm i.d.	72
Honey	Fructose, glucose, and sucrose/ amino acid proline	Dilution in water	CZE-UV	10 mM sodium benzoate (pH 12.4) and 1.5 mM CTAB	25 °C, + 25 kV, hydrodynamic injection: 0.5 psi for 3 s Capillary: 58 cm, 50 µm i.d.	73

(Table 4) cont.....

Sample	Analytes	Sample Preparation	CE Mode/Detection	Background Electrolyte	Instrumental Conditions	Ref
Grapevine	Glucose, fructose, and sucrose	Extraction with ethanolic solution, centrifugation, and filtration	CZE-UV	7 mM potassium sorbate, 50 mM NaOH (pH 12.8), 0.50 mM CTAB and 10% (v/v) methanol	20 ºC, -20 kV, injection: 30 mbar for 5 s Capillary: 55 cm, 75 µm i.d.	74
UHT milk	Lactulose	Solubilization in 50% (v/v) methanol	CZE-UV	20 mM PCD, 50 Mm NaOH (pH 12.5) and 0.5 mM CTAB	20 °C, -15 kV, hydrodynamic injection: 10 mbar for 10 s Capillary: 58.5 cm, 75 µm i.d.	75
Honey and milk	Mannitol, sucrose, lactose, glucose, and fructose	Solubilization and dilution in 75mM NaOH	CZE-AD	75 mM NaOH aqueous solution.	12 kV, electrokinetic injection: 12 kV for 6 s, Capillary 40 cm, 25 µm i.d.	76
Model mixture containing the analytes	Mannitol, sucrose, lactose, glucose, and fructose	Solubilization and dilution in 75mM NaOH	CZE-AD	75 mM NaOH aqueous solution.	9 kV, electrokinetic injection: 9 kV for 6 s, Capillary 40 cm, 25 µm i.d.	77
Honey	Glucose and fructose	Dilution in BGE	MCE-AD	20 mM NaOH and 10mM H_3BO_3 (pH 12)	+15 kV, injection: +15 kV for 5s Glass-plate microchip: 85 mm channel	78
Model mixture containing the analytes	Fructose, galactose, glucose, lactose, and sucrose	Solubilization in water and diluted in 10% BGE	MCE-C^{4}D	75 mM NaOH and 15 mM Na_3PO_4	+12 kV, injection: 0.5 kV for 10 s. Glass microchip: 85 mm separation channel	79

UHT = Ultra high temperature β-CD = β-cyclodextrin PCD = 2,6-pyridinedicarboxylic acid; CTAB = cetyltrimethylammonium bromide; HDB = hexadimethrine bromide; SDS = sodium dodecyl sulphate; M = mol/L; i.d. = internal diameter

The CZE mode was the most employed for carbohydrates determination in the last years [63 - 77]. This is due to the sufficient solubility of carbohydrates in the aqueous buffer medium. The determination of carbohydrates by CZE depends on the compound's degree of dissociation and, therefore, an electrolyte capable of ionizing carbohydrates must be used. According to the carbohydrates pKa values (usually above 11.9), the solution of the running buffer should be strongly alkaline (pH $\geq$ 12.0) to ensure analyte ionization. Contrary to lipids, where compounds separation always occurred with analytes as anions in the counter electroosmotic mode (cathodic way), carbohydrates separation was performed in some studies in the reversed electroosmotic flow (anodic way). In some methods, some cationic agents, such as hexadimethrine bromide (HDB) or cetyltrimethylammonium bromide (CTAB), were used as electroosmotic flow modifiers to reverse the electroosmotic flow [69, 72, 74, 75].

MEKC also demonstrated to be an excellent alternative to analyze compounds of difficult separation, although only one method employed this CE mode for carbohydrates separation [62]. Furthermore, the potential of miniaturized systems was also demonstrated in the analytical methods which employed MCE for carbohydrates separation [78, 79].

When considering detection systems, derivatization reactions during sample pretreatment may be required in systems using direct detection since carbohydrates lack a strong UV chromophore. However, the use of indirect absorbance detection offers an alternative without the need of derivatization steps [61]. Most methods developed employed UV detection [62 - 75] for carbohydrates identification. Nevertheless, the use of C^4D [79] and AD [76 - 78] were also demonstrated. These detection systems are very useful in the determination of non-UV-absorbing analytes, as they do not rely on the presence of chromophores, unlike spectrophotometric methods. However, commercial instrumentation for CE is generally not available coupled with these detection systems.

Among the methods which employed direct UV detection for carbohydrates determination, Biluca *et al.* [69], determined fructose, glucose, and sucrose in thirteen bee honey samples. Analyses were carried out by CZE-UV under 254 nm, before and after thermal treatment. The carbohydrates determination was performed, according to Rizelio *et al.* [80]. BGE was composed of sorbic acid and CTAB adjusted in pH 12.2 with NaOH, and an uncoated fused silica capillary with 60.0 cm total length was employed. Lu *et al.* [66] also analyzed honey samples by CZE-UV at 245 nm. They developed a method capable of separating ten monosaccharides and a disaccharide (maltose).

Analyses were carried out in a fused silica capillary with 58.5 cm total length and a 200 mM borate buffer containing 4% methanol as BGE.

Sirén *et al.* [63] used CE-UV, liquid chromatography, and inductively coupled plasma atom spectrometry to analyze carbohydrates, organics acids, aldehydes, anthocyanins, phenolic compounds, inorganic anions and metals in *Pinot noir* red wines. The carbohydrates levels of saccharose, galactose, glucose, rhamnose, fructose, and xylose were analyzed by CZE-UV at 270 nm, according to the method developed by Rovio *et al.* [81]. In 2016, Zhao *et al.* [65] also optimized the CZE method developed by Rovio and collaborators to quantify carbohydrates in grape tissues. In this work, the BGE used was 130 mM NaOH and 36 mM Na_2HPO_4, but the pH was adjusted up to 13, and the separation voltage was 10 kV.

Still using direct UV absorption as detection mode, a CZE method was developed by Toutounji *et al.* [64] to analyze sucrose, maltose, glucose, and fructose in commercial breakfast cereals with no sample preparation or derivatization step, since direct UV absorption detection was chosen. In the same year, Monti *et al.*

[62] applied Bao's CZE method [82] to determine sialylated oligosaccharides in cow, goat, and equine milk. The results established the high variability of the sialylated oligosaccharides compounds in the different milk species.

A CZE-UV method under 185 nm was also used to analyze lactose, galactose, and glucose in caprine whey by Oliveira *et al.* [70] in 2014. A year later, Navarro-Pascual-Ahuir *et al.* [71] developed a CZE-UV method to determine sugars and cyclamate contents in commercial juices. Run time was less than 6 minutes. The separation method allied to linear discriminant analysis could distinguish the juice by the fruit type. Fig. (**2**) shows the electropherograms for apple juice (A), grape juice (B), orange juice (C), and orange nectar (D) analysis.

Jiang *et al.* [72] used a dynamically coating capillary to analyze 32 carbohydrates in milk, juice, honey, and candied jujube in less than 12 minutes. The analysis was carried out using hexadimethrine bromide as an electroosmotic flow modifier and indirect UV detection at 275 nm. Dominguez *et al.* [73] developed a CZE-UV method, monitored at 224 nm, for simultaneous determination of fructose, glucose, sucrose, and the amino acid proline in honey samples. All the analytes were separated within 5 minutes employing only a single sample dilution in water. To evaluate the matrix effects in the quantification of glucose, fructose and sucrose in grapevines, Moreno *et al.* [74] optimized the Lumex protocol for sugar analysis (Sugar Analysis Chemical Kit, Beverages, version 20.01.2014, Lumex Ltd, St Petersburg, Russia) which used a BGE of 7 mM potassium sorbate, 0.5 mM hexadecyltrimethylammonium bromide, 50 mM NaOH and methanol 10%

(v/v) at pH 12.8 in a CZE-UV method. Finally, Neves *et al.* [75] also proposed a CZE-UV method, monitored at 275 nm, to determine lactulose in UHT milk without lactose interference.

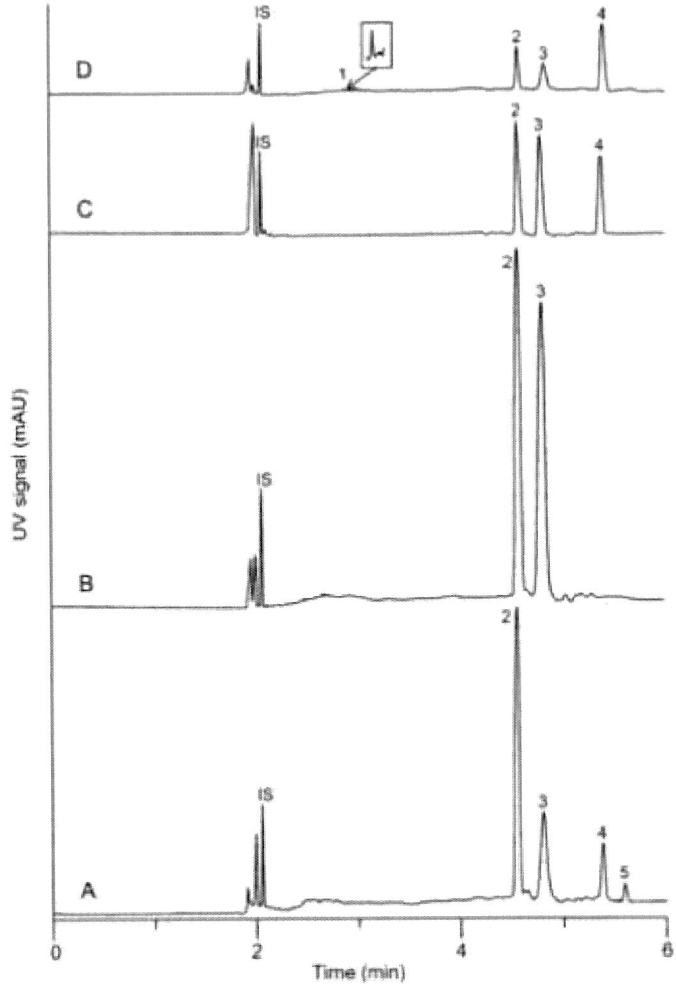

Fig. (2). Representative electropherograms of apple juice **(A)**, grape juice **(B)**, orange juice **(C)**, and orange nectar **(D)**. Peak identification: IS, Internal Standard; 1, cyclamate; 2, fructose; 3, glucose; 4, sucrose; 5, sorbitol. CZE conditions: BGE composed of 10 mM PDC and 0.5 mM CTAB at pH 12.1; hydrodynamic injection, 50 mbar for 6 s; separation voltage, −25 kV at 15 °C. Reprinted with permission from [71].

Because sensitive and selective detection is highly demanded in CE, AD is a promising detection system for CE with the advantages of high sensitivity and selectivity, low-power demands, and high integration of detector and control instrumentation [77]. The potential of this detection system was demonstrated by

Liang *et al.* [76], Mao *et al.* [77] and García, and Escarpa [78]. In Ling's method, authors have developed a graphene-cobalt microsphere hybrid paste electrode for the determination of carbohydrates in honey and milk in 2016. The performance of the electrodes was demonstrated by detecting mannitol, sucrose, lactose, glucose, and fructose after CE separation. The five analytes were well separated within 9 min in a 40 cm long capillary at a separation voltage of 12 kV. The electrodes exhibited pronounced electrocatalytic activity, lower detection potentials, enhanced signal-to-noise characteristics, and higher reproducibility. Two years later, the same analytes were studied in a model mixture by Mao *et al.* [77]. In this study, the authors developed a magnet-assembled alignment device for AD. The carbohydrates were well separated within 13 min in a 40 cm long capillary at a separation voltage of 9 kV using only a 75 mM NaOH aqueous solution as BGE, same as Liang *et al.*

In the study of García and Escarpa [78], copper nanowires (CuNWs) were coupled with MCE for the selective analysis of monosaccharides in honey samples. The approach allowed the separation of glucose and fructose in less than 250 s. MCE was also employed by Duarte-Júnior *et al.* [79]. A method was developed using this technique coupled with C^4D as a detection system for the separation of fructose, galactose, glucose, lactose, and sucrose. The five carbohydrates were separated within 180 s with excellent repeatability (RSD <1%).

PROTEINS

Proteins are highly complex polymers made up of 21 different amino acids that play a central role in biological systems. Although the information on cell evolution and biological organization is contained in DNA, enzymes are the molecules responsible for catalyzing the biochemical processes that sustain the life of an organism/cell [83].

At the elemental level, proteins contain on wt/wt basis 50-55% carbon, 6-7% hydrogen, 20-23% oxygen, 12-19% nitrogen and 0.2-3.0% sulfur. According to their biological function, they can be classified as enzyme catalysts, structural, contractile proteins, antibodies, storage proteins, and protective proteins. Food proteins have functional properties due to their physicochemical and structural characteristics, which can be defined as those proteins that are easily digested, nontoxic, nutritionally adequate, functionally usable in food products, available in abundance, and agriculturally sustainable [84]. The functional role of various proteins in different food products, according to Kinsella *et al.* [85], can be observed in Table **5**.

Table 5. Functional roles of food proteins.

Functional Roles of Food Proteins in Food Systems			
Function	**Mechanism**	**Food**	**Protein Type**
Solubility	Hydrophilicity	Beverages	Whey proteins
Viscosity	Water binding, hydrodynamic size, and shape	Soups, gravies, and salad dressings, desserts	Gelatin
Water binding	Hydrogen bonding, ionic hydration	Meat sausages, cakes, and breads	Muscle proteins, egg proteins
Gelation	Water entrapment and immobilization, network formation	Meats, gels, cakes, bakeries, cheese	Muscle proteins, egg proteins, milk proteins
Cohesion - adhesion	Hydrophobic, ionic and hydrogen bonding	Meats, sausages, pasta, baked goods	Muscle proteins, egg proteins, whey proteins
Elasticity	Hydrophobic bonding, disulfide crosslinks	Meats, bakery	Muscle proteins, cereal proteins
Emulsification	Adsorption and film formation at interfaces	Sausages, bologna, soup, cakes, dressings	Muscle proteins, egg proteins, milk proteins
Foaming	Interfacial adsorption and film formation	Whipped toppings, ice cream, cakes, desserts	Egg proteins, milk proteins
Fat and flavor binding	Hydrophobic bonding, entrapment	Low - fat bakery products, doughnuts	Milk proteins, egg proteins, cereal proteins

Proteins are nutrients needed in the human diet and essential for the proper growth and maintenance of the human body. Thus, the development of new analytical methods for the separation and determination of these biomolecules has been one of the major challenges for the food industry and regulatory agencies, aiming at providing food quality assurance [86]. CE is perfectly adequate to determine proteins and amino acids in food. It provides high efficiency, short analysis time, low solvent consumption, and is very versatile. The technique also offers the possibility of enantiomeric separation using different CE modes [87 - 89].

Table **6**, **7**, **8** and **9** provide an overview of the scientific publication over the past five years covering CZE, MEKC, MCE, and other CE modes, respectively, employed for protein identification and quantification in food samples using CE. Analytical methods revised within each CE mode were organized in chronological order.

The separation of proteins employing CE can be performed by a variety of modes since these compounds differ in electric charge, relative molecular mass, conformation, hydrophobicity, and even specific binding capability. CZE and Isotachophoresis (ITP) can be used for protein separation as a result of differences in electrophoretic mobility. Separation in sieving media such as gels and entangled polymeric networks can be performed considering the different sizes of compounds. Isoelectric focusing (IEF) or EKC modes can be applied in order to achieve separation based on proteins charge. Besides, MEKC can be used when considering hydrophobicity of compounds, and separation by bioaffinity electrophoresis may be an alternative for separation, approaching specific integration with other biomolecules [130].

Table 6. Description of analytical methods developed for protein analysis employing CZE in the last five years.

Sample	Analytes	Sample Preparation	CE Mode/Detection	Background Electrolyte	Instrumental Conditions	Ref
Cheese	Lysozyme	Extraction with acetic acid 1 M and filtration	CZE-MS	100 mM formic acid	+20 kV, hydrodynamic injection: 100 mbar for 5 s.	90
Rye	Rye HMW secalin subunit	Extraction, centrifugation, and alkylation followed by protein precipitation New centrifugation, dissolution in a solution of 40% ACN + 0.1% TFA and 2 M urea. Final centrifugation	CZE-UV	75 mM iminodiacetic acid, 0.15% poly(ethylene oxide), 26 mM N-dodecyl-N,N-dimethyl-3-amm-nio-1-propanesulfonate, and 15% acetonitrile	40 ºC, +10 kV, hydrodynamic injection: 3.45 Pa for 10 s; Capillary: 31.2 cm, 50 i.d.	91
Olive leaves and pulps	Protein profiles	Lyophilization. Extraction followed by sonication. Centrifugation and dilution with BGE.	CZE-UV	50 mM phosphate, 50 mM tetraborate and 0.1% poly(vinyl)alcohol	25 ºC, +10 kV, hydrodynamic injection: 50 mbar for 3 s; Capillary: 33.5 cm, 50 i.d.	92
Milk	Protein profile	Centrifugation and filtration followed by dilution	CZE-UV	0.19 M monohydrate citric acid, 0.02 M trisodium citrate dehydrate, 6 M urea and w/w 0,05% methylhydroxyethylcellulose 3000	45 ºC, 25 kV, hydrodynamic injection: 5 kPa for 7 s; Capillary: 40 cm, 50 µm i.d	93
Milk	Casein	Dilution in buffer and centrifugation	CZE/UV	0.19 M monohydrate citric acid, 0.02 M trisodium citrate dehydrate, 6 M urea and w/w 0.05% methylhydroxyethylcellulose 3000	45 ºC, 25 kV, hydrodynamic injection: 50 mbar for 7 s; Capillary: 40 cm, 50 µm i.d.	94
Milk powder	Melamine	Dilution with acetic acid centrifugation and filtration	CZE-UV	30 mM Tris-H_3PO_4	25 ºC, +20 kV, hydrodynamic injection: 0.5 psi for 3 s; Capillary: 40 cm x 50 i.d.	95
Cheese	Casein	Dissolution in 10 mM sodium phosphate containing 8 M urea and 10 mM DTT and filtration	CZE- UV	10 mM sodium phosphate containing 6 M urea and 0.05% HPMC	23 ºC, +18.5 kV, hydrodynamic injection: 3447 Pa for 5 s. Capillary: 57 cm, 75 µm i.d.	96

(Table 6) cont.....

Cheese	Peptide profile (pH 4.6-insoluble fraction)	Dissolution in 10 mM sodium phosphate, 8 M urea, and 10 mM dithiothreitol and filtration	CZE- UV	10 mM sodium phosphate, 6 M urea, and 0.05% hydroxypropyl methylcellulose	23 °C, +18.5 kV, hydrodynamic injection: 0.5 psi for 5 s. Capillary: 57 cm, 75 µm i.d.	97
Milk	Protein profile	Raw milk: Centrifugation, dilution with distilled water, and filtration. Whey proteins obtention: precipitation, centrifugation, and filtration. Dissolution with the BGE	CZE- UV	150 mM borate	+7.8 kV, hydrodynamic injection: 200 Pa for 1 s. Capillary: 17 cm, 25 µm i.d.	98
Flour	Monomeric prolamin	Pre-extraction with 0.4 M NaCl and 0.067 M $HKNaPO_4$, followed by extraction with ethanol	CZE- UV	0.06 M iminodiacetic acid, 0.1% (w/v) polyvinylpyrrolidone with a molecular weight of 40,000, and 20% (v/v) acetonitrile	40 °C, hydrodynamic injection: 3.45 Pa for 6 s. Capillary: 31.2 cm, 50 µm i.d.	99
Sport nutritional supplements	Leucine, isoleucine, and valine	Acid hydrolysis with HCl, neutalization with NaOH, centrifugation, dilution with purified water and filtration	CD-CZE- UV	40 mM sodium tetraborate and 40 mM β-cyclodextrin (for direct UV detection); 2 mM Na_2HPO_4, 10 mM PAS, 40 mM β-cyclodextrin (for indirect UV detection)	25 °C, +15 kV, hydrodynamic injection: 0.5 psi for 5 s. Capillary: 60 cm, 75 µm i.d.	100
Human milk	Human milk proteins	Extraction with calcium phosphate	CZE-UV	50 mM iminodiacetic acid, 0.5% hydroxyethylcellulose and 6 M urea (pH 3.1)	25 °C, 15 kV, hydrodynamic injection: 50 mbar for 5 s; Capillary: 33.5 cm, 50 µm i.d.	101
Bovine lactoferrin	Peptides	Enzymatic hydrolysis followed by solid-phase extraction with formic acid and methanol. Lyophilization and filtration	CZE-MS	A 10% acetic acid	+20 kV, hydrodynamic injection: 20 psi for 60 s. Capillary: 91 cm, 30 µm i.d.	102
Fish	Parvalbumin	Protein extraction, and centrifugation. Column extraction with linear saline gradient buffer (0.1 M NaCl to 0.5 M NaCl) and filtration	CZE- UV	25 mM borateborax	25 ºC, +15 kV, hydrodynamic injection: 0.5 psi for 5 s Capillary: 30.2 cm, 75 µm i. d.	103
Nutraceuticals	Antihypertensive peptides from a bovine milk protein	Dissolution in 1 mL of warm water, centrifugation, and filtration	CZE-MS	1 M acetic acid	25 ºC, +10 kV, hydrodynamic injection: 34 mbar for 15 s Capillary: 72 cm, 75 i.d.	104
Cheese	Intact Caseins and Long Peptides	Dissolution and centrifugation.Mixture with 10 M urea, 0.83 g/L methyl-hydroxyethylcellulose, 167 mM Tris, 67 mM disodium EDTA, 42 mM 3-(--morpholino)propanesulfonic acid, and 17 mM dithiothreitol	CZE- UV	20 mM trisodium citrate, 190 mM citric acid, 0.5 g/L of methylhydroxyethylcellulose and 6 M urea	45 °C, linear voltage gradient from 0 to +25 kV in 3 min, followed by a constant voltage of +25 kV hydrodynamic injection: 34 mbar for 20 s. Capillary: 56 cm, 50 µm i.d.	105

(Table 6) cont.....

Milk	Casein	Dissolution in 10 mol/L urea buffer, dilution with the same buffer and filtration	CZE- UV	10 mM urea buffer	45° C, +30 kV, hydrodynamic injection: 0.5 psi for 20 s. Capillary: 50 cm	106
Energy drinks, fresh milk, powdered infant formula, and probiotic green rice	Taurine and choline	Powder samples: sonication, filtration, and dilution with deionized water Liquid samples: filtration and dilution with deionized water	CZE-C[4]D	150 mM tris/lactic acid for taurine and 150 mM tris/acetic acid for choline	-10 kV, hydrodynamic injection: 30s. Capillary: 60 cm, 50 µm i.d.	107

ACN = acetonitrile, DTT = Dithiothreitol, EDTA = ethylenediamine-tetracetic acid, HMW = high molecular weight, HPMC = hydroxypropyl methylcellulose, PAS = p-aminosalicylic acid, TFA = trifluoroacetic acid, Tris = Tris (hydroxymethyl)aminomethane, M = mol/L;i.d. = internal diameter.

Table 7. Description of analytical methods developed for protein analysis employing MEKC in the last five years.

Sample	Analytes	Sample Preparation	CE Mode/Detection	Background Electrolyte	Instrumental Conditions	Ref
Milk and yogurt	Estriol, zearalanol, estradiol, zearalenone, ethinylestradiol, zearalanol, methoxyestradiol, and estrone	Precipitation followed by microextraction	MEKC-MS	45 mM ammonium perfluorooctanoate (pH 9) with 10% (v/v) methanol	25 ° C, +25 kV, hydrodynamic injection: 0.5 psi for 25 s. Capillary: 60 cm, 50 µm i.d.	108
Egg	Ovalbumin, conalbumin, ovomucoid, and lysozyme	Extraction followed by precipitation and dilution in buffer	MEKC- UV	300 mM borate buffer (pH 9.2) and 25 mM SDS	25 °C, +10 kV, hydrodynamic injection: 40 mbar for 5 s Capillary: 31.5 cm, 50 µm i.d	109
Cheese and pasteurize d milk	Nisin	Dilution in water followed by centrifugation	MEKC-UV	50 mM phosphate (pH 3.95) and 80 mM SDS	30 °C, -13 kV. hydrodynamic injection 50 mbar for 10 s Capillary: 59 cm, 50 µm i.d.	110
Plastron	Amino acid profile of plastron	Alkali digestion followed by derivatization procedure	MEKC-UV	20 mM borax (pH 8.74) and 0.1 M SDS with 6% (v/v) methanol	25 °C, 25 kV, electrokinetic injection: 9 kV for 9 s; Capillary: 60.2 cm, 50 µm i.d.	111

SDS = sodium dodecyl sulfate; M = mol/L; i.d. = internal diameter

Table 8. Description of analytical methods developed for protein analysis employing MCE in the last five years.

Sample	Analytes	Sample Preparation	CE Mode/Detection	Background Electrolyte	Instrumental Conditions	Ref
Infant milk formula	Lactoferrin, immunoglobulin G, α-lactalbumin, β-lactoglobulin A, β -lactoglobulin B, bovine serum albumin, α-casein	Dissolution in IEF buffer, centrifugation, and filtration	MCE/UV based on IEF-IT--CZE	IEF buffer: 2% pharmalyte, 8 M urea, 1% Triton X100, 0.1 g/L PEO, 0.5 g/L HEMC, 0.25% TEMED. Anolyte: 0.05 M HAc, 0.1 g/L PEO, 0.5 g/L HEMC (pH 3.0). Catholyte: 0.04 M NaOH, 0.1 g/L PEO, 0.5 g/L HEMC (pH 10.0). Terminating electrolyte: 0.05 M HAc, 0.1 g/L PEO, 0.5 g/L HEMC (pH 3.0). Leading electrolyte: 0.05 M Tris–HCl, 0.1 g/L PEO, 0.5 g/L HEMC, same as BGE	See article for channel lengths, detailed voltages and times used for filling BGE, loading protein samples and IEF, ITP and CZE run	112
Dairy products	cytochrome C, myoglobin, carbonic anhydrase II, α-lactalbumin, albumin from bovine serum.	Dissolution in IEF buffer, centrifugation, and filtration	MCE/UV based on IEF-IT--CZE	The same reported in reference [112]	See article for channel lengths, detailed voltages and times used for filling BGE, loading protein samples and IEF, ITP and CZE run	113
Poultry meat	Carnosine and anserine	Homogenization of the grounded sample with water, centrifugation, filtration, deproteinization. Remotion of precipitated proteins and particulates, centrifugation. Supernatant was evaporated, supplemented with 0.5 mL of water, and sonicated.	MCE-C^{4}D	200 mM acetic acid, 10 mM 2-hydroxyisobutyric acid and 0.3 mM potassium acetate	Separation channel: 87 mm. Electrokinetic injection: +1.20 kV for 21 s, separation voltage: +3.50 kV.	114

(Table 8) cont.....

Chicken, beef, horse and rabbit meat	Carnosine and anserine	The same procedure described in reference [114]	MCE-C⁴D	250 mM acetic acid, 10 mM, 2-hydroxy-2-methylpropanoic acid, and 0.3 mM rubidium acetate (pH 2.6)	Separation channel: 87 mm. Electrokinetic injection: +0.9 kV for 16 s, separation voltage: +3.30 kV	115

HAc = acetic acid; PEO = polyethylene oxide; HEMC = hydroxyethyl methyl cellulose; TEMED = tetramethylethylenediamine; M = mol/L;

Table 9. Description of analytical methods developed for protein analysis employing other CE modes in the last five years.

Sample	Analytes	Sample Preparation	CE Mode/Detection	Background Electrolyte	Instrumental Conditions	Ref
Chicken egg and milk	α, β, κ caseins, α lactalbumin; β lactoglobulin, albumin from bovine serum (milk); lysozyme, ovalbumin, conalbumin (eggs)	Chicken egg: dilution in BGE and filtration. Milk: fat removal, dilution in BGE, and filtration.	CEC-UV	20 mM phosphate-citrate (pH 3.0)	25 ºC, -20 kV, hydrodynamic injection: 0.5 psi for 5 s; Capillary: polydopamine-coated open tubular column; 39.2 cm, 50 µm i.d.	116
Commercial ovalbumin products	Ovalbumin	Dilution with water and filtration.	CIEF-UV	Anolyte: 200 mM H_3PO_4 Catholyte: 300 mM NaOH Spacers: 10.7 mM L-Arginine (cathodic) and 1.8 mM iminodiacetic acid (anodic)	20 ºC; hydrodynamic injection: 900 mbar for 200 s; Capillary: 31.4 – 32.6 cm; 50 µm i.d. CIEF was performed with anolyte in CIEF gel and catholyte applying +25.0 kV for 15.0 min. For chemical mobilization, the catholyte was replaced by 50 mM L- glutamic acid (pH 10.50), and +30.0 kV was applied. Samples focusing: 15 min at +25.0 kV.	117

(Table 9) cont.....

Sample	Analytes	Sample Preparation	CE Mode/Detection	Background Electrolyte	Instrumental Conditions	Ref
Skim milk powder and infant formulas	Casein and whey proteins	Dilution with water	SDS-CGE-UV	100 mM Tris–HCl, (pH 9.0); with 1% SDS	25 °C, - 497 V/cm, electrokinetic injection: - 5 kV for 20 s; Capillary: 20 cm; 50 μm i.d.	118

SDS = sodium dodecyl sulphate; M = mol/L; i.d. = internal diameter

In ITP mode, not approached in this chapter yet, samples are introduced between a leading electrolyte (fast zone) and a terminating electrolyte (slow zone). Charged analytes are separated based on their ionic mobility, as in the CZE mode. However, the use of two electrolytes makes possible analytes focusing, which can be very useful to detect compounds present at very low levels [119]. On the other hand, in the IEF mode, or the capillary isoelectric focusing (CIEF) mode, proteins and peptides are separated according to their isoelectric point values. In this case, the capillary is filled with a combination of sample and ampholytes, and the capillary ends are submerged in anolyte and catholyte solutions. The practice of isoelectric focusing in the capillary provides superior resolving power of conventional gel IEF [120].

When considering macronutrients, proteins are the most analyzed compounds by CE [29]. Amorim and Oliveira reported that the number of mentionings of CE and proteins was > 4000 in the Web of Science database considering the 2010s decade. The number of mentions for CE and carbohydrates, and CE and lipids were < 400 and < 300, respectively. Thus, this chapter will approach only a few main representative analytical methods for protein determination in food samples by CE.

It can be observed in Table **6** that CZE has been identified as the most widely used CE mode for protein analysis, coupled to different detection systems (UV, C^4D, MS) [90 - 107]. Besides, milk and dairy products proteins were the most studied employing CZE. However, other samples, such as milk products, fish, grains, fruits, and nutritional supplements, were also approached.

Casein is one of the most studied proteins in dairy samples, and two works can be highlighted considering the determination of this compound using CZE. Vries *et al.* evaluated the effect of shortening or omitting the dry period of dairy cows on milk casein composition employing CZE-UV [94], and Gustavsson *et al.* investigated the relationships between casein genetic variants and detailed protein composition in Swedish and Danish dairy milk, also using CZE-UV [93]. Specific conditions of these methods can be observed in detail in Table **6**.

Besides, much work has been reported approaching the detection of microorganisms in dairy samples, since the microbial community of milk is complex, variable, and depends on various conditions [121]. Employing CZE-UV, a recent study evaluated *Pseudomonas fluorescens, Ps. poae,* and *Chryseobacterium joostei* strains in raw milk [106]. The presence of these strains in milk is a challenge for UHT milk manufacturers and authors have proposed to evaluate the presence of C-terminal casein-macropeptide as a useful indicator of milk susceptibility to gelation, irrespective of the responsible species. The proposed methodology has much potential to be used in quality control for routine evaluation of raw milk before processing, considering the simplicity of the CE methodology developed.

CE coupled to MS has also proved to be an important tool in the identification of dairy product peptides. Chen *et al.* [102], for example, identified bovine lactoferrin digested peptides as functional hydrolysates by CE-Q-TOF-MS. 102 peptides were characterized in terms of molecular weight, mass/charge ratio, and isoelectric point.

The work of Benavente *et al.* [104] also employed CZE-MS in combination with advanced chemometric tools for the identification of antihypertensive peptides in a nutraceutical derived from a bovine milk protein hydrolyzate. The proposed methodology was successfully employed in nutraceuticals analyses and demonstrated potential to obtain a characteristic and activity-related fingerprint for quality control and authentication of antihypertensive nutraceuticals. Bioactive peptides from food proteins exert several beneficial effects on human health, and some of them present antihypertensive properties. Thus, much research has also been carried out to detect and quantify these compounds in foods [122].

Another innovative study using CE in tandem with mass spectrometry (MS/MS) was developed by Taivosalo *et al.* (2017) [105]. Authors determined the composition of water-insoluble and soluble proteinaceous fractions in cheese and studied in detail the degradation of these compounds during eight months of ripening. More than 3000 small peptides were detected, which represents a full casein peptidome in cheese. Proteins in cheese samples were also determined by CZE coupled to MS detector in the work of Kondeková *et al.* [90]. In this work, a method was developed for the detection of lysozyme, a protein considered as an allergenic agent.

Other studies employing CZE to obtain the protein profiles in cheese have also been reported employing a UV detector. Merheb-Dini *et al.* explored the use of a new coagulant for the manufacture of a variety of highly cooked starter-free cheeses. Authors evaluated their physicochemical and functional characteristics compared with cheeses made with a traditional commercial coagulant using a CZE-UV method [96]. Besides, cheese samples were also analyzed in the work of Baptista *et al.* [97] employing CZE-UV. Authors described the characterization of the peptide profile in commercial Prato cheese by MALDI and CE. They reported that even commercial cheeses produced with different raw material and processing conditions showed very similar peptide profiles when assessed at the molecular level. Moreover, only nine peptides were responsible for discrimination of cheeses.

CZE methods have also been widely used in milk adulteration studies such as contamination of sheep's milk with cow's milk. This fraud has been observed in the dairy industry due to the lower price and availability of cow's milk over the year. In this context, an interesting method by CZE for the quantification of sheep's milk in sheep/bovine milk mixtures allowed for quick and cheap recognition of sheep's milk and cow's milk adulteration. Under the optimized experimental conditions, sheep and bovine milk exhibited distinct and recognizable CE protein profiles, with a specific sheep peak showing a reproducible migration zone in the sheep/bovine mixtures. Based on the sheep-specific CE peak, a method for quantifying sheep's milk in sheep/bovine skimmed milk mixtures was developed [98]. It showed good linearity, precision, accuracy, and a minimum amount of detectable fraudulent cow's milk equal to 5% was reported.

The work of Zhang *et al.* can also be cited considering the studies in food adulteration approaching proteins. Authors developed a CZE-UV method for the monitoring of melamine in milk products [95] since melamine fraud is common to increase the apparent protein content in milk and dairy products. Low limits of detection and quantification were reported, as well as good recoveries and repeatability. The developed methodology provided a simple and relatively sensitive method for the determination of melamine in milk powder.

Concerning the sample preparation for milk analysis, some authors addressed the difficulty of analyzing milk proteins without the precipitation of casein. The milk protein fractionation is generally performed at pH 4.6 (the isoelectric pH). However, in this pH, some whey proteins may co-precipitate with the caseins and vice versa [123]. Given this, Ten-Doménech *et al.* [101] proposed a simple sample pretreatment method for the fractionation of human milk proteins (whey and caseins) and evaluated method efficiency by CZE-UV. The developed method is very promising due to its simplicity in the treatment of milk samples. Besides, the CE fraction collection off-line coupled to LC-MS/MS has proved to be useful in the identification of minor proteins present in human milk.

Although most of the studies approached considered the analysis of milk and dairy products, recent methods for protein determination in other food matrices were also found in the literature. A simple analytical method using ion-exchange chromatography coupled with CZE was developed for the detection and quantification of parvalbumin, a major allergen, in fish [103]. Besides, the fractional protein composition of triticale flour and gluten was also studied using CZE-UV [99]. Salmanowicz *et al.* [91] also demonstrated the potential of CZE-UV to a fast and accurate characterization of rye storage proteins. Moreover, a simple, fast, and environmentally friendly CZE-UV method has been described by Vergara-Barberan *et al.* [92] for obtaining the protein profiles of olive leaves and pulp. In this work, discriminant analysis of the CZE data enabled to correctly classify the leaf and pulp samples belonging to nine cultivars from different Spanish regions with excellent resolution across all categories.

When considering nutritional supplements, Nguyen *et al.* employed CZE coupled with C^4D for a rapid determination of choline and taurine in energy drinks and dietary supplements [107]. Qiu *et al.* performed the quantitation of underivatized branched-chain amino acids in sport nutritional supplements by CD-CZE-UV [100]. Both studies demonstrated the potential of CZE coupled with different detection systems for the rapid analysis of nutritional supplements.

The MEKC mode was applied in some studies found in the literature for the determination of proteins in food samples, employing different detection systems, such as UV, C^4D, and MS, as depicted in Table **7** [108 - 111]. The work of Drevinskas *et al.* [111] demonstrated the use of a MEKC-C^4D method for nisin separation and determination in fermented dairy products. The developed method required a simple sample preparation (only sample dilution in water) and centrifugation. Besides, it permitted quantification down to 0.02 g/mL of nisin.

A method applying MEKC-UV to evaluate the authenticity of functional amino acid profiled plastron-derived foods was also recently published. The fraud with plastron-derived food is very well known since plastron is a nutritive and superior functional food. Due to its limited supply and enormous demands, some functional foods supposed to contain plastron may be forged with other substitutes. The method developed by Li *et al.* [109] detected 18 amino acids in diverse food matrices such as chicken and pork, fish, and eggs. It has shown great potential to detect food fraud in these products.

Uysal *et al.* [110] employed MEKC-UV for evaluating the effect of heat treatment parameters in liquid whole egg. Electropherograms of samples demonstrated the differences in protein contents in eggs before and after heat treatment. Authors reported that conalbumin and lysozyme were influenced by the heating, while ovalbumin and ovomucoid were not affected. Moreover, authors combined CE with multivariate analysis and obtained a powerful methodology for untreated and treated egg samples classification.

When considering MS detection, D'Orazio *et al.* [108], have adapted a MEKC method for the analysis of a group of active estrogen in dairy samples. After adequate deproteinization and degreasing of the whole cow, skimmed and semi-skimmed cow's milk, and natural whole cow's yogurt, the methodology was adequate for the analysis of the target compounds in 18 samples currently marketed in Spain. The method allowed good selectivity and sensitivity for the products analyzed.

MCE also proved to be a useful tool for protein determination in food samples. In addition to the advantages of conventional CE, such as high efficiency, low consumption of chemicals and samples, MCE presents the unique benefits of miniaturization, and the ability to perform fast and real-time analysis [124. Analytical methods employing MCE were described in Table **8** [112 - 115].

Wu *et al.* for example, developed an MCE method for determining low levels of functional proteins in a matrix simulating infant milk formula [125]. A multi-dimension strategy was adopted in a 2D MCE device to isolate major proteins on-chip and enrich minor proteins in the capillary before separation employing IEF and ITP. The results using a standard protein mixture showed a complete isolation of high abundance proteins by a 2-min 1D IEF run. The subsequent t-ITP/CZE run was able to enrich the isolated protein fractions 60 times. The authors also developed a similar approach to evaluate soy protein adulteration in a subsequent work [113]. The analysis time was less than 10 min, and adulteration with 0.1% soy protein in dairy products could be detected using the developed method.

Additionally, MCE coupled to C^4D has successfully been applied to the separation of pharmacologically important dipeptides that have important bioactivities in diseases such as diabetes, cataracts, ischemia, Alzheimer's, and Parkinson's: carnosine and anserine. In the method proposed by Jozanović *et al.* [114], electropherograms exhibited well-resolved symmetrical peaks and provided satisfactory quantification of carnosine and anserine as single dipeptides and as a single mixture. This method was successfully used to determine both compounds in the thigh and breast muscle of poultry meat. Results were similar to those obtained by a commercial MCE-C^4D device and those obtained by high-performance liquid chromatography (HPLC).

A subsequent study was also reported for carnosine and anserine determination in chicken, beef, horse, and rabbit meat. In this study, an improved MCE device was developed, coupled with C^4D detection [115]. The developed methodology was shown to be a cheap, fast, and a simpler alternative to the more complicated and expensive existing HPLC methods. It does not require optical detectors and sample derivatization, which are unavoidable in HPLC methods. This method has been shown to be promising for animal species determination in unknown meat samples using the carnosine/anserine ratio, and it has potential to be used for food fraud detection.

Advances in electrophoretic studies using microchips are also evidenced by some comprehensive reviews published recently. A critical review showing MCE applications in biotechnology for the analysis of genetically modified organisms is available in the literature. Authors approached the studies of substantial equivalence, discrimination between transgenic and non-transgenic cultivars, unintended effects caused by genetic modification, or its modification response to various stressful situations or conditions (*e.g.*, environmental, climatic, infections) [126]. Another relevant review showed the quantification of proteins/peptides and other metabolites in wine samples focusing on the different methodologies employed to properly perform CE in wine analysis informing about wine chemistry concerning polyphenol analysis and protein profile [127]. Furthermore, the work of Dawod *et al.* [128] can be referred to for recent advances in protein analysis by capillary and MCE.

Considering the potential of other CE modes for protein determination in food samples, some studies approached the use of CEC [116], CIEF [117], and Sodium Dodecyl Sulfate-Capillary Gel Electrophoresis (SDS-CGE) [118] recently. Analytical conditions employed in these methods were summarized in Table **9** [116 - 118].

CEC mode, which combines chromatography and electrophoretic separation principles, is a powerful alternative for improving protein separation in diverse matrices. It can be useful to overcome some problems in protein analysis by CE, such as peak widening and tailing, poor reproducibility between runs, low peak efficiency, and low protein recovery due to strong electrostatic interaction and/or hydrogen bonding between proteins and the capillary silanol groups [129]. Xiao *et al.* [116], developed a simple and economical method for the preparation of polydopamine-coated open tubular column. The column performance was evaluated on several parameters and validated for the separation of several proteins in chicken egg and milk samples. The developed methodology provided interesting results for proteins determination with great precision. The relative standard deviations of migration times of proteins considering run-to-run, day-t--day, and column-to-column experiments were less than 3.5%.

CIEF was employed by Kristl and Stutz [117]' also aiming ovalbumin detection, a major food allergen present in eggs and some industrialized products. In this study, authors compared pressure and chemical mobilization modes and observed that the mobilization with an acidic zwitterion provided the best selectivity results for ovalbumin detection. The application of the developed methodology revealed differences in the constituent profile of three commercial ovalbumin products and proved the suitability of the method for ovalbumin detection in food samples.

In the work of Feng *et al.* [118], quantification of whey protein content in skim milk powder and infant formulas was performed employing SDS-CGE. In this CE mode, proteins separation is achieved in a gel *via* capillary columns filled with BGE. The use of this CE mode was demonstrated to be a powerful new analytical method for the separation of protein molecules based on their molecular mass. The developed methodology was capable of accurately determining the ratio of whey to casein in products manufactured using different whey ingredients. Furthermore, the method was validated in terms of linearity, accuracy, precision, specificity, limit of quantification, and range. The impressive results obtained showed the potential of the method to be applied in a routine quality control environment.

Overall, the papers presented here show the trends and new electrophoretic approaches for protein analysis in food samples, highlighting the importance of CE as a potential tool for Proteomics in science and food technology. Further information about protein analysis by CE can be obtained in the comprehensive chapter of Huck and Bonn [130]. The review of our research group considering the analysis of amino acids, proteins, carbohydrates, and lipids by CE can also be consulted for older applications in food samples [61].

CONCLUDING REMARKS

This chapter summarized recent systematic information from literature involving the application of CE methodologies for the analyses of lipids, proteins, and carbohydrates in food matrices considering the past five years. Different CE modes, such as CZE, CD-CZE, MEKC, MEEKC, NACE, and MCE, and diversified detection systems, such as UV, LIF, AD, C^4D, and MS have been used. The text content has been written in an attractive and precise manner, and the main analytical methods data were compiled in extensive tables throughout the document. The potential association of CE with multivariate analysis aiming for fast classification of food samples was also demonstrated in some studies. This chapter can serve both interested lay readers to gather the essential information making possible the minimum understanding of each method and more experienced readers who want to go deeper into the topic, by consulting the other reviews and book chapters referred to. Within this context, we believe that the present chapter has as a descriptive focus an extensive portfolio of information tangent to the state- of- the art of CE methodologies developed in the last five years, within the theme here suggested.

CONSENT FOR PUBLICATION

Not applicable.

CONFLICT OF INTEREST

The author declares no conflict of interest, financial or otherwise.

ACKNOWLEDGEMENTS

The authors aknowledge Coordenação de Aperfeiçoamento de Nível Superior - CAPES (Finance Code 001 and Project PNPD 23071022702/2018-43), Conselho Nacional de Desenvolvimento Científico e Tecnológico - CNPq (Projects 303355/2017-4 and 424032/2018-0), Instituto Nacional de Ciência e Tecnologia de Bioanalítica – INCTBio (Projects FAPESP 2014/50867-3 and CNPq 465389/2014- 7), Financiadora de Inovação e Projetos – FINEP (Project CT-INFRA 01/2013- REF 0633/13), Rede Mineira de Química - RQ-MG (Project CEX.RED- 00010-14) and Fundação de Amparo a Pesquisa do Estado de Minas Gerais - FAPEMIG for fellowships and financial support.

REFERENCES

[1] Solomons, T.; Fryhle, C. *Organic Chemistry,* 10th ed; Wiley & Sons: New York, **2009**.

[2] Lehninger, A.; Nelson, D.L.; Cox, M.M. *Biochemistry Principles,* 3rd ed; Sarvier: São Paulo, **2000**.

[3] Gunstone, F.D.; Norris, F.A. *Lipids in Food Chemistry, Biochemistry and Technology,* 1st ed; Pergamon Press, **1983**.

[4] *FAO. Fats and Fatty Acids in Human Nutrition*; World and Health Organization: Geneva, **2010**.

[5] Buettner, R.; Schölmerich, J.; Bollheimer, L.C. High-fat diets: modeling the metabolic disorders of human obesity in rodents. *Obesity (Silver Spring),* **2007**, *15*(4), 798-808. [http://dx.doi.org/10.1038/oby.2007.608] [PMID: 17426312]

[6] Zhao, Y.; Xiang, L.; Liu, Y.; Niu, M.; Yuan, J.; Chen, H. Atherosclerosis Induced by a High-Cholesterol and High-Fat Diet in the Inbred Strain of the Wuzhishan Miniature Pig. *Anim. Biotechnol.,* **2018**, *29*(2), 110-118. [http://dx.doi.org/10.1080/10495398.2017.1322974] [PMID: 28636491]

[7] de Oliveira, M.A.L.; Porto, B.L.S.; Faria, I.D.L.; de Oliveira, P.L.; de Castro Barra, P.M.; de Jesus Coelho Castro, R.; Sato, R.T. 20 years of fatty acid analysis by capillary electrophoresis. *Molecules,* **2014**, *19*(9), 14094-14113. [http://dx.doi.org/10.3390/molecules190914094] [PMID: 25207715]

[8] Fahy, E.; Subramaniam, S.; Brown, H.A.; Glass, C.K.; Merrill, A.H., Jr; Murphy, R.C.; Raetz, C.R.H.; Russell, D.W.; Seyama, Y.; Shaw, W.; Shimizu, T.; Spener, F.; van Meer, G.; VanNieuwenhze, M.S.; White, S.H.; Witztum, J.L.; Dennis, E.A. A comprehensive classification system for lipids. *J. Lipid*

Res., **2005**, *46*(5), 839-861.
[http://dx.doi.org/10.1194/jlr.E400004-JLR200] [PMID: 15722563]

[9] Chow, C.K. *Fatty Acids in Foods and Their Health Implications,* 3rd ed; CRC Press, **2008**.

[10] Ahmad, M.U. *Fatty Acids. Chemistry, Synthesis, and Applications*; AOCS Press, **2017**.

[11] Wu, J.; Ge, Y.; Qin, W. Combination of running-buffer-mediated extraction and polyamidoamine-dendrimer-assisted capillary electrophoresis for rapid and sensitive determination of free fatty acids in edible oils. *J. Agric. Food Chem.,* **2014**, *62*(18), 4104-4111.
[http://dx.doi.org/10.1021/jf4056105] [PMID: 24758299]

[12] Sato, R.T.; de Jesus Coelho Castro, R.; de Castro Barra, P.M.; de Oliveira, M.A.L. Rapid separation of free fatty acids in vegetable oils by capillary zone electrophoresis. *Phytochem. Anal.,* **2014**, *25*(3), 241-246.
[http://dx.doi.org/10.1002/pca.2499] [PMID: 24453169]

[13] Böckel, W.J.; Da Silva, Y.P.; Mendonça, C.R.B.; Simó-Alfonso, E.F.; Ramis-Ramos, G.; Piatnicki, C.M.S. Direct Determination of Oleic Acid in Soybean Oil by Capacitively Coupled Contactless Conductivity Detection Capillary Electrophoresis in an Oil-Miscible KOH/1-Propanol/Methanol Medium. *J. Braz. Chem. Soc.,* **2014**, *25*(9), 1662-1666.
[http://dx.doi.org/10.5935/0103-5053.20140155]

[14] Porto, B.L.S.; Faria, I.D.L.; Oliveira Mendes, T.; Oliveira, M.A.L. Fast Screening Method for the Analysis of Trans Fatty Acids in Processed Food by CZE-UV with Direct Detection. *Food Control,* **2015**, *55*, 230-235.
[http://dx.doi.org/10.1016/j.foodcont.2015.02.027]

[15] de Oliveira, L.N.; de Jesus Coelho Castro, R.; de Oliveira, M.A.; de Oliveira, L.F. Lipid Characterization of White, Dark, and Milk Chocolates by FT-Raman Spectroscopy and Capillary Zone Electrophoresis. *J. AOAC Int.,* **2015**, *98*(6), 1598-1607.
[http://dx.doi.org/10.5740/jaoacint.15-083] [PMID: 26651573]

[16] de Oliveira Mendes, T.; Porto, B.L.S.; Bell, M.J.V.; Perrone, Í.T.; de Oliveira, M.A.L. Capillary zone electrophoresis for fatty acids with chemometrics for the determination of milk adulteration by whey addition. *Food Chem.,* **2016**, *213*, 647-653.
[http://dx.doi.org/10.1016/j.foodchem.2016.07.035] [PMID: 27451230]

[17] Mimura, A.M.S.; Faza, L.P.; Marques, R.; Le Hyaric, M.; De Oliveira, M.A.L.; Silva, J.C.J. Determination of Cu, Fe, Mn, Zn and Free Fatty Acids in Pequi Oil. *Quim. Nova,* **2016**, *39*(5), 621-626.
[http://dx.doi.org/10.5935/0100-4042.20160067]

[18] Lee, J.H.; Kim, S.J.; Lee, S.; Rhee, J.K.; Lee, S.Y.; Na, Y.C. Saturated fatty acid determination method using paired ion electrospray ionization mass spectrometry coupled with capillary electrophoresis. *Anal. Chim. Acta,* **2017**, *984*, 223-231.
[http://dx.doi.org/10.1016/j.aca.2017.06.052] [PMID: 28843567]

[19] Prado, T.L.A.; Porto, B.L.S.; de Oliveira, M.A.L. Method Optimization for Trans Fatty Acid Determination by CZE-UV under Direct Detection with a Simple Sample Preparation. *Anal. Methods,* **2017**, *9*(6), 958-965.
[http://dx.doi.org/10.1039/C6AY02906J]

[20] Amorim, T.L.; Duarte, L.M.; Dos Santos, H.F.; de Oliveira, M.A.L. Screening method for simultaneous detection of elaidic and vaccenic trans fatty acid isomers by capillary zone electrophoresis. *Anal. Chim. Acta,* **2019**, *1048*(7), 212-220.
[http://dx.doi.org/10.1016/j.aca.2018.10.057] [PMID: 30598152]

[21] Amorim, T.L.; Duarte, L.M.; Chellini, P.R.; de Oliveira, M.A.L. A Validated Capillary Electrophoresis Method for Fatty Acid Determination in Encapsulated Vegetable Oils Supplements. *Lebensm. Wiss. Technol.,* **2019**, *114*, 108380.
[http://dx.doi.org/10.1016/j.lwt.2019.108380]

[22] Amorim, T. Pena, M. G.; Costa, F. F.; Oliveira, M. A.; Chellini, P. R., A Fast and Validated Capillary Zone Electrophoresis Method for the Determination of Selected Fatty Acids Applied to Food and Cosmetic Purposes. *Anal. Methods,* **2019**, *11*, 5607-5612.
[http://dx.doi.org/10.1039/C9AY01917K]

[23] Duarte, L.M.; Amorim, T.L.; Adriano, L.H.C.; de Oliveira, M.A. A CZE-UV Method for Saturated and Unsaturated Fatty Acids Determination in Hops. *J. Am. Soc. Brew. Chem.,* **2019**, •••, 32-42.

[24] Amorim, T.L.; Duarte, L.M.; de la Fuente, M.A.; de Oliveira, M.A.L.; Gómez-Cortés, P. Fast capillary electrophoresis method for determination of docosahexaenoic and eicosapentaenoic acids in marine oils omega-3 supplements. *J. Chromatogr. A,* **2020**, *1613*, 460641.
[http://dx.doi.org/10.1016/j.chroma.2019.460641] [PMID: 31677769]

[25] Hjertén, S. Free zone electrophoresis. *Chromatogr. Rev.,* **1967**, *9*(2), 122-219, 147-219.
[http://dx.doi.org/10.1016/0009-5907(67)80003-6] [PMID: 4883163]

[26] Mofaddel, N.; Krajian, H.; Villemin, D.; Desbène, P.L. Enantioseparation of binaphthol and its monoderivatives by cyclodextrin-modified capillary zone electrophoresis: a mathematical approach. *Talanta,* **2009**, *78*(2), 631-637.
[http://dx.doi.org/10.1016/j.talanta.2008.12.027] [PMID: 19203636]

[27] Hancu, G.; Tero-Vescan, A.; Filip, C.; Rusu, A. Capillary Electrophoresis in the Analysis of Polyunsaturated Fatty Acids. *Acta Med. Marisiensis,* **2015**, *61*(4), 378-381.
[http://dx.doi.org/10.1515/amma-2015-0103]

[28] Prado, T.L.A.; de Oliveira, M.A.L. Trans Fatty Acid Determination by Capillary Zone Electrophoresis the State of the Art and Applications. *Anal. Methods,* **2017**, *9*(17), 2483-2494.
[http://dx.doi.org/10.1039/C7AY00193B]

[29] Amorim, T.L.; de Oliveira, M.A.L. Advances in Lipid Capillary Electromigration Methods to Food Analysis Within the 2010s Decade. *Food Anal. Methods,* **2020**, *13*, 1503-1522.
[http://dx.doi.org/10.1007/s12161-020-01772-w]

[30] De Oliveira, M.A.L.; Vaz, F.A. Duarte, L.; Chellini, P. R.; Marques, R.; Sato, R. T.; Prado, T. L. A., Employment of the Chemometric Methods in Capillary Electrophoresis.*Chemometrics: Methods, Applications and New Research*; Nova Science Publishers, **2017**.

[31] de Oliveira, M.A.L.; Micke, G.A.; Bruns, R.E.; Tavares, M.F.M. Factorial design of electrolyte systems for the separation of fatty acids by capillary electrophoresis. *J. Chromatogr. A,* **2001**, *924*(1-2), 533-539.
[http://dx.doi.org/10.1016/S0021-9673(01)00940-2] [PMID: 11521907]

[32] Vaz, F.A.S.; Oliveira, C.L.M.; Oliveira, M.A.L. Fundamentos de Eletroforese Capilar: Uma Abordagem Por Animações. *Quim. Nova,* **2015**, *38*(5), 732-737.

[33] Stolz, A.; Jooß, K.; Höcker, O.; Römer, J.; Schlecht, J.; Neusüß, C. Recent advances in capillary electrophoresis-mass spectrometry: Instrumentation, methodology and applications. *Electrophoresis,* **2019**, *40*(1), 79-112.
[http://dx.doi.org/10.1002/elps.201800331] [PMID: 30260009]

[34] Bittman, R. Glycerolipids: Chemistry.*Encyclopedia of Biophysics*; Roberts, G.C.K., Ed.; Springer: Berlin, Heidelberg, **2013**, pp. 907-914.
[http://dx.doi.org/10.1007/978-3-642-16712-6_527]

[35] Karantonis, H.C.; Nomikos, T.; Demopoulos, C.A. Triacylglycerol metabolism. *Curr. Drug Targets,* **2009**, *10*(4), 302-319.
[http://dx.doi.org/10.2174/138945009787846443] [PMID: 19355855]

[36] Montealegre, C.; Sánchez-Hernández, L.; Crego, A.L.; Marina, M.L. Determination and characterization of glycerophospholipids in olive fruit and oil by nonaqueous capillary electrophoresis with electrospray-mass spectrometric detection. *J. Agric. Food Chem.,* **2013**, *61*(8), 1823-1832.
[http://dx.doi.org/10.1021/jf304357e] [PMID: 23379923]

[37] Fung, F.M.; Su, M.; Feng, H.T.; Li, S.F.Y. Extraction, separation and characterization of endotoxins in water samples using solid phase extraction and capillary electrophoresis-laser induced fluorescence. *Sci. Rep.*, **2017**, *7*(1), 10774.
[http://dx.doi.org/10.1038/s41598-017-11232-x] [PMID: 28883632]

[38] Xiao, M.W.; Bai, X.L.; Liu, Y.M.; Yang, L.; Liao, X. Simultaneous determination of trace Aflatoxin B_1 and Ochratoxin A by aptamer-based microchip capillary electrophoresis in food samples. *J. Chromatogr. A*, **2018**, *1569*, 222-228.
[http://dx.doi.org/10.1016/j.chroma.2018.07.051] [PMID: 30037541]

[39] Gao, J.; Wang, J.; Wu, C.; Hou, F.; Chang, S.; Wang, Z.; Pu, Q.; Guo, D.; Fu, H. Fast screening of aflatoxins in dairy cattle feeds with CE-LIF method combined with preconcentration technique of vortex assisted low density solvent-microextraction. *Electrophoresis*, **2019**, *40*(4), 499-507.
[http://dx.doi.org/10.1002/elps.201800339] [PMID: 30467879]

[40] Wu, X.; Xu, Z.; Huang, Z.; Shao, C. Large volume sample stacking of cationic tetracycline antibiotics toward 10 ppb level analysis by capillary electrophoresis with UV detection. *Electrophoresis*, **2016**, *37*(22), 2963-2969.
[http://dx.doi.org/10.1002/elps.201600189] [PMID: 27494422]

[41] Hong, Y.Q.; Guo, X.; Chen, G.H.; Zhou, J.W.; Zou, X.M.; Liao, X.; Hou, T. Determination of Five Macrolide Antibiotic Residues in Milk by Micellar Electrokinetic Capillary Chromatography with Field Amplified Sample Stacking. *J. Food Saf.*, **2018**, *38*(1), 1-7.
[http://dx.doi.org/10.1111/jfs.12382]

[42] Vesper, H.; Schmelz, E.M.; Nikolova-Karakashian, M.N.; Dillehay, D.L.; Lynch, D.V.; Merrill, A.H., Jr Sphingolipids in food and the emerging importance of sphingolipids to nutrition. *J. Nutr.*, **1999**, *129*(7), 1239-1250.
[http://dx.doi.org/10.1093/jn/129.7.1239] [PMID: 10395583]

[43] Montealegre, C.; Verardo, V.; Luisa Marina, M.; Caboni, M.F. Analysis of glycerophospho- and sphingolipids by CE. *Electrophoresis*, **2014**, *35*(6), 779-792.
[http://dx.doi.org/10.1002/elps.201300534] [PMID: 24301713]

[44] D. A. A.; Wooldridge, K. G., Bacterial Pathogenicity.*Medical Microbiology*; Churchill Livingstone, **2012**, pp. 156-167.
[http://dx.doi.org/10.1016/B978-0-7020-4089-4.00028-7]

[45] Wassenaar, T.M.; Zimmermann, K. Lipopolysaccharides in Food, Food Supplements, and Probiotics: Should We be Worried? *Eur. J. Microbiol. Immunol. (Bp.)*, **2018**, *8*(3), 63-69.
[http://dx.doi.org/10.1556/1886.2018.00017] [PMID: 30345085]

[46] Li, J.; Richards, J.C. Application of capillary electrophoresis mass spectrometry to the characterization of bacterial lipopolysaccharides. *Mass Spectrom. Rev.*, **2007**, *26*(1), 35-50.
[http://dx.doi.org/10.1002/mas.20105] [PMID: 16967446]

[47] O'Hagan, D. Biosynthesis of polyketide metabolites. *Nat. Prod. Rep.*, **1992**, *9*(5), 447-479.
[http://dx.doi.org/10.1039/np9920900447] [PMID: 1436739]

[48] Hanson, J.R. Polyketides from Fungi.*The Chemistry of Fungi*; Hanson, J.R., Ed.; Royal Society of Chemistry, **2008**, pp. 47-72.

[49] Morita, H.; Abe, I.; Noguchi, H.; Plant Type, P.K.S., III *Comprehensive Natural Products II*; Liu, H-W.; Mander, L., Eds.; Elsevier Ltd, **2010**, pp. 171-225.
[http://dx.doi.org/10.1016/B978-008045382-8.00022-8]

[50] World Health Organization. *Aflatoxins. In: Food Safety Digest*; , **2018**, pp. 1-5.

[51] Sáenz, J.P.; Sezgin, E.; Schwille, P.; Simons, K. Functional convergence of hopanoids and sterols in membrane ordering. *Proc. Natl. Acad. Sci. USA*, **2012**, *109*(35), 14236-14240.
[http://dx.doi.org/10.1073/pnas.1212141109] [PMID: 22893685]

[52] Gañán, J.; Morante-Zarcero, S.; Gallego-Picó, A.; Garcinuño, R.M.; Fernández-Hernando, P.; Sierra, I. Evaluation of a molecularly imprinted polymer for determination of steroids in goat milk by matrix solid phase dispersion. *Talanta,* **2014**, *126*, 157-162.
[http://dx.doi.org/10.1016/j.talanta.2014.03.041] [PMID: 24881547]

[53] El Fellah, S.; Duporté, G.; Sirén, H. Steroid Hormones, Inorganic Ions and Botrydial in Drinking Water. Determination with Capillary Electrophoresis and Liquid Chromatography-Orbitrap High Resolution Mass Spectrometry. *Microchem. J.,* **2017**, *133*, 126-136.
[http://dx.doi.org/10.1016/j.microc.2017.03.016]

[54] Sirén, H.; Vesanen, S.; Suomi, J. Separation of steroids using vegetable oils in microemulsion electrokinetic capillary chromatography. *J. Chromatogr. B Analyt. Technol. Biomed. Life Sci.,* **2014**, *945-946*, 199-206.
[http://dx.doi.org/10.1016/j.jchromb.2013.11.052] [PMID: 24355214]

[55] Kuzuyama, T.; Seto, H. Diversity of the biosynthesis of the isoprene units. *Nat. Prod. Rep.,* **2003**, *20*(2), 171-183.
[http://dx.doi.org/10.1039/b109860h] [PMID: 12735695]

[56] Wang, X.; Li, K.; Yao, L.; Wang, C.; Van Schepdael, A. Recent advances in vitamins analysis by capillary electrophoresis. *J. Pharm. Biomed. Anal.,* **2018**, *147*, 278-287.
[http://dx.doi.org/10.1016/j.jpba.2017.07.030] [PMID: 28784252]

[57] Lu, D.; Yang, Y.; Li, Y.; Sun, C. Analysis of Tocopherols and Tocotrienols in Pharmaceuticals and Foods: A Critical Review. *Curr. Pharm. Anal.,* **2014**, *11*(1), 66-78.
[http://dx.doi.org/10.2174/1573412910666140630170055]

[58] Butnariu, M. Methods of Analysis (Extraction, Separation, Identification and Quantification) of Carotenoids from Natural Products. *J. Ecosyst. Ecography,* **2016**, *6*(2)
[http://dx.doi.org/10.4172/2157-7625.1000193]

[59] Poinsot, V.; Ta, H.Y.; Meang, V.O.; Perquis, L.; Gavard, P.; Pipy, B.; Couderc, F. A digest of capillary electrophoretic methods applied to lipid analyzes. *Electrophoresis,* **2019**, *40*(1), 190-211.
[http://dx.doi.org/10.1002/elps.201800264] [PMID: 30073670]

[60] Otieno, A.C.; Mwongela, S.M. Capillary electrophoresis-based methods for the determination of lipids--a review. *Anal. Chim. Acta,* **2008**, *624*(2), 163-174.
[http://dx.doi.org/10.1016/j.aca.2008.06.026] [PMID: 18706322]

[61] De Oliveira, M.A.L.; Porto, B.L.S.; De Bastos, C.A.; Sabarense, C.M.; Vaz, F.A.S.; Neves, L.N.O.; Duarte, L.M.; Da Campos, N.S.; Chellini, P.R.; Da Silva, P.H.F.; de Sousa, R.A.; Marques, R.; Sato, R.T.; Grazul, R.M.; Lisboa, T.P.; Mendes, T.O.; Rios, V.C. Analysis of Amino Acids, Proteins, Carbohydrates and Lipids in Food by Capillary Electromigration Methods: A Review Analysis of Amino Acids, Proteins, Carbohydrates and Lipids in Food by Capillary Electromigration Methods: A Review. *Anal. Methods,* **2016**, *8*(18), 3649-3680.
[http://dx.doi.org/10.1039/C5AY02736E]

[62] Monti, L.; Cattaneo, T.M.P.; Orlandi, M.; Curadi, M.C. Capillary electrophoresis of sialylated oligosaccharides in milk from different species. *J. Chromatogr. A,* **2015**, *1409*, 288-291.
[http://dx.doi.org/10.1016/j.chroma.2015.07.076] [PMID: 26228851]

[63] Sirén, H.; Sirén, K.; Sirén, J. Evaluation of Organic and Inorganic Compounds Levels of Red Wines Processed from Pinot Noir Grapes. *Anal. Chem. Res.,* **2015**, *3*, 26-36.
[http://dx.doi.org/10.1016/j.ancr.2014.10.002]

[64] Toutounji, M.R.; Van Leeuwen, M.P.; Oliver, J.D.; Shrestha, A.K.; Castignolles, P.; Gaborieau, M. Quantification of sugars in breakfast cereals using capillary electrophoresis. *Carbohydr. Res.,* **2015**, *408*, 134-141.
[http://dx.doi.org/10.1016/j.carres.2015.03.008] [PMID: 25841355]

[65] Zhao, L.; Chanon, A.M.; Chattopadhyay, N.; Dami, I.E.; Blakeslee, J.J. Quantification of

Carbohydrates in Grape Tissues Using Capillary Zone Electrophoresis. *Front. Plant Sci.*, **2016**, *7*, 818. [http://dx.doi.org/10.3389/fpls.2016.00818] [PMID: 27379118]

[66] Lu, Y.; Hu, Y.; Wang, T.; Yang, X.; Zhao, Y. Rapid Determination and Quantitation of Compositional Carbohydrates to Identify Honey by Capillary Zone Electrophoresis. *CYTA J. Food,* **2017**, *15*(4), 531-537. [http://dx.doi.org/10.1080/19476337.2017.1308970]

[67] Lechtenberg, M.; Hensel, A. Determination of Glucosinolates in Broccoli-Based Dietary Supplements by Cyclodextrin-Mediated Capillary Zone Electrophoresis. *J. Food Compos. Anal.*, **2019**, *78*, 138-149. [http://dx.doi.org/10.1016/j.jfca.2019.02.007]

[68] Surapureddi, S.R.K.; Kunta, R.; Sameer Kumar, G.S.; Sappidi, S.R.; Dadke, S. A sensitive and high throughput method for the analysis of d-psicose by capillary electrophoresis. *Food Chem.*, **2019**, *281*, 36-40. [http://dx.doi.org/10.1016/j.foodchem.2018.12.081] [PMID: 30658762]

[69] Biluca, F.C.; Della Betta, F.; de Oliveira, G.P.; Pereira, L.M.; Gonzaga, L.V.; Costa, A.C.O.; Fett, R. 5-HMF and carbohydrates content in stingless bee honey by CE before and after thermal treatment. *Food Chem.*, **2014**, *159*, 244-249. [http://dx.doi.org/10.1016/j.foodchem.2014.03.016] [PMID: 24767051]

[70] Oliveira, D.L.; Wilbey, R.A.; Grandison, A.S.; Roseiro, L.B. Natural Caprine Whey Oligosaccharides Separated by Membrane Technology and Profile Evaluation by Capillary Electrophoresis. *Food Bioprocess Technol.*, **2014**, *7*(3), 915-920. [http://dx.doi.org/10.1007/s11947-013-1153-1]

[71] Navarro Pascual-Ahuir, M.; Lerma-García, M.J.; Simó-Alfonso, E.F.; Herrero-Martínez, J.M. Rapid differentiation of commercial juices and blends by using sugar profiles obtained by capillary zone electrophoresis with indirect UV detection. *J. Agric. Food Chem.*, **2015**, *63*(10), 2639-2646. [http://dx.doi.org/10.1021/acs.jafc.5b00122] [PMID: 25719749]

[72] Jiang, T.F.; Chong, L.; Yue, M.E.; Wang, Y.H.; Lv, Z.H. Separation and Determination of Carbohydrates in Food Samples by Capillary Electrophoresis Using Dynamically Coating the Capillary with Indirect UV Detection. *Food Anal. Methods,* **2015**, *8*(10), 2588-2594. [http://dx.doi.org/10.1007/s12161-015-0157-z]

[73] Dominguez, M.A.; Jacksén, J.; Emmer, Å.; Centurión, M.E. Capillary Electrophoresis Method for the Simultaneous Determination of Carbohydrates and Proline in Honey Samples. *Microchem. J.*, **2016**, *129*, 1-4. [http://dx.doi.org/10.1016/j.microc.2016.05.017]

[74] Moreno, D.; Berli, F.; Bottini, R.; Piccoli, P.N.; Silva, M.F. Grapevine tissues and phenology differentially affect soluble carbohydrates determination by capillary electrophoresis. *Plant Physiol. Biochem.*, **2017**, *118*, 394-399. [http://dx.doi.org/10.1016/j.plaphy.2017.07.010] [PMID: 28711788]

[75] de Oliveira Neves, L.N.; Marques, R.; da Silva, P.H.F.; de Oliveira, M.A.L. Lactulose determination in UHT milk by CZE-UV with indirect detection. *Food Chem.*, **2018**, *258*, 337-342. [http://dx.doi.org/10.1016/j.foodchem.2018.03.069] [PMID: 29655742]

[76] Liang, P.; Sun, M.; He, P.; Zhang, L.; Chen, G. Determination of carbohydrates in honey and milk by capillary electrophoresis in combination with graphene-cobalt microsphere hybrid paste electrodes. *Food Chem.*, **2016**, *190*, 64-70. [http://dx.doi.org/10.1016/j.foodchem.2015.05.059] [PMID: 26212942]

[77] Mao, H.; Tian, Z.; Zhang, L.; Chen, G. Fabrication of a Magnet-Assembled Alignment Device for the Amperometric Detection of Carbohydrates in Combination with CE. *Chromatographia,* **2018**, *81*(10), 1421-1430. [http://dx.doi.org/10.1007/s10337-018-3589-y]

[78] García, M.; Escarpa, A. Microchip electrophoresis-copper nanowires for fast and reliable

determination of monossacharides in honey samples. *Electrophoresis,* **2014**, *35*(2-3), 425-432. [http://dx.doi.org/10.1002/elps.201300458] [PMID: 24115078]

[79] Duarte-Junior, G.F.; Lobo-Júnior, E.O.; Medeiros Junior, Í.; da Silva, J.A.F.; do Lago, C.L.; Coltro, W.K.T. Separation of carbohydrates on electrophoresis microchips with controlled electrolysis. *Electrophoresis,* **2019**, *40*(5), 693-698. [http://dx.doi.org/10.1002/elps.201800354] [PMID: 30597587]

[80] Rizelio, V.M.; Tenfen, L.; da Silveira, R.; Gonzaga, L.V.; Costa, A.C.O.; Fett, R. Development of a fast capillary electrophoresis method for determination of carbohydrates in honey samples. *Talanta,* **2012**, *93*, 62-66. [http://dx.doi.org/10.1016/j.talanta.2012.01.034] [PMID: 22483877]

[81] Rovio, S.; Yli-Kauhaluoma, J.; Sirén, H. Determination of neutral carbohydrates by CZE with direct UV detection. *Electrophoresis,* **2007**, *28*(17), 3129-3135. [http://dx.doi.org/10.1002/elps.200600783] [PMID: 17661315]

[82] Bao, Y.; Zhu, L.; Newburg, D.S. Simultaneous quantification of sialyloligosaccharides from human milk by capillary electrophoresis. *Anal. Biochem.,* **2007**, *370*(2), 206-214. [http://dx.doi.org/10.1016/j.ab.2007.07.004] [PMID: 17761135]

[83] Damodaran, S.; Parkin, K.L. *Fennema's food chemistry,* 5th ed; CRC press, **2017**, pp. 235-356.

[84] Damodaran, S.; Parkin, K.L.; Fennema, O.R. *Fennema's Food Chemistry,* 4th ed; CRC press, **2007**, pp. 217-330.

[85] Kinsella, J.E.; Damodaran, S.; German, B. *New Protein Food: Seed Storage Proteins*; Altshul, A.M.; Wilcke, H.L., Eds.; Academic Press: London, **1985**, Vol. 5, pp. 107-179. [http://dx.doi.org/10.1016/B978-0-12-054805-7.50012-9]

[86] Acunha, T.; Ibáñez, C.; García-Cañas, V.; Simó, C.; Cifuentes, A. Recent advances in the application of capillary electromigration methods for food analysis and Foodomics. *Electrophoresis,* **2016**, *37*(1), 111-141. [http://dx.doi.org/10.1002/elps.201500291] [PMID: 26256797]

[87] García-Ruiz, C.; Marina, M.L. *Analysis and Detection by Capillary Electrophoresis*; Marina, M.L.; Ríos, A.; Valcárcel, M., Eds.; Elsevier Science: Amsterdam, **2005**, Vol. 45, pp. 617-701. [http://dx.doi.org/10.1016/S0166-526X(05)45013-8]

[88] Pérez-Miguez, R.; Castro-Puyana, M.; Marina, M.L. *Advances in food analysis research*; Haynes, A., Ed.; Nova Science Publishers, Incorporated: New York, **2015**, pp. 89-120.

[89] Stuknytė, M.; Cattaneo, S.; Masotti, F.; De Noni, I. Occurrence and fate of ACE-inhibitor peptides in cheeses and in their digestates following *in vitro* static gastrointestinal digestion. *Food Chem.,* **2015**, *168*, 27-33. [http://dx.doi.org/10.1016/j.foodchem.2014.07.045] [PMID: 25172679]

[90] Kondeková, M.; Maier, V.; Ginterová, P.; Marák, J.; Sevčík, J. Analysis of lysozyme in cheese samples by on-line combination of capillary zone electrophoresis and mass spectrometry. *Food Chem.,* **2014**, *153*, 398-404. [http://dx.doi.org/10.1016/j.foodchem.2013.12.078] [PMID: 24491746]

[91] Salmanowicz, B.P.; Langner, M.; Kubicka-Matusiewicz, H. Variation of high-molecular-weight secalin subunit composition in Rye (Secale cereale L.) inbred lines. *J. Agric. Food Chem.,* **2014**, *62*(43), 10535-10541. [http://dx.doi.org/10.1021/jf502926q] [PMID: 25307487]

[92] Vergara-Barberán, M.; Lerma-García, M.J.; Herrero-Martínez, J.M.; Simó-Alfonso, E.F. Use of protein profiles established by CZE to predict the cultivar of olive leaves and pulps. *Electrophoresis,* **2014**, *35*, 1652-1659. [http://dx.doi.org/10.1002/elps.201300530]

[93] Gustavsson, F.; Buitenhuis, A.J.; Johansson, M.; Bertelsen, H.P.; Glantz, M.; Poulsen, N.A.; Lindmark

Månsson, H.; Stålhammar, H.; Larsen, L.B.; Bendixen, C.; Paulsson, M.; Andrén, A. Effects of breed and casein genetic variants on protein profile in milk from Swedish Red, Danish Holstein, and Danish Jersey cows. *J. Dairy Sci.*, **2014**, *97*(6), 3866-3877. [http://dx.doi.org/10.3168/jds.2013-7312] [PMID: 24704225]

[94] de Vries, R.; van Knegsel, A.; Johansson, M.; Lindmark-Månsson, H.; van Hooijdonk, T.; Holtenius, K.; Hettinga, K. Effect of shortening or omitting the dry period of Holstein-Friesian cows on casein composition of milk. *J. Dairy Sci.*, **2015**, *98*(12), 8678-8687. [http://dx.doi.org/10.3168/jds.2015-9544] [PMID: 26454294]

[95] Zhang, Y.; Chen, L.; Zhang, C.; Liu, S.; Zhu, H.; Wang, Y. Polydopamine-assisted partial hydrolyzed poly(2-methyl-2-oxazolinze) as coating for determination of melamine in milk by capillary electrophoresis. *Talanta,* **2016**, *150*, 375-387. [http://dx.doi.org/10.1016/j.talanta.2015.12.054] [PMID: 26838421]

[96] Merheb-Dini, C.; Chaves, K.S.; Gomes, E.; da Silva, R.; Gigante, M.L. Coalho Cheese Made with Protease from Thermomucor indicae-seudaticae N31: Technological Potential of the New Coagulant for the Production of High-Cooked Cheese. *J. Food Sci.*, **2016**, *81*(3), C563-C568. [http://dx.doi.org/10.1111/1750-3841.13217] [PMID: 26799929]

[97] Baptista, D.P.; Araújo, F.D.D.S.; Eberlin, M.N.; Gigante, M.L. A Survey of the Peptide Profile in Prato Cheese as Measured by MALDI-MS and Capillary Electrophoresis. *J. Food Sci.*, **2017**, *82*(2), 386-393. [http://dx.doi.org/10.1111/1750-3841.13618] [PMID: 28098943]

[98] Trimboli, F.; Morittu, V.M.; Cicino, C.; Palmieri, C.; Britti, D. Rapid capillary electrophoresis approach for the quantification of ewe milk adulteration with cow milk. *J. Chromatogr. A,* **2017**, *1519*, 131-136. [http://dx.doi.org/10.1016/j.chroma.2017.08.075] [PMID: 28888682]

[99] Pruska-Kędzior, A.; Makowska, A.; Kędzior, Z.; Salmanowicz, B.P. Rheological characterisation of gluten from triticale (x Triticosecale Wittmack). *J. Sci. Food Agric.*, **2017**, *97*(14), 5043-5052. [http://dx.doi.org/10.1002/jsfa.8555] [PMID: 28722133]

[100] Qiu, J.; Wang, J.; Xu, Z.; Liu, H.; Ren, J. Quantitation of underivatized branched-chain amino acids in sport nutritional supplements by capillary electrophoresis with direct or indirect UV absorbance detection. *PLoS One,* **2017**, *12*(6), e0179892. [http://dx.doi.org/10.1371/journal.pone.0179892] [PMID: 28640882]

[101] Ten-Doménech, I.; Simó-Alfonso, E.F.; Herrero-Martínez, J.M. Improving Fractionation of Human Milk Proteins through Calcium Phosphate Coprecipitation and Their Rapid Characterization by Capillary Electrophoresis. *J. Proteome Res.*, **2018**, *17*(10), 3557-3564. [http://dx.doi.org/10.1021/acs.jproteome.8b00526] [PMID: 30156851]

[102] Chen, H.; Shi, P.; Fan, F.; Tu, M.; Xu, Z.; Xu, X.; Du, M. Complementation of UPLC-Q-TOF-MS and CESI-Q-TOF-MS on identification and determination of peptides from bovine lactoferrin. *J. Chromatogr. B Analyt. Technol. Biomed. Life Sci.*, **2018**, *1084*, 150-157. [http://dx.doi.org/10.1016/j.jchromb.2018.03.022] [PMID: 29604613]

[103] Fu, L.; Zhou, J.; Wang, C.; Zhang, Y.; Ma, A.; Wang, Y. Determination of a major allergen in fish samples by simple and effective label-free capillary electrophoretic analysis after background suppression in ion-exchange chromatography. *Food Chem.*, **2018**, *261*, 124-130. [http://dx.doi.org/10.1016/j.foodchem.2018.04.037] [PMID: 29739572]

[104] Benavente, F.; Pero-Gascon, R.; Pont, L.; Jaumot, J.; Barbosa, J.; Sanz-Nebot, V. Identification of antihypertensive peptides in nutraceuticals by capillary electrophoresis-mass spectrometry. *J. Chromatogr. B Analyt. Technol. Biomed. Life Sci.*, **2018**, *1579*, 129-137. [http://dx.doi.org/10.1016/j.chroma.2018.10.018] [PMID: 30361036]

[105] Taivosalo, A.; Kriščiunaite, T.; Seiman, A.; Part, N.; Stulova, I.; Vilu, R. Comprehensive analysis of proteolysis during 8 months of ripening of high-cooked Old Saare cheese. *J. Dairy Sci.*, **2018**, *101*(2),

944-967.
[http://dx.doi.org/10.3168/jds.2017-12944] [PMID: 29174156]

[106] D'Incecco, P.; Brasca, M.; Rosi, V.; Morandi, S.; Ferranti, P.; Picariello, G.; Pellegrino, L. Bacterial proteolysis of casein leading to UHT milk gelation: An applicative study. *Food Chem.,* **2019**, *292*, 217-226.
[http://dx.doi.org/10.1016/j.foodchem.2019.04.066] [PMID: 31054668]

[107] Nguyen, T.D.; Nguyen, M.H.; Vu, M.T.; Duong, H.A.; Pham, H.V.; Mai, T.D. Dual-channeled capillary electrophoresis coupled with contactless conductivity detection for rapid determination of choline and taurine in energy drinks and dietary supplements. *Talanta,* **2019**, *193*, 168-175.
[http://dx.doi.org/10.1016/j.talanta.2018.10.002] [PMID: 30368287]

[108] D'Orazio, G.; Asensio-Ramos, M.; Hernández-Borges, J.; Rodríguez-Delgado, M.Á.; Fanali, S. Evaluation of the combination of a dispersive liquid-liquid microextraction method with micellar electrokinetic chromatography coupled to mass spectrometry for the determination of estrogenic compounds in milk and yogurt. *Electrophoresis,* **2015**, *36*(4), 615-625.
[http://dx.doi.org/10.1002/elps.201400452] [PMID: 25394185]

[109] Li, L.Q.; Baibado, J.T.; Shen, Q.; Cheung, H.Y. Determination of the authenticity of plastron-derived functional foods based on amino acid profiles analysed by MEKC. *J. Chromatogr. B Analyt. Technol. Biomed. Life Sci.,* **2017**, *1070*, 23-30.
[http://dx.doi.org/10.1016/j.jchromb.2017.10.036] [PMID: 29107730]

[110] Uysal, R.S.; Boyacı, I.H.; Soykut, E.A.; Ertaş, N. Effects of heat treatment parameters on liquid whole egg proteins. *Food Chem.,* **2017**, *216*, 201-208.
[http://dx.doi.org/10.1016/j.foodchem.2016.08.050] [PMID: 27596410]

[111] Drevinskas, T.; Stankevičius, M.; Bimbiraitė-Survilienė, K.; Naujokaitytė, G.; Kaškonienė, V.; Kondrotienė, K.; Šernienė, L.; Kornyšova, O.; Malakauskas, M.; Maruška, A. Optimization of a capillary zone electrophoresis-contactless conductivity detection method for the determination of nisin. *Electrophoresis,* **2018**, *39*(19), 2425-2430.
[http://dx.doi.org/10.1002/elps.201800143] [PMID: 29935020]

[112] Wu, R.; Wang, Z.; Zhao, W.; Yeung, W.S.B.; Fung, Y.S. Multi-dimension microchip-capillary electrophoresis device for determination of functional proteins in infant milk formula. *J. Chromatogr. A,* **2013**, *1304*, 220-226.
[http://dx.doi.org/10.1016/j.chroma.2013.06.073] [PMID: 23870546]

[113] Wu, R.; Wang, Z.; Fung, Y.S.; Seah, D.Y.; Yeung, W.S.B. Assessment of adulteration of soybean proteins in dairy products by 2D microchip-CE device. *Electrophoresis,* **2014**, *35*(11), 1728-1734.
[http://dx.doi.org/10.1002/elps.201300559] [PMID: 25025095]

[114] Jozanović, M.; Hajduković, M.; Galović, O.; Kralik, G.; Kralik, Z.; Sakač, N.; Medvidović-Kosanović, M.; Sak-Bosnar, M. Determination of anti-oxidative histidine dipeptides in poultry by microchip capillary electrophoresis with contactless conductivity detection. *Food Chem.,* **2017**, *221*, 1658-1665.
[http://dx.doi.org/10.1016/j.foodchem.2016.10.122] [PMID: 27979143]

[115] Jozanović, M.; Sakač, N.; Sak-Bosnar, M.; Carrilho, E. A simple and reliable new microchip electrophoresis method for fast measurements of imidazole dipeptides in meat from different animal species. *Anal. Bioanal. Chem.,* **2018**, *410*(18), 4359-4369.
[http://dx.doi.org/10.1007/s00216-018-1087-6] [PMID: 29707755]

[116] Xiao, X.; Wang, W.; Chen, J.; Jia, L. Polydopamine-coated open tubular column for the separation of proteins by capillary electrochromatography. *J. Sep. Sci.,* **2015**, *38*(16), 2893-2899.
[http://dx.doi.org/10.1002/jssc.201500315] [PMID: 26017540]

[117] Kristl, T.; Stutz, H. Comparison of different mobilization strategies for capillary isoelectric focusing of ovalbumin variants. *J. Sep. Sci.,* **2015**, *38*(1), 148-156.
[http://dx.doi.org/10.1002/jssc.201400890] [PMID: 25358859]

[118] Feng, P.; Fuerer, C.; McMahon, A. Quantification of whey protein content in infant formulas by

sodium dodecyl sulfate-capillary gel electrophoresis (SDS-CGE): single-laboratory validation, first action 2016.15. *J. AOAC Int.*, **2017**, *100*(2), 510-521.
[http://dx.doi.org/10.5740/jaoacint.16-0344] [PMID: 28088252]

[119] Malá, Z.; Gebauer, P.; Boček, P. Recent progress in analytical capillary isotachophoresis. *Electrophoresis,* **2015**, *36*(1), 2-14.
[http://dx.doi.org/10.1002/elps.201400337] [PMID: 25130397]

[120] Wehr, T. Capillary isoelectric focusing. *Sep. Sci. Technol.,* **2005**, 181-210.

[121] Vithanage, N.R.; Dissanayake, M.; Bolge, G.; Palombo, E.A.; Yeager, T.R.; Datta, N. Biodiversity of culturable psychrotrophic microbiota in raw milk attributable to refrigeration conditions, seasonality and their spoilage potential. *Int. Dairy J.,* **2016**, *57*, 80-90.
[http://dx.doi.org/10.1016/j.idairyj.2016.02.042]

[122] Lee, S.Y.; Hur, S.J. Antihypertensive peptides from animal products, marine organisms, and plants. *Food Chem.,* **2017**, *228*, 506-517.
[http://dx.doi.org/10.1016/j.foodchem.2017.02.039] [PMID: 28317757]

[123] Liao, Y.; Alvarado, R.; Phinney, B.; Lönnerdal, B. Proteomic characterization of human milk whey proteins during a twelve-month lactation period. *J. Proteome Res.,* **2011**, *10*(4), 1746-1754.
[http://dx.doi.org/10.1021/pr101028k] [PMID: 21361340]

[124] Álvarez, G.; Montero, L.; Llorens, L.; Castro-Puyana, M.; Cifuentes, A. Recent advances in the application of capillary electromigration methods for food analysis and Foodomics. *Electrophoresis,* **2018**, *39*(1), 136-159.
[http://dx.doi.org/10.1002/elps.201700321] [PMID: 28975648]

[125] Wu, R.; Wang, Z.; Fung, Y.S. Multidimensional microchip-capillary electrophoresis device for determination of functional proteins in infant milk formula. *Methods Mol. Biol.,* **2015**, *1274*, 111-118.
[http://dx.doi.org/10.1007/978-1-4939-2353-3_10] [PMID: 25673487]

[126] Domínguez Vega, E.; Marina, M.L. Characterization and study of transgenic cultivars by capillary and microchip electrophoresis. *Int. J. Mol. Sci.,* **2014**, *15*(12), 23851-23877.
[http://dx.doi.org/10.3390/ijms151223851] [PMID: 25535077]

[127] Coelho, C.; Bagala, F.; Gougeon, R.D.; Schmitt-Kopplin, P. Capillary Electrophoresis in wine science. *Methods Mol. Biol.,* **2016**, *1483*, 509-523.
[http://dx.doi.org/10.1007/978-1-4939-6403-1_23] [PMID: 27645750]

[128] Dawod, M.; Arvin, N.E.; Kennedy, R.T. Recent advances in protein analysis by capillary and microchip electrophoresis. *Analyst (Lond.),* **2017**, *142*(11), 1847-1866.
[http://dx.doi.org/10.1039/C7AN00198C] [PMID: 28470231]

[129] Wang, C.; Lucy, C.A. Mixed cationic/anionic surfactants for semipermanent wall coatings in capillary electrophoresis. *Electrophoresis,* **2004**, *25*(6), 825-832.
[http://dx.doi.org/10.1002/elps.200305760] [PMID: 15004842]

[130] Huck, C.W.; Bonn, G.K. *Methods in Molecular Biology*; Walker, J.M., Ed.; Humana Press, **2008**, pp. 507-540.

CHAPTER 4

Analysis of Peptides by Capillary Electromigration Methods

Sille Štěpánová[1] and **Václav Kašička**[1,*]

[1] *Institute of Organic Chemistry and Biochemistry of the Czech Academy of Sciences, Flemingovo n. 2, 166 10 Prague 6, Czechia*

Abstract: These peptides themselves and especially as products of enzymatic or chemical cleavage of parental proteins, belong to the important components of foodstuffs. They significantly influence their nutritional, biological, technological, and functional properties. Some of these peptides were found to have effects on human health and nutrition, *e.g.*, by affecting human digestive, endocrine, cardiovascular, immune, and nervous systems. Hence, qualitative and quantitative analysis of peptides in foods is of great importance. For the separation and quantification of peptides in foods, capillary electromigration methods represent one of the most suitable analytical methods. This chapter presents a comprehensive overview of the developments and applications of high performance capillary and microchip electromigration methods (zone electrophoresis, isotachophoresis, isoelectric focusing, affinity electrophoresis, electrokinetic chromatography and electrochromatography) for separation and analysis of peptides in foods and food products in the time period since 2010 up to the middle of 2020. Various aspects of the application of capillary electromigration methods for peptide analysis in foods, such as sample preparation, peptide preseparation, preconcentration, derivatization, adsorption suppression, and detection, are described and discussed. Several particular applications of capillary electromigration methods for separation and analysis of peptides in various food samples of animal, plant, and microbial origin are demonstrated.

Keywords: Bioactive peptides, Capillary electrophoresis, Electromigration methods, Food analysis, Peptides in foods.

INTRODUCTION

Peptides themselves and especially as products of enzymatic (proteolytic) and chemical hydrolysis of parental proteins belong to the important components of foodstuffs. They possess significant effect on their nutritional, biological, functional and technological properties. Food proteins are a major source of biologically active peptides.

* **Corresponding author Václav Kašička:**, Institute of Organic Chemistry and Biochemistry of the Czech Academy of Sciences, Flemingovo n. 2, 166 10 Prague 6, Czechia; E-mail: kasicka@uochb.cas.cz

María Castro-Puyana, Miguel Herrero and María Luisa Marina (Eds.)
Capillary Electrophoresis in Food Analysis

Food-derived biopeptides are mostly released from animal, plant or microbial proteins by enzymatic hydrolysis under *in-vivo* or *in-vitro* conditions using variable proteases. The peptides remain biologically inactive within the parent proteins until they are released during gastrointestinal digestion, during the processing of food or by using proteolytic enzymes or microbial fermentation [1 - 3]. Bioactive peptides exhibit many vitally important activities in all living organisms. They act as hormones, neurotransmitters, enzyme substrates and inhibitors, co-enzymes, ionophores, antibiotics, and drugs. Peptides in foods possess many biological effects. Based on their mode of action, they can be classified as antioxidant, antimicrobial, antifungal, antithrombotic, anticoagulant, antihypertensive, antidiabetic, opioid, hypo-cholesterolemic, immunomodulatory, mineral binding, and other [3 - 5]. Antihypertensive peptide inhibitors of angiotensin converting enzyme have been derived from milk, corn and fish proteins. Opioid peptides were found to be released by cleavage of wheat gluten or milk casein by pepsin. Immunomodulatory peptides were derived from hydrolysates of rice and soybean proteins by trypsin. Antioxidant properties were observed for peptides originating from milk proteins. Both hydrophilic and lipophilic antioxidant activity were found at caseinophosphopeptides obtained by a tryptic digestion of casein. Because of the possible health-influencing effects of peptides by affecting the human digestive, endocrine, cardiovascular, immune, and nervous systems, the separation and qualitative and quantitative analysis of bioactive peptides in foods is a subject of great interest and importance and a challenge for analytical methods.

Besides the most spread separation methods, different modes of (ultra)high-performance liquid chromatography ((U)HPLC), capillary electromigration (CE) techniques represent alternative and/or complementary separation and analytical tools for food analysis and foodomics, including analysis of peptides in foods and food products [6 - 8].

CE methods enable the separation of a wide set of compounds ranging from small molecules, such as metal ions, inorganic and organic acids and bases, and amino acids, through the medium sized peptides and oligonucleotides up to the large biopolymers, such as proteins, polysaccharides and nucleic acids (see the recent comprehensive review [9] and book [10] on the latest developments and applications of CE methods for the analysis of various types of (bio)molecules and (bio)particles). Based on the separation principles, CE methods can be divided into two major groups. The first one includes purely electrophoretic modes, such as zone electrophoresis in a free solution (CZE) or in gel and sieving media (CGE and CSE), Isotachophoresis (CITP), and Isoelectric Focusing (CIEF), Affinity Electrophoresis (ACE), which separate the soluble charged compounds according to their electrophoretic mobilities governed by the charge/size (charge/mass) ratio of their molecules. The second group of CE methods contains the combined electro- chromatographic methods, particularly, the Capillary Electrokinetic Chromatography (EKC), especially its most spread micellar mode, Micellar EKC (MEKC), and Capillary Electrochromatography (CEC). These methods can separate both charged and non-charged compounds and their separation is based mainly on their different interactions with stationary (CEC) or pseudostationary (MEKC) phases.

The advantages of all the above CE methods include high separation efficiency, short analysis time, relatively simple method development, low sample and solvent consumption, and low running costs. Their main drawback is their relatively low concentration sensitivity due to the short optical path of the most frequently used UV-Vis absorption detector.

Peptide separations and analyses by CE methods in general have been regularly reviewed within a two-year period in the last 20 years (see the last review of this series [11] and the references cited therein). CE methods combined with variable detection modes, UV-absorption, fluorescence, contactless conductivity, and especially mass spectrometry (MS), are frequently employed in the fields of peptidomics and proteomics [12 - 14].

This chapter presents a comprehensive overview of the developments and applications of CE methods for the separation and analysis of peptides in foods and food products from 2010 until the middle of 2020. Various aspects of application of CE methods for peptide analysis in foods, such as sample preparation, peptide preseparation, preconcentration, derivatization, adsorption suppression, and detection, are described and discussed. Moreover, several particular applications of CE methods for separation and analysis of peptides in various food samples of animal, plant and microbial origin are demonstrated.

METHODOLOGY

Sample Preparation

One major problem in the analysis of peptides in foods in general is that the food matrix is usually very complex. Various food matrix components, additives, and food treatment during processing and storage can interfere with peptide determinations in foodstuffs. Additionally, the sensitivity of the employed detection methods or the separation power of the CE technique might be not sufficient for the separation of complex peptide mixtures and for direct analysis of peptides present at low concentrations. For these reasons, the sample preparation, comprising peptide preseparation and/or preconcentration, is often an important and necessary step in the CE analysis of peptides in foods.

Preseparation and preconcentration techniques using both CE and LC principles for sample preparation prior to CE analysis of biomolecules, including peptides, in complex matrices have been regularly reviewed in the last ten years (see the last review of this series [15] and the references cited therein and some other recent reviews [16 - 20]). Other papers provided an overview of specific preseparation or preconcentration methods, such as Solid Phase Extraction (SPE) [21, 22], pH mediated SPE [23], immunoaffinity SPE [24], Solid State Microextraction (SPME) [25], Electromembrane Extraction (EME) [26] or electrophoretic sample stacking [27]. The electrophoretic stacking methods embrace Field Amplified Sample Stacking (FASS) [28], Field Enhanced or Field Amplified Sample Injection (FASI or FESI) [29], Large Volume Sample Stacking (LVSS) [30], Transient- Isotachophoresis (t-ITP) [31, 32], dynamic pH junction [33, 34], sweeping [35, 36], and electrodialysis with ultrafiltration membrane [37].

The strategies for sample preparation and other aspects of peptide analysis in peptidomic studies, such as purification, identification and quantitation, have been discussed in the comprehensive review of current peptidomics [38]. Sample preparation typically involves sampling, extraction, clean up and concentration and it is followed by the CE separation. Peptides differ in several parameters, such as size, effective (net) charge, hydrophobicity, and binding capabilities of their molecules. Their physicochemical and functional variety is further increased by posttranslational modifications, *e.g.*, glycosylation, phosphorylation, oxidation, and acetylation. Due to the natural diversity of peptides, it is often the case that multiple peptide purification methods are used sequentially on the same sample to achieve a complete separation/isolation of peptides from the complex matrix [38].

The first step of sample pre-treatment is typically a physical homogenization (*e.g.* grinding, pulverization, ultrasonication, increased pressure, and freezing/thawing). The food samples in peptide analysis are usually homogenized in water or aqueous buffers. Sometimes, homogenization, centrifugation and filtration are sufficient for the sample clean-up.

Nevertheless, it is oftentimes necessary to separate high-molecular-mass biopolymers, proteins, nucleic acids, and polysaccharides, from the low- or middle- molecular-mass peptides. An effective method for isolation of specific peptide mass/size ranges is size-exclusion chromatography. In addition, molecular mass cut-off membrane (ultra)filtration can be utilized but it does not allow for the complete separation of a specific mass/size range without partial losses of peptides and/or partial contamination from the undesired fractions. Proteins can also be removed by selective precipitation with the aid of organic solvents (*e.g.* acetonitrile, methanol, acetone), acids (*e.g.*, trichloroacetic acid, trifluoroacetic acid) or their combinations. Nevertheless, during protein precipitation, some peptides may aggregate and can be lost in the precipitate. Additionally, to avoid uncontrolled proteolytic activity of proteases during and after sample collection, protease inhibitors (*e.g.*, phenylmethylsulfonyl fluoride) may be added, or, for protein denaturation, solvents or acids can be used, or microwave irradiation can be applied. Described procedures are usually followed by centrifugation and/or (ultra)filtration. Sometimes samples can be further purified by SPE and/or preparative chromatography.

In proteomic studies, where tryptic peptide fragments of proteins are separated and analyzed, the strategies for sample preparation are somewhat more complicated than for peptide analysis or in peptidomic studies. Useful information can be found in a recent comprehensive review of sample preparation in foodomics [39] where the procedures for sample preparation in genomics, metabolomics, proteomics, and peptidomics are described. Sample preparation for proteomics and peptidomics also involves homogenization and protein/polypeptide extraction step which is followed by their solubilization by addition of buffers, detergents (*e.g.*, anionic sodium dodecyl sulfate (SDS), cationic cetyltrimethylammonium bromide (CTAB) or amphoteric 3-[(--cholamidopropyl)dimethylammonio]-1-propanesulfonate (CHAPS), and/or chaotropic agents, *e.g.*, urea. Uncontrolled enzymatic degradation of proteins and polypeptides should be avoided. Frequently, various depletion, prefractionation and enrichment strategies are applied in proteomics and peptidomics. The bottom-up approach involves the digestion of proteins and polypeptides by trypsin or other site-specific proteases or chemical cleavage into peptides.

Sometimes, in order to change the analytes separation and detection characteristics as part of the sample preparation, derivatization is performed [40]. Peptides derivatization involves labelling with fluorophore in order to enable their detection by (laser induced) fluorescence (LIF). Fluorescein isothiocyanate (FITC) was utilized as the precolumn derivatization agent to label fluorophore on five β- casomorphins (β-CMs) in the development of their microfluidic CE separation and analysis in cheese samples [41]. In order to attain the greatest fluorescence intensity, factors like BGE composition, concentration, and pH (15 mM boric acid- sodium borate, pH 10.5), and derivatization conditions (FITC:β-CMs 533.3:1 (molar ratio), 5 mmol/L sodium borate buffer, pH 9.0; reaction time 12 h at 25°C) were optimized. The achieved LODs of β-CMs were in the range 18.7 – 75.1 nmol/L. The CE-LIF electropherograms of the analysis of cheese samples where two FITC labeled β-CMs were identified by spiking the sample with synthetic peptides with defined sequences are shown in Fig. (**1**).

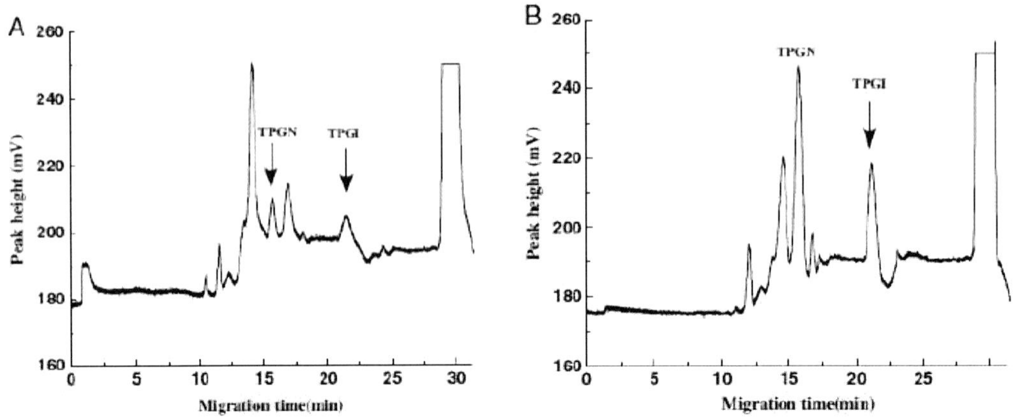

Fig. (1). CE-LIF electropherogram of analysis of FITC labeled β-casomorphins (β-CMs) in (A) real cheese sample and (B) real cheese sample spiked with 1 μmol/g of FITC labeled peptides, TPGN (Tyr-Pro-Phe--ro-Gly-NH_2) and TPGI (Tyr-Pro-Phe-Pro-Gly-Pro-Ile). Derivatization conditions: FITC:β-CMs 533.3:1 (molar ratio), 5 mmol/L sodium borate buffer, pH 9.0; reaction time 12 h at 25°C. BGE: 15 mM boric acid-sodium borate, pH 10.5. Glass chip with separation channel (width×depth×length) 100 μm × 25 μm × 18 cm, separation voltage +1.8 kV. Reprinted with permission from [41].

Another detection method, Inductively Coupled Plasma Mass Spectrometry (ICP-MS) requires labelling of peptides and proteins with metallic ions. The metal chelate labelling method where diethylenetriamine-*N,N,N',N",N"*-penta-acetic acid (DTPA) was employed as chelating reagent and Eu^{3+} as labelling element, was utilized to tag three β-CMs with Eu^{3+} before their CZE-ICP-MS analysis [42]. The Eu^{3+}-tagged peptides were separated by CZE and determined with ICP-MS by monitoring ^{151}Eu and ^{153}Eu. Mass LODs of β-CMs were very low, 1.79-2.13 amol.

Selection of CE Separation Conditions

For CE analysis of peptides, general rules for selection of proper separation conditions as well as specific peptide properties should be taken into account. Principles for choosing suitable background electrolytes (BGE) for CE and MCE have been summarized in several general reviews [43 - 45] as well as in specialized review on CE of peptides [46 - 48]. In brief, BGEs must contain suitable amount of ions to conduct the electric current but not to cause too high current and Joule heating. At least one component of the BGE must have sufficient buffering capacity at chosen pH; the pH of the BGE should be close to the acidity constant (pK_a) of the buffering component of the BGE, *i.e.* within the range $pK_a \pm 1$. A co-ion of the BGE should have mobility similar to those of major peptide sample components, in order to minimize electromigration dispersion and to obtain sharp and symmetrical peaks of these peptides. The composition of the BGE should be kept as simple as possible. The use of multiple co-ions that causes the formation of system peaks should be avoided unless it is inevitable for indirect

UV-absorption detection or other purposes, *e.g.* specific interactions with some analytes.

However, in the CE analysis of peptides in foods, it is sometimes necessary to use various additives of BGE or employ coated capillaries in order to avoid adsorption of peptides on the bare fused-silica (FS) capillary wall and to prevent peak tailing, reduction of separation efficiency and low repeatability of the migration times. With covalently attached of physically adsorbed permanent capillary coatings, no further additives to the BGE are required [19]. However, for the dynamic coatings, the BGE must contain the coating reagents (*e.g.* oligo- or polyamines, polysaccharides, surfactants) during the CE separation.

In the selection of BGE, also the detection method should be taken into account. For example, when performing CE separations of peptides with MS detection, sensitivity of analysis considerably decreases if the high buffer concentrations, non- volatile constituents and/or surfactants that may cause ion suppression, are utilized. Additionally, application of non-volatile BGE components may result in ion source contamination and high background signals.

In the selection of suitable separation conditions, the specific properties of peptides resulting from their amino acid composition and the sequence and secondary structure should also be considered. Peptides are amphoteric polyelectrolytes composed of amino acids (AAs) comprising various ionogenic groups bringing both positive and negative charges to peptides molecules. Carboxyl groups of C- terminal AAs and carboxyl groups of acidic AAs, aspartic and glutamic acids contribute to negative charge of peptides whereas α-amino groups of N-terminal AAs and ε-amino group of lysine, imidazolium group of histidine and guanidinium group of arginine bring positive charges to peptide chains. The ionogenic groups of peptides and the range of their acidity (ionization) constants (pK_a) are listed in Table **1**.

Table 1. Ionogenic groups of amino acid residues (AARs) in peptides and approximate range of their acidity (ionization) constants (pK_a).

Ionogenic group	Amino acid residue	pK_a
α-COOH	C-terminal AAR	1.8-3.5
β-COOH	Aspartic acid	3.5-4.5
γ-COOH	Glutamic acid	4.0-4.5
Imidazolium	Histidine	5.6-6.9
α-NH3 +	N-terminal AAR	7.5-8.6
-SH	Cysteine	9.0-10.5

(Table 1) cont.....

Ionogenic group	Amino acid residue	pK_a
+ ε-NH3	Lysine	9.0-10.8
Phenol	Tyrosine	9.8-11.0
Guanidinium	Arginine	10.0-12.5

The effective (net) charge of peptides is equal to the sum of charges (including their signs) of all ionogenic groups present in peptide molecules. The effective charge of peptides is strongly dependent on pH and acidity constants (pK_a) of ionogenic groups of AA residues of peptide chain and can be calculated from the AA composition or sequence of a given peptide if the p$K_a$values of the present ionogenic groups are known or can be estimated. Actually, pK_a values depend not only on the amino acid residue itself but also on the neighboring amino acid residues and on the steric arrangement of the polypeptide chain. The electrophoretic mobility of peptides is directly proportional to their effective charge and indirectly proportional to their size (molecular mass). Additionally, mobility of peptides is influenced by the shape (conformation) of their molecules. Thc important characteristic of peptides is their isoelectric point (pI), *i.e.* pH at which their effective charged and effective mobility are equal to zero. Ionogenic peptides can be analyzed by CZE as cations at (usually acidic) pH at least 1-2 pH units below their pI or as anions at (usually neutral or alkaline) pH at least 1-2 pH units above pI. Non-ionogenic peptides with blocked or derivatized ionogenic groups can be separated by EKC or CEC methods. Due to their versatile amino acid composition and sequence, peptides differ by their solubility in water and other solvents. Solubilization of peptides in deionized and/or distilled water is advantageous because of the possibility to use the same sample solution for CE separations in different BGEs, and the FASS effect as well. For peptides insoluble in water, acidified or alkalized aqueous buffer solutions or mixed hydro-organic solvents (*e.g.* water/methanol, ethanol, propan-1-ol, propan-2-ol or acetonitrile) or the above organic solvent themselves have to be used. In addition, solubilization of peptides can be achieved by adding detergents, anionic SDS or cationic CTAB or chaotropic agents (urea) to the sample solvent solution. On the other hand, employing BGE solution as the sample solvent ensures constant separation conditions during the whole CE experiment which is good especially for exact measurements of effective electrophoretic mobilities of peptides.

Peptide biological activity, chemical and temperature lability as well as the complexity of some peptide mixtures isolated from food matrices are also important issues to be considered when selecting proper CE separation conditions. Thus, parameters like applied separation voltage, temperature of the capillary coolant or capillary dimensions should also be optimized.

Detection

Because of the absorption of short wavelength UV-radiation (185-220 nm range) by peptide bond and the absorption of UV-radiation around the 280 nm wavelength by the aromatic amino acids, tyrosine, tryptophan and phenylalanine, the detection based on the absorption of UV radiation is the most used in CE analysis of peptides. CE combined with UV-detector offers a good mass sensitivity but relatively low concentration sensitivity, typically in the micromolar concentration range with minimal limits of detection (LODs) in the range 10^{-5} – 10^{-7} M. For the particular application of UV-absorption detection for peptide analysis in foods, see the below Tables **2** and **3**.

One of the most sensitive detection modes employed for CE separations of peptides in different matrices, including foods, is the laser induced fluorescence (LIF). The LODs of peptides determined with LIF detectors are generally in nanomolar range [49, 50]. However, only peptides with aromatic AA residues of tryptophan, tyrosine and phenylalanine exhibit native fluorescence in the region 200-300 nm. For their excitation, UV-lasers or multiphoton excitation are necessary. Commonly, pre- column or post-column derivatization of peptides with a fluorescent label before fluorescence detection is performed. The disadvantage of derivatization is that peptides usually contain more than one derivatization site which leads to multiple labelling and generation of several derivatives with different electrophoretic mobilities. Then, multiple peaks for originally single peptide are observed in the electropherograms. For the particular application of LIF for peptide detection, see the above analysis of FITC-labeled β-CMs in Fig. (**1**) and other examples presented in Tables **2** and **3** below.

Table 2. CE analyses of peptides in foods of animal origin.

Peptide(s)	Food or Food Products	Sample Preparation	CE Method BGE Composition, pH Capillary Type and Dimensions Sep. Voltage, Temperature	Detection LOD LOQ Analysis Time	Significant Findings	Refs.
Anserine, carnosine	Poultry meat (thigh and breast muscle)	Grounding, homogenization, centrifugation, filtration, protein precipitation, centrifugation, evaporation, dissolution, ultrasonication	CZE 200 mM acetic acid, 10 mM HIBA,0.3 mM potassium acetate, pH 2.7 PMMA chip with separation channel (width×depth×length): 50 μm × 50 μm × 87 mm, +3.50 kV	C^4D 0.16/ 0.10 μM 0.53/0.33 μM 120 s	With home-made instrument the measured dipeptides concentrations were comparable to the values obtained by commercial MCE/C^4D device or HPLC.	[57]
Anserine, carnosine	Meat: chicken, beef, horse, rabbit	Homogenization, extraction, deproteinization, ultrasonication	CZE 300 mM acetic acid, 7.5 mM 2-hydroxy-2-methylpropanoic acid, 0.2 mM rubidium acetate, pH 2.6 ± 0.1 PMMA chip with separation channel (width×depth×length): 50 μm × 50 μm × 87 mm, +3.3 kV	C^4D 11.5/8.6 mg/L 31.2/24.3 mg/L 100 s	Analysis of real biological samples had satisfactory agreement with HPLC method and could be used in food fraud prevention.	[58]

(Table 2) cont.....

Peptide(s)	Food or Food Products	Sample Preparation	CE Method BGE Composition, pH Capillary Type and Dimensions Sep. Voltage, Temperature	Detection LOD LOQ Analysis Time	Significant Findings	Refs.
Anserine, carnosine	Meat, meat products: chicken, pork, turkey and beef meats, hams, wursts, sausages and salami	Heating and pressure treatment, homogenization, extraction, deproteinization, centrifugation, filtration	CZE 100 mM phosphate buffer, pH 2.5 FS capillary, 50 µm id, 45 cm L_{det} +15 kV, 38°C	UV@200 nm 1.64/1.58 mg/L 5.47/5.27 mg/L 10 min	Content of these imidazole dipeptides was independent of heating and pressure treating.	[77]
Anserine, carnosine, glutathione	Salted herring meat	Homogenization (with 6.5% TCA), centrifugation, dilution, (SPE)	CZE 0.1 M phosphate, pH 2.5 FS capillary, 50 µm id, 32 cm L_{det}, 20 kV, 35°C	UV@200 (260, 280) nm n.p. n.p. 25 min	Method may be suitable for fast monitoring of the salted herring ripening process.	[79]
Antihypertensive peptides	Nutraceuticalderived from a bovine milk protein hydrolysate	Grounding to fine powder, dilution, centrifugation, filtration	CZE 1 M acetic acid, pH 2.3 FS capillary, 75 µm id, 72 cm L_{tot} +10 kV, 25°C	MS; MS/MS n.p. n.p. 25 min	17 antihypertensive peptides were identified.	[88]
Bioactive peptides	Hypoallergenic infant milk formulas (Hydrolysate of bovine milk proteins)	Dissolution, dilution, extraction, centrifugation, filtration, SPE	CZE 1 M acetic acid FS capillary, 50 µm id, 57 cm L_{tot}, +15 kV, 25°C	ESI-IT-MS, ESI-TOF-MS n.p. n.p. 20 min	5 peptides identified are angiotensin converting enzyme inhibitors and one of them has antimicrobial activity.	[81]
Bioactive peptides	Hypoallergenic infant milk formulas (Hydrolysate of bovine milk proteins)	Dissolution, dilution, extraction, centrifugation, filtration, SPE	CZE 1 M acetic acid, pH 2,3 FS capillary, 50 µm id, 57 cm L_{tot}, +15 kV, 25°C	ESI-TOF-MS n.p. n.p. n.p.	24, 30, and 38 bioactive peptides were determined in 3 different infant milk formulas. Many of them are angiotensin converting enzyme inhibitors.	[82]
β-casomorphins	Cheese	Enzymatic degradation, ultrasonication, 50° 2 h, 100°C 10 min, centrifugation, filtration, derivatization with FITC	CZE 15 mM boric acid-sodium borate, pH 10.5 Glass chip with separation channel (width×depth×length): 100 µm × 25 µm × 18 cm, +1.8 kV	LIF λ_{ex} = 473 nm λ_{em} =525 nm 18.7-75.1 nM n.p. 30 min	In cheese, β-casomorphins, Tyr-Pro-Phe-Pro-Gly-Pro-Ile (0.162 µM/g) and Tyr-Pr- -Phe-Pro-Gly-NH_2 (0.169 µM/g) were determined.	[41]
β-casomorphins	Cheese	Enzymatic degradation (pepsin), ultrasonication, 50°C 2h, 100°C 10 min, centrifugation, filtration, lyophilization, labeling with DTPA and Eu^{3+}	CZE 20 mM sodium phosphate- 5 mM sodium tetraborate, pH 6.25 75 µ id, 70 cm L_{tot}, +18 kV	ICP-MS 1.79-2.13 amol n.p. 15 min	In cheese samples, β-casomorphins, Pro-Phe-P-o-Gly-Pro-Ile (0.056-0.148 µM) and Tyr-Pro-Phe--ro-Gly-Pro (0.256-0.345 µM) were determined.	[42]
Carnosine	Cured beef meat (bresaola)	Homogenization, adjustment of pH to 2.5, centrifugation, filtration, boiling, filtration, dilution, centrifugation	CZE 110 mM phosphate, pH 2.5 FS capillary, 50 µm id, 51 cm L_{det}, +15 kV, 38°C	UV@200 nm n.p. n.p. n.p.	Carnosine content in two undigested bresaola samples was 628 and 671 mg/100g.	[78]
Inhibitory peptides of angioconverting enzyme (putative)	Fermented skim milk	Extraction, centrifugation, ultrafiltration, dissolution	CZE Cation buffer solution (H3301-1001, Human Metabolome Technologies, Tsuruoka, Japan) FS capillary, 50 µm id, 80 cm +27 kV	ESI-TOF-MS n.p. n.p. n.p.	In γ-aminobutyric acid rich fermented milk, the levels of 6 putative inhibitory peptides of angiotensin converting enzyme were higher than in control milk.	[89]

(Table 2) cont.....

Peptide(s)	Food or Food Products	Sample Preparation	CE Method BGE Composition, pH Capillary Type and Dimensions Sep. Voltage, Temperature	Detection LOD LOQ Analysis Time	Significant Findings	Refs.
Nisin (Polycyclic antimicrobial peptide from bacterium *Lactococcus lactis)*	Milk (2%), whipping cream, homogenized milk, plain yogurt, yogurt drink, cheese, salad dressing, canned tomatoes, pre-drop beer, wine	Dilution and homogenization (plain yogurt, yogurt drink, processed cheese, salad dressings, canned tomatoes), filtration, centrifugation	MEKC 50 mM sodium phosphate, 80 mM SDS, pH 3.75 FS capillary, 50 µm id, 50 cm L_{tot}, -20kV, 25°C	UV@214 nm 0.3-0.8 mg/L 1.0-2.8 mg/L 8 min	Fast and sensitive method for identification and quantification of nisin in various food products.	[74]
Nisin (Polycyclic antimicrobial peptide from bacterium *Lactococcus lactis)*	Cheese	Extraction/dilution, centrifugation	CZE 0.5 M acetic acid Polyacrylamide coated capillary, 50 µm id, 47 cm L_{det}, 59 cm L_{tot}, +13kV, 40°C	C^4D n.p. 0.02 µg/mL 20 min	The values of different nisin forms were ranging from 0.056 ± 0.003 µg/mL to 9.307 ± 0.437 µg/mL.	[90]
Peptide biomarker panel for intramammary infection	Milk	Addition of 0.1% PMSF, extraction, filtration, desalting, centrifugation	CZE	MS n.p. n.p. n.p.	The developed peptide biomarker panel for intramammary infection comprising 77 peptides allowed to differentiate infected from non-infected milk. It could be possibly used as biomarker panel of bovine mastitis.	[91]
Peptides from *Mytilus edulis*	Blue mussel	Homogenization, enzymatic hydrolysis, centrifugation, freeze-drying, 24 h in 75% alcohol followed by centrifugation, ultrafiltration, filtration, preparative chromatography (AKTA avant 25, GE Healthcare)	CZE 10% acetic acid, pH 2.3, FS capillary, 30 µm id, 91 cm L_{tot}, +20 kV	ESI-Q-TOF-MS	Identification of novel peptide YPRKDETGAERT from neutral proteinase enzymatic hydrolysate of <3 kDa after alcohol precipitation of *Mytilus edulis* with osteogenic activity in mouse MC3T3-E1 preosteoblast cells.	[80]
Peptides from α_{s1}-CN proteolysis	Cheese (from goat milk)	Protein solubilization, dilution, filtration	CZE Coated FS capillary (DB-WAX; J & W Agilent Technologies Inc., Santa Clara, CA)	UV@214 nm	The formation and the progressive increase of peptides originating from α_{s1}-casein proteolysis can be used to monitor the proteolytic phenomena occurring during cheese aging.	[85]
Prato cheese peptide profile	Cheese	Homogenization, acidification (pH 4.6), centrifugation, filtration; insoluble fraction dissolved in buffer (10 mM sodium phosphate, 8 M urea, 10 mM dithiothreitol, pH 8.0)	CZE 10 mM sodium phosphate, 6 M urea, 0.05% HPMC, pH 3.0 FS capillary, 75 µm id, 50 cm L_{det}, 57 cm L_{tot}, 18.5 kV, 23°C	UV@214 nm 60 min In addition independent MALDI-MS analysis of insoluble peptide fractions.	Commercial cheeses produced from different raw materials and possibly different processing conditions have similar casein hydrolysis profiles; only 9 peptides were responsible for the discrimination of cheeses.	[86]
Tryptic protein fragments	Milk powder (bovine)	Dilution, extraction, digestion, lyophilization	t-ITP Neutrally (polyacrylamide) coated capillary LE, ammonium acetate, pH 4.0 +25 kV, 5515.8 Pa for 60 min	ESI-IT-MS Less than 2.5 ng	12 phosphopeptides could be determined, including two di-phosphorylated peptides.	[68]
Aspartame	Soft fruit drinks, tabletop sweetener formulations	Degasification by sonication, dilution	CZE 100 mM Tris, 10 mM His, pH 9.4 FS capillary, 75 µm id, 60 cm L_{det}, 70 cm L_{tot} +30 kV	C^4D 4.2 mg/L 14.1 mg/L 6 min	Simple sample preparation and fast analysis.	[59]

(Table 2) cont.....

Peptide(s)	Food or Food Products	Sample Preparation	CE Method BGE Composition, pH Capillary Type and Dimensions Sep. Voltage, Temperature	Detection LOD LOQ Analysis Time	Significant Findings	Refs.
Aspartame	Lemon tea	Dissolution, vacuum filtration	CZE 20 mM sodium tetraborate, 15 mM Tris, 7.5 mM benzoic acid, pH 9.15 FS capillary, 50 μm id, 40 cm L_{det}, 48.5 cm L_{tot} +20 kV	UV@215 nm 6.8 mg/L 22.0 mg/L 6 min	Simple sample preparation and short analysis time.	[97]
Aspartame	Liquid/solid sweeteners, peach/lemon tea, syrup	Degasification by sonication, dilution	CZE 45 mM 2-Amino-2-(hydroxymethyl)-1,3-propanediol, 15 mM benzoic acid, pH 8.4 FS capillary, 50 μm id, 8.5 cm L_{det}, 48.5 cm L_{tot} +30 kV, 25°C	UV@220nm 5.1 mg/L 9.4 mg/L 1 min	Fast determination with simple sample preparation; determination of pK_a and ionic mobilities.	[98]
Glutathione	Orange (juice)	Filtration	CZE 1 M formic acid, pH 2.4 (pH adjusted with ammonium hydroxide) Cationic capillary, coated with Poly(TEDETAMA-co-HPMA) (50:50) -20 kV, 25°C	ESI-TOF-MS n.p. n.p. 15 min	Detection of 87 metabolites in orange juice.	[65]
Japanese sake peptide profile	Rice wine	Dilution, centrifugation, filtration	CZE 1 M formic acid FS capillary, 50 μm id, 100 cm L_{tot}, +30kV, 20°C	ESI-TOF-MS n.p. n.p. n.p.	In unpasteurized sake, several individual peptide concentrations increased moderately, while they remained almost constant in pasteurized sake.	[100]

DTPA, diethylenetriamine-*N,N,N*',*N*'',*N*''-pentaacetic acid; HPMC, hydroxypropyl methyl cellulose; IT-MS, ion-trap MS; n.p., not presented; PMMA, polymethyl methacrylate; Poly(TEDETAMA-co-HPMA), poly-(*N,N,N*',*N*'-tetraethyldiethylenetriamine), N-(2-hydroxypropyl) methacrylamide copolymer; PMSF, phenylmethylsulfonyl fluoride; TCA, trichloroacetic acid; t-ITP, transient isotachophoresis; Tris, tris(hydroxymethyl)aminomethane

Table 3. CE analyses of peptides in foods of plant origin.

Peptide(s)	Foods or Food Products	Sample Preparation	CE Method BGE Composition, pH Capillary Type and Dimensions Sep. Voltage, Temperature	Detection LOD LOQ Analysis Time	Significant Findings	Refs.
Aspartame	Soft fruit drinks, tabletop sweetener formulations	Degasification by sonication, dilution	CZE 100 mM Tris, 10 mM His, pH 9.4 FS capillary, 75 μm id, 60 cm *L*det, 70 cm *L*tot +30 kV	C^4D 4.2 mg/L 14.1 mg/L 6 min	Simple sample preparation and fast analysis.	[59]
Aspartame	Lemon tea	Dissolution, vacuum filtration	CZE 20 mM sodium tetraborate, 15 mM Tris, 7.5 mM benzoic acid, pH 9.15 FS capillary, 50 μm id, 40 cm *L*det, 48.5 cm *L*tot +20 kV	UV@215 nm 6.8 mg/L 22.0 mg/L 6 min	Simple sample preparation and short analysis time.	[97]

(Table 3) cont.....

Peptide(s)	Foods or Food Products	Sample Preparation	CE Method BGE Composition, pH Capillary Type and Dimensions Sep. Voltage, Temperature	Detection LOD LOQ Analysis Time	Significant Findings	Refs.
Aspartame	Liquid/solid sweeteners, peach/lemon tea, syrup	Degasification by sonication, dilution	CZE 45 mM 2-Amino-2-(hydroxymethyl)-1,3-propanediol, 15 mM benzoic acid, pH 8.4 FS capillary, 50 µm id, 8.5 cm *L*det, 48.5 cm *L*tot +30 kV, 25°C	UV@220nm 5.1 mg/L 9.4 mg/L 1 min	Fast determination with simple sample preparation; determination of pK_a and ionic mobilities.	[98]
Glutathione	Orange (juice)	Filtration	CZE 1 M formic acid, pH 2.4 (pH adjusted with ammonium hydroxide) Cationic capillary, coated with Poly(TEDETAMA-co-HPMA) (50:50) -20 kV, 25°C	ESI-TOF-MS n.p. n.p. 15 min	Detection of 87 metabolites in orange juice.	[65]
Japanese sake peptide profile	Rice wine	Dilution, centrifugation, filtration	CZE 1 M formic acid FS capillary, 50 µm id, 100 cm *L*tot, +30kV, 20°C	ESI-TOF-MS n.p. n.p. n.p.	In unpasteurized sake, several individual peptide concentrations increased moderately, while they remained almost constant in pasteurized sake.	[100]
Systemin peptides	Tomato plant (stem and leaves)	Homogenization, dilution, centrifugation, filtration	CZE 25 mM phosphate, pH 2.5 FS capillary, 75 µm id, $^{50\ cm}L$det, $^{60\ cm}L$tot 20 kV, 25°C	UV@214 nm n.p. n.p. 20 min	Synthetic systemin peptides could be directly detected in the crude plant matrixes; they were effectively spread throughout tomato stem and leaves.	[93]

(Table 3) cont.....

Peptide(s)	Foods or Food Products	Sample Preparation	CE Method BGE Composition, pH Capillary Type and Dimensions Sep. Voltage, Temperature	Detection LOD LOQ Analysis Time	Significant Findings	Refs.
Tryptic protein fragments	Maize flour	Grinding, extraction, centrifugation, filtration, digestion	CZE 200 mM iminodiacetic acid, pH 2.26 FS capillary, 50 µm id, $^{20\ cm}L$det· $^{31\ cm}L$tot· +10 kV, 22-25°C	UV@206 nm n.p. n.p. 8 min	Some significant relative qualitative and quantitative differences in CZE-UV profiling of tryptic proteins from transgenic and two non-transgenic maize species were found.	[95]
Tryptic protein fragments	Spinach chloroplasts	Grinding, extraction, MeOH/chloroform precipitation, centrifugation, digestion	CZE 8/70/22% v/v FA/MeOH/H2O HPC coated FS capillary, 50 µm id, 100 cm +30 kV	ESI-MS/MS n.p. n.p. 90 min	3579 peptides and 1141 proteins were identified.	[28]
Tryptic protein fragments	Soybean seeds	Grinding, extraction, sonication, centrifugation, dilution, digestion	CZE 0.5 M Formic acid, FS capillary, 50 µm id, 90 cm *L*tot, 25 kV, 25°C	MS-TOF n.p. n.p. 30 min	Based on the protein fraction studied, no significant differences were found between the conventional and the transgenic soyabean.	[96]

FA, formic acid; HPMC, hydroxypropyl methyl cellulose; IT-MS, ion-trap MS; MeOH, methanol; n.p., not presented; PMMA, polymethyl methacrylate; poly(TEDETAMA-co-HPMA), poly-(*N*,*N*,*N*',*N*'-tetraethyldiethylenetriamine), N-(2-hydroxypropyl) methacrylamide copolymer; PMSF, phenylmethylsulfonyl fluoride; TCA, trichloroacetic acid; t-ITP, transient isotachophoresis; Tris, tris(hydroxymethyl)aminomethane

Thanks to its universality, selectivity, sensitivity (LODs in the range $10^{-8} - 10^{-10}$ M) and high identification power, mass spectrometry (MS) has become the most powerful detection mode in CE methods [51, 52]. The importance of MS as an analytical and structure elucidation method has significantly increased in the last years due to its key role in the identification and characterization of peptides, proteins and metabolites in proteomic, peptidomic and metabolomic studies [53,

54]. CE can be coupled to different ionization sources of MS detectors but it is usually combined with electrospray ionization (ESI) or matrix assisted laser desorption ionization (MALDI) [55]. As mass analyzers, mostly ion trap (IT), Orbitrap, quadrupole (Q) or time of flight (TOF) analyzer are employed in MS detectors.

For peptide analysis, oftentimes ESI is the ionization technique mostly used. In CE- ESI MS analyses, the volatile BGE constituents and additives should be utilized. With MALDI-MS, CE is mostly coupled off-line. The off-line arrangement is more tolerant to salts and enables to store samples/analytes. In addition, off-line combination allows for independent optimization of CE separation and MS detection. Several examples of some specific applications of MS detection for peptide analysis in foods are provided in the figures below and Tables **2** and **3**.

Besides the above detection modes, a universal capacitively coupled contactless conductivity detection (C^4D) based on the on-line measurement of electric conductivity of the BGE solution inside the on-column capillary detection cell can be used for the CE analysis of peptides in different matrices including foods. Instrumental developments as well as applications of CE-C^4D in the period 2010-2020 have been described in a series of review articles (see the last review of this series [56] and the references cited therein). In the area of CE analysis of peptides in foods, C^4D was used *e.g.* for detection of anti-oxidative histidine dipeptides, anserine and carnosine, in poultry, beef, horse, and rabbit meats [57, 58] and for determination of dipeptide sweetener aspartame in soft drinks [59], see the figures below and Tables **2** and **3**.

Suppression of Wall Adsorption and Control of EOF

Adsorption of peptides onto the inner wall of the FS capillaries significantly decreases the separation efficiency and migration time repeatability in CE methods. Especially hydrophobic and basic peptides tend to interact with bare inner wall of the FS capillary. Different approaches have been proposed to suppress peptide adsorption during CE analysis. Adsorption of peptides on the capillary surface can be reduced by using either very acidic or basic BGEs and/or by employing BGEs with high ionic strength. The negative aspect of employing BGEs with extreme pH is that it can cause problems with peptide stability and solubility. On the other hand, usage of the BGEs with high ionic strength leads to high electric current and high Joule heating resulting in peak distortion and in a higher noise and instability of the detection baseline. Another possibility to reduce or eliminate adsorption and also alter the magnitude and/or direction of EOF is to modify or coat the separation capillary or separation channel. By modifying or coating their inner walls, the adsorption can be suppressed or eliminated and the magnitude and direction of EOF can be controlled as well. Different kinds of coatings are available, such as dynamic coatings, covalently bound coatings and physically adsorbed coatings. Survey of column coatings utilized in CE separations of proteins, which are appropriate also for the CE analysis of peptides can be found in recent reviews [60, 61].

Various reagents, such as oligo- and polyamines, polysaccharides and other polymers, surfactants (at sub-micellar concentrations) or ionic liquids are used as dynamic coating materials added to the BGE. Using dynamic coatings is simple but this type of coating is not suitable for CE separations on-line coupled with MS detection because ionization suppression or contamination of the ion source by the above additives can occur [62].

Covalent coatings that are bound to the silica surface by chemical reactions with the silanol groups are suitable for CE-MS analysis. However, the stability of these coatings is mostly limited to the pH range 2-8. Moreover, covalent coating procedures can be laborious, time-consuming and expensive. For physically permanently adsorbed coatings, neutral (poly(ethylene oxide) (PEO), poly(vinyl alcohol) (PVA)), cationic (chitosan, linear or branched polyethylenimine (PEI), polybrene (PB), poly(diallyldimethylammonium chloride) (PDADMAC)) or anionic polymers (dextran sulfate (DS), poly(vinyl sulfonate) (PVS)) are employed. It is not necessary to add them to the BGE, the capillaries are just flushed with a solution of these polymers before CE analyses and the coatings can be renewed by repeated rinsing of the capillary by solutions of the above polymers. Nevertheless, very alkaline BGEs and/or high content of organic solvent in the BGE may significantly decrease the stability of these coatings. A high stability of the coatings and an efficient suppression of peptide and protein sorption to the inner capillary wall is achieved by the so-called successive multiple ionic polymer layer (SMIL) coatings. For example, three-layer coating, PB-DS-PB, effectively prevented sorption of polypeptides and proteins during the investigation of their interactions by affinity CE [63].

Besides preventing the adsorption on the inner capillary surface, various capillary coatings are utilized also for tuning of electroosmotic flow (EOF) to maximize the resolution of analytes with similar electrophoretic mobilities [64]. Poly-(*N*,*N*,*N*',*N*'- tetraethyldiethylenetriamine,*N*-(2-hydroxypropyl) methacry lamide, TEDETAMA- co-HPMA (50:50) copolymer was applied as capillary coating material for metabolite (including peptide) profiling of orange juice and wine by CE-MS [65]. The coating, which reversed EOF, was generated by flushing the FS capillary for 20 min with 1 mg/mL of polymer solution. Between runs, the capillary was sequentially washed with 0.1 mg/mL solution of cationic polymer and the separation buffer. Anionic metabolites were analyzed in these capillaries with reversed polarity of the separation voltage (cathode at the injection capillary end) using 1 M formic acid (pH 2.4) as BGE. Negative ion mode was used also for the ESI source. Performed separations were fast, had high efficiency and good repeatability of migration times and peak areas. The method was utilized for the CE separation of metabolites in orange juice and red wine. Among the separated metabolites, a peptide, particularly the tripeptide glutathione was also identified.

Peptide Separations by Different CE Methods

Capillary Zone Electrophoresis

Capillary Zone Electrophoresis (CZE) is the simplest and most frequently applied method for the separation and analysis of peptides in various matrices including foods. When further in the text, the CE method is mentioned without further specification of the separation mode, CZE separation principle is supposed to be used in the particular application. In CZE, diverse BGEs of variable composition and a wide pH range are utilized (see Tables **2** and **3**). In addition to the buffering constituents, oftentimes non-buffering additives (*e.g.* complexing agents, organic solvents, surfactants) are also added to the BGE in order to enhance solubility, improve separation selectivity and resolution or to diminish adsorption of peptides on the capillary wall of bare FS capillary. CZE is mostly performed in aqueous BGEs but for water insoluble or poorly soluble peptides, non-aqueous CZE/CE (NACE) performed in pure organic solvents, *e.g.* methanol or acetonitrile, or CZE/CE in mixed hydro-organic solvents, *e.g.* water/methanol, water/ACN, can be used. Change of the solvent can also improve the selectivity of the CE separation of peptides with identical or close to identical mobilities (charge/size ratios) in the aqueous media.

Isotachophoresis

Capillary Isotachophoresis (ITP) is mainly used as a preconcentration transient-ITP (t-ITP) method for peptides present at low concentrations and/or in complex matrices with subsequent separation and analysis of concentrated peptides by CZE. For recent developments and applications of CITP itself and for on-lie coupling of t-ITP and CE, see the comprehensive reviews [27, 66]. In addition, the CITP is a suitable method for determination of anionic and cationic counterions of strongly basic or strongly acidic peptides [67].

Concerning the application of ITP to analysis of peptides in foods, the following example can be mentioned here. The t-ITP preconcentration followed by the sheathless-ESI-MS detection was employed for the determination of phosphopeptides in bovine milk digest [68]. In this study, the effect of ultra-low flow ESI interface for the MS detection of phosphopeptides was investigated. It was shown that the detection sensitivity for all phosphorylated peptides is strongly enhanced when working at flow rates below 30 nL/ min. For the t-ITP, neutrally coated capillaries were utilized (see Fig. (**2**)). The flow required for ESI was produced by the application of a slight pressure at the capillary inlet. Small amount 5 ng of milk digest (0.25 μL volume of a sample at 20 ng/μL concentration) was injected. With this strategy, twelve phosphopeptides could be determined, including two di-phosphorylated peptides. ITP-ESI-MS system was compared to nano-LC-MS setup where ten times more material, 50 ng of milk digest (2.5 μL volume of a sample at 20 ng/ μL concentration) was loaded. With nano-LC- MS, nine monophosphorylated peptides were identified. With ITP-ES- -MS five and with nano-LC-MS three phosphopeptides unique for their separation technique were determined indicating complementarity of the techniques.

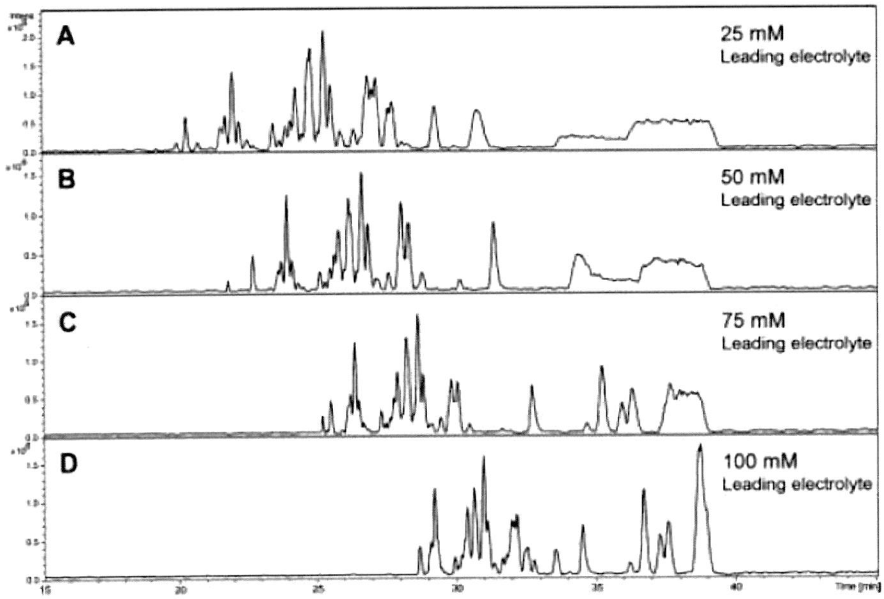

Fig. (2). Base peak electropherogram of t-ITP-MS analysis of 2.5 ng of milk protein digest in neutral (polyacrylamide) coated FS capillary (37% capillary fill, 250 nL, 30 psi for 60 s). Sample dissolved in leading electrolyte (LE) composed of 10% acetic acid and ammonium acetate, pH 4.0, at various ionic strength: **(A)** 25 mM, **(B)** 50 mM, **(C)** 75 mM, and **(D)** 100 mM). Analyses performed in capillary with porous tip (30 μm id, 150 μm od, total !ength 100 cm) at +25 kV. MS conditions: ESI voltage of −1050 V; detection range 300–2900 m/z. Reprinted with permission from [68].

Isoelectric Focusing

Capillary Isoelectric Focusing (CIEF) is another CE method with inherent concentration and self-sharpening effects frequently used for analysis of peptides and proteins. For basic aspects of this method, such as selection of carrier ampholytes and pI markers, sample application, and hydrodynamic or electro-driven mobilization of the focused peptides/proteins, and for the recent developments and applications of this technique, see the highly informative tutorial and review articles [48, 69].

Affinity Electrophoresis

Affinity Capillary Electrophoresis (ACE) combines the electrophoretic movement of charged analytes with their (bio)specific interactions with component(s) of the BGE. This way, the mobility of the sample components interacting with the affinity ligand in the BGE is selectively influenced, which can be aimed to their separation or isolation from the given complex matrix. ACE is often applied for the separation of enantiomers and other stereoisomers of (bio)molecules using BGEs with the addition of stereoselectors, *e.g.* cyclodextrins, crown ethers, macrocyclic antibiotics and others. Additionally, different modes of ACE are used to study non- covalent interactions of peptides with other (bio)molecules [70].

Electrokinetic Chromatography

Capillary Electrokinetic Chromatography (CEKC) and more specifically its most spread mode, Micellar EKC (MEKC), is especially suitable for the separation and analysis of non-charged species including non-ionogenic peptides (*i.e.* peptides with blocked or derivatized ionogenic groups) or peptides with the same or very similar charge-to-mass ratio (mobility) but different hydrophobicity. Separation is based on the distribution of analytes between aqueous mobile phase and micellar pseudostationary phase of anionic (*e.g.* SDS), cationic (*e.g.* CTAB) or amphoteric (*e.g.* CHAPS) detergents. For the basis of MEKC, see the review article by the inventor of this method, S. Terabe [71], while later developments and applications of MEKC and micro-emulsion EKC (MEEKC) are described in ref [72, 73].

In the field of peptide analysis in foods, MEKC using SDS-based micellar pseudophase and mobile phase composed of 50 mM sodium phosphate, pH 3.75, was applied for the fast (6 min) and sensitive (LODs 0.3-0.8 mg/L) analysis of nisin, a polycyclic antibacterial peptide containing 34 amino acid residues including the uncommon amino acids lanthionine (Lan), methyllanthionine (MeLan), didehydroalanine (Dha), and didehydroaminobutyric acid.

It is produced by bacteria of milk fermentation (*Lactococcus lactis)* and is used as food preservatives in several foods or beverages, *e.g.* milk, yogurt, cheese, beer, and wine [74].

Capillary Electrochromatography

Capillary Electrochromatography (CEC) is a hybrid method utilizing both electrokinetic and chromatographic separation principles. Recent developments and application of CEC for separation and analysis of both charged and non-charged analytes including peptides are described in detailed reviews [75, 76]. Despite inherent advantages of CEC (combination of high separation efficiency of CE and high selectivity of chromatographic stationary phases), it did not find a wide application neither in analysis of peptides in general nor in the analysis of peptides in foods.

APPLICATIONS

High separation power of CE methods and a high concentration and/or mass sensitivity of MS, LIF, UV-absorption, and C^4D detection modes, potentially enhanced by on-line sample preconcentration procedures, allow their application for separation and determination of peptides present at low concentration levels in complex biomatrices including foods. According to food origin, applications will be sorted to CE analysis of peptides in foods of animal, plant and microbial origin.

CE Analysis of Peptides in Foods of Animal Origin

In the samples of animal origin, peptides were mostly separated and analyzed in different types of meats and meat products or milks and milk products.

CZE with UV-absorption detection at low UV wavelength 200 nm, using 100 mM phosphate buffer, pH 2.5, as BGE in the FS capillary of 50 um id, 45 cm effective (to the detector) length, and 15 kV separation voltage at 38°C was applied for the determination of imidazole (histidine) antioxidative dipeptides present in muscles, carnosine (β-alanyl-L-histidine) and anserine (β-alanyl-1-L-methyl-histidine), in thirty-two samples of meats and meat products, such as pork, beef, chicken and turkey meats, hams, wursts, sausages, and salami [77]. The measured values were in accordance with the data given on food labels. Additionally, no significant changes in dipeptide contents after heat and pressure treatment of pork muscle samples were observed.

Anserine and carnosine were determined in meat samples of different animal species also by using a home-made portable microchip device coupled with C^4D [58], see Fig. (**3**). The analysis was carried out in a poly(methyl methacrylate) microchip with 50 μm wide x 50 μm deep, 87 mm long separation channel and 10 mm long injection channel at 377.1 V/cm electric field intensity. BGE was composed of 250 mM acetic acid, 10 mM 2-hydroxy-2-methylpropanoic acid, and

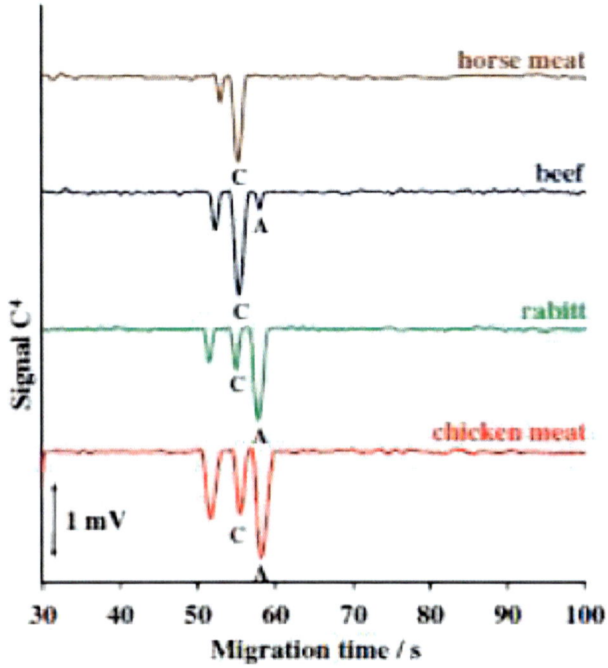

Fig. (3). CE-C⁴D electropherogram of analysis of anserine and carnosine in chicken, rabbit, beef, and horse meat extract. BGE: 300 mM acetic acid, 7.5 mM 2-hydroxy-2-methylpropanoic acid, 0.2 mM rubidium acetate, pH 2.6. Analyses were performed in poly(methylacrylamide) microfluidic chip with separation channel with the width×depth×length dimensions: 50 µm × 50 µm × 87 mm, at separation voltage +3.3 kV. Reprinted with permission from [58].

0.3 mM rubidium acetate, pH 2.6. The method was linear in 0-20 mg/L range. The measured values of concentrations of these peptides showed a satisfactory agreement with those obtained by standard HPLC method. The method was applied for animal species determination in unknown meat samples using the carnosine/anserine ratio and was suggested to be used in food fraud prevention. Content of carnosine in cured beef meat was determined by CZE with UV-absorption detection at 200 nm using BGE composed of 110 mM sodium phosphate, pH 2.5, FS capillary of 50 µm id and 51 cm effective length, and separation voltage +15 kV at 38°C [78].

Identification of peptides isolated from the trichloroacetic acid (TCA) extracts of salted herring meat and analyzed by CZE-UV method was performed by comparing the theoretically predicted migration times of amino acids and peptides based on the Offord's surface charge (q)/mass (M) model ($q/M^{2/3}$) and the experimentally measured migration times of the standard peptides [79]. Standards were separated individually, in the mixture, and with the sample (at increasing concentrations). On the electropherogram of salted herring meat, 36 peaks were observed from which among other compounds peptides, anserine, carnosine, and glutathione were also identified.

Health benefit peptides and proteins are provided also by seafood. Twenty-three peptides from the enzymatic hydrolysate of water soluble protein of Blue mussel (*Mytilus edulis*) were identified by CE-ESI-Q-TOF [80]. Among them, a novel dodecapeptide (YPRKDETGAERT) was found to have the highest osteogenic activity in mouse preosteoblast cells.

CE-ESI-IT-MS and CE-ESI-TOF-MS in 1 M acetic acid BGE were employed for the separation and identification of bioactive peptides in three commercial hypoallergenic infant milk formulas [81, 82]. A sample clean-up pretreatment with a citrate buffer containing dithiothreitol and urea followed by solid-phase extraction (SPE) with different reversed-phase commercial cartridges allowed for optimum detection sensitivity of both CE-MS setups (see Fig. (**4**). In this study, classical semiempirical relationship (classical polymer model) between electrophoretic mobility and charge-to-mass ratio ($q/M^{1/2}$) that describes peptide migration behavior, helped to confirm the identities of the proposed sequences. Finally, 24, 30 and 38 bioactive peptides were identified in each of the three infant milk formulas.

Digested peptides of bovine lactoferrin as the functional hydrolysates of this milk protein were identified by ESI-Q-TOF tandem MS coupled to UPLC and CE [83]. The former (UPLC) method identified 106 peptides while the latter (CE) characterized 102 peptides after comparison of their relative molecular masses, mass to charge (m/z) ratio and isoelectric point (pI). CE separations were performed in bare FS capillary (30 µm id, 91 cm total length) using 10% acetic acid as BGE, pH 2.3, and 20 kV separation voltage. The same CE-ESI-Q-TOF system enabled the identification of hundreds of novel peptides from lactoferrin in the gastrointestinal tract of mice [84].

The changes in chemical and microbiological properties of fresh goat milk cheese were investigated also by CZE monitoring of the degradation of the main casein fractions and the formation of new peptides [85]. The results have shown that rennet played the main role in the proteolysis of this cheese. It was suggested that the formation and the progressive increase in peptides originating from α_{s1}-casein proteolysis could be used to monitor the proteolytic phenomena occurring during cheese aging.

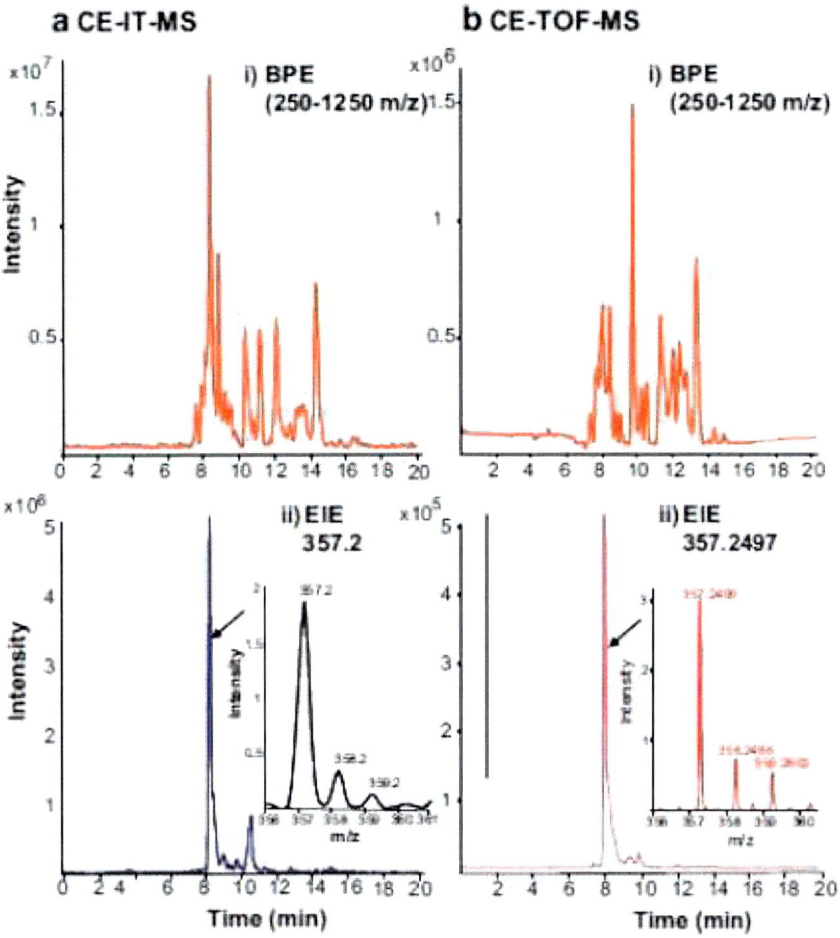

Fig. (4). CE-IT-MS (a) and CE-TOF-MS (b) base peak electropherograms (i) (250–1250 m/z) and extracted ion electropherograms (ii) (CE-IT-MS 357.2±0.5 m/z and CE-TOF-MS 357.2497±0.0005 m/z) of analyses of a fraction of hypoallergenic infant milk formulas after concentration with C18 cartridges. A piece of the mass spectra of the electrophoretic peak corresponding to the bioactive peptide is also shown. CE separations were performed in a 57 cm×50 μm id FS capillary, in 1 M acetic acid BGE at separation voltage +15 kV. The sheath liquid solution consisted of a hydro-organic mixture isopropylalcohol:water (60:40, v/v), with 0.05% v/v of formic acid. Reprinted with permission from [81].

Peptide profiling of the extract of commercial Prato cheese carried out by CZE in the BGE containing 10 mM sodium phosphate, 6 mM urea, and 0.05% hydroxypropylmethylcellulose, pH 3.0, provided characteristic pattern of hydrolysis with formation of α-, β- and para-κ-caseins [86].

Other examples include i) separation and determination of five FITC derivatized β- casomorphins in real cheese samples with LODs in 18.7–75.1 nM range by CE in glass microfluidic chip with LIF detection at 473/525 nm excitation/emission wavelengths in 15 mM sodium borate BGE, pH 10.5 [41], see above Fig. (**1**); ii) determination of twelve phosphopeptides in small amount of 2.5ng of bovine milk digest by t-ITP-ESI-MS [81], see above Fig. (**2**); and iii) ultrasensitive determination of three europium chelate labeled β-casomorphins in pepsin digested cheese samples by CE with ICP-MS detection [87]. This method is based on casomorphins labeling with Eu^{3+} diethylenetriamine-N,N,N',N",N"-pen-a-acetic acid and on their CE separation in the 75 µm id, 70 cm long FS capillary, in the BGE composed of 20 mM sodium phosphate, 5 mM sodium tetraborate, pH 6.25, at +18 kV and 23-25°C. Using on-line ICP-MS detection, the ultralow LODs, 1.79-2.13 amol (17.9-21.3 fmol/L for 100 nL of injected sample volume) were achieved.

For the detailed experimental conditions of the above and other examples of CE analysis of peptides in foods of animal origin, see Table **2**.

CE Analysis of Peptides in Foods of Plant Origin

As shown in the review of recent trends and analytical challenges in the separation and identification of bioactive peptides in plants, CE is also a suitable method for the analysis of plant bioactive peptides [92]. A case in point is a CE study of the spreading of systemin (18 amino acid residues containing plant peptide hormone involved in the regulation of plant's defensive response) and its derivatives in tomato plant [93]. CE analyses were performed in 75 µm id, 60 cm total length, 50 cm effective length capillary, at 20 kV separation voltage, 25°C, 214 nm detection wavelength and using 25 mM phosphate buffer as BGE, see Fig. (**5**). This CE study has shown that systemin peptides are spread throughout tomato stem and leaves and can be directly detected in the crude plant matrices.

The combination of dynamic pH junction with a CE-ESI-MS method allowed determination of trace amounts of systemins, plant peptide hormones participating in the regulation of plant defensive responses, in the extracts from tomato and tobacco [94]. With the online enrichment factor 90-127 of the pH junction preconcentration and qTOF MS detection, the LODs lower than 0.5 nM and linear range 3-120 ng/mL were achieved.

Tryptic fragments of envelope proteins extracted from spinach chloroplasts were analyzed by high-organic content CE-MS/MS (HOCE-MS) using neutral hydroxypropyl cellulose (HPC) coated capillary, and applying on-line stacking technique, FASS [28]. 70% methanol, 8% formic acid and 22% H_2O (v/v/v) was used as BGE, and 0.03% formic acid in ACN/H_2O (67/33, v/v) was employed as low conductive sample matrix. Final sample concentration was 1 µg/µL, and sample was injected at 1 psi for 60 s (150 ng of sample was loaded). Application of FASS technique increased peak heights of peptides 58-fold.

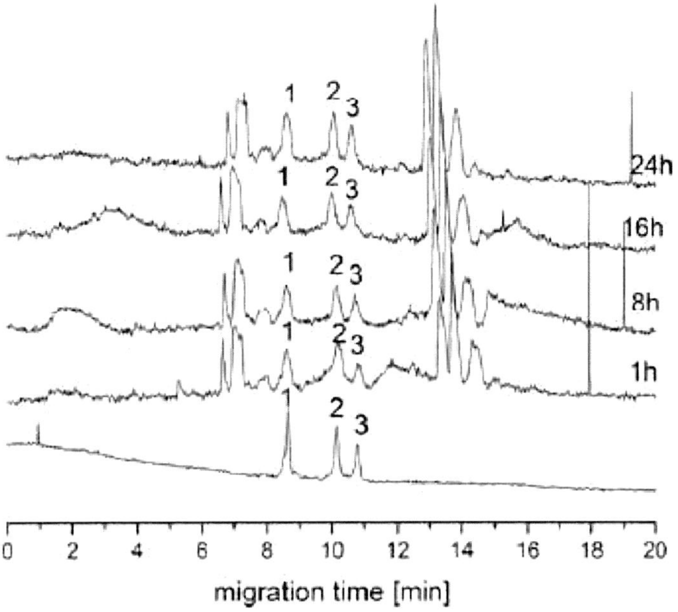

Fig. (5). CE monitoring of stability of systemin (1) and its derivatives, *N*-propiolyl systemin (2) and 3'-azido- 2',3'-dideoxythymidine systemin (3) in tomato stem within 1-24 hours. Experimental conditions: bare FS capillary (60/50 cm total/effective length, 75 µm id); BGE 25 mM phosphate buffer, pH 2.5, +20 kV, 25°C, electrokinetic sample injection at 10 kV for 5 s, detection at 214 nm. Reprinted with permission from plant [93].

In Fig. (**6**), comparison of separations of tryptic peptide fragments of envelope proteins extracted from spinach chloroplasts in different types of capillaries is demonstrated. Finally, neutral HPC capillary that eliminated both electrostatic peptide-wall interactions and EOF was selected. With the applied high-organic content CE-MS (HOCE) more peptides and proteins were identified compared to the conventional aqueous CE-MS (BGE: 10% acetic acid in water). Even though higher numbers of peptides and proteins were identified with nano-LC-MS method, HOCE-MS outperformed it at detecting peptides with molecular mass below 2 kDa. From the full set of peptides using the two techniques, 19% were identified only by HOCE-MS.

Tryptic peptide maps of water soluble proteins extracted from maize flours of *Bacillus thuringiensis* (Bt)-transgenic (Aristis-Bt) and two native non-transgenic (Aristis and Coventry) maize varieties obtained by CZE-UV profiling of tryptic hydrolysates of these proteins in three acid classical BGEs (50/100 mM Tris-phosphate, pH 2.25; 500 mM acetic acid, pH 2.54; 200 mM formic acid, 200 mM acetic acid, pH 2.05) and one acid isoelectric BGE (200 mM iminodiacetic acid, pH 2.26) showed significant qualitative and quantitative differences in the composition of these complex mixtures [95]. The differences in these profiles (see Fig. (7)) could be potentially used to differentiate the transgenic Aristis Bt and non-transgenic Aristis varieties or two native non-transgenic varieties, Aristis and Coventry.

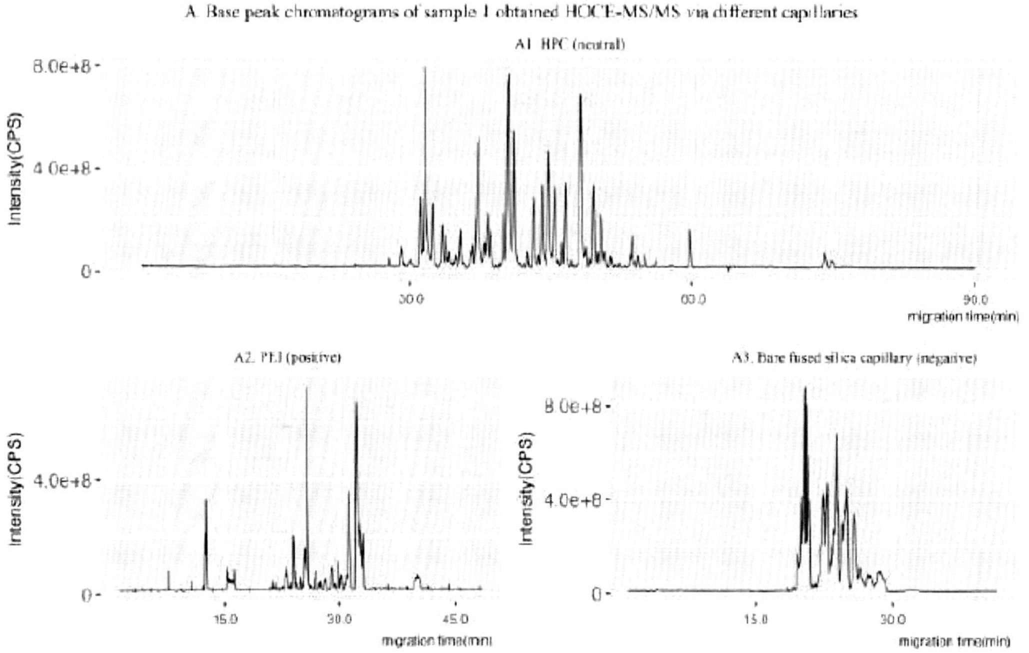

Fig. (6). Base peak electropherograms of tryptic peptide fragments of envelope proteins extracted from spinach leaves obtained by CE-MS/MS using coated and uncoated FS capillaries. A1, capillary with neutral coating by hydroxypropyl cellulose (HPC); A2, capillary with positively charged polyethylenimine (PEI) coating; A3, negatively charged bare FS capillary. Experimental conditions: BGE: 8/70/22% v/v formic acid/MeOH/water; capillary: 50 µm id, 100 cm total length; Separation voltage: +30 kV. Reprinted with permission from [28].

A shotgun proteomic approach based on tryptic digestion and CZE-ESI-TOF-MS separation and detection has been developed by Cifuentes *et al.* for peptide mapping of complex protein hydrolysates from native and genetically modified soybeans [96]. Optimized separation and mass characterization of more than 150 peptide fragments was achieved within 30 min in 50 µm id, 90 cm long bare FS capillary using volatile 0.5 M formic acid, pH 2.02, as BGE and propan--ol/water, 50/50 v/v, as sheath liquid at a 3 µL/min flow rate, see Fig. (**8**).

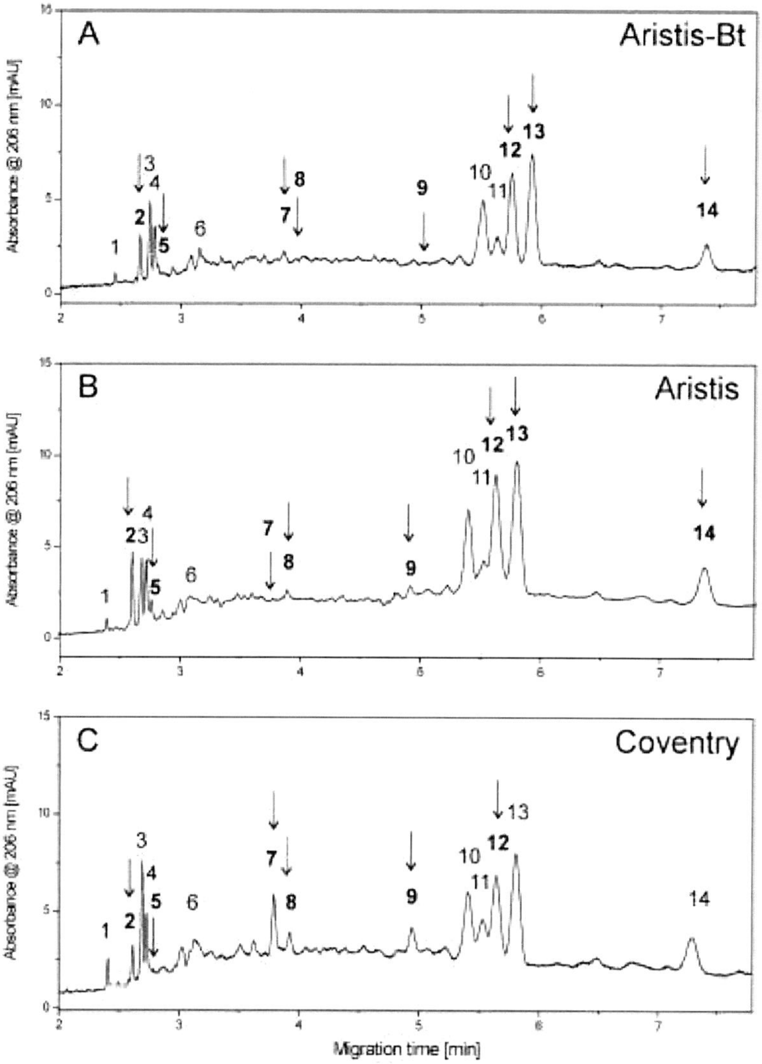

Fig. (7). CZE-UV profiling of tryptic peptide fragments of water soluble proteins extracted from maize flour of transgenic Aristis-Bt **(A)**, and non-transgenic Aristis **(B)** and Coventry **(C)** maize varieties in the isoelectric BGE (200 mM iminodiacetic acid, pH 2.26) in the bare FS capillary (50/375 μm id/od, 31/20 cm total/effective length) at 10 kV separation voltage and 206 nm detection wavelength. The bold printed numbers and arrows indicate the peaks of the significant differences (at least 100% differences in the corrected (migration times normalized) peak areas) between transgenic Aristis Bt and non-transgenic Aristis species or between two non-transgenic species, Aristis and Coventry, respectively. Reprinted with permission from [95].

CZE combined with C^4D [59] was applied for the simultaneous determination of synthetic dipeptide aspartame (Asp-Phe methyl ester) and other artificial sweeteners (cyclamate, acesulfam-K, and saccharine) in soft drinks and tabletop sweetener formulations. Fig. (**9**) shows CE separation of these artificial sweeteners in two beverages, cola and flavored guarana; in 75 μm id, 70/60 cm total/effective length FS capillary using BGE composed of 100 mM Tris, 10 mM His, pH 9.4, and 30 kV separation voltage [59]. Small amount of aspartame was detected only in the cola sample.

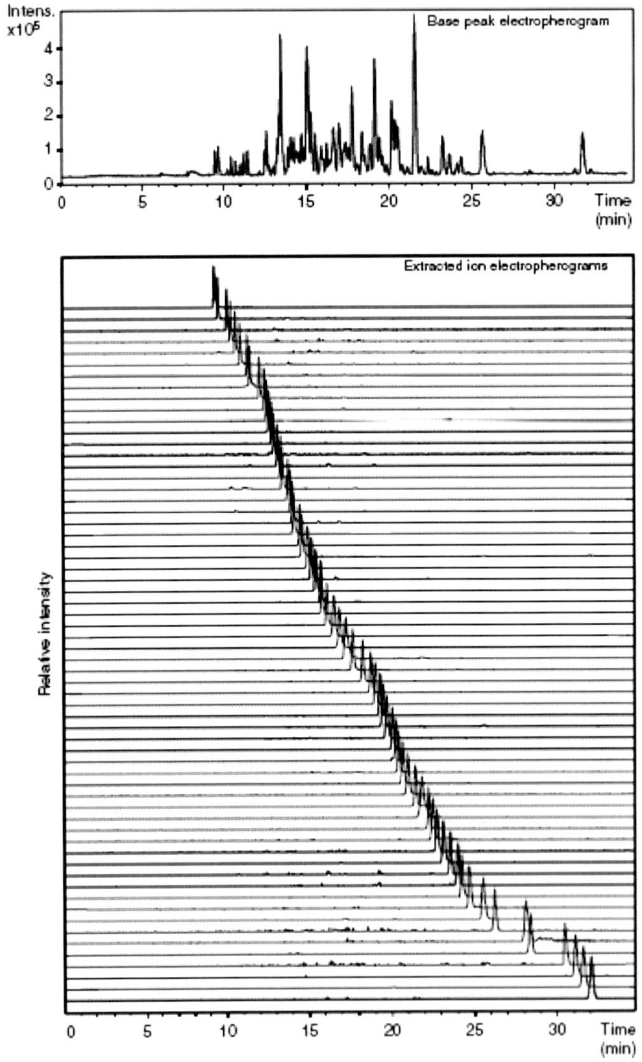

Fig. (8). CE-TOF MS base peak electropherogram of the digested soybean protein isolate extract and their corresponding extracted ion electropherogram. Experimental conditions: bare FS capillary (50 μm id, 90 cm); BGE: 0.5 M formic acid; injection time: 20 s at 34.5 mbar; separation voltage: 25 kV; sheath liquid: propan-2- ol/water (50:50 v/v) at a 3 μL/min flow rate; nebulizer gas: 0.4 bar; drying gas: 4 L/min nitrogen at 200°C; MS in positive ion mode; scan: *m/z* 50–3000. Reprinted with permission from [96].

CZE with UV-detection at 215 nm was used for determination of these sweeteners in lemon tea sachet samples using 20 mM sodium tetraborate, 15 mM Tris, 7.5 mM benzoic acid, pH 9.15, as the BGE, bare FS capillary 50 μm id, 48.5/40 cm total/effective length, and 20 kV separation voltage. The achieved LODs were in the range 0.5-12.0 mg/L [97].

Ultrafast (sub-minute) simultaneous determination of the aspartame and the above- referenced sweeteners in soft drinks, syrupy, analgesic suspensions and solid and liquid sweeteners was performed by CZE with UV detection at 220 nm, in 48.5 cm long 50 μm id capillary with short effective length (8.5 cm) using the short end injection, 30 kV separation voltage over the whole 48.5 cm long capillary and at 25°C temperature, using BGE composed of 45 mM 2-amino-2-(hydroxymethyl)-1,3- propanediol and 15 mM benzoic acid, pH 8.4 [98]. This method provided good linearity (R^2 > 0.9972) and LODs in the range 3.3-6.4 mg/mL.

CE-MS applications include determination of a toxic dipeptide γ-glutamyl-S-ethenyl- cysteine in the extract of the legume plant *Vicia narbonensis* seeds by CE-ESI-IT- MS/MS using a target-based approach [99]. For other examples of CE analysis of peptides in foods of plant origin, see Table **3**.

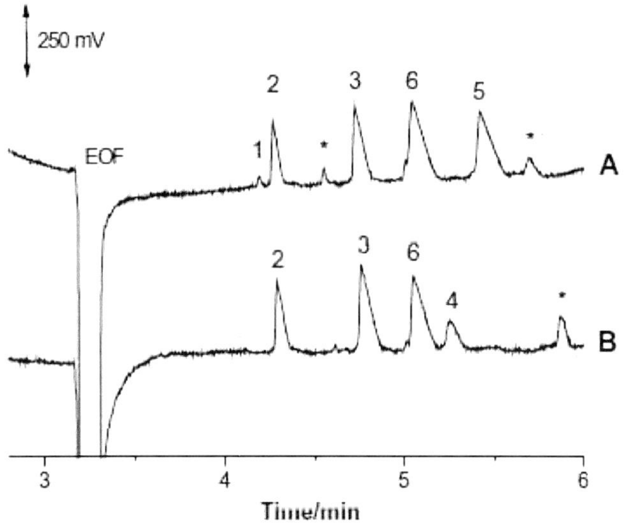

Fig. (9). CE-C⁴D electropherograms of soft drink samples: (A) cola and (B) guarana flavored. HEPES (4-(--hydroxyethyl)-1-piperazineethanesulfonic acid) was added as internal standard. Peaks: (1) aspartame; (2) HEPES; (3) cyclamate; (4) saccharin; (5) acesulfam-K; (6) benzoate; (*) unidentified peaks. Experimental conditions: BGE: 100 mM Tris, 10 mM His, pH 9.4, FS capillary, 75 μm id, 60 cm L_{det}, 70 cm L_{tot}, +30 kV. Reprinted with permission [59].

CE Analysis of Peptides in Foods of Microbial Origin

CE-ESI-TOF-MS was employed to measure peptides in different lots of the inactivated probiotic formulations, a bacterial lysate of heat-inactivated *Escherichia coli* and *Enterococcus faecalis* [101]. Over 5000 peptides were detected in five different lots of the bacterial lysate and in a sample of culture medium. 71 to 75% of the total peptide content was identical in all lots; the percentage increased to 87-89% when the absence of a peptide in one of the five samples was allowed. The results, based on over 2000 peptides, showed high similarity of the five different lots. With peptide sequence analysis, peptides originated from both *E. coli* and *E. faecalis* as well as peptides originating from the culture medium were identified.

CONCLUDING REMARKS

As demonstrated in many of the examples provided above, peptides are important components of foods and food products and CE methods represent powerful tools for their qualitative and quantitative analysis and characterization. CE analyses of peptides can be employed for the evaluation of quality and safety of foods, for identification of new (bioactive) peptides present in the extracts of foods of animal, plant or microbial origin, for determination of levels of particular synthetic or natural peptides in food products, or for peptide profiling in foods with the aim to distinguish foods originating from different animal or plant species including native or genetically modified organisms. Frequently, CE methods are used as complements and/or counterparts of the most commonly used LC techniques, especially when applied in the fields of peptidomic, metabolomic, and proteomic analyses of foods, collectively called foodomics. In these applications, several relevant advantages of CE methods are utilized, such as high separation efficiency, high sensitivity, high speed, ultra-small sample volume per single analysis, high flexibility, small reagents and solvents consumption, low environmental pollution, and low running costs. For peptide analysis in foods, mostly the CZE and MEKC modes of CE methods are employed, on-line coupled with various detection schemes, especially with MS detection due to its universality and selectivity, high sensitivity and high identification power, and UV-absorption detection for its simplicity and low cost. Contactless conductivity or fluorescence detection modes are applied less frequently.

Peptides belong to vitally important biomolecules. Many of them and many of their functions are known but many peptides including those in foods, still have to be revealed and their functions have to be recognized. Hence, separation, analysis, identification and structure elucidation of peptides in foods will remain one of the challenging tasks of CE methods in the future. Concerning CE methods themselves, their further developments will lead to further miniaturization, implementation on microfluidic devices, and integration into micro-total analytical systems, online coupled with highly sensitive MS or LIF detectors.

CONSENT FOR PUBLICATION

Not applicable.

CONFLICT OF INTEREST

The author declares no conflict of interest, financial or otherwise.

ACKNOWLEDGEMENTS

Some of the results presented have been achieved with the support of the Czech Science Foundation, grant no. 18-02597S, and of the Czech Academy of Sciences, project no. RVO 61388963.

REFERENCES

[1] Kitts, D.D.; Weiler, K. Bioactive proteins and peptides from food sources. Applications of bioprocesses used in isolation and recovery. *Curr. Pharm. Des.*, **2003**, *9*(16), 1309-1323. [http://dx.doi.org/10.2174/1381612033454883] [PMID: 12769739]

[2] Di Bernardini, R.; Harnedy, P.; Bolton, D.; Kerry, J.; O'Neill, E.; Mullen, A.M.; Hayes, M. Antioxidant and antimicrobial peptidic hydrolysates from muscle protein sources and by-products. *Food Chem.*, **2011**, *124*, 1296-1307. [http://dx.doi.org/10.1016/j.foodchem.2010.07.004]

[3] Wang, Y.; Huang, Q.; Kong, D.; Xu, P. Production and Functionality of Food-derived Bioactive Peptides: A Review. *Mini Rev. Med. Chem.*, **2018**, *18*(18), 1524-1535. [http://dx.doi.org/10.2174/1389557518666180424110754] [PMID: 29692240]

[4] Sanchez, A.; Vazquez, A. Bioactive peptides: A review. *Food Qual. Saf.*, **2017**, *1*, 29-46. [http://dx.doi.org/10.1093/fqs/fyx006]

[5] Arroume, N.; Froidevaux, R.; Kapel, R.; Cudennec, B.; Ravallec, R.; Flahaut, C.; Bazinet, L.; Jacques, P.; Dhulster, P. Food peptides: purification, identification and role in the metabolism. *Curr. Opin. Food Sci.*, **2016**, *7*, 101-107. [http://dx.doi.org/10.1016/j.cofs.2016.02.005]

[6] Acunha, T.; Ibáñez, C.; García-Cañas, V.; Simó, C.; Cifuentes, A. Recent advances in the application of capillary electromigration methods for food analysis and Foodomics. *Electrophoresis,* **2016**, *37*(1), 111-141. [http://dx.doi.org/10.1002/elps.201500291] [PMID: 26256797]

[7] Álvarez, G.; Montero, L.; Llorens, L.; Castro-Puyana, M.; Cifuentes, A. Recent advances in the application of capillary electromigration methods for food analysis and Foodomics. *Electrophoresis,* **2018**, *39*(1), 136-159. [http://dx.doi.org/10.1002/elps.201700321] [PMID: 28975648]

[8] Aydogan, C. Nanoscale separations based on LC and CE for food analysis: A review. *TrAC. Trends Analyt. Chem.,* **2019**, *121*, 115693. [http://dx.doi.org/10.1016/j.trac.2019.115693]

[9] Voeten, R.L.C.; Ventouri, I.K.; Haselberg, R.; Somsen, G.W. Capillary Electrophoresis: Trends and

Recent Advances. *Anal. Chem.*, **2018**, *90*(3), 1464-1481.
[http://dx.doi.org/10.1021/acs.analchem.8b00015] [PMID: 29298038]

[10] Poole, C.F. *Capillary Electromigration Separation Methods*; Elsevier Inc.: Amsterdam - Oxford - Cambridge (US), **2018**, pp. 1-606.

[11] Kašička, V. Recent developments in capillary and microchip electroseparations of peptides (2017-mid 2019). *Electrophoresis,* **2020**, *41*(1-2), 10-35.
[http://dx.doi.org/10.1002/elps.201900269] [PMID: 31657477]

[12] Štěpánová, S.; Kašička, V. Recent developments and applications of capillary and microchip electrophoresis in proteomics and peptidomics (2015-mid 2018). *J. Sep. Sci.*, **2019**, *42*(1), 398-414.
[http://dx.doi.org/10.1002/jssc.201801090] [PMID: 30457207]

[13] Latosinska, A.; Siwy, J.; Mischak, H.; Frantzi, M. Peptidomics and proteomics based on CE-MS as a robust tool in clinical application: The past, the present, and the future. *Electrophoresis,* **2019**, *40*(18-19), 2294-2308.
[http://dx.doi.org/10.1002/elps.201900091] [PMID: 31054149]

[14] Zhang, Z.B.; Qu, Y.Y.; Dovichi, N.J. Capillary zone electrophoresis-mass spectrometry for bottom-up proteomics. *TrAC. Trends Analyt. Chem.*, **2018**, *108*, 23-37.
[http://dx.doi.org/10.1016/j.trac.2018.08.008]

[15] Breadmore, M.C.; Grochocki, W.; Kalsoom, U.; Alves, M.N.; Phung, S.C.; Rokh, M.T.; Cabot, J.M.; Ghiasvand, A.; Li, F.; Shallan, A.I.; Keyon, A.S.A.; Alhusban, A.A.; See, H.H.; Wuethrich, A.; Dawod, M.; Quirino, J.P. Recent advances in enhancing the sensitivity of electrophoresis and electrochromatography in capillaries and microchips (2016-2018). *Electrophoresis,* **2019**, *40*(1), 17-39.
[http://dx.doi.org/10.1002/elps.201800384] [PMID: 30362581]

[16] Kohler, I.; Schappler, J.; Rudaz, S. Microextraction techniques combined with capillary electrophoresis in bioanalysis. *Anal. Bioanal. Chem.*, **2013**, *405*(1), 125-141.
[http://dx.doi.org/10.1007/s00216-012-6367-y] [PMID: 22965532]

[17] Kitagawa, F.; Otsuka, K. Recent applications of on-line sample preconcentration techniques in capillary electrophoresis. *J. Chromatogr. A,* **2014**, *1335*, 43-60.
[http://dx.doi.org/10.1016/j.chroma.2013.10.066] [PMID: 24210300]

[18] Kawai, T. Recent Studies on Online Sample Preconcentration Methods in Capillary Electrophoresis Coupled with Mass Spectrometry. *Chromatography (Basel),* **2017**, *38*, 1-8.
[http://dx.doi.org/10.15583/jpchrom.2017.001]

[19] Ramos-Payán, M.; Ocaña-Gonzalez, J.A.; Fernández-Torres, R.M.; Llobera, A.; Bello-López, M.A. Recent trends in capillary electrophoresis for complex samples analysis: A review. *Electrophoresis,* **2018**, *39*(1), 111-125.
[http://dx.doi.org/10.1002/elps.201700269] [PMID: 28791719]

[20] Jarvas, G.; Guttman, A.; Miekus, N.; Baczek, T.; Jeong, S.; Chung, D.S.; Patoprsty, V.; Masar, M.; Hutta, M.; Datinska, V.; Foret, F. Practical sample pretreatment techniques coupled with capillary electrophoresis for real samples in complex matrices. *TrAC. Trends Analyt. Chem.*, **2020**, *122*, 115702.
[http://dx.doi.org/10.1016/j.trac.2019.115702]

[21] Ramautar, R.; Somsen, G.W.; de Jong, G.J. Developments in coupled solid-phase extraction-capillary electrophoresis 2013-2015. *Electrophoresis,* **2016**, *37*(1), 35-44.
[http://dx.doi.org/10.1002/elps.201500401] [PMID: 26464172]

[22] Dziomba, S.; Araya-Farias, M.; Smadja, C.; Taverna, M.; Carbonnier, B.; Tran, N.T. Solid supports for extraction and preconcentration of proteins and peptides in microfluidic devices: A review. *Anal. Chim. Acta,* **2017**, *955*, 1-26.
[http://dx.doi.org/10.1016/j.aca.2016.12.017] [PMID: 28088276]

[23] Sonker, M.; Knob, R.; Sahore, V.; Woolley,. A.T. Integrated electrokinetically driven microfluidic

devices with pH-mediated solid-phase extraction coupled to microchip electrophoresis for preterm birth biomarkers. *Electrophoresis,* **2017**, *38*(13-14), 1743-1754.
[http://dx.doi.org/10.1002/elps.201700054] [PMID: 28272749]

[24] Kubota, K.; Kobayashi, N.; Yabuta, M.; Ohara, M.; Naito, T.; Kubo, T.; Otsuka, K. Identification and characterization of a thermally cleaved fragment of monoclonal antibody-A detected by sodium dodecyl sulfate-capillary gel electrophoresis. *J. Pharm. Biomed. Anal.,* **2017**, *140*, 98-104.
[http://dx.doi.org/10.1016/j.jpba.2017.03.027] [PMID: 28346883]

[25] Boyacı, E.; Rodríguez-Lafuente, Á.; Gorynski, K.; Mirnaghi, F.; Souza-Silva, E.A.; Hein, D.; Pawliszyn, J. Sample preparation with solid phase microextraction and exhaustive extraction approaches: Comparison for challenging cases. *Anal. Chim. Acta,* **2015**, *873*, 14-30.
[http://dx.doi.org/10.1016/j.aca.2014.12.051] [PMID: 25911426]

[26] Huang, C.; Seip, K.F.; Gjelstad, A.; Pedersen-Bjergaard, S. Electromembrane extraction for pharmaceutical and biomedical analysis - Quo vadis. *J. Pharm. Biomed. Anal.,* **2015**, *113*, 97-107.
[http://dx.doi.org/10.1016/j.jpba.2015.01.038] [PMID: 25669728]

[27] Šlampová, A.; Malá, Z.; Gebauer, P. Recent progress of sample stacking in capillary electrophoresis (2016-2018). *Electrophoresis,* **2019**, *40*(1), 40-54.
[http://dx.doi.org/10.1002/elps.201800261] [PMID: 30073675]

[28] Cheng, J.; Morin, G.B.; Chen, D.D.Y. Bottom-up proteomics of envelope proteins extracted from spinach chloroplast *via* high organic content CE-MS. *Electrophoresis,* **2020**, *41*(5-6), 370-378.
[http://dx.doi.org/10.1002/elps.201900452] [PMID: 31994203]

[29] Kowalski, P.; Olędzka, I.; Plenis, A.; Miękus, N.; Pieckowski, M.; Bączek, T. Combination of field amplified sample injection and hydrophobic interaction electrokinetic chromatography (FASI-HIEKC) as a signal amplification method for the determination of selected macrocyclic antibiotics. *Anal. Chim. Acta,* **2019**, *1046*, 192-198.
[http://dx.doi.org/10.1016/j.aca.2018.09.047] [PMID: 30482299]

[30] DeLaney, K.; Li, L. Capillary electrophoresis coupled to MALDI mass spectrometry imaging with large volume sample stacking injection for improved coverage of C. borealis neuropeptidome. *Analyst (Lond.),* **2019**, *145*(1), 61-69.
[http://dx.doi.org/10.1039/C9AN01883B] [PMID: 31723949]

[31] Heemskerk, A.A.M.; Wuhrer, M.; Busnel, J.M.; Koeleman, C.A.M.; Selman, M.H.J.; Vidarsson, G.; Kapur, R.; Schoenmaker, B.; Derks, R.J.E.; Deelder, A.M.; Mayboroda, O.A. Coupling porous sheathless interface MS with transient-ITP in neutral capillaries for improved sensitivity in glycopeptide analysis. *Electrophoresis,* **2013**, *34*(3), 383-387.
[http://dx.doi.org/10.1002/elps.201200357] [PMID: 23161657]

[32] Ivanov, A.V.; Nikiforova, K.A.; Bulgakova, P.O.; Virus, E.D.; Kubatiev, A.A. Determination of Blood Plasma Aminothiols Using Derivatization-enhanced Capillary Transient Isotachophoresis. *Anal. Sci.,* **2018**, *34*(4), 505-508.
[http://dx.doi.org/10.2116/analsci.17N036] [PMID: 29643316]

[33] Wang, L.; MacDonald, D.; Huang, X.; Chen, D.D.Y. Capture efficiency of dynamic pH junction focusing in capillary electrophoresis. *Electrophoresis,* **2016**, *37*(9), 1143-1150.
[http://dx.doi.org/10.1002/elps.201600008] [PMID: 26949078]

[34] Zhu, G.; Sun, L.; Dovichi, N.J. Dynamic pH junction preconcentration in capillary electrophoresis-electrospray ionization-mass spectrometry for proteomics analysis. *Analyst (Lond.),* **2016**, *141*(18), 5216-5220.
[http://dx.doi.org/10.1039/C6AN01140C] [PMID: 27460877]

[35] Pyell, U.; Rageh, A.H.; El-Awady, M. The Concept of Stationary and Moving Boundaries Modelled as Accelerating or Decelerating Planes in the Understanding of Sweeping Processes Employed for Online Focusing in Capillary Zone Electrophoresis and Electrokinetic Chromatography. *Chromatographia,* **2017**, *80*, 359-382.

[http://dx.doi.org/10.1007/s10337-017-3261-y]

[36] Chen, Y.; Zhang, L.; Cai, Z.; Chen, G. Dynamic pH junction-sweeping for on-line focusing of dipeptides in capillary electrophoresis with laser-induced fluorescence detection. *Analyst (Lond.),* **2011**, *136*(9), 1852-1858.
[http://dx.doi.org/10.1039/c0an00744g] [PMID: 21394345]

[37] Dlask, O.; Vaclavikova, N. Electrodialysis with ultrafiltration membranes for peptide separation. *Chem. Pap.,* **2018**, *72*, 261-271.
[http://dx.doi.org/10.1007/s11696-017-0293-6]

[38] Dallas, D.C.; Guerrero, A.; Parker, E.A.; Robinson, R.C.; Gan, J.; German, J.B.; Barile, D.; Lebrilla, C.B. Current peptidomics: applications, purification, identification, quantification, and functional analysis. *Proteomics,* **2015**, *15*(5-6), 1026-1038.
[http://dx.doi.org/10.1002/pmic.201400310] [PMID: 25429922]

[39] Martinović, T.; Šrajer Gajdošik, M.; Josić, D. Sample preparation in foodomic analyses. *Electrophoresis,* **2018**, *39*(13), 1527-1542.
[http://dx.doi.org/10.1002/elps.201800029] [PMID: 29660150]

[40] Wuethrich, A.; Quirino, J.P. Derivatisation for separation and detection in capillary electrophoresis (2015-2017). *Electrophoresis,* **2018**, *39*(1), 82-96.
[http://dx.doi.org/10.1002/elps.201700252] [PMID: 28758685]

[41] Wang, Z.; Wang, W.; Wang, W.; Xu, L.; Chen, G.; Fu, F. Separation and determination of β-casomorphins by using glass microfluidic chip electrophoresis together with laser-induced fluorescence detection. *J. Sep. Sci.,* **2011**, *34*(2), 196-201.
[http://dx.doi.org/10.1002/jssc.201000634] [PMID: 21246725]

[42] Yang, M.W.; Wang, Z.W.; Fang, L.; Zheng, J.P.; Xu, L.J.; Fu, F.F. Simultaneous and ultra-sensitive quantification of multiple peptides by using europium chelate labeling and capillary electrophoresis-inductively coupled plasma mass spectrometry. *J. Anal. At. Spectrom.,* **2012**, *27*, 946-951.
[http://dx.doi.org/10.1039/c2ja30065f]

[43] Beckers, J.L.; Bocek, P. The preparation of background electrolytes in capillary zone electrophoresis: golden rules and pitfalls. *Electrophoresis,* **2003**, *24*(3), 518-535.
[http://dx.doi.org/10.1002/elps.200390060] [PMID: 12569542]

[44] Persat, A.; Chambers, R.D.; Santiago, J.G. Basic principles of electrolyte chemistry for microfluidic electrokinetics. Part I: Acid-base equilibria and pH buffers. *Lab Chip,* **2009**, *9*(17), 2437-2453.
[http://dx.doi.org/10.1039/b906465f] [PMID: 19680570]

[45] Persat, A.; Suss, M.E.; Santiago, J.G. Basic principles of electrolyte chemistry for microfluidic electrokinetics. Part II: Coupling between ion mobility, electrolysis, and acid-base equilibria. *Lab Chip,* **2009**, *9*(17), 2454-2469.
[http://dx.doi.org/10.1039/b906468k] [PMID: 19680571]

[46] Kasicka, V.; Prusík, Z. Application of capillary isotachophoresis in peptide analysis. *J. Chromatogr. A,* **1991**, *569*(1-2), 123-174.
[http://dx.doi.org/10.1016/0378-4347(91)80228-5] [PMID: 1939486]

[47] Janini, G.M.; Issaq, H.J. Selection of buffers in capillary zone electrophoresis: Application to peptide and protein analysis. *Chromatographia,* **2001**, *53*, S18-S26.

[48] Righetti, P.G.; Sebastiano, R.; Citterio, A. Capillary electrophoresis and isoelectric focusing in peptide and protein analysis. *Proteomics,* **2013**, *13*(2), 325-340.
[http://dx.doi.org/10.1002/pmic.201200378] [PMID: 23180512]

[49] Galievsky, V.A.; Stasheuski, A.S.; Krylov, S.N. "Getting the best sensitivity from on-capillary fluorescence detection in capillary electrophoresis" - A tutorial. *Anal. Chim. Acta,* **2016**, *935*, 58-81.
[http://dx.doi.org/10.1016/j.aca.2016.06.015] [PMID: 27543015]

[50] Ban, E.; Song, E.J. Recent developments and applications of capillary electrophoresis with laser-

induced fluorescence detection in biological samples. *J. Chromatogr. B Analyt. Technol. Biomed. Life Sci.,* **2013**, *929*, 180-186.
[http://dx.doi.org/10.1016/j.jchromb.2013.04.028] [PMID: 23685428]

[51] Týčová, A.; Ledvina, V.; Klepárník, K. Recent advances in CE-MS coupling: Instrumentation, methodology, and applications. *Electrophoresis,* **2017**, *38*(1), 115-134.
[http://dx.doi.org/10.1002/elps.201600366] [PMID: 27783411]

[52] Stolz, A.; Jooß, K.; Höcker, O.; Römer, J.; Schlecht, J.; Neusüß, C. Recent advances in capillary electrophoresis-mass spectrometry: Instrumentation, methodology and applications. *Electrophoresis,* **2019**, *40*(1), 79-112.
[http://dx.doi.org/10.1002/elps.201800331] [PMID: 30260009]

[53] Mikšík, I. Coupling of CE-MS for protein and peptide analysis. *J. Sep. Sci.,* **2019**, *42*(1), 385-397.
[http://dx.doi.org/10.1002/jssc.201800817] [PMID: 30238606]

[54] Ramautar, R.; Somsen, G.W.; de Jong, G.J. CE-MS for metabolomics: Developments and applications in the period 2016-2018. *Electrophoresis,* **2019**, *40*(1), 165-179.
[http://dx.doi.org/10.1002/elps.201800323] [PMID: 30232802]

[55] Zhong, X.; Zhang, Z.; Jiang, S.; Li, L. Recent advances in coupling capillary electrophoresis-based separation techniques to ESI and MALDI-MS. *Electrophoresis,* **2014**, *35*(9), 1214-1225.
[http://dx.doi.org/10.1002/elps.201300451] [PMID: 24170529]

[56] Kubáň, P.; Hauser, P.C. Contactless conductivity detection for analytical techniques: Developments from 2016 to 2018. *Electrophoresis,* **2019**, *40*(1), 124-139.
[http://dx.doi.org/10.1002/elps.201800248] [PMID: 30010203]

[57] Jozanović, M.; Hajduković, M.; Galović, O.; Kralik, G.; Kralik, Z.; Sakač, N.; Medvidović-Kosanović, M.; Sak-Bosnar, M. Determination of anti-oxidative histidine dipeptides in poultry by microchip capillary electrophoresis with contactless conductivity detection. *Food Chem.,* **2017**, *221*, 1658-1665.
[http://dx.doi.org/10.1016/j.foodchem.2016.10.122] [PMID: 27979143]

[58] Jozanović, M.; Sakač, N.; Sak-Bosnar, M.; Carrilho, E. A simple and reliable new microchip electrophoresis method for fast measurements of imidazole dipeptides in meat from different animal species. *Anal. Bioanal. Chem.,* **2018**, *410*(18), 4359-4369.
[http://dx.doi.org/10.1007/s00216-018-1087-6] [PMID: 29707755]

[59] Bergamo, A.B.; da Silva, J.A.F.; de Jesus, D.P. Simultaneous determination of aspartame, cyclamate, saccharin and acesulfame-K in soft drinks and tabletop sweetener formulations by capillary electrophoresis with capacitively coupled contactless conductivity detection. *Food Chem.,* **2011**, *124*, 1714-1717.
[http://dx.doi.org/10.1016/j.foodchem.2010.07.107]

[60] Štěpánová, S.; Kašička, V. Recent applications of capillary electromigration methods to separation and analysis of proteins. *Anal. Chim. Acta,* **2016**, *933*, 23-42.
[http://dx.doi.org/10.1016/j.aca.2016.06.006] [PMID: 27496994]

[61] Hajba, L.; Guttman, A. Recent advances in column coatings for capillary electrophoresis of proteins. *TrAC. Trends Analyt. Chem.,* **2017**, *90*, 38-44.
[http://dx.doi.org/10.1016/j.trac.2017.02.013]

[62] Huhn, C.; Ramautar, R.; Wuhrer, M.; Somsen, G.W. Relevance and use of capillary coatings in capillary electrophoresis-mass spectrometry. *Anal. Bioanal. Chem.,* **2010**, *396*(1), 297-314.
[http://dx.doi.org/10.1007/s00216-009-3193-y] [PMID: 19838682]

[63] Domínguez-Vega, E.; Haselberg, R.; Somsen, G.W.; de Jong, G.J. Simultaneous assessment of protein heterogeneity and affinity by capillary electrophoresis-mass spectrometry. *Anal. Chem.,* **2015**, *87*(17), 8781-8788.
[http://dx.doi.org/10.1021/acs.analchem.5b01701] [PMID: 26214104]

[64] MacDonald, A.M.; Bahnasy, M.F.; Lucy, C.A. A modified supported bilayer/diblock polymer-

-working towards a tunable coating for capillary electrophoresis. *J. Chromatogr. A,* **2011**, *1218*(1), 178-184.
[http://dx.doi.org/10.1016/j.chroma.2010.10.111] [PMID: 21122870]

[65] Acunha, T.; Simó, C.; Ibáñez, C.; Gallardo, A.; Cifuentes, A. Anionic metabolite profiling by capillary electrophoresis-mass spectrometry using a noncovalent polymeric coating. Orange juice and wine as case studies. *J. Chromatogr. A,* **2016**, *1428*, 326-335.
[http://dx.doi.org/10.1016/j.chroma.2015.08.001] [PMID: 26296988]

[66] Bahga, S.S.; Santiago, J.G. Coupling isotachophoresis and capillary electrophoresis: a review and comparison of methods. *Analyst (Lond.),* **2013**, *138*(3), 735-754.
[http://dx.doi.org/10.1039/C2AN36150G] [PMID: 23232502]

[67] Stěpánová, S.; Kašička, V. Determination of impurities and counterions of pharmaceuticals by capillary electromigration methods. *J. Sep. Sci.,* **2014**, *37*(15), 2039-2055.
[http://dx.doi.org/10.1002/jssc.201400266] [PMID: 24838601]

[68] Heemskerk, A.A.M.; Busnel, J.M.; Schoenmaker, B.; Derks, R.J.E.; Klychnikov, O.; Hensbergen, P.J.; Deelder, A.M.; Mayboroda, O.A. Ultra-low flow electrospray ionization-mass spectrometry for improved ionization efficiency in phosphoproteomics. *Anal. Chem.,* **2012**, *84*(10), 4552-4559.
[http://dx.doi.org/10.1021/ac300641x] [PMID: 22494114]

[69] Pergande, M.R.; Cologna, S.M. Isoelectric Point Separations of Peptides and Proteins. *Proteomes,* **2017**, *5*(1), 4-14.
[http://dx.doi.org/10.3390/proteomes5010004] [PMID: 28248255]

[70] Štěpánová, S.; Kašička, V. Capillary electrophoretic methods applied to the investigation of peptide complexes. *J. Sep. Sci.,* **2015**, *38*(15), 2708-2721.
[http://dx.doi.org/10.1002/jssc.201500399] [PMID: 25982955]

[71] Terabe, S. Capillary separation: micellar electrokinetic chromatography. *Annu. Rev. Anal. Chem. (Palo Alto, Calif.),* **2009**, *2*, 99-120.
[http://dx.doi.org/10.1146/annurev.anchem.1.031207.113005] [PMID: 20636055]

[72] Silva, M. Micellar electrokinetic chromatography: a review of methodological and instrumental innovations focusing on practical aspects. *Electrophoresis,* **2013**, *34*(1), 141-158.
[http://dx.doi.org/10.1002/elps.201200349] [PMID: 23161127]

[73] Ryan, R.; Altria, K.; McEvoy, E.; Donegan, S.; Power, J. A review of developments in the methodology and application of microemulsion electrokinetic chromatography. *Electrophoresis,* **2013**, *34*(1), 159-177.
[http://dx.doi.org/10.1002/elps.201200375] [PMID: 23161220]

[74] Soliman, L.C.; Donkor, K.K. Method development for sensitive determination of nisin in food products by micellar electrokinetic chromatography. *Food Chem.,* **2010**, *119*, 801-805.
[http://dx.doi.org/10.1016/j.foodchem.2009.06.062]

[75] Ahmed, M.A.; Felisilda, B.M.B.; Quirino, J.P. Recent advancements in open-tubular liquid chromatography and capillary electrochromatography during 2014-2018. *Anal. Chim. Acta,* **2019**, *1088*, 20-34.
[http://dx.doi.org/10.1016/j.aca.2019.08.016] [PMID: 31623713]

[76] Mikšík, I. Capillary electrochromatography of proteins and peptides (2006-2015). *J. Sep. Sci.,* **2017**, *40*(1), 251-271.
[http://dx.doi.org/10.1002/jssc.201600908] [PMID: 27731543]

[77] Szerdahelyi, E.; Csehi, B.; Takacs, K.; Korompai, E.; Nagy, A.; Gelencser, E.; Friedrich, L. Monitoring of imidazole dipeptides in meat products by capillary zone electrophoresis. *Czech J. Food Sci.,* **2020**, *38*, 36-42.
[http://dx.doi.org/10.17221/192/2019-CJFS]

[78] Marcolini, E.; Babini, E.; Bordoni, A.; Di Nunzio, M.; Laghi, L.; Maczó, A.; Picone, G.; Szerdahelyi,

E.; Valli, V.; Capozzi, F. Bioaccessibility of the Bioactive Peptide Carnosine during *in vitro* Digestion of Cured Beef Meat. *J. Agric. Food Chem.*, **2015**, *63*(20), 4973-4978.
[http://dx.doi.org/10.1021/acs.jafc.5b01157] [PMID: 25966609]

[79] Felisiak, K.; Szymczak, M.; Kolakowski, E. Identification of non-protein nitrogen compounds separated by CZE without derivatization from TCA extract from salted herring meat. *J. Food Compos. Anal.*, **2019**, *77*, 108-114.
[http://dx.doi.org/10.1016/j.jfca.2019.01.010]

[80] Xu, Z.; Chen, H.; Wang, Z.; Fan, F.; Shi, P.; Tu, M.; Du, M. Isolation and Characterization of Peptides from Mytilus edulis with Osteogenic Activity in Mouse MC3T3-E1 Preosteoblast Cells. *J. Agric. Food Chem.*, **2019**, *67*(5), 1572-1584.
[http://dx.doi.org/10.1021/acs.jafc.8b06530] [PMID: 30614690]

[81] Català-Clariana, S.; Benavente, F.; Giménez, E.; Barbosa, J.; Sanz-Nebot, V. Identification of bioactive peptides in hypoallergenic infant milk formulas by capillary electrophoresis-mass spectrometry. *Anal. Chim. Acta,* **2010**, *683*(1), 119-125.
[http://dx.doi.org/10.1016/j.aca.2010.10.002] [PMID: 21094390]

[82] Català-Clariana, S.; Benavente, F.; Giménez, E.; Barbosa, J.; Sanz-Nebot, V. Identification of bioactive peptides in hypoallergenic infant milk formulas by CE-TOF-MS assisted by semiempirical model of electromigration behavior. *Electrophoresis,* **2013**, *34*(13), 1886-1894.
[http://dx.doi.org/10.1002/elps.201200547] [PMID: 23564639]

[83] Chen, H.; Shi, P.; Fan, F.; Tu, M.; Xu, Z.; Xu, X.; Du, M. Complementation of UPLC-Q-TOF-MS and CESI-Q-TOF-MS on identification and determination of peptides from bovine lactoferrin. *J. Chromatogr. B Analyt. Technol. Biomed. Life Sci.*, **2018**, *1084*, 150-157.
[http://dx.doi.org/10.1016/j.jchromb.2018.03.022] [PMID: 29604613]

[84] Fan, F.; Shi, P.; Chen, H.; Tu, M.; Wang, Z.; Lu, W.; Du, M. Identification and availability of peptides from lactoferrin in the gastrointestinal tract of mice. *Food Funct.*, **2019**, *10*(2), 879-885.
[http://dx.doi.org/10.1039/C8FO01998C] [PMID: 30693932]

[85] Masotti, F.; Battelli, G.; De Noni, I. The evolution of chemical and microbiological properties of fresh goat milk cheese during its shelf life. *J. Dairy Sci.*, **2012**, *95*(9), 4760-4767.
[http://dx.doi.org/10.3168/jds.2011-5039] [PMID: 22916879]

[86] Baptista, D.P.; Araújo, F.D.D.; Eberlin, M.N.; Gigante, M.L. A Survey of the Peptide Profile in Prato Cheese as Measured by MALDI-MS and Capillary Electrophoresis. *J. Food Sci.*, **2017**, *82*(2), 386-393.
[http://dx.doi.org/10.1111/1750-3841.13618] [PMID: 28098943]

[87] Johannesen, S.A.; Beeren, S.R.; Blank, D.; Yang, B.Y.; Geyer, R.; Hindsgaul, O. Glycan analysis *via* derivatization with a fluorogenic pyrylium dye. *Carbohydr. Res.*, **2012**, *352*, 94-100.
[http://dx.doi.org/10.1016/j.carres.2012.02.016] [PMID: 22425443]

[88] Benavente, F.; Pero-Gascon, R.; Pont, L.; Jaumot, J.; Barbosa, J.; Sanz-Nebot, V. Identification of antihypertensive peptides in nutraceuticals by capillary electrophoresis-mass spectrometry. *J. Chromatogr. A,* **2018**, *1579*, 129-137.
[http://dx.doi.org/10.1016/j.chroma.2018.10.018] [PMID: 30361036]

[89] Hagi, T.; Kobayashi, M.; Nomura, M. Metabolome analysis of milk fermented by γ-aminobutyric acid-producing Lactococcus lactis. *J. Dairy Sci.*, **2016**, *99*(2), 994-1001.
[http://dx.doi.org/10.3168/jds.2015-9945] [PMID: 26686724]

[90] Drevinskas, T.; Stankevičius, M.; Bimbiraitė-Survilienė, K.; Naujokaitytė, G.; Kaškonienė, V.; Kondrotienė, K.; Šernienė, L.; Kornyšova, O.; Malakauskas, M.; Maruška, A. Optimization of a capillary zone electrophoresis-contactless conductivity detection method for the determination of nisin. *Electrophoresis,* **2018**, *39*(19), 2425-2430.
[http://dx.doi.org/10.1002/elps.201800143] [PMID: 29935020]

[91] Thomas, F.C.; Mullen, W.; Tassi, R.; Ramírez-Torres, A.; Mudaliar, M.; McNeilly, T.N.; Zadoks,

R.N.; Burchmore, R.; David Eckersall, P. Mastitomics, the integrated omics of bovine milk in an experimental model of Streptococcus uberis mastitis: 1. High abundance proteins, acute phase proteins and peptidomics. *Mol. Biosyst.*, **2016**, *12*(9), 2735-2747.
[http://dx.doi.org/10.1039/C6MB00239K] [PMID: 27412456]

[92] Piovesana, S.; Capriotti, A.L.; Cavaliere, C.; La Barbera, G.; Montone, C.M.; Zenezini Chiozzi, R.; Laganà, A. Recent trends and analytical challenges in plant bioactive peptide separation, identification and validation. *Anal. Bioanal. Chem.*, **2018**, *410*(15), 3425-3444.
[http://dx.doi.org/10.1007/s00216-018-0852-x] [PMID: 29353433]

[93] Mucha, P.; Ruczynski, J.; Dobkowski, M.; Backtrog, E.; Rekowski, P. Capillary electrophoresis study of systemin peptides spreading in tomato plant. *Electrophoresis*, **2019**, *40*(2), 336-342.
[http://dx.doi.org/10.1002/elps.201800206] [PMID: 30259532]

[94] Bai, Y.; Chang, C.; Du, F.; Tan, Z.; Bai, Y.; Liu, H. Combination of dynamic pH junction with capillary electrophoresis-mass spectrometry for the determination of systemins in plant samples. *Electrophoresis*, **2014**, *35*(14), 1984-1988.
[http://dx.doi.org/10.1002/elps.201300434] [PMID: 24668451]

[95] Sázelová, P.; Kašička, V.; Leon, C.; Ibáñez, E.; Cifuentes, A. Capillary electrophoretic profiling of tryptic digests of water soluble proteins from Bacillus thuringiensis-transgenic and non-transgenic maize species. *Food Chem.*, **2012**, *134*(3), 1607-1615.
[http://dx.doi.org/10.1016/j.foodchem.2012.02.220] [PMID: 25005989]

[96] Simó, C.; Domínguez-Vega, E.; Marina, M.L.; García, M.C.; Dinelli, G.; Cifuentes, A. CE-TOF MS analysis of complex protein hydrolyzates from genetically modified soybeans--a tool for foodomics. *Electrophoresis*, **2010**, *31*(7), 1175-1183.
[http://dx.doi.org/10.1002/elps.200900448] [PMID: 20209569]

[97] Fernandes, V.N.O.; Fernandes, L.B.; Vasconcellos, J.P.; Jager, A.V.; Tonin, F.G.; de Oliveira, M.A.L. Simultaneous analysis of aspartame, cyclamate, saccharin and acesulfame-K by CZE under UV detection. *Anal. Methods*, **2013**, *5*, 1524-1532.
[http://dx.doi.org/10.1039/c3ay26187e]

[98] Vistuba, J.P.; Dolzan, M.D.; Vitali, L.; de Oliveira, M.A.L.; Micke, G.A. Sub-minute method for simultaneous determination of aspartame, cyclamate, acesulfame-K and saccharin in food and pharmaceutical samples by capillary zone electrophoresis. *J. Chromatogr. A*, **2015**, *1396*, 148-152.
[http://dx.doi.org/10.1016/j.chroma.2015.03.070] [PMID: 25895731]

[99] Simo, C.; Cifuentes, A.; Kasicka, V. , **2013**.

[100] Sugimoto, M.; Kaneko, M.; Onuma, H.; Sakaguchi, Y.; Mori, M.; Abe, S.; Soga, T.; Tomita, M. Changes in the charged metabolite and sugar profiles of pasteurized and unpasteurized Japanese sake with storage. *J. Agric. Food Chem.*, **2012**, *60*(10), 2586-2593.
[http://dx.doi.org/10.1021/jf2048993] [PMID: 22352923]

[101] Klein, G.; Schanstra, J.P.; Hoffmann, J.; Mischak, H.; Siwy, J.; Zimmermann, K. Proteomics as a Quality Control Tool of Pharmaceutical Probiotic Bacterial Lysate Products. *PLoS One*, **2013**, *8*(6), e66682.
[http://dx.doi.org/10.1371/journal.pone.0066682] [PMID: 23840518]

CHAPTER 5

Amino Acid Analysis by Capillary Electromigration Methods

María Castro-Puyana[1,2,*] and **María Luisa Marina**[1,2]

[1] *Departamento de Química Analítica, Química Física e Ingeniería Química, Universidad de Alcalá, Ctra. Madrid-Barcelona Km. 33.600. 28871 Alcalá de Henares (Madrid), Spain*

[2] *Instituto de Investigación Química Andrés M. del Río (IQAR), Universidad de Alcalá, Ctra. Madrid-Barcelona Km. 33.600. 28871 Alcalá de Henares (Madrid), Spain*

Abstract: The relevance of amino acids analysis is widely recognized in different fields. In Food Science, the determination of amino acids is of special interest since it can provide valuable information related to the nutritional, quality, and safety properties of food samples. For this reason, the development of robust, efficient, sensitive and cost-effective analytical methodologies is essential. Among the different analytical techniques, capillary electrophoresis has shown great potential in the last decades as a powerful tool to carry out the analysis of amino acids in food samples. This chapter aims at providing a comprehensive overview of the most recent applications of capillary electrophoresis for the analysis of protein and nonprotein amino acids in foodstuffs. The main experimental conditions concerning the separation and detection of amino acids are discussed and given in tables.

Keywords: Capillary electrophoresis, Food, Nonprotein amino acids, Protein amino acids.

INTRODUCTION

Amino acids are organic compounds which contain in their structure an amino group, a carboxylic group, and a side chain (R group) linked to a carbon atom. The R group is specific for each amino acid so that it determines not only the amino acid identity but also its properties. The wide variety of side chains that may be part of the amino acid structure makes possible the existence of hundreds of differen amino acids. Among them, just twenty have specific codons in the genetic code and are the protein building blocks in living organisms.

* **Corresponding autor María Castro-Puyana:** Departamento de Química Analítica, Química Física e Ingeniería Química, Universidad de Alcalá, Ctra. Madrid-Barcelona Km. 33.600, 28871 Alcalá de Henares (Madrid), Spain; E-mail: maria.castrop@uah.es

María Castro-Puyana, Miguel Herrero and María Luisa Marina (Eds.)
Capillary Electrophoresis in Food Analysis

Protein amino acids have a relevant role since they act as key compounds in the regulation of metabolic pathways to improve the health, survival, growth, development, and reproduction of organisms [1, 2]. These amino acids have been traditionally classified as "*essential*" or "*nonessential*" based on the need (or not) to be provided by the diet for nitrogen balance or growth [3]. The human body is able to produce nonessential amino acids in quantities needed for healthy functioning [4]. In contrast, those amino acids that cannot be produced by the human body must be ingested from the diet. These are the group of essential amino acids where phenylalanine (Phe), isoleucine (Ile), leucine (Leu), lysine (Lys), methionine (Met), threonine (Thr), tryptophan (Trp) and valine (Val), are included. In some cases, dietary supplementation with one or more amino acids can be used to treat some health problems. Considering that elevated levels of amino acids and their products are pathogenic factors for neurological disorders, oxidative stress, and cardiovascular disease, it is crucial to maintain an optimum balance among amino acids in the diet and circulation to guarantee body homeostasis [3]. Along with the twenty protein amino acids, there are hundreds of other amino acids that are not found in the main protein chain either for lack of a specific transfer RNA and codon triplet or because they do not arise from protein amino acids by post-translational modification [5]. This kind of nonprotein amino acids mainly exist in foods as products formed during food processing, as metabolic intermediates, or as additives to increase the nutritional and functional properties of foodstuffs [6].

In the food analysis framework, amino acids are responsible for nutritional and organoleptic properties. Technological processes (fermentation, aging, etc.) may affect their concentration or even generate new amino acids. In this way, amino acids and their relative concentration can be considered as an important marker of authenticity, quality, and safety of food [7]. For this reason, the determination of protein and nonprotein amino acids has been exploited for years in the food field. Consequently, the relevance of high-throughput and sensitive analytical techniques to determine these components in foodstuffs is unquestionable.

Since its introduction over thirty years ago, the popularity and attractiveness of Capillary Electrophoresis (CE) have considerably increased owing to its intrinsic advantages (short analysis time, high sensitivity, and low consumption of solvents). Nowadays, CE is considered an excellent and reliable analytical tool. The high interest in the development and application of CE approaches to perform amino acid analysis in different matrices including foodstuffs can be deduced from the review articles published on this topic during the last ten years and collected in Table **1**.

Although not included in this table, other interesting reviews have also been published on the use of capillary electromigration methods for the analysis of a large variety of food-related molecules with different chemical properties including amino acids [17 - 22].

Table 1. Representative reviews published in the last ten years on amino acid analysis in food by capillary electromigration methodologies.

Year	Title	Ref.
2010	MEKC: A powerful tool for the determination of amino acids in a variety of biomatrices.	8
2012	Recent novel MEKC applications to analyze free amino acids in different biomatrices: 2009-2010.	9
2012	Recent advances in amino acid analysis by capillary electrophoresis.	10
2013	Recent trends in the analysis of amino acids in fruits and derived foodstuffs.	11
2014	Recent advances in amino acid analysis by capillary electromigration methods, 2011–2013.	12
2016	Recent advances in amino acid analysis by capillary electromigration methods, 2013-2015.	13
2016	Capillary electrophoresis determination of non-protein amino acids as quality markers in foods	6
2016	Analysis of amino acids, proteins, carbohydrates and lipids in food by capillary electromigration methods: a review.	14
2018	Recent advances in amino acid analysis by capillary electromigration methods: June 2015–May 2017.	15
2019	Advances in the determination of nonprotein amino acids in foods and biological samples by capillary electrophoresis.	16

Among the different CE modes, Capillary Zone Electrophoresis (CZE) and Micellar Electrokinetic Chromatography (MEKC) are the preferred options to carry out the amino acid analysis in food samples. Other modes, Capillary Electrochromatography (CEC) and Capillary Isotachophoresis (ITP) have been used in a lesser extent. Also, Microchip Capillary Electrophoresis (MCE) can result in a very attractive tool for rapid amino acid analysis. Regarding detection systems, UV, fluorescence, conductivity and mass spectrometry (MS) have been employed in combination with the different CE modes. Among them, UV and laser-induced fluorescence (LIF) are the most popular for the determination of amino acids in food samples. However, the absence of chromophore or fluorophore groups in most of the amino acid structures makes that a derivatization step be often essential in order to enhance their detectability using optical detection. Derivatization procedures also provide merits for electrochemical detection or MS. In the former, the amino acid reacts with an electroactive group to improve its redox properties enabling an adequate derivative for detection, and in the latter, the change in mass- to-charge ratio may lead to a better analyte identification [23]. With the aim of providing suitable structural features to increase the detection sensitivity of amino acids in CE, a high number of derivatizing reagents have been successfully employed. Ideally, a derivatizing reagent should fulfill several requirements such as be stable and react rapidly and quantitatively with the amino acid to give rise to sufficiently stable derivatives without interfering side products [24, 25]. Among the broad variety of labeling agents, 9-fluoroenylmethylchloroformate (FMOC-Cl), fluorescein isothiocyanate (FITC), dansyl chloride (DNS-Cl), naphthalene-2,3- dicarboxaldehyde (NDA) and o-phthaldialdehyde (OPA) are the most employed for the derivatization of protein and nonprotein amino acids. To carry out the derivatization reactions in CE, three different strategies can be followed: precapillary (which can be performed either off-line (manual) or at-line (automated)), in-capillary, and postcapillary, which take place before, during, or after the electrophoretic separation, respectively. Selecting one way or another will depend on different parameters such as the number of samples to be analyzed or the properties of the analyte and the reagent, among others [26]. Even though precapillary derivatization has some drawbacks (the possible formation of side products, a loss in sensitivity due to the dilution of the sample, and the fact that it is often time consuming), this derivatization strategy is oftentimes the most used due to high flexibility in the reaction conditions and the wide availability of reagents that can be employed. However, the current trend is to establish procedures to adapt the derivatization step to the in-capillary mode. For more information on derivatizing reagents, reaction conditions, spectral properties, and derivatization schemes, readers are referred to different interesting publications [23 - 28].

A relevant point that should be mentioned is the chirality of a great number of protein and nonprotein amino acids. Except glycine, all protein amino acids have an asymmetric carbon and exhibit optical activity so that they are chiral molecules that can exist (at least) as one pair of enantiomers. As indicated by Armstrong *et al.* in 1990, enantioselective separations can be used in food and beverage studies for identifying adulterated samples, to control and to monitor fermentation processes and products, to evaluate the effects of age, treatment and storage or to evaluate flavor and fragrance components, among others [29]. Since this book includes a chapter specifically dedicated to chiral analysis, this aspect is not discussed herein. Readers are encouraged to read Chapter 9 to obtain a more in-depth knowledge about the potential of CE to achieve the enantiomeric analysis of amino acids and other chiral food components.

The purpose of the present chapter is to provide the reader with a comprehensive overview on the most recent applications of CE for the analysis of protein and nonprotein amino acids in food matrices covering the literature published in the last five years. The main advantages and drawbacks of these procedures are discussed with the aim of showing the potential of CE to perform the analysis of these food components. Experimental conditions of the developed methodologies will be detailed in tables and some representative examples will be illustrated.

DETERMINATION OF PROTEIN AMINO ACIDS

As previously mentioned previously, the determination of amino acids in food matrices provides relevant information related to nutritional, quality, and safety properties. Table **2** summarizes the characteristics of different CE methodologies developed from 2015 till now for the analysis of protein amino acids in a great variety of food matrices, including fruit juices, beers, cereal products, soy sauce, vinegars, or microalgae, among others. The main objectives of these CE approaches were to determine the amino acid content in a variety of foodstuffs and to perform their quality control through authenticity assessment and the detection of adulterations. From the information presented in Table **2**, it can be deduced that CZE and MEKC continue to be the most popular CE modes in the analysis of protein amino acids.

Table 2. Characteristics of the CE approaches devoted to the determination of protein amino acids in foods published in the last years (2015-June 2020)

Protein amino acids	Sample	Mode/detection	Experimental Conditions	LOD	Ref.
Ala, Asn, Gln, Pro, Ser, Val	Sudanese foods	CZE-UV (475 nm)	100 mM borate (pH 9.7), capillary, 50 µm x 30 cm, 25 kV, 25°C	0.32-0.56 mg/L	**30**
Trp	Soy sauce	CZE-UV (280 nm)	10 mM borate (pH 10.0) capillary, 50 µm x 61.5 cm, 20 kV, temperature, not indicated	-	**31**
Lys, Arg, His, Gly, Ala, Leu, Ile, Trp, Thr, Met, Tyr, Phe, Glu, Pro, Cys, Asp *(Separated from other twenty small charged molecules and small peptides)*	Salted herring meat	CZE-UV (200, 260 and 280 nm)	0.1 M phosphate (pH 2.5), capillary, 50 µm x 32 cm, 20 kV, 35°C	-	**32**

(Table 2) cont.....

Protein amino acids	Sample	Mode/detection	Experimental Conditions	LOD	Ref.
Glu *(Separated from benzoic acid and sorbic acid)*	Oyster, sweet and sour chili sauce	CZE-UV (230 nm)	5 mM OPA + 6 mM 3-MPA in 25 mM borax, capillary, 50 µm x 42 cm, 25 kV, 25°C	0.46 mg/L	**33**
Lys, His, Asp, Cys, Trp, Ser, Phe, Gly, Val, Ala, Leu, Ile, Try, Met, Pro, Glu, Thr	Beer	CZE-UV (254 nm)	50 mM $CuSO_4$ (pH 4.4), capillary, 50 µm x 65 cm; 22.5 kV, temperature, not indicated	0.13-0.25 µg/mL	**34**
Lys, His, Asp, Cys, Trp, Ser, Phe, Gly, Val, Ala, Leu, Ile, Try, Met, Pro, Glu, Thr	Vinegar	CZE-UV (254 nm)	50 mM $CuSO_4$ (pH 4.4), capillary, 50 µm x 65 cm; 22.5 kV, temperature, not indicated	0.13-0.25 µg/mL	**35**
Glu	Soy sauce	CZE-C^4D	5 M acetic acid (pH 2.0), capillary, 50 µm x 30 cm, 20 kV, 25°C	59.2 µM	**36**
Glu	Soy sauce	CZE-C^4D	6% PEG (w/v) in 5 M acetic acid (pH 2.0) capillary, 50 µm x 40 cm, 18.5 kV, 20°C	-	**37**
Lys, Ala, Pro, Glu, Asp *(Separated from 10-hydroxy-2-decenoic acid)*	Royal jelly based dietary supplement	CE-C^4D	2 M lactic acid, capillary, 50 µm x 48 cm, -16 kV, temperature, not indicated	0.039-0.090 mg/g	**38**
Ile, Gln, Asn, β-Ala, and α-Ala	*Dunaliella salina*	CZE-LIF (λ_{ex} 488 nm; λ_{em} 520 nm)	10 mM β-CD in 100 mM tetraborate (pH 9.4), capillary, 50 µm x 50 cm, 25 kV, 25°C	LOQ: 3-30 nM	**39**
Trp, Tyr *(Separated from tryptamine and tyramine)*	Beer	CZE-AD	50 mM borate (pH 9.35), capillary, 25 µm x 19.5 cm, 3 kV, 25°C	Trp: 0.45 µM Tyr: 0.52 µM	**40**
Phe *(Separated from Trp, His, Tyr)*	Cereal products	MEKC-UV (200 nm)	100 mM SDS + 25% methanol in 30 mM phosphoric acid (pH 1.9); capillary, 50 µm x 8.5 cm, -20 kV, 25°C	1.0 mg/L	**41**
20 protein amino acids *(Separated from cystine)*	Passion fruit juices	MEKC-UV (265 nm)	30 mM SDS + 5% methanol in 60 mM borate (pH 10.1); capillary, 50 µm x 72 cm; 25 kV, 23°C	0.01-0.7 mM	**42**

(Table 2) cont.....

Protein amino acids	Sample	Mode/detection	Experimental Conditions	LOD	Ref.
Ala, Arg, Asn, Cys, Gly, His, Lys, Met, Phe,Trp, Tyr, Val. *(Separated from lactic acid, pyroglutamic acid, taurine, organic acids, cystine and tryptamine)*	Fermentation broth of tapioca starch and yeast extract	MEKC-UV (200 nm)	50 mM SDS in 25 mM tetraborate (pH 9.1); capillary, 50 μm x 56 cm; 30 kV, 35°C	-	**43**
Trp, Tyr *(Separated from 7 biogenic amines)*	Wine and beer	MEKC-UV (200 nm)	20 mM SDS in 5 mM phosphate (pH 3.7), capillary, 75 μm x 40 cm, -15 kV, temperature, not indicated	Trp: 0.48 μM Tyr: 0.13 μM	**44**
Ser, Asn, Thr, Gln, His, Gly, Glu, Ala, Asp, Tyr, Val, Met, Ile, Trp, Leu, Phe, Arg *(Separated from γ-Aminobutyric acid, citrulline, phosphoryl ethanolamine and ethanolamine)*	Beers	MEKC-LIF (λ_{ex} 442 nm; λ_{em} 485 nm)	60 mM SDS + 2 mM HP-β-CD in 40 mM tetraborate (pH 9.2); capillary, 10 μm x 10 cm; -25 kV, 25°C	2.0-5.0 nM	**45**
Arg, Lys, Leu, Ile, His, Gln, Val, Phe, The, Asn, Pro, Ser, Gly, Glu, Asp *(Separated from tyramine)*	Yeast (*Saccharomyces cerevisiae*)	MEKC-LIF (λ_{ex} 488 nm; λ_{em} 520 nm)	30 mM SDS + 20 mM 18-crown-6-ether in 50 mM tetraborate (pH 9.0), capillary, 50 μm x 50 cm, 30 kV, 15°C	LOQ: 62.5 pM-6.25 nM	**46**
Leu, Ile, Val	Food supplement	ITP-CD	Leading electrolyte: 0.01 M HCl, 0,05% hydroxyethyl cellulose, + β-alanine (pH 3.6), terminating electrolyte: 0.01 M valeric acid with sodium hydroxide (pH 7.24)	0.62-0.66 mg/L	**47**

*Capillary dimensions expressed as internal diameter x effective length (cm to the detector), and LODs units expressed as in the original work.

CZE

UV detector is the most common detection approach used in combination with CZE for the determination of protein amino acids in foods, although a derivatization step is often required to enhance amino acids detectability.

As it can be seen in Table **2**, several CZE-UV methodologies were proposed in the period of time covered in this chapter.

A simple CZE-UV method was proposed by Omar *et al.* to carry out the analysis of six free amino acids (alanine (Ala), asparagine (Asn), glutamine (Gln), proline (Pro), serine (Ser), Val) in five different Sudanese food samples (potato, eggplant, chickpeas, soft wheat flour, sorghum durra flour) [30]. To achieve a high UV sensitivity, the influence of different derivatization parameters (pH, temperature, reaction time, labeling concentration) on the derivatization reaction between 4-chloro-7-nitro-2,1,3-benzoxadiazole (NBD-Cl) and the selected amino acids were investigated. The optimized conditions were based on the use of 40 mM NBD-Cl in borate buffer (pH 8.5) at 70ºC for 60 min. The derivatized amino acids were separated in borate buffer at pH 9.7 ranging the LODs achieved between 0.32 and 0.56 mg/L. The results obtained in this work for Asn concentration in food raw material were correlated with those obtained in a previous work [48] for acrylamide concentration in final food product. However, the relationship between them were not clear so that more studies are needed. Borate buffer was also used in a different work to perform the analysis of Trp in soy samples [31]. In this case, the CE-UV method was employed to validate the data obtained by fluorescence resonance energy transfer in a flow-injection system. Taking into account that this essential protein amino acid decomposed during manufacturing of non-fermented soy sauce by an acid-hydrolysis procedure, the analytical method was used to discriminate pure fermented soy sauces, adulterated soy sauces and chemical soy sauces. Other work reported the identification of a broad variety of compounds including sixteen amino acids (see Table **2**) and other twenty small charged molecules and small peptides, in trichloroacetic extracts from salted herring meat by CZE-UV [32]. This methodology showed great potential for the fast monitoring of the salted herring ripening process in which the proteolysis of herring meat takes place. This gives rise to the release of high amounts of free amino acids and peptides responsible for the typical and distinctive taste and aroma of mature meat [49, 50]. Therefore, the degree of hydrolysis provides information related to the sensory quality and nutritive value of a food product.

As an alternative to precapillary derivatization, Aung and Pyell developed a strategy by combining the determination of in-column derivatized glutamic acid (Glu) together with the determination of underivatized benzoic and sorbic acids within one CZE method (using the same background electrolyte) at a fixed UV-wavelength [33]. The method was based on the use of 25 mM borax, 5 mM o-phthalaldehyde (OPA) and 6 mM 3-mercaptopropionic acid (3-MPA) as separation medium and the in-capillary derivatization of Glu with OPA (in presence of 3- MPA). Once the experimental parameters were optimized following a fundamental investigation to minimize the reaction band broadening, the developed methodology was validated in terms of LOD, LOQ, linearity, recovery and precision. Then, to verify the feasibility of this analytical system, canned food and other processed food (namely, soy, fish, oyster, and sweet and sour chili sauces) were analyzed after application of appropriate clean-up steps. Glu was detected in all the samples without interferences from other matrix components. While it was present at trace levels in sweet and sour chili sauce, a relative high concentration was found in the fish sauce.

AD, amperometric detection; CD, conductivity detector; C^4D, capacitively coupled contactless conductivity; HP-β-CD, 2-hydroxypropyl-β-cyclodextrin; LIF, laser-induced fluorescence; 3-MPA, 3-mercaptopropionic acid; OPA, o-phthalaldehyde; PEG, polyethylene glycol; SDS, sodium dodecyl sulfate

An interesting CZE-UV methodology was developed by Luo *et al.*, to achieve the sensitive analysis of underivatized amino acids [34, 35]. To make possible the UV amino acid detection, the methodology was based on the coordination interaction between amino acids and copper ions since the complexes formed have a strong absorption at 254 nm. In this way, a derivatization step is not necessary. Since the presence of copper ions (which present UV absorption) in the separation buffer reduced sensitivity, the sweeping technique was used to improve it. Once the separation conditions were optimized, the methodology was applied to the direct determination of seventeen free amino acids in beer samples (see Fig. **1**) [34]. Then, using the United Nations Food Agriculture Organization/World Health Organization (FAO/WHO) model of essential amino acid pattern, the quality and flavor of three types of beer were evaluated. The results obtained showed that the contents of Phe, Pro, Ser and Ile were relatively high in the samples analyzed which had a great influence on flavor. Subsequently, the authors employed the same methodology to evaluate vinegars according to the ratios essential amino acids/total amino acids and essential amino acids/nonessential amino acids stipulated by FAO/WHO, and principal component analysis [35].

During the period of time covered in this chapter, three different works demonstrated the applicability of capacitively coupled contactless conductivity detector (C^4D) in combination with CZE for the analysis of protein amino acids. Campos *et al.* presented a systematic investigation focused on the changes induced by organic solvents on the migration properties of Glu, in which experimental and mathematical simulated data were compared to determine the parameters affecting the defocusing (peak splitting) of glutamate zones [36]. This artefact peaks can be observed in the analysis of ampholytic compounds and depend on the composition of the separation medium. Authors demonstrated in this study that the addition of acetonitrile prevents peak splitting of Glu and that this addition is more effective if it is made in the sample instead of in the separation medium. To show the applicability of their methodology, the authors applied the CZE-C[4]D method to the quantitative analysis of Glu in soy samples from different suppliers. Subsequently, the same authors developed other CZE-C[4]D strategy based, in this case, on the coupling of an on-line liquid-liquid extraction with CZE which allowed an efficient analysis in samples with high sodium content [37].

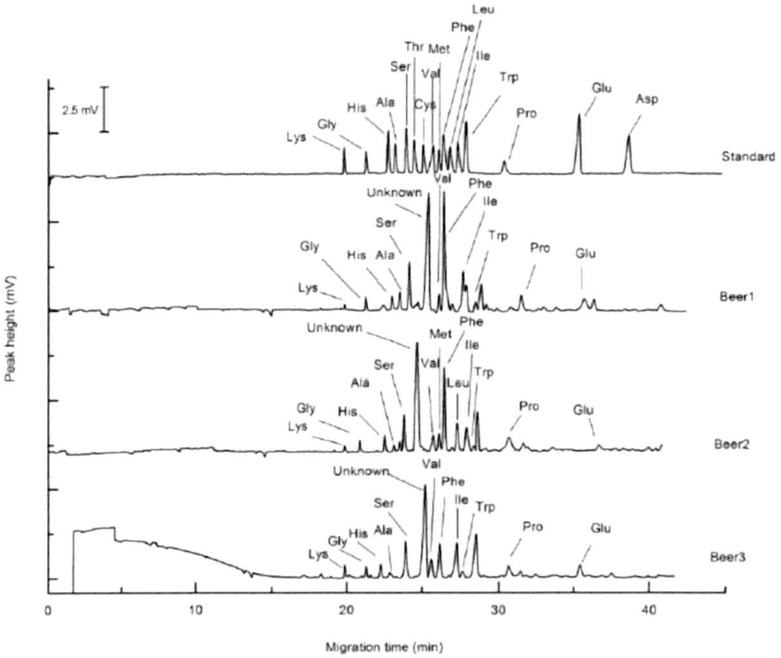

Fig. (1). Electrophregrams obtained in the analysis of several protein amino acids in beer samples. Experimental conditions: Capillary, 50 µm x 65 cm (effective length); buffer, 50 mM CuSO4 (pH 4.4); voltage, 22.5 kV. Reprinted with permission from [34].

In fact, samples with high sodium content (soy sauce contains around 17-19% w/v of sodium) lead to the deterioration of the efficiency and reproducibility. An electrically driven extraction of the target compounds was achieved by using an aqueous two-phase system which gave rise to the sodium depletion. To do that, the sample was prepared in Dextran- rich phase and the acetic acid buffer contained 6% w/v of polyethylene glycol. As a proof of concept, the developed strategy was successfully applied to the determination of Glu in soy sauce. More recently, Dong *et al.,* employed, for the first time, a purpose-made dual-channeled CE-C[4]D instrument for the fast determination of 10-hydroxy-2-decenoic acid (10-HAD) and five protein amino acids (Lys, Ala, Pro, Glu, aspartic acid (Asp)) in royal jelly supplements [38]. By using this approach, it was possible to perform the analysis of compounds of different categories in a single run since the dual-channel setup allows independent optimization of the working conditions for each category. The reliability of the data was demonstrated by the good agreement of the results obtained with the proposed method and those obtained with a standard confirmation method (liquid chromatography-UV (LC-UV)). This CE-C^4D strategy presents relevant advantages including faster analysis time for higher throughput, lower operation cost and low consumption of reagents demonstrating to be an inexpensive solution for quality control of food products.

Other detection systems such as LIF and amperometric detection (AD) were also employed in combination with CZE for the determination of protein amino acids in foodstuffs. Thus, a CZE-LIF method was used for the sensitive analysis of several amino acids (see Table **2**) extracted by electroporation from *Dunaliella salina* microalgae [39]. With this aim, amino acids were labeled with FITC employing microwave-assisted derivatization at 80°C (680 W) during 150 s. Satisfactory electrophoretic resolutions were obtained using 100 mM borate buffer (pH 9.4) containing 10 mM β-cyclodextrin (10 mM) as separation medium, and limits of quantification at nanomolar level (3-30 nM) were reached. To evaluate its efficiency the developed analytical strategy was applied to investigate the effect of cultivation conditions on the amino acid expression in the microalgae. Results obtained showed that the production of amino acids was induced under stress conditions and confirmed the accumulation of Ala in response to exposure to a variation of stress conditions which can be considered as a universal first stress signal expressed by cells. Regarding to AD, it was employed as detection system in a miniaturized CE equipment for the determination of two amino acids (Trp and tyrosine (Tyr)) and two bioamines (tryptamine and tyramine) [40]. Bioamines are the products of microbial amino acid decarboxylation whose presence in food is of high importance from the safety and quality viewpoints, given their relation with toxic incidents [44]. The use of a gold-nanoparticles and reduced graphene oxide composite electrode combined with a borate separation buffer made possible the separation and detection of the four target compounds in 7 min with a good sensitivity (LODs of 0.45 and 0.52 μM for Trp and Tyr, respectively) and selectivity [40]. The CZE-AD approach was satisfactorily applied to analyze beer samples demonstrating to be an alternative tool for the rapid assessment of food safety.

MEKC

The determination of the amount of some amino acids in foods is, in certain occasions, of special relevance for human health. For instance, the essential amino acid Phe is involved in an autosomal recessive disease caused by a mutation in gene of the enzyme phenylalanine hydroxylase (which is responsible for transformation of Phe to tyrosine). Deficiency of this enzyme results in intolerance to the dietary intake of this essential amino acid and produces different disorders including phenylketonuria and hyperphenylalaninemia [51]. Considering that many food products have this amino acid in their composition, it is relevant to inform people suffering the above-mentioned disorders on the concentration levels of Phe in foods. In this sense, Camargo Valese *et al.* developed a rapid MEKC methodology with UV detection for the determination of Phe in cereals products [41]. To find the most adequate experimental conditions, the effect of different variables such as voltage, temperature, percentage of methanol, and concentration of phosphoric acid and sodium dodecyl sulfate (SDS) was evaluated employing a factorial experimental design. In this study, other amino acids such as Trp, Tyr and histidine (His) were also considered since they are present in cereals and could interfere in the determination of Phe. Using 8.5 cm as capillary effective length and under optimal separation conditions (see Table **2**), the amino acids separation was achieved in just 2.2 min with a LOD for Phe of 1.0 mg/L (which corresponds to 20 mg/g sample). After validating different parameters according to the guidelines for method validation, the MEKC-UV method was successfully applied to evaluate the amount of Phe in commercial breakfast cereals products. Statistical analysis was performed to compare the data obtained by the proposed MEKC method and a LC- MS method. Results obtained in the *t-test* showed that the contents of Phe obtained by both methods in the cereal samples analyzed were statistically equal at a confidence level of 95%. This fact demonstrates the potential of the fast MEKC methodology for the routine monitoring of Phe in food products based on cereals.

MEKC-UV was also used as analytical platform for the quantification of the twenty protein amino acids along with cystine (and using hydroxyproline (Hyp) as internal standard) in passion fruit juices [42]. Using FMOC as derivatizing reagent and 30 mM SDS + 5% methanol in 60 mM borate (pH 10.1) as separation buffer, it was possible to achieve the separation of the twenty-two compounds. After validation, the methodology was applied on eight brands of industrial processed juices plus one fresh natural juice. In these samples, twelve amino acids (Gln, Ala, Ser, Pro, Glu, Asp, Arg, Val, Ile, Leu, Phe, and Lys) were detected. The statistical analysis of the data obtained from the quantitative analysis of six of these amino acids (namely, Pro, Glu, Ile, Leu, Phe, and Arg) enabled the characterization of different types of juices and showed potential to detect possible adulterations on industrial juices. Another MEKC-UV methodology based on the use of a borate buffer containing SDS was proposed by Laube *et al.*, to perform the analysis of lactic acid and different organic impurities including twelve protein amino acids (see Table **2**), cysteine, taurine, tryptamine, pyroglutamic acid, and ten organic acids in the fermentation broth of tapioca starch and yeast extract [43]. The developed methodology proved to be a rapid and reliable approach for the analysis of lactic acid and the impurity monitoring. It could be used as a tool for the benchmarking of the methods to produce pure lactic acid from fermentation solutions. Huang *et al.* also demonstrated the potential of a MEKC-UV method for the simultaneous determination of two amino acids (Trp and Tyr) and seven biogenic amines [44]. Under optimal separation conditions (see Table **2**), the nine studied compounds were separated in 11 min with LODs of 0.48 and 0.13 μM for Trp and Tyr, respectively. This method was successfully applied to the analysis of these analytes in wine and beer samples. Trp was only detected in of Tyr was detected both in beer and in white wines.

As shown in Table **2**, MEKC methodologies with LIF detection were also developed in the last five years for the analysis of protein amino acids in food. On the one hand, Zhu *et al.* developed a methodology enabling the simultaneous separation of seventeen protein amino acids, γ-aminobutyric acid (GABA), citrulline, phosphoryl ethanolamine and ethanolamine in beers [45]. The reported method was based on a flow-gated CE system coupled with alternate electrokinetic injections carried out by a microfabricated switch (see Fig. **2A**). By using this strategy, standards and samples, previously derivatized with NDA in presence of cyanide, were alternatively delivered to the flow gate which made possible to use standards as side-by-side calibration and references for quantitation. Employing a capillary effective length of 10 cm and an optimized separation medium consisting of 60 mM SDS and 2 mM 2-hydroxypropyl-β-cyclodextrin (HP-β-CD) in 40 mM tetraborate (pH 9.2), the separation of all the compounds studied was achieved in just 90 s. Once the method was validated, it was applied to the analysis of amino acids in eight different brands of beer. Fig. (**2B**) displays the electropherograms obtained for a mixture of standard amino acids, a beer sample and a blank sample (obtained mixing NDA and cyanide without amino acids). The results obtained showed that the most abundant amino acids were Ala, Val, and GABA whereas citrulline, Gln, and Met were the less abundant. Although this methodology is a fast and accurate approach to quantify amino acids, it can be only applied in those samples in which matrix effect is negligible. MEKC with LIF detection was also used as analytical platform to study amino acid uptake by yeast during beer fermentation [46]. After an extensive study of the most adequate conditions to perform the derivatization with Oregon green™ 488 carboxylic acid, succinimidyl ester (OG-SE, a fluorinated analogue of fluorescein) and the separation medium composition (see Table **2**), the MEKC-LIF method enabled to carry out the simultaneous monitoring, identification and quantification of amino acids. This methodology enabled to achieve quantification limits as low as 62.5 pM which correspond to below attomole level and provides a system of high relevance for the optimization of process parameters and cultivation performance.

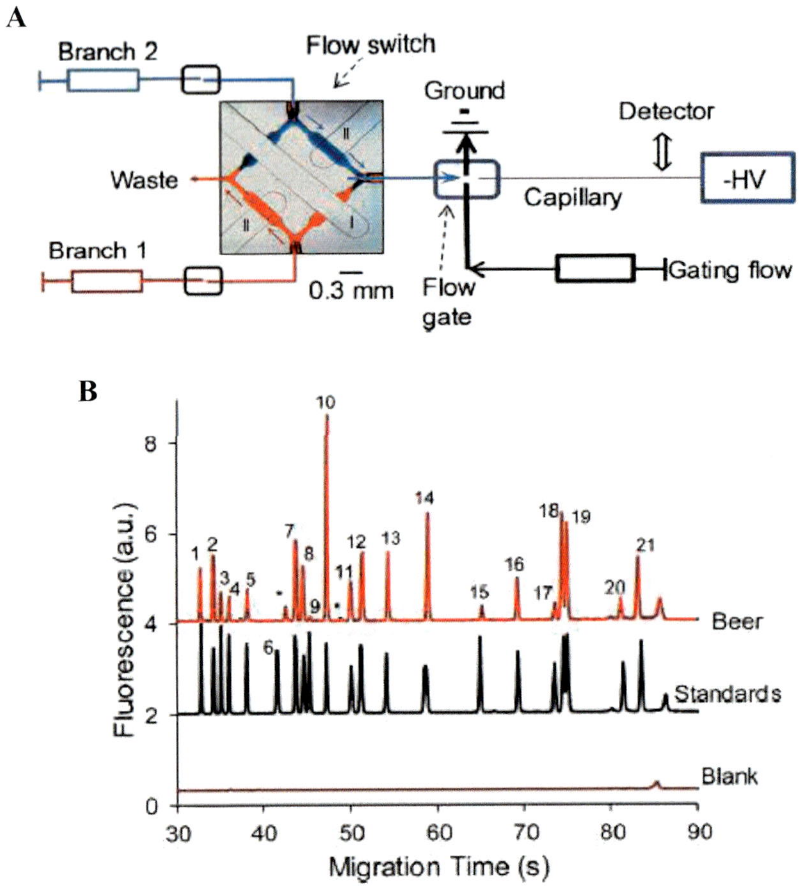

Fig. (2). **A)** Schematic diagram of the flow-gated CE system. The flow switch shows that Control I was depressed, and Control II was released. Solid arrows indicate flow directions. **B)** Electropherograms of standard amino acids (10 mM each) and the beer sample (diluted by 50-fold). Blank was obtained by mixing NDA and cyanide without AAs. "*" represents unknown peaks. Experimental conditions: Capillary, 10 μm x 10 cm (effective length); buffer, 60 mM SDS + 2.0 mM HP-β-CD in 40 mM tetraborate (pH 9.2); injection, -5.0 kV for 0.3 s; voltage, -25 kV, temperature, 25°C. Peak identification: 1 Ser, 2 Asn, 3 Thr, 4 Gln, 5 His, 6 phosphoryl ethanolamine, 7 Gly, 8 Glu, 9 Citrulline, 10 Ala, 11 Asp, 12 Tyr, 13 GABA, 14 Val, 15 Met, 16 Ile, 17 Trp, 18 Leu, 19 Phe, 20 ethanolamine and 21 Arg. Reprinted with permission from [45].

ITP

During the period of time covered in this chapter, ITP was also employed as CE mode to develop a methodology for the determination of three essential amino acids in food supplements [47]. The method, developed by Korarová *et al.*, employed for the first time the conversion of amino groups to diazo groups which are hydrolyzed to form hydroxy acids. Subsequently, the hydroxy acids were analyzed by ITP coupled to a conductivity detector. Applying the optimal experimental conditions (see Table **2**), the feasibility of this methodology was tested in the analysis of food supplements containing Val, Leu, and Ile. Good agreement between the measured content and the declared one was observed.

DETERMINATION OF NONPROTEIN AMINO ACIDS

The different CE methodologies applied from 2005 to present, to the determination of nonprotein amino acids in food matrices were mainly focused on the determination of amino acids levels which is the most relevant aspect in the case that these compounds have toxic effects. As it can be deduced from the information presented in Table **3**, during the last years, CZE has been the preferred CE mode to perform the analysis of nonprotein amino acids in a broad variety of matrices, although other modes such as MEKC, CEC and MCE have also demonstrated to be suitable analytical platforms.

Table 3. Characteristics of the CE approaches devoted to the determination of nonprotein amino acids in foods published in the last years (2015-June 2020).

Nonprotein Amino Acids	Sample	Mode/Detection	Experimental Conditions	LOD	Ref.
β-N-Oxalyl-L-α,β-diaminopropionic acid Homoarginine	Legumes (*Lathyrus* species)	CZE-UV (195 nm)	25 mM borate (pH 9.2); capillary, 75 μm x 50 cm; 21 kV, 22°C	-	**52**
Mycosporine-like amino acids	Macroalgae (*Palmaria palmata, Porphyra umbilicalis, Porphyra sp.*)	CZE-UV (320 nm)	30 mM tetraborate (pH 10.3), capillary, 75 μm x 80 cm; 25 kV, 25°C	4.3-4.8 μg/mL	**53**
β-N-Methylamino-L-alanine	Cycad, mussel and lobster samples	CZE-UV (192 nm) CZE(QqQ)MS/MS	250 mM phosphate (pH 3.0); capillary, 50 μm x 46 cm; 25 kV, 17°C 5 M formic acid + 10% acetonitrile (v/v) (pH 1.55); capillary, 50 μm x 100 cm; 20 kV, 17°C	0.25 μg/mL 0.8 ng/mL	**54**

(Table 3) cont.....

Nonprotein Amino Acids	Sample	Mode/Detection	Experimental Conditions	LOD	Ref.
Carnitine Acetyl-L-carnitine	Milk	CZE-indirect UV (200 nm)	3.0 mM melamine + 10% methanol (pH 2.1); capillary, 75 µm x 39 cm; 10 kV, 20°C	L-carnitine: 3.0 µM Acetyl-L-carnitine: 5.0 µM	**55**
Taurine *(Separated from Citrate and Carbonate)*	Energy drink	CZE-C^{4}D	Instrument with a coaxial flow-gating interface: 20 mM CHES + 10 mM NaOH (pH 9.5); capillary, 25 µm x 18 cm; 20 kV, 25°C	14.4 mg/L	**56**
Taurine *(Separated from choline)*	Energy drink and food supplements	CZE-C^{4}D	150 mM Tris/lactic acid (pH 8.96), capillary, 50 µm x 49 cm; -10 kV, temperature, not indicated	0.27 mg/L	**57**
Hydroxyproline Hydroxylysine *(Separated from 18 protein amino acids)*	Plastron-derived functional foods	MEKC-UV (214 nm)	0.1 M SDS + 6% methanol in 20 mM tetraborate and phosphate (pH 8.7); capillary, 50 µm x 50 cm; 25 kV, 25°C	Hydroxyproline: 1.57 x 10^{-4} mg/mL Hydroxylysine: 6.65 x 10^{-4} mg/mL	**58**
Domoic acid	Shellfish samples (oyster, *mytilus edulis* and *ruditapes*)	pCEC-LIF (λ_{ex} 473 nm; λ_{em} 530 nm)	5 mM phosphate (pH 2.5):acetonitrile (40:60 v/v) at 1.2 µL/min; packed capillary column, 100 µm x 55 cm (total length of which 20 cm was packed with ODS particles); 6 kV, 25°C; supplementary pressure 7.2 MPa	10 ng/mL	**59**

(Table 3) cont.....

Nonprotein Amino Acids	Sample	Mode/Detection	Experimental Conditions	LOD	Ref.
Domoic acid	Shellfish tissues	MCE-LIF (λ_{ex} 475 nm; λ_{em} 535 nm)	5 mM tetraborate (pH 9.2); glass microchip cross-channel design; 27 mm from injection to the detector; 400 V/cm; temperature, not indicated	0.28 nM	**60**
Taurine *(Separated from Lysine and Vitamin B_3)*	Functional drinks	MCE-LIF (λ_{ex} 635 nm; λ_{em} 495 nm)	100 mM borate (pH 9.9); glass microchip with a simple cross channel design; 45 mm from injection to the detector; applied potential and temperature, not indicated	0.50 nM	**61**

*Capillary dimensions expressed as internal diameter x effective length (cm to the detector), and LODs units expressed as in the original work. C^4D, capacitively coupled contactless conductivity; CHES, N-cyclohexy--2-aminoethanesulfonic acid; MCE, microchip capillary electrophoresis; ODS, octadecyl silica; pCEC, pressurized capillary electrochromatography; (QqQ)MS/MS, triple quadrupole tandem mass spectrometry; SDS, sodium dodecyl sulfate; Tris, tris(hydroxymethyl)aminomethane.

The dietary importance of legume seeds as source of proteins represents an alternative to consumption of animal food sources. In this context, *Lathyrus* species have been cultivated as protein-rich legumes for human food and as grain, forage and green manure crops. However, this species (mainly *L. sativus)* are related to lathyrism, a disease leading to paralysis in humans and animals [62]. The causative agent of this disease is the nonprotein amino acid β-N-Oxaly--L-α,β- diaminopropionic acid. On the other hand, consumption of *Lathyrus* species also produces a high amount of the nonprotein amino acid homoarginine which is considered as a positive factor since it can be converted into Lys. For all these reasons, the determination and quantification of both compounds in *Lathyrus* seeds are very important from a human and animal nutrition viewpoint. With this objective in mind, Sacristán *et al.,* developed a CZE-UV method, based on the use of 25 mM borate buffer (pH 9.2), allowing the simultaneous analysis of both nonprotein amino acids in two different *Lathyrus* species (*L. sativus* and *L. cicera*) [52]. To achieve the extraction of the target compounds from *Lathyrus* seeds, ethanol:water (60:40 v/v) was employed as extracting solvent and a Ultra-Turrax homogenizer was used. This method enabled the determination of homoarginine (from 8.08 mg/g to 12.44 mg/g) and β-N-Oxalyl-L-α,β-diaminopropionic acid (from 0.79 to 5.05 mg/g) contents in the samples analyzed. The levels of β-N- Oxalyl-L-α,β-diaminopropionic acid obtained were below the recommended ones in *L. cicera* samples but in the case of *L. sativus* samples the levels found exceeded those recommended.

CZE-UV was also employed by Harmann *et al.* to perform for the first time the analysis of mycosporine-like amino acids (which have photo-protective potential) in marine algae [53]. The baseline separation of five different mycosporine-like amino acids was achieved in 27 min using borate at pH 10.3 as separation medium. LODs obtained with the proposed methodology were from 4.3 to 4.8 μg/mL. Once fully validated, the method was applied to the quantitative determination of mycosporine-like amino acids in three different marine macroalgae (*Palmaria palmata, Porphyra umbilicalis*, and *Porphyra sp.*), as well as the lichen *Lichina pygmaea.* This CZE-UV methodology has raised much interest since it may contribute to enhancing the knowledge on this group of natural products which could have great potential in medicinal and cosmetic applications.

Another interesting research work was based on the use of CZE to carry out the determination of the neurotoxin β-N-methylamino-L-alanine which is related to many neurodegenerative pathologies such as Parkinson or Alzheimer, among others [64]. The widespread occurrence of this nonprotein amino acid in cyanobacteria and seafood has raised concern due to its negative effects on public health. The determination of β-N-methylamino-L-alanine is usually performed by LC-MS/MS although it does not provide the complete separation of isomeric and isobaric compounds. Matrix suppression and conjugated forms are also other limitations. For this reason, Kerrin *et al.* developed a CZE-MS/MS method as an alternative for the quantitative determination of the neurotoxin in cycad, mussel and lobster samples [54]. First, the authors optimized a CZE-UV methodology (see conditions in Table **3**) which, although enabled the separation of β--methylamino-L-alanine and its isomers, provided a high LOD. Afterwards, they developed a CZE-MS/MS approach to improve the sensitivity. The coupling CE-MS was made using a custom interface with a straight tube enclosing the CE capillary which eliminated plugging problems. 5 M formic acid containing 10% acetonitrile (v/v) at pH 1.55 was used as separation buffer and a preconcentration by field amplified stacking after injection was achieved by lowering sample ionic strength with a cation-exchange cleanup procedure. Under these conditions, a LOD of 0.8 ng/mL (16 ng/g dry mass) was attained which is comparable to those obtained by LC-MS/MS. In addition, CE-MS/MS enabled to achieve an excellent resolution of isomers and no interferences from matrix components were observed.

Indirect UV detection in combination with CZE also demonstrated to be a reliable tool to carry out the sensitive determination of L-carnitine and acetyl-L-carnitine [55]. These are nonprotein amino acids with multiple functions such as transporting long-fatty acids for oxidation or reducing the intracellular accumulation of toxic metabolites under ischemic conditions [64, 65]. Even though carnitine can be produced in humans, its levels are very low so that it may be taken from diet. This fact makes relevant its quantitative determination in food. Using an orthogonal experimental design, the adequate parameters to perform the quantitative analysis of both compounds by a CZE-indirect UV strategy were selected. Thus, using 3.0 mM melamine and 10% methanol (pH 2.1) as separation medium, the studied nonprotein amino acids could be detected in 6 min with LODs of 3.0 and 5.0 μM for carnitine and acetyl-L-carnitine, respectively. Subsequently, fourteen types of milks samples were analyzed by this developed method showing carnitine contents from 43.6 to 121.5 mM and acetyl-L-carnitine contents from 17.5 to 68.5 mM.

Another suitable strategy for the analysis of nonprotein amino acids in food samples is the use of CZE with C^4D. In fact, two different methodologies based on CZE- C^4D were described in recent years for the determination of taurine in energy drinks and food supplements [56, 57]. On the one hand, a coaxial flow-gating interface in which the separation capillary passes through the sampling capillary was used. This system enabled reliable and repeatable sample injections. A comparison between the results obtained using the developed interface and an Agilent commercial equipment demonstrated that a lower analysis time and similar LODs were achieved with the proposed interface. Both methods were able to determine lower percentages of taurine in energy drinks than the declared value [56]. On the other hand, a dual-channeled CZE-C^4D methodology was employed for the rapid concurrent determination of taurine [57]. A central composite design in the response surface methodology (RSM-CCD) was employed to optimize several parameters prior to experiment implementation. Under the best conditions, taurine was analyzed in the first CE channel using 150 mM Tris/lactic acid (pH 8.96) whereas choline was analyzed in the second one using a different separation medium. To verify the feasibility of this analytical system, energy drinks and food supplements were tested. Good agreement between the results obtained with the CE-C^4D proposed method and those corresponding to the reference method (LC-UV and LC MS/MS) was achieved, demonstrating the potential of this approach as a tool for the screening of compounds with different characteristics in a single run which could be relevant from a food control viewpoint.

Only one MEKC-UV methodology was described in the time covered in this chapter for the analysis of nonprotein amino acids in food matrices. It was proposed by Li *et al.,* to evaluate the authenticity of plastron-derived functional foods since the high price of these samples in the market gives rise to possible adulteration with some low-price collagen-containing materials [58]. To compare the amino acid profiles of plastron and its possible substitutes, Li *et al.* developed a methodology based on the use of 0.1 M SDS + 6% methanol in 20 mM tetraborate and phosphate (pH 8.7) to achieve the separation of Hyp and hydroxylysine (Hyl) along with 18 protein amino acids (all previously derivatized with DNS-Cl) and the by-products formed during derivatization. The absence of Hyp and Hyl in chicken, egg, fish, milk, pork, nail and hair samples made possible to use both compounds as markers to distinguish easily authentic plastron from adulterated samples. For other collagen-tissue samples (pigskin, fish skin, calf tendon, and chicken tendon), a principal component analysis (PCA) was conducted and plastron was successfully distinguished. The application of the proposed methodology to the authentication of turtle shell glue enabled to detect some fake samples.

Other CE modes such as CEC and MCE were also successfully employed for the determination of domoic acid or taurine in shellfish samples or functional drinks, respectively. On the one hand, Chen *et al.* developed and applied a pressurized CEC (pCEC) methodology with LIF detection for the quantification of domoic acid (the principal causative toxin of amnesic shellfish poisoning) in different shellfish samples [59]. A solid-liquid extraction followed by a clean-up procedure and a derivatization step with NBD-F was performed, and the CEC analysis was conducted using 5mM phosphate (pH 2.5):acetonitrile (40:60 v/v) as mobile phase. As Fig. (**3A**) shows, two driving forces including hydraulic pressure and electroosmotic flow (EOF) could be employed in pCEC. Then, electrophoretic migration and retention in a packed stationary phase could be employed for the separation. Sensitive analysis of domoic acid was achieved with LOD in shellfish tissue extracts as low as 10 ng/mL. Applied to shellfish samples (oyster, *mytilus edulis* and *ruditapes*), the recoveries were satisfactory and the quantification was acceptable when compared with a LC-MS/MS method. Fig. (**3B**) displays pCEC analysis of domoic acid in shellfish tissue extracts. This nonprotein amino acid was also determined in shellfish tissues using MCE with LIF detection [60]. Using FITC as derivatizing reagent and 5 mM tetraborate (pH 9.2), Chen *et al.* achieved the determination of domoic acid within 60 s with a LOD low enough (0.28 nM) to detect the nonprotein amino acid in the naturally-contaminated samples for regulatory purposes. This methodology can be considered a powerful alternative approach for toxin detection. On the other hand, the potential of MCE-LIF was also demonstrated by successful application to determine taurine, lysine and vitamin B_3 in functional drinks [61]. The utilization of an on-line preconcentration strategy based on the use of field-amplified sample stacking combined with reverse-field stacking enabled improvement in the sensitivity and the separation efficiency of a conventional MCE-LIF method.

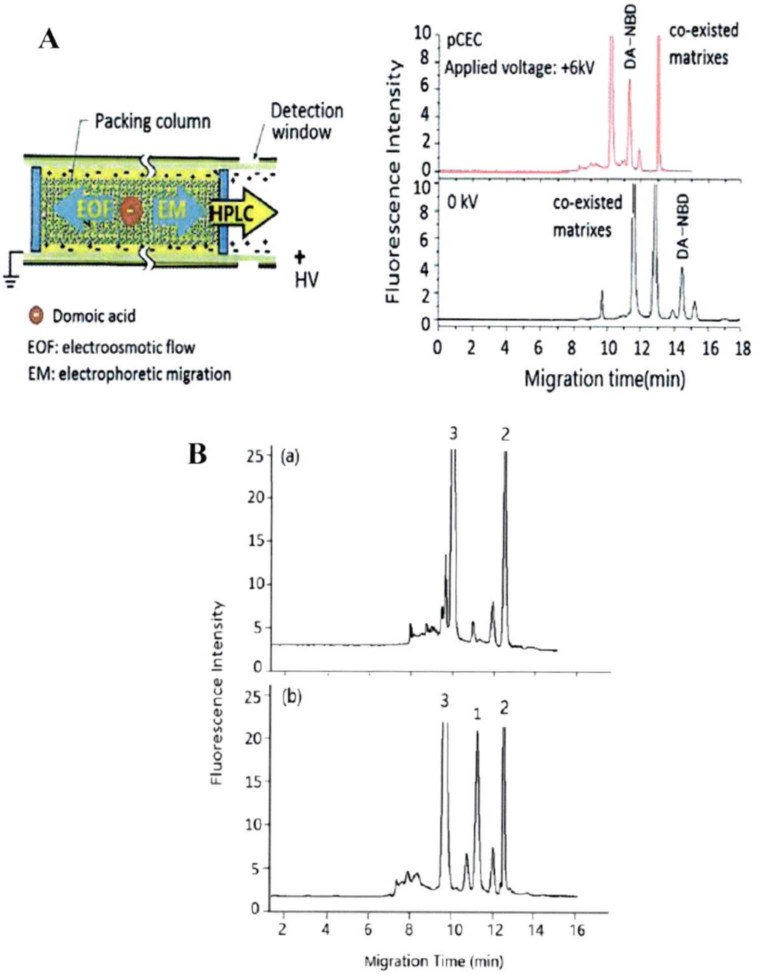

Fig. (3). **A)** Scheme of multiple separation interactions in pCEC for domoic acid detection. **B)** pCEC analysis of domoic acid in shellfish tissue extracts. Experimental conditions: Mobile phase, 5 mM phosphate (pH 2.5): acetonitrile (40:60 v/v) at 1.2 μL/min; voltage: 6 kV, supplementary pressure: 1000 psi. (a) Blank sample, (b) spiked with 1 μg/mL domoic acid standard solution. Peaks: 1, NBD-domoic acid; 2, NBD-F hydrolysis products; 3, co-eluted compounds. Reprinted from [59].

Under optimal conditions, *i.e.* the use of 100 mM borate (pH 9.88) as running buffer and sulfoindocyanine succinimidyl ester (Cy5) as derivatizing reagent, the analytes were simultaneous and completely separated within 4 min. Then, the approach was satisfactorily used to analyze the three functional ingredients in eight functional drinks. The concentrations of taurine, lysine and vitamin B_3detected met the standard of the values claimed of all drink samples analyzed except in one of them in which the content found for taurine was about half of the labeled value.

CONCLUDING REMARKS

CE has shown to be a powerful tool for the analysis of amino acids in food. CZE, MEKC and ITP were the CE separation modes employed in the last five years to achieve the determination of protein amino acids in food matrices, while CZE, MEKC, CEC and MCE were applied for nonprotein amino acids analysis in food. Although UV has been the most frequently used detection system, indirect UV detection, LIF, conductivity detection, C^4D, amperometric detection and MS were also employed to a lesser extent in the period of time covered in this chapter. Precapillary and in-capillary derivatization strategies using different derivatizing reagents (NBD-Cl, NBD-F, OPA, FTIC, FMOC, NDA, DNS-Cl, OG-SE, Cy5) were applied to increase detection sensitivity although complexation of amino acids with metal ions such as Cu(II) enabled the determination of underivatized amino acids. Off-line (solid-liquid extraction, liquid-liquid extraction, extraction by electroporation) and in-capillary preconcentration strategies (field amplified stacking, reverse-field stacking, sweeping) were also applied in some cases to decrease limits of detection. The developed methodologies were applied to the analysis of a wide variety of food samples and drinks (including salted herring meat, sauces, beers, vinegar, wines, dietary supplements, fruit juices, microalgae and macroalgae, functional drinks, shellfish samples, legumes, milk, energy drinks, among others). The developed methods by CZE-LIF made possible the quantitation of protein amino acids at the nM level (using extraction by electroporation and precapillary derivatization with FTIC) while MEKC-LIF enabled the detection of amino acids at nM levels and their quantitation at around 60 pM levels. Most sensitive strategies for nonprotein amino acids detection included the only work reporting MS detection (LOD 0.8 ng/mL) and MCE-LIF enabling LODs ranging from 0.28 to 0.50 nM. The implementation of MCE-LIF together with miniaturized CE systems with amperometric detection for amino acids analysis in food samples, within the current tendency to miniaturize analytical processes, gave rise to short run times and powerful strategies for rapid food quality control. Undoubtedly, the analysis of amino acids by CE will continue being used as a powerful tool for food quality and safety assessment and more developments in CE systems miniaturization and in the use of MS detection can be expected.

CONSENT FOR PUBLICATION

Not applicable.

CONFLICT OF INTEREST

The author declares no conflict of interest, financial or otherwise.

ACKNOWLEDGEMENTS

Authors thank the Comunidad of Madrid (Spain) and European funding from FSE and FEDER Programs (project S2018/BAA-4393, AVANSECAL-II-CM). M.C.P. thanks the Spanish Ministry of Economy and Competitiveness for her "Ramón y Cajal" research contract (RYC-2013-12688).

REFERENCES

[1] Wu, G. Functional amino acids in growth, reproduction, and health. *Adv. Nutr.*, **2010**, *1*(1), 31-37. [http://dx.doi.org/10.3945/an.110.1008] [PMID: 22043449]

[2] Wu, G. Functional amino acids in nutrition and health. *Amino Acids*, **2013**, *45*(3), 407-411. [http://dx.doi.org/10.1007/s00726-013-1500-6] [PMID: 23595206]

[3] Wu, G. Amino acids: metabolism, functions, and nutrition. *Amino Acids*, **2009**, *37*(1), 1-17. [http://dx.doi.org/10.1007/s00726-009-0269-0] [PMID: 19301095]

[4] Sarkadi, L.S. Amino acids and biogenic amines as food quality factors. *Pure Appl. Chem.*, **2019**, *91*, 289-300. [http://dx.doi.org/10.1515/pac-2018-0709]

[5] Hunt, S. The Non-Protein amino acids.*Chemistry and Biotechnology of amino acids*; Barret, G. C., Ed.; Chaptman and Hall: USA, **1985**, p. 55. [http://dx.doi.org/10.1007/978-94-009-4832-7_4]

[6] Pérez-Míguez, R.; Marina, M.L.; Castro-Puyana, M. Capillary electrophoresis determination of non-protein amino acids as quality markers in foods. *J. Chromatogr. A*, **2016**, *1428*, 97-114. [http://dx.doi.org/10.1016/j.chroma.2015.07.078] [PMID: 26233255]

[7] Papetti, A.; Colombo, R. High-performance capillary electrophoresis for food quality evaluation.*Evaluation Technologies for Food Quality*; Zhong, J.; Wang, X., Eds.; Woodhead Publishing, **2019**, pp. 301-377. [http://dx.doi.org/10.1016/B978-0-12-814217-2.00014-7]

[8] Viglio, S.; Fumagalli, M.; Ferrari, F.; Iadarola, P. MEKC: a powerful tool for the determination of amino acids in a variety of biomatrices. *Electrophoresis*, **2010**, *31*(1), 93-104. [http://dx.doi.org/10.1002/elps.200900366] [PMID: 20014052]

[9] Viglio, S.; Fumagalli, M.; Ferrari, F.; Bardoni, A.; Salvini, R.; Giuliano, S.; Iadarola, P. Recent novel MEKC applications to analyze free amino acids in different biomatrices: 2009-2010. *Electrophoresis*, **2012**, *33*(1), 36-47. [http://dx.doi.org/10.1002/elps.201100336] [PMID: 22127995]

[10] Poinsot, V.; Carpéné, M-A.; Bouajila, J.; Gavard, P.; Feurer, B.; Couderc, F. Recent advances in amino acid analysis by capillary electrophoresis. *Electrophoresis*, **2012**, *33*(1), 14-35. [http://dx.doi.org/10.1002/elps.201100360] [PMID: 22213525]

[11] Mandrioli, R.; Mercolini, L.; Raggi, M.A. Recent trends in the analysis of amino acids in fruits and derived foodstuffs. *Anal. Bioanal. Chem.*, **2013**, *405*(25), 7941-7956. [http://dx.doi.org/10.1007/s00216-013-7025-8] [PMID: 23686004]

[12] Poinsot, V.; Ong-Meang, V.; Gavard, P.; Couderc, F.; Couderc, F. Recent advances in amino acid analysis by capillary electromigration methods, 2011-2013. *Electrophoresis*, **2014**, *35*(1), 50-68. [http://dx.doi.org/10.1002/elps.201300306] [PMID: 24174162]

[13] Poinsot, V.; Ong-Meang, V.; Gavard, P.; Couderc, F.; Couderc, F. Recent advances in amino acid analysis by capillary electromigration methods, 2013-2015. *Electrophoresis*, **2016**, *37*(1), 142-161. [http://dx.doi.org/10.1002/elps.201500302] [PMID: 26434399]

[14] De Oliveira, M.A.L.; Porto, B.L.S.; Bastos, C.A.; Sabarense, C.M.; Vaz, F.A.S.; Neves, L.N.O.; Duarte, L.M.; Campos, N.S.; Chellini, P.R.; Da Silva, P.H.F.; De Sousa, R.A.; Marques, R.; Sato, R.T.; Grazul, R.M.; Lisboa, T.P.; Mendes, T.O.; Rios, V.C. Analysis of amino acids, proteins, carbohydrates and lipids in food by capillary electromigration methods: a review. *Anal. Methods,* **2016**, *8*, 3649-3680.
[http://dx.doi.org/10.1039/C5AY02736E]

[15] Poinsot, V.; Ong-Meang, V.; Ric, A.; Gavard, P.; Perquis, L.; Couderc, F. Recent advances in amino acid analysis by capillary electromigration methods: June 2015-May 2017. *Electrophoresis,* **2018**, *39*(1), 190-208.
[http://dx.doi.org/10.1002/elps.201700270] [PMID: 28805963]

[16] Pérez-Míguez, R.; Salido-Fortuna, S.; Castro-Puyana, M.; Marina, M.L. Advances in the determination of nonprotein amino acids in foods and biological samples by capillary electrophoresis. *Crit. Rev. Anal. Chem.,* **2019**, *49*(5), 459-475.
[http://dx.doi.org/10.1080/10408347.2018.1546113] [PMID: 30963784]

[17] Herrero, M.; García-Cañas, V.; Simó, C.; Cifuentes, A. Recent advances in the application of capillary electromigration methods for food analysis and Foodomics. *Electrophoresis,* **2010**, *31*(1), 205-228.
[http://dx.doi.org/10.1002/elps.200900365] [PMID: 19967713]

[18] Castro-Puyana, M.; García-Cañas, V.; Simó, C.; Cifuentes, A. Recent advances in the application of capillary electromigration methods for food analysis and Foodomics. *Electrophoresis,* **2012**, *33*(1), 147-167.
[http://dx.doi.org/10.1002/elps.201100385] [PMID: 22147337]

[19] García-Cañas, V.; Simó, C.; Castro-Puyana, M.; Cifuentes, A. Recent advances in the application of capillary electromigration methods for food analysis and Foodomics. *Electrophoresis,* **2014**, *35*(1), 147-169.
[http://dx.doi.org/10.1002/elps.201300315] [PMID: 24114758]

[20] Acunha, T.; Ibáñez, C.; García-Cañas, V.; Simó, C.; Cifuentes, A. Recent advances in the application of capillary electromigration methods for food analysis and Foodomics. *Electrophoresis,* **2016**, *37*(1), 111-141.
[http://dx.doi.org/10.1002/elps.201500291] [PMID: 26256797]

[21] Álvarez, G.; Montero, L.; Llorens, L.; Castro-Puyana, M.; Cifuentes, A. Recent advances in the application of capillary electromigration methods for food analysis and Foodomics. *Electrophoresis,* **2018**, *39*(1), 136-159.
[http://dx.doi.org/10.1002/elps.201700321] [PMID: 28975648]

[22] Rocco, A.; Donati, E.; Touloupakis, E.; Aturki, Z. Miniaturized separation techniques as analytical methods to ensure quality and safety of dietary supplements. *TrAC. Trends Analyt. Chem.,* **2018**, *103*, 156-183.
[http://dx.doi.org/10.1016/j.trac.2018.04.004]

[23] Wuethrich, A.; Quirino, J.P. Derivatisation for separation and detection in capillary electrophoresis (2015-2017). *Electrophoresis,* **2018**, *39*(1), 82-96.
[http://dx.doi.org/10.1002/elps.201700252] [PMID: 28758685]

[24] Wan, H.; Blomberg, L.G. Chiral separation of amino acids and peptides by capillary electrophoresis. *J. Chromatogr. A,* **2000**, *875*(1-2), 43-88.
[http://dx.doi.org/10.1016/S0021-9673(99)01209-1] [PMID: 10839139]

[25] Marina, M.L.; Castro-Puyana, M. Derivatization in capillary electrophoresis.*Capillary Electrophoresis: Methods and Protocols,* 2nd ed; Schmitt-Kopplin, P., Ed.; Humana Press, **2016**, pp. 37-52.
[http://dx.doi.org/10.1007/978-1-4939-6403-1_3]

[26] Bardelmeijer, H.A.; Lingeman, H.; de Ruiter, C.; Underberg, W.J. Derivatization in capillary electrophoresis. *J. Chromatogr. A,* **1998**, *807*(1), 3-26.

[http://dx.doi.org/10.1016/S0021-9673(98)00230-1] [PMID: 9646486]

[27] Underberg, W.J.M.; Waterval, J.C.M. Derivatization trends in capillary electrophoresis: an update. *Electrophoresis,* **2002**, *23*(22-23), 3922-3933.
[http://dx.doi.org/10.1002/elps.200290010] [PMID: 12481285]

[28] Fukushima, T.; Usui, N.; Santa, T.; Imai, K. Recent progress in derivatization methods for LC and CE analysis. *J. Pharm. Biomed. Anal.,* **2003**, *30*(6), 1655-1687.
[http://dx.doi.org/10.1016/S0731-7085(02)00511-3] [PMID: 12485710]

[29] Armstrong, D.W.; Chang, C-D.; Li, W.Y. Relevance of enantiomeric separations in food and beverage analyses. *J. Agric. Food Chem.,* **1990**, *38*, 1674-1677.
[http://dx.doi.org/10.1021/jf00098a010]

[30] Omar, M.M.A.; Elbashir, A.A.; Schmitz, O.J. Capillary electrophoresis method with UV-detection for analysis of free amino acids concentrations in food. *Food Chem.,* **2017**, *214*, 300-307.
[http://dx.doi.org/10.1016/j.foodchem.2016.07.060] [PMID: 27507479]

[31] Gao, Y.S.; Hsieh, B.C.; Cheng, T.J.; Chen, R.L. Judgment of pure fermented soy sauce by fluorescence resonance energy transfer of OPA-tryptophan adduct. *Food Chem.,* **2015**, *178*, 122-127.
[http://dx.doi.org/10.1016/j.foodchem.2015.01.028] [PMID: 25704692]

[32] Felisiak, K.; Szymczak, M.; Kołakowski, E. Identification of non-protein nitrogen compounds separated by CZE without derivatization from TCA extract from salted herring meat. *J. Food Compos. Anal.,* **2019**, *77*, 108-114.
[http://dx.doi.org/10.1016/j.jfca.2019.01.010]

[33] Aung, H-P.; Pyell, U. In-capillary derivatization with o-phthalaldehyde in the presence of 3-mercaptopropionic acid for the simultaneous determination of monosodium glutamate, benzoic acid, and sorbic acid in food samples *via* capillary electrophoresis with ultraviolet detection. *J. Chromatogr. A,* **2016**, *1449*, 156-165.
[http://dx.doi.org/10.1016/j.chroma.2016.04.033] [PMID: 27156753]

[34] Luo, T.; Ke, J.; Xie, Y.; Dong, Y. Determination of underivatized amino acids to evaluate quality of beer by capillary electrophoresis with online sweeping technique. *J. Food Drug Anal.,* **2017**, *25*(4), 789-797.
[http://dx.doi.org/10.1016/j.jfda.2017.03.003] [PMID: 28987354]

[35] Luo, T.; Dong, Y.; He, S.; Yang, J.; Dong, Y. Establishing a sensitive capillary electrophoresis-UV method for direct determination of amino acids to evaluate vinegar quality. *Electrophoresis,* **2018**, *39*(11), 1410-1416.
[http://dx.doi.org/10.1002/elps.201800001] [PMID: 29464729]

[36] Campos, C.D.M.; de Campos Braga, P.A.; Reyes, F.G.R.; da Silva, J.A.F. Elimination of the artefact peaks in capillary electrophoresis determination of glutamate by using organic solvents in sample preparation. *J. Sep. Sci.,* **2015**, *38*(21), 3781-3787.
[http://dx.doi.org/10.1002/jssc.201500601] [PMID: 26332708]

[37] Campos, C.D.M.; Reyes, F.G.R.; Manz, A.; da Silva, J.A.F. On-line electroextraction in capillary electrophoresis: Application on the determination of glutamic acid in soy sauces. *Electrophoresis,* **2019**, *40*(2), 322-329.
[http://dx.doi.org/10.1002/elps.201800203] [PMID: 30246879]

[38] Duong, H.A.; Vu, M.T.; Nguyen, T.D., Ngyyen, M.H. Determination of 10-hydroxy-2-decenoic acid and free amino acids in royal jelly supplements with purpose-made capillary electrophoresis coupled with contactless conductivity detection. *J. Food Compos. Anal.,* **2020**, *87*, 103422.
[http://dx.doi.org/10.1016/j.jfca.2020.103422]

[39] Nehmé, R.; Atieh, C.; Fayad, S.; Claude, B.; Chartier, A.; Tannoury, M.; Elleuch, F.; Abdelkafi, S.; Pichon, C.; Morin, P. Microalgae amino acid extraction and analysis at nanomolar level using electroporation and capillary electrophoresis with laser-induced fluorescence detection. *J. Sep. Sci.,* **2017**, *40*(2), 558-566.

[http://dx.doi.org/10.1002/jssc.201601005] [PMID: 27813309]

[40] Luo, D.; Zhao, X.; Zheng, Y.; Ma, Y.; Cun, F.; Ye, J.; Chu, Q. Fast determination of main bioamines and precursor amino acids in beer by miniaturized electrophoresis using gold nanoparticle composite electrode. *Food Anal. Methods,* **2019**, *12*, 644-651.
[http://dx.doi.org/10.1007/s12161-018-1395-7]

[41] Valese, A.C.; Vitali, L.; Fett, R.; Micke, G.A.; de Oliveira, M.A.; Oliveira Costa, A.C. A rapid method for analysis of phenylalanine in cereal products by MEKC-UV using LC/MS/MS as a comparative method. *J. AOAC Int.,* **2015**, *98*(6), 1632-1639.
[http://dx.doi.org/10.5740/jaoacint.15-052] [PMID: 26651576]

[42] Passos, H.M.; Cieslarova, Z.; Simionato, A.V. CE-UV for the characterization of passion fruit juices provenance by amino acids profile with the aid of chemometric tools. *Electrophoresis,* **2016**, *37*(13), 1923-1929.
[http://dx.doi.org/10.1002/elps.201500483] [PMID: 26800985]

[43] Laube, H.; Boden, J.; Schneider, R. Capillary electrophoresis method for the analysis of organic acids and amino acids in the presence of strongly alternating concentrations of aqueous lactic *acid. Bioprocess Biosyst. Eng.,* **2017**, *40*(7), 981-988.
[http://dx.doi.org/10.1007/s00449-017-1761-7] [PMID: 28382460]

[44] Huang, C-Y.; Wang, Y-X.; Wang, X-Z.; Hu, C-C.; Chiu, T-C. Analysis of seven biogenic amines and two amino acids in wines using micellar electrokinetic chromatography. *Appl. Sci. (Basel),* **2019**, *9*, 1193.
[http://dx.doi.org/10.3390/app9061193]

[45] Zhu, Q.; Zhang, N.; Gong, M. Rapid amino acid analysis of beers using flow-gated capillary electrophoresis coupled with side-by-side calibration. *Anal. Methods,* **2017**, *9*, 4520-4256.
[http://dx.doi.org/10.1039/C7AY01267E]

[46] Turkia, H.; Sirén, H.; Penttilä, M.; Pitkänen, J-P. Capillary electrophoresis with laser-induced fluorescence detection for studying amino acid uptake by yeast during beer fermentation. *Talanta,* **2015**, *131*, 366-371.
[http://dx.doi.org/10.1016/j.talanta.2014.07.101] [PMID: 25281116]

[47] Kovarová, M.; Bartos, M.; Mikysek, T.; Hlouskova, D.; Hyhlikova, B.; Vladykova, J. Isotachophoretic determination of amino acids after their conversion to hydroxy acids. *Monatshefte für Chemie - Chemical Monthly,* **2018**, *149*, 1709-1718.

[48] Omar, M.M.A.; Elbashir, A.A.; Schmitz, O.J. Determination of acrylamide in Sudanese food by high performance liquid chromatography coupled with LTQ Orbitrap mass spectrometry. *Food Chem.,* **2015**, *176*, 342-349.
[http://dx.doi.org/10.1016/j.foodchem.2014.12.091] [PMID: 25624242]

[49] Szymczak, M.; Kołakowski, E.; Felisiak, K. Influence of salt concentration on properties of marinated meat from fresh and frozen herring (Clupea harengus L.). *Int. J. Food Sci. Technol.,* **2012**, *47*, 282-289.
[http://dx.doi.org/10.1111/j.1365-2621.2011.02837.x]

[50] Szymczak, M.; Kołakowski, E.; Felisiak, K. Effect of addition of different acetic acid concentrations on the quality of marinated herring. *J. Aquat. Food Prod. Technol.,* **2015**, *24*, 566-581.
[http://dx.doi.org/10.1080/10498850.2013.796504]

[51] Mitchell, J.J.; Trakadis, Y.J.; Scriver, C.R. Phenylalanine hydroxylase deficiency. *Genet. Med.,* **2011**, *13*(8), 697-707.
[http://dx.doi.org/10.1097/GIM.0b013e3182141b48] [PMID: 21555948]

[52] Sacristán, M.; Varela, A.; Pedrosa, M.M.; Burbano, C.; Cuadrado, C.; Legaz, M.E.; Muzquiz, M. Determination of β-N-oxalyl-L-α,β-diaminopropionic acid and homoarginine in *Lathyrus sativus* and *Lathyrus cicera* by capillary zone electrophoresis. *J. Sci. Food Agric.,* **2015**, *95*(7), 1414-1420.
[http://dx.doi.org/10.1002/jsfa.6792] [PMID: 24954190]

[53] Hartmann, A.; Murauer, A.; Ganzera, M. Quantitative analysis of mycosporine-like amino acids in marine algae by capillary electrophoresis with diode-array detection. *J. Pharm. Biomed. Anal.*, **2017**, *138*, 153-157.
[http://dx.doi.org/10.1016/j.jpba.2017.01.053] [PMID: 28213175]

[54] Kerrin, E.S.; White, R.L.; Quilliam, M.A. Quantitative determination of the neurotoxin β--methylamino-L-alanine (BMAA) by capillary electrophoresis-tandem mass spectrometry. *Anal. Bioanal. Chem.*, **2017**, *409*(6), 1481-1491.
[http://dx.doi.org/10.1007/s00216-016-0091-y] [PMID: 27909777]

[55] Kong, Y.; Yang, G.; Chen, S.; Hou, Z.; Du, X.; Li, H.; Kong, L. rapid and sensitive determination of L-carnitine and acetyl-L-carnitine in liquid milk samples with capillary zone electrophoresis using indirect uv detection. *Food Anal. Methods*, **2018**, *11*, 170-177.
[http://dx.doi.org/10.1007/s12161-017-0987-y]

[56] Opekar, F.; Tůma, P. Coaxial flow-gating interface for capillary electrophoresis. *J. Sep. Sci.*, **2017**, *40*(15), 3138-3143.
[http://dx.doi.org/10.1002/jssc.201700412] [PMID: 28581040]

[57] Nguyen, T.D.; Nguyen, M.H.; Vu, M.T.; Duong, H.A.; Pham, H.V.; Mai, T.D. Dual-channeled capillary electrophoresis coupled with contactless conductivity detection for rapid determination of choline and taurine in energy drinks and dietary supplements. *Talanta*, **2019**, *193*, 168-175.
[http://dx.doi.org/10.1016/j.talanta.2018.10.002] [PMID: 30368287]

[58] Li, L.Q.; Baibado, J.T.; Shen, Q.; Cheung, H.Y. Determination of the authenticity of plastron-derived functional foods based on amino acid profiles analysed by MEKC. *J. Chromatogr. B Analyt. Technol. Biomed. Life Sci.*, **2017**, *1070*, 23-30.
[http://dx.doi.org/10.1016/j.jchromb.2017.10.036] [PMID: 29107730]

[59] Chen, Q.; Deng, L.; Chi, J.; Liu, M.; Lin, X.; Xie, Z. Sensitive Profiling of Trace Neurotoxin Domoic Acid by Pressurized Capillary Electrochromatography with Laser-Induced Fluorescence Detection. *RSC Advances*, **2017**, *7*, 53778-53784.
[http://dx.doi.org/10.1039/C7RA10662A]

[60] Cheng, Y.; Guo, C.; Zhao, B.; Yang, L. Fast analysis of domoic acid using microchip electrophoresis with laser-induced fluorescence detection. *J. Sep. Sci.*, **2017**, *40*(7), 1583-1588.
[http://dx.doi.org/10.1002/jssc.201600982] [PMID: 28160435]

[61] Wu, M.; Gao, F.; Zhang, Y.; Wang, Q.; Li, H. Sensitive analysis of amino acids and vitamin B3 in functional drinks *via* field-amplified stacking with reversed-field stacking in microchip electrophoresis. *Talanta*, **2015**, *131*, 624-631.
[http://dx.doi.org/10.1016/j.talanta.2014.08.051] [PMID: 25281150]

[62] Hugon, J.; Ludolph, A.C.; Spencer, P.S. N-Oxalylamino-L-alanine.*Experimental and Clinical Neurotoxicology*; Spencer, P.S.; Schaumburg, H., Eds.; Oxford University Press: New York, **2000**, pp. 925-938.

[63] Zhou, X.; Escala, W.; Papapetropoulos, S.; Zhai, R.G. *β*-N-methylamino-L-alanine induces neurological deficits and shortened life span in Drosophila. *Toxins (Basel)*, **2010**, *2*(11), 2663-2679.
[http://dx.doi.org/10.3390/toxins2112663] [PMID: 22069570]

[64] Longo, N.; Frigeni, M.; Pasquali, M. Carnitine transport and fatty acid oxidation. *Biochim. Biophys. Acta*, **2016**, *1863*(10), 2422-2435.
[http://dx.doi.org/10.1016/j.bbamcr.2016.01.023] [PMID: 26828774]

[65] Kelly, J.G.; Hunt, S.; Doyle, G.D.; Laher, M.S.; Carmody, M.; Marzo, A.; Arrigoni Martelli, E. Pharmacokinetics of oral acetyl-L-carnitine in renal impairment. *Eur. J. Clin. Pharmacol.*, **1990**, *38*(3), 309-312.
[http://dx.doi.org/10.1007/BF00315038] [PMID: 2340852]

CHAPTER 6

Vitamins Analysis by Capillary Electrophoresis

Xuan Liu[1], Jinhui Li[1], Fei Zhao[1], Zhuoting Liu[1], Ann Van Schepdael[2,*] and Xu Wang[1,*]

[1] *State Key Laboratory of Food Nutrition and Safety, College of Food Science and Engineering, Tianjin University of Science and Technology, Tianjin, 300457, People Republic of China*

[2] *Department of Pharmaceutical and Pharmacological Sciences, Pharmaceutical Analysis, KU Leuven - University of Leuven, Leuven, Belgium*

Abstract: Vitamins are a series of trace compounds obtained from food that play important roles in human health. Vitamin analysis is essential for nutritional assessment and food production. A comprehensive overview of capillary electrophoresis for vitamins analysis is given. This chapter includes papers published since 1996 and can be seen as a guidance note of vitamins analysis using capillary electrophoresis. The analyses are discussed for water-soluble vitamins and fat-soluble vitamins according to different separation modes. In addition, various sample pretreatment methods avoiding matrix interferences with the analysis of vitamins are also described. Articles pertaining to different vitamins from a variety of food and beverages, dietary supplements, and pharmaceutical samples are included. This chapter highlights the unique performance of capillary electrophoresis for the qualitative analysis of vitamins in food.

Keywords: Capillary electrophoresis, Vitamins, Food.

INTRODUCTION

As one of the seven major nutrients, vitamins are important for human nutrition. They can be subdivided into water-soluble and fat-soluble vitamins [1, 2]. Vitamins cannot be produced within the body with the exception of vitamin D. They can be obtained either from food products or *via* pharmaceutical formulations. The dietary sources of these vitamins are represented in Table **1** [3, 4].

* **Corresponding authors Xu Wang and Ann Van Schepdael**: State Key Laboratory of Food Nutrition and Safety, Tianjin University of Science and Technology, Tianjin 300457, P.R. China; and Department of Pharmaceutical and Pharmacological Sciences, Pharmaceutical Analysis, KU Leuven-University of Leuven, Leuven, Belgium; E-mail: xuwang@tust.edu.cn, ann.vanschepdael@kuleuven.be

María Castro-Puyana, Miguel Herrero and María Luisa Marina (Eds.)
Capillary Electrophoresis in Food Analysis

Table 1. Food sources of vitamins.

Nutrient	Food Sources
Thiamin (Vitamin B_1)	Nuts and seeds, legumes, wholegrain/enriched cereals and breads, pork
Riboflavin (Vitamin B_2)	Organic meats, milk, bread products, fortified cereals
Niacin (Vitamin B_3)	Meat, fish, poultry, enriched and whole-grain breads and bread products, fortified cereals
Pantothenic Acid (Vitamin B_5)	Chicken, beef, potatoes, oats, cereals, tomato products, liver, kidney, yeast, egg yolk, broccoli, whole grains
Biotin (Vitamin B_7)	Liver and smaller amounts in fruits and meats
Folic Acid (Vitamin B_9)	Enriched cereal grains, dark leafy vegetables, liver, orange juice, seeds, legumes, leafy green vegetables
Cobalamin (Vitamin B_{12})	Milk, milk products, eggs, seafood, fish, poultry, meat. It is not present in plant foods.
Ascorbic Acid (Vitamin C)	Found in vegetables and fruits, especially citrus fruits, tomatoes, tomato juice, potatoes, brussels sprouts, cauliflower, broccoli, strawberries, cabbage, spinach
Vitamin A	Liver, dairy products, fish, darkly colored fruits and leafy vegetables
Vitamin D	Fatty fish, fortified milk products, fortified cereals
Vitamin E	Vegetable oils, unprocessed cereal grains, nuts, fruits, vegetables, meats
Vitamin K	Green vegetables, brussels sprouts, cabbage, plant oils, margarine

Some vitamins can also be used as biomarkers for certain diseases [5]. Therefore, is necessary to have fast methods for vitamins analysis in different sample matrices.

High Performance Liquid Chromatography (HPLC) is one of the most common techniques for the analysis of vitamins [6 - 8]. However, it usually requires a longer analysis time and larger sample volume with much more solvent consumption. As an alternative tool to HPLC, Capillary Electrophoresis (CE) offers unique advantages due to its fast speed, less sample consumption and low reagent cost [9].

The separation of CE in different modes is based on the differential migration in an electric field [10]. Capillary Zone Electrophoresis (CZE) is based on differences in the charge-to-mass ratio enabling to analyze charged analytes [11]. Water-soluble vitamins can be easily charged and can be separated using CZE [12]. Micellar Electrokinetic Chromatography (MEKC) is another mode of CE whereby the separation of neutral compounds was achieved [13]. MEKC was mostly used for the simultaneous analysis of water-soluble vitamins [14, 15]. In addition, other separation modes were also used to analyze vitamins in foods, such as Microchip Electrophoresis (MCE), Nonaqueous Capillary Electrophoresis (NACE),

Microemulsion Electrokinetic Chromatography (MEEKC) and Capillary Electrochromatography (CEC).

Currently, a variety of detection techniques have been used for CE in different food analyses, including UV, laser-induced fluorescence (LIF), chemiluminescence (CL), electrochemical (EC) and MS detection [16]. Most CE-UV methods are mainly used for high concentration analyte analysis or simultaneous determination of several vitamins [17]. More sensitive detectors and/or on-line preconcentration techniques are developed to improve CE sensitivity [18]. LIF could provide 1000 times more sensitivity than the UV detector [19]. However, it is only suitable for analytes containing a fluorophore. EC detection coupled with CE is attractive due to its good selectivity towards electroactive analytes. MS detectors have a higher sensitivity but the instrumentation required is expensive [20]. A stable CE current and high ionization efficiency were achieved by the developed interfaces [21].

This chapter provides an overview of CE method developments in vitamins analysis from 1996 to 2019. The different types of samples together with separation mode, buffer, detection type and obtained sensitivity are listed in **Tables 2-6**.

Table 2. Summary of vitamin B analysis by CE.

Analyte	Sample	Separation Mode	Detection	Buffer	LOD	Ref
Analyte	**Sample**	**Separation Mode**	**Detection**	**Buffer**	**LOD**	**Ref**
Thiamine(B_1)			UV	65 mM boric acid,	2×10^{-4} mM,	
TH, THP,	Tablets	CZE	200 nm	8 mM sodium tetraborate,	1×10^{-4} mM,	[38]
THPP				pH 8.24	1×10^{-4} mM	

(Table 2) cont.....

Analyte	Sample	Separation Mode	Detection	Buffer	LOD	Ref
Thiamine(B_1)	Pork & beef	MEKC	UV	0.05 M SDS,	NA	[39]
			254 nm	0.1 M sodium tetraborate,		
				0.05 M sodium phosphate and 10% isopropanol		
				pH 7.6		
Thiamine(B_1)	Food supplement	NACE	MS	Acetonitrile buffer containing 2 M HAc and 4 mM NH_4Ac	1.3 μM	[40]
Riboflavin(B_2)	Wines	CZE	LIF	30 mM phosphate	0.5 μg/L	[43]
			EX: 442 nm	pH 9.8		
			EM: 515 nm			
Riboflavin(B_2)	Milk & yoghurt & raw egg (white and yolk)	CZE	LIF	30 mM phosphate pH 9.8	0.5 μg/L,	[44]
RF			EX: 442 nm		4 μg/L,	
FMN			EM: 515 nm		6 μg/L	
FAD						
Riboflavin(B_2)	Beverage	CZE	LIF	10 mM borate	3 nM	[47]
	green tea		EX: 474 nm	pH 9.6		
			EM: 515 nm			
Riboflavin(B_2)	Spinach &	CZE	LIF	50 mM borate	1.2 nM,	[48]
RF	lettuce leaves		EX: 450 nm	pH 8.5	7.0 nM,	
FMN			EM: 530 nm		7.0 nM	
FAD						
Riboflavin(B_2)	Beer	MEKC	UV LED-	10 mM sodium tetraborate	200	[50]
			induced fluorescence detection:	60 mM SDS,	ng/mL	
			380 nm	pH 9.3		
		Dynamic pH-		140 mM sodium tetraborate,	5 ng/mL	
		junction		pH 8.5		

(Table 2) cont.....

Analyte	Sample	Separation Mode	Detection	Buffer	LOD	Ref
Riboflavin(B_2)	Beer	MEKC	Blue LED-	10 mM sodium tetraborate;	480 ng/mL	[51]
			induced fluorescence detection:	80 mM SDS,		
			467 nm	pH 9.3		
		Dynamic pH-		100 mM sodium tetraborate,	1 ng/mL	
		junction		pH 8.2		
Riboflavin(B_2)	Barbeque sauce	MCE	SERS	20 mM 3-(N-morpholino) propane sulfonic acid,	1 μM	[53]
			473 nm	pH 6.5		
Niacin	Yeast	CZE	UV	0.02 M disodium	NA	[55]
(B_3)	spreads		254 nm	hydrogen orthophosphate and 0.02 M sodium tetraborate,		
				pH 9.2		
Niacin	Meat & fish & chicken & nuts & sunflower seeds	CZE	UV	15% acetonitrile, 85% of 0.02 M potassium dihydrogen orthophosphate, 0.02 M disodium hydrogen orthophosphate	0.5 mg/100g	[56]
(B_3)			254 nm	pH 7.0		
Niacin	Raw & cooked meat	CZE	UV	7.5% acetonitrile, 92.5 % of 0.02 M potassium dihydrogen orthophosphate, 0.02 M disodium hydrogen orthophosphate	0.5 mg/100g	[57]
(B_3)			254 nm	pH 7.0		
Niacin	Lentils &	MEKC	UV	20 mM sodium tetraborate, 15 mM SDS and 20% isopropylalcohol solution,	NA	[58]
(B_3)	faba beans &		254 nm	pH 9.2		
	peas					

(Table 2) cont.....

Analyte	Sample	Separation Mode	Detection	Buffer	LOD	Ref
Niacin (B_3)	Propolis	MEKC	UV 254 nm	30 mM borax, 12 mM SDS 5% EtOH. pH 9.0	0.036 mg/ml	[59]
Niacin (B_3)	Functional drinks	MCE	LIF EX: 635 nm	100 mM borate pH 9.88	0.2 nM	[60]
Folic acid(B_9)	Citrus products	CZE	PDA 290 nm	55 mM sodium tetraborate with 5% acetonitrile	3.09 μM	[64]
Folic acid (B_9)	Instant fried noodles	CZE	UV 214 nm	8 mM phosphate, 12 mM borate with 5% MeOH, pH 9.5	2.2 mg/L	[17]
Folic acid(B_9)	Apple juice	CZE	CL	35 mM sodium borate pH 9.4	2.0×10^{-8} M	[65]
Cobalamin(B_{12})	Nutritive supplement	MEKC	ICP/MS	25 mM Tris buffer pH 9.0; 15 mM SDS	0.3 ng/mL	[66]
B_1, B_2, B_3, B_5, B_6, B_{12}	Soft drinks & vitamin supplements	MEKC	UV 214 nm	50 mM borate, 25 mM SDS pH 8.5	0.06 -2.00 mg/mL	[67]
B_1, B_2	Vitamin enriched drink	MEKC	DAD	135 mM SDS in borate buffer	0.1 μg/mL,	[68]
B_3, B_6			270 nm	pH 8.0	1 μg/mL, 1 μg/mL, 1.5 μg/mL	
B_2, B_3 B_9	Health drinks	CZE	LIF EX: 488 nm EM: 520 nm	25 mM borate buffer pH 9.85	$2.59.8^3$ nM, 1.5 nM, 1 nM	[34]
B_1, B_2, B_3, B_5, B_6, B_7, B_9, B_{12}	Pharmaceuticals & dietary supplements	CZE	MS/MS	50 mmol/L formic acid pH 2.05	0.0852 - 0.2252 μg/mL	[69]

VITAMIN B

Vitamin B complex mainly contains vitamin B1 (thiamine), vitamin B_2 (riboflavin), vitamin B_3 (niacin), vitamin B_5 (pantothenic acid), vitamin B_6 (pyridoxal), vitamin B_7 (biotin), vitamin B_9 (folic acid) and vitamin B_{12} (cyanocobalamin) [22]. As indispensable vitamin in the human body, vitamin B is an essential molecule in the metabolism of carbohydrates, fat and proteins [23 - 31]. Therefore, deficiencies in any of these B vitamins can cause serious diseases [32].

It is clear that sensitive analytical methods for vitamin B are of high importance. In recent years, several CE methods coupled with different detectors have appeared, including UV, LIF, CL and MS, being UV and LIF the preferred [33]. As far as UV detection is concerned, it is mostly used for detecting vitamin B at a detection wavelength of 200-290 nm. The analysis of vitamin B2 can be directly done by LIF, while other vitamins must be detected through a derivatization method [34].

Additionally, in order to improve separation efficiency, CE with different separation modes including CZE, MEKC, NACE and MCE have been employed to study vitamin B in different food matrices. Therefore, vitamin B analysis by CE was summarized according to different separation modes (Table **2**).

Vitamin B_1

Thiamin is commonly known as vitamin B_1. In food, endogenous thiamine exists as four vitamers: thiamine (Th), thiamine monophosphate (ThP), thiamine pyrophosphate (ThPP) and thiamine triphosphate (ThPPP) [35, 36]. Thiamine predominates in cereals and plants, while the esters, in particular thiamine pyrophosphate, are the principal forms in animal products. Much of the endogenous thiamine in food is bound to protein [37].

CZE

A simultaneous analysis of free thiamin and its phosphate esters was achieved using CZE with a UV at 200 nm. The separation was obtained with 65 mM boric acid and 8 mM sodium tetraborate as running buffer at pH 8.24 and a voltage of 30 kV. The LODs of Th, ThP and ThPP were $2\times10-4$ mM, $1\times10-4$ mM and $1\times10-4$ mM, respectively. Tablets samples were dissolved and sonicated for about 4 min in 25 mL buffer before being injected into the CZE instrument. The method was applied to the analysis of thiamin in commercial samples [38].

MEKC

The determination of thiamine in pork and beef samples was achieved by MEKC using a fused-silica capillary (70 cm×75 μm) and a buffer consisting of 50 mM SDS, 100 mM sodium tetraborate, 50 mM sodium phosphate (pH 7.6) and 10% IPO [39]. The analysis was performed at a temperature of 50°C and a voltage of 15 kV by UV detection at 254 nm. The sample was hydrolyzed by hydrochloric acid and taka-diastase, then the extract was purified using ethanol and ion exchange. The protein was precipitated by adding ethanol. The supernatant was neutralised with 4 M sodium acetate after centrifugation. The solution was then passed through Amberlite CG-50 weak cation exchange resin. The thiamine was removed from the column with hydrochloric acid. The eluent was evaporated to dryness and dissolved in Milli-Q water. Meanwhile, the results obtained by MEKC were comparable to those obtained by HPLC. The concentration of thiamine was 0.38 mg/100g for pork and 0.39 mg/100g for beef.

NACE

The organic solvents used in NACE provide a variety of physicochemical properties [40]. A method implementing NACE-C4D-MS was applied for the determination of thiamine in food supplement [41]. The acetonitrile-based buffer containing 2 M HAc and 4 mM NH4Ac was used as BGE. Tablets were ground and dissolved in 10 mL of 2 M HAc/ACN: H_2O (95:5). The thiamine could be also detectable by the C4D with an LOD of 18.8 μM, which was compared with MS having an LOD of 1.3 μM. Therefore, MS was selected to quantify thiamine.

Vitamin B_2

Riboflavin (RF), commonly called vitamin B_2, is an essential part of many dietary supplements with a recommended daily intake of 1.0-1.4 mg [42].

CZE

Cataldi *et al.* [43] have developed CZE with LIF for RF analysis in a variety of wines (white, rosé, red). Because RF is susceptible to degradation, all samples were protected from light. Sample pretreatment was simply achieved by dilution. The separation was performed at a temperature of 15°C and a voltage of 30 kV using 30 mM phosphate buffer pH 9.8 with a hydrodynamic injection of 10 s. The results indicate that the average value of RF in rosé wine was a little higher than in white wines, while a significantly higher value was found in red wines (Fig. **1**).

Later on, researchers from the same group used similar CZE-LIF conditions to quantitate vitamin B_2 in a variety of foods [44]. Vitamin B_2 can not only exist naturally as RF, but also in the form of flavin mononucleotide and flavin adenine dinucleotide. In this study, samples were extracted by NH_4Ac (pH 6). The LODs of the method were in the range of 0.5-6 μg/L. The result showed that the compounds were all detected in milk and yoghurt. The average content of RF was higher than FMN and FAD. For raw egg yolk and white samples, the mean content of RF was 310 and 380 mg/100 g, respectively, while FAD and FMN were not detectable.

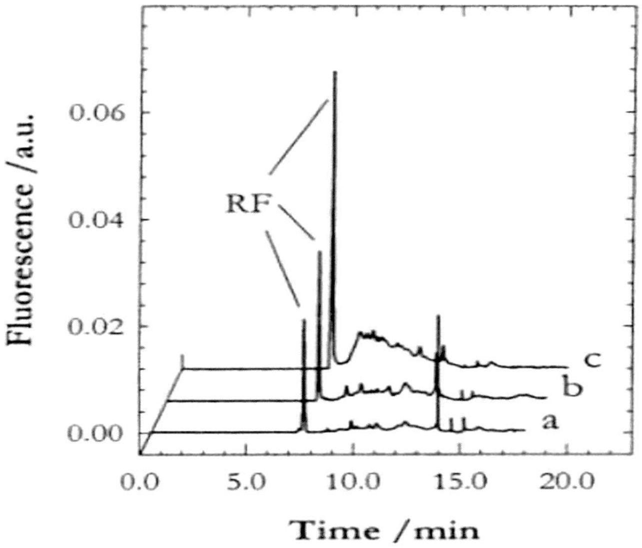

Fig. (1). Typical separations by CZE-LIF of three samples of white **(a)**, rosé **(b)**, and red **(c)** wine. Reprinted with permission from [43].

High background noise and low detection sensitivity were sometimes found in LIF detection [45, 46]. To avoid these shortcomings, a new method was proposed by CE-LIF by in-column optical fiber detection for RF [47]. In this study, an optical fiber was inserted directly into the tail end of the capillary for about 2 cm, with a diode-pumped solid-state blue laser source using an excitation wavelength at 474 nm. The other end of the fiber was inserted into the separation capillary to introduce the exciting light into the detection window with an emission wavelength of 515 nm. The method was linear (r2 = 0.998) for riboflavin in the range of 0.05 to 20 μM with a LOD of 3.0 nM. The separation was performed with 10 mM borate buffer at pH 9.6 and at a voltage of 18 kV. The sample was injected into the capillary by hydrodynamic flow at a height difference of 20 cm for 10 s. The mean content of RF was 2.81 and 1.58 μM for beverage and green tea samples, respectively. The content of RF found with the current method was consistent with that obtained by the AOAC method.

Yu *et al.* [48] developed a wall-free detection method based on liquid junction in a capillary gap. This wall-free CE-LIF detection approach was applied to the analysis of RF, FMN and FAD in spinach and lettuce leaves. All samples were treated with CH_3OH/CH_2Cl_2 and NH_4Ac (pH 6.0). Sample solutions were injected at 10 cm height difference for 10 s. The separation was achieved with 50 mM borate buffer (pH 8.5) and a voltage of 15 kV. The sensitivity of the method was improved several times compared with the on-column cell. The result showed that RF, FAD and FMN were all detected in spinach and lettuce leaves, where the average content of FAD was higher than FMN and RF to a large extent.

MEKC

The beam intensity of a Light-emitting diodes (LED) is not superior to that of a laser [49]. Chang *et al.* [50] proposed a MEKC method using UV LED source for the analysis of RF. The optimized conditions were achieved with a 10 mM sodium tetraborate buffer (pH 9.3) containing 60 mM SDS. A simple method using UV LED source with pH junction techniques was used to improve sensitivity. The optimum analysis of riboflavin was performed using 140 mM sodium tetraborate (pH 8.5), a 16 cm column length and a voltage of 16 kV, with excitation at 380 nm. The LOD of RF was 5 ng/mL which improved 40 times compared with normal MEKC mode. The concentration of RF in beer was 240 ng/mL. Researchers from the same group used blue LED-induced fluorescence detection combined with a dynamic pH junction technique in beers when excited at 467 nm [51]. The LOD of RF was 1 ng/mL which improved 5 times compared with UV-LED. The result showed that riboflavin was found in all 12 commercial beer samples in the range of 130 to 280 ng/mL.

MCE

Microchip Electrophoresis (MCE) is becoming a new tool in micro-fluidics for food analysis. MCE has the advantage of miniaturization, less sample consumption and short analysis time [52]. The sample consumption of microchips (typically picoliters) is much less than that of CE. Tycova *et al.* [53] combined MCE and surface enhanced Raman spectroscopy (SERS) with 473 nm laser for the determination of RF in a barbecue sauce. The barbecue sample was dispersed in 1 mL of 5 mM NaOH, followed by ultrasonication. The dispersion was subjected to centrifugation at 13 000 rpm for 7 min and diluted three fold using deionized water. The separation was operated in 20 mM 3-(N-morpholino) propanesulfonic acid (MOPS) at pH 6.5. The migration time was less than 60 s and the RSD was determined to be below 0.5%. The approach allowed the analysis of RF above 1 μM.

Vitamin B_3

The niacin vitamers in foods include the free form of nicotinic acid and nicotinamide, as well as bound form of nicotinamide adenine dinucleotide (NAD) and nicotinamide adenine dinucleotide phosphate (NADP). In general, the niacin vitamers in cereal grains and other seeds are principally nicotinic acid forms, whereas those in meat and fish are primarily the nicotinamide forms. Bound forms of niacin are not metabolized by the human gut, so they are considered "non-available niacin" [54].

CZE

Ward *et al.* [55] proposed a CE method for the analysis of nicotinic acid in three different yeast spreads. Alkaline extraction was used to liberate the niacin from the food by autoclave treatment for 2 h at 121°C (104 kPa). In order to improve sensitivity the supernatant was passed through a C18 and SCX cation exchange column in series after centrifugation at 2000 rpm. Then, the nicotinic acid was eluted with a 2% ammonia in methanol. The nicotinic acid was separated using a fused-silica capillary (75 cm×75 μm) with a 1: 1 mixed buffer (pH 9.2) of 20 mM sodium tetraborate, 20 mM disodium orthophosphate, and a voltage of 20 kV at 254 nm. The result showed that nicotinic acid was found in the range of 100-126 mg/100 g. The method was then extended to the analysis of niacin in foods, *e.g.* meat, fish, chicken, nuts and sunflower seeds [56]. The sample extraction was the same as described for yeast extraction. A buffer containing 15% acetonitrile, 85% of a 1:1 mixture (pH 7) of 20 mM potassium dihydrogen orthophosphate, 20 mM disodium hydrogen orthophosphate, was selected for all samples analysis. The LOD was 5 mg/kg for all foods. Nicotinic acid was in the range of 27-170 mg/kg in meat and 13-125 mg/kg in beans.

It is believed that an acid hydrolysis of niacin is preferred for its bioavailability. Windahl *et al.* [57] developed a CZE method to determine niacin in meat. Niacin is liberated from the sample matrix by autoclaving with 0.8 M H_2SO_4 at 121°C for 2 h. SPE (C18 and SCX cartridges) was used to clean up the nicotinic acid. CE separation was performed with capillary (64.5 cm×50 μm) using a buffer (pH 7) consisting of 7.5% acetonitrile, 92.5% of 20 mM potassium dihydrogen orthophosphate, 20 mM disodium hydrogen orthophosphate. The LOD for the method was 5 mg/kg. The content of niacin in cooked meat was higher than in raw meat. In addition, the level determined by the acid extraction was similar to the alkaline extraction procedure.

Slightly higher levels were recorded for meat samples with the acid extraction, and for cereal samples with the alkaline extraction.

MEKC

Vidal-Valverde [58] determined the bioavailable and total niacin content in lentils, faba beans and peas by MEKC. Bioavailable niacin was obtained using hydrolysis with HCl followed by taka-diastase and total niacin was extracted using an alkaline extraction. The extracts were subsequently purified using a strong anion exchange resin (Dowex-1-X8) and the retained vitamin was eluted with HCl. The separation buffer consisted of a 20 mM sodium tetraborate, 15 mM SDS and 20% isopropyl alcohol solution (pH 9.2). The analysis was carried out on a fused-silica capillary column (70 cm×75 μm ID) with a voltage of 30 kV and UV detection at 254 nm. The niacin content obtained by CE was similar to the levels obtained by HPLC.

MEKC has also been used to determine hesperetin, cinnamic acid and nicotinic acid in propolis by Lu *et al.* [59]. The optimized separation buffer contained 30 mM borax (pH 9.0), 12 mM SDS and 5% (v/v) EtOH. The three compounds could be well separated in 20 min with a UV detector at 214 nm and an LOD of 0.036 mg/mL. The concentration of nicotinic acid in propolis was in the range of 0.10-1.40 mg/mL.

MCE

A MCE method with LIF detection was established for the analysis of niacinamide in functional drinks [60]. The sample reacted with the same volume of 100 μM sulfoindocyanine succinimidyl ester (Cy5) at room temperature in the dark. The maximum yield was obtained in 4 h using 20 mM borate buffer (pH 8.6) as the dilution buffer. 100 mM borate solution (pH 9.85) was used as separation buffer and the LIF detector applied an excitation at 635 nm. The analysis was achieved within 4 min and the LOD for niacinamide was 0.2 nM which improved 236-fold in comparison with the method without concentration technique. Niacinamide was found in the range of 0.003-0.038 μM for 8 functional drink samples.

Vitamin B_9

Folic acid (FA) is also called vitamin B_9. The major source of the total folate content is 5-methyl tetrahydrofolic acid (5mTHF), which could be used in fortified foods [61 - 63]. Matella [64] developed a method to determine 5mTHF in citrus. 50 mL of not-from-concentrate were centrifuged (10000 g for 10 min at 0°C) and adjusted to pH 7. The supernatants were concentrated 100-fold using combined folate-affinity chromatography and C18 extraction. The samples were analyzed under the following conditions: 55 mM sodium tetraborate buffer with 5% acetonitrile, 18 kV voltage, fused-silica column (50 cm × 50 μm), 25°C, UV absorbance at 290 nm. The method has a recovery of 82% and an LOD of 3.09 μM for folate. The results confirmed that the commercial samples contained 1 nmol/mL of folate.

Cheung *et al.* [17] developed a CZE method with UV detection at 214 nm for the detection of FA in instant noodles. The method involved 8 mM phosphate-12 mM borate with 5% methanol at pH 9.5 as running buffer. In instant noodles, the extraction of the folic acid was performed using α-amylase without the need for either protease or conjugase. Sample extraction was homogenized with running buffer for 1h and the mixture was adjusted to pH 7.0 before the addition of the enzyme solution and 1 h incubation at 65°C. The limit of detection was 2.2 mg/L. The results obtained indicated that FA in instant fried noodles was stable.

Zhao *et al.* [65] developed a CE method with CL detection for the analysis of folic acid in apple juice. The running buffer was made with a 35 mM sodium borate (pH 9.4) containing 0.8 mM luminol and 1.6 mM NaBrO in 100 mM Na_2CO_3 solution (pH 12.0). Prior to injection, samples were mixed with 35 mM borate buffer and heated at 80°C for 5 min. The analysis of FA was achieved in less than 20 min, and the LOD was 20×nM. The results indicated that the folic acid were in the range of 0.33-0.43 mg/L for three kinds of apple juice.

Vitamin B_{12}

Cobalamin, also called vitamin B_{12}, contains a transition metal. A CE coupled with ICP-MS method was developed for the analysis of cobalamin in nutritive supplement [66]. A microwave-assisted extraction method was used for the extraction of the compounds. Approximately 0.25 g of powder samples was weighed into 15 mL polyethylene centrifuge tubes, and 5 mL of 5% v/v HNO_3 solution was added. After microwave heating, the cooled samples were centrifuged for 5 min at 3500 rpm and the supernatant was diluted by water before CE separation. The separation was performed with a 25 mM Tris buffer (pH 9.0) containing 15 mM SDS as and a voltage of +20 kV. Recovery was in the range of 94-105% for various cobalt species. The LOD of the method was 0.3 ng/mL for vitamin B_{12} and the cyanocobalamin content of the nutritive supplement was 189 ng/mL.

Multivitamin B analysis

Schreiner *et al.* [67] developed a MEKC method for separating six water-soluble vitamins in some soft drinks and vitamin supplements. These compounds were separated using a buffer consisting of 50 mM borate and 25 mM SDS (pH 8.5) with UV absorbance at 214 nm. Soft drinks were diluted with the same volume of internal standard (40 mg/L Na-saccharin) and vitamin supplements were dissolved in 1 L of water. The RSDs of the intra-day and inter-day precision were in the range of 1.08-3.68% and 1.26-3.35%, respectively. The recoveries for pantothenate in vitamin supplements were in the range of 41% to 61% due to weak spectral properties at 214 nm. The LOD of detected vitamins was in the range of 0.06-2.00 mg/mL.

Okamoto *et al.* [68] developed a MEKC method for determining riboflavin phosphate based on its conversion to riboflavin with alkaline phosphatase using in- capillary enzymatic reaction. Meanwhile, three other water-soluble vitamins including thiamin, niacin and pyridoxine were simultaneously quantified in the presence of RF. In this method, sandwich plugs of solution without SDS were injected into the capillary on both sides of the reaction zone to protect the enzyme reaction. A weak electric field was used to proceed the reaction. The compounds were separated using a running buffer of 135 mM SDS in borate (pH 8.0) and a UV detection at 270 nm. The linearity was excellent with correlation coefficients of 0.9999. Recoveries were obtained in the range of 99.3-101.8% with RSD of 0.1- 2.5%, respectively. The LODs of riboflavin, thiamine, pyridoxine and niacin were in the range of 0.1-1.5 μg/mL.

Zhao *et al.* [34] developed a CE-LIF method for the determination of riboflavin (RF), folic acid (FA) and niacinamide (NA) in health drinks. The drink samples were shaken to expel CO_2 and diluted with 50 mM borate solution. NA and FA were derivatized by mixing vitamin solution with the same volume of 5 mM Fluorescein isothiocyanate (FITC) solution at pH 9.3. The reaction was performed at 43°C for 2 h and then at room temperature for 6 h in the dark. Three B vitamins were separated with a voltage of 22 kV using 25 mM borate buffer at pH 9.85. The obtained RSDs were in the range of 0.4 to 1.1% for migration times and 1.9 to 4.4% for peak areas. The LODs of RF, FA and NA were 2.5 × 103 nM, 1.5 nM, 1 nM, respectively.

In order to enhance sensitivity and selectivity, a method based on CE-ESI-QqQ was applied to simultaneously identify and determine the content of nine B vitamins [69]. The method was achieved in multiple reaction monitoring (MRM) mode (Fig. **2**). A tablet dissolved in 500 μL of 26% ammonia solution and aq formic acid solution (pH 7.00) was transferred into a 50 mL volumetric flask and the mixtures were ultrasonicated for 5 min before injection into the CE equipment. 50 mM aqueous formic acid with pH 2.05 was used as the running buffer and a sheath liquid consisted of 0.1% formic acid in 50% methanol at a flow rate of 8 μL/min. The obtained LODs were in the range of 0.0852-0.2252 μg/mL.

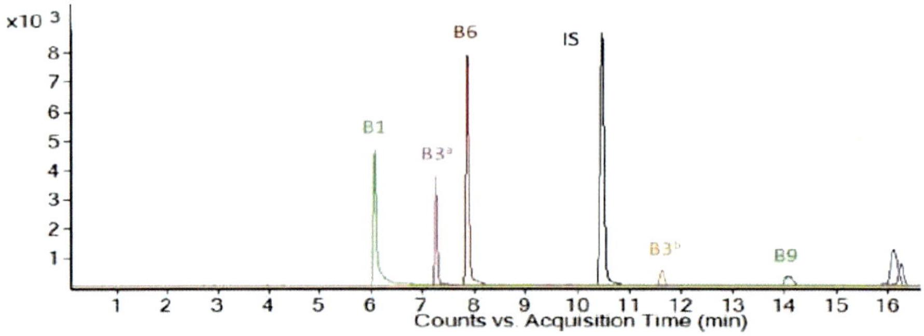

Fig. (2). Electropherogram from the simultaneous CZE-ESI-QqQ determination of vitamins B in a mixed standard solution. Adapted from reference [69].

VITAMIN C

Ascorbic acid (AA or L-AA), is a reducing carbohydrate and participates in redox processes needed for a number of functions in humans [70 - 73]. In the aspect of cellular and biochemical processes, ascorbic acid is an important antioxidant and it can suppress the development of liver cancer cells and prevent gastric cancer with strong antioxidant characteristics [74 - 76]. Additionally, ascorbic acid has obtained much attention due to its neuroprotective and neuromodulating properties [77]. Ascorbic acid cannot be synthesized in the human body. Therefore, it should be obtained from fruits, vegetables and various pharmaceuticals [78]. The acceptable daily intake of AA is 75 and 90 mg/d for women and men, respectively [79]. However, in high doses (>2000 mg/d) it could cause diarrhea and upset stomach [80]. Thus, the determination of AA has been of interest [81].

From an analytical aspect, the most important physicochemical property of ascorbic acid is lack of stability [37]. Vitamin C is highly sensitive to light, temperature, heat, salt, pH, oxygen, enzymes, *etc* [82], and the fast degradation of vitamin C makes its quantification difficult [83]. CE-UV methods were developed for the analysis of AA in different samples including pharmaceuticals, drinking juices and milks, due to its relatively inexpensive and simple detection [84]. However, the application of CE-UV was limited owing to its unsatisfactory sensitivity. Electrochemical detection (ECD) coupled with CE enabled the detection of ascorbic acid as electroactive substance [85]. It was reported that single cell analysis can be performed from femto- to attomole levels by CE-ECD

[86]. It must be isolated from the CE separation in order to adapt the ECD potentials (mV) with much larger CE voltage (kV), which can be solved by end-column approach [87].

Furthermore, CE with different separation modes including capillary CEC, CZE, MEKC and MCE has been successfully employed to study ascorbic acid (Table **3**).

Table 3. Summary of vitamin C analysis by CE.

Sample	Separation Mode	Detection	Buffer	LOD	Ref
Juices and spinach	CZE	UV 265nm	0.2 M borate buffer, pH 9.0	1 μg/mL	[83]
Apricot juices	CZE	UV 265nm	50 mM tricine buffer, pH 8.8	1.6 mg/L	[90]
Effervescent tablets	CZE	UV 270nm	0.1 M tricine buffer, pH 8.8	0.8 mg/L	[87]
Tablets, drinking juices and milk samples	CZE	UV 265nm	12.5 mM phosphate buffer, pH 7.0	0.56 μg/mL	[84]
Food samples and medicinal plants	CZE	UV 254nm	borate buffer, pH 8.0	0.03 mg/g	[95]
Grapefruit juice	CZE	ECD	60 mM borate buffer, pH 9.0	1×10^{-6} g/mL	[96]
Beverages	CZE	ECD	30 mM sodium hydroxide, pH 12.5	4.9×10^{-7} mol/L	[98]
Oolong tea infusion	CZE	ECD	Phosphate-borate (30/40 mmol/L, pH 8.5)	2.39 μmol/L	[99]
Beverage samples	MCE	ECD	5 mM phosphate buffer (pH 6.9) containing 2 mM TTAB	9.7 μM	[108]
Medicine sample	MCE	ECD	10 mM phosphate at pH 6.5 and 200 μM CTAB	3.1 μM	[109]
Soft drinks and vitamin C tablets	MCE	ECD	histidine/tartrate buffer (pH 6.5) containing 0.025% HP-β-CD and 0.1 mM CTAB	10 mg/L	[112]
Fresh tea leaves and tea liquor	MEKC	UV 200nm	100 mM SDS, 25 mM phosphate, methanol, pH 7	20 μg/mL	[113]
Beverage and fruit samples	CEC	UV 200nm	30 mM phosphate buffer, pH 8.0	10 μg/L	[115]

Capillary Zone Electrophoresis (CZE)

CZE-UV

AA is a polar compound that could be dissociated in neutral or basic solution to a large extent, so CZE is a suitable mode for separation [88]. L-Ascorbic acid is the major form, whereas its epimer D-isoascorbic acid (D-IAA) does not naturally exist in foods, which presents the opportunity to utilize D-IAA as an internal standard for quantitative determination of L-ascorbic acid. To minimize the inaccuracy from the sample introduction, in 2001, Liao *et al.* [83] used D-IAA as the internal standard to quantify L-AA in spinach and fruits by CZE. Since L-AA is prone to degradation, its stability was maintained by the addition of metaphosphoric acid and EDTA during sample preparation. In practice, 2 mL juice samples were mixed well with 6 mL 1 mM EDTA/3.0% metaphosphoric acid and then the mixtures were centrifuged. The 20 g spinach sample (without roots) was homogenized with 20 mL 1 mM EDTA/0.2 M KH2PO4/0.474 M H3PO4 (pH 2.0). Then, 6 g homogenates were added to 24 mL 1 mM EDTA/3.0% metaphosphoric acid. The supernatants were collected to be analyzed by CE. The method utilized a fused-silica capillary (50 cm×75 μm) with 0.2 M borate buffer (pH 9.0), and avoltage of 30 kV. Analyte was detected with UV at a wavelength of 265 nm. The detection limit of L-AA was found to be 1 μg/mL. Juice from lemon contains about 360 μg/mL L-AA which was four times higher than spinach (about 88 μg/mL). The recoveries of L-AA were between 95 and 105%.

D-isoascorbic acid (D-IAA) has about 5% activity of L-AA [89]. Therefore, there is an ongoing interest in the simultaneous determination of L-AA and D-IAA. In 2004, Versari *et al.* [90] presented a method for the determination of L-AA and D- IAA in fruit juice using CE with UV detection at 265 nm. Apricot juices were diluted 2-times (1:1, v/v) with 10% metaphosphoric acid and centrifuged before analysis. The method used a fused-silica capillary (50 cm×75 μm) with 50 mM tricine buffer at pH 8.8, and a voltage of 11 kV. The LOD was 1.6 and 1.1 mg/L for L-AA and D-IAA, respectively. The content of AA in juices was in the range of n.d. to 486 mg/L.

Neuberger *et al.* [88] developed a CZE-UV method for the determination of AA and its related substances (mono- and diacetylated AA). The tablets were dissolved in 250 mL 70 mg/L 2-naphthalenesulfonic acid (internal standard). The separation was performed using a fused-silica capillary (50 cm×50 μm) with a 0.1 M tricine buffer (pH 8.8) and a voltage of 20 kV. AA was quantified with a DAD at 270 nm. The linearity for AA was in the range of 25-2500 mg/L (R2 = 0.999) with an LOD of about 0.8 mg/L. When the tablets were exposed to an environment of defined humidity (75%) and temperature (30°C) for 72 h, it was found that more than 80% of AA was degraded.

The instability of AA in aqueous solutions is very complex and contains a variety of redox and intermolecular rearrangement reactions [91, 92]. Macromolecules could combine with cationic surfactant to form complexes protecting AA from degradation in aqueous media [93]. In 2018, Albishri *et al.* [84] evaluated β-cyclodextrin (β-CD) combined with cetyltrimethyl ammonium bromide (CTAB) as stabilization solution of AA in different samples. The samples were dissolved in 0.44 mM β-CD mixed with 10.0 mM CTAB. Moreover, the CZE-UV method was developed using a fused-silica capillary (56 cm×50 µm) with a 12.5 mM phosphate (pH 7.0) as running buffer and an operating voltage of 20 kV for the analysis of AA at 265 nm. The linearity for AA was determined in the range of 1-1000 µg/mL (= 0.9959) with an LOD of about 0.56 µg/mL. Furthermore, the recoveries of AA in the commercial samples were found to be in the range of 89.6-102.5% within the first 11 d. Therefore, reliable analysis could be obtained by β-CD-CTAB complexes.

Direct analysis for solid sample is a challenging task for analytical chemists [94]. In 2014, Falkova *et al.* [95] developed an automated procedure based on multi-commutated stepwise injection analysis (MCSWIA) connected with CZE which provided the possibility to carry out the at-line extraction and detection of L-AA in food samples and medicinal plants. The method utilized a fused-silica capillary column (60 cm×75 µm) with borate buffer (pH 8.0) and a voltage of 20 kV. The borate buffer solution was prepared by mixing 50 mL of 50 mM sodium tetraborate, 25 mL of 0.2 mol/L HCl and 25 mL of 0.4 mol/L EDTA. The linearity of the method was in the range of 0.1-5.0 mg/g for AA with an LOD of 0.03 mg/g at 254 nm. The content found of AA was 0.6-2.5 mg/g in food samples and medicinal plants.

CZE-ECD

The direct detection by UV absorbance is not possible for all compounds. In recent years, alternative electrochemical detection techniques have been performed by lots of scientists. Wu *et al.* [96] developed a CE-ECD method for the analysis of AA in grapefruit juice. The separation was carried out using a fused-silica capillary column (75 cm×50 µm i.d) with a 60 mM borate buffer (pH 9.0) and a voltage of 12 kV. In addition, the juice was centrifuged and 80 µL sample was diluted with running buffer to 1 mL and then it was electrokinetically injected 6 s at 12 kV for analysis. The method has a linear range of 0.005-0.1mg/mL for AA with an LOD of 0.001mg/mL. The content of AA was 979.17×10-6 g/mL in grapefruit juice. The robustness in the analysis of AA was demonstrated by the contents of AA in several samples and the results were consistent with the results from a website database.

In general, chemically modified electrodes have extended the applicability of CE-ECD [97]. In 2007, Dong *et al.* [98] used a kind of novel carbon paste electrode modified with double modifiers-polyethylene glycol (PEG) and Cu_2O (PEG/Cu_2O CPME) in CZE-ECD to detect AA in beverages. The separation was performed in a 70 cm×25 μm i.d. fused-silica capillary at a voltage of 17 kV with 30 mM NaOH (pH 12.5). Samples were electrokinetically injected at 17 kV for 10 s. The linearity was in the range of 1-50×10 mM with the LOD of 0.49×nM. The beverage samples were successfully analyzed with recoveries of 93% and RSDs of 2.9%.

Yang *et al.* [99] proposed a CZE-AD method to determine the analytes in oolong tea infusion. 2 g tea leaves was infused with 30 mL hot water (100°C) for 30 min and the filtrate was diluted fivefold before CE analysis. The separation was conducted using a fused-silica capillary (70 cm×25 μm i.d) at a voltage of 18 kV with a solution of phosphate-borate (30/40 mmol/L, pH 8.5) as BGE. The detection cell was filled with 0.1 mol/L NaOH solution to immerse the surface of the Cu disc electrode. All samples were electrokinetically injected at 18 kV for 8 s. The linearity of the method for AA was in the range of 0.01~10 mmol/L with an LOD of 2.39 μmol/L. Under the selected conditions, AA was not detected in the oolong tea infusion.

Microchip Electrophoresis (MCE)

MCE-ECD

In comparation with traditional CE, MCE offer several advantages like shorter analysis times, lower sample consumption, higher throughput capability and portability [100, 101]. Moreover, MCE devices can be fabricated in many different substrate materials, including glass, quartz and polymers [102 - 105]. In addition, ECD was compatible with microfabrication technologies allowing the integration of detection electrodes with microfluidic channels toward portable μTASs [106, 107]. In 2014, Lucca *et al.* [108] developed a method using polydimethylsiloxane (PDMS) MCE coupled with amperometric detection for the analysis of AA in beverage samples. The samples were diluted in ultrapure water and injected into the chip without any pretreatment step. The method utilized a PDMS microchip with 5 mM phosphate buffer (pH 6.9) containing 2 mM tetradecyl trimethyl ammonium bromide (TTAB), with electrokinetic injection at 1400 V for 5 s and a voltage of 1600 V. The detection potential was 1.0 V at carbon fiber electrode *versus* Ag/AgCl. The linearity of method was in the range of 10-110 μM with an LOD of 9.7 μM. Moreover, the recovery of 98%~110%

indicates a satisfactory accuracy of the PDMS microchip amperometric detection system.

Graphene can be used to fabricate the detection electrodes in amperometric detection. Lucca *et al.* [109] developed a MCE method with amperometric detection for the analysis of AA. The method applied a detection potential of +0.80 V, a separation voltage of +800 V and an injection voltage at −600 V for 7 s. The separation buffer consisted of 10 mM phosphate at pH 6.5 and 200 μM CTAB surfactant. The linearity of method was in the range of 10 to 110 μM with an LOD of 3.1 μM. The RSD values for peak currents and migration time were 9.8% and 3.0%, respectively.

Conductometric detection (CD) is applicable to ionic substances while, contact CD has the possible drawback of electrode contamination [110]. The development of contactless CD could prevent degradation of the electrode surface [111]. In 2005, Law *et al.* [112] presented a method using MCE with capacitively coupled contactless conductivity detection (C4D) for the separation and detection of vitamin C in soft drinks and vitamin C tablets. The method utilized external metallic film electrodes of the C4D cell, which were affixed to the cover of a poly (methyl methacrylate) (PMMA) microchip of 90×16 mm dimensions. The excitation frequency and the amplitude were set to 600 kHz and 20 Vpp (peak to peak), respectively. The separation was carried out with a histidine/tartrate buffer (pH 6.5) containing 0.025% HP-β-CD and 0.1 mM CTAB, with electrokinetic injection at - 1 kV for 3 s and a voltage of -5 kV, 20±1°C.

The method has a linear range of 20-150 mg/L and LOD of 10 mg/L. Moreover, the RSDs for the migration times and peak areas were in the range of 0.2-0.4% and 0.5-3.5%, respectively.

Micellar Electrokinetic Chromatography (MEKC)

J.P. Aucamp *et al.* [113] developed a MEKC method enabling the detection of AA in tea. The method was based on the use of an uncoated fused-silica capillary (50 cm×50 μm) with a solution of 25 mM phosphate containing 100 mM SDS and 6% (v/v) methanol (pH 7) as BGE for tea leaf. The analysis was performed with a separation voltage of 14 kV and a UV absorbance at 200 nm. Tea liquors were prepared from dried tea leaves samples as 1% (w/v) tea solution. The water extracts of tea samples were shaken in a flask with boiling hot water for 10 min then diluted as 0.1% (w/v) solution before injection. The linearity of the method was in the range of 50-500 μg/mL (R2 = 0.994) with an LOD of about 20 μg/mL. The content of AA was about 225 μg/mL in bottled black tea but it was not detected in the black tea and dried fresh tea leaf samples.

Capillary Electrochromatography (CEC)

CEC offers the possibility to combine electrophoretic with chromatographic separation mechanisms [114]. Liu *et al.* [115] developed a method using CE with diamino moiety functionalized silica nanoparticles (NPs) for the analysis of AA. Beverage samples were diluted tenfold with phosphate buffer solution before injection. Fruit samples were pulped and the pulp was centrifuged at 10000 r/min for min. The method was developed based on a fused-silica capillary (27 cm×75 μm) using a 30 mM phosphate as running buffer (pH 8.0) with a separation voltage at ±8 kV and an electrokinetic injection at -7 kV for 3 s. The UV detection wavelength of AA was set at 200 nm. The linearity for AA was in the range of 30-200 μg/L (R2 = 0.9912) with an LOD of about 10 μg/L. The RSD for the precision for five replicate separations of AA was 2.71% for peak area and 1.27% for migration time. The content of AA was found in the range of ND-551 mg/L for beverages and 241-495 mg/g for fruits.

SIMULTANEOUS DETERMINATION OF WATER-SOLUBLE VITAMINS MEKC

MEKC was an effective tool in the determination of water-soluble vitamins. In 2013, Serni *et al.* [116] developed a MEKC method for the analysis of five Vit C, $VitB_1$, $VitB_2$, nicotinamide and nicotinic acid (B_3). The separation was performed using a fused-silica capillary (50 cm × 50 μm i.d.) with running buffer consisting of 40 mM boric acid and 60 mM SDS (pH 8.2). The analysis was performed at a voltage of 30 kV and a wavelength of 260 nm. 1.0 g dietary supplements was dissolved in 8.0 mL of 0.1 M HCl with 5% acetonitrile and heated for 5 min at 90°C with magnetic stirring. The artichoke sample (100 g) was homogenized with 0.1 M HCl (100 mL) by grinding in liquid nitrogen for 10 min and then subjected to acid hydrolysis with 150 mL of 0.1 M HCl for 3 h. The suspension was cooled at room temperature and centrifuged at 10000 rpm for 15 min; the resulting pellets were separated and washed with 0.1 M HCl (50 mL). The supernatants were collected at -20°C and freeze-dried for 72 h afterwards. The resulting solid was reduced to powder and stored in a desiccator before analysis. The linearity of method was in the range of 10-100 μg/mL (R2 >0.99) with an LOD of 5 μg/mL. Results were close to the label claim on the package. Table **4** gives an overview of the analysis of water-soluble vitamins by CE.

In 2016, Navarro-Pascual-Ahuir *et al.* [117] developed a simple MEKC method for the analysis of eight water-soluble vitamins including Vit C, $VitB_1$, $VitB_2$, nicotinamide and nicotinic acid (B_3), $VitB_5$, $VitB_6$, Vit B_{12} in energy and sport beverages. The separation used an uncoated fused-silica capillary (50 cm×75 μm i.d.) with 40 mM borate containing 40 mM of SDS and 5% MeOH at pH 8.5, as the running buffer. The analysis was carried out at a voltage of 20 kV with UV detection at 265 nm for Vit C and 214 nm for the others. The LODs and LOQs were in the range of 0.10-1.20 mg/mL and 0.33-3.96 mg/mL for eight water-soluble vitamins. The results showed that vitamin C was below the values mentioned on the labels of energy drinks and fortified vitamin nectars.

Table 4. Summary of water-soluble vitamins analysis by CE.

Analyte	Sample	Separation Mode	Detection	Buffer	LOD	Ref
C, B_1, B_2, B_3	Multivitamin dietary supplements and artichoke extracts	MEKC	UV 260 nm	40 mM boric acid and 60 mM SDS running buffer, pH 8.2	5 μg/mL	[116]
C, B_1, B_2, B_3, B_5, B_6, B_{12}	Energy and sport beverages	MEKC	UV 214 nm, except for Vit. C at 265 nm	40 mM borate at pH 8.5, 40 mM SDS and 5% MeOH	0.10-1.20 mg/mL	[117]
B_6, B_9 and C	Tablets	MCE	ECD	10 mM phosphate buffer containing 1 M nitric acid, pH 6	15-28 μM	[118]
B_6, B_9 and C	Tablets	MCE	ECD	10 mM phosphate buffer containing 1 M nitric acid, pH 7	8-11 μM	[119]
B_1, B_2, NA, B_3, CP, B_6-H, B_6-OH, B_6-NH_2, B_9, B_7, B_{12} and C	Corn samples	CEC	UV 210 nm	4 mM phosphate (pH 2.5) in methanol-water (26/74, v/v)	0.02-0.1 μg/mL	[120]

MCE

Crevillén *et al.* [118] demonstrated microchip CE with amperometric detection to determine water-soluble vitamins including Vit B_6, fol Vit B_9 and Vit C in tablets. The method utilized 10 mM phosphate buffer (pH 6) containing 1 M nitric acid. In a single-channel glass microchip, and an electrokinetic injection at 2 kV for 3 s, a separation voltage at 2 kV and the detection potential of 1.2 V were applied. The time of separation was achieved within 130 s. 50 mg of tablet was dissolved in 25 mL running buffer and analyzed within 24 h. The method had LODs of 15-28 μM and LOQs of 20-50 μM for three water-soluble vitamins. The RSD values for migration times and peak areas were less than 0.7% and 6%, respectively.

Crevillén *et al.* [119] demonstrated the better performance of screen-printed electrode (SPE) modified with multiwalled carbon nanotubes (MWCNTs) for the vitamins (B_6, B_9 and C) analysis compared with glassy carbon electrode (GCE) in pharmaceutical preparations. 10 mM phosphate buffer (pH 7) containing 1 M nitric acid was used in this method. In the CE microchip, an electrokinetic injection at 2 kV for 20 s, a separation voltage at 2 kV and a detection potential of 1.2 V was applied. 50 mg of tablet powders and capsules were dissolved with 25 mL of running buffer (pH 7) and ultrasonicated for 30 min at 40°C, followed by centrifugation at 3500 rpm for 10 min. The method has increased the sensitivity and showed an improved electrode stability. The LODs and LOQs for three water-soluble vitamins were in the range of 8-11 μM and 25-37 μM, respectively. Excellent recoveries ranging from 91% to 97% were obtained in all cases examined.

CEC

Jia *et al.* [120] developed a pressurized capillary electrochromatography (pCEC) for the analysis of water-soluble vitamins including Vit B_1, Vit B_2, nicotinic acid (NA), Vit B_3, calcium pantothenate (CP), pyridoxal hydrochloride (B_6-H), pyridoxine hydrochloride (B_6-OH), pyridoxamine dihydrochloride (B_6-NH2), Vit B_9, VitB_7, Vit $B_1$2 and Vit C. The CEC method used a separation voltage at -12 kV and a 250 mm×0.2 mm monolithic silica-ODS column. The mobile phase was made of 4 mM phosphate (pH 2.5) in methanol-water (26/74, v/v) with a flow rate of 20 μL/min. The analysis was performed with detection wavelength at 210 nm. The corn (120 mg) was extracted with 4 mL of cold methanol for 30 min and then the solution was centrifuged at 5000 rpm for 25 min. The supernatants were evaporated and the dried sample was dissolved in 1 mL of water. The LODs was in the range of 0.02-0.1 μg m/L for eleven water-soluble vitamins. B_6-OH, NA, $B_1$2, FA and B_2 were detected and NA was not quantified due to the insufficient resolution with matrix interferences in the corn sample.

VITAMIN E

Vitamin E (VE) as natural antioxidant is important for human normal growth [121]. The antioxidative effect of vitamin E is to eliminate free radicals reactions that can cause cell membrane damage and organ injury in the body [122]. Their positive effect to inhibit the progress of precancerous lesions has also been demonstrated [123]. Because of physiological functions of Vitamin E, its lack or excess can cause health problems. However, it is ingested only from foods. According to dietary recommendations, the estimated average requirement (EAR) and the recommended dietary allowance (RDA) of VE (2R-α-tocopherol) is 12 and 15 mg, respectively [124].

Vitamin E consists of eight isomers known as α-, β-, γ-, δ-tocopherols and α-, β-, γ-, δ-tocotrienols. The structure of tocopherols and tocotrienols is similar with the difference of three trans-double bonds in the hydrocarbon tail in tocotrienols (Fig. **3**). Meanwhile, tocopherols have a different number of methyl groups on the chromanol ring [125]. These isomers differ in antioxidative properties, with α-tocopherol as the most active isomer. Due to their important nutritional value and health benefits, the analysis of vitamin E has been of much interest. The separation of these isomers has become a challenge so far. As they are fat-soluble, it is not possible to separate them in an aqueous CE system. CEC and NACE methods have been developed to improve separation efficiencies.

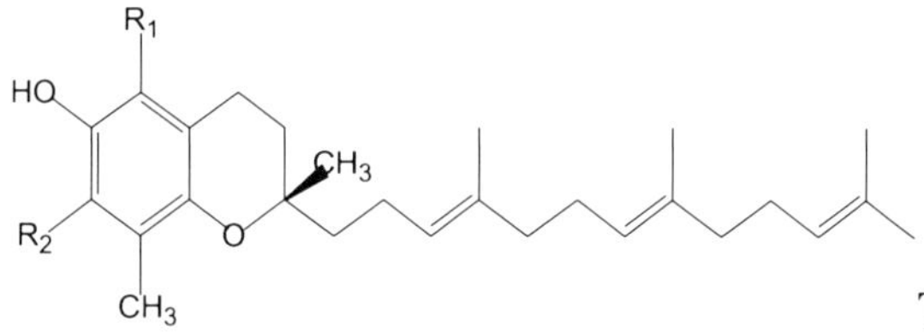

Tocotrienol

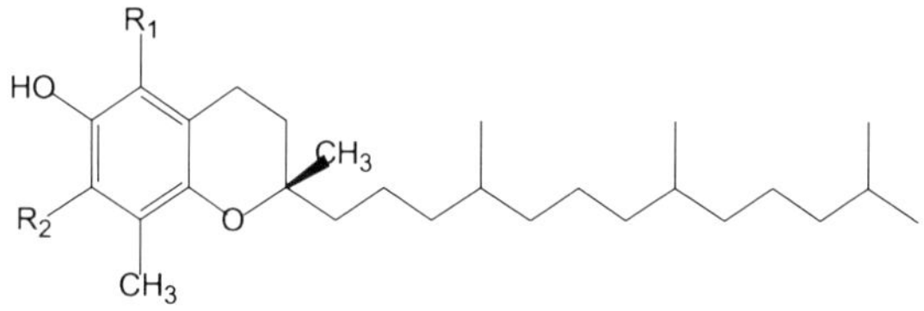

Tocopherol

Isomer	**R_1**	**R_2**
α-tocopherol/tocotrienol	CH_3	CH_3
β- tocopherol/tocotrienol	CH_3	H
γ- tocopherol/tocotrienol	H	CH_3
δ- tocopherol/tocotrienol	H	H

Fig. (3). Chemical structures of the eight isomers of tocopherol and tocotrienol.

CEC

RP18 silica particles can be simultaneously used for the analysis of neutral compounds and polar compounds [126, 127]. Fanali *et al.* [128] developed a CEC method for the analysis of the mixture of α-, γ-, δ-tocopherols and the antioxidant

BHT. The separation was performed in an octadecyl silica (ODS) stationary phase packed in a fused-silica capillary (100 μm I.D., 365 μm O.D. ×33 cm of total length and 24.6 or 8.4 cm effective length). The mobile phase was composed of methanol- acetonitrile (30:70) containing 0.01% (w/v) of ammonium acetate. The LOD and LOQ were reported at 10 μg/mL and 25 μg/mL, respectively (see Table 5).

Table 5. Summary of fat-soluble vitamins analysis by CE.

Analytes	Sample	Separation mode	Detection	Buffer	LOD	Ref
α-,γ-,δ-tocopherols and the antioxidant BHT	-	CEC ODS stationary phase	DAD	70:30ACN/MeOH containing 0.01% ammonium acetate	10 μg/mL	[128]
α-,γ-,δ-tocopherols , α-tocopherol acetate and BHT	Vegetable oils	CEC C18 stationary phase	UV 205nm	50/50 v/v methanol-acetonitrile containing 0.01% ammonium acetate	1.2 μg/mL	[130]
α-, β-, γ- and δ- **TOHs**	-	CEC ULTIMA C18	UV 205 nm	95:5 v/v methanol/water (5 mmol/L Tris, final concentration)	NA	[138]
α-, β, γ-tocopherols and α-tocopherol acetate	Vitamin E supplement capsule	CEC polymeric and monomeric C_{30}	UV 214 and 254 nm	1% water/methanol	NA	[132]
Tocopherols, tocotrienols	Vegetable oils	CEC: methacrylate ester-based monolithic columns	UV 205 nm 295nm	93:7 v/v MeOH–aqueous buffer	1.5-2.3 μg/mL	[142]
α-,β,γ-,δ-tocopherols and α-tocopherol acetate	Capsule	CEC PEDAS-EDMA monolithic column	UV 200 nm	3:10:87 v/v/v 100 mM Tris buffer (pH 9.3)/methanol/ACN	1 μg/mL	[144]
α-, (β + γ)- and δ-tocopherol	Vegetable oils	NACE	DAD 220, 240, 290 and 330 nm	40 mM KOH in a methanol/1-propanol (PrOH) mixture (15:85 v/v)	3-5 μg/mL	[147]

(Table 5) cont.....

Analytes	Sample	Separation mode	Detection	Buffer	LOD	Ref
α-, (β+γ)-, and δ-tocopherol	Vegetable oil	NACE	DAD 200 nm FLD Excitation 297 nm; emission 320 nm	12 mM borate (3 mM sodium tetraborate), 60 mM sodium cholate, and 12 mM NaOH in methanol	0.18-0.56 µg /mL (FLD)	[148]
vitamin D_2 and vitamin D_3	——	MEKC	UV: 280 nm	25 mM silver nitrate added. pH 10 CAPS buffer, 25 mM SDS	NA	[152]
retinal, retinol and retinoic acid	——	MEKC	UV: 340 nm	0.02 M sodium tetraborate, containing 0.1 M sodium cholate and 20% acetonitrile	NA	[156]
vitamin A palmitate, E acetate, and D_3	Commercial formulations	EKC	DAD: 238 nm	acetonitrile–water (80:20, v/v) with 80 mM TDA bromide	NA	[157]
vitamins A palmitate, E acetate and D_3	Tablet	EKC	DAD: 235 nm	80 mM TDAB and 4 mM borate in acetonitrile water (80/20 v/v).	0.07 mg/mL	[158]
vitamin D_2, vitamin E and vitamin A	Multivitamin tablets and vitamin E soft capsules	MEKC	UV	80 mM borate-60 mM sodium deoxycholate-20 mM dioctyl sulfosuccinate sodium salt-2 mM Brij 35 (pH 8.5)	0.028-0.17 µg/mL	[162]
vitamins A palmitate, E acetate and D	Commercial vitamin tablets	MEEKC	DAD: 300 nm (VA) 200 nm (VE) 260 nm (VD)	66.6% (w/w) 25 mM phosphate buffer pH 2.5, 20.0% (w/w) 2-propanol, 6.6% (w/w) 1-butanol, 6.0% (w/w) sodium lauryl sulphate (SLS), and 0.8% (w/w) n-octane	Vitamin A palmitate: 6 µg/mL Vitamin E acetate: 2 µg/mL Vitamin D: 3 µg/mL	[168]

(Table 5) cont.....

Analytes	Sample	Separation mode	Detection	Buffer	LOD	Ref
vitamins A, D, E and K	Animal feed	MEEKC	UV: 280 nm	25 mmol/L phosphate buffer (pH 2.5), 2-propanol, 1-butanol, sodium dodecyl sulfate and octane	vitamins A, D and K : 7 mg/kg vitamin E: 25 mg/kg	[169]

The main sources of vitamin E are vegetable oils, fats, nuts and seeds [129]. Aturki *et al.* [130] developed a method using C18 stationary phase for the analysis of vitamin E in vegetable oils. Due to tocopherols autoxidation, an antioxidant compound (BHT) was added during the extraction process. A fused-silica capillary partially packed with 3 μm C18 stationary phase (75 μm ID, 33 cm total length, 7 cm effective and packed lengths) was utilized in the study. The optimum mobile phase was a mixture of methanol-acetonitrile (50:50, v/v) containing 0.01% ammonium acetate. The separation of the tocopherols and the BHT was achieved within 2.5 min using a voltage of -25 kV by the short end injection. Vitamin E isomers in 1 g oil were extracted using 2.5 mL methanol and the supernatant was collected after centrifugation (3500 g/min) for 10 min. The residue was extracted again with two 2.5 mL portions of a mixture of methanol/isopropanol (8:2, v/v). The combined extracts were dried using a rotary pump at 40°C and were reconstituted in 1 mL of ethanol. The result showed that soybean oil has the higher content of γ-tocopherol and the quantity of δ-tocopherol is greater than the amount of α-tocopherol. Sunflower, hazelnut and virgin oils contain a high amount of α- tocopherol and δ-tocopherol is found only in sunflower oil in very small amount. The method could determine adulteration of the oil, but it cannot separate β- and γ- isomers.

It is difficult for vitamin E isomers to be separated on C8 or C18 phases [128, 130 - 134]. The shape selectivity for alkyl stationary phases increases with increased alkyl chain length in the stationary phase according to the Sander and Wise group [135 - 137]. Charles *et al.* [132] compared two different modes of alkyl phase including polymeric and monomeric C30 using CEC with UV detection at wavelengths of 214 and 254 nm. The separation of the tocopherol isomers was achieved using the monomerically bound C30 stationary phase and 1% water/methanol mobile phase. The method was applied to detect the tocopherols in a vitamin E capsule with simple dissolution of sample in methanol.

Carabias-Martínez *et al.* [138] developed a method for the separation of β-, γ-tocopherols using ULTIMA C18. The ULTIMA can improve selectivity for the separation of β-, γ-tocopherols. In addition, the ULTIMA C18 is very similar in length of the alkyl chain to the normal C18, but it has a similar selectivity to C30.

Monolithic capillary columns can be manufactured with an excellent reproducibility. The methacrylate ester-based monolithic columns have several advantages including an adjustable polarity, control of pore characteristics, and high stability under extreme pH conditions [139 - 141]. Lerma-García *et al.* [142] established a method to separate tocopherols in vegetable oils. 4 g oil samples were extracted twice with 10 mL of methanol containing 0.1% w/v BHT, followed by 10 mL of a methanol/isopropanol mixture (8:2, v/v). All extractions were performed by a 30 s shaking followed by centrifugation at 8000 *g* for 10 min and the extracts were evaporated to dryness. The residue was dissolved in 1 mL of methanol and stored at -20°C before analysis.

Although the PEDAS monolith was successful for the separation of all the tocopherol compounds, the addition of EDMA as copolymer could improve the separation efficiency. It was noted that the preparation of PEDAS-EDMA monolith can be made according to Okanda and El Rassi [143]. Chaisuwan *et al.* [144] developed a method for separating α-, β-, γ-, δ-tocopherols and α-tocopherol acetate (TAc) with PEDAS-EDMA monolithic column. The separation was achieved within 25 min using 100 mM Tris buffer (pH 9.3)/methanol/acetonitrile (3:10:87, v/v/v) as the mobile phase. The LODs for all four tocopherols and TAc were below 1 ppm with recoveries from 93 to 99%. The method was successfully applied to a capsule analysis after dissolving sample in 10.00 mL methanol. Results indicated that the capsule contains four isomers, including α-, β-, γ-, δ-tocopherols at 262±5.2 mg, 1.51±0.01 mg, 1.65±0.02 mg and 0.57±0.01 mg, respectively. The detected amounts were practically identical to the claimed amounts.

NACE

NACE is suitable for the analysis of weak water-soluble compounds [145, 146]. Mendonça *et al.* [147] developed a NACE method for VE isomers analysis in vegetable oils. The method has the advantage of direct injection for vegetable oil samples without pretreatment. The oil-miscible background electrolyte consisting of 40 mM KOH in PrOH (15:85, v/v) was selected to support the direct injection. Samples were only diluted with methanol/1-propanol before analysis. The concentrations found in refined soybean oil were 29, 269 and 96 mg/mL for α-, (β+γ)- and δ-tocopherol, respectively.

Galeano-Díaz *et al.* [148] developed a NACE method with serial UV and fluorescence detection for the analysis of tocopherols in vegetable oil. The buffer of 12 mM borate buffer (3 mM sodium tetraborate), 60 mM sodium cholate, and

12 mM sodium hydroxide in methanol was selected for separation. Tocopherol can be easily detected with fluorescence detection but not with UV detection. For the sample extraction, 0.1~0.6 g oil sample was dissolved in 4 mL of hexane and the solution passed through a Sep-Pak Plus Silica (690 mg) cartridge and eluted with 2 mL of diethyl ether. The eluate was evaporated to dryness and the residue was dissolved with BGE-hexane-ethanol (60:20:20, v/v/v) before analysis.

VITAMIN D

Vitamin D is a steroid derivative. The most important members of the vitamin D group are ergocalcitol (vitamin D2) and cholecalcitol (vitamin D3). Vitamin D3 is the major dietary source present mostly in foods of animal origin. Vitamin D2 is the form mostly used in supplements and fortified foods [149]. 25(OH)D is a derivative of vitamin D3 and vitamin D2 by the action of 25-hydroxylase and it is the indicator to reflect the nutritional status of Vitamin D in the body. It is necessary to meet the required daily intake of vitamin D [150]. Meanwhile vitamin D3 is related to calcium absorption and its deficiency causes osteomalacia [151].

The addition of Ag(I) in MEKC can be used for the analysis of compounds that show selectivity toward Ag(I) complexation, like vitamin D. Wright and Dorsey [152] developed a MEKC method to separate vitamin D2 and vitamin D3. The separation was performed by using 3-cyclohexylamino-1-propanesulfonic acid (CAPS) buffer at pH 10, through the addition of 25 mM silver nitrate to the buffer containing SDS with UV detection at 280 nm.

VITAMIN A

Vitamin A components, retinol and retinoic acid, are fat-soluble vitamins and are two physiologically important retinoids [153]. Vitamin A deficiency causes many diseases [154]. It has been reported that more than one million children become permanently blind each year due to vitamin A deficiency [155].

S. Profumo *et al.* [156] developed a MEKC method to separate the three vitamin A derivatives including retinal, retinol and retinoic acid. The derivatives were separated within 25 min and detected at 340 nm. The capillary used was an unmodified fused-silica capillary of 75 μm ID with effective length 50 cm. The optimum buffer condition is 20 mM sodium tetraborate, containing 100 mM sodium cholate and 20% acetonitrile. Samples were injected into the capillary by 0.5 psi pressure for 1.5 s.

SIMULTANEOUS DETERMINATION OF FAT-SOLUBLE VITAMINS EKC

Pedersen-Bjergaard *et al.* [157] developed an electrokinetic chromatography (EKC) method for the analysis of vitamin A palmitate, E acetate and D3. Acetonitrile–water (8:2, v/v) with 80 mM tetradecylammonium bromide (TDAB) was selected as separation electrolyte. Sample was injected by 0.5 psi pressure for 1 s and three vitamins were separated within 6 min. The analysis was performed with an uncoated fused-silica capillary (47 cm×50 µm ID) and a voltage of 10 kV at UV 238 nm. The solid product was extracted with 10 mL of tetrahydrofuran (THF) and centrifuged for 5 min. After another extraction for 10 min with n-hexane and 5 min of centrifugation, the liquid product was evaporated and dissolved in 0.4 mL THF before analysis. The LOD and LOQ for vitamin E acetate were 0.07 mg/mL and 0.20 mg/mL, respectively. The samples were determined to contain 97% of the labeled amount [158].

MEKC

The sensitivity of the fat-soluble vitamins in conventional MEKC methods is low and the detection limit ranges from 5.7 to 40 µg/mL [159 - 161]. Liu *et al.* [162] developed a MEKC method for the analysis of three fat-soluble vitamins including vitamin D2, α-tocopherol and retinol palmitate with UV detection and on-line sample concentration. The pressure-assisted field-enhanced sample injection (PA- FESI) provided a little advancement in the concentration efficiency compared with the sweeping and field-enhanced sample injection. The separation of the three compounds was obtained by using 80 mM borate-60 mM sodium deoxycholate-20 mM SDS salt-2 mM Brij 35 (pH 8.5) as BGE. The best concentration efficiency was obtained by using 80 mM borate solution (pH 8.5) containing 1% methanol as the sample matrix when applying a pressure of 10.35 kPa and an injection of -7 kV for 36 s. In PA-FESI, the LODs of the vitamins were in the range of 0.028-0.17 µg/mL, which enhanced 42 to 103-fold compared with conventional injection. For tablets analysis, one tablet was dissolved in 500 mL ice-cold methanol followed by ultrasonication for 30 min and centrifugation. The upper layer was taken for evaporation at 30°C, reconstituted in 100 µL methanol and diluted to 10 mL with 80 mM borate solution. For capsules analysis, 10 mL of ice-cold methanol was added to 10% of the content of one capsule and the sample solution was ultrasonicated for 30 min. The contents determined of VE in multivitamin tablets and capsules are consistent with the labeled contents.

MEEKC

In general, alkaline buffers were used for fat-soluble vitamins analysis in MEEKC [163 - 165]. The utility of acidic MEEKC systems has received limitted attention

[163, 166, 167]. Pedersen-Bjergaard *et al.* [168] developed a MEEKC method for the analysis of A palmitate, E acetate and D. These vitamins were separated with 66.6% (w/w) 25 mM phosphate buffer pH 2.5, 20.0% (w/w) 2-propanol, 6.6% (w/w) 1-butanol, 6.0% (w/w) sodium lauryl sulphate (SLS) and 0.8% (w/w) n-octane as the separation buffer. The analysis was performed in uncoated fused-silica capillaries with 30 cm of effective length with a voltage of 20 kV at a wavelength of 205 nm. The vitamin E tablets were subsequently extracted by ultra- sonication for 15 min with a microemulsion. The tablet on analysis was found to contain 48.5 mg of VE per tablet.

Oledzka *et al.* [169] also developed a MEEKC method for the analysis of vitamins A, D, E and K in animal feed with UV detection. The separation of four fat-soluble vitamins was achieved with buffer consisting of 25 mM phosphate buffer (pH 2.5), 2-propanol, 1-butanol, SDS and octane (Fig. **4**). Samples were introduced into the capillary (I.D. 50 µm, total length 47 cm, effective length 40 cm) by a pressure of 0.5 psi for 2 s. The polarity of the separation voltage was reversed in view of the strongly suppressed electro-osmotic flow, and the applied voltage was set at 12 kV. The LODs were 7 mg/kg for vitamins A, D and K and 25 mg/kg for vitamin E. For feed sample analysis, 1 g of sample was spiked with 0.1 g of vitamin C as antioxidant and 300 mg/kg methyltestosterone as IS. Basic ethanol was used to extract the vitamins and the supernatant was mixed with n-hexane. Subsequently,

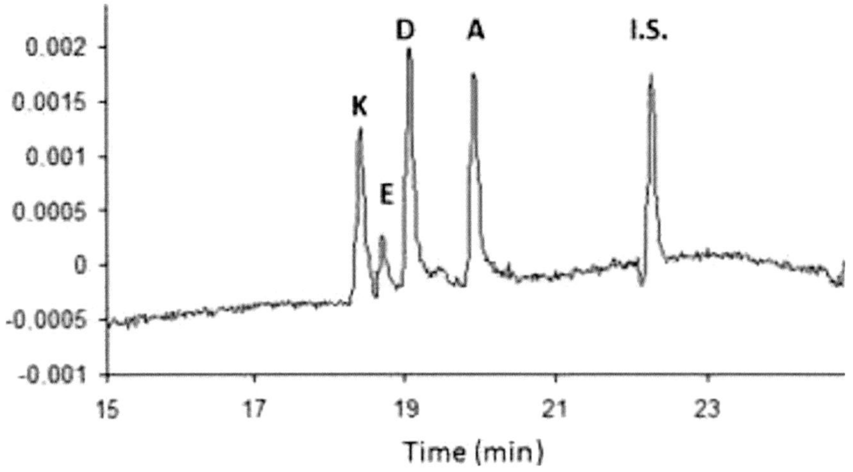

Fig. (4). Typical electropherogram of feed sample spiked with vitamins of interest and I.S. Adapted from reference [169].

the hexane phase was evaporated to dryness in a water bath at 45°C and the residue was dissolved in 100 μL of microemulsion for MEEKC analysis.

SIMULTANEOUS DETERMINATION OF WATER-SOLUBLE AND FAT-SOLUBLE VITAMINS

EKC

Delgado-Zamarreno *et al.* [161] developed a EKC method with UV detector at 200 nm for the analysis of eight vitamin B_1, B_2, B_6 and B_7 and K3, A and E. The separation of eight vitamins was accomplished with 30 mM bis (2-ethylhexyl) sodium sulfosuccinate (AOT) in a water-acetonitrile (20:80, v/v) solution with 2% methanol in the presence of 20 mM boric acid-sodium borate buffer (pH 8.4). The analysis was performed in fused-silica capillary (57 cm×75 μm I.D.) at a voltage of 24 kV at 25°C. After being ground and homogenized, vitamin tablets were dissolved in 5 mL of the separation buffer and sonicated for 5 min with the pH adjusted to 8.4. The content of vitamin B_1, Vitamin B_2, Vitamin B_6 and Vitamin E was 122.1%, 102.5%, 136.5% and 125.0% of the labeled amount, respectively (Table **6**).

Table 6. Simultaneous determination of water-soluble and fat-soluble vitamins by CE.

Analytes	Sample	Separation Mode	Detection	Buffer	LOD	references
B_1, B_2, B_6, B_7 and K_3, A and E	Vitamin tablets	EKC	UV 200 nm	30 mM AOT, acetonitrile-20 mM boric acid-sodium borate buffer (20:80 v/v), pH 8.4 with 2% methanol	NA	[161]
B_1, B_2, B_3, B_6, B_{12} and C and vitamins A palmitate, E acetate, D and K	Multi-vitamin formulation	MEEKC	UV 254 nm	40 mM borate buffer, 80 mM SDS, 5% butanol, 0.8% octane, 15% propanol	NA	[170]
B_{12}, PP, B_2, B_6, B_1, B_3, B_9, B_7, C, B_5 and retinol, tocopherol, Vitamin D_3	Commercial multivitamin tablet	MEEKC	DAD 205 nm	1.2% (w/w) SDS, 21% (v/v) 1-butanol, 18% (v/v) acetonitrile, 0.8% (w/w) n-hexane, 20 mM borax (pH 8.7)	0.2-12 μg/mL	[171]

(Table 6) cont.....

Analytes	Sample	Separation Mode	Detection	Buffer	LOD	references
B_2, D_3 and B_1	Drug tablet	MEEKC	UV 272nm ESI-MS: B_1 *m/z* 265.15, B_2 *m/z* 377.19, D_3 *m/z* 385.39	37.5% water, 43.75% ethanol, 18.75% 1-octanol (w/w/w) and 150 mmol/kg ammonium acetate	B_1: 0.53 μg/mL B_2: 7.54 μg/mL, D_3: 2.89 μg/mL	[174]
B_1, B_2, B_3, PP, B_6, C and retinol acetate, E, D_3 and K_3	Commercial multivitamin tablet	CEC: methacrylate-based capillary monolithic column	UV 210 nm	1 M formic acid (pH 1.9)/ACN (30:70, v/v) containing 10 mM ammonium formate	NA	[151]

MEEKC

In 2002, Sanchez *et al.* [170] compared MEKC and MEEKC for the separation of water-(B_1, B_2, B_3, B_6, B_12 and C) and fat-soluble (A palmitate, E acetate, D and K) vitamins. In conclusion, MEKC is only useful in the analysis of water-soluble vitamins when SDS is used as the surfactant. MEEKC gave better resolution than MEKC in the separation of fat-soluble vitamins. The microemulsion was prepared with SDS as the surfactant, octane as the non-polar modifier, butanol as the co-surfactant, and propanol as the second co-surfactant. The analysis was obtained within 60 min when a voltage of 15 kV and an injection time of 10 s were applied with UV detection at 254 nm.

Most MEEKC experiments were carried out with a common microemulsion [165]. A higher concentration of SDS led to a longer migration time. A rapid MEEKC method was developed to separate 10 water-soluble and 3 fat-soluble vitamins [171]. In this study, lower SDS concentration with the addition of 1-butanol was established. The result indicated that the internal lipophilic phase had a minor impact on the separation efficiency in MEEKC [172, 173]. The optimum microemulsion buffer consisted of 1.2% (w/w) SDS, 21% (v/v) 1-butanol, 18% (v/v) acetonitrile, 0.8% (w/w) n-hexane, 20 mM borax (pH 8.7). The analysis was accomplished within 30 min using a voltage of 25 kV and a temperature of 25°C. One tablet of commercial multivitamin was extracted with 50 mL microemulsion and the contents found were approximately consistent with those declared on the label for vitamins B_1, PP, B_2, E, B_6, B_3, C and B_9.

A surfactant-free microemulsion electrokinetic chromatography (SF-MEEKC) method [174] was developed to separate vitamins B_2, D3 and vitamin B_1 using UV at 272 nm. Since ammonium acetate was compatible with MS, instead of SDS-based microemulsions, a SFME consisting of 37.5% water, 43.75% ethanol, 18.75% 1-octanol (w/w/w) and 150 mmol/kg ammonium acetate was used to couple with ESI-MS. Positive ion mode with -4 kV of electrospray ionization voltage was carried out in SF-MEEKC-ESI-TOF-MS. Vitamin B_1 *m/z* 265.15, vitamin B_2 *m/z* 377.19, vitamin D3 *m/z* 385.39 and 25-hydroxycholecalciferol (IS) *m/z* 401.40 were selected as the respective quantifiers for each analyte. LODs were determined to be 0.53 μg/mL for vitamin B_1, 7.54 μg/mL for vitamin B_2 and 2.89 μg/mL for vitamin D3. The tablet dissolved with 1 mL SFME solution was vortexed for 5 min and centrifuged for 10 min. The contents of vitamin D3 determined in samples was in the range of 95-113 μg, which is in good agreement to the declared mass of 100 μg.

CEC

A method for determination of water- and fat-soluble vitamins was achieved using pressure-assisted CEC with a methacrylate-based capillary monolithic column at 210 nm [151]. A mixture of six water-soluble and four fat-soluble vitamins was separated within 20 min using isocratic elution at 15 kV. 1 M formic acid (pH 1.9)/acetonitrile (30:70, v/v) containing 10 mM ammonium formate was used as mobile phase. The column was a poly (BMA-*co*-EDMA-*co*-AMPS5%) monolithic column with a length of 22.5 cm and 0.1 MPa He gas pressure was applied. The commercial pharmaceutical tablets were extracted with ethanol. However, the developed method did not enable to determine all the vitamins due to insufficient sensitivity.

CONCLUDING REMARKS

Recent research has demonstrated a robust performance of CE for the analysis of vitamins in a number of food, dietary supplements and pharmaceutical samples. Various sample preparation procedures could be helpful for more efficient analysis and to avoid matrix interference. High sensitivity of vitamins analysis could be achieved by fluorescence detection, where the electrochemical detection was attractive for an electroactive substance. Most CE methods for the simultaneous determination of vitamins have been performed using CE-UV. The mode of MEKC was mainly used in the separation of water-soluble vitamins. Promising areas of future growth are the use of MCE offering several advantages in the analysis of water-soluble vitamins. MEEKC or CEC are more selective for fat-soluble vitamins and could be used in the determination of water-soluble and fat-soluble vitamins.

CONSENT FOR PUBLICATION

Not applicable.

CONFLICT OF INTEREST

The author declares no conflict of interest, financial or otherwise.

ACKNOWLEDGEMENTS

The work was supported by Open Fund from the State Key Lab of Food Nutrition and Safety, Tianjin University of Science and Technology (grant number SKLFNS- KF-201820).

REFERENCES

[1] Analysis of Food Constituents, in: J. L. Multon (Ed.), In: *Wiley-VCH*; Multon, J.L., Ed. New York, **1997**.

[2] Lukaski, H.C. Vitamin and mineral status: effects on physical performance. *Nutrition,* **2004**, *20*(7-8), 632-644. [http://dx.doi.org/10.1016/j.nut.2004.04.001] [PMID: 15212745]

[3] Miller, S. Vitamins and minerals. *J. Sch. Health,* **2010**, *4*(2a), 4.

[4] Tyśkiewicz, K.; Dębczak, A.; Gieysztor, R.; Szymczak, T.; Rój, E. Determination of fat- and water-soluble vitamins by supercritical fluid chromatography: A review. *J. Sep. Sci.,* **2018**, *41*(1), 336-350. [http://dx.doi.org/10.1002/jssc.201700598] [PMID: 28771927]

[5] Wiedemann, B.; Paul, K.D.; Stern, M.; Wagner, T.O.; Hirche, T.O. Evaluation of body mass index percentiles for assessment of malnutrition in children with cystic fibrosis. *Eur. J. Clin. Nutr.,* **2007**, *61*(6), 759-768. [http://dx.doi.org/10.1038/sj.ejcn.1602582] [PMID: 17213872]

[6] Mendiola, J.A.; Marin, F.R.; Señoráns, F.J.; Reglero, G.; Martín, P.J.; Cifuentes, A.; Ibáñez, E. Profiling of different bioactive compounds in functional drinks by high-performance liquid chromatography. *J. Chromatogr. A,* **2008**, *1188*(2), 234-241. [http://dx.doi.org/10.1016/j.chroma.2008.02.054] [PMID: 18325525]

[7] Petteys, B.J.; Frank, E.L. Rapid determination of vitamin B_2 (riboflavin) in plasma by HPLC. *Clin. Chim. Acta,* **2011**, *412*(1-2), 38-43. [http://dx.doi.org/10.1016/j.cca.2010.08.037] [PMID: 20816949]

[8] Ciulu, M.; Solinas, S.; Floris, I.; Panzanelli, A.; Pilo, M.I.; Piu, P.C.; Spano, N.; Sanna, G. RP-HPLC determination of water-soluble vitamins in honey. *Talanta,* **2011**, *83*(3), 924-929. [http://dx.doi.org/10.1016/j.talanta.2010.10.059] [PMID: 21147338]

[9] Revermann, T.; Götz, S.; Künnemeyer, J.; Karst, U. Quantitative analysis by microchip capillary electrophoresis: current limitations and problem-solving strategies. *Analyst (Lond.),* **2008**, *133*(2), 167-174. [http://dx.doi.org/10.1039/B711165G] [PMID: 18227937]

[10] Ouimet, C. M.; D'Amico, C. I.; Kennedy, R. T. Advances in Capillary Electrophoresis and the Implications for Drug Discovery *Expert Op. Drug Discovery,* **2016**, *12*(2) [http://dx.doi.org/17460441.2017.1268121.]

[11] Wall, M.M.; Bosland, P.W. Analytical methods for color and pungency of chiles (capsicums). *Dev. Food Sci.,* **1998**, *39*(98), 347-373.

[http://dx.doi.org/10.1016/S0167-4501(98)80014-9]

[12] Fotsing, L.; Fillet, M.; Bechet, I.; Hubert, P.; Crommen, J. Determination of six water-soluble vitamins in a pharmaceutical formulation by capillary electrophoresis. *J. Pharm. Biomed. Anal.*, **1997**, *15*(8), 1113-1123.
[http://dx.doi.org/10.1016/S0731-7085(96)02010-9] [PMID: 9215964]

[13] Lindeberg, J. Capillary electrophoresis in food analysis. *Food Chem.*, **1996**, *55*(1), 73-94.
[http://dx.doi.org/10.1016/0308-8146(96)81729-8]

[14] Honoré Hansen, S.; Gabel□Jensen, C.; Pedersen□Bjergaard, S. Comparison of microemulsion electrokinetic chromatography and solvent□modified micellar electrokinetic chromatography. *J. Sep. Sci.*, **2001**, *24*(8), 643-650.
[http://dx.doi.org/10.1002/1615-9314(20010801)24:8<643::AID-JSSC643>3.0.CO;2-U]

[15] Ong, C.P.; Ng, C.L.; Lee, H.K.; Li, S.F.Y. Separation of water- and fat-soluble vitamins by micellar electrokinetic chromatography. *J. Chromatogr. A*, **1991**, *547*(01), 419-428.
[http://dx.doi.org/10.1016/S0021-9673(01)88665-9]

[16] Alegre, M.; Luisa, M.; Puyana, C. María; Míguez, P.; Raquel, Capillary electrophoresis determination of non-protein amino acids as quality markers in foods. *J. Chromatogr. A*, **2007**, *1428*, 97-114.

[17] Hau Fung Cheung, R.; Morrison, P.D.; Small, D.M.; Marriott, P.J. Investigation of folic acid stability in fortified instant noodles by use of capillary electrophoresis and reversed-phase high performance liquid chromatography. *J. Chromatogr. A*, **2008**, *1213*(1), 93-99.
[http://dx.doi.org/10.1016/j.chroma.2008.09.098] [PMID: 18930464]

[18] Hempel, G. Strategies to improve the sensitivity in capillary electrophoresis for the analysis of drugs in biological fluids. *Electrophoresis*, **2000**, *21*(4), 691-698.
[http://dx.doi.org/10.1002/(SICI)1522-2683(20000301)21:4<691::AID-ELPS691>3.0.CO;2-U] [PMID: 10733208]

[19] Han, F.; Huynh, B.H.; Shi, H.; Lin, B.; Ma, Y. Pteridine analysis in urine by capillary electrophoresis using laser-induced fluorescence detection. *Anal. Chem.*, **1999**, *71*(7), 1265-1269.
[http://dx.doi.org/10.1021/ac981218v] [PMID: 10204031]

[20] Týčová, A.; Ledvina, V.; Klepárník, K. Recent advances in CE-MS coupling: Instrumentation, methodology, and applications. *Electrophoresis*, **2017**, *38*(1), 115-134.
[http://dx.doi.org/10.1002/elps.201600366] [PMID: 27783411]

[21] Simó, C.; Barbas, C.; Cifuentes, A. Capillary electrophoresis-mass spectrometry in food analysis. *Electrophoresis*, **2005**, *26*(7-8), 1306-1318.
[http://dx.doi.org/10.1002/elps.200410108] [PMID: 15761921]

[22] Kopruszinski, C.M.; Reis, R.C.; Bressan, E.; Reeh, P.W.; Chichorro, J.G. Vitamin B complex attenuated heat hyperalgesia following infraorbital nerve constriction in rats and reduced capsaicin *in vivo* and *in vitro* effects. *Eur. J. Pharm.*, **2015**, *762*, 326-332.
[http://dx.doi.org/10.1016/j.ejphar.2015.05.063] [PMID: 26048309]

[23] Kjer-Nielsen, L.; Patel, O.; Corbett, A.J.; Le Nours, J.; Meehan, B.; Liu, L.; Bhati, M.; Chen, Z.; Kostenko, L.; Reantragoon, R.; Williamson, N.A.; Purcell, A.W.; Dudek, N.L.; McConville, M.J.; O'Hair, R.A.; Khairallah, G.N.; Godfrey, D.I.; Fairlie, D.P.; Rossjohn, J.; McCluskey, J. MR1 presents microbial vitamin B metabolites to MAIT cells. *Nature*, **2012**, *491*(7426), 717-723.
[http://dx.doi.org/10.1038/nature11605] [PMID: 23051753]

[24] Patel, O.; Kjer-Nielsen, L.; Le Nours, J.; Eckle, S.B.; Birkinshaw, R.; Beddoe, T.; Corbett, A.J.; Liu, L.; Miles, J.J.; Meehan, B.; Reantragoon, R.; Sandoval-Romero, M.L.; Sullivan, L.C.; Brooks, A.G.; Chen, Z.; Fairlie, D.P.; McCluskey, J.; Rossjohn, J. Recognition of vitamin B metabolites by mucosal-associated invariant T cells. *Nat. Commun.*, **2013**, *4*, 2142.
[http://dx.doi.org/10.1038/ncomms3142] [PMID: 23846752]

[25] Muroyama, K.; Murosaki, S.; Yamamoto, Y.; Ishijima, A.; Toh, Y. Effects of intake of a mixture of

thiamin, arginine, caffeine, and citric acid on adiposity in healthy subjects with high percent body fat. *Biosci. Biotechnol. Biochem.*, **2003**, *67*(11), 2325-2333. [http://dx.doi.org/10.1271/bbb.67.2325] [PMID: 14646190]

[26] Kozik, A.; Korytowski, W.; Sarna, T.; Bloom, A.S. Interactions of flavins with melanin. Studies on equilibrium binding of riboflavin to dopa-melanin and some spectroscopic characteristics of flavin-melanin complex. *Biophys. Chem.*, **1990**, *38*(1-2), 39-48. [http://dx.doi.org/10.1016/0301-4622(90)80038-9] [PMID: 2128193]

[27] Goldberg, A.C. A meta-analysis of randomized controlled studies on the effects of extended-release niacin in women. *Am. J. Cardiol.*, **2004**, *94*(1), 121-124. [http://dx.doi.org/10.1016/j.amjcard.2004.03.042] [PMID: 15219522]

[28] Sonmez, A.; Lurie, D.; Chuong, C.J. Effects of pantothenic acid on postoperative adhesion formation in a rat uterine horn model. *Arch. Gynecol. Obstet.*, **2000**, *263*(4), 164-167. [http://dx.doi.org/10.1007/s004040050274] [PMID: 10834323]

[29] Morrow, L.E.; Grimsley, E.W. Long-term diuretic therapy in hypertensive patients: effects on serum homocysteine, vitamin B6, vitamin B12, and red blood cell folate concentrations. *South. Med. J.*, **1999**, *92*(9), 866-870. [http://dx.doi.org/10.1097/00007611-199909000-00003] [PMID: 10498160]

[30] Báez-Saldaña, A.; Zendejas-Ruiz, I.; Revilla-Monsalve, C.; Islas-Andrade, S.; Cárdenas, A.; Rojas-Ochoa, A.; Vilches, A.; Fernandez-Mejia, C. Effects of biotin on pyruvate carboxylase, acetyl-CoA carboxylase, propionyl-CoA carboxylase, and markers for glucose and lipid homeostasis in type 2 diabetic patients and nondiabetic subjects. *Am. J. Clin. Nutr.*, **2004**, *79*(2), 238-243. [http://dx.doi.org/10.1093/ajcn/79.2.238] [PMID: 14749229]

[31] Clements, F.W. Nutrition 7. Vitamin and mineral supplementation. *Med. J. Aust.*, **1975**, *1*(19), 595-599. [http://dx.doi.org/10.5694/j.1326-5377.1975.tb111601.x] [PMID: 1152710]

[32] Jacobs, P.; Wood, L. Hematology of malnutrition. Part two. *Dis. Mon.*, **2003**, *49*(11), 624-690. [http://dx.doi.org/10.1016/j.disamonth.2003.09.002] [PMID: 14614434]

[33] Castro-Puyana, M.; Crego, A.L.; Marina, M.L.; García-Ruiz, C. CE methods for the determination of non-protein amino acids in foods. *Electrophoresis*, **2007**, *28*(22), 4031-4045. [http://dx.doi.org/10.1002/elps.200700169] [PMID: 17960856]

[34] Zhao, D.; Lu, M.; Cai, Z. Separation and determination of B vitamins and essential amino acids in health drinks by CE-LIF with simultaneous derivatization. *Electrophoresis*, **2012**, *33*(15), 2424-2432. [http://dx.doi.org/10.1002/elps.201200040] [PMID: 22887164]

[35] Tallaksen, C.M.E.; Bøhmer, T.; Karlsen, J.; Bell, H. Determination of thiamin and its phosphate esters in human blood, plasma, and urine. *Methods Enzymol.*, **1997**, *279*(279), 67-74. [http://dx.doi.org/10.1016/S0076-6879(97)79010-4] [PMID: 9211258]

[36] Brunnekreeft, J.W.I.; Eidhof, H.; Gerrits, J. Optimized determination of thiochrome derivatives of thiamine and thiamine phosphates in whole blood by reversed-phase liquid chromatography with precolumn derivatization. *J. Chromatogr. A*, **1989**, *491*(1), 89-96. [http://dx.doi.org/10.1016/S0378-4347(00)82822-4] [PMID: 2793986]

[37] Trenerry, V.C. The application of capillary electrophoresis to the analysis of vitamins in food and beverages. *Electrophoresis*, **2001**, *22*(8), 1468-1478. [http://dx.doi.org/10.1002/1522-2683(200105)22:8<1468::AID-ELPS1468>3.0.CO;2-Q] [PMID: 11386658]

[38] Shabangi, M.; Sutton, J.A. Separation of thiamin and its phosphate esters by capillary zone electrophoresis and its application to the analysis of water-soluble vitamins. *J. Pharm. Biomed. Anal.*, **2005**, *38*(1), 66-71. [http://dx.doi.org/10.1016/j.jpba.2004.11.061] [PMID: 15907621]

[39] Vidal-Valverde, C.; Diaz-Pollán, C. Optimization analysis by capillary electrophoresis of thiamine in meat: comparison with high perfomance liquid chromatography. *Eur. Food Res. Technol.,* **1999**, *209*(5), 355-359.
[http://dx.doi.org/10.1007/s002170050509]

[40] Matysik, F-M.; Marggraf, D.; Gläser, P.; Broekaert, J.A.C. Nonaqueous capillary electrophoresis with indirect electrochemical detection. *Electrophoresis,* **2002**, *23*(21), 3711-3717.
[http://dx.doi.org/10.1002/1522-2683(200211)23:21<3711::AID-ELPS3711>3.0.CO;2-Z] [PMID: 12432533]

[41] Beutner, A.; Scherer, B.; Matysik, F.M. Dual detection for non-aqueous capillary electrophoresis combining contactless conductivity detection and mass spectrometry. *Talanta,* **2018**, *183*, 33-38.
[http://dx.doi.org/10.1016/j.talanta.2018.02.012] [PMID: 29567184]

[42] Gropper, S.A.S.; Smith, J.L.; Groff, J., Eds. , **2013**. *Advanced nutrition and human metabolism.* in:

[43] Cataldi, T.R.I.; Nardiello, D.; Scrano, L.; Scopa, A. Assay of riboflavin in sample wines by capillary zone electrophoresis and laser-induced fluorescence detection. *J. Agric. Food Chem.,* **2002**, *50*(23), 6643-6647.
[http://dx.doi.org/10.1021/jf020212a] [PMID: 12405754]

[44] Catald, T.R.; Nardiello, D.; De Benedetto, G.E.; Bufo, S.A. Optimizing separation conditions for riboflavin, flavin mononucleotide and flavin adenine dinucleotide in capillary zone electrophoresis with laser-induced fluorescence detection. *J. Chromatogr. A,* **2002**, *968*(1-2), 229-239.
[http://dx.doi.org/10.1016/S0021-9673(02)00953-6] [PMID: 12236507]

[45] Yang, X.; Yuan, H.; Wang, C.; Zhao, S.; Xiao, D.; Choi, M.M.F. In-column fiber-optic laser-induced fluorescence detection for CE. *Electrophoresis,* **2007**, *28*(17), 3105-3114.
[http://dx.doi.org/10.1002/elps.200600815] [PMID: 17674418]

[46] Zhao, S.; Yuan, H.; Xiao, D. Optical fiber light-emitting diode-induced fluorescence detection for capillary electrophoresis. *Electrophoresis,* **2006**, *27*(2), 461-467.
[http://dx.doi.org/10.1002/elps.200500300] [PMID: 16358357]

[47] Hu, L.; Yang, X.; Wang, C.; Yuan, H.; Xiao, D. Determination of riboflavin in urine and beverages by capillary electrophoresis with in-column optical fiber laser-induced fluorescence detection. *J. Chromatogr. B Analyt. Technol. Biomed. Life Sci.,* **2007**, *856*(1-2), 245-251.
[http://dx.doi.org/10.1016/j.jchromb.2007.06.011] [PMID: 17625992]

[48] Yu, C.Z.; He, Y.Z.; Xie, H.Y.; Gao, Y.; Gan, W.E.; Li, J. On-line wall-free cell for laser-induced fluorescence detection in capillary electrophoresis. *J. Chromatogr. A,* **2009**, *1216*(20), 4504-4509
[http://dx.doi.org/10.1016/j.chroma.2009.03.038] [PMID: 19329123]

[49] Su, A.K.; Lin, C.H. Determination of riboflavin in urine by capillary electrophoresis-blue light emitting diode-induced fluorescence detection combined with a stacking technique. *J. Chromatogr. B Analyt. Technol. Biomed. Life Sci.,* **2003**, *785*(1), 39-46.
[http://dx.doi.org/10.1016/S1570-0232(02)00849-8] [PMID: 12535836]

[50] Chang, Y-S.; Shih, C-M.; Lin, C-H. UV light-emitting diode-induced fluorescence detection combined with online sample concentration techniques for capillary electrophoresis. *Anal. Sci.,* **2006**, *22*(2), 235-240.
[http://dx.doi.org/10.2116/analsci.22.235] [PMID: 16512414]

[51] Su, A.K.; Chang, Y.S.; Lin, C.H. Analysis of riboflavin in beer by capillary electrophoresis/blue light emitting diode (LED)-induced fluorescence detection combined with a dynamic pH junction technique. *Talanta,* **2004**, *64*(4), 970-974.
[http://dx.doi.org/10.1016/j.talanta.2004.04.011] [PMID: 18969698]

[52] Liu, S.; Ren, H.; Gao, Q.; Roach, D.J.; Jr, R.T.L.; Armstrong, T.M.; Mao, Q.; Blaga, I.; Barker, D.L.; Jovanovich, S.B. Parallel DNA Sequencing on Microfabricated Electrophoresis Chips. *Proc. Natl. Acad. Sci. USA,* **2000**, *97*(10), 5369-5374.

[http://dx.doi.org/10.1073/pnas.100113197] [PMID: 10792056]

[53] Tycova, A.; Gerhardt, R.F.; Belder, D. Surface enhanced Raman spectroscopy in microchip electrophoresis. *J. Chromatogr. A,* **2018**, *1541*, 39-46. [http://dx.doi.org/10.1016/j.chroma.2018.02.014] [PMID: 29433820]

[54] Prodanov, M.; Sierra, I.; Vidal-Valverde, C. Effect of germination on the thiamine, riboflavin and niacin contents in legumes *Zeitschrift für Lebensmitteluntersuchung und -Forschung A,* **1997**, *205*(1), 48-52.

[55] Ward, C.M.; Trenerry, V.C.; Pant, I. The application of capillary electrophoresis to the determination of total niacin in concentrated yeast spreads. *Food Chem.,* **1997**, *58*(3), 185-192. [http://dx.doi.org/10.1016/S0308-8146(96)00241-5]

[56] Ward, C.M.; Trenerry, V.C. The determination of niacin in cereals, meat and selected foods by capillary electrophoresis and high performance liquid chromatography. *Food Chem.,* **1997**, *60*(4), 667-674. [http://dx.doi.org/10.1016/S0308-8146(97)00022-8]

[57] Windahl, K.L.; Trenerry, V.C.; Ward, C.M. The determination of niacin in selected foods by capillary electrophoresis and high performance liquid chromatography: acid extraction. *Food Chem.,* **1999**, *65*(2), 263-270. [http://dx.doi.org/10.1016/S0308-8146(98)00225-8]

[58] Vidal-Valverde, C.; Sierra, I.; Díaz-Pollán, C.; Blázquez, I. Determination by capillary electrophoresis of total and available niacin in different development stage of raw and processed legumes: comparison with high-performance liquid chromatography. *Electrophoresis,* **2001**, *22*(8), 1479-1483. [http://dx.doi.org/10.1002/1522-2683(200105)22:8<1479::AID-ELPS1479>3.0.CO;2-I] [PMID: 11386659]

[59] Lu, Y.; Wu, C.; Yuan, Z. Determination of hesperetin, cinnamic acid and nicotinic acid in propolis with micellar electrokinetic capillary chromatography. *Fitoterapia,* **2004**, *75*(3-4), 267-276. [http://dx.doi.org/10.1016/j.fitote.2003.12.026] [PMID: 15158983]

[60] Wu, M.; Gao, F.; Zhang, Y.; Wang, Q.; Li, H. Sensitive analysis of amino acids and vitamin B3 in functional drinks *via* field-amplified stacking with reversed-field stacking in microchip electrophoresis. *Talanta,* **2015**, *131*, 624-631. [http://dx.doi.org/10.1016/j.talanta.2014.08.051] [PMID: 25281150]

[61] Buttriss, J. Strategies designed to increase awareness about folates and health, and to increase folate intake: A review. *Trends Food Sci. Technol.,* **2005**, *16*(6), 246-252. [http://dx.doi.org/10.1016/j.tifs.2005.03.012]

[62] Pawlosky, R.J.; Hertrampf, E.; Flanagan, V.P.; Thomas, P.M. Mass spectral determinations of the folic acid content of fortified breads from Chile. *J. Food Compos. Anal.,* **2003**, *16*(3), 281-286. [http://dx.doi.org/10.1016/S0889-1575(03)00033-4]

[63] Dong, F.M.; Oace, S.M. Folate distribution in fruit juices. *J. Am. Diet. Assoc.,* **1973**, *62*(2), 162-166. [http://dx.doi.org/10.1016/S0002-8223(21)08797-6] [PMID: 4631389]

[64] Matella, N.J.; Braddock, R.J.; Gregory, J.F., III; Goodrich, R.M. Capillary electrophoresis and high-performance liquid chromatography determination of polyglutamyl 5-methyltetrahydrofolate forms in citrus products. *J. Agric. Food Chem.,* **2005**, *53*(6), 2268-2274. [http://dx.doi.org/10.1021/jf048285c] [PMID: 15769167]

[65] Zhao, S.; Yuan, H.; Xie, C.; Xiao, D. Determination of folic acid by capillary electrophoresis with chemiluminescence detection. *J. Chromatogr. A,* **2006**, *1107*(1-2), 290-293. [http://dx.doi.org/10.1016/j.chroma.2005.11.052] [PMID: 16413558]

[66] Chen, J-H.; Jiang, S-J. Determination of cobalamin in nutritive supplements and chlorella foods by capillary electrophoresis-inductively coupled plasma mass spectrometry. *J. Agric. Food Chem.,* **2008**, *56*(4), 1210-1215.

[http://dx.doi.org/10.1021/jf073213h] [PMID: 18247530]

[67] Schreiner; Razzazi, M.; Luf, E., Determination of water-soluble vitamins in soft drinks and vitamin supplements using capillary electrophoresis. *Mol. Nutr. Food Res.*, **2003**, *47*(4), 243-247.

[68] Okamoto, H.; Nakajima, T.; Ito, Y. Simultaneous determination of water-soluble vitamins in a vitamin-enriched drink by an in-capillary enzyme reaction method. *J. Chromatogr. A*, **2003**, *986*(1), 153-161.
[http://dx.doi.org/10.1016/S0021-9673(02)01916-7] [PMID: 12585333]

[69] Maráková, K.; Piestanský, J.; Havránek, E.; Mikus, P. Simultaneous analysis of vitamins B in pharmaceuticals and dietary supplements by capillary electrophoresis hyphenated with triple quadrupole mass spectrometry. *Pharmazie*, **2014**, *69*(9), 663-668.
[PMID: 25272936]

[70] Scheffler, J.; Bork, K.; Bezold, V.; Rosenstock, P.; Gnanapragassam, V.S.; Horstkorte, R. Ascorbic acid leads to glycation and interferes with neurite outgrowth. *Exp. Gerontol.*, **2019**, *117*, 25-30.
[http://dx.doi.org/10.1016/j.exger.2018.08.005] [PMID: 30149062]

[71] Bazel, Y.; Riabukhina, T.; Tirpak, J. Spectrophotometric determination of ascorbic acid in foods with the use of vortex-assisted liquid-liquid microextraction. *Microchem. J.*, **2018**, *143*, 160-165.
[http://dx.doi.org/10.1016/j.microc.2018.08.003]

[72] Block, G. Vitamin C and cancer prevention: the epidemiologic evidence. *Am. J. Clin. Nutr.*, **1991**, *53*(1) Suppl., 270S-282S.
[http://dx.doi.org/10.1093/ajcn/53.1.270S] [PMID: 1985398]

[73] Holst, A.; Frölich, T. Experimental Studies Relating to "Ship-beri-beri" and Scurvy. *J. Hyg. (Lond.)*, **1907**, *7*(5), 634-671.
[PMID: 20474337]

[74] Duran, B.O.S.; Góes, G.A.; Zanella, B.T.T.; Freire, P.P.; Valente, J.S.; Salomão, R.A.S.; Fernandes, A.; Mareco, E.A.; Carvalho, R.F.; Dal-Pai-Silva, M. Ascorbic acid stimulates the *in vitro* myoblast proliferation and migration of pacu (Piaractus mesopotamicus). *Sci. Rep.*, **2019**, *9*(1), 2229.
[http://dx.doi.org/10.1038/s41598-019-38536-4] [PMID: 30778153]

[75] Pohle, T.; Brzozowski, T.; Becker, J.C.; Van der Voort, I.R.; Markmann, A.; Konturek, S.J.; Moniczewski, A.; Domschke, W.; Konturek, J.W. Role of reactive oxygen metabolites in aspirin-induced gastric damage in humans: gastroprotection by vitamin C. *Aliment. Pharmacol. Ther.*, **2001**, *15*(5), 677-687.
[http://dx.doi.org/10.1046/j.1365-2036.2001.00975.x] [PMID: 11328262]

[76] Neuberger, S.; Jooß, K.; Flottmann, D.; Scriba, G.; Neusüß, C. Raman spectroscopy and capillary zone electrophoresis for the analysis of degradation processes in commercial effervescent tablets containing acetylsalicylic acid and ascorbic acid. *J. Pharm. Biomed. Anal.*, **2017**, *134*, 122-129.
[http://dx.doi.org/10.1016/j.jpba.2016.11.020] [PMID: 27907854]

[77] Cheng, H.; Li, L.; Zhang, M.; Jiang, Y.; Yu, P.; Ma, F.; Mao, L. Recent advances on *in vivo* analysis of ascorbic acid in brain functions. *TrAC. Trends Analyt. Chem.*, **2018**, *109*, 247-259.
[http://dx.doi.org/10.1016/j.trac.2018.10.017]

[78] Matei, N.; Birghila, S.; Popescu, V.; Dobrinas, S.; Soceanu, A.; Oprea, C.; Magearu, V. Kinetic study of vitamin C degradation from pharmaceutical products. *Rom. J. Phys.*, **2008**, *53*(1-2), 343-351.

[79] Padayatty, S.J.; Levine, M. New insights into the physiology and pharmacology of vitamin C. *CMAJ*, **2001**, *164*(3), 353-355.
[PMID: 11232136]

[80] Hathcock, J.N.; Azzi, A.; Blumberg, J.; Bray, T.; Dickinson, A.; Frei, B.; Jialal, I.; Johnston, C.S.; Kelly, F.J.; Kraemer, K.; Packer, L.; Parthasarathy, S.; Sies, H.; Traber, M.G. Vitamins E and C are safe across a broad range of intakes. *Am. J. Clin. Nutr.*, **2005**, *81*(4), 736-745.
[http://dx.doi.org/10.1093/ajcn/81.4.736] [PMID: 15817846]

[81] Boonpangrak, S.; Lalitmanat, S.; Suwanwong, Y.; Prachayasittikul, S.; Prachayasittikul, V. Analysis of Ascorbic Acid and Isoascorbic Acid in Orange and Guava Fruit Juices Distributed in Thailand by LC-IT-MS/MS. *Food Anal. Methods,* **2016**, *9*(6), 1616-1626. [http://dx.doi.org/10.1007/s12161-015-0337-x]

[82] Finholt, P.; Paulssen, R.B.; Higuchi, T. Rate of anaerobic degradation of ascorbic acid in aqueous solution. *J. Pharm. Sci.,* **1963**, *52*(10), 948-954. [http://dx.doi.org/10.1002/jps.2600521008] [PMID: 14076502]

[83] Liao, T.; Jiang, C.M.; Wu, M.C.; Hwang, J.Y.; Chang, H.M. Quantification of L-ascorbic acid and total ascorbic acid in fruits and spinach by capillary zone electrophoresis. *Electrophoresis,* **2001**, *22*(8), 1484-1488. [http://dx.doi.org/10.1002/1522-2683(200105)22:8<1484::AID-ELPS1484>3.0.CO;2-9] [PMID: 11386660]

[84] Albishri, H.M.; Almalawi, A.M.; Alshitari, W.; Abd El-Hady, D. Use of β-cyclodextrin inclusion concurrent with cationic surfactant shielding for the enhancement of ascorbic acid stability followed by ultra-high performance liquid chromatography and online preconcentration capillary electrophoresis. *J. Liq. Chromatogr. Relat. Technol.,* **2018**, *41*(11), 732-739. [http://dx.doi.org/10.1080/10826076.2018.1511996]

[85] Wang, X.; Li, K.; Yao, L.; Wang, C.; Van Schepdael, A. Recent advances in vitamins analysis by capillary electrophoresis. *J. Pharm. Biomed. Anal.,* **2018**, *147*, 278-287. [http://dx.doi.org/10.1016/j.jpba.2017.07.030] [PMID: 28784252]

[86] Fischer, D.J.; Hulvey, M.K.; Regel, A.R.; Lunte, S.M. Amperometric detection in microchip electrophoresis devices: effect of electrode material and alignment on analytical performance. *Electrophoresis,* **2009**, *30*(19), 3324-3333. [http://dx.doi.org/10.1002/elps.200900317] [PMID: 19802847]

[87] Holland, L.A.; Leigh, A.M. Amperometric and voltammetric detection for capillary electrophoresis. *Electrophoresis,* **2002**, *23*(21), 3649-3658. [http://dx.doi.org/10.1002/1522-2683(200211)23:21<3649::AID-ELPS3649>3.0.CO;2-O] [PMID: 12432525]

[88] Neuberger, S.; Jooß, K.; Ressel, C.; Neusüß, C. Quantification of ascorbic acid and acetylsalicylic acid in effervescent tablets by CZE-UV and identification of related degradation products by heart-cut CZE-CZE-MS. *Anal. Bioanal. Chem.,* **2016**, *408*(30), 8701-8712. [http://dx.doi.org/10.1007/s00216-016-9734-2] [PMID: 27405873]

[89] Fabianek, J.; Herp, A. Antiscorbutic activity of D-araboascorbic acid. *Proceedings of the Society for Experimental Biology and Medicine,* **1967**, , pp. 462-5.

[90] Versari, A.; Mattioli, A.; Parpinello, G.P.; Galassi, S. Rapid analysis of ascorbic and isoascorbic acids in fruit juice by capillary electrophoresis. *Food Control,* **2004**, *15*(5), 355-358. [http://dx.doi.org/10.1016/S0956-7135(03)00097-5]

[91] Garnero, C.; Longhi, M. Development of HPLC and UV spectrophotometric methods for the determination of ascorbic acid using hydroxypropyl-beta-cyclodextrin and triethanolamine as photostabilizing agents. *Anal. Chim. Acta,* **2010**, *659*(1-2), 159-166. [http://dx.doi.org/10.1016/j.aca.2009.11.038] [PMID: 20103119]

[92] Johnston, C.S.; Bowling, D.L. Stability of ascorbic acid in commercially available orange juices. *J. Am. Diet. Assoc.,* **2002**, *102*(4), 525-529. [http://dx.doi.org/10.1016/S0002-8223(02)90119-7] [PMID: 11985409]

[93] Garnero, C.; Longhi, M. Study of ascorbic acid interaction with hydroxypropyl-beta-cyclodextrin and triethanolamine, separately and in combination. *J. Pharm. Biomed. Anal.,* **2007**, *45*(4), 536-545. [http://dx.doi.org/10.1016/j.jpba.2007.07.030] [PMID: 17851014]

[94] Pavlovic, D.M.; Babic, S.; Horvat, A.J.M.; Kastelan-Macan, M. Sample preparation in analysis of

pharmaceuticals. *TrAC. Trends Analyt. Chem.*, **2007**, *26*(11), 1062-1075. [http://dx.doi.org/10.1016/j.trac.2007.09.010]

[95] Falkova, M.T.; Bulatov, A.V.; Pushina, M.O.; Ekimov, A.A.; Alekseeva, G.M.; Moskvin, L.N. Multicommutated stepwise injection determination of ascorbic acid in medicinal plants and food samples by capillary zone electrophoresis ultraviolet detection. *Talanta,* **2015**, *133*, 82-87. [http://dx.doi.org/10.1016/j.talanta.2014.04.092] [PMID: 25435231]

[96] Wu, T.; Guan, Y.; Ye, J. Determination of flavonoids and ascorbic acid in grapefruit peel and juice by capillary electrophoresis with electrochemical detection. *Food Chem.,* **2007**, *100*(4), 1573-1579. [http://dx.doi.org/10.1016/j.foodchem.2005.12.042]

[97] Voegel, P.D.; Zhou, W.; Baldwin, R.P. Integrated capillary electrophoresis/electrochemical detection with metal film electrodes directly deposited onto the capillary tip. *Anal. Chem.,* **1997**, *69*(5), 951-957. [http://dx.doi.org/10.1021/ac960748t] [PMID: 21639232]

[98] Dong, S.; Zhang, S.; Cheng, X.; He, P.; Wang, Q.; Fang, Y. Simultaneous determination of sugars and ascorbic acid by capillary zone electrophoresis with amperometric detection at a carbon paste electrode modified with polyethylene glycol and Cu(2)O. *J. Chromatogr. A,* **2007**, *1161*(1-2), 327 - 333. [http://dx.doi.org/10.1016/j.chroma.2007.05.077] [PMID: 17574259]

[99] Yang, Z.; Li, Z.; Zhu, J.; Wang, Q.; He, P.; Fang, Y. Use of different buffers for detection and separation in determination of physio-active components in oolong tea infusion by CZE with amperometric detection. *J. Sep. Sci.,* **2010**, *33*(9), 1312-1318. [http://dx.doi.org/10.1002/jssc.200900586] [PMID: 20309896]

[100] Jacobson, S.C.; Culbertson, C.T.; Daler, J.E.; Ramsey, J.M. Microchip structures for submillisecond electrophoresis. *Anal. Chem.,* **1998**, *70*(16), 3476-3480. [http://dx.doi.org/10.1021/ac980349t]

[101] Xu, Z.; Oleschuk, R.D. A fluorous porous polymer monolith photo-patterned chromatographic column for the separation of a flourous/fluorescently labeled peptide within a microchip. *Electrophoresis,* **2014**, *35*(2-3), 441-449. [http://dx.doi.org/10.1002/elps.201300365] [PMID: 24170603]

[102] Pessoa de Santana, P.; Segato, T.P.; Carrilho, E.; Lima, R.S.; Dossi, N.; Kamogawa, M.Y.; Gobbi, A.L.; Piazzeta, M.H.; Piccin, E. Fabrication of glass microchannels by xurography for electrophoresis applications. *Analyst (Lond.),* **2013**, *138*(6), 1660-1664. [http://dx.doi.org/10.1039/c3an36540a] [PMID: 23392529]

[103] Li, R.; Wang, L.; Gao, X.; Du, G.; Zhai, H.; Wang, X.; Guo, G.; Pu, Q. Rapid separation and sensitive determination of banned aromatic amines with plastic microchip electrophoresis. *J. Hazard. Mater.,* **2013**, *248-249*, 268-275. [http://dx.doi.org/10.1016/j.jhazmat.2013.01.010] [PMID: 23385207]

[104] Coltro, W.K.; Lunte, S.M.; Carrilho, E. Comparison of the analytical performance of electrophoresis microchannels fabricated in PDMS, glass, and polyester-toner. *Electrophoresis,* **2008**, *29*(24), 4928-4937. [http://dx.doi.org/10.1002/elps.200700897] [PMID: 19025869]

[105] Yap, Y.C.; Guijt, R.M.; Dickson, T.C.; King, A.E.; Breadmore, M.C. Stainless steel pinholes for fast fabrication of high-performance microchip electrophoresis devices by CO2 laser ablation. *Anal. Chem.,* **2013**, *85*(21), 10051-10056. [http://dx.doi.org/10.1021/ac402631g] [PMID: 24063252]

[106] Vandaveer, W.R., IV; Pasas, S.A.; Martin, R.S.; Lunte, S.M. Recent developments in amperometric detection for microchip capillary electrophoresis. *Electrophoresis,* **2002**, *23*(21), 3667-3677. [http://dx.doi.org/10.1002/1522-2683(200211)23:21<3667::AID-ELPS3667>3.0.CO;2-W] [PMID: 12432527]

[107] Wang, J. Electrochemical detection for microscale analytical systems: a review. *Talanta,* **2002**, *56*(2),

223-231.
[http://dx.doi.org/10.1016/S0039-9140(01)00592-6] [PMID: 18968498]

[108] Lucca, B.G.; Lunte, S.M.; Tomazelli Coltro, W.K.; Ferreira, V.S. Separation of natural antioxidants using PDMS electrophoresis microchips coupled with amperometric detection and reverse polarity. *Electrophoresis,* **2014**, *35*(23), 3363-3370.
[http://dx.doi.org/10.1002/elps.201400359] [PMID: 25224541]

[109] Lucca, B.G.; de Lima, F.; Coltro, W.K.; Ferreira, V.S. Electrodeposition of reduced graphene oxide on a Pt electrode and its use as amperometric sensor in microchip electrophoresis. *Electrophoresis,* **2015**, *36*(16), 1886-1893.
[http://dx.doi.org/10.1002/elps.201500092] [PMID: 25884327]

[110] Klampfl, C.W. Determination of organic acids by CE and CEC methods. *Electrophoresis,* **2007**, *28*(19), 3362-3378.
[http://dx.doi.org/10.1002/elps.200700159] [PMID: 17806127]

[111] Solínová, V.; Kasicka, V. Recent applications of conductivity detection in capillary and chip electrophoresis. *J. Sep. Sci.,* **2006**, *29*(12), 1743-1762.
[http://dx.doi.org/10.1002/jssc.200600167] [PMID: 16970183]

[112] Law, W.S.; Kubán, P.; Zhao, J.H.; Li, S.F.; Hauser, P.C. Determination of vitamin C and preservatives in beverages by conventional capillary electrophoresis and microchip electrophoresis with capacitively coupled contactless conductivity detection. *Electrophoresis,* **2005**, *26*(24), 4648-4655.
[http://dx.doi.org/10.1002/elps.200500437] [PMID: 16278922]

[113] Aucamp, J.P.; Hara, Y.; Apostolides, Z. Simultaneous analysis of tea catechins, caffeine, gallic acid, theanine and ascorbic acid by micellar electrokinetic capillary chromatography. *J. Chromatogr. A,* **2000**, *876*(1-2), 235-242.
[http://dx.doi.org/10.1016/S0021-9673(00)00145-X] [PMID: 10823519]

[114] Breadmore, M.C.; Hilder, E.F.; Macka, M.; Haddad, P.R. Determination of inorganic anions by capillary electrochromatography. *TrAC. Trends Analyt. Chem.,* **2001**, *20*(6-7), 355-364.
[http://dx.doi.org/10.1016/S0165-9936(01)00073-5]

[115] Liu, F.J.; Ding, G.S.; Tang, A.N. Simultaneous separation and determination of five organic acids in beverages and fruits by capillary electrophoresis using diamino moiety functionalized silica nanoparticles as pseudostationary phase. *Food Chem.,* **2014**, *145*, 109-114.
[http://dx.doi.org/10.1016/j.foodchem.2013.08.035] [PMID: 24128456]

[116] Serni, E.; Audino, V.; Del Carlo, S.; Manera, C.; Saccomanni, G.; Macchia, M. Determination of water-soluble vitamins in multivitamin dietary supplements and in artichokes by micellar electrokinetic chromatography. *Nat. Prod. Res.,* **2013**, *27*(23), 2212-2215.
[http://dx.doi.org/10.1080/14786419.2013.800977] [PMID: 23731489]

[117] Navarro-Pascual-Ahuir, M.; Lerma-García, M.J.; Simó-Alfonso, E.F.; Herrero-Martínez, J.M. Determination of water-soluble vitamins in energy and sport drinks by micellar electrokinetic capillary chromatography. *Food Control,* **2016**, *63*, 110-116.
[http://dx.doi.org/10.1016/j.foodcont.2015.11.033]

[118] Crevillén, A.G.; Blasco, A.J.; González, M.C.; Escarpa, A. A fast and reliable route integrating calibration and analysis protocols for water-soluble vitamin determination on microchip-electrochemistry platforms. *Electrophoresis,* **2006**, *27*(24), 5110-5118.
[http://dx.doi.org/10.1002/elps.200600213] [PMID: 17161002]

[119] Crevillén, A.G.; Pumera, M.; González, M.C.; Escarpa, A. Carbon nanotube disposable detectors in microchip capillary electrophoresis for water-soluble vitamin determination: analytical possibilities in pharmaceutical quality control. *Electrophoresis,* **2008**, *29*(14), 2997-3004.
[http://dx.doi.org/10.1002/elps.200700947] [PMID: 18576364]

[120] Jia, L.; Liu, Y.; Du, Y.; Xing, D. Pressurized capillary electrochromatographic analysis of water-soluble vitamins by combining with on-line concentration technique. *J. Chromatogr. A,* **2007**,

1154(1-2), 416-422.
[http://dx.doi.org/10.1016/j.chroma.2007.03.080] [PMID: 17442323]

[121] Kamal-Eldin, A.; Appelqvist, L.A. The chemistry and antioxidant properties of tocopherols and tocotrienols. *Lipids,* **1996**, *31*(7), 671-701.
[http://dx.doi.org/10.1007/BF02522884] [PMID: 8827691]

[122] Palozza, P.; Krinsky, N.I. The inhibition of radical-initiated peroxidation of microsomal lipids by both α-tocopherol and β-carotene. *Free Radic. Biol. Med.,* **1991**, *11*(4), 407-414.
[http://dx.doi.org/10.1016/0891-5849(91)90158-Y] [PMID: 1797626]

[123] Hidiroglou, M.; Farnworth, E.; Butler, G. Effects of vitamin E and fat supplementation on concentration of vitamin E in plasma and milk of sows and in plasma of piglets. *Int. J. Vitam. Nutr. Res.,* **1993**, *63*(3), 180-187.
[PMID: 8300328]

[124] Lampi, A.M.; Nurmi, T.; Ollilainen, V.; Piironen, V. Tocopherols and tocotrienols in wheat genotypes in the HEALTHGRAIN Diversity Screen. *J. Agric. Food Chem.,* **2008**, *56*(21), 9716-9721.
[http://dx.doi.org/10.1021/jf801092a] [PMID: 18921970]

[125] Sun, C.; Li, Y.; Yi, Y.; Dan, L. Analysis of Tocopherols and Tocotrienols in Pharmaceuticals and Foods: A Critical Review. *Curr. Pharm. Anal.,* **2015**, *11*(1), 66-78.

[126] Sandra, P.; Dermaux, A.; Ferraz, V.; Dittmann, M.M.; Rozing, G. Analysis of triglycerides by capillary electrochromatography. *J. Microcolumn Sep.,* **2015**, *9*(5), 409-419.
[http://dx.doi.org/10.1002/(SICI)1520-667X(1997)9:5<409::AID-MCS7>3.0.CO;2-5]

[127] Desiderio, C.; Fanali, S. Capillary electrochromatography and capillary electrochromatography-electrospray mass spectrometry for the separation of non-steroidal anti-inflammatory drugs. *J. Chromatogr. A,* **2000**, *895*(1-2), 123-132.
[http://dx.doi.org/10.1016/S0021-9673(00)00658-0] [PMID: 11105854]

[128] Fanali, S.; Catarcini, P.; Quaglia, M.G.; Camera, E.; Rinaldi, M.; Picardo, M. Separation of δ-, γ- and α-tocopherols by CEC. *J. Pharm. Biomed. Anal.,* **2002**, *29*(6), 973-979.
[http://dx.doi.org/10.1016/S0731-7085(02)00137-1] [PMID: 12110381]

[129] Slover, H.T. Tocopherols in foods and fats. *Lipids,* **1971**, *6*(5), 291-296.
[http://dx.doi.org/10.1007/BF02531818]

[130] Aturki, Z.; D'Orazio, G.; Fanali, S. Rapid assay of vitamin E in vegetable oils by reversed-phase capillary electrochromatography. *Electrophoresis,* **2005**, *26*(4-5), 798-803.
[http://dx.doi.org/10.1002/elps.200410181] [PMID: 15669010]

[131] Gliszczyńska-Swigło, A.; Sikorska, E. Simple reversed-phase liquid chromatography method for determination of tocopherols in edible plant oils. *J. Chromatogr. A,* **2004**, *1048*(2), 195-198.
[http://dx.doi.org/10.1016/j.chroma.2004.07.051] [PMID: 15481256]

[132] Henry, C.W., III; Fortier, C.A.; Warner, I.M. Separation of tocopherol isomers using capillary electrochromatography: comparison of monomeric and polymeric C30 stationary phases. *Anal. Chem.,* **2001**, *73*(24), 6077-6082.
[http://dx.doi.org/10.1021/ac011018a] [PMID: 11791583]

[133] Abidi, S.L.; Rennick, K.A. Capillary electrochromatographic evaluation of vitamin E-active oil constituents: tocopherols and tocotrienols. *J. Chromatogr. A,* **2001**, *913*(1-2), 379-386.
[http://dx.doi.org/10.1016/S0021-9673(00)01068-2] [PMID: 11355836]

[134] Abidi, S.L.; Mounts, T.L. Reversed-phase high-performance liquid chromatographic separations of tocopherols. *J. Chromatogr. A,* **1997**, *782*(1), 25-32.
[http://dx.doi.org/10.1016/S0021-9673(97)00471-8]

[135] Sander, L.C.; Wise, S.A. Effect of phase length on column selectivity for the separation of polycyclic aromatic hydrocarbons by reversed-phase liquid chromatography. *Anal. Chem.,* **1987**, *59*(18), 2309-2313.

[http://dx.doi.org/10.1021/ac00145a020] [PMID: 3674444]

[136] Bell, C.M.; Sander, L.C.; Fetzer, J.C.; Wise, S.A. Synthesis and characterization of extended length alkyl stationary phases for liquid chromatography with application to the separation of carotenoid isomers. *J. Chromatogr. A,* **1996**, *753*(1), 37-45.
[http://dx.doi.org/10.1016/S0021-9673(96)00562-6] [PMID: 8997743]

[137] Sander, L.C.; Lippa, K.A.; Wise, S.A. Order and disorder in alkyl stationary phases. *Anal. Bioanal. Chem.,* **2005**, *382*(3), 646-668.
[http://dx.doi.org/10.1007/s00216-005-3127-2] [PMID: 15827723]

[138] Carabias-Martínez, R.; Rodríguez-Gonzalo, E.; Smith, N.W.; Ruano-Miguel, L. Use of a polar-embedded stationary phase for the separation of tocopherols by CEC. *Electrophoresis,* **2006**, *27*(22), 4423-4430.
[http://dx.doi.org/10.1002/elps.200600223] [PMID: 17058307]

[139] Svec, F.; Fréchet, J.M.J. New designs of macroporous polymers and supports: from separation to biocatalysis. *Science,* **1996**, *273*(5272), 205-211.
[http://dx.doi.org/10.1126/science.273.5272.205] [PMID: 8662498]

[140] Peters, E.C.; Petro, M.; Svec, F.; Fréchet, J.M.J. Molded rigid polymer monoliths as separation media for capillary electrochromatography. 1. Fine control of porous properties and surface chemistry. *Anal. Chem.,* **1998**, *70*(11), 2288-2295.
[http://dx.doi.org/10.1021/ac9713518] [PMID: 9624900]

[141] Peters, E.C.; Petro, M.; Svec, F.; Fréchet, J.M.J. Molded rigid polymer monoliths as separation media for capillary electrochromatography. *Anal. Chem.,* **1997**, *69*(17), 3646-3649.
[http://dx.doi.org/10.1021/ac970377w] [PMID: 9286168]

[142] Lerma-García, M.J.; Simó-Alfonso, E.F.; Ramis-Ramos, G.; Herrero-Martínez, J.M. Determination of tocopherols in vegetable oils by CEC using methacrylate ester-based monolithic columns. *Electrophoresis,* **2007**, *28*(22), 4128-4135.
[http://dx.doi.org/10.1002/elps.200700195] [PMID: 17948270]

[143] Okanda, F.M.; El Rassi, Z. Capillary electrochromatography with monolithic stationary phases. 4. Preparation of neutral stearyl-acrylate monoliths and their evaluation in capillary electrochromatography of neutral and charged small species as well as peptides and proteins. *Electrophoresis,* **2005**, *26*(10), 1988-1995.
[http://dx.doi.org/10.1002/elps.200500073] [PMID: 15841500]

[144] Chaisuwan, P.; Nacapricha, D.; Wilairat, P.; Jiang, Z.; Smith, N.W. Separation of alpha-, beta-, gamma-, delta-tocopherols and alpha-tocopherol acetate on a pentaerythritol diacrylate monostearate-ethylene dimethacrylate monolith by capillary electrochromatography. *Electrophoresis,* **2008**, *29*(11), 2301-2309.
[http://dx.doi.org/10.1002/elps.200700689] [PMID: 18548461]

[145] Benson, L.M.; Tomlinson, A.J.; Reid, J.M.; Walker, D.L.; Ames, M.M.; Naylor, S. Study of *in vivo* pyrazoloacridine metabolism by capillary electrophoresis using isotachophoresis preconcentration in non□aqueous separation buffer. *J. Sep. Sci.,* **1993**, *16*(5), 324-326.

[146] Tomlinson, A.J.; Benson, L.M.; Naylor, S. Nonaqueous solvents in the on-line capillary electrophoresis-mass spectrometry analysis of drug metabolites. *J. Sep. Sci.,* **1994**, *17*(3), 175-177.

[147] Mendonça, C.R.B.; Bica, C.I.D.; Piatnicki, C.M.S.; Simó-Alfonso, E.F.; Ramis-Ramos, G. Characterization of hydroxyaromatic compounds in vegetable oils by capillary electrophoresis with direct injection in an oil-miscible KOH/propanol/methanol medium. *Electrophoresis,* **2005**, *26*(17), 3307-3314.
[http://dx.doi.org/10.1002/elps.200410394] [PMID: 16097022]

[148] Galeano-Diaz, T.; Isabel Acedo-Valenzuela, M.; Silva-Rodriguez, A. Determination of tocopherols in vegetable oil samples by non-aqueous capillary electrophoresis (NACE) with fluorimetric detection. *J. Food Compos. Anal.,* **2012**, *25*(1), 24-30.

[http://dx.doi.org/10.1016/j.jfca.2011.04.014]

[149] Borel, P.; Caillaud, D.; Cano, N.J.; Vitamin, D. Vitamin D bioavailability: state of the art. *Crit. Rev. Food Sci. Nutr.*, **2015**, *55*(9), 1193-1205.
[http://dx.doi.org/10.1080/10408398.2012.688897] [PMID: 24915331]

[150] Wimalawansa, S.J. Vitamin D in the new millennium. *Curr. Osteoporos. Rep.*, **2012**, *10*(1), 4-15.
[http://dx.doi.org/10.1007/s11914-011-0094-8] [PMID: 22249582]

[151] Yamada, H.; Kitagawa, S.; Ohtani, H. Simultaneous separation of water- and fat-soluble vitamins in isocratic pressure-assisted capillary electrochromatography using a methacrylate-based monolithic column. *J. Sep. Sci.*, **2013**, *36*(12), 1980-1985.
[http://dx.doi.org/10.1002/jssc.201201191] [PMID: 23559554]

[152] Wright, P.B.; Dorsey, J.G. Silver(I)-mediated separations by capillary zone electrophoresis and micellar electrokinetic chromatography: argentation electrophoresis. *Anal. Chem.*, **1996**, *68*(3), 415-424.
[http://dx.doi.org/10.1021/ac950802q] [PMID: 8712354]

[153] Eskild, W.; Troen, G.; Blaner, W.S.; Nilsson, A.; Hansson, V. Evidence for independent control at the mRNA and protein levels of cellular retinol binding protein 1 in rat Sertoli cells. *J. Reprod. Fertil.*, **2000**, *119*(1), 101-109.
[http://dx.doi.org/10.1530/jrf.0.1190101] [PMID: 10864819]

[154] Shi, H.; Ma, Y.; Humphrey, J.H.; Craft, N.E. Determination of vitamin A in dried human blood spots by high-performance capillary electrophoresis with laser-excited fluorescence detection. *J. Chromatogr. B Biomed. Appl.*, **1995**, *665*(1), 89-96.
[http://dx.doi.org/10.1016/0378-4347(94)00520-F] [PMID: 7795805]

[155] Reddy, V. Control of vitamin A deficiency and blindness. *Acta Paediatr. Scand. Suppl.*, **1991**, *374*(s374), 30-37.
[http://dx.doi.org/10.1111/j.1651-2227.1991.tb12004.x] [PMID: 1957625]

[156] Profumo, A.; Profumo, V.; Vidali, G. Micellar electrokinetic capillary chromatography of three natural vitamin A derivatives. *Electrophoresis,* **1996**, *17*(10), 1617-1621.
[http://dx.doi.org/10.1002/elps.1150171022] [PMID: 8957192]

[157] Pedersen-Bjergaard, S.; Rasmussen, K.E.; Tilander, T. Separation of fat-soluble vitamins by hydrophobic interaction electrokinetic chromatography with tetradecylammonium ions as pseudostationary phase. *J. Chromatogr. A,* **1998**, *807*(2), 285-295.
[http://dx.doi.org/10.1016/S0021-9673(98)00087-9]

[158] Naess, O.; Tilander, T.; Pedersen-Bjergaard, S.; Rasmussen, K.E. Analysis of vitamin formulations by electrokinetic chromatography utilizing tetradecylammonium ions as the pseudostationary phase. *Electrophoresis,* **1998**, *19*(16-17), 2912-2917.
[http://dx.doi.org/10.1002/elps.1150191619] [PMID: 9870388]

[159] Le, T.H.; Kraak, J.C.; Kok, W.T. Application of micellar electrokinetic chromatography for the separation of retinoids. *J. Pharm. Biomed. Anal.*, **2000**, *22*(5), 879-885.
[http://dx.doi.org/10.1016/S0731-7085(00)00298-3] [PMID: 10815729]

[160] Hsieh, Y.Z.; Kuo, K.L. Separation of retinoids by micellar electrokinetic capillary chromatography. *J. Chromatogr. A,* **1997**, *761*(1-2), 307-313.
[http://dx.doi.org/10.1016/S0021-9673(96)00820-5] [PMID: 9084156]

[161] Delgado-Amarreño, M.M.; González-Maza, I.; Sánchez-Pérez, A.; Carabias-Martinez, R. Separation and simultaneous determination of water-soluble and fat-soluble vitamins by electrokinetic capillary chromatography. *J. Chromatogr. A,* **2002**, *953*(1-2), 257-262.
[http://dx.doi.org/10.1016/S0021-9673(02)00130-9] [PMID: 12058939]

[162] Liu, Q.; Jia, L.; Hu, C. On-Line Concentration Methods for Analysis of Fat-Soluble Vitamins by MEKC. *Chromatographia,* **2010**, *72*(1-2), 95-100.

[http://dx.doi.org/10.1365/s10337-010-1608-8]

[163] Bächmann, K.; Göttlicher, B. New particles as pseudostationary phase for electrokinetic chromatography. *Chromatographia,* **1997**, *45*(1), 249-254.
[http://dx.doi.org/10.1007/BF02505567]

[164] Altria, K.D. Highly efficient and selective separations of a wide range of analytes obtained by an optimised microemulsion electrokinetic chromatography method. *Chromatographia,* **1999**, *49*(7-8), 457-464.
[http://dx.doi.org/10.1007/BF02467624]

[165] Miola, M.F.; Snowden, M.J.; Altria, K.D. The use of microemulsion electrokinetic chromatography in pharmaceutical analysis. *J. Pharm. Biomed. Anal.,* **1998**, *18*(4-5), 785-797.
[http://dx.doi.org/10.1016/S0731-7085(98)00217-9] [PMID: 9919981]

[166] Quirino, J.P.; Terabe, S.; Otsuka, K.; Vincent, J.B.; Vigh, G. Sample concentration by sample stacking and sweeping using a microemulsion and a single-isomer sulfated β-cyclodextrin as pseudostationary phases in electrokinetic chromatography. *J. Chromatogr. A,* **1999**, *838*(1), 3-10.
[http://dx.doi.org/10.1016/S0021-9673(99)00235-6]

[167] Gluck, S. J. Benk?, M. H.; Hallberg, R. K.; Steele, K. P., Indirect determination of octanol-water partition coefficients by microemulsion electrokinetic chromatography *J. Chromatogr. A,* **1996**, *744*(s 1-2), 141-146.

[168] Pedersen-Bjergaard, S.; Naess, O.; Moestue, S.; Rasmussen, K.E. Microemulsion electrokinetic chromatography in suppressed electroosmotic flow environment. Separation of fat-soluble vitamins. *J. Chromatogr. A,* **2000**, *876*(1-2), 201-211.
[http://dx.doi.org/10.1016/S0021-9673(00)00117-5] [PMID: 10823515]

[169] Olędzka, I.; Kowalski, P.; Bałuch, A.; Bączek, T.; Paradziej-Łukowicz, J.; Taciak, M.; Pastuszewska, B. Quantification of the level of fat-soluble vitamins in feed based on the novel microemulsion electrokinetic chromatography (MEEKC) method. *J. Sci. Food Agric.,* **2014**, *94*(3), 544-551.
[http://dx.doi.org/10.1002/jsfa.6290] [PMID: 23801189]

[170] Sánchez, J.M.; Salvadó, V. Comparison of micellar and microemulsion electrokinetic chromatography for the analysis of water- and fat-soluble vitamins. *J. Chromatogr. A,* **2002**, *950*(1-2), 241-247.
[http://dx.doi.org/10.1016/S0021-9673(02)00026-2] [PMID: 11990998]

[171] Yin, C.; Cao, Y.; Ding, S.; Wang, Y. Rapid determination of water- and fat-soluble vitamins with microemulsion electrokinetic chromatography. *J. Chromatogr. A,* **2008**, *1193*(1-2), 172-177.
[http://dx.doi.org/10.1016/j.chroma.2008.04.016] [PMID: 18440539]

[172] Pedersen-Bjergaard, S.; Gabel-Jensen, C.; Honoré Hansen, S. Selectivity in microemulsion electrokinetic chromatography. *J. Chromatogr. A,* **2000**, *897*(1-2), 375-381.
[http://dx.doi.org/10.1016/S0021-9673(00)00791-3] [PMID: 11128221]

[173] Terabe, S.; Matsubara, N.; Ishihama, Y.; Okada, Y. Microemulsion electrokinetic chromatography: comparison with micellar electrokinetic chromatography. *J. Chromatogr. A,* **1992**, *608*(1), 23-29.
[http://dx.doi.org/10.1016/0021-9673(92)87102-E]

[174] Mohorič, U.; Beutner, A.; Krickl, S.; Touraud, D.; Kunz, W.; Matysik, F-M. Surfactant-free microemulsion electrokinetic chromatography (SF-MEEKC) with UV and MS detection - a novel approach for the separation and ESI-MS detection of neutral compounds. *Anal. Bioanal. Chem.,* **2016**, *408*(30), 8681-8689.
[http://dx.doi.org/10.1007/s00216-016-0057-0] [PMID: 27866256]

CHAPTER 7

Application of Capillary Electrophoresis to the Determination of Polyphenols in Food Samples

Merichel Plaza[1,2,*], **Andrea Martin-Ortiz**[1] and **María Luisa Marina**[1,2]

[1] *Departamento de Química Analítica, Química Física e Ingeniería Química, Universidad de Alcalá, Ctra. Madrid-Barcelona Km. 33.600, 28871 Alcalá de Henares (Madrid), Spain*

[2] *Instituto de Investigación Química Andrés M. del Río (IQAR), Universidad de Alcalá, Ctra. Madrid-Barcelona. Km. 33.600, 28871 Alcalá de Henares (Madrid), Spain*

Abstract: Polyphenols are naturally occurring compounds found in fruits, vegetables, cereals and beverages. Nowadays, there is a high interest in these compounds because of their potential health benefits associated with the protection against the development and progression of many degenerative diseases due to their antioxidant capacity. However, their composition changes both qualitatively and quantitatively depending on the natural source. Thus, the determination of these compounds is not straightforward. Among the different techniques employed for their analysis, capillary electrophoresis is a very interesting alternative due to its high separation efficiency, high resolution power, short analysis time and low consumption of samples and reagents. This chapter presents an overview of the recent developments and applications of capillary electrophoresis for the analysis of phenolic compounds from food samples, including articles published since 2010 to date. In addition, the characteristics of the most relevant developed methodologies using different separation modes are broadly discussed.

Keywords: Capillary electrophoresis, Food samples, Phenolic compounds, Polyphenols.

INTRODUCTION

Polyphenols are a broad family of secondary natural metabolites which are products of the pentose phosphate, shikimate and phenylpropanoid pathways in plants [1]. Structurally, they contain aromatic rings, with one or more hydroxyl substituents, and range from simple phenolic molecules to highly polymeric compounds (see Fig. **1**) [2].

* **Corresponding author Merichel Plaza:** Departamento de Química Analítica, Química Física e Ingeniería Química, Universidad de Alcalá, Ctra. Madrid-Barcelona Km. 33.600, 28871 Alcalá de Henares (Madrid), Spain; E-mail: merichel.plaza@uah.es

María Castro-Puyana, Miguel Herrero and María Luisa Marina (Eds.)
Capillary Electrophoresis in Food Analysis

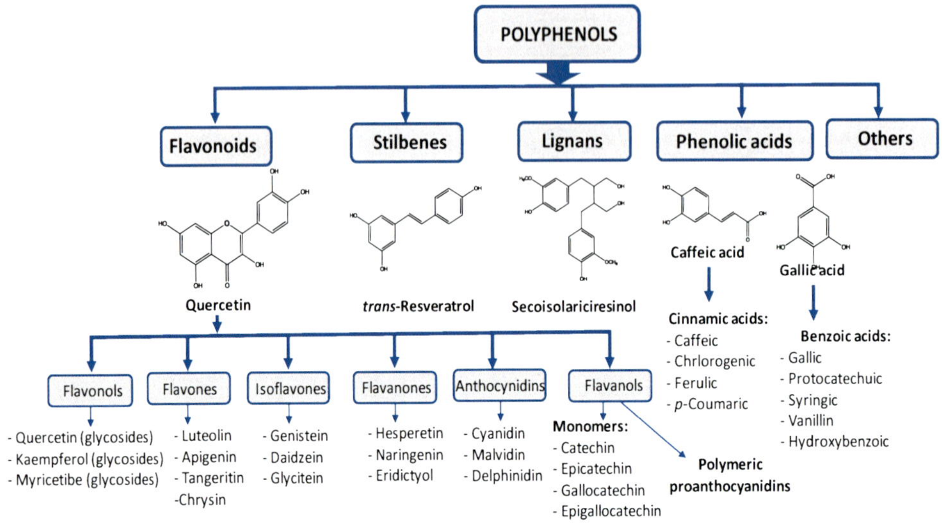

Fig. (1). Classes of polyphenols based on the number of phenol rings and their structural elements.

The majority of polyphenols in nature are found as conjugates with mono- and polysaccharides or as functional derivatives like esters and methyl esters [3]. They can be classified into different classes according to the chemical structures of the aglycones such as flavonoids (flavonols, flavones, isoflavones, flavanones, anthocyanidins, and flavanols), stilbenes, phenolic acids, lignans, and others, as shown in Fig. (**1**) which represents a classification based on the work of D'Archivio *et al.* [4]. Plant polyphenols provide protection against pathogens and predators, playing an important role in growth and reproduction. They are also responsible for the colour and the sensory characteristics of fruits and vegetables [5]. Moreover, polyphenols present several health beneficial properties such as anti- inflammatory, anticarcinogenic, antimicrobial, antithrombotic, cardioprotective, vasodilatory, and anti-neurodegenerative effects [6]. Many of these properties are due to their antioxidant capacity [7, 8]. Their antioxidant properties are mainly due to their redox properties which allow them to act as reducing agents, hydrogen donors, and singlet oxygen quenchers [9].

Polyphenols are common constituents of plants and foods of plant origin. The polyphenol composition from one to another natural source changes both qualitatively and quantitatively. For instance, some phenolic compounds are extensively spread while others are limited to specific natural sources (*i.e.* isoflavones in legumes). That is why the determination of polyphenols is a challenging analytical assignment. Generally, two different approaches are used in order to achieve their determination. One is the quantification of plant polyphenols through spectrophotometric methods. For example, Folin-Ciocalteau and Folin- Denis methods have been widely employed as spectrophotometric methods to measure total phenolic compounds in plants for many years. The main disadvantage of these methods is that they only allow for an estimation of the phenolic content because they are not specific for phenolic compounds. Furthermore, these methods are not selective and robust being highly affected by matrix effects [10]. Moreover, spectrophotometric methods do not provide quantitative measurement of individual compounds because they do not allow an analytical separation of the compounds. Therefore, the second approach comprises the separation of polyphenols before their detection. With this purpose, chromatographic techniques have mainly been employed although Capillary Electrophoresis (CE) has also demonstrated its potential in the separation of phenolic compounds from plant extracts. Regarding chromatographic techniques, High-Performance Liquid Chromatography (HPLC) is by far the preferred technique for the separation and quantification of polyphenols from natural sources. Gas Chromatography (GC) has also been employed with this aim although to a lesser extent due to the need for a derivatization step of compounds [11]. In the last years, supercritical fluid chromatography (SFC) has also been used in the separation and identification of phenolic compounds. SFC is more versatile than HPLC, more cost-effective and environmentally friendly, with higher output, better resolution and faster analysis times than general liquid chromatographic methods [12]. On the other hand, CE has already been exhibited as an attractive alternative to HPLC or GC for the analysis of phenolic compounds. CE is a versatile analytical tool for the routine determination of a wide variety of phenolic compounds in different types of samples due to its high separation efficiency, high resolution power, short analysis times and low consumption of samples and reagents [13, 14]. CE methods developed for the analysis of phenolic compounds were generally based on the use of UV-Vis detection while detection by mass spectrometry (MS) is less frequent [15 - 17].

This chapter presents an overview of the recent developments and applications of CE to the determination of phenolic compounds in different food matrices covering the last ten years (from 2010 until the present). The characteristics of the developed methodologies using different CE separation modes are detailed in Tables and the most recent and relevant applications of CE in this field are discussed.

ANALYSIS OF PHENOLIC COMPOUNDS

Different separation modes were employed in the analysis of phenolic compounds by CE. The most employed CE mode was Capillary Zone Electrophoresis (CZE) although Electrokinetic Chromatography (EKC) using cyclodextrins (CDs) in the separation buffer, Micellar Electrokinetic Chromatography (MEKC) mainly using sodium dodecyl sulfate (SDS) as micellar system, Non-Aqueous CE (NACE) and Microchip Capillary Electrophoresis (MCE) were also used to analyze phenolic compounds in food. Table **1** groups the characteristics of the developed methodologies for the analysis of phenolic compounds by CZE whereas the characteristics of the methods developed by EKC, MEKC, NACE, and MCE are detailed in Table **2**.

Table 1. CZE methodologies developed in the last ten years for the analysis of phenolic compounds in food samples.

Phenolic Compounds	Food Samples	Separation Conditions	Detection Conditions	LOD	Remarks	Ref.
Narirutin, epicatechin, catechin, rutin, kaempferol, myricetin, quercetin, morin, trans-resveratrol, cinnamic acid, ferulic acid, p-coumaric acid, vanillic acid, caffeic acid, gallic acid, 3,4-dihydroxybenzoic acid	Red, white and rose wines	BGE: 175 mM boric acid (pH 9.0) Injection 50 mbar for 30 s Run voltage (+30 kV) Temp: 25 °C Run time: 20 min Capillary: 50 µm i.d. x 60 cm L_{eff}	UV at 217 nm	0.06-0.41 mg/L	On-line preconcentration by LVSS	[18]
2-(4-hydroxyphenyl)ethanol, resveratrol, (−)-epicatechin, (+)-catechin, veratric acid, homovanillic acid, vanillin, trans-cinnamic acid, sinapic acid, quercitrin, homogentistic acid, syringic acid, ferulic acid, fisetin, p-coumaric acid, quercetin, 4-hydroxybenzoic acid, caffeic acid, gallic acid, 3,4-dihydroxybenzoic acid	Red wines	BGE: 30 mM sodium tretraborate buffer (pH 9.2) with 5% isopropanol Injection 3.5 kPa for 10 s Run voltage (+25 kV) Temp: 25 °C Run time: 18 min Capillary: 75 µm i.d. x 50 cm L_{eff}	UV-Vis (DAD) at 280, 310, and 370 nm	0.3-2.6 mg/L	PCA allowed wines to be clustered according to their origins	[19]
Syringic acid, kaempferol, myricetin, quercetin, caffeic acid, gallic acid	Red wine	BGE: 20 mM sodium tetraborate (pH 9.0) with 10% methanol (v/v) Injection 50 mbar for 5 s Run voltage (+25 kV) Temp: 25 °C Run time: 12 min Capillary: 50 µm i.d. x 40 cm L_{eff}	UV-Vis at 280 nm	1.59-2.24 mg/L	Sample preparation by LLE	[20]

(Table 1) cont.....

Phenolic Compounds	Food Samples	Separation Conditions	Detection Conditions	LOD	Remarks	Ref.
Gallic acid, vanillic acid, syringic acid, ellagic acid, sinapinic acid, cinnamic acid, quercetin	Red, white and rose wines	BGE: 30 mM sodium tetraborate (pH 9.1) with 10% methanol (v/v) Injection 10 mbar for 3 s Run voltage (+20 kV) Temp: 20 °C Run time: 14 min Capillary: 75 µm i.d. x 50 cm L_{eff}	UV at 254 nm	0.045 mg/L	Home-made tip SPE technique	[21]
Tyrosol, hydroxytyrosol, oleuropein glycoside, ferrulic acid, *p*-coumaric acid, cinnamic acid, *p*-hydroxybenzoic acid, gallic acid, caffeic acid, luteolin, apigenin, vanillic acid, and 3,4-dihydroxybenzoic acid	Extra-virgin olive oil	BGE: 50 mM boric acid (pH 10.2) Injection 50 mbar for 25 s Run voltage (+30 kV) Temperature: 25 °C Run time: 12 min Capillary: 50 µm i.d. x 60 cm L_{eff}	UV-Vis (DAD) at 210 nm	0.04-0.85 mg/L	Central composite design (boric acid concentration, pH, and voltage)	[22]
Caffeic, gallic, vanillic, syringic, cinnamic, *p*-coumaric acids and oleuropein, apigenin, luteolin, 3-hydroxytyrosol, and tyrosol	Virgin olive oil	BGE: 30 mM boric acid (pH 9.5) Injection 30 mbar for 3 s Run voltage (+25 kV) Temp: 30 °C Run time: 7 min Capillary: 75 µm i.d. x 55 cm L_{eff}	UV at 200 nm	0.004-0.253 mg/kg	DLLME with an on-line preconcentration stacking step	[23]
Tyrosol, vinylphenol, oleuropein, hydroxytyrosol, rutin, catechin, naringenin, cinnamic acid, chlorogenic acid, syringic acid, luteolin, apigenin, vanillic acid, quercetin, caffeic acid	Virgin olive oil	BGE: 30 mM boric acid buffer (pH 9.5) Injection 30 mbar for 2 s Run voltage (+25 kV) Temp: 25 °C Run time: 8 min Capillary: 75 µm i.d. x 50 cm L_{eff}	UV at 240 nm	0.078-1.094 µg/mL	Sample preparation by SPE. Simple phenols and cinnamic acid derivatives showed a higher correlation with the antioxidant capacity	[24]

(Table 1) cont.....

Phenolic Compounds	Food Samples	Separation Conditions	Detection Conditions	LOD	Remarks	Ref.
Oleacein, oleacanthal, tyrosol, hydroxytyrosol	Olive oil	BGE: 45 mM borax buffer (pH 9.5) Injection 50 mbar for 3 s Run voltage (+25 kV) Temp: 30 °C Run time: 18 min Capillary: 50 µm i.d. x 52 cm L_{eff}	UV-Vis (DAD) at 200 nm	0.49-1.61 µg/mL	Oleocanthal showed high anti-inflammatory effects	[25]
Vanillic, homovanillic, isovanillic, ferulic, 4-hydroxybenzoic, benzoic, trans-cinnamic, syringic, caffeic, gallic, gentisic, *p*-coumaric, *m*-coumaric, *o*-coumaric, protocatechuic, *o*-pyrocatechuic, sinapic, chlorogenic, α-resorcylic, β-resorcylic, and γ-resorcylic acids	Avocado	BGE: 40 mM sodium tetraborate (pH 9.4) Injection 0.5 psi for 5 s Run voltage (+30 kV) Temp: 25 °C Run time: 30 min Capillary: 50 µm i.d. x 50 cm L_{eff}	UV at 254 nm	9-145 µg/L	Hydroxycinnamic acid showed the highest influence on data variability. Hydroxybenzoic acid composition seemed to be more cultivar-related.	[26]
Caffeic, gallic, cinnamic, ferulic, chlorogenic, syringic, vanillic, benzoic, *p*-hydroxybenzoic, 2,4-dihydroxybenzoic, *o*-coumaric, *m*-coumaric, and *p*-coumaric acids	Vegetable oil	BGE: 25 mM sodium tetraborate with 5% methanol (pH 9.15) Injection 50 mbar for 4 s Run voltage (+30 kV) Temp: 25 °C Run time: 10 min Capillary: 50 µm i.d. x 40 cm L_{tot}	UV at 200 nm	0.017–0.16 mg/L	Sample preparation by in-vial LLME	[27]
Arbutin, bergenin	Bergeniae Rhizoma	BGE: 50 mM borate buffer (pH 9.2) Injection 12 kV for 6 s Run voltage (+12 kV) Run time: 10 min Capillary: 25 µm i.d. x 45 cm L_{tot}	Amperometric detection, 0.8V (*vs.* Ag/AgCl electrode)	0.21-0.23 µM	Use of a novel CNT-epoxy composite electrode as a sensitive amperometric detector	[28]

(Table 1) cont.....

Phenolic Compounds	Food Samples	Separation Conditions	Detection Conditions	LOD	Remarks	Ref.
Vanillin, ferulic acid, vanillic acid, caffeic acid, protocatechuic acid	*Rhizoma Chuanxiong*	BGE: 50 mM borate buffer (pH 9.2) Injection 15 kV for 6 s Run voltage (+15 kV) Run time: 13 min Capillary: 25 µm i.d. x 45 cm L_{tot}	Amperometric detection, 0.8V (*vs.* Ag/AgCl electrode)	0.15-0.34 µM	Use of a CNT-polydimethylsiloxane composite electrode	[29]
Isoferulic, ferulic, and caffeic acids	*Cimicifugae rhizoma*	BGE: 50 mM borate buffer (pH 9.2) Injection 15 kV for 6 s Run voltage (+15 kV) Run time: 10 min Capillary: 25 µm i.d. x 40 cm L_{tot}	Amperometric detection, 0.8V (*vs.* Ag/AgCl electrode)	0.18-0.25 µM	Use of a novel graphene–phenolic resin composite electrode	[30]
Caffeic acid hexose, catechin-glucoside, (Epi)afzelchin-(Epi)catechin Isomer A, (Epi)afzelchin-(Epi)catechin Isomer B, catechin, procyanidin B2, 2-hydroxy---O-b-D-glucopyranosil-benzoic acid, epiafzelchin-epicatechin-O-di methyl gallate, epicatechin, procyanidin B2 dimethyl gallate, epiafzelchin-epicatechi--O-methyl gallate, epicatechin-O-3-4-dimethylgallate, swertiamacroside isomer A, rutin, swertiamacroside isomer B, hyperin, dihydroxy-trimethoxyisoflavan, 5,7,4´-trimethoxyflavan	Buckwheat	BGE: 50 mM ammonium acetate/NH_3 with 2.5% methanol (pH 9.2) Injection 0.5 psi for 20 s Run voltage (+28 kV) Temp: 26 °C Run time: 18 min Capillary: 50 µm i.d. x 90 cm L_{tot}	MS (micro ToF)	0.7-0.52 µg/mL	Swertiamacroside and 2-hydroxy-3---ß-D-glucopyranosil-benzoic acid were identified for the first time.	[15]
Sinapic, ferulic, syringic, p-coumaric, vanillic, and p-hydroxybenzoic acids	Common wheat (*Triticum aestivum* L.)	BGE: 50 mM sodium phosphate (pH 6.0) Injection 50 mbar for 5 s Run voltage (+27 kV) Temp: 30 °C Run time: 12 min Capillary: 50 µm i.d. x 30 cm L_{eff}	UV-Vis at 200, 220 and 300 nm	0.3-0.4 µg/mL	Significant differences on the quantitative phenolic acids profile between the modern and the ancient genotypes.	[31]

(Table 1) cont.....

Phenolic Compounds	Food Samples	Separation Conditions	Detection Conditions	LOD	Remarks	Ref.
Phenolic compounds/profile	Fruits, fruit-based products and commercial cranberry-based products	BGE: 35 mM sodium tetraborate (pH 9.4) with 5% methanol Injection 3.5 KPa for 10 s Run voltage (+25 kV) Temp: 25 °C Run time: 40 min Capillary: 75 µm i.d. x 50 cm L_{eff}	UV-Vis at 280 nm		A method developed for the authentication of commercial natural and pharmaceutical products	[32]
Phenolic compounds/profile	Carob kibbles	BGE: 15 mM sodium tetraborate decahydrate with 8% methanol (pH 9) Injection 50 mbar for 5 s Run voltage (+30 kV) Temp: 30 °C Run time: 20 min Capillary: 50 µm i.d. x 56 cm L_{eff}	UV-Vis (DAD) at 220, 280, 320 and 375 nm		Potential use as an antioxidant source	[33]
Phenolic compounds/profile (ferulic acid, *p*-coumaric acid, tricin, vanillin, syringaldehyde, hydroxybenzaldehyde).	Rice straw	BGE: 15 mM borate buffer in 10% methanol (pH 9.1) Injection 50 mbar for 6 s Run voltage (+30 kV) Temp: 25 °C Run time: 14 min Capillary: 50 µm i.d. x 56 cm L_{eff}	UV-Vis (DAD) at 220 and 280 nm		Hydrolysis pre-treatment and organosolv delignification were successful to separate and recover hemi-celluloses and lignin-derived compounds	[34]
Phenolic compounds/profile	Lemon balm leaves and rosemary leaves	BGE: 50 mM ammonium acetate (pH 9.1) Injection 50 mbar for 5 s Run voltage (+23 kV) Temp: 25 °C Run time: 20 min Capillary: 50 µm i.d. x 73 cm L_{tot}	UV-Vis (DAD) at 190-600 nm and MS (ESI-QToF)		Addition of DPPH to the sheath-liquid for MS analyses of antioxidant phenolic compounds	[17]

(Table 1) cont.....

Phenolic Compounds	Food Samples	Separation Conditions	Detection Conditions	LOD	Remarks	Ref.
Rosmarinic and carnosic acids	Salvia species	BGE: 20 mM borate buffer (pH 9.6) Injection 50 mbar, 6 s Run voltage (+28 kV) Temp: 25 °C Run time: 10 min Capillary: 50 μm i.d. x 59 cm L_{eff}	UV at 210 nm	1.72-1.86 μg/mL	Strong correlations between rosmarinic acid contents and bioactivities	[35]
Naringin, rutin, carnosic acid, apigenin, quercetin, morin, chicoric acid	Fruit and herbal samples	BGE: 20 mM borate buffer (pH 9.2) Injection 100 mbar for 20 s Run voltage (+16 kV) Temp: 25 °C Run time: 26 min Capillary: 50 μm i.d. x 56 cm Leff	UV-Vis (DAD) at 214 nm	0.012-0.086 μg/mL	Central composite design (concentration of borate buffer, pH, voltage) LVSS	[36]
Pyrogallol, homogentisic acid, ferulic acid, cinnamic acid, gentisic acid, *p*-coumaric acid, chlorogenic acid, *p*-hydroxybenzoic acid, caffeic acid, gallic acid, 3,4 -dihydroxybenzoic acid	Edible mushrooms	BGE: 175 mM boric acid (pH 8.5) Injection 50 mbar for 5 s Run voltage (+30 kV) Temp: 25 °C Run time: 20 min Capillary: 50 μm i.d. x 72 cm Ltot	UV-Vis (DAD) at 210 nm	0.12-0.80 mg/L	Hydrolysis of phenolic acids pre-treatment	[37]
Taxifolin-7-O-glucoside, flavanomarein, quercetagetin 7-O-glucoside, okanin 4′-- -glucoside, okanin, chlorogenic acid	*Coreopsis tinctoria* Nutt.	BGE: 50 mM borax buffer with 15% acetonitrile (pH 9.0) Injection 0.5 psi for 5 s. Run voltage (+25 kV) Temp: 25 °C Run time: 18 min Capillary: 75 μm i.d. x 50 cm Leff	UV-Vis (DAD) at 280 nm	0.58-3.23 μg/mL	Qualitative and quantitative differences of flavonoids in different parts and origin	[38]

(Table 1) cont.....

Phenolic Compounds	Food Samples	Separation Conditions	Detection Conditions	LOD	Remarks	Ref.
Chlorogenic, ferulic, *p*-coumaric and caffeic acids	Green tea, red fruit juice and honey samples	BGE: 25 mM borate buffer (pH 9.2) Injection 50 mbar for 1 s Run voltage (+25 kV) Temp: 25 °C Run time: 10 min Capillary: 75 µm i.d x 45.1 cm Leff	UV-Vis (DAD) at 214, 280 and 320 nm	18.4-81.8 µg/L	Bar adsorptive microextraction coated with a mixed-mode anion exchange/RP followed by liquid desorption to decrease LODs	[39]
Resveratrol, 3-acetylcoumarin, 6-hydroxycoumarin, catechin, rutin, ferulic acid, quercitrin, kaempferol, fisetin, myricetin, quercetin, caffeic acid, gallic acid, 4-hydroxycinnamic acid.	*Connarus perrottetii* var. *Angustifolius*	BGE: 20 mM borate buffer (pH 9.2) with 15% methanol (v/v) Injection 20 cm for 60 s Run voltage (-20 kV) Temp: 25 °C Run time: 26 min Capillary: 75 µm i.d. x 56 cm Leff	UV at 230 nm	0.09-2.96 mg/L	Phenolic antioxidants present in these plant species documented for the first time	[40]
Luteolin-7-O-glucoside, phenylpyruvic acid-2-*O*-glucoside, nothofagin, aspalathin, ferulic acid, phloretin-3′,5′-di-*C*-glucoside, iriflophenone-3-*C*-glucoside-4-*O*-glucoside, mangiferin, isomangiferin, iriflophenone-3-*C*-glucoside, isovitexin, isoorientin, vitexin, orientin, chrysoeriol, luteolin, scolymoside, hesperidin, eriocitrin, quercetin, isoquercitrin, hyperoside, rutin	Rooibos (*Aspalathus linearis*) and honeybush (*Cyclopia* spp.) (herbal teas)	BGE: 200 mM borate buffer (pH 8.80 for rooibos and 9.25 for honeybush) Injection 500 mbar/s Run voltage (+30 kV) Temp: 20 °C for rooibos and 30 °C for honeybush Run time: 21 min for rooibos and 32 min for honeybush Capillary: 50 µm i.d. x 80 cm Leff	UV-Vis at 244, 283, 330, and 384 nm	0.73–14.9 µg/mL	Differences in the phenolic profiles of fermented and unfermented teas	[41]

(Table 1) cont.....

Phenolic Compounds	Food Samples	Separation Conditions	Detection Conditions	LOD	Remarks	Ref.
Catechin, isorhamnetin, kaempferol, rutin, isoquercitrin, hyperoside, quercetin, quercetin-3-O-glucuronide	Lotus leaves (*Nelumbo nucifera)*	BGE: 30 mM borax solution (pH 8.0) with 15% of acetonitrile (v/v) Run voltage (+30 kV) Injection: 50 mbar pressure for 5 s Temp: 30 °C Run time: 10 min Capillary: 50 µm i.d. x 62 cm L_{eff}	UV at 210 nm	0.50-1.69 µg/mL	qualitative and quantitative differences were observed depending on growing area and season	[42]
Rutin, caffeic acid, 3,4-dihydoxybenzoic acid	Mate herb (*Ilex paraguarensis*)	BGE: 100 mM boric acid (pH 9.0) Injection 50 mbar for 30 s Run voltage (+30 kV) Temp: 25 °C Run time: 4 min Capillary: 50 µm i.d. x 40 cm Leff	UV-Vis (DAD) at 217 nm	0.05-0.14 mg/L	Acid hydrolysis extraction. Before hydrolysis: rutin was detected. After hydrolysis: 3,4-dihydroxybenzoic acid and caffeic acid were detected	[43]
Four flavonoid glycosides (quercetin and kaempferol glycosides) and chlorogenic acid	*Urceola rosea* leaves	BGE: 25 mM borate buffer (pH 8.5) Injection 50 mbar for 4 s Run voltage (+25 kV) Temp: 40 °C Run time: 8 min Capillary: 50 µm i.d. x 58.5 cm Leff	UV-Vis (DAD) at 254 nm	3.6-7.0 µg/mL	Higher LODs, lower analysis time and an extremely economic and environmentally friendly operation compared to HPLC	[44]

ABTS, 2,2'-azino-bis(3-ethylbenzothiazoline-6-sulfonic acid; BGE, background electrolyte; CNT, carbon nanotube; DAD, diode array detection; DLLME, dispersive liquid-liquid microextraction; DPPH, 2,2-diphenyl-1-picrylhydrazyl; ESI, electrospray ionization; L_{eff}, effective length; L_{tot}, total length; LLE, liquid-liquid extraction; LLME, liquid-liquid microextraction; LOD, limit of detection; LVSS, large volume sample stacking; PCA, principal component analysis; QToF, quadrupole-time of flight; RP, reversed phase; SPE, solid-phase extraction; ToF, time of flight.

Table 2. Developed methodologies by EKC, MEKC, NACE and MCE in the last ten years for the analysis of phenolic compounds in food samples.

CE Mode	Phenolic Compounds	Food Sample	Separation Conditions	Detection Conditions	LOD	Remarks	Ref.
EKC	Caffeic, cinnamic, p-coumaric, ferulic, protocatechuic, syringic, and vanilic acids	Ethanolic extract from *Epilobium parviflorum*,	BGE: 200 mM borate with 37.5% methanol, 0.001% hexadimethrine bromide, and 15 mM 2-HP-β-CD (pH 9.2) Run voltage (-30 kV) Temp: 20 °C Run time: 22 min Capillary: 50 μm i.d. x 70.5 cm L_{eff}	UV-Vis (DAD) at 214 nm	11-31 ng/mL	Injection of large plugs of low concentration samples without overloading column capacity and low detection limits with preserved high separation efficiency	[45]
EKC	Protocatechuic aldehyde, salvianolic acid C, rosmarinic acid, salvianolic acid A, danshensu, salvianolic acid B, protocatechuic acid	Danshen plant	BGE: 20 mM sodium tetraborate with 12 mM β-CD (pH 9.0) Injection 3447.4 Pa for 10 s Run voltage (+20 kV) Temp: 25 °C Run time: 17 min Capillary: 75 μm i.d. x 50 cm L_{eff}	UV-Vis (DAD) at 280 nm	0.14-0.36 μg/mL	Study on the effect of β-CD concentration. 12 mM β-CD was chosen because it originated the shortest migration time with baseline separation of analytes	[46]
MEKC	Flavonoids, phenolic acids, saponins	Qishenyiqi dropping pills	BGE: 10 mM borate buffer (pH 9.0) with 100 mM SDS, 6% propanol and 4 μg/mL ILs-MWNTs Injection 50 mbar for 3 s Run voltage (+28 kV) Temp: 30 °C Run time: 11 min Capillary: 50 μm i.d. x 41.5 cm L_{eff}	UV-Vis (DAD) at 200 nm	1.01-76.32 μg/mL	The addition of MWNTs changed the electrophoretic behaviour of tested compounds and enhanced their resolutions	[47]
MEKC	Flavonoids, caffeic acids, leontopodic acid	Edelweiss	BGE: 60 mM sodium tetraborate solution with 25 mM SDS and 35% acetonitrile (pH 6.75) Injection 50 mbar for 4 s Run voltage (+30 kV) Temp: 40 °C Run time: 22 min Capillary: 50 μm i.d. x 62 cm L_{eff}	UV-Vis (DAD) at 254 nm	1.2-7.2 μg/mL	Correlation between the content of phenolic compound and the growth status of the plant	[48]
MEKC	Cinnamic, sinapic, ferulic, *p*-coumaric, benzoic, caffeic, syringic, 2,4-dihydroxybenzoic, vanillic, protocatechuic acids	Fruit juices	BGE: 80 mM sodium phosphate buffer containing 175 mM SDS (pH 2.0) Injection -10 kV for 900 s Run voltage (-20 kV) Run time: 7 min Capillary: 50 μm i.d. x 30 cm L_{eff}	UV-Vis (DAD) at 214 nm	0.01-2.5 ng/mL	A sensitivity enhancement of four orders of magnitude was achieved with MECK in comparison to conventional CZE methods.	[49]
MEKC	Epigallocatechin gallate, gallocatechin gallate, epicatechin, epicatechin gallate, epigallocatechin, hydroquinone, pyrocatechol, resorcinol.	Green tea and red wine	BGE: 25 mM borate buffer (pH 9.2) with 25 mM SDS Injection 30 mbar, 5 s Run voltage (+25 kV) Run time: 15 min Capillary: 75 μm i.d. x 60 cm L_{eff}	UV at 230 nm	2.3-3.5 μg/mL	MEKC improves the selectivity of the separation of catechols, compared to conventional CZE	[50]

(Table 2) cont.....

CE Mode	Phenolic Compounds	Food Sample	Separation Conditions	Detection Conditions	LOD	Remarks	Ref.
MEKC	Protocatechuic aldehyde, salvianolic acid A, rosmarinic acid, salvianolic acid B, lithospermicacid, caffeic acid, sodium danshensu, salvianolic acid D	Danshen Dripping Pills (traditional Chinese herbal formulation)	BGE: 10 mM sodium borate (pH 9.2), 0.70% (w/w) 1-butyl-3-methylimidazolium tetrafluoroborate, and 8.1% (w/w) Tween 20 Injection 50 mbar for 3 s Run voltage (+25 kV) Temp: 30 °C Run time: 10 min Capillary: 50 µm i.d. x 36.5 cm L_{eff}	UV-Vis (DAD) at 205 nm	1.24-4.60 µg/mL	The addition of ILs to the BGE improved the resolution of analytes	[51]
MEKC	Caffeic, *p*-coumaric, ferulic and chlorogenic acids, kaempferol, quercetin, myricetin, naringenin, rutin.	Tomato	BGE: 11.3 mM borax and 11.2 mM SDS (pH 8.5) Injection 7500 Pa for 20 s Run voltage (+25 kV) Temp: 15 °C Run time: 20 min Capillary: 50 µm i.d. x 60 cm L_{eff}	UV-Vis (DAD) at 270, 320 and 345 nm	0.8 to 3.8 mg/kg	D-optimal design showed SDS and acetonitrile concentrations benefit the separation of phenolic compounds in an antagonistic way	[52]
NACE	Hydroxytyrosol, oleuropein, tyrosol, caffeic acid, vanillic acid, quercetin, cinnamic acid, *o*-coumaric acid, gentisic acid	Extra virgin olive oil	BGE: 25 mM boric acid and 18 mM KOH in 74:26 (v/v) 1-propanol/methanol, (pH 11.2) Injection -30 mbar for 6 s Run voltage (+20 kV) Temp: 35 °C Run time: 25 min Capillary: 75 µm i.d. x 49 cm L_{eff}	UV at 220 nm and fluorescence $\lambda_{exc}/\lambda_{em}$ =297/320 nm	0.21-3.4 µg/mL	Box-Behnken experimental design demonstrated the ability of NACE with UV and fluorescence detection to carry out a rapid and simple separation of several phenolic compounds	[53]
NACE	Tyrosol, hydroxytyrosol	Virgin olive oil	BGE: 25 mM boric acid and 18 mM KOH in 74:26 v/v 1-propanol/methanol (pH 11.2) Injection −30 mbar for 6 s Run voltage (+20 kV) Temp: 35 °C Run time: 25 min Capillary: 75 µm i.d. x 49 cm L_{eff}	UV-Vis (DAD) at 220 nm	0.09-0.3 µg/g	LLE to preconcentrate phenolic compounds	[54]
NACE	*p*-Coumaric, caffeic, ferulic, 3,4-dihydroxybenzoic, vanillic, and 4-hydroxyphenylacetic acids	Virgin olive oil	BGE: 25 mM boric acid and 18 mM KOH in 74:26 v/v 1-propanol:methanol (pH 11.2) Injection −30 mbar for 6 s Run voltage (+20 kV) Temp: 35 °C Run time: 10 min Capillary: 75 µm i.d. x 49 cm L_{eff}	UV-Vis (DAD) at 220 nm	0.4-0.9 µg/mL	Implementation of MCR-ALS to resolve standard mixtures of phenolic acids	[55]
NACE	Epicatechin, catechin, kaempferol, quercetin, naringenin, ferulic acid, p-coumaric acid	Rice	BGE: 20 mM borate buffer (pH 9.0) and 5% acetonitrile in methanol Injection: 50 mbar for 30 s Run voltage (-20 kV) Temp: 30 °C Run time: 9 min Capillary: 50 µm i.d. x 68 cm L_{eff}	UV-Vis (DAD) at 217 nm	NACE. 0.33-2.0 µg/L NACE-LVSS: 0.06-0.19 µg/L	LVSS improved around 10 55 times the detectability of phenolic compounds	[56]

(Table 2) cont.....

CE Mode	Phenolic Compounds	Food Sample	Separation Conditions	Detection Conditions	LOD	Remarks	Ref.
NACE	Quercetin, luteolin, kaempferol	*Semen Plantaginis*	BGE: 50 mM 1-ethyl-3-methylimidazolium ILs-hydrogen sulfate salt in acetonitrile Injection: 0.5 psi for 5 s Run voltage (+20 kV) Temp: 25 °C Run time: 8 min Capillary: 50 µm i.d. x 60.2 cm L_{eff}	UV-Vis (DAD) at 254 nm	0.3-0.5 µg/mL	Three different 1-ethyl-3-methylimidazolium ILs were tried: chloride, hydrogen sulfate and tetrafluoroborate salts. The better results were achieved with 5 mM hydrogen sulfate	[57]
MCE	Epigallocatechin gallate, gallocatechin gallate, epicatechin, epicatechin gallate, epigallocatechin, hydroquinone, pyrocatechol, resorcinol.	Green tea and red wine	BGE: 20 mM borate buffer (pH 9.2) with 1 mM SDS Chamber pressure: 200 Pa Run time: 4 min Microfluidic chip analyzer of PDMS	Amperometric detection, 120 V (*vs.* Ag/AgCl electrode)		Short analysis time	[50]

2-HP-β-CD, 2-hydroxypropyl-β-cyclodextrin; BGE, background electrolyte; EKS, electrokinetic supercharging; ILs, ionic liquids; LLE, liquid-liquid extraction; LOD, limit of detection; LVSS, large volume sample stacking; MCR-ALC, multivariate curve resolution-alternating least-squares; MWNTs, ionic liquids coated multi-walled carbon nanotubes; PDMS, polydimethylsiloxane; SDS, sodium dodecyl sulfate; tITP, transient isotachophoresis.

Capillary Zone Electrophoresis (CZE)

As shown in Table **1**, CZE has been broadly employed for the analysis of phenolic compounds in food samples. One of the most important factors to perform a good separation of phenolic compounds by CZE is the buffer pH which is considered crucial for the dissociation of analytes [58].

The optimization of the best pH value is a challenge because each phenolic compound has a specific behaviour at a given pH. In general, the pKa value of phenolic compounds is around 9. Therefore, if the pH is higher than 9 the dissociation of phenolic OH groups happens and the compounds are negatively charged, presenting better separations under counterelectroosmotic flow [59]. Borate buffers are mainly used, but in certain cases, the addition of around 2.5-15% (v/v) of an organic solvent (such as acetonitrile, methanol or isopropanol) to the background electrolyte (BGE) was needed to enhance the separation. Basic pH values (8.0-9.5) were generally employed. For instance, using a 175 mM borate buffer at pH 8.5 allowed the analysis within 20 min of 11 phenolic acids in mushrooms after the previous hydrolysis during aqueous extraction [37]. This CZE method was successfully applied to the analysis of 42 samples of edible mushrooms showing that homogentisic acid, cinnamic acid and *p*-coumaric acid were detected in samples of edible mushrooms grown in Brazil, with the prevalence of cinnamic acid.

Although basic pHs between 8.0 and 9.5 and borate buffers are the most commonly employed, Gotti *et al.* used a BGE composed of sodium phosphate solution (50 mM) in water/methanol 80:20 (v/v) at pH 6.0 to perform the quantitation of six of the most important phenolic acids (sinapic, ferulic, syringic, *p*-coumaric, vanillic, and *p*-hydroxybenzoic acids) in whole grain [31]. The optimized and validated CZE-UV method allowed the analysis of ten varieties of soft wheat observing significant differences in the quantitative phenolic acids profile between the modern and the ancient genotypes, with the latter showing higher amount of the main representative phenolic acids [31].

Regarding detection systems, although UV-Vis detection was the most frequently used, other alternative detection systems such as mass spectrometry and amperometric detectors were also employed. Verardo *et al.* developed a CZE methodology coupled to MS to investigate the phenolic fraction of buckwheat [15]. 19 different compounds could be identified including phenolic acids, procyanidins, and galloylated properlargonidins with an optimized MS-friendly BGE (50 mM ammonium acetate/NH3 with 2.5% methanol and pH 9.2) in less than 18 min. The presence of swertiamacroside and 2-hydroxy-3-- -ß-D-glucopyranosil-benzoic acid was unequivocally identified for the first time in buckwheat whereas 5,7,4′- trimethoxyflavan and dihydroxy-trimethoxyisoflavan were tentatively identified for the first time [15]. In another work, extracts from lemon balm and rosemary leaves were characterized by using a CZE□MS method [17]. In this work, a MS- friendly BGE composed of 50 mM ammonium acetate (pH 9.1) was applied. The two major components observed in the rosemary extract were carnosic acid and rosmarinic acid while for lemon balm leaves extracts the major compound was rosmarinic acid. Also, the addition of the free radical 2,2-diphenyl 1-picrylhydrazyl (DPPH) to the sheath-liquid for MS analysis was studied and the abundance of most molecular ions of phenolic compounds was reduced due to their antioxidant capacity. Using MS/MS fragmentation, a total of 15 and 13 phenolic compounds could be determined by CZE□MS in rosemary and lemon balm leaves extracts, respectively.

On the other hand, composite electrodes of carbon nanotube (CNT)-polydimethylsiloxane (PDMS) and CNT-epoxy were developed for the CZE determination of different phenolic acids in *Rhizoma Chuanxiong* and Bergeniae Rhizoma, respectively [28, 29]. The CNT-based amperometric detectors presented high sensitivity using 50 mM borate buffer (pH 9.2). In addition, the same authors tried successfully a graphene-phenolic resin composite electrode with CZE for the analysis of 3 phenolic acids (isoferulic, ferulic, and caffeic acids) from *Cimicifugae rhizome* [30].

One interesting application of CZE was the analysis of phenolic acids and flavonoids in wine samples (see Table **1**) [18 - 21]. These compounds are very important in the quality assessment of wines regarding taste and health effects. For instance, a CZE-UV method was employed for the analysis in 12 min of six phenolic compounds in red wine from Brazil employing 20 mM sodium tetraborate (pH 9.0) and 10% methanol (v/v) as BGE [20]. As can be seen in Table **1**, this method provided LODs similar to the ones obtained in another work reporting the simultaneous analysis of 20 phenolic compounds in 49 Spanish red wines in less than 25 min by CZE-UV (see Fig. **2a**) [19]. In this work, principal component analysis (PCA) of the phenolic profile and the peaks areas of the most abundant compounds was employed to determine the significant features that can contribute to wine discrimination according to the geographical areas of production. Results from PCA proved that such compositional data allowed wines to be clustered according to their origins (see Fig. **2b**). Tyrosol, gallic, protecatechuic, *p*- coumaric, and caffeic acids were some of the representative discriminant compounds. In order to decrease the LODs in the determination of phenolic compounds from wine by CE, a preconcentration step can be used. For instance, on-line preconcentration by large-volume sample stacking (LVSS) and CZE-UV were used to optimize the simultaneous separation of 16 phenolic compounds present in 23 samples of red, white and rose wines in 19 min [18]. This method exhibited high analytical reliability and avoided exhaustive extractions. In addition, miniaturized solid-phase extraction (SPE) prior to CE-UV allowed the determination of ellagic acid and other phenolic compounds (7 in total) in thirty Argentinian wines from red, white and rose grapes employing 30 mM sodium tetraborate (pH 9.1) and 10% methanol (v/v) within 10 min [21]. The extraction was performed using a home-made tip SPE which decreased the conditioning/sample/washing/elution volumes and increased the sensitivity [21].

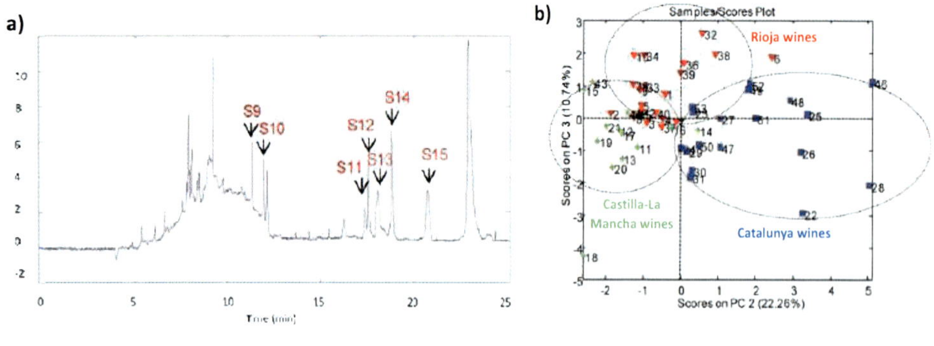

Fig. (2). a) Electropherogram of a wine sample recorded at 370 nm. **b)** PCA results (score plot) using selected peak areas as analytical data (adapted from Franquet-Griell *et al.* [19]).

In order to determine the phenolic compounds present in olive oils, several CZE methods were developed (see Table **1**). For instance, a central composite design, response surface analysis and the simultaneous optimization method of Derringer and Suich were employed to optimize a method for the simultaneous separation of 13 phenolic compounds from extra virgin olive oil in 12 min. These multivariate procedures were efficient in determining the optimal separation conditions using five peak pair resolutions and runtime as responses. Additionally, an online stacking pre-concentration method was applied increasing 6-fold area and 4-fold peak height. This method satisfied the official requirements for the quantification of phenolic compounds in samples of extra-virgin olive oil [22].

Dispersive liquid-liquid microextraction together with an on-line stacking preconcentration and CZE-UV were applied for the analysis of 11 phenolic compounds in virgin olive oil [23]. This method allowed the selective determination of phenolic compounds in virgin olive oils with satisfactory sensitivities, recoveries, and relative standards deviations (RSDs), compatible with levels present in olive oil samples [23]. The same authors studied the relationship between 15 phenolic compounds quantified by SPE-CZE-UV, the corresponding Folin Ciocalteau results, and the total antioxidant capacity. Simple phenols and cinnamic acid derivatives showed higher correlation with the antioxidant capacity [24]. In addition, Vulcano *et al.* developed the first validated CZE-method suitable for the simultaneous quantitative determination of oleocanthal and oleacein as well as their precursors tyrosol and hydroxytyrosol in olive oil [25]. These compounds are responsible for the burning pungent sensation of certain olive oils. Furthermore, oleocanthal showed potent anti-inflammatory effects [25].

Phenolic acids were also analyzed by CZE in other oily matrices such as avocado and vegetable oils (see Table **1**). 21 phenolic acids were separated by CZE-UV using 40 mM sodium tetraborate at pH 9.4 as BGE [26]. From 8 to 14 phenolic acids were quantified in 13 varieties of avocado fruit at 2 ripening stages according to the variety. Phenolic profiles were found to be characteristic from each variety. For instance, gallic acid was only found in Sir Prize variety and *trans*-cinnamic, *m*-coumaric, benzoic, *γ*-resorcylic, *p*-coumaric, and protoca techuic acids were detected in the ten genetically related samples out of the total 13 varieties analysed. PCA was applied to highlight the differences between varieties and ripening degrees. The statistical studies highlighted the fact that hydroxycinnamic acids, particularly *p*-coumaric acid (its concentration increased during ripening) was the variable with the highest influence on data variability while hydroxybenzoic acid composition was more cultivar-related [26]. On the other hand, an in-vial liquid- liquid microextraction (in-vial LLME)-CZE method was developed for the simultaneous determination of 13 phenolic acids in vegetable oils. A BGE composed of 25 mM sodium tetraborate with 5% methanol (pH 9.15) showed a successful separation in less than 10 min. The validated in-vial LLME-CZE method was applied to the analysis of different vegetable oil samples (extra virgin olive oil, virgin olive oil, pure olive oil, walnut oil and grapeseed oil) showing good sensitivity [27].

CZE with UV-Vis detection enabled to obtain the phenolic profile of carob kibbles which are a by-product of the carob bean gum production. The phenolic profile and its potential as an added-value antioxidant-rich product were evaluated showing a promising use of this biomass as an antioxidant source [33]. Furthermore, CZE-UV has been used to obtain the phenolic profile of other by-products such rice-straw [34]. This by-product was pre-treated by hydrolysis at 210 ºC (autohydrolysis) followed by an organosolv delignification under mild conditions previously to its analysis. The identified lignin-derived products, tricin, vanillin, ferulic acid, *p*- coumaric acid, seringaldehyde and hydroxybezaldehyde, are added-value compounds due to their antioxidant and surfactant properties.

CZE and HPLC techniques combined with PCA were employed to carry out the authentication of fruit products based on the phenolic profile [32]. Different types of samples were analyzed including fruits (cranberry, blueberry, grapes, and raisins), fruit□based products (grape□juice and cranberry□juice) as well as commercial cranberry□based products (pharmaceutical natural extracts, powder capsules, syrup, and sachets). After data analysis, the results showed that the combination of CZE, HPLC and PCA is a good strategy for the characterization of these products according to their fruit of origin and therefore it might be proposed for the authentication of commercial natural and pharmaceutical products [32].

A CZE-LVSS method was developed in order to increase sensitivity in the simultaneous determination of naringin, rutin, carnosic acid, apigenin, quercetin, morin and chicoric acid from a number of food samples [36]. Optimal CZE parameters were achieved by central composite design showing a good separation using 20 mM borate buffer at pH 9.2. This method provided a 3 to 5-fold improvement in the LODs for the determination of phenolic compounds from fruit and herbal samples [36]. In another work, a method based in bar adsorptive microextraction coated with mixed-mode anion exchange/reverse phase followed by liquid desorption combined with CZE-UV-Vis was used to decrease the LOD of phenolic acids (chlorogenic, ferulic, *p*-coumaric, and caffeic acids) from different food matrices [39]. The application of the method to the analysis of food matrices such as green tea, red fruit juice, and honey allowed very good performances for the determination of minor amounts of phenolic acids since this method presented low LODs (18.4-81.8 µg/L) (see Table **1**).

Rosmarinic and carnosic acids have also been determined by CZE-UV with a simple method (using 20 mM borate buffer (pH 9.6)) in 14 *Salvia* species in order to correlate their content with *in vitro* antioxidant capacities and antidiabetic activity. In this study, strong correlations between rosmarinic acid contents and bioactivities of *Salvia* samples were found [35]. Polar phenolic compounds (kaempherol and quercetin glycosides and chlorogenic acid) from *Urceola rosea* leaves were determined and quantified by CZE with 25 mM borate buffer (pH 8.5) in less than 8 min [44]. Rutin, caffeic acid and 3,4-dihydroxybenzoic acid were determined in mate herb by CZE-UV with 100 mM boric acid (pH 9.0) in less than 4 min after hydrolysis treatment [43]. The phenolic content found in mate herbs hydrolyzates indicated that this matrix is a great source of these compounds [43]. Moreover, a CZE-UV-Vis method was successfully developed for the analysis of rooibos and honeybush phenolics (23 different phenolic compounds) with 200 mM borate buffer at pH 8.20 and 9.25 within 21 and 32 min, respectively [41]. Clear differences in the phenolic profiles of fermented and unfermented rooibos tea samples were observed. Similar results were obtained for honeybush tea samples, where important additional variations in phenolic contents of *C. subternata* and *C. maculata* samples were noticed [41].

In order to analyze phenolic compounds with different polarities, the addition of organic solvents was needed because many flavonoids are not dissolved in pure water systems (Table **1**). For instance, 8 flavonoids (catechin, isorhamnetin, kaempferol, rutin, isoquercitrin, hyperoside, quercetin, and quercetin-3-O-glucuronide), some of which are aglycones, and other glycosides, were determined within 10 min in lotus leaves collected in different places of Vietnam employing a BGE composed of 30 mM borax solution at pH 8.0 containing 15% of acetonitrile (v/v) [42]. These flavonoids are responsible for the anti-inflammatory properties of lotus leaves. The main flavonoid found was quercetin-3-*O*-glucuronide followed by hyperoside and isoquercitrin. Nonetheless, significant qualitative and quantitative differences were observed depending on the growing area and season [42]. Other example was the analysis of five flavonoids and one phenolic acid in different parts and origins of *Coreopsis tinctoria* and its related species (13 samples) [38]. Analyses were performed within 18 min by using 50 mM borax buffer containing 15% acetonitrile as a modifier (pH 9.0). The qualitative and quantitative differences of flavonoids in different parts and origins of *C. tinctoria* demonstrated that this method can aid to the quality control of this material [38]. Furthermore, a CZE-UV method was developed for the simultaneous determination of 14 phenolic compounds extracted from *Connarus perrottetii* var. *angustifolius* (a medicinal plant native to Brazilian Amazonia) within 26 min using a 20 mM borate buffer (pH 9.2) containing 15% methanol (v/v) as BGE [40]. This method allowed the quantification of catechin and rutin in the extracts.

EKC is a separation mode enabling to expand the applicability of CE to the separation of neutral compounds but it is also employed in the separation of charged compounds. EKC is an analytical separation method based on the use of the CZE technique in combination with the principle of chromatography. In EKC, a new component, the pseudophase, is added to the separation buffer employed in CZE.

In order to carry out the simultaneous determination of seven phenolic acids in Danshen and related medicinal plants, a running buffer composed of 20 mM sodium tetraborate adjusted to pH 9.0, and containing 12 mM β-cyclodextrin was employed [46]. High resolution and relatively short migration times (17 min) were obtained with the addition of β-cyclodextrin to the running buffer. Under optimal conditions, samples of Danshen and related medicinal plants were analyzed using the developed method with high separation efficiency and short analysis time with good accuracy and satisfactory repeatability.

In order to pre-concentrate and determine seven phenolic acids in plant extracts, a method based on the combination of an on-line transient isotachophoretic (tITP) preconcentration and EKC was developed [45]. The EKC separation was carried out with a BGE composed of 200 mM borate buffer (pH 9.2) containing 37.5% methanol, 0.001% hexadimethrine bromide, and 15 mM 2-hydroxypropyl-β-cyclodextrin in less than 20 min. The tITP-EKC method showed low LODs ranging from 11 ng/mL to 31 μg/mL being a suitable method for performing the quality control of different supplementary products and medicinal plants.

Micellar Electrokinetic Chromatography (MEKC)

MEKC with SDS, the most common anionic surfactant used in MEKC, has been mainly used to separate flavonoids and phenolic acids (see Table **2**). MEKC is a hybrid technique that combines chromatographic and electrophoretic separation principles and extends the applicability of CE to the analysis of neutral compounds [60]. In MEKC, the surfactants are added to the separation buffer in concentrations above their critical micellar concentration to develop micelles which perform as a pseudostationary phase. Thus, the separation of the compounds is achieved by their distribution between hydrophobic micelles and a hydrophilic buffer system [60].

As can be seen in Table **2**, borax buffers at pH higher than 6.75 are used in MEKC analysis of flavonoids and phenolic acids because they provided good results [47, 48, 50 - 52]. Borate ions form complexes with flavonoids and facilitate their separation because of enhanced selectivity [61]. This fact was also noticed by Ganzera *et al.* [48] when they tried different buffers (phosphate, citric acid/phosphate, borate/phosphate and ammonium acetate/acetic acid) at pH-range between 6.50 and 7.25 for analyzing 12 relevant phenolic compounds (flavonoids, cafeic acids and leontopodic acid) in the aerial parts of *Leontopodium alpinum.* Borax buffer at pH 6.75 allowed the best separation of all phenolic compounds with resolution values higher than 3.0. The addition of SDS and an organic modifier to the buffer was needed because without surfactant several phenolic compounds co- eluted and without the organic modifier some flavonoids were not dissolved in the buffer. These analytes were well resolved in less than 25 min by using a borax buffer (60 mM) containing SDS (25 mM) and 35% acetonitrile (v/v). This work showed a correlation between the content of phenolic compounds and the growth status of the plant (*i.e.* flavonoids were not found in early stages but all phenolics reached high levels during blooming) [48].

Therefore, besides the selection of the buffer in the BGE composition, the buffer concentration, surfactant concentration and the use of organic modifier in the separation of phenolic compounds is very important. That is why, the BGE composition (borax, acetonitrile, and SDS concentration) and pH was optimized by response surface methodology with a D-optimal design to achieve the analysis of flavonoids and phenolic acids from tomato [52]. Methanol and acetonitrile were considered as modifiers in order to improve the resolution of phenolic compounds. Acetonitrile provided better resolution than methanol and it was selected as organic modifier for further optimization. The design pointed out that SDS and acetonitrile concentrations benefited from the separation of phenolic compounds in an antagonistic way. Accordingly, the optimum included SDS in the absence (or very low concentration) of acetonitrile or vice versa. The optimum for pH was identified in the extremes (8.5 and 9.5) with lower resolution for the pH values between them. A desirability function was employed to obtain the optimal conditions for the separation of phenolic compounds within 20 min using 11.3 mM borax buffer and 11.2 mM SDS adjusted to pH 8.5 as BGE. This developed method can be used to determine tomato polyphenols in quality control procedures.

A faster method within 5 min was achieved by MEKC to determine and quantify flavonoids from *Eruka* and *Diplotaxis* rocket salad species [62]. The optimum separation was achieved using 40 mM sodium tetraborate solution containing SDS (15 mM) as a surfactant because the peak resolution increased (see Fig. **3**). The effect of pH was tested from 8.0 to 10.0 and the best peak resolution was obtained using a buffer with a pH value of 9.0. In addition, when the applied voltage and temperature were increased, peak migration times decreased as a result of the positive effect of these two parameters on electrophoretic mobility selecting a voltage of 25 kV and a temperature of 28 °C. The results obtained by this optimized MEKC method were positively correlated with those achieved by an HPLC method (R^2 = 0.9887, $p < 0.001$) confirming the good agreement between results obtained by these two techniques. *Eruka* and *Diplotaxis* samples contained mostly kaempferol and quercetin derivatives, respectively [62].

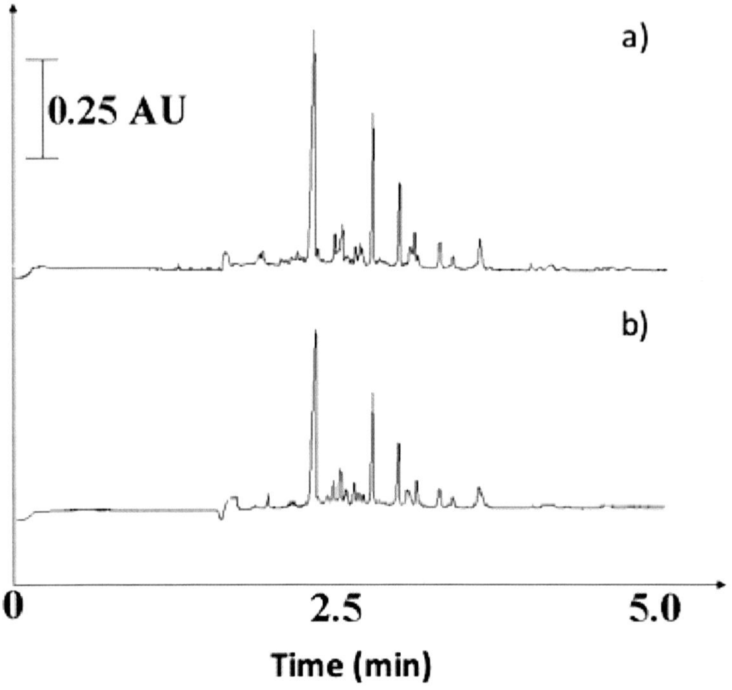

Fig. (3). Effect of the presence of SDS on the CE separation of *Eruca* phenolic compounds with a 40 mM sodium tetraborate buffer (pH 9.0), a) 0 mM SDS or b) 15 mM SDS (adapted from Verardo *et al.* [62]).

On the other hand, the addition of ionic liquids (ILs) to the BGE could improve the separation performance of MEKC. ILs are a kind of green solvent consisting of an organic cation and an inorganic or organic anion, and which remain in liquid state at room temperature [63]. Due to their non-volatile character and high conductivity, ILs were also introduced as an alternative to organic modifiers in MEKC and are receiving increased attention [64, 65]. An ILs-modified MEKC method was developed for the separation and determination of eight phenolic acids in a Chinese herbal formulation (Compound Danshen Dripping Pills) [51]. In order to increase the separation efficiency and selectivity, the micelle system consisting of aqueous mixtures of ILs, polyoxyethylene sorbitan monolaurate (Tween 20) and borate was optimized using a D-optimal design. The results indicated that the addition of ILs in the BGE could significantly alter the electrophoretic behaviour and improve the resolution of target analytes. A satisfactory separation of all analytes was achieved in 10 min with optimized buffer composed of 0.70% (w/w) 1-butyl-3- methylimidazolium tetrafluoroborate, 8.1% (w/w) Tween 20 and 10 mM sodium borate at pH 9.2 [51].

On the other hand, Cao *et al.* developed a new MEKC system using SDS and ILs coated multi-walled carbon nanotubes (ILs-MWNTs) for the simultaneous determination of four flavonoids, four phenolic acids, and two saponins [47]. Several parameters affecting the separation were studied including the choice of ILs, ILs-MWNTs concentration, the respective use of ILs and MWNTs, buffer pH, SDS concentration and borate content. Results revealed that the addition of IL- MWNTs in running electrolytes enhanced the separation of target compounds compared to conventional SDS micellar solution because the surface of carbon nanotubes interacted favorably with the analytes. The separation was carried out in a 10 mM borate buffer (pH 9.0) containing 100 mM SDS, 6% propanol and 4 μg/mL IL (1-dodecyl-3-methylimidazolium choride)-MWNTs within 10 min. The method was successfully applied to the determination of different phenolic compounds in Qishenyiqi dropping pills being this method suitable for the determination of complex samples because it reduces interferences of compounds from the matrix [47].

In order to increase sensitivity in the analysis of different phenolic acids in fruit juice, a rapid (less than 7 min) and efficient MEKC method with electrokinetic supercharging (EKS) and sweeping for the on-line preconcentration of analytes was developed [49]. The EKS-sweeping parameters were optimized obtaining LODs for the 10 phenolic acids in the range 0.01-2.5 ng/mL. As can be seen in Table **2**, these LODs values were the lowest obtained by MEKC for phenolic compounds. Also, compared with conventional CZE method, the sensitivity for the analytes can increase four orders of magnitude [49]. Moreover, this method was successfully applied to the analysis of phenolic acids in juice samples.

Non-Aqueous Capillary Electrophoresis (NACE)

CZE can be performed in aqueous as well as organic buffer solutions. When organic buffer solutions are used, it is called Non-Aqueous Capillary Electrophoresis (NACE). According to the properties of the analytes, different organic solvents (such as methanol, ethanol, acetonitrile, 1-propanol, 2-propanol, among others) and their mixtures can be employed in NACE. NACE mode is usually selected for the analysis of hydrophobic compounds because of their solubility in the organic solvents. This CE mode presents enhanced selectivity and reduced electrophoretic currents and Joule heating [66]. As can be observed in Table **2**, the separation selectivity of phenolic compounds from olive oil by NACE has been demonstrated [53 - 55]. For instance, Godoy-Caballero *et al.* [53]

employed a Box-Behnken experimental design to optimize several injection and separation variables to analyze various phenolic compounds from olive oil samples by NACE with UV and fluorescence detection. The hydrophobicity of BGE (25 mM boric acid and 18 mM KOH in a mixture of 74:26 (v/v) 1-propanol/methanol with an apparent pH of 11.2) allowed its miscibility with the olive oil sample so that the phenolic characterization in this sample was possible without any pre-treatment. This method showed a rapid and simple separation of nine phenolic compounds from olive oil within 25 min with LODs in the order of micrograms per millilitre (0.21- 3.4 µg/mL). To increase sensitivity the authors also developed a simple liquid-liquid extraction method based on a relationship 5:1 (w/v) olive oil/ethanol to achieve the necessary preconcentration for the quantitation of phenolic compounds present at lower concentrations [54]. Moreover, they combined the NACE method with UV detection with multivariate curve resolution-alternating least-squares (MCR-ALS) algorithm to carry out the resolution and quantitation of six phenolic acids (*p*-coumaric, caffeic, ferulic, 3,4-dihydroxyphenylacetic, vanillic and 4- hydroxyphenilacetic acids) in virgin olive oils simplifying the previous sample pre- treatment stages [55].

Additionally, NACE can be used for other types of samples different from olive oil (see Table **2**). For instance, a rapid NACE method for the separation and determination of epicatechin, catechin, kaempferol, quercetin, naringenin, ferulic acid and *p*-coumaric acid was developed and applied in different rice varieties [56]. The seven compounds were separated employing 20 mM borate buffer at pH 9.0 and 5% acetonitrile in methanol within 9 min. The authors optimized the LVSS technique to preconcentrate the sample and to analyze non-detectable phenolic compounds. The NACE and NACE-LVSS methods presented high sensitivity with low LODs (0.33 to 2.0 µg/L and 0.06 to 0.19 µg/L, respectively). LVSS technique improved around 10-55 times the detectability of phenolic compounds in rice samples [56].

ILs based on 1-ethyl-3-methylimidazolium have been employed as BGE in NACE to analyze flavonoids (quercetin, luteolin, kaempferol) in *Semen Plantaginis* [57]. Three different 1-ethyl-3-methylimidazolium ILs, namely chloride, hydrogen sulfate and tetrafluoroborate salts using acetonitrile as solvent were studied. According of the IL, the three flavonoids can be separated under positive voltage with a low IL concentration and negative voltage with a high IL concentration. The LODs ranged from 0.3 to 0.5 µg/mL being lower than those obtained in aqueous CE for these compounds (see Table **2**). The best results on the determination of these three flavonoids in *Semen Plantaginis* were attained with low concentration (5 mM) of the hydrogen sulfate ionic liquid.

Microchip Electrophoresis (MCE)

Nowadays, miniaturization is a trend in science and technology. That is why CE has been adapted to this tendency through the use of microfluidic chips. Samples can be analyzed in short time with negligible sample and solvent consumption and practically without waste production [67]. The development of point-of-analysis (*in-situ* testing) analysis systems using microchips can be very interesting in food control which represents an area of research of great potential and interest [67]. However, in the last ten years, MCE has scarcely been used for the analysis of phenolic compounds from food matrices (see Table **2**). One example is the development of a microfluidic system of CE with electrochemical detection for the determination of antioxidant phenolic compounds in green tea and red wine [50]. The optimum BGE composition for the separation of phenolics by MEKC on the chip was studied presenting the best separation with 25 mM borate buffer (pH 9.2) with the addition of 25 mM SDS. This method allowed the analysis of phenolic compounds within 4 min and was applied to the rapid analysis of antioxidant phenolic compounds in natural samples.

CONCLUDING REMARKS

CE is an attractive technique for the analysis of phenolic compounds in very different types of food matrices. CZE has been the most employed CE separation mode for the analysis of different classes of phenolic compounds in a variety of food samples and beverages. However, other CE separation modes such as EKC, MEKC, and NACE were also used with this aim. ILs have been considered an interesting alternative to organic modifiers in MEKC and NACE enabling improved separations of target compounds. UV-Vis was by far the most frequently used detector although fluorescence, amperometric (with novel CNT-composite electrodes) and MS detection systems were also successfully employed in some cases. Different on-line preconcentration strategies based on electrophoretic principles were implemented to increase method sensitivity. An application of MCE to the analysis of phenolic compounds was also reported showing its potential to provide fast analysis of food samples and usefulness for quality control with very low consumption of reagents. Based on the results obtained in the articles reviewed in this chapter, it is expected that CE continues being a powerful tool for the analysis of phenolic compounds in food samples. New developments can also be expected in the near future based on MCE and an increased use of MS detectors, which allow the structural elucidation of many phenolic compounds commercially unavailable as standards.

CONSENT FOR PUBLICATION

Not applicable.

CONFLICT OF INTEREST

The author declares no conflict of interest, financial or otherwise.

ACKNOWLEDGEMENTS

This work was supported by the Comunidad of Madrid (Spain) and European funding from FSE and FEDER Programs (project S2018/BAA-4393, AVANSECAL-II-CM).

REFERENCES

[1] Randhir, R.; Lin, Y.T.; Shetty, K. Stimulation of phenolics, antioxidant and antimicrobial activities in dark germinated mung bean sprouts in response to peptide and phytochemical elicitors. *Process Biochem.*, **2004**, *39*, 637-646. [http://dx.doi.org/10.1016/S0032-9592(03)00197-3]

[2] Lin, D.; Xiao, M.; Zhao, J.; Li, Z.; Xing, B.; Li, X.; Kong, M.; Li, L.; Zhang, Q.; Liu, Y.; Chen, H.; Qin, W.; Wu, H.; Chen, S. An Overview of Plant Phenolic Compounds and Their Importance in Human Nutrition and Management of Type 2 Diabetes. *Molecules,* **2016**, *21*(10), 1374. [http://dx.doi.org/10.3390/molecules21101374] [PMID: 27754463]

[3] Pandey, K.B.; Rizvi, S.I. Plant polyphenols as dietary antioxidants in human health and disease. *Oxid. Med. Cell. Longev.*, **2009**, *2*(5), 270-278. [http://dx.doi.org/10.4161/oxim.2.5.9498] [PMID: 20716914]

[4] D'Archivio, M.; Filesi, C.; Di Benedetto, R.; Gargiulo, R.; Giovannini, C.; Masella, R. Polyphenols, dietary sources and bioavailability. *Ann. Ist. Super. Sanita,* **2007**, *43*(4), 348-361. [PMID: 18209268]

[5] Lattanzio, V. Bioactive Polyphenols: Their Role in Quality and Storability of Fruit and Vegetables. *J. Appl. Bot. Food Qual.*, **2003**, *77*, 128-146.

[6] Ajila, C.M.; Brar, S.K.; Verma, M.; Tyagi, R.D.; Godbout, S.; Valéro, J.R. Extraction and analysis of polyphenols: recent trends. *Crit. Rev. Biotechnol.*, **2011**, *31*(3), 227-249. [http://dx.doi.org/10.3109/07388551.2010.513677] [PMID: 21073258]

[7] Abbas, M.; Saeed, F.; Anjum, F.M.; Afzaal, M.; Tufail, T.; Bashir, M.S.; Ishtiaq, A.; Hussain, S.; Suleria, H.A.R. Natural polyphenols: an overview. *Int. J. Food Prop.*, **2017**, *20*, 1689-1699. [http://dx.doi.org/10.1080/10942912.2016.1220393]

[8] Szwajgier, D.; Borowiec, K.; Pustelniak, K. The neuroprotective effects of phenolic acids: molecular mechanism of action. *Nutrients,* **2017**, *9*(5), 477. [http://dx.doi.org/10.3390/nu9050477] [PMID: 28489058]

[9] Liang, T.; Yue, W.; Li, Q. Comparison of the phenolic content and antioxidant activities of Apocynum venetum L. (Luo-Bu-Ma) and two of its alternative species. *Int. J. Mol. Sci.*, **2010**, *11*(11), 4452-4464. [http://dx.doi.org/10.3390/ijms11114452] [PMID: 21151449]

[10] Smeraglia, J.; Baldrey, S.; Watson, D. Matrix effects and selectivity issues in LC-MS-MS. *Chromatographia,* **2002**, *55*, S95-S99. [http://dx.doi.org/10.1007/BF02493363]

[11] Proestos, C.; Sereli, D.; Komaitis, M. Determination of phenolic compounds in aromatic plants by RP-HPLC and GC-MS. *Food Chem.*, **2006**, *95*, 44-52. [http://dx.doi.org/10.1016/j.foodchem.2004.12.016]

[12] Silva, C.G.A.; Collins, C.H. Super/subcritical fluid chromatography with packed columns: state of the

art and applications. Quím. *Nova, São Paulo,* **2014**, *37*(6), 1047-1057.
[http://dx.doi.org/10.5935/0100-4042.20140158]

[13] Gomez, F.J.V.; Monasterio, R.P.; Vargas, V.C.S.; Silva, M.F. Analytical characterization of wine and its precursors by capillary electrophoresis. *Electrophoresis,* **2012**, *33*(15), 2240-2252.
[http://dx.doi.org/10.1002/elps.201100595] [PMID: 22887148]

[14] Anderson, H.E.; Santos, I.C.; Hildenbrand, Z.L.; Schug, K.A. A review of the analytical methods used for beer ingredient and finished product analysis and quality control. *Anal. Chim. Acta,* **2019**, *1085*, 1-20.
[http://dx.doi.org/10.1016/j.aca.2019.07.061] [PMID: 31522723]

[15] Verardo, V.; Gómez-Caravaca, A.M.; Segura-Carretero, A.; Caboni, M.F.; Fernández-Gutiérrez, A. Development of a CE-ESI-microTOF-MS method for a rapid identification of phenolic compounds in buckwheat. *Electrophoresis,* **2011**, *32*(6-7), 669-673.
[http://dx.doi.org/10.1002/elps.201000609] [PMID: 21365659]

[16] Contreras-Gutiérrez, P.K.; Hurtado-Fernández, E.; Gómez-Romero, M.; Ignacio Hormaza, J.; Carrasco-Pancorbo, A.; Fernández-Gutiérrez, A. Determination of changes in the metabolic profile of avocado fruits (Persea americana) by two CE-MS approaches (targeted and non-targeted). *Electrophoresis,* **2013**, *34*(19), 2928-2942.
[http://dx.doi.org/10.1002/elps.201200676] [PMID: 24228266]

[17] Maringer, L.; Ibáñez, E.; Buchberger, W.; Klampfl, C.W.; Causon, T.J. Using sheath-liquid reagents for capillary electrophoresis-mass spectrometry: application to the analysis of phenolic plant extracts. *Electrophoresis,* **2015**, *36*(2), 348-354.
[http://dx.doi.org/10.1002/elps.201400429] [PMID: 25308871]

[18] Ballus, C.A.; Dillenburg Meinhart, A.; Grando de Oliveira, R.; Teixeira Godoy, H. Optimization of capillary zone electrophoresis separation and on-line preconcentration of 16 phenolic compounds from wines produced in South America. *Food Res. Int.,* **2012**, *45*, 136-144.
[http://dx.doi.org/10.1016/j.foodres.2011.09.029]

[19] Franquet-Griell, H.; Checa, A.; Núñez, O.; Saurina, J.; Hernández-Cassou, S.; Puignou, L. Determination of polyphenols in Spanish wines by capillary zone electrophoresis. Application to wine characterization by using chemometrics. *J. Agric. Food Chem.,* **2012**, *60*(34), 8340-8349.
[http://dx.doi.org/10.1021/jf302078j] [PMID: 22866993]

[20] Souza Dias, F.; Klassen, A.; Tavares, M.F.; Mauricio, J. Fast Determination of Phenolic Compounds in Brazilian Wines from Vale do Sao Francisco Region by CE. *Chromatographia,* **2013**, *76*, 559-563.
[http://dx.doi.org/10.1007/s10337-013-2399-5]

[21] Spisso, A.; Gomez, F.J.V.; Fernanda Silva, M. Determination of ellagic acid by capillary electrophoresis in Argentinian wines. *Electrophoresis,* **2018**, *39*(13), 1621-1627.
[http://dx.doi.org/10.1002/elps.201700487] [PMID: 29572874]

[22] Ballus, C.A.; Meinhart, A.D.; Bruns, R.E.; Godoy, H.T. Use of multivariate statistical techniques to optimize the simultaneous separation of 13 phenolic compounds from extra-virgin olive oil by capillary electrophoresis. *Talanta,* **2011**, *83*(4), 1181-1187.
[http://dx.doi.org/10.1016/j.talanta.2010.07.013] [PMID: 21215853]

[23] Monasterio, R.P.; de Los Angeles Fernández, M.; Silva, M.F. High-throughput determination of phenolic compounds in virgin olive oil using dispersive liquid-liquid microextraction- capillary zone electrophoresis. *Electrophoresis,* **2013**, *34*(12), 1836-1843.
[http://dx.doi.org/10.1002/elps.201300117] [PMID: 23576183]

[24] Fernandez, M.A.; Soto Vargas, V.C.; Siva, M.F. Phenolic Compounds and Antioxidant Capacity of Monovarietal Olive Oils Produced in Argentina. *J. Am. Oil Chem. Soc.,* **2014**, *91*, 2021-2033.
[http://dx.doi.org/10.1007/s11746-014-2558-3]

[25] Vulcano, I.; Halabalaki, M.; Skaltsounis, L.; Ganzera, M. Quantitative analysis of pungent and anti-inflammatory phenolic compounds in olive oil by capillary electrophoresis. *Food Chem.,* **2015**, *169*,

381-386.
[http://dx.doi.org/10.1016/j.foodchem.2014.08.007] [PMID: 25236241]

[26] Hurtado-Fernández, E.; Contreras-Gutiérrez, P.K.; Cuadros-Rodríguez, L.; Carrasco-Pancorbo, A.; Fernández-Gutiérrez, A. Merging a sensitive capillary electrophoresis-ultraviolet detection method with chemometric exploratory data analysis for the determination of phenolic acids and subsequent characterization of avocado fruit. *Food Chem.*, **2013**, *141*(4), 3492-3503.
[http://dx.doi.org/10.1016/j.foodchem.2013.06.007] [PMID: 23993512]

[27] Abu Bakar, N.B.; Makahleh, A.; Saad, B. In-vial liquid-liquid microextraction-capillary electrophoresis method for the determination of phenolic acids in vegetable oils. *Anal. Chim. Acta,* **2012**, *742*, 59-66.
[http://dx.doi.org/10.1016/j.aca.2012.02.045] [PMID: 22884208]

[28] Zhang, L.; Zhang, W.; Chen, W.; Chen, G. Simultaneous determination of five bioactive constituents in Rhizoma Chuanxiong by capillary electrophoresis with a carbon nanotube-polydimethylsiloxane composite electrode. *J. Pharm. Biomed. Anal.,* **2016**, *131*, 107-112.
[http://dx.doi.org/10.1016/j.jpba.2016.08.031] [PMID: 27589027]

[29] Zhang, L.; Zhang, W.; Chen, G. Determination of arbutin and bergenin in Bergeniae Rhizoma by capillary electrophoresis with a carbon nanotube-epoxy composite electrode. *J. Pharm. Biomed. Anal.,* **2015**, *115*, 323-329.
[http://dx.doi.org/10.1016/j.jpba.2015.07.044] [PMID: 26263060]

[30] Mao, H.; Zhang, Y.; Chen, G. Determination of three phenolic acids in Cimicifugae rhizoma by capillary electrophoresis with a graphene–phenolic resin composite electrode. *Anal. Methods,* **2019**, *11*, 303.
[http://dx.doi.org/10.1039/C8AY01942H]

[31] Gotti, R.; Amadesi, E.; Fiori, J.; Bosi, S.; Bregola, V.; Marotti, I.; Dinelli, G. Differentiation of modern and ancient varieties of common wheat by quantitative capillary electrophoretic profile of phenolic acids. *J. Chromatogr. A,* **2018**, *1532*, 208-215.
[http://dx.doi.org/10.1016/j.chroma.2017.11.058] [PMID: 29195662]

[32] Navarro, M.; Núñez, O.; Saurina, J.; Hernández-Cassou, S.; Puignou, L. Characterization of fruit products by capillary zone electrophoresis and liquid chromatography using the compositional profiles of polyphenols: application to authentication of natural extracts. *J. Agric. Food Chem.,* **2014**, *62*(5), 1038-1046.
[http://dx.doi.org/10.1021/jf404776d] [PMID: 24432703]

[33] Roseiro, L.B.; Tavares, C.T.; Roseiro, J.C.; Rauter, A.P. Antioxidants from aqueous decoction of carob pods biomass (Ceretonia siliqua L.): Optimisation using response surface methodology and phenolic profile by capillary electrophoresis. *Ind. Crops Prod.,* **2013**, *44*, 119-126.
[http://dx.doi.org/10.1016/j.indcrop.2012.11.006]

[34] Moniz, P.; Serralheiro, C.; Matos, C.T.; Boeriu, C.G.; Frissen, A.E.; Duarte, L.C.; Roseiro, L.B.; Pereira, H.; Carvalheiro, F. Membrane separation and characterisation of lignin and its derived products obtained by a mild ethanol organosolv treatment of rice straw. *Process Biochem.,* **2018**, *65*, 136-145.
[http://dx.doi.org/10.1016/j.procbio.2017.11.012]

[35] Adımcılar, V.; Kalaycıoğlu, Z.; Aydoğdu, N.; Dirmenci, T.; Kahraman, A.; Erim, F.B. Rosmarinic and carnosic acid contents and correlated antioxidant and antidiabetic activities of 14 Salvia species from Anatolia. *J. Pharm. Biomed. Anal.,* **2019**, *175*, 112763.
[http://dx.doi.org/10.1016/j.jpba.2019.07.011] [PMID: 31330278]

[36] Memon, A.F.; Palabiyik, I.M.; Solangi, A.R.; Memon, S.Q.; Mallah, A.B. Large Volume Sample Stacking (LVSS) in Capillary Electrophoresis (CE) with Response Surface Methodology (RSM) for the Determination of Phenolics in Food Samples. *Anal. Lett.,* **2019**, *52*, 2853-2867.
[http://dx.doi.org/10.1080/00032719.2019.1624371]

[37] de Souza Campos Junior, F.A.; Petrarca, M.H.; Meinhart, A.D.; de Jesus Filho, M.; Godoy, H.T. Multivariate optimization of extraction and validation of phenolic acids in edible mushrooms by capillary electrophoresis. *Food Res. Int.*, **2019**, *126*, 108685. [http://dx.doi.org/10.1016/j.foodres.2019.108685] [PMID: 31732036]

[38] Deng, Y.; Lam, S.C.; Zhao, J.; Li, S.P. Quantitative analysis of flavonoids and phenolic acid in Coreopsis tinctoria Nutt. by capillary zone electrophoresis. *Electrophoresis,* **2017**, *38*(20), 2654-2661. [http://dx.doi.org/10.1002/elps.201700129] [PMID: 28681383]

[39] da Rosa Neng, N.; Sequeiros, R.C.P.; Florêncio Nogueira, J.M. Combining bar adsorptive microextraction with capillary electrophoresis--application for the determination of phenolic acids in food matrices. *Electrophoresis,* **2014**, *35*(17), 2488-2494. [http://dx.doi.org/10.1002/elps.201400101] [PMID: 24789174]

[40] Sabo Muller, L.; Domingos da Silveira, G.; Dal Pra, V.; Lameira, O.; Viana, V.; Machado de Carvalho, L. Investigation of phenolic antioxidants as chemical markers in extracts of Connarus perrottetii var. Angustifolius Radlk by capillary zone electrophoresis. *J. Liq. Chrom. Relat. Tech.*, **2016**, *39*, 13-20. [http://dx.doi.org/10.1080/10826076.2015.1115768]

[41] Arries, W.J.; Tredoux, A.G.J.; de Beer, D.; Joubert, E.; de Villiers, A. Evaluation of capillary electrophoresis for the analysis of rooibos and honeybush tea phenolics. *Electrophoresis,* **2017**, *38*(6), 897-905. [http://dx.doi.org/10.1002/elps.201600349] [PMID: 27921291]

[42] Do, T.C.; Nguyen, T.D.; Tran, H.; Stuppner, H.; Ganzera, M. Quantitative determination of phenolic compounds in lotus (Nelumbo nucifera) leaves by capillary zone electrophoresis. *Planta Med.,* **2012**, *78*(16), 1796-1799. [http://dx.doi.org/10.1055/s-0032-1315258] [PMID: 22923198]

[43] Schaper Bizzotto, C.; Dillenburg Meinhart, A.; Poloni Rybka, A.C.; Sobrinho, M.R.; Bogusz, S., Junior; Ballus, C.A.; Teixera Godoy, H. Quantification of phenolic compounds by capillary zone electrophoresis in extracts of four commercial types of mate herb before and after acid hydrolysis. *Food Res. Int.*, **2012**, *48*, 763-768. [http://dx.doi.org/10.1016/j.foodres.2012.06.020]

[44] Gufler, V.; Ngoc, H.N.; Stuppner, H.; Ganzera, M. Capillary electrophoresis as a fast and efficient alternative for the analysis of Urceola rosea leaf extracts. *Fitoterapia,* **2018**, *125*, 1-5. [http://dx.doi.org/10.1016/j.fitote.2017.12.007] [PMID: 29229419]

[45] Honegr, J.; Pospíšilová, M. Determination of phenolic acids in plant extracts using CZE with on-line transient isotachophoretic preconcentration. *J. Sep. Sci.*, **2013**, *36*(4), 729-735. [http://dx.doi.org/10.1002/jssc.201200806] [PMID: 23401390]

[46] Cao, J.; Wei, J.; Tian, K.; Su, H.; Wan, J.; Li, P. Simultaneous determination of seven phenolic acids in three Salvia species by capillary zone electrophoresis with β-cyclodextrin as modifier. *J. Sep. Sci.*, **2014**, *37*(24), 3738-3744. [http://dx.doi.org/10.1002/jssc.201400845] [PMID: 25296843]

[47] Cao, J.; Li, P.; Yi, L. Ionic liquids coated multi-walled carbon nanotubes as a novel pseudostationary phase in electrokinetic chromatography. *J. Chromatogr. A,* **2011**, *1218*(52), 9428-9434. [http://dx.doi.org/10.1016/j.chroma.2011.11.013] [PMID: 22119676]

[48] Ganzera, M.; Greifeneder, V.; Schwaiger, S.; Stuppner, H. Chemical profiling of Edelweiss (Leontopodium alpinum Cass.) extracts by micellar electrokinetic capillary chromatography. *Fitoterapia,* **2012**, *83*(8), 1680-1686. [http://dx.doi.org/10.1016/j.fitote.2012.09.023] [PMID: 23041520]

[49] Lin, Y.H.; Huang, H.C.; Hsu, W.L. Application of capillary electrophoresis with electrokinetic supercharging and sweeping for the on-line preconcentration of phenolic acids. *Anal. Bioanal. Chem.*, **2015**, *407*(23), 7093-7100.

[http://dx.doi.org/10.1007/s00216-015-8869-x] [PMID: 26159571]

[50] Nikovaev, A.V.; Kartsova, L.A.; Filimonov, V.V. A microfluidic chip for the determination of polyphenolic antioxidants. *J. Anal. Chem.*, **2015**, *70*, 863-869. [http://dx.doi.org/10.1134/S1061934815060118]

[51] Liu, L.; Wu, B.; Liu, K.; Li, C.R.; Zhou, X.; Li, P.; Yang, H. Multivariable optimization of the micellar system for the ionic liquid-modified MEKC separation of phenolic acids. *J. Pharm. Biomed. Anal.*, **2016**, *126*, 1-8. [http://dx.doi.org/10.1016/j.jpba.2016.04.027] [PMID: 27136281]

[52] Martí, R.; Valcárcel, M.; Herrero-Martínez, J.M.; Cebolla-Cornejo, J.; Roselló, S. Simultaneous determination of main phenolic acids and flavonoids in tomato by micellar electrokinetic capillary electrophoresis. *Food Chem.*, **2017**, *221*, 439-446. [http://dx.doi.org/10.1016/j.foodchem.2016.10.105] [PMID: 27979225]

[53] Godoy-Caballero, M.P.; Acedo-Valenzuela, M.I.; Durán-Merás, I.; Galeano-Díaz, T. Development of a non-aqueous capillary electrophoresis method with UV-visible and fluorescence detection for phenolics compounds in olive oil. *Anal. Bioanal. Chem.*, **2012**, *403*(1), 279-290. [http://dx.doi.org/10.1007/s00216-012-5799-8] [PMID: 22349335]

[54] Godoy-Caballero, Mdel.P.; Galeano-Díaz, T.; Isabel Acedo-Valenzuela, M. Simple and fast determination of phenolic compounds from different varieties of olive oil by nonaqueous capillary electrophoresis with UV-visible and fluorescence detection. *J. Sep. Sci.*, **2012**, *35*(24), 3529-3539. [http://dx.doi.org/10.1002/jssc.201200696] [PMID: 23225718]

[55] Godoy Caballero, Mdel.P.; Culzoni, M.J.; Galeano-Díaz, T.; Acedo-Valenzuela, M.I. Novel combination of non-aqueous capillary electrophoresis and multivariate curve resolution-alternating least squares to determine phenolic acids in virgin olive oil. *Anal. Chim. Acta,* **2013**, *763*, 11-19. [http://dx.doi.org/10.1016/j.aca.2012.12.014] [PMID: 23340281]

[56] Parveen, S.; Memon, S.Q.; Siyal, A.N.; Memon, N.; Khuhawar, M.Y. Large-Volume Sample Staking of Rice Polyphenols Prior to Their Determination by Non-aqueous Capillary Electrophoresis. *Food Anal. Methods,* **2016**, *9*, 2152-2160. [http://dx.doi.org/10.1007/s12161-015-0394-1]

[57] Lu, Y.; Jia, C.; Yao, Q.; Zhong, H.; Breadmore, M.C. Analysis of flavonoids by non-aqueous capillary electrophoresis with 1-ethyl-3-methylimidazolium ionic-liquids as background electrolytes. *J. Chromatogr. A,* **2013**, *1319*, 160-165. [http://dx.doi.org/10.1016/j.chroma.2013.10.014] [PMID: 24188995]

[58] Rabanes, H.R.; Guidote, A.M., Jr; Quirino, J.P. Capillary electrophoresis of natural products: highlights of the last five years (2006-2010). *Electrophoresis,* **2012**, *33*(1), 180-195. [http://dx.doi.org/10.1002/elps.201100223] [PMID: 22147686]

[59] Novaes Fonseca, F.; Kato, M.J. Oliveira Jr.; L., Pinto Neto, N.; Franco Maggi Tavares, M., Critical assessment of electrolyte system for the capillary electrophoresis analysis of phenolic compounds in herbal extracts. *J. Microcolumn Sep.,* **2001**, *13*, 227-235. [http://dx.doi.org/10.1002/mcs.1047]

[60] Rabanes, H.R.; Aranas, A.T.; Benbow, N.L.; Quirino, J.P. Synergistic effect of field enhanced sample injection on micelle to solvent stacking in capillary electrophoresis. *J. Chromatogr. A,* **2012**, *1267*, 74-79. [http://dx.doi.org/10.1016/j.chroma.2012.08.073] [PMID: 22995193]

[61] Zehrmann, N.; Zidorn, C.; Ganzera, M. Analysis of rare flavonoid C-glycosides in Celtis australis L. by micellar electrokinetic chromatography. *J. Pharm. Biomed. Anal.,* **2010**, *51*(5), 1165-1168. [http://dx.doi.org/10.1016/j.jpba.2009.11.028] [PMID: 20022443]

[62] Verardo, V.; Riciputi, Y.; Pasini, F.; Filippo D'Antuono, L.; Caboni, M.F. Determination of flavonoids in Eruca and Diplotaxis rocket salad species. *Ital. J. Food Sci.,* **2011**, *23*, 352-359.

[63] Weber, C.C.; Masters, A.F.; Maschmeyer, T. Structural features of ionic liquids:consequences for material preparation and organic reactivity. *Green Chem.,* **2013**, *15*, 2655-2679. [http://dx.doi.org/10.1039/c3gc41313f]

[64] Abd El-Hady, D.; Albishri, H.M. Simultaneous determination of acrylamide, asparagine and glucose in food using short chain methyl imidazolium ionic liquid based ultrasonic assisted extraction coupled with analyte focusing by ionic liquid micelle collapse capillary electrophoresis. *Food Chem.,* **2015**, *188*, 551-558. [http://dx.doi.org/10.1016/j.foodchem.2015.05.047] [PMID: 26041230]

[65] Ni, X.; Yu, M.; Cao, Y.; Cao, G. Microstructure of microemulsion modified with ionic liquids in microemulsion electrokinetic chromatography and analysis of seven corticosteroids. *Electrophoresis,* **2013**, *34*(17), 2568-2576. [http://dx.doi.org/10.1002/elps.201300168] [PMID: 23801369]

[66] Teng, H.; Yuan, B.Q.; You, T.Y. Recent advanced in application of nonaqueous capillary electrophoresis. *Chin. J. Anal. Chem.,* **2010**, *38*, 1670-1677. [http://dx.doi.org/10.1016/S1872-2040(09)60078-5]

[67] Gómez, F.J.V.; Silva, M.F. Microchip electrophoresis for wine analysis. *Anal. Bioanal. Chem.,* **2016**, *408*(30), 8643-8653. [http://dx.doi.org/10.1007/s00216-016-9841-0] [PMID: 27526092]

CHAPTER 8

Analysis of Food Additives by Capillary Electrophoresis

Samah Lahouidak[1,2], Mohammed Zougagh[2,3] and Ángel Ríos[1,2,*]

[1] *Analytical Chemistry and Food Technology Department, University of Castilla-La Mancha, Camilo José Cela Avenue, E-13005, Ciudad Real, Spain*

[2] *Regional Institute for Applied Scientific Research, IRICA, Camilo José Cela Avenue, E-13005, Ciudad Real, Spain*

[3] *Analytical Chemistry and Food Technology Department, Faculty of Pharmacy, University of Castilla-La Mancha, Albacete, Spain*

Abstract: Electrophoretic approaches are increasingly used for the determination of food additives in real samples based on the easy operation modes and enhanced separation efficiency of Capillary Electrophoresis (CE). This chapter presents a summary of recent breakthroughs related to the development of different analytical strategies focused on enhancing the study of food samples and reviews the determination of food additives by CE, including some promising approaches. The effectiveness of these strategies to solve alimentary issues is also discussed.

Keywords: Capillary electrophoresis, Detection, Food additives, Microchips, Nanomaterials, Preconcentration, Screening.

INTRODUCTION

Food additives have become more and more noticeable for regulatory and consumer agencies. These kinds of chemicals are deliberately added to food products for different reasons [1]. However, potential health risks to humans have increased particularly when these chemicals are ingested in excess. This fact is leading governments and controlling authorities to regulate the permitted levels of food additives [2]. To avoid food fraud and protect consumers there is need for developing rapid analytical methods suitable for the simultaneous determination of many component additives mixtures with high performance, appropriate reproducibility, and good sensitivity.

* **Corresponding author Ángel Ríos:** Analytical Chemistry and Food Technology Department, University of Castilla-La Mancha, Camilo José Cela Avenue, E-13005, Ciudad Real, Spain; E-mail: angel.rios@uclm.es

María Castro-Puyana, Miguel Herrero and María Luisa Marina (Eds.)
Capillary Electrophoresis in Food Analysis

Electrophoretic analytical separation techniques have been used to determine food additives in food. A comparison of LC and GC with the CE technique reveals some interesting facts (see Table **1**).

Table 1. Comparison of capillary electrophoresis (CE), liquid chromatography (LC) and gas chromatography (GC) techniques for the determination of food additives in food.

	HPLC	GC	CE
INVESTMENT	High	Medium	Low
SOLVENT CONSUMPTION	High	High	Low
SKILLS NEEDED	Medium	High	Medium
SAMPLE PRETREATMENT	Medium	High	Low
SENSITIVITY	High	Medium	Low

CE is an interesting alternative because of its low investment costs and the possibility of performing multiple fast separations and detection with little solvent and sample consumption; however, it has a low concentration sensitivity with optical detection. CE separation is achieved by the application of an electric field through a narrow capillary [3]. This technique has become an important approach and a good alternative separation technique for the determination of a large number of food additives allowing the simultaneous determination of diverse chemicals and with the ability to analyze charged and neutral species [4]. The main advantages of CE for daily analysis are small injection volume and low waste, reduced reagent and sample consumption, shorter analysis times than those commonly obtained by LC, and the fact that it enables the separation of analytes exhibiting a poor behavior in GC [5]. In addition, complex samples can be injected into the CE system after an easy and simplified sample pre-treatment step without a detriment in the separation which makes CE more attractive for the analysis of diverse food matrices [6]. CE also provides the possibility to achieve analyses in miniaturized format which has demonstrated to be an extraordinary alternative for microchip technology because volumes in the region of nanoliters can be easily handled, it requires fixed parts and provides quick separations with a high resolution. Microchip-based CE has some other additional advantages namely fast analysis time, small size and low cost. These features are the main objectives of many analytical methods [7].

CE is generally used with a low cost and easy UV-Vis detector. However, one of the limitations of this detection device is the low concentration sensitivity because of the short length of the detection path and the low sample volumes commonly injected. To improve sensitivity many approaches have been developed using various types of detectors (electrochemical (ECD), laser-induced fluorescence

(LIF), or mass spectrometry (MS)), chemical reactions (chemiluminescence, derivatization and nanomaterials) and sample treatment units for preconcentration and clean-up (Fig. **1**). These sample treatment units can be used off-line in batch mode or incorporated on-line by coupling or integrating them with the CE system.

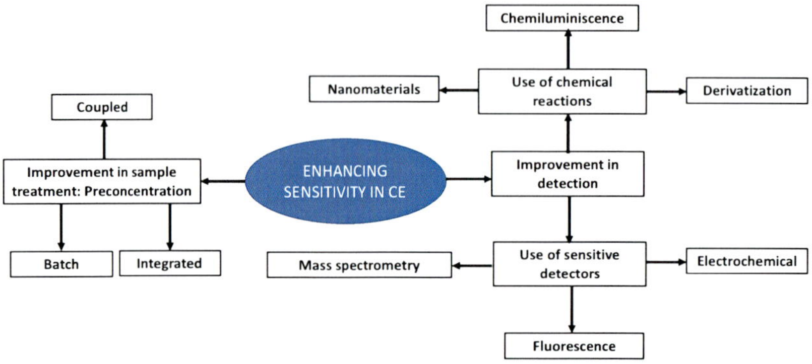

Fig. (1). Analytical strategies applied to enhance sensitivity in CE.

CE has become a promising separation technique based on its versatility given that it permits to separate very important molecules by using new and different coatings; it may be miniaturized into a microfluid-based device suited for several techniques such as Mass Spectrometry, and it has also proved its on-line coupling with sample pretreatments approaches for on-line analysis and applicability for the analysis of complex samples [8].

The aim of this chapter is to introduce food additives, their importance and uses, and discuss the common analytical strategies developed for the determination of food additives by CE including preconcentration, separation, screening, detection and the use of microchips. Some recent applications for monitoring additives in food samples by electrophoretic methods are included.

FOOD ADDITIVES: IMPORTANCE AND USES

Food additives are chemicals incorporated during food production for several purposes, for example, to increase the safety of food by preventing the undesirable effects of microbial growth and also to enhance food color or make food odor, flavor, and texture more attractive [9] but above all, these additives are employed to facilitate food preparation.

The use of food additives is not novel. The need and importance of preserving food already existed in ancient times, when plain salt was used to preserve fish or to make meat last longer. Similarly, the resins added to Greek or Roman wine acted as mild preservatives. Different authorities have defined food additives differently. According to the European Commission on regulation of food safety, a food additive is "*Any substance not normally consumed as a food in itself and not normally used as characteristic ingredient of food, whether or not it has nutritive value, the intentional addition of which to a food for a technological purpose in the manufacture, processing, preparation, treatment, packaging, transport or storage of such food results, or may be reasonably expected to result, in it or its by-products becoming directly or indirectly a component of such food*" [10].

Food additives are nowadays present in the majority of finished food products because they are directly or indirectly included in different stages of industrial food production [11]. Direct food additives are usually intentionally added by the manufacturer to food products for a specific purpose and indirect food additives are substances that drift into food during food production. Although the difference seems clear, in some cases there is some confusion between ingredients and additives. Moreover, some substances are considered additives in some cases and sometimes they are not. Caffeine is a good case in point since it is added to many food samples as a stimulant and regarded as an additive; however, it is not an additive in food in which it occurs naturally [2].

Food additives can be divided into two types. Raw ingredients extracted from natural sources, for example corn and soybeans release lecithin responsible for maintaining consistency, beets produce a powder useful to color food and some other components are not likely to be found in the natural world and thus they must be synthesized as artificial constituents. There are also natural ingredients that may be artificially and economically manufactured with even higher purity and better quality than their natural counterparts. For example, ascorbic acid or vitamin C may derive from citrus fruit or be synthesized in a laboratory [12].

New chemical substances that may be used to preserve and color foods were developed during the nineteenth and early twentieth centuries. The necessity to combat commercial fraud and overuse led to France's first regulation of chemical food additives in 1912, giving rise to the notion of a "positive list" of chemicals authorized to be added to foods. For the first time, there was an express restriction on the adding of chemicals or colorants to foods that had not previously been approved by law. The notion of positive additive lists has survived to the present day and is being used in current international regulation [13].

Besides the strict legal requirements of food risk assessments, regulations are also very complelling concerning food safety and this is why the usefulness and purpose of additives must be indicated on the packaging labels of drinks and foods where they are added. Labels must indicate the additive function and name or number. Food additive identification using a designated number is an alternative way to name additives with their specific nomination which is normally a complex and long task due to the structure of the chemical. In 1960, the European Union assigned traditional numbers to all food additives to facilitate their identification. Each substance was distinguished with a sole number of three or four figures, preceded by the letter E. This initial specifies that the additives have been previously assessed and approved by the European Food Safety Authority (EFSA) for their use in the whole EU. In the Codex Alimentarius (CA) framework, the Codex Committee on Food Additives and Contaminants (CCFAC) launched the International Numbering System (INS) in 1989.This system was established to achieve a universal accord on identifying additives by number and establishing an order list of additives including the number of every single one and its technological role. The European Commission and the Codex Alimentarius have come to an agreement on using the same identification number although the INS does not consider the E prefix exclusive for EU [13].

There are various methods to classify additives: chemical role, alphabetical order or numbering each chemical within a compounds collection. The method most oftentimes used though is the classification of additives by functional categories. CA, EFSA, and Food and Drug Administration (FDA) have grouped additives based on their function categories which have changed over time. Nowadays EFSA has settled 26 groups whereas CA has one more, therefore totaling 27 categories of food additives [12]. Considering the broader FDA definition, grouping additives as anything taking part of a food item, it only divides additives in 18 sorts. Whereas CA and EFSA establish similar classifications and names for the different additives, FDA seems to arrange these chemicals according to distinct terminologies as detailed in Table **2**. Even with this number of discriminating classes, some additives may be included in two or more categories. Six main categories of food additives are shown below together with their roles and some practical examples.

Table 2. Representation of the different classes of additives among the EFSA, CA and FDA.

EFSA	Codex Alimentarius	FDA
Acidity regulators	Acidity regulators	Anti-caking agents
Acids	Anti-caking agent	Color additives
Anti-caking agents	Anti-foaming agent	Dough strengtheners and conditioners
Anti-foaming agents	Antioxidant	emulsifiers
Antioxidants	Bleaching agent	Enzyme preparations
Bulking agents	Bulking agent	Fat replacers

(Table 2) cont.....

EFSA	Codex Alimentarius	FDA
Carries	Carbonating agent	Firming agents
colors	Carrier	Flavors and spices
Emulsifiers	Color	Flavor enhancers
Emulsifying salts	Color retention agent	Gases
Firming agents	Emulsifier	Humectants
Flavor enhancers	Emulsifying salt	Leavening agents
Flour treatment agents	Firming agent	Nutrients
Foaming agents	Flavor enhancer	pH control
Gelling agents	Flour treatment	Preservatives
Glazing agents	Foaming agent	Stabilizers
Humectants	Gelling agent	Sweeteners
Modified starches	Glazing agent	Yeast
Packaging gases	Humectant	-
Preservatives	Packaging gas	-
Propellants	Preservative	-
Raising agents	Propellant	-
Sequestrants	Raising agent	-
Stabilizers	Sequestrant	-
Sweeteners	Stabilizer	-
thickeners	Sweetener	-
	thickener	-

Preservatives

Preservatives are perhaps the single most relevant additive type, allowing food to slow or even stop down spoilage (edibility, nutritive value, or loss of quality) provoked or speeded up by micro-organisms. There are methods using benign bacteria, fungi, or yeasts to add unique qualities and to preserve food (*e.g.* wine, cheese) [14]. Keeping or creating nutritional value, flavor and texture is of importance when preserving its value as a food item.

Preservation typically encompasses prevention from bacteria growth, as well as fungi and other microorganism apart from inhibiting fat oxidation, responsible for rancidity. It also involves processes used to stop natural ageing and discoloration likely to happen during food preparation, such as Maillard reaction in apples after being cut [15].

Antioxidants

Antioxidants are compounds that inhibit oxidation of foodstuffs keeping their taste and color and remaining edible over a long time. Fats, oil, flavoring vitamins, substances, and colors can all spontaneously oxidize because of exposure to air. The deterioration rate may considerably vary and is influenced by the existence of natural antioxidants and other components, oxygen available, and how sensitive the substance is to oxidation, light, and temperature [16]. Oxidation may be minimized or delayed by several means namely air replacement by inert gases, oxygen removal with glucose oxidase, addition of a chemical absorbing UV light in transparent materials for packaging, cooling, and use of isolating agents. This might not be likely in all scenarios or sufficient to guarantee an appropriate shelf-life for food products. Therefore, antioxidants are exploited to delay decay provoked by oxygen and lengthen shelf-life. There are antioxidants responsible for oxygen removal by self-oxidation such as ascorbic acid while others intervene in the oxidation mechanism as gallic acid esters, tocopherols, BHT and BHA. All antioxidants show unique features, their applicability being more effective in some cases. The mixture of two or more antioxidants is oftentimes more efficient than when they are used separately due to synergy effects [17].

Sweeteners

Sugar is by far the most relevant flavoring substance. Its action causes the unique sweetness perception and it also constitutes an energy source. Nonetheless, an excess of sugar ingestion is associated with serious health conditions namely diabetes, obesity, and tooth decay. In this regard, alternative compounds to conventional sugar have been proposed and developed with the aim of providing sweetness with minimum energy content. Saccharin was first found in 1878 but it was not until the 1950's that customers took an interest in low-calorie foods, and sweeteners became of importance. Sweeteners may be arranged in two principal categories: intense and bulk, they are often much sweeter than conventional sugar and both types are frequently used at very low concentration levels to mimic the sweetness commonly provided by sugar. Intense sweeteners, such as sucralose, saccharin, aspartame and acesulfame-K possess, as suggested by their name, a significantly high sweetening power, variable depending on type but typically several times greater than sucrose. Bulk sweeteners, mostly polyols, enclose sorbitol, erythritol, lactitol and isomalt. They are less sweet but provoke a mouthfeel enhancement. Owing to the lower sweetening power exhibited by polyols, it is likely to combine them with other polyols or even with intense sweeteners to enhance taste quality and sweetness due to synergy effects [18]. Another asset of polyols is that they hide the undesirable metallic, bitter aftertaste of intense sweeteners. Some combinations frequently used are cyclamate with saccharin, aspartame with acesulfame-K and erythritol with acesulfame-K amongst others [9].

Coloring

Color additives, namely pigments and food dyes, are chemicals from both animal, plant, mineral sources or even synthetic [19]. Their aim is to enhance food color, add color to otherwise colorless food products, modify natural color variations, make 'funny' foods namely cake decorations and help identify flavors (for example, yellow for lemon or purple for grape flavor). Sometimes this sort of additives is known as food dyes. The guidelines on colorants are discussed in 3 groups, colorants whose ADI values are defined and the use of which is authorized, colorants authorized to be used for a particular end, for example aluminium and silver colours are only permitted to be used as surface finishing, and colorants that can only be used in specific foods like titanium dioxide, vegetable carbon and red beet [20]. The use of dyes outside this classification is prohibited.

Nanomaterials

Nanotechnology is truly a scientific revolution on continuous growth and development due to its suitability and applicability to a wide range of products and nanomaterials, its enormous interest in economic and technological issues and its impact on many industries [21]. Certain approved additives are used (i) to improve food appearance, *e.g.* by changing the color, flavor, consistency, texture of food products, (ii) develop new tastes, (iii) monitor the release of nutrients and flavors maximizing the absorption of nutraceuticals and nutrients inside the body, and minimizing fat amount, color, preservatives to favor consumption of healthy foods, and (iv) remove undesired chemicals with nanofiltration incorporating nanosensors in food packaging for food safety and traceability. Nanomaterials used in food products can be divided into two groups: inorganic compounds such as silicon dioxide (E551), titanium dioxide (E171) or silver (E174) and organic compounds based on proteins, fats, or sugars [22]. In the European Union, such additives may only be used after their official approval according to Regulation (EC) No. 1333/2008 with a labelling obligation for packaged and bulk foodstuffs in the form of E-numbers [23].

ANALYTICAL STRATEGIES FOR THE DETERMINATION OF ADDITIVES IN FOOD SAMPLES BY CE

Concerning the determination of additives in food samples by CE, four aspects can be distinguished as illustrated in Fig. (**2**).

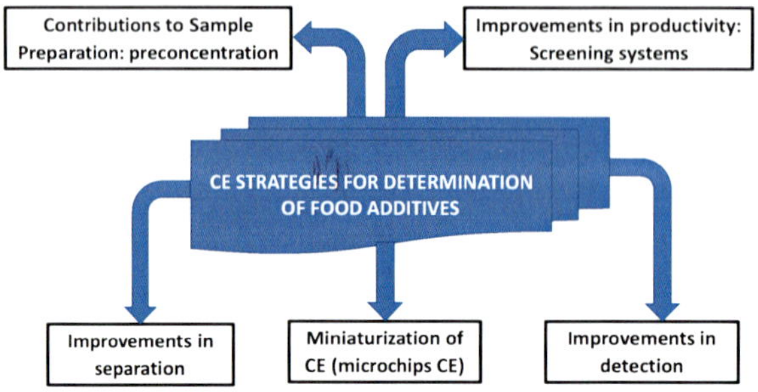

Fig. (2). Main analytical strategies to be considered for the determination of additives in food samples by CE.

Thus, some aspects can be distinguished: (i) aspects related with sample preparation particularly preconcentration; (ii) improvements in separation; (iii) aspects associated to productivity requirements (screening); (iv) miniaturization of CE (microchips CE); and (v) improvements in detection (sensitivity). These aspects are briefly described bellow.

Aspects Related with Sample Preparation: preconcentration

On-line sample preconcentration strategies with an electrokinetic basis and large volumes constitute an effective, elegant, and simple way to enhance sensitivity for detection.

Recently many online sample preconcentration approaches in CE have been described in order to increase sample amount to be injected into the system [24]. The online preconcentration enrichment caused by amendment of chemical features of the samples and/or BGS in order to handle analytes electrophoretic mobility and/or to reach partitions of the analyte into a pseudo- or a stationary phase to alter preconcentration [25]. There are four major online preconcentration techniques commonly used in CE: (i) field amplification, (ii) dynamic pH junction and (iii) transient isotachophoresis (tr-ITP).

The simplest technique aimed at stacking sample is field-amplified sample stacking (FASS). The basis of this method is the electrophoretic migration of ions from a solution with a low conductivity to another one with a high conductivity used as sample zone and background respectively. The strength of the electric field will increase across the sample zone with a low conductivity in comparison to BGE. Consequently, analytes velocity will be higher in the sample zone until reaching the buffer interface. As simpler ions enter BGE they sharply slow down and get stacked on a narrow zone [26].

Large-volume sample stacking (LVSS) is a method developed to increase sample volume to be injected when using FASS by an aqueous sample solution and hydrodynamically filling 1/2-1/3 of the capillary with sample solution, applying a reversed polarity. The electrophoretic mobility of analytes must oppose that of EOF to make possible the stacking on the boundary between the sample and the electrolyte [27]. Lee *et al.* [2012] developed a quick CE approach to analyze flavonoids in broccoli aliquots. Two key flavonols, quercetin and kaempferol were enriched on-line and separated in a short analysis time (8 min) by means of an optimized LVSS/CZE method [28]. This stacking strategy allowed for both a sensitive detection and a reproducible quantification of low flavonoid concentrations in broccoli with no need for costly detectors.

Field-amplified sample injection (FASI) is another method to increase the amount of charged species into the capillary using electrokinetic injection according to FASS mechanism, with the possibility of introducing a large analyte amount in relation with FASS [29]. Wei and al [30]. described a CE method using FASI with capacitively coupled contactless conductivity detection (FASI-CE-C4D) for sensitive determination of sorbic and benzoic acids in soy sauce. Benzoic acid and sorbic acid were adequately separated in 10 min, LODs being 0.08 μM (9.8 μg/L) and 0.05 μM (5.6 μg/L), respectively.

Isotachophoretic (ITP) stacking is a method aimed at preconcentrating but it is also useful as a purification approach, and therefore appropriate for the analysis of samples with a high ionic strength [25]. In this method, the sample ions are compressed and led to a terminating electrolyte. This leading electrolyte is composed of a high amount of an ion with a high mobility of a similar charge to the analytes and the terminating electrolyte includes an ion with a lower mobility compared to the analytes. Usually, anions or cations may run in one direction with exceptions. Regarding cation analysis, the leading buffer is placed on the cathodic edge of the sample zone, while the terminating buffer is applied at the end of the sample zone. When a steady state is reached, all ions migrate at constant velocity [31]. A. Jastrzbska [32], reported a rapid and simple Capillary Isotachophoretic strategy for the concurrent determination of nitrate and nitrite ions in meat. The determination limit of elaborated procedure was sufficiently low, whereas quantification limits ranged from 4.4 to 11.9 mg N/L (pre-separation column) and 0.7 to 1.0 mg N/L (analytical column). The correlation coefficients indicated good precision of linear calibration curves for tested ions ($R^2 \leq 0.998$). Michaela Herrmannová *et al.* [33] developed a direct simultaneous identification of eight different sweeteners by capillary ITP in chewing gum and candies. In a single capillary with no sample pretreatment the method enables fast and reliable monitoring of individual sweeteners with sufficient sensitivity and reproducibility.

Sweeping is a very popular technique, utilizing changes in analytes migration velocity because of the interaction with a pseudostationary phase in electrokinetic chromatography [34]. Cao and co-workers [35], reported flavonoids sweeping by micro-emulsion electrokinetic chromatography. In comparison with normal sample injection, an enhancement of 185-508 folds in sensitivity was obtained in terms of detection limit.

Dynamic pH junction is increasingly growing in popularity as a practical approach for preconcentration in CE. Aebersold and Morrison first proposed the employment of a sample solution exhibiting a different pH for the background electrolyte to achieve stacking [36]. If analytes get concentrated at a limited pH range owing to an alteration in the analyte velocity or mobility, the stacking strategy is dynamic pH junction [25]. This technique resembles the effect of capillary isoelectric focusing on either amphoteric analytes or analytes having a certain pI value. Within a pH range, analytes can migrate throughout the capillary until reaching a pH region equal to their pI value. The analytes tend to accumulate around this region because of their neutral feature, given that they do not have a value for electrophoretic mobility. A low-cost, efficient and easy online focusing method combining dynamic pH junction and capillary electrophoresis sweeping with polymers was optimized and developed for simultaneous identification of sorbic acid (SA) and benzoic acid (BA) [37]. Under optimal conditions, the whole analytical procedure was achieved in 7 min and an enhancement of 900 folds in sensitivity was achieved, detection limits (S/N=3) falling down to 6.1 and 8.2 nM for SA and BA, respectively.

SPE has also been incorporated in-line with CE for determination of additives in food samples in order to enhance sensitivity, employing a capillary column containing reverse phase at the beginning of the capillary, or covalent material associated with the capillary inner wall [38]. SA and BA were determined by flow analysis-ion pair solid phase extraction-CZE [39]. Due to the many bubbles caused by C18-bonded silica during elution, affecting sampling in CZE and detection by spectrophotometry, C8 was chosen as an adsorbent for the SPE column. Tetrabutylammonium bromide (TAB) was selected as an ion-pair reagent to improve preservative retention on the sorbent based on C8-silica.

Improvements in Separation

Among CE features, one of the most important applications is the wide compound range with diverse characteristics due to the various CE separation modalities, broadening the applicability of these techniques to diverse fields. In this part we will discuss the most commonly modes reported for food additives analysis in real samples including Micellar Electrokinetic Chromatography (MEKC), Capillary

Zone Electrophoresis (CZE), Capillary Electrochromatography (CEC), Microemulsion Electrokinetic Chromatography (MEEKC), and Capillary Isotachophoresis (cITP).

CZE is by far the simplest, basic, and most universal CE separation mode [40 - 45]. In CZE, the capillary is filled with the background electrolyte prior to use. The different electrophoretic mobilities allow the separation of analyte according density of charge at a certain pH value. Solutes velocity and direction are defined by both EOF and electrophoretic mobility [46].

An appropriate separation of five artificial food colorants in commercial fruit was achieved within 10 min by an uncoated capillary made of fused silica, filled with a running buffer composed of 2 mM hydroxypropyl-β-cyclodextrin and 25 mM acetic acid and (pH =3.5), separation voltage was set at −15.0 kV, and electrokinetic injection at −12.0 kV×10 s [40]. After removing the complex matrix interferences by dispersive solid phase extraction (dSPE), and using separation by electrophoretic methods and FASI conditions, detection limits for the aforementioned colorants were in the region of 0.035-0.055 mg/kg.

CZE was frequently used for flavonoids detection in wine. Ballus and co-workers [41] developed a CZE approach for detection of 16 phenolic compounds in 23 samples of rose, white, and red wine. All experiments were carried out in a 60-cm capillary (50 μm × 68 cm) filled with an electrolyte containing 175 mM boric acid (pH=9.0) and applying a 50-mbar hydrodynamic injection during 30 s. Time analysis was 19 min.

Quercetin, Rutin, Naringin, and Naringenin have also been analyzed by CZE in different fruits using photodiode array (PDA) detector [42]. The effects of CE variables including buffer pH and concentration, injection time and voltage were subject to optimization. Under optimal conditions, flavonoids were appropriately determined in a buffer containing 10 mM (pH=8.5) within 10 min and voltage value being 25 kV. Detection limits were 0.406, 0.314, 0.582, and 0.333 μg/mL, for Quercetin, Rutin, Naringin, and Naringenin respectively.

MEKC is a hybrid technique between chromatography and electrophoresis. Running buffer in MEKC contains one or more micelles in the CZE buffer solution. If concentration is higher than critical micelle concentration, the micelle is formed and charged micelles start to move throughout the EOF. Separation implies the dual role of capillary electrophoresis electric migration and micellar solubilization. This strategy may be used for separation of ions or neutral molecules. The latter migrate under the influence of an electric percolation in the electric field [46, 47]. The most used substance is lauryl sodium sulfate (SDS).

Flavonoids and tomato phenolic acid have been determined by MEKC combined with a DAD device [48]. Polyphenols separation was obtained at 15 °C within 20 min using 11.2 mM SDS and 11.3 mM borax, as micellar BGE with no organic modifier. Under optimal conditions the LOD ranged from 0.8 to 3.8 μg/mL.

Guan and co-workers [49] developed a MEKC method for determining synthetic phenolic antioxidants in different matrix foods including vegetable oil, mushroom cream, and fish soup. The compounds were separated in a 75-cm length and 25μm i.d and 360 μm o.d. fused-silica capillary using 20 mM borate (pH 7.4), 25 mM SDS.

MEEKC is a type of electrokinetic chromatography derived from MEKC and developed as a technique in the 1990s [50]. The micro-emulsion is stable and the transparent liquid for the ultrasonic treatment mixture is composed of buffer, auxiliary surfactants, surfactant and n-alkanes. The nanoregion of micro-emulsion was sprinkled as a stationary phase in the buffer. The analyte distributes between pseudo-stationary phases and mobile phase. MEEKC is suitable for a simultaneous separation of water-soluble, fat-soluble, uncharged or charged compounds. The separated compounds may have a wide polarity range. Between MEKC and MEEKC, the main difference is that the large sample capacity of the former [49]. Bordagaray determined by MEEKC four color additives commonly used. The addition of SDS in borax buffer was not enough to separate all the relevant peaks, therefore the addition of β-CD was necessary. The best separation was obtained using 25kV, 25°C, and 12mm borax, 6mm SDS, AND 5mm β-CD. LOD values were between 0.24 and 1.21 mg/L. Another successful MEEKC approach was implemented to discriminate seven preservatives in food products such as wines, soy sauces and soft drinks after a C8 solid phase extraction [51].

CEC is another CE mode with applicability to separate neutral or charged compounds. The technique makes use of a stationary phase in a similar way to HPLC, together with a certain voltage and with no pump constituting hybrid separation procedure of interest [52]. The displacement of mobile phase is achieved by electric field while the electroosmotic flow is used to transport solvent, then having an absence of column back pressure. A small volume of sample and solvent is required, with a positive impact on net analysis cost. Because of this reason, CEC has attracted a certain potential, especially in biomedicine and pharmaceutical fields. Nevertheless, food analysis has attracted attention of a great deal of researchers focusing on this direction. Bottoli and co-workers [53] described a CEC approach for determining parabens in sweeteners. Commercial C18 silica capillary column was used as separation phase. An appropriate parabens separation was accomplished within 18.5 minutes with a running electrolyte containing 50:50 v/v acetonitrile and tris(hydroxymethyl) aminomethane (pH=8.0). This method was successfully implemented to the quantitative determination of parabens in sweeteners with direct injection.

cITP was recently described as a new approach for food additives analysis. It is an electrophoretic technique based on moving boundary, and two background electrolytes are combined while these zones move at the same speed. The zones remain enclosed between terminating and leading electrolytes and separation is regulated by the running electrolyte [54]. The principal benefit is the chance to perform a fast, concurrent detection of a great deal of anions. This method has been recently applied to determination of functional additives, food preservatives and selected ions. Orthophosphates, pyrophosphates, tripolyphosphates and nitrites were identified and quantified by Capillary Isotachophoresis in various food products [55].

Improvements in Productivity: Screening Systems

Detection methods for food additives are commonly arranged into two categories, confirmatory and screening. Confirmatory methods are usually based on quantification of analyte concentration. However, their application is limited by factors such as the laboriousness of the process, costly sample analysis and laboratory instruments maintenance, the requirement for qualified and experienced staff and the difficulty of the sample preparation process [56]. Screening methods are intended to decide whether a sample contains an additive or not. If a positive result is obtained, selecting the appropriate method to correctly identify, verify and quantify food additives is essential. Screening methods permit to identify an analyte or a family of analytes at a certain concentration level, and normally provide semiquantitative results. The main advantages of a screening assay are the low rate of samples giving false-positive results, simplicity, high throughput, timesaving, low cost and high selectivity [57].

Moreno and co-workers [58] developed a screening confirmation method for distinction and characterization of titanium dioxide TiO_2 between nano- and micro- fraction in sugary foods. Fig. **3** shows the procedure followed in this work. Firstly, a portable Raman spectrometer was used to confirm the presence of TiO_2 in the samples and then CE was used to characterize and separate the different size particles present in the samples. The results obtained in this study demonstrated that the combined use of Raman spectroscopy and CE techniques can be a valuable analytical approach to screen specific micro/nanoparticles in food items.

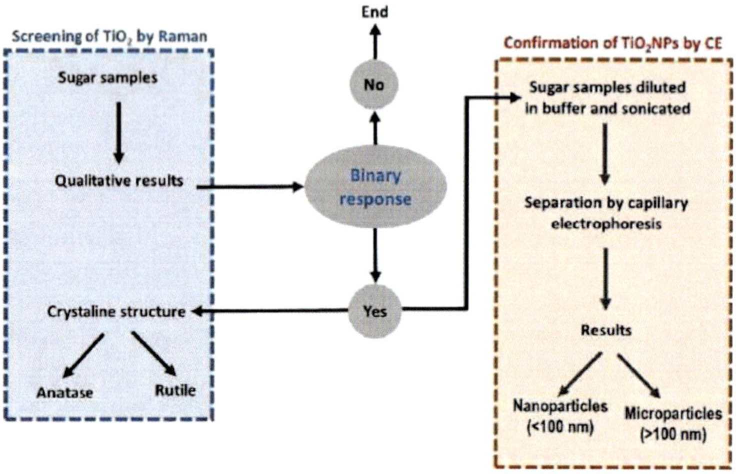

Fig. (3). General analytical nanometrological approach for the screening-confirmation of titanium dioxide nano/micro-particles in sugary food samples based on Raman spectroscopy- capillary electrophoresis. Adapted from ref [58]. and reproduced with permission.

Ávila and co-workers [59], described a strategy based on a quick separation of fingerprint markers from Vanilla planifolia extracts and vanilla-related specimens by microfluidic-electrochemistry chip making use of an amperometric detector. Microfluidic chip worked simultaneously as screening and confirmatory mode simplifying the analytical process. The detection of fingerprint Vanilla planifolia markers and ethyl vanillin allowed for a quick and simple confirmation of possible frauds in vanilla-related food sample. Hao and co-workers [57] developed a screening method followed by determination of food additives by means of capillary electrophoresis hyphenated with a contactless conductivity detection. The method developed is simple and inexpensive, making these activities feasible for food control even in local laboratories with modest funding and limited qualified personnel. There was a consistency between figures obtained with CE-C^4D instrument and the ones obtained with the confirmation method (HPLC-PDA).

Miniaturization of CE (Microchips CE)

Microchip is a miniaturization of analytical and bioanalytical instruments resulting in an extremely rapid separation consuming only sample volume at pL level and low consumption reagents, giving the possibility of combining both preparation and analysis of samples with a single device with high efficiency, high throughput and easy operation [60]. A microchip application to alkyl gallates separation and nordihydroguaiaretic acid [61] present the microchip potential for separation and routinary analysis of antioxidants in food samples, offering a low expensive portable and customized tool for analyzing small samples including phenolic antioxidants. Two artificial food dyes, patent blue and Green S, were separated by capillary electrophoresis microsystem with in-channel setting for amperometric detection significantly improving the performance of electrochemical detection for microchip CE, because of the employment of a ground-floating electrically isolated potentiostat [62]. Moreover, the proposed method is particularly appealing for separation and detection of electroactive additives in food.

Improvements in Detection

Several methods of detection were recently reported for CE analysis including UV detection, mass spectrometry, contactless conductivity detection, chemiluminescence detection, electrochemical detector, and laser-induced fluorescence detection. In approximately 80 percent of the cases, a UV detection system was used for the separation of food additives by CE [63 - 73]. UV detection suffers from poor concentration sensitivity due to the short path length given by the narrow separation capillaries [74]. Thus, a sensitivity improvement was necessary.

Sensitive CE with a mass spectrometry device for detection is frequently used [75, 76] because of its appropriate selectivity, high sensitivity, and information of molecular structure to identify unknown compounds [77]. Surowiec and co-workers [78] made a comparison between capillary electrophoresis with ultraviolet and mass spectrometry detection. Detection limits that were achieved by means of MS were even lower than in the case of UV-Vis. Micellar electrokinetic chromatography-MS/MS using a volatile surfactant was used to determine Sudan dyes in chili [79].

The applicability of capacitively coupled contactless conductivity detection (C^4D) has been tested for CE methods (CE-C^4D) with the aim of determining food additives, in particular those exhibiting low UV absorption [80, 43]. A simple analytical approach based on CE hyphenated with C4D was proposed for determination of xylitol, maltitol, erythritol, and sorbitol in chocolate samples free of sugar [81]. Saccharin, cyclamate, aspartame and acesulfame-k were also determined using CE-C4D. The LODs and LOQs obtained were regarded to be better than values commonly obtained by photometric detection hyphenated with CE [82].

Chemiluminescent devices have also been used for additives detection. Quercetin and rutin were detected with this systems. The post column chemiluminescent reagent for detection was K3Fe (CN)6 in alkali. The detection limits were 1.51×10^{-3} μg/mL and 6.10×10^{-4} μg/mL, respectively [83].

Detection based on electrochemistry was another promising approach. Amperometry-based detection was the most extensively used among electrochemical strategies. It is advantageous due to low cost and easy operation. CZE with electrochemical devices [74] and MEKC with electrochemical detection [84] have also been reported. Qian described CE with electrochemical detection. Limits of detection were between 0.27 and 0.88 µg/mL [73].

CE in combination with amperometric detection was also employed for separation of phenolic acid and flavonoids samples of Chinese herbal tea. Limits of detection fell down to 10 µg/mL [75]. CE with amperometric detection was also used for determination of Folium Eriobotryae, limits of detection down to 0.83-1.36 µg/mL for all analytes [85]. A home-made system combining CE and amperometric detection was employed to quantify flavonoids in traditional Chinese drugs. Limits of detection ranged from 0.15 to 0.26 µg/mL [86].

Capillary electrophoresis in combination with laser-induced fluorescence detection was also described for separation of food additives in real samples. Laser-induced fluorescence detection is adequate for both fluorescent chemicals and derivatized compounds [87]. CZE, when combined with laser-induced fluorescence detection, was applied to separate and determine red colorants, the limits of detections obtained were about one-order lower compared to results obtained with CE-UV/Vis [88], and limits of detection for all compounds ranged from 0,2 to 0.5 µg/mL [89]. MEEKC hyphenated with laser-induced fluorescence detection was important to quantify seven flavonoids. LODs for all the compounds ranged from 0.07 to 17.99.

RECENT APPLICATIONS FOR MONITORING ADDITIVES IN FOOD SAMPLES BY ELECTROPHORETIC METHODS

Preservatives

Preservatives are food additives that protect against the action of microorganisms to prolong self-life of foodstuff. Commonly used preservatives, organic acids, namely benzoic acid (E210), sorbic acid (E200) and propanoic acid (E280), are also preservatives extensively used in low pH foods. Relevant information relating to the determination of preservatives by CE is presented in Table **3**.

Table 3. Determination of preservatives in food.

Food application	Analytes	Method	Separation conditions	Detection/LOD	Ref.
Lemon sauce, pickle, fruit juice, ketchup, olive past, and cake	benzoic acid sorbic acid	CZE	20 mM borate buffer at 9.3 pH	UV detection, 220 nm, 260 nm. 0.405 µg/mL, 0.415 µg/mL	[15]
Bread	Propionate, sorbate	CZE	10 mM benzoic acid, 15 mM L-histidine and 0.2mM of cetyltrimethylammonium bromide (CTAB)	UV, 235 and 250 nm. 4.3 and 1.5 mg/kg	[64]
Fish	Nitrates, nitrites	Sample Stacking-CE	30 mM formic acid and 30 mM sodium sulfate at a pH of 4.0	UV detection, 210 nm. 0.55 and 0.82 µmol/L	[65]
Tea	Nitrites	Multiple-Injection CZE	4.0 g/L β-alanine and 1.5 g/L perchloric acid (pH 3.79)	UV-Vis diode- array detector, 200 nm. 0.324 µg/mL	[66]
Yoghurt	Natamycin	CZE	20 mM Na2HPO4, pH 9.2	UV detection 306 nm. 0.065µg/mL	[77]

In this sense, Öztekin N [15]. presented a simple, rapid CE method for the simultaneous determination of sorbic acid and benzoic acid in a range of foodstuffs including lemon sauce, pickle, fruit juice, ketchup, olive paste, and cake. The ionization degree of an organic acid depends on the selected buffer. Its pH has an influence on ionization degree and separation. Simultaneous migration of cinnamic acid, sorbic acid and benzoic acid was assayed in three diverse background electrolytes with different pH range. When borate buffer at 9.30 pH was used as separation media, also the effect of borate concentration was studied, optimal peak area and peak symmetry were achieved with a 20 mM borate buffer. A value of 28 kV was chosen as the suitable voltage as a compromise between migration time and resolution. Under optimal conditions, there was no matrix effect in the samples that were highly diluted. Sorbic and benzoic acid were detected at different wavelengths, 220 nm and 260 nm respectively, migration time was 3.1 min and the LODs were found to be 0.405 µg/mL for benzoic acid and 0.415 µg/mL in the case of sorbic acid. The LOQs were 1.35 µg/mL and 1.38 µg/mL, respectively.

De Jesus and co-workers [64] described a new method for the simultaneous determination of sorbate and propionate in bread. The sample pre-treatment was achieved by simply dilution and sonication before filtering by means of a polyvinylidene difluoride (PVDF) membrane filter (0.2 μm). Acetate was set as an internal standard. Lactic acid (as lactate) frequently present in bread to control its acidity, shows a similar mobility to propionate. Thus, the BGE with 5.8 was chosen to minimize co-migration of these particular ions. As acetate and propionate do not possess a strong chromophore substituent, benzoic acid was added in the BGE composition to enhance UV absorption, thus allowing for an indirect detection and obtaining negative peaks for these chemicals at 235nm. By contrast, sorbate was detected at 250 nm. A good resolution was achieved in less than 5 min and the LOQs were 1.5 and 4.3 mg/kg for sorbate and propionate, respectively.

Nitrates and nitrites are small anions commonly applied to inhibit *Clostridium botulinum* in foods containing raw meat such as fish, sausage, ham, bacon, and salamis. To promote a faster migration for smaller anions, EOF is decreased by working with buffers at a low pH value and sample injection is performed with an inverse polarity, that is to say, from the cathode [91].

In Zeynep Kalaycıoğlu and co-workers' study [65], a formate/formic acid buffer was chosen as low pH medium and EOF reduced. Under optimal conditions regarding resolution and migration times, only the content of nitrate would be detected. For this reason, a sample stacking method was described to enhance the detection sensitivity of both ions for real sample analysis and the separation buffer conductivity was augmented with the aid of sodium sulfate to this end. Acetonitrile is also added to fish specimens to remove proteins and improve the stacking effect due to its low conductivity. The staking mode resulted in an improvement of in detection sensitivity of 30 folds for both anions compared with small-volume injection. Also, Ana p. Ruani co-workers [66] reported the utilization of CE with multiple injection as a quick approach to determine antinitrosating capacity of nine commercially available tea types by means of nitrite quantification. This methodology consisted of injecting the sample prior to injection of the control solution, permitting determination of the antinitrosating capacity (AC) in a sole run.

Nisin (E234) and natamycin (E235) are antibiotics used to inhibit fungi and bacteria growth in various foods. Xiaobin Li co-workers [67] presented a CZE method to determine Natamycin in food items and a comparison with chromatographic approaches. Both methodologies for analyzing Natamycin originated good results. LOD obtained with HPLC method was lower than in the case of CZE. Nevertheless, CZE is more advantageous given that reagent consumption is smaller and detection time is shorter.

Antioxidants

Antioxidants, both synthetic and of natural origin, are included in the food composition to avoid or decrease the velocity of oxidation reactions that cause food to go rancid. Artificial antioxidants are polyphenols with a relative liposoluble degree and they must not be noxious. Additionally, it is advisable not to confer different tastes to the food items where they are added [49]. The most broadly used primary antioxidants are butylated hydroxytoluene (BHT), tert-butylhydroquinone (BHQ), butylated hydroxyanisole (BHA), and gallic-acid esters. Natural antioxidants frequently added to food encompass isoascorbic acid, ascorbic acid, and a variety of tocopherols. Several analytical strategies have been addressed and validated for antioxidants analysis by HPLC and GC. The representative contributions of CE in the determination of antioxidants in food and the most relevant aspects of each application are reported in Table **4**. Guan and co-workers [49] implemented a MEKC approach (25 mM SDS in 20 mM borate, pH=7.4) for separating BHQ, BHA, PG and BHT. With the aim of enhancing sensitivity, they made use of amperometry-based detectors and reported 1 mM as LOD. This methodology was useful to determine BHQ and PG in vegetable soup and oil, their levels being below the maximum permitted values. Ding and co-workers [61] also used amperometric detection in combination with MEKC (SDS being present in the pseudostationary phase) to separate DG, PG, OG and nordihydroguaiaretic acid. They accomplished separation on a microchip, showing a separation in solely 2 min and similar sensitivity values (2-4 mM) to Guan and co-workers [49]. Darji and co- workers [92] described a useful method for the identification of authorized synthetic antioxidants using a typical MEEKC method with in-capillary concentration of BHT by sweeping. Firstly, the effect of the pH was studied. When using the high pH buffer five out of seven antioxidants were separated in less than 20 min although α-tocopherol and highly non-polar BHT comigrated. Since at the low pH value analysis time was shorter (less than 5 min), only four peaks were observed. Non-polar antioxidants (α-tocopherols, DG, OG and BHT) strongly interacted with the microemulsion and it was not possible to separate them. To change the elution order or improve the resolution, the buffer polarity was altered by including an organic solvent. The addition of propanol---ol was useful to resolve all antioxidants in only 10 min. The buffer consisted of 0.6 g of octane, 6.6 g of butan-1-ol, 3.3 g of SDS, 69.3 g of phosphate buffer (25 mM) and 20 g of propan- 2-ol. Detection was performed at 214 nm and LODs ranged from 0.9 to 44 μg/L.

Table 4. Determination of antioxidants in food.

Food application-	Analytes	Method	Separation conditions	Detection/ LOD	Ref.
Food sample	PG, OG, LG, NDGA	Microchip MEKC	30 mM borate+30 mM SDS at pH 9.7	pulsed amperometric detection 1.4-4.1 μM	[61]
Fruit juices	ascorbic acid, sorbic acid	CZE	80 mM boric acid 5 mM borax	UV detection 1.70 and 0.54 mg/L	[68]
Vegetable oil, mushroom cream, fish soup	PG, TBHQ, BHA, BHT	MEKC	20 mM borate, 25 mM SDS pH 7.4	electrochemical detection $2,9.10^{-7}$ to $2,7.10^{-6}$ mol/L	[49]
Tomato	Resveratrol, naringenin, rutin, ascorbic acid chlorogenic, acid myricetin.	CE	50 mM borate buffer (pH 8.7)	Electrochemical detection 1.10^{-8} - 1.10^{-6} g/mL	[93]
Wine	BHQ, PG BHA, OG BHT, DG tocopherol AA, IA	Mixed Micellar Electrokinetic Capillary Chromatography	sodium cholate (40 mM), sodium dodecyl sulfate (15 mM), 10% methanol, and 10 mM borate at pH 9.3.	UV detection 1-10 g/mL	[92]
Edible oils	BHA BHT DG	L-L extraction-MECK	20% ACN, 20 mM boric-borate at pH 9.2 2 mM bis-(2-ethylhexyl) sodium sulfosuccinate surfactant	UV detection at 280 nm 0.45, 1.47 and 0.27 ppm	[63]

Applicability of CE to the determination of artificial antioxidants, especially regarding food analysis, is limited. By contrast, several CE studies have focused on the separation and detection natural antioxidants as ascorbic acid and other vitamins with antioxidant activity.

Tang and co-workers [68] described a rapid approach for the simultaneous determination of sorbic acid and ascorbic acid in fruit juices by CZE. Both chemicals were successfully separated in a buffer containing 20 mM boric acid and 5 mM borax (pH=8.0) in 5 min at 270 nm as the detection wavelength. Under optimal conditions the detection limits were 1.7 mg/L in the case of ascorbic acid and 0.54 mg/L for sorbic acid. The determination of ascorbic acid and phenols in different tomato fractions was successfully developed by Peng and co-workers [93] using CE hyphenated with electrochemical detection to analyze ascorbic acid and phenolics. A 300-µm diameter carbon disk electrode was selected as the working electrode showing a good response at +0.90 V (against saturated calomel electrode) for the compounds. Under optimal conditions, the analytes were separated at baseline in 20 min using a buffer composed of 50 mM borate (pH=8.7). Notably, an extraordinary linear range was achieved with limits of detection between 1×10^{-8} and 2×10^{-7} g/mL for all analytes. This methodology has been implemented to follow phenols and ascorbic acid content in actual specimens, with rewarding results.

Sweeteners

Synthetic high-intensity sweeteners (also known as non-nutritive sweeteners) constitute a relevant sort of food additives frequently employed as a low-calorie alternative to sugar. This kind of compounds are prescribed for individuals needing to minimize their sugar intake for health reasons [80]. Another advantage of artificial sweeteners is that there is no fermentation in teeth microflora, therefore making teeth healthier. Nonetheless, because of their potential toxicity (*e.g.*, saccharin may enclose a carcinogenic effect) these compounds are allowed to be added in foods by a specific legislation intended to guarantee consumer safety. For this reason, there is need for developing analytical approaches to achieve a sensitive analysis of synthetic sweeteners in food items. In this regard, a CZE-C^4D-based approach was developed for the simultaneous determination of four common sweeteners (aspartame, cyclamate, acesulfame-K and saccharin) (see Table **5**). Separation was performed in less than 3 min with superimposed hydrodynamic pumping in the capillary in order to improve separation time, using a 150 mM 2- (cyclohexylamino) ethanesulfonic acid (CHES) and 400 mM tris(hydroxymethyl)aminomethane (Tris) (pH=9.1) in order to render all analytes in the fully deprotonated anionic form. Good sensitivity was obtained by adjusting the injected amount. The LOD obtained by the method developed were (3.8, 4.0, 5.0 and 6.5 µmol/L for acesulfame-K, saccharin, cyclamate and aspartame, respectively [82]. Xia and co-workers [80] reported an analytical method for quantification of three sulfanilamides in candied fruits and beverage by CE-C_4D with porous aromatic frameworks (PAFs) enhancing SPE, as an adequate sample handling, to minimize the matrix interference and achieve an excellent LOD. The BGE consisted of 20 mM Tris, 20 mM $Na_2B_4O_7$ and 50 µM CTAB (pH=9.74) and LODs were 0.09 µM, 0.12 µM and 0.22 µM, respectively.

Table 5. Determination of colorants in food.

Food applications	Analytes	Method	Separation conditions	LOD	Ref.
Beverage samples	aspartame acesulfame-K, saccharin, and cyclamate	CZE-C⁴D	150 mM CHES 400 mM Tris at pH 9.1	6.5, 5.0, 4.0, 3.8 μmol/L	[82]
Beverage samples, candied fruits	acesulfame-K, saccharin, and cyclamate	SPE–CE–C⁴D	20 mmol/L Tris, 20 mmol/L $Na_2B_4O_7$ and 50 μmol/L CTAB at pH 9.74	0.09, 0.12 and 0.22 μM	[80]
Sugar-free chocolate	erythritol, maltitol, xylitol, sorbitol	CZE-C⁴D	25 mM borate at pH 8.5	2.7–4.8 μg/g	[81]
Beverages	acesulfame-K, saccharin, and cyclamate	FASI-CE-C⁴D	20 mmol/L HAc	4.4-8.8 μg/L	[95]
Free-flowing sweetener, Sweetener tablets Liquid, sweetener Jupík fruit beverage, Staropramen fruit beer.	Rebaudioside A, Stevioside.	CE-C⁴D	170 mM H3BO3/ LiOH at pH 9.0	0.3 mg/L	[43]

Polyols or sugar alcohols, namely sorbitol, erythritol, xylitol and maltitol are extensively utilized as sweeteners to substitute sucrose in low-calorie foods. The determination of these compounds in sugar-free or foods with a low-sugar content is relevant regarding quality control and nutritional information. Polyols separation by CE may be carried out with a strong alkaline BGE for the ionizing these compounds, owing to high pKa values (above 12). Detection of polyols by UV is challenging because of the lack of chromophore groups in these compounds. C^4D might be a feasible alternative for polyols determination using CE separation. Erythritol, sorbitol, xylitol and maltitol were simultaneously analyzed in sugar-free chocolates by means of a CZE-C^4D system by Coelho and co-workers [81]. To achieve this objective, polyols were preconcentrated from the actual samples by an easy procedure consisting in the use of water and ultrasonic energy. A 25 mM borate (pH 8.5) buffer caused the formation of anionic borate esters (formed by the existing interaction of polyols with borate ions), separated in less than 6 min. Method validation showed LODs in the region of 2.7-4.8 μg/g and performance values between 70 and 116% (RSD ranging from 0.2 to 19%). The applicability of the proposed methodology to actual items demonstrated that the majority contained maltitol as principal sweetener. Online preconcentration may be considered as a major breakthrough in CE, especially to avoid the sensitivity limitation in the determination of artificial sweeteners in food samples.

A sensitive method for simultaneous determination of three synthetic sweeteners by CE with field- amplified sample injection (FASI) and C^4D was also developed [95]. Under optimal conditions, separation was carried out in 10 min with the detection limits in the range of 4.4-8.8 µg/L showing a great potential method compared with the CE-C^4D method.

Colorants

Regarding food technology concerns, food colorants are chemicals added to food to balance the loss of natural color typically occurring during storage and to retain the desirable color appearance of food products. Based on the increasing demand for more and more attractive products, several artificial food colorants have been synthesized for their application to food production. As it is well known, artificial food colorants might be pathogenic and may have adverse effects on consumers health. Various methods have been reported for colorant analysis in food such as gas chromatography (GC), high-performance liquid chromatography (HPLC) and thin-layer chromatography (TLC). HPLC is mostly used because of its adequate sensitivity when coupled with a classical UV-Vis detector. Nevertheless, the majority of HPLC approaches already described for colorant analysis show a scarce separation degree and a lot of time is spent for conditioning the column, especially if a great deal of food colorants is needed to be determined. Several CE strategies have been described as attractive and valuable alternatives for the determination of food colorants in foodstuff especially to resolve mixtures of closely related dyes standards and mixtures of dyes typically present in selected foods. Most synthetic dyes contain carboxyl acid or sulfonic acid groups, forming colored anions at a high pH.

When charged, food dyes are suitable analytes for CE separation according to electrophoretic mobilities and although there is a good prospective for CE in dye analysis, the low detection sensitivity present a major disadvantage. Many approaches have been described to enhance CE sensitivity for detection of colorants in different food matrices by employing a more sensitive detection method or including a step for sample preconcentration prior to separation (see Table **6**).

Table 6. Determination of colorants in food.

Food application	Analytes	Method	Separation conditions	Detection/ LOD	Ref.
Soft drinks, milk beverages, jellies	Brilliant blue FCF, carminic acid, Allura red AC indigo carmine Sunset yellow FCF Tartrazine Fast green FCF New coccine	CE -LVSS	20 mM sodium hydroxide, 15 mM borax, 7 mM β-Cyclodextrin at 10 pH	UV–visible detection 0.1–0.5 mg/ml	[44]
Fruit juice	Carminic acid Tartrazine Sunset yellow Brilliant blue	dSPME-CE	30 mM phosphate buffer (pH 6.0) containing 2 mM β-CD	UV absorption at 220 nm. 0.030-0.36mg/L	[40]
Vinegar, Soy Sauce	5-hydroxymethylfurfural (HMF)	UALLME-CE	ND	UV detection at 280 nm 0.10 mg/L	[71]
Solid dyes (for paella) Energy drinks, Non- alcoholic drinks, Ice-pops, Liquid jellies Drug.	Tartrazine, Sunset Yellow, Red Allure, Blue Brilliant	MEEKC	12.5mM Borax, 6mM SDS and 5mM β-CD	UV detection 0.80-4.03 mg/L	[56]
Chili products	Sudan I, Sudan II, Sudan III, Sudan IV	MEKC	60:20:10:10 (v/v/v/v) APFO solution /ACN/ MeOH /THF	MS/MS 22 μg/kg	[79]

Huang and co-workers [44] described a strategy consisting in an on-line capillary concentration namely large-volume sample stacking (LVSS) and efficient CE separation to detect and analyze Brilliant blue FCF, carminic acid, Allura red AC, indigo carmine, Sunst yellow FCF, Tartrazine, Fast green and new coccine in food items such as jellies, milk beverages and soft drinks. All samples used in LVSS were treated in a low-conductivity matrix and injected hydrodynamically into the capillary electrophoresis. The concentration of colorants was 50 μg/mL in the case of classical CE and 1 μg/mL for LVSS. The complex matrices of the food samples examined make difficult to choose a particular colorant to be enriched due to the simultaneous concentration of most components in these samples during the staking procedure which resulted in the appearance of unknown spikes in overlapping peaks. To minimize the matrix effect during the separation process, SPE with polyamide was chosen as the pre-handling step. The developed technique provided various limits of detection in the region of 0.10-0.50 mg/L.

Another method was studied by Yi and co-workers [40]. They reported a high-sensitive detection of artificial food colorants in preserved fruits by FASI coupled with CE with a contactless conductivity detector after dispersive solid-phase extraction (dSPE) sample clean-up. The complexity of matrix interferents, *i.e.*, biomolecules and inorganic salts co-existing with food colorants at a trace level in preserved fruits were eliminated after cleanup and there was a sensitivity enhancement in FASI. Under optimal conditions, five artificial food colorants were properly separated within 10 min under CE conditions and LOD values ranged from 0.035 to 0.055 mg/kg.

Over the last decade, nanoparticles (NPs) have been employed as CE components. NPs can take part in pseudostationary phases (PSPs) to improve column efficiency and separation selectivity. Moreover, NPs have been appealing due to their elevated specific surface area when exploited as sorbents [96]. Liu and co-workers [70] addressed a novel methodology for food colorants determination in beverages using silica nanoparticles functionalized with diamino moiety both being good sorbents in pseudo-stationary phases for CE and in dispersive solid-phase micro-extraction. Detection limits for four food colorants fell in the region of 0.030-0.36 mg/L.

Liquid-Liquid Extraction (LLE), is one of the most popular techniques in food colorants determination after solid-phase extraction. LLE deals with compounds separation according to solubility in two immiscible liquids. Typically, solvents used for extracting artificial colorants from food items are methanol, ethanol, ammoniac ethanol, ethyl acetate, isopropyl alcohol, ammonia, tetra-n-butyl ammonium phosphate, cyclohexane or even water [97]. Wu and co-workers [71] reported an extraction approach with a basis on Ultrasonic Assisted Liquid-Liquid Micro-Extraction (UALLME) in two steps and coupled with a CE-UV system for analysis of 5-hydroxymethylfurfural (HMF) in high ion-strength samples (soy sauce and vinegar). Under optimal conditions, LOD and LOQ for HMF were 0.03 and 0.10 mg/L, respectively.

Sudan I, II, III and IV are industrial dyes commonly used as coloring plastics and other synthetic materials. Due to their low price and intense red color, these compounds have been illegally added as a dye in some food preparations, mainly in curry, chili, and some paprika powders, palm oil, to intensify their color [74]. Sudan dyes are weak acids and their pKa do not exceed 11.5. Therefore, they may be regarded as neutral chemicals in a broad pH range and MEKC is a suitable technique to accomplish full separation of these substances. A recent MEKC-MS/MS hyphenation has been developed by González and co-workers [79] to determine four Sudan dyes in chili products using ammonium perfluorooctanoate (APFO) as volatile surfactant, coupling MEKC to AJS-ESI-MS/MS. Under optimal conditions, LOQ value was no higher than 22 μg/kg.

Nanomaterials

The advent of nanotechnology has meant a true revolution in science, and it has completely changed several aspects of our daily lives. For this reason, the interest for this area is increasing more and more owing to the advantages in many research lines and its applicability to numerous fields such as agriculture, aerospace, chemistry, food, electronics, pharmacy, industry, transportation and even medical goods. According to the European legislation, silica nanoparticles SiO_2NPs, silver nanoparticles AgNPs, titanium dioxide nanoparticles TiO_2NPs, gold nanoparticles AuNPs were arranged in the list of authorized food additives being coded as E551, E174, E171, E175 respectively, with maximum permitted levels for each of them. The size separation and characterization of NPs are one of the most relevant trends in recent years, given that their chemical and physical features depend on their size. CE has recently become one of the most powerful analytical techniques for separation, and it can be used for separation purposes or analysis of various materials with different sizes. This fact makes CE an appealing technique for the identification and determination of NPs. Table **7** reports some selected applications. The most common CE detectors are UV/Vis absorbance, fluorescence conductivity or MS. The advantage when using CE in comparison with the rest of separation techniques is the high resolution power that may be obtained and the ability to analyze both NPS and ionic species. However, there are three principal challenges limiting the usefulness of CE in the analysis of NPs, which are associated to the nanoparticles′ own features, run-to-run variation and capillary heating. The determination of nanoparticles in food matrix by CE is still challenging which makes this kind of applications rarely available. Adelantado *et al* [98] described a simple and low-cost CE method coupled with an evaporative light scattering detector (ELSD) to discriminate and separate SiO_2NPs of different sizes. This application is based on the on-line coupling of ELSD with CE system using a new and adapted interface. Fig. **4** shows a schematic diagram of the CE-ELSD system used for the separation and determination of SiO_2NPs of 20, 50 and 100 nm [99]. Salt and sweeteners were chosen as real samples for determination of silica. In this case, it was proven that there is matrix influence on nanoparticles peaks when using CE in combination with ELSD. Palomero *et al* [100] have also used CE for determination of silver nanoparticles in food products including mussels and orange juice as complex matrices. To remove the effect of matrix, a simple and efficient dispersive micro solid extraction was developed using a sorbent such as sulfonated nanocellulose. Haiou Qu *et al.* [101] established an analytical strategy with high resolution composed of a CE interface with ICP-MS to characterize and separate multimode gold nanoparticles mixtures including different sizes with slights differences (5, 10, 15, 20, 30nm). The common surfactant used in CE for separation of AuNPS is SDS, but in this case, no full separation was accomplished in SDS-bearing buffers due to the small size difference of adjacent nanoparticles fraction (<10 nm). The good resolution was reached using SDBS as surfactant in the running electrolyte. The method was applied to a variety of dietary supplements, providing a characterization and speciation with high resolution for metal nanoparticles.

Table 7. Some selected determinations of nanoparticles in food samples.

Food application	Analytes	Method	Separation conditions	Detection	Ref.
Salt, Sweeteners	SiO2NPs	CE	Aqueous mixture of NH4Ac 3 mM 1% in methanol (pH 6.9)	ELSD	[99]
Mussels, orange juice	AgNPs	CE	40 mM SDS and 10 mM CAPS adjusted to pH 10 with NaOH	UV detection	[100]
Sugary food	TiO2NPs	Raman - CE	2.5 mM Na2HPO4 adjusted at pH 9.5	UV detection	[58]
dietary supplements	AuNPs	CE	Mixture of cyclohexylaminoethane sulfonic acid 10 mM and sodium dodecylbenzenesulfonate 70 mM (pH 10)	ICP-MS	[101]

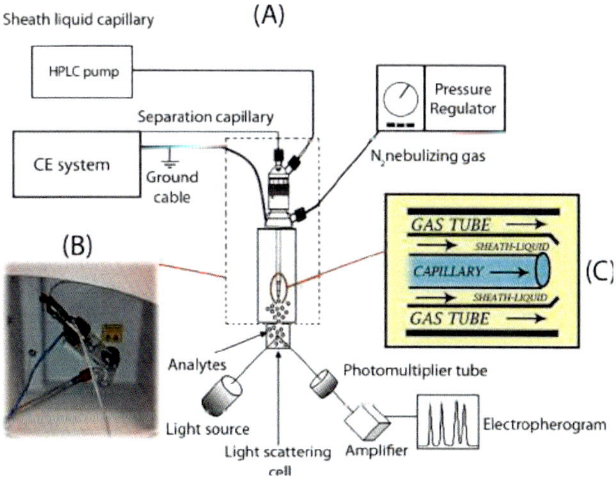

Fig. (4). (**A**) Scheme of the interface for coupling the CE equipment with the ELSD detector *via* a stainless-steel triple-tube nebulizer. (**B**) Picture of the nebulizer accommodated in a customized nebulizer chamber and scheme of the different connections. (**C**) Details of the flow tubes to produce an efficient nebulization. Adapted from Ref. 99 and reproduced with permission.

Other Additives

Vanillin(4-hydroxy-3-methoxybenzaldehyde) (VAN) is an extensively-used flavoring compound as a component in tobacco, pharmacy, cosmetics, beverage and food industry. VAN is an organic chemical, the formula being $C_8H_8O_3$. It is the main component of the vanilla bean extract. It may also be prepared from cheap raw materials namely lignin, eugenol and 2-methoxyphenol [102]. Vanillin may trigger migraine in susceptible individuals. In China, vanillin is forbidden in milk powder for 0 to 6-month-old babies [103].

Shu and co-workers [72] reported a new methodology for detection and quantification of vanillin in milk powder by means of CE in combination with Dispersive Liquid-Liquid Microextraction (DLLME). Analytical conditions for CZE and DLLME parameters were subject to optimization, the running buffer being a 10 mM borate solution (pH=8.0), and 4-Hydroxybenzaldehyde was chosen as the internal standard (see Table **8**). Optimal microextraction conditions were: a volume of 5.0 mL for sample solution (pH=5.5), an 8% (m/v) of NaCl was added, 120 of μL $CHCl_3$ (extraction solvent), and 0.8 mL of tetrahydrofuran (THF) (disperser solvent). After both extraction and detection conditions were optimized, vanillin was identified in 5 min, detection limit being 0.02 μg/mL and quantification limit ranged from 0.05 to 10 μg/mL. This methodology was adequately applied to vanillin analysis in milk powder for infants. Recovery values obtained ranged from 89.7 to 95.1%, the relative standard deviation (RSD) not exceeding 2.5%

Table 8. Determination of other additives in food.

Food application	Analytes	Methods	Separation conditions	Detection/ LOD	Ref.
Milk powder	Vanillin	DLLME-CZE	10 mM borate buffer at pH 8.0	UV detection 0.02 μg/mL	[72]
Vanilla extract products	COUM, EVA, PHB, PHBA, VAN, VANA, VOH	CE	sodium tetraborate anhydrous (50 mM, pH = 9.50) ethanol 20%	UV detection 0.4-2.6 μg/mL	[45]
Energy drinks	Caffeine, Taurine	MEKC	16.20 mM/L of benzoic acid and 39.90 mM/L of SDS, a pH value of 7.26	Diode-array detection 2 mg/L and 7.5 mg/L	[104]

(Table 8) cont.....

Food application	Analytes	Methods	Separation conditions	Detection/ LOD	Ref.
Fresh milk, Probiotic, green rice, Milk powder, Energy drink.	Choline, Taurine.	Dual-channelled capillary electrophoresis	150mM tris(hydroxymethyl) aminomethane/ lactic acid (pH8.96) for Taurine .150mM tris(hydroxymethyl) aminomethane/acetic acid (pH 9.5) for Choline.	C^4D 0.27 and 0.45 mg/L	[105]

Vanilla bean extracts are composed of vanillin and a broad range of phenylpropanoid-related (C6-C3) compounds (notably VANA, PHBA and PHB), all of them combined to confer the unique delicacy of natural vanilla. There are two sorts of synthetic vanilla flavorings, containing vanillin prepared from low expensive raw materials namely eugenol, guaiacol or lignin, a polymeric substance naturally found in wood; and/or addition of ethyl vanillin, this being another synthetic vanilla compound which is three times more flavoring than vanillin. There are vanilla extracts that may be adulterated with coumarin, a type of a phytochemical, to increment the perception of vanilla flavor since it possess a sweet smell and it is used as a fragrance enhancer. Coumarin provokes hepatoxicity in animals and its use as food additive has been banned in U.S. since 1956, especially since it was considered to be genotoxic and carcinogenic.

Lahouidak and co-workers [45] developed a CE approach to discriminate between artificial and natural vanilla flavour to prevent food-based frauds. Vanilla markers (COUM, EVA, PHB, PHBA, VAN, VANA, and VOH) were first detected and then separated in a CE device coupled with a diode array detector (DAD) in 15 min. The carrier electrolyte was 50 mM sodium tetraborate anhydrous (pH=9.50), with 20% ethanol and this solution was chosen as the running buffer. CE temperature and voltage were set at 20°C and 25kV, respectively, and detection was performed at both 200 and 275 nm. Detection limits were in the range 0.4-2.6 µg/mL. For the discrimination between artificial and natural vanilla flavour, four selected real extract products were analysed. First, if analyses indicated the existence of VAN combined with secondary compounds derived from natural vanilla (PHB, PHBA or VANA, these samples clearly came from a natural origin. The presence of EVA means that the sample is not natural. The detection of COUM demonstrate that the samples were adulterated. When VAN is initially detected but no secondary components are found, the concentration of these compounds need to be studied in the samples. Caffein and similar compounds naturally occur in food or are added during food processing. Taurine is a free amino acid playing an essential role in the human body because it acts as an antioxidant, absorbs fats and protects the heart [104]. The difficulty in developing methods for a simultaneous analysis of taurine and caffeine is to ascertain what detection system needs to be used. Alves *et al* developed an approach for the concurrent analysis of taurine and caffeine in energy drinks by MEKC and DAD, employing indirect and direct detection concomitantly [105]. Once optimized, the method made use of a capillary tube with a 50- µm i.d. x 72-cm bulb, a running buffer composed of 39.90 mmol/L SDS and 16.20 mmol/L benzoic acid (pH=7.26), a positive voltage of 30 kV, caffeine was directly detected at 274 nm and taurine was indirectly detected at 230 nm with detection limits of 2 mg/L in the case of caffeine and 7.5 mg/L for taurine.

CONCLUDING REMARKS

The bibliographic sources commented above reflect that CE has become a valuable alternative in relation with chromatography-based strategies for the determination of food additives in food products. This technique has significant operational advantages regarding high-separation efficiency, quickness of analysis and method development, small solvent volume used and low cost. However, the insufficient sensitivity of CE with optical detection requires that new methods continue to be developed to overcome this drawback.

Many applications have been developed incorporating very suitable techniques to solve the limited sensitivity of CE including off-column and on-column trace-enrichment and more sensitive detection systems.

CONSENT FOR PUBLICATION

Not applicable.

CONFLICT OF INTEREST

The author declares no conflict of interest, financial or otherwise.

ACKNOWLEDGEMENTS

Financial supports from the Spanish Ministry of Science and Innovation (PID2019- 104381GB-I00), Junta de Comunidades de Castilla-La Mancha (SBPLY/17/180501/000262) and University of Castilla-La Mancha (2020-GRIN-28882) are gratefully acknowledged.

REFERENCES

[1] Boyce, M.C. Determination of additives and organic contaminants in food by CE and CEC. *Electrophoresis,* **2007**, *28*(22), 4046-4062.
[http://dx.doi.org/10.1002/elps.200700280] [PMID: 17948272]

[2] Boyce, M.C. Determination of additives in food by capillary electrophoresis. *Electrophoresis,* **2001**, *22*(8), 1447-1459.
[http://dx.doi.org/10.1002/1522-2683(200105)22:8<1447::AID-ELPS1447>3.0.CO;2-#] [PMID:

11386656]

[3] Tagliaro, F.; Turrina, S.; Smith, F.P. Capillary electrophoresis: principles and applications in illicit drug analysis. *Forensic Sci. Int.*, **1996**, *77*(3), 211-229. [http://dx.doi.org/10.1016/0379-0738(95)01863-8] [PMID: 8819995]

[4] García-Cañas, V.; Simó, C.; Castro-Puyana, M.; Cifuentes, A. Recent advances in the application of capillary electromigration methods for food analysis and Foodomics. *Electrophoresis,* **2014**, *35*(1), 147-169. [http://dx.doi.org/10.1002/elps.201300315] [PMID: 24114758]

[5] Anastos, N.; Barnett, N.W.; Lewis, S.W. Capillary electrophoresis for forensic drug analysis: A review. *Talanta,* **2005**, *67*(2), 269-279. [http://dx.doi.org/10.1016/j.talanta.2005.03.038] [PMID: 18970167]

[6] Soga, T.; Serwe, M. Determination of carbohydrates in food samples by capillary electrophoresis with indirect UV detection. *Food Chem.*, **2000**, *69*, 339-344. [http://dx.doi.org/10.1016/S0308-8146(00)00044-3]

[7] Henry, C.S. Microchip capillary electrophoresis: an introduction. *Methods Mol. Biol.*, **2006**, *339*, 1-10. [PMID: 16790862]

[8] Sastre Toraño, J.; Ramautar, R.; de Jong, G. Advances in capillary electrophoresis for the life sciences. *J. Chromatogr. B Analyt. Technol. Biomed. Life Sci.*, **2019**, *1118-1119*, 116-136. [http://dx.doi.org/10.1016/j.jchromb.2019.04.020] [PMID: 31035134]

[9] Saltmarsh, M.; Insall, L. Food additives and why they are used. In: *Essential guide to food additives, The Royal Society of Chemistry*; Saltmarsh, M., Ed.; , **2013**. chapter 1.

[10] European Commission. Regulation (EC) No 1333/2008 of the European Parliament and of the Council, of 16 December 2008, on food additives. **2008**.

[11] Paritoch, S.; Akshay, S.; Ashok, S. Food additives from an organic chemistry perspective. *MOJ Biorg Org Chem.*, **2017**, *1*, 70-79.

[12] Carocho, M.; Barreiro, M.F.; Morales, P.; Ferreira, I.C.F.R. Adding Molecules to Food, Pros and Cons: A Review on Synthetic and Natural Food Additives. *Compr. Rev. Food Sci. Food Saf.*, **2014**, *13*(4), 377-399. [http://dx.doi.org/10.1111/1541-4337.12065] [PMID: 33412697]

[13] Ferreira, I.C.F.R. Food Additives: Classification, Regulation and Analysis. In: *Encyclopedia of Analytical Science*; , **2005**; 2, pp. 217-225.

[14] Silva, M.M.; Lidon, F.C. Food preservatives-An overview on applications and side effects. *Emir. J. Food Agric.*, **2016**, *28*, 366-373. [http://dx.doi.org/10.9755/ejfa.2016-04-351]

[15] Öztekin, N. Simultaneous determination of benzoic acid and sorbic acid in food products by capillary electrophoresis. *Food and Health,* **2018**, *4*, 176-182. [http://dx.doi.org/10.3153/FH18018]

[16] Shahidi, F.; Zhong, Y. Lipid oxidation: Measurement methods. *Bailey's industrial oil and fat products,* **2005**, *1*, 357-386.

[17] Shahidi, F.; Zhong, Y. Measurement of antioxidant activity. *J. Funct. Foods,* **2015**, *18*, 757-781. [http://dx.doi.org/10.1016/j.jff.2015.01.047]

[18] Chattopadhyay, S.; Raychaudhuri, U.; Chakraborty, R. Artificial sweeteners - a review. *J. Food Sci. Technol.*, **2014**, *51*(4), 611-621. [http://dx.doi.org/10.1007/s13197-011-0571-1] [PMID: 24741154]

[19] Zelman, K. Color Additives. *foodandnutrition.org,* **2017**.

[20] Sezgin, A.C.; Ayyıldız, S. Food Additives: Colorants, Science within Food: Up-to-date Advances on

Research and Educational Ideas. *Shipin Kexue,* **2017**, *1*, 87-94.

[21] Sekhon, B.S. Food nanotechnology - an overview. *Nanotechnol. Sci. Appl.,* **2010**, *3*, 1-15. [PMID: 24198465]

[22] Anastasi, E.; Riviere, G.; Teste, B. *Nanomaterials in Food - Prioritisation & Assessment*; FSA Journal, **2019**, p. 17.

[23] Contado, C. Nanomaterials in consumer products: a challenging analytical problem. *Front Chem.,* **2015**, *3*, 48. [http://dx.doi.org/10.3389/fchem.2015.00048] [PMID: 26301216]

[24] Chen, H.; Terabe, S. On-line sample preconcentration techniques in capillary electrophoresis. Analysis and detection by capillary electrophoresis. XLV, **2005**; pp. 135-172. [http://dx.doi.org/10.1016/S0166-526X(05)45003-5]

[25] Kitagawa, F. Otsuka. K. *J. Chromatogr. A,* **2004**, *1335*, 43-60. [http://dx.doi.org/10.1016/j.chroma.2013.10.066] [PMID: 24210300]

[26] Osbourn, D.M.; Weiss, D.J.; Lunte, C.E. On-line preconcentration methods for capillary electrophoresis. *Electrophoresis,* **2000**, *21*(14), 2768-2779. [http://dx.doi.org/10.1002/1522-2683(20000801)21:14<2768::AID-ELPS2768>3.0.CO;2-P] [PMID: 11001283]

[27] Chien, R.L.; Burgi, D.S. Sample stacking of an extremely large injection volume in high-performance capillary electrophoresis. *Anal. Chem.,* **1992**, *64*, 1046-1050. [http://dx.doi.org/10.1021/ac00033a015]

[28] Lee, S.L.; Boyce, M.C.; Michael, C. Extraction, and on-line concentration of flavonoids in Brassica oleracea by capillary electrophoresis using large volume sample stacking. *Food Chem.,* **2012**, *133*, 205-211. [http://dx.doi.org/10.1016/j.foodchem.2012.01.006]

[29] Wu, Y.W.; Liu, J.F.; Xiao, T.X.; Han, D.Y.; Zhang, H.L.; Pan, J.C. Field-amplified sample injection for the determination of albumin and transferrin in human urines by MEKC. *Electrophoresis,* **2009**, *30*(4), 668-673. [http://dx.doi.org/10.1002/elps.200800684] [PMID: 19180545]

[30] Wei, R.; Li, W.; Yang, L.; Jiang, Y.; Xie, T. Online preconcentration in capillary electrophoresis with contactless conductivity detection for sensitive determination of sorbic and benzoic acids in soy sauce. *Talanta,* **2011**, *83*(5), 1487-1490. [http://dx.doi.org/10.1016/j.talanta.2010.11.036] [PMID: 21238741]

[31] Breadmore, M.C.; Haddad, P.R. Approaches to enhancing the sensitivity of capillary electrophoresis methods for the determination of inorganic and small organic anions. *Electrophoresis,* **2001**, *22*(12), 2464-2489. [http://dx.doi.org/10.1002/1522-2683(200107)22:12<2464::AID-ELPS2464>3.0.CO;2-U] [PMID: 11519951]

[32] Jastrzbska, A. Application of Capillary Isotachophoretic Method to the Determination of Nitrate and Nitrite Ions in Meat Products. *J. Anal. Chem.,* **2010**, *65*, 1170-1175. [http://dx.doi.org/10.1134/S1061934810110134]

[33] Herrmannová, M.; Krivánková, L.; Bartos, M.; Vytras, K. Direct simultaneous determination of eight sweeteners in foods by capillary isotachophoresis. *J. Sep. Sci.,* **2006**, *29*(8), 1132-1137. [http://dx.doi.org/10.1002/jssc.200500441] [PMID: 16830728]

[34] Quirino, J.P.; Kim, J.B.; Terabe, S. Sweeping: concentration mechanism and applications to high-sensitivity analysis in capillary electrophoresis. *J. Chromatogr. A,* **2002**, *965*(1-2), 357-373. [http://dx.doi.org/10.1016/S0021-9673(01)01499-6] [PMID: 12236536]

[35] Cao, J.; Dun, W.L. Separation and sweeping of flavonoids by microemulsion electrokinetic chromatography using mixed anionic and cationic surfactants. *Talanta,* **2011**, *84*(1), 155-159.

[http://dx.doi.org/10.1016/j.talanta.2010.12.037] [PMID: 21315913]

[36] Anderson, J.; Berthod, A.; Pino, V. Modern injection modes (stacking for CE). *Analytical Separation Science,* **2016**, *5*, 563.

[37] Hsu, S.H.; Hu, C.C.; Chiu, T.C. Online dynamic pH junction-sweeping for the determination of benzoic and sorbic acids in food products by capillary electrophoresis. *Anal. Bioanal. Chem.,* **2014**, *406*(2), 635-641.
[http://dx.doi.org/10.1007/s00216-013-7481-1] [PMID: 24247553]

[38] Ramautar, R.; Somsen, G.W.; de Jong, G.J. Recent developments in coupled SPE-CE. *Electrophoresis,* **2010**, *31*(1), 44-54.
[http://dx.doi.org/10.1002/elps.200900510] [PMID: 20014054]

[39] Han, F.; He, Y.Z.; Li, L.; Fu, G.N.; Xie, H.Y.; Gan, W.E. Determination of benzoic acid and sorbic acid in food products using electrokinetic flow analysis-ion pair solid phase extraction-capillary zone electrophoresis. *Anal. Chim. Acta,* **2008**, *618*(1), 79-85.
[http://dx.doi.org/10.1016/j.aca.2008.04.041] [PMID: 18501248]

[40] Yi, J.; Zeng, L.; Wu, Q.; Yang, L.; Xie, T. Sensitive Simultaneous Determination of Synthetic Food Colorants in Preserved Fruit Samples by Capillary Electrophoresis with Contactless Conductivity Detection. *Food Anal. Methods,* **2018**, *11*, 1608-1618.
[http://dx.doi.org/10.1007/s12161-017-1141-6]

[41] Ballus, A.; Meinhart, A.D.; Oliveira, R.; Godoy, H.T. Optimization of capillary zone electrophoresis separation and on-line preconcentration of 16 phenolic compounds from wines produced in South America. *Food Res. Int.,* **2012**, *45*, 136-144.
[http://dx.doi.org/10.1016/j.foodres.2011.09.029]

[42] Memon, F.; Solangi, R.; Memon, Q.; Mallah, A.; Memon, N.; Ayaz, A. Simultaneous Determination of Quercetin, Rutin, Naringin, and Naringenin in Different Fruits by Capillary Zone Electrophoresis. *Food Anal. Methods,* **2017**, *10*, 83-91.
[http://dx.doi.org/10.1007/s12161-016-0552-0]

[43] Pavlíček, V.; Tůma, P. The use of capillary electrophoresis with contactless conductivity detection for sensitive determination of stevioside and rebaudioside A in foods and beverages. *Food Chem.,* **2017**, *219*, 193-198.
[http://dx.doi.org/10.1016/j.foodchem.2016.09.135] [PMID: 27765216]

[44] Huang, H.Y.; Chiu, C.W.; Sue, S.L.; Cheng, C.F. Analysis of food colorants by capillary electrophoresis with large-volume sample stacking. *J. Chromatogr. A,* **2003**, *995*(1-2), 29-36
[http://dx.doi.org/10.1016/S0021-9673(03)00530-2] [PMID: 12800920]

[45] Lahouidak, S.; Salghi, R.; Zougagh, M.; Ríos, A. Capillary electrophoresis method for the discrimination between natural and artificial vanilla flavor for controlling food frauds. *Electrophoresis,* **2018**, *39*(13), 1628-1633.
[http://dx.doi.org/10.1002/elps.201700480] [PMID: 29508413]

[46] García, M.Á.; Marina, M.L.; Ríos, A. Valcárcel. M., Separation modes capillary electrophoresis.*Analysis and detection by capillary electrophoresis*; Marina, M.L.; Ríos, A.; Valcarcel, M., Eds.; Elsevier, **2005**, pp. 31-134.
[http://dx.doi.org/10.1016/S0166-526X(05)45002-3]

[47] Pyell, U. Micellar electrokinetic chromatography--from theoretical concepts to real samples (review). *Fresenius J. Anal. Chem.,* **2001**, *371*(6), 691-703. [Review].
[http://dx.doi.org/10.1007/s002160100981] [PMID: 11768454]

[48] Martí, R.; Valcárcel, M.; Herrero-Martínez, J.M.; Cebolla-Cornejo, J.; Roselló, S. Simultaneous determination of main phenolic acids and flavonoids in tomato by micellar electrokinetic capillary electrophoresis. *Food Chem.,* **2017**, *221*, 439-446.
[http://dx.doi.org/10.1016/j.foodchem.2016.10.105] [PMID: 27979225]

[49] Guan, Y.; Chu, Q.; Fu, L.; Wu, T.; Ye, J. Determination of phenolic antioxidants by micellar electrokinetic capillary chromatography with electrochemical detection. *Food Chem.*, **2006**, *94*, 157-162.
[http://dx.doi.org/10.1016/j.foodchem.2005.01.015]

[50] Hansen, S.; Gabel-Jensen, C.; El-Sherbiny, D. Microemulsion electrokinetic chromatography or solvent-modified micellar electrokinetic chromatography? *TrAc. Trends Analyt. Chem.*, **2001**, *20*, 11.
[http://dx.doi.org/10.1016/S0165-9936(01)00127-3]

[51] Bordagaray, A.; Garcia-Arrona, R.; Vidal, M.; Ostra, M. Determination of food colorants in a wide variety of food matrices by microemulsion electrokinetic capillary chromatography. Considerations on the found concentrations and regulated consumption limits. *Food Chem.*, **2018**, *262*, 129-133.
[http://dx.doi.org/10.1016/j.foodchem.2018.04.086] [PMID: 29751899]

[52] Altria, K.D. Overview of capillary electrophoresis and capillary electrochromatography. *J. Chromatogr. A,* **1999**, *856*(1-2), 443-463.
[http://dx.doi.org/10.1016/S0021-9673(99)00830-4] [PMID: 10526799]

[53] Bottoli, C.; Ponce, G.; Aguiar, V.; De Aquino, W. Determination of parabens in sweeteners by capillary electrochromatography. *Braz. J. Pharm. Sci.,* **2011**, *47*, 4.
[http://dx.doi.org/10.1590/S1984-82502011000400015]

[54] Wen, Y.; Li, J.; Ma, J.; Chen, L. Recent advances in enrichment techniques for trace analysis in capillary electrophoresis. *Electrophoresis,* **2012**, *33*(19-20), 2933-2952.
[http://dx.doi.org/10.1002/elps.201200240] [PMID: 23019127]

[55] Jastrzębska, A. Capillary Isotachophoretic as rapid method for determination of orthophosphates, pyrophosphates, tripolyphosphates and nitrites in food samples. *J Food Compos. Anal.*, **2011**, *24*, 1049-1056.
[http://dx.doi.org/10.1016/j.jfca.2010.12.007]

[56] Lerma-García, M.; Zougagh, M.; Ríos, A. The Applied Side of Capillary Electrophoresis: A Critical View. *Curr. Anal. Chem.*, **2014**, *10*, 184-196.
[http://dx.doi.org/10.2174/15734110113099990005]

[57] Le, H. T.; Nguyen, T.; Tran, C.; Vu, T., Screening determination of food additives using capillary electrophoresis coupled with contactless conductivity detection: A case study in Vietnam. *Food Control,* **2017**, *77*, 281-289.
[http://dx.doi.org/10.1016/j.foodcont.2017.02.020]

[58] Moreno, V.; Zougagh, M.; Ríos, Á. Analytical nanometrological approach for screening and confirmation of titanium dioxide nano/micro-particles in sugary samples based on Raman spectroscopy - Capillary electrophoresis. *Anal. Chim. Acta,* **2019**, *1050*, 169-175.
[http://dx.doi.org/10.1016/j.aca.2018.10.067] [PMID: 30661586]

[59] Ávila, M.; Zougagh, M.; Escarpa, A.; Ríos, A. Fast single run of vanilla fingerprint markers on microfluidic-electrochemistry chip for confirmation of common frauds. *Electrophoresis,* **2009**, *30*(19), 3413-3418.
[http://dx.doi.org/10.1002/elps.200900238] [PMID: 19739138]

[60] Breadmore, M.C. Capillary and microchip electrophoresis: challenging the common conceptions. *J. Chromatogr. A,* **2012**, *1221*, 42-55.
[http://dx.doi.org/10.1016/j.chroma.2011.09.062] [PMID: 22000781]

[61] Ding, Y.; Mora, F.; Garcia, D. Analysis of alkyl gallates and nordihydroguaiaretic acid using plastic capillary electrophoresis – microchips. *Anal. Chim. Acta,* **2006**, *561*, 126-132.
[http://dx.doi.org/10.1016/j.aca.2006.01.013]

[62] Dossi, N.; Toniolo, R.; Pizzariello, A.; Susmel, S.; Perennes, F.; Bontempelli, G. A capillary electrophoresis microsystem for the rapid in-channel amperometric detection of synthetic dyes in food. *J. Electroanal. Chem. (Lausanne),* **2007**, *601*, 1-7.

[http://dx.doi.org/10.1016/j.jelechem.2006.10.019]

[63] Delgado, M.; González-Maza, I.; Pérez, S.; Martínez, R. Analysis of synthetic phenolic antioxidants in edible oils by micellar electrokinetic capillary chromatography. *Food Chem.,* **2007**, *100*, 1722-1727. [http://dx.doi.org/10.1016/j.foodchem.2005.10.018]

[64] de Jesus, F.; Coelho, A.; Pallos, L.; da Silva, J.; Daniel, D.; de Jesus, D. Simple and fast method for simultaneous determination of propionate and sorbate in bread by capillary electrophoresis with UV spectrophotometric detection. *J. Food Compos. Anal.,* **2018**, *72*, 93-96. [http://dx.doi.org/10.1016/j.jfca.2018.06.010]

[65] Kalaycıoğlu, Z.; Erim, F.D. Simultaneous Determination of Nitrate and Nitrite in Fish Products with Improved Sensitivity by Sample Stacking-Capillary Electrophoresis. *Food Anal. Methods,* **2016**, *9*, 706-711. [http://dx.doi.org/10.1007/s12161-015-0241-4]

[66] Ruani, A.P.; Vitalia, L.; Micke, G.A. Multiple-Injection Capillary Zone Electrophoresis as a Fast Strategy to Determine Antinitrosating Capacity of Commercial Teas. *J. Braz. Chem,* **2018**, *10*, 2150-2157. [http://dx.doi.org/10.21577/0103-5053.20180090]

[67] Li, X.; Gao, F.; Liu, H.; Gao, Y. Development of a capillary zone electrophoresis method to determine natamycin in food and a comparison with a liquid chromatography method. *J. Sci. Food Agric.,* **2020**, *100*(2), 811-816. [http://dx.doi.org/10.1002/jsfa.10089] [PMID: 31617212]

[68] Tang, Y.; Wu, M. A quick method for the simultaneous determination of ascorbic acid and sorbic acid in fruit juices by capillary zone electrophoresis. *Talanta,* **2005**, *65*(3), 794-798. [http://dx.doi.org/10.1016/j.talanta.2004.08.013] [PMID: 18969870]

[69] Boyce, M.C.; Spickett, E.E. Separation of food grade antioxidants (synthetic and natural) using mixed micellar electrokinetic capillary chromatography. *J. Agric. Food Chem.,* **1999**, *47*(5), 1970-1975. [http://dx.doi.org/10.1021/jf980959u] [PMID: 10552480]

[70] Liu, F.J.; Liu, C.T.; Li, W.; Tang, A.N. Dispersive solid-phase microextraction and capillary electrophoresis separation of food colorants in beverages using diamino moiety functionalized silica nanoparticles as both extractant and pseudostationary phase. *Talanta,* **2015**, *132*, 366-372. [http://dx.doi.org/10.1016/j.talanta.2014.09.014] [PMID: 25476319]

[71] Wu, J.; Feng, T.; Liang, X.; Yu, C.; Dong, L.; Du, M. Determination of HMF in Vinegar and Soy Sauce Using Two-Step Ultrasonic Assisted Liquid-Liquid Micro-Extraction Coupled with Capillary Electrophoresis-Ultraviolet Detection. *Food Anal. Methods,* **2018**, *11*, 479-485. [http://dx.doi.org/10.1007/s12161-017-1018-8]

[72] Shu, M.; Man, Y.; Ma, H.; Luan, F.; Liu, H.; Gao, Y. Determination of Vanillin in Milk Powder by Capillary Electrophoresis Combined with Dispersive Liquid-Liquid Microextraction. *Food Anal. Methods,* **2016**, *9*, 1706-1712. [http://dx.doi.org/10.1007/s12161-015-0347-8]

[73] Qian, X.; Zhang, Q.; Zhang, Y.; Tu, Y. Separation/determination of flavonoids and ascorbic acid in rat serum and excrement by capillary electrophoresis with electrochemical detection. *Anal. Sci.,* **2010**, *26*(5), 557-560. [http://dx.doi.org/10.2116/analsci.26.557] [PMID: 20467130]

[74] Chi, L.Z.; Li, Z.C.; Dong, S.Q.; He, P.G.; Wang, Q.J.; Fang, Y.Z. Simultaneous Determination of Flavonoids and Phenolic Acids in Chinese Herbal Tea by Beta-Cyclodextrin Based Capillary Zone Electrophoresis. *Mikrochim. Acta,* **2009**, *167*, 179-185. [http://dx.doi.org/10.1007/s00604-009-0229-0]

[75] Bustamante-Rangel, M.; Delgado-Zamarreño, M.M.; Carabias-Martínez, R.; Domínguez-Álvarez, J. Analysis of isoflavones in soy drink by capillary zone electrophoresis coupled with electrospray ionization mass spectrometry. *Anal. Chim. Acta,* **2012**, *709*, 113-119.

[http://dx.doi.org/10.1016/j.aca.2011.10.015] [PMID: 22122939]

[76] Zhang, Y.; Chen, Z. Separation of isomeric bavachin and isobavachalcone in the fructus Psoraleae by capillary electrophoresis-mass spectrometry. *J. Sep. Sci.*, **2012**, *35*(13), 1644-1650. [http://dx.doi.org/10.1002/jssc.201200173] [PMID: 22761143]

[77] Wang, S.F.; Fang, H.Y.; Qu, H.B. Optimization of micellar electrokinetic capillary chromatography method using central composite design for the analysis of components in Yangwei granule. *J. Zhejiang Univ. Sci. B*, **2011**, *12*(3), 193-200. [http://dx.doi.org/10.1631/jzus.B1000135] [PMID: 21370504]

[78] Surowiec, I.; Pawelec, K.; Rezeli, M.; Kilar, F.; Trojanowicz, M. Capillary electrophoretic determination of main components of natural dyes with MS detection. *J. Sep. Sci.*, **2008**, *31*(13), 2457-2462. [http://dx.doi.org/10.1002/jssc.200800032] [PMID: 18604840]

[79] Moreno-González, D.; Jáč, P.; Švec, F.; Nováková, L. Determination of Sudan dyes in chili products by micellar electrokinetic chromatography-MS/MS using a volatile surfactant. *Food Chem.*, **2020**, *310*, 125963. [http://dx.doi.org/10.1016/j.foodchem.2019.125963] [PMID: 31838374]

[80] Xia, S.; Yin, D.; Chen, Y.; Yang, Z.; Miao, Y.; Zhang, W.; Chen, S.; Zhao, W.; Zhang, S. Simultaneous determination of three sulfanilamide artificial sweeteners in foodstuffs by capillary electrophoresis coupled with contactless conductivity detection based on porous aromatic frameworks enhanced solid phase extraction. *Can. J. Chem.*, **2019**, *97*, 344-351. [http://dx.doi.org/10.1139/cjc-2018-0410]

[81] Coelho, A.G.; de Jesús, D.P. A simple method for determination of erythritol, maltitol, xylitol, and sorbitol in sugar-free chocolates by capillary electrophoresis with capacitively coupled contactless conductivity detection. *Electrophoresis*, **2016**, *37*(22), 2986-2991. [http://dx.doi.org/10.1002/elps.201600263] [PMID: 27520494]

[82] Stojkovic, M.; Mai, T.D.; Hauser, P.C. Determination of artificial sweeteners by capillary electrophoresis with contactless conductivity detection optimized by hydrodynamic pumping. *Anal. Chim. Acta*, **2013**, *787*, 254-259. [http://dx.doi.org/10.1016/j.aca.2013.05.039] [PMID: 23830447]

[83] Wang, J.; Han, S. Capillary Electrophoresis Chemiluminescence for the Analysis of Flavonoids in Pharmaceuticals and Human Plasma. *Chromatographia*, **2013**, *76*, 715-718. [http://dx.doi.org/10.1007/s10337-013-2439-1]

[84] Yu, L.S.; Xu, X.Q.; Huang, L.; Ling, J.M.; Chen, G.N. Separation and determination of flavonoids using microemulsion EKC with electrochemical detection. *Electrophoresis*, **2008**, *29*(3), 726-733. [http://dx.doi.org/10.1002/elps.200700140] [PMID: 18186530]

[85] Chen, Z.; Zhang, L.; Chen, G. Microwave-assisted extraction followed by capillary electrophoresis-amperometric detection for the determination of antioxidant constituents in Folium Eriobotryae. *J. Chromatogr. A*, **2008**, *1193*(1-2), 178-181. [http://dx.doi.org/10.1016/j.chroma.2008.04.032] [PMID: 18462740]

[86] Wang, W.; Lin, P.; Ma, L.; Xu, K.; Lin, X. Separation and determination of flavonoids in three traditional chinese medicines by capillary electrophoresis with amperometric detection. *J. Sep. Sci.*, **2016**, *39*(7), 1357-1362. [http://dx.doi.org/10.1002/jssc.201501287] [PMID: 26829244]

[87] Wu, T.; Yu, C.; Li, R. Li, J., Minireview: Recent advances in the determination of flavonoids by capillary electrophoresis. *Instrum. Sci. Technol.*, **2018**, **•••**, 1525-6030.

[88] Ryvolová, M.; Táborský, P.; Vrábel, P.; Krásenský, P.; Preisler, J. Sensitive determination of erythrosine and other red food colorants using capillary electrophoresis with laser-induced fluorescence detection. *J. Chromatogr. A*, **2007**, *1141*(2), 206-211. [http://dx.doi.org/10.1016/j.chroma.2006.12.018] [PMID: 17188280]

[89] Xu, J.J.; An, M.; Yang, R.; Cao, J.; Ye, L.H.; Peng, L.Q. Trace amounts of poly-β-cyclodextrin wrapped carbon nanotubes for the microextraction of flavonoids in honey samples by capillary electrophoresis with light-emitting diode induced fluorescence detection. *Electrophoresis,* **2016**, *37*(13), 1891-1901.
[http://dx.doi.org/10.1002/elps.201600016] [PMID: 27060460]

[90] Nair, M.S.; Nair, D.; Johny, A.; Venkitanarayanan, K. *Use of food preservatives and additives in meat and their detection techniques*; Meat Quality Analysis, **2020**, pp. 187-213.

[91] Liu, H.; Wen, Y.; Luan, F.; Gao, Y. Analysis of Food Additives by Capillary Electrophoresis. *Acta Chromatogr.,* **2008**, *20*, 239-246.
[http://dx.doi.org/10.1556/AChrom.20.2008.2.8]

[92] Darji, V.; Boyce, M.C.; Bennett, I.; Breadmore, M.C.; Quirino, J. Determination of food grade antioxidants using microemulsion electrokinetic chromatography. *Electrophoresis,* **2010**, *31*(13), 2267-2271.
[http://dx.doi.org/10.1002/elps.201000050] [PMID: 20533329]

[93] Peng, Y.; Zhang, Y.; Ye, J. Determination of phenolic compounds and ascorbic acid in different fractions of tomato by capillary electrophoresis with electrochemical detection. *J. Agric. Food Chem.,* **2008**, *56*(6), 1838-1844.
[http://dx.doi.org/10.1021/jf0727544] [PMID: 18284201]

[94] Vasconcelos, M.A.; Orsolin, P.C.; Silva-Oliveira, R.G.; Nepomuceno, J.C.; Spanó, M.A. Assessment of the carcinogenic potential of high intense-sweeteners through the test for detection of epithelial tumor clones (warts) in Drosophila melanogaster. *Food Chem. Toxicol.,* **2017**, *101*, 1-7.
[http://dx.doi.org/10.1016/j.fct.2016.12.028] [PMID: 28027980]

[95] Medrano, L.; Flores-Aguilar, J.; Islas, G.; Rodríguez, A.; Ibarra, S. Solid-Phase Extraction and Large-Volume Sample Stacking-Capillary Electrophoresis for Determination of Artificial Sweeteners in Water Samples. *Food Anal. Methods,* **2019**, *12*, 526-533.
[http://dx.doi.org/10.1007/s12161-018-1383-y]

[96] Nilsson, C.; Viberg, P.; Spégel, P.; Jörntén-Karlsson, M.; Petersson, P.; Nilsson, S. Nanoparticle-based continuous full filling capillary electrochromatography/electrospray ionization-mass spectrometry for separation of neutral compounds. *Anal. Chem.,* **2006**, *78*(17), 6088-6095.
[http://dx.doi.org/10.1021/ac060526n] [PMID: 16944888]

[97] Ntrallou, K.; Gika, H.; Tsochatzis, E. Analytical and Sample Preparation Techniques for the Determination of Food Colorants in Food Matrices. *Foods,* **2020**, *9*(1), 58.
[http://dx.doi.org/10.3390/foods9010058] [PMID: 31936025]

[98] Adelantado, C.; Rodríguez-Fariñas, N.; Rodríguez Martín-Doimeadios, R.C.; Zougagh, M.; Ríos, Á. Analysis of silica nanoparticles by capillary electrophoresis coupled to an evaporative light scattering detector. *Anal. Chim. Acta,* **2016**, *923*, 82-88.
[http://dx.doi.org/10.1016/j.aca.2016.03.055] [PMID: 27155305]

[99] Bouri, M.; Salghi, R.; Zougagh, M.; Ríos, A. Design and adaptation of an interface for commercial capillary electrophoresis-evaporative light scattering detection coupling. *Anal. Chem.,* **2013**, *85*(10), 4858-4862.
[http://dx.doi.org/10.1021/ac400370f] [PMID: 23611663]

[100] Ruiz-Palomero, C.; Soriano, M.L.; Valcárcel, M. Sulfonated nanocellulose for the efficient dispersive micro solid-phase extraction and determination of silver nanoparticles in food products. *J. Chromatogr. A,* **2016**, *1428*, 352-358.
[http://dx.doi.org/10.1016/j.chroma.2015.06.023] [PMID: 26116191]

[101] Qu, H.; Linder, S.W.; Mudalige, T.K. Surface coating and matrix effect on the electrophoretic mobility of gold nanoparticles: a capillary electrophoresis-inductively coupled plasma mass spectrometry study. *Anal. Bioanal. Chem.,* **2017**, *409*(4), 979-988.
[http://dx.doi.org/10.1007/s00216-016-0012-0] [PMID: 27783123]

[102] Sinha, A.K.; Sharma, U.K.; Sharma, N. A comprehensive review on vanilla flavor: extraction, isolation and quantification of vanillin and others constituents. *Int. J. Food Sci. Nutr.*, **2008**, *59*(4), 299-326.
[http://dx.doi.org/10.1080/09687630701539350] [PMID: 17886091]

[103] Standardization Administration of China. National food safty standards. *Standards of using food additives,* **2014**, 120.

[104] Nguyen, T.D.; Nguyen, M.H.; Vu, M.T.; Duong, H.A.; Pham, H.V.; Mai, T.D. Dual-channeled capillary electrophoresis coupled with contactless conductivity detection for rapid determination of choline and taurine in energy drinks and dietary supplements. *Talanta,* **2019**, *193*, 168-175.
[http://dx.doi.org/10.1016/j.talanta.2018.10.002] [PMID: 30368287]

[105] Alves, A.C.; Meinhart, A.D.; Filho, J.T.; Godoy, H.T. Development of a method for simultaneous analysis of caffeine and taurine in energy drinks by micellar electrokinetic chromatography with diode-array detector. Food Sci. Techn, **2019**; pp. 1-10.

CHAPTER 9

Chiral Capillary Electrophoresis in Food Analysis

Samuel Bernardo-Bermejo[1], **Elena Sánchez-López**[2], **María Castro-Puyana**[1,3] and **María Luisa Marina**[1,3,*]

[1] *Departamento de Química Analítica, Química Física e Ingeniería Química, Universidad de Alcalá, Ctra. Madrid-Barcelona Km. 33.600,28871 Alcalá de Henares (Madrid), Spain*

[2] *Department of Human Genetics, Leiden University Medical Center, Leiden, The Netherlands*

[3] *Instituto de Investigación Química Andrés M. del Río (IQAR), Universidad de Alcalá, Ctra. Madrid-Barcelona Km. 33.600, 28871 Alcalá de Henares (Madrid), Spain*

Abstract: Chiral analysis is a powerful tool in Food Science for quality and safety assessment since the enantiomeric composition of food samples can reveal adulterations, the effects of processing and storage or give valuable information on the bioactivity, traceability or even toxicity of foods. This chapter describes the potential of Capillary Electrophoresis in the chiral analysis of food and beverages. The separation modes used in CE for the chiral analysis of food samples are described, including different strategies for sample preparation and sensitivity enhancement. The most relevant applications developed in the period from 2010 to the present are depicted and the main conclusions and future prospects are outlined.

Keywords: Amino acids, Capillary electrochromatography, Catechins, Chiral capillary electrophoresis, Chiral separation, Enantiomers, Electrokinetic chromatography, Food samples, Organic acids, Sweeteners.

INTRODUCTION

Chiral analysis is nowadays an important area in different fields due to the different biological activities that enantiomers may have. In fact, the enantiomeric determination of chiral compounds has become paramount in a big variety of samples such as pharmaceutical, biomedical, agrochemical, environmental or food matrices, among others.

In food science, chiral analysis is a powerful tool for food quality and safety assessment.

* **Corresponding author María Luisa Marina:**, Departamento de Química Analítica, Química Física e Ingeniería Química, Universidad de Alcalá, Ctra. Madrid-Barcelona Km 33.600, 28871 Alcalá de Henares (Madrid), Spain and Instituto de Investigación Química Andrés M. del Río (IQAR), Universidad de Alcalá, Ctra. Madrid-Barcelona Km. 33.600, 28871 Alcalá de Henares (Madrid), Spain E-mail: mluisa.marina@uah.es

María Castro-Puyana, Miguel Herrero and María Luisa Marina (Eds.)
Capillary Electrophoresis in Food Analysis

This is due to the fact that numerous food components are chiral (amino acids, organic acids, vitamins, bioactive compounds or sweeteners, among others) but also different toxic compounds potentially present in food are chiral. As a consequence, the enantiomeric composition of food samples can reveal adulterations, alterations due to the effects of processing or storage (including microbial racemization of the enantiomers naturally present in food or beverages), can explain changes in organoleptic properties or can give valuable information on the bioactivity or traceability of food samples.

Aimed to individually determine the enantiomers of chiral compounds, different analytical techniques have been employed. Separation techniques such as High Performance Liquid Chromatography (HPLC) or capillary LC, Gas Chromatography (GC), Supercritical Fluid Chromatography (SFC) and electromigration techniques such as Capillary Electrophoresis (CE) or Capillary Electrochromatography (CEC) have been the most used and their potential to achieve chiral analysis in food science has been reviewed in several relevant articles [1 - 3]. However, the use of other chiral techniques such as countercurrent chromatography, sensors, biosensors, and direct mass spectrometry (MS) in food science has also been recently reviewed [3]. Among chromatographic and electromigration techniques, HPLC has been the most employed for enantiomer determinations due to its applicability for preparative and analytical purposes, easy operation mode, wide range of chiral stationary phases commercially available, and the possibility of working in two dimensional chromatography, although some drawbacks can also be mentioned such as the use of organic solvents and expensive chiral columns [3]. In this context, CE has demonstrated in the last years very interesting characteristics and even some advantages for chiral analysis. Among these advantages, the high efficiency that can be obtained and the easy change of the chiral selector for a given enantiomeric separation (by simply adding the selector to the separation medium) can be highlighted [4, 5]. Moreover, the fact that the use of a chiral chromatographic column is not necessary (although it can be employed in CEC if desired) together with the low consumption of reagents required to work in CE, confers this technique a special attraction from economic and environmental points of view. In addition to these interesting features, the coupling of CE with MS has increased the potential of this technique for chiral analysis [6 - 8], although the use of some strategies was necessary in order to avoid the entrance of non-volatile chiral selectors into the ionization source of the mass spectrometer. Some of these strategies were partial filling techniques, counter migration of chiral selector or employing the indirect separation mode for chiral analysis consisting of using pure chiral reagents to originate diastereomers having different electrophoretic mobilities in CE with achiral separation media.

Although several articles have reviewed the main advances in the application of capillary electromigration methods for food analysis and foodomics in the last decade [9 - 13], and some others were focused on the analysis of relevant food components such as non-protein amino acids [14, 15], just one review article has been published in the last decade specifically devoted to chiral capillary electrophoresis in food analysis [16].

This chapter presents the potential of CE in the chiral analysis of food samples. With this aim, the CE separation modes enabling the separation of enantiomers are described, including specific features for sample preparation and sensitivity enhancement, and the most recent and relevant applications of chiral CE for food analysis are discussed. The characteristics of the methodologies developed are detailed and presented in Tables and the main conclusions and future prospects are outlined.

CHIRAL SEPARATION MODES IN CE

Chiral separations are definitely one of the most challenging tasks in the Separation Sciences. The separation of two species that have the same physicochemical properties and that are mirror images of one another can only be achieved under a chiral environment, *i.e.*, when interacting with an appropriate chiral selector. CE is a microscale analytical separation technique valuable in the chiral analysis of numerous application fields. There are two possibilities that enable chiral separations in CE based on how a given pair of enantiomers interact with this chiral environment. Firstly, these can be classified into indirect and direct modes. The indirect mode relies on the covalent reaction of the enantiomers with an enantiomerically pure derivatizing agent. This turns the separation of a pair of enantiomers into a separation of a pair of diastereoisomers which now possess different physicochemical properties and, thus, can be separated under achiral conditions. As will be further detailed in the *Applications* section of this Chapter, indirect procedures have been used in previous years but this approach is not as popular as the direct strategies. This is because of the additional experimental time and costs derived from having enantiomerically pure chemicals that enable the formation of the purest diastereoisomers. On the contrary, the direct mode is based on the direct interaction of a chiral selector and the given enantiomers to form temporary diastereoisomers that ideally will have different electrophoretic mobilities and consequent chiral separation. The direct mode is the gold standard approach in CE that can be further divided into two modes depending on whether the chiral selector is solubilized in the background electrolyte (Electrokinetic Chromatography, EKC) or forms part of the capillary column (CEC).

EKC is definitely the most used CE mode to perform chiral separations. In EKC, a chiral selector is added to the background electrolyte (BGE), forming a pseudostationary phase. This chiral selector will interact with the enantiomers and, if this interaction is stronger with one of the enantiomers than with the other, the chiral separation will take place. Due to the high peak efficiency in CE, *i.e.* narrower peaks when compared to liquid chromatography, regarding differences in interactions for each enantiomer do not have to be that different. One of the reasons for EKC being such a popular and versatile mode is the diversity of the chiral selectors that can be used for maximising its chance of success. The only prerequisite for these molecules is that they must be chiral, soluble in the BGE and, ideally, affordable. The most commonly used chiral selectors are cyclodextrins (CDs) and surfactants, followed by proteins, macrocyclic antibiotics, chiral ligands, polysaccharides, crown ether, and many more. For a very insightful revision on recently used chiral selectors, readers are redirected to a recent published review article on this matter [17].

CDs are a group of sugar-based macromolecules consisting of 6, 7 or 8 glucopyranose units linked by α-(1-4) bonds and are named α-, β-, and γ-CD, respectively, all of them being referred to as native CDs. Chemical modification of these native CDs by adding, for instance, hydroxyl-, methyl-, or sulfo- groups, to name a few, further extend their possibilities in chiral separation [18]. This is because the host-guest complexes formed between the chiral selector and the target enantiomers are based on non-covalent interactions such as dipole, electrostatic, hydrogen bonding and steric interactions [17] and these groups might enhance these interactions.

Surfactants are also widely used in chiral separations and when used above the critical micelle concentration form micelles. These spherical-like aggregates form a hydrophilic exterior whereas the center is hydrophobic. The mechanism of enantioseparation is based on the electrophoretic mobility of the charged analytes and the quick exchange of the target enantiomers between the free and bound states with the micelle. This particular setup receives the name of Micellar Electrokinetic Chromatography (MEKC) and can be used to separate both neutral and charged analytes. MEKC further increases the possibilities in chiral separations and different modes are known. In the chiral MEKC approach, the micelle acts as a chiral selector where sodium cholate, sodium deoxycholate or sodium taurocholate are some of the most popular ones [19]. It is also possible for a CD to be added to the system acting as a chiral selector, which is known as CD-modified MEKC or CD-MEKC. In this case, the micelle used might be achiral such as anionic sodium dodecyl sulfate (SDS). Both the chiral analyte and the CD will interact with the micellar phase.

In CEC, the chiral selector is attached to a stationary phase or even to the capillary walls. Thus, the separation principle of CEC is considered as a hybrid between LC and CE, maintaining the advantages of both strategies. This approach combines the high selectivity of LC and the high efficiency of CE. In the last decades, CEC has experienced a major development in column technology in the form of packed (PC), monolithic (MC) or open-tubular columns (OTC), especially in the latter two [20 - 22].

All of the above supports the flexibility of CE in the chiral analysis field as different approaches, be it direct or indirect strategies, and in the case of the direct ones, the applicability of EKC, MEKC or CEC might be a possibility to enable a desired chiral separation.

SAMPLE PREPARATION AND ENHANCING OF THE SENSITIVITY

Without a doubt, in CE, the chiral selector plays a very important role in the separation of chiral molecules. However, the sample preparation step should not be overlooked. A proper sample preparation step will also contribute towards a successful chiral separation because cleaner and interference-free matrices will improve the interaction between chiral selector-analyte, enhancing the chiral separation, among others [2]. On the other hand, an ideal sample preparation step should also improve the sensitivity. Although CE is a very powerful eco-friendly technique, it also possesses some drawbacks. The main limitation of CE is its somehow low sensitivity that derives from its inherent small dimension. This is the use of low injection samples (usually in the nL range), although beneficiary from the point of view of sample availability and reduced costs, it has the counter-reaction of displaying limited sensitivity. This combined with the most employed detection system in CE, *i.e.*, UV-Vis detection, which is characterized by a small in-capillary built detection window of just a few μm, only worsens this scenario. It is important to note that this lower sensitivity is also reinforced in the chiral analysis since in many cases some enantiomers are present in low concentrations.

Fortunately, the flexibility of CE enables that there are numerous useful strategies to enhance the sensitivity. Using a more sensitive detection system such as MS or LIF would, in principle, alleviate this issue, but since not everybody has access to those more expensive setups, there are several available options that are compatible with UV-Vis detection. They can be sorted in off-line and in-capillary sample treatments.

Off-line sample treatments are focused on reducing matrix interferences and isolating the compounds of interest to further enhance not only the sensitivity but also the selectivity. This is especially relevant in food analysis since typical foodstuffs are complex matrices with presence of different classes of compounds, thus, extraction methods acquire high importance. The most used extraction methods include traditional Liquid-Liquid Extraction (LLE) and Solid Phase Extraction (SPE). LLE is based on the extraction from the sample phase to another liquid phase. Once the analyte is in the new solvent it can be evaporated to dryness and then reconstituted into a lesser volume to enable the desired preconcentration. SPE consists of passing the liquid sample through a cartridge containing a stationary phase (most of the times based on C_{18}) so that analytes of interest are retained. Then, a second step includes elution of the analytes using a specific solvent. As in the LLE, this solvent can be further evaporated and after consequent reconstitution the preconcentration and isolation of specific analytes or class of analytes can be achieved.

Off-line derivatization of the sample is also a possibility to increase selectivity and sensitivity. In the case of amino acids, this is twice as important because derivatization is paramount so that they can be detected using UV or LIF detectors. Even for MS detection, derivatization of amino acids is useful as it increases their size and thus their pseudomolecular ion what improves their sensitivity by being able to fractionate them in tandem MS experiments. Derivatization also improves the chances to obtain a chiral separation given the larger size of the molecule to be separated. One example is the enantioseparation of amino acids derivatized with 9- fluorenylmethoxycarbonyl chloride (FMOC) [23].

In-capillary sample treatment approaches include those strategies focused on enhancing the sensitivity in-capillary. These strategies might involve coupling CE to other than electrophoretic techniques such as membrane techniques, or chromatographic ones to improve sensitivity [24]. SPE is a popular low-cost technique that can be combined directly to CE *via* connecting column [25]. In this setup it is possible to wash the SPE column and preconcentrate the analytes independently *via* a flow-switching interface so that only the concentrated fraction goes to the CE system [26]. However, one of the major limitations of this approach is to lose some of the sensitivity gained due to the dead volume of the system [24], but strategies to overcome this issue have been proposed [27].

Regarding in-capillary approaches based on electrokinetic approaches or stacking methods, these are focused towards modifying the electrophoretic mobility of the analytes such as by changing the pH, conductivity or incorporating additives [25]. Field Amplified Sample Stacking (FASS) is one of the most used stacking methods. In FASS, the sample is solved in a low conductivity solvent and then is hydrodynamically injected into the capillary which contains a BGE whose conductivity is much higher. When applying the voltage, the sample zone will have a higher electric field. Analytes in the sample zone will move faster until they reach the BGE band where they will be quickly slowed down in this boundary and therefore being stacked. Other examples of sample stacking include Large Volume Sample Stacking (LVSS), Isotachophoresis (ITP) or dynamic pH junction. For a very comprehensive and detailed description of these and other preconcentration strategies, readers are redirected elsewhere [26]. Sweeping techniques involve the accumulation of the analytes by the pseudostationary phase, *e.g.* micelles or charged CDs, that enter the sample zone [24]. Unlike in FASS conditions, sweeping is characterized by a homogeneous electric field strength that is assumed throughout the whole capillary. Note that combination of two of these approaches can be used for a more efficient improvement of the sensitivity. For example, combination of sweeping and stacking enabled to obtain a sensitivity improvement of up 220 times in nine pairs of FMOC-labelled amino acids analyzed in beer samples [28].

In-capillary derivatization is another attractive strategy that allows for low reagent and solvent consumption and often brings better repeatability and reproducibility than off-line derivatization [29]. This procedure needs to be carefully optimized. Examples of in-capillary derivatization include derivatization of amino acids using (-)-1-(9-fluorenyl)ethyl chloroformate in a MEKC-UV system [30], and using 6- aminoquinolyl-N-hydroxysuccinimidyl carbamate to derivatize amino acids in wine and food supplements in an EKC-UV system [31].

APPLICATIONS

This section is devoted to illustrate the different possibilities of CE in chiral food analysis through an in-depth overview of the most representative applications published during the period 2010-2020. As shown in the results obtained from a search in the database "ISI Web of Knowledge" (using as keywords "capillary electrophoresis" and "chiral or enantio*" combined with "food" or "beverages"), the research work developed in the last ten years in this field has been mainly focused on the enantioselective determination of amino acids, organic acids, catechins, sweeteners and other chiral compounds (such as aliphatic primary amines, organosulfur compounds, lipids or pesticides) in a great variety of food matrices including food supplements, fruit juices, tea, wines, milk, beer, algae, *etc.* (Fig. **1A**).

The main characteristics of the methodologies developed for the enantioseparation of all these food components by CE are summarized in Table **1** - **4**. From a global perspective, the most employed CE mode has been CD-MEKC followed by EKC with CDs as chiral selectors and Ligand-Exchange Capillary Electrophoresis (LECE) (Fig. **1B**).

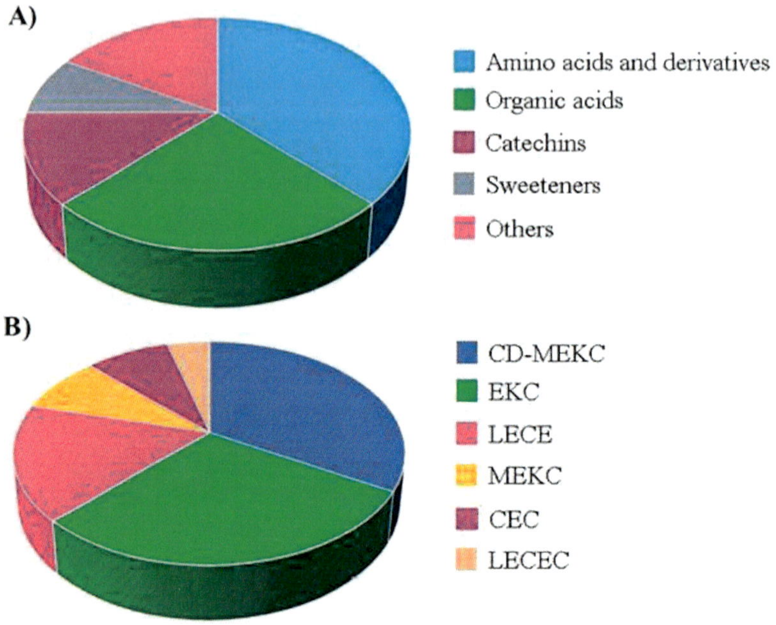

Fig. (1). **A)** Compounds analyzed in food matrices by chiral CE during the period 2010-2020. **B)** Most employed CE mode in chiral food analysis during the period 2010-2020. (Results obtained from the search in the database "ISI Web of Knowledge" using as keywords "capillary electrophoresis" and "chiral or enantio*" combined with "food" or "beverages".

Table 1. Representative applications published during the period 2010-2020 dealing with the chiral analysis of protein and non-protein amino acids.

Analytes	Food Sample	Sample Preparation	Mode/Detection	Experimental Conditions	LOD	Application	Ref.
Glu, Asp	Rice wine	Dilution in acetonitrile (1:1 v/v), centrifugation, and derivatization with FMOC	CD-MEKC/UV (λ= 254 nm)	35 mM β-CD + 6 mM HP-β-CD + 25 mM D-fructose + 30 mM SDS + 15% (v/v) isopropanol in 100 mM borate (pH 9.5)	2.5 μM	Study of the fermentation process by the determination of D-Glu and D-Asp	[32]
Arg, Pro, Ala, Asp	Fruit juices	Derivatization with FITC	CD-MEKC/LIF λ_{ex}= 488 nm	10 mM β-CD + 5 mM SDBS in 50 mM borate (pH 9.6)	0.85-3.99 nM	L-Asn as potential marker for the adulteration of pomegranate juices with apple juice	[33]

(Table 1) cont.....

Asp	Soymilk and beer	Dilution with water (50-fold) and derivatization with NDA	CD-MEKC/LEDIF λ_{em}= 410 nm	60 mM HP-β-CD + 150 mM SDS in 150 mM Tris-borate (pH 9.0) *In capillary preconcentration*: PEO-based stacking and sweeping	2.5×10^{-10} M (L-Asp) 2.4×10^{-10} M (D-Asp)	Analysis of Asp enantiomers	[34]
Trp, Phe, Glu	Beers	Derivatization with FMOC	CD-MEKC/UV (λ= 260 nm)	35 mM β-CD + 35 mM STDC + 12.5% (v/v) isopropanol in 1.5 mM tris-borate (pH 8.5) *In capillary preconcentration*: PEO-based stacking and sweeping	40.1-40.4 nM	Sensitive determination of Trp, Phe, and Glu enantiomers	[28]
Cit	Food supplements	*Capsules*: Dissolution in water, ultrasonication, centrifugation, filtration, and derivatization with FMOC	CEC/UV (λ= 210 and 260 nm)	*Stationary phase*: cellulose tris(3-chloro-4-methylphenylcarbamate) *BGE*: 0.5 M formate/water/acetonitrile (1/19/80, v/v/v) (pH 2.5)	7.5×10^{-7} M (D and L-Cit)	Determination of Cit enantiomers	[35]
Cit	Food supplements	*Capsules*: Dissolution in water, ultrasonication, centrifugation, filtration, and derivatization with FMOC	EKC/UV (λ= 210 nm)	10 mM S-γ-CD in 100 mM formate (pH 3.0)	0.21 μM (D-Cit) 0.18 μM (L-Cit)	Analysis of D-and L-Cit	[36]
17 pairs of protein amino acids	Vinegars	Filtration through 3kDa cutoff filters using centrifugation and dilution in water	EKC/MS	30 mM ($18C_6H_4$) in 1M formate	0.07-1.03 μg/mL	Determination of L- and D-amino acids	[37]
Carn	Food supplements	*Drinks*: Dilution in water and derivatization with FMOC *Biscuits, tablets, and capsules*: Dissolution in water, centrifugation, dilution in water (1/10), and derivatization with FMOC	EKC/MS^2	0.2% (w/v) Succ-γ-CD in 0.5 M ammonium formate (pH 2.5)	10 ng/mL (L-Carn)	Identification and quantification of L- and D-Carn	[38]

(Table 1) cont.....

SeMet	Food supplements	*Capsules:* Dissolution in borate buffer (26 mM, pH 9.2), ultrasonication, centrifugation, filtration, and derivatization with FLEC	MEKC/UV (λ= 210 nm) *Indirect mode*	100 mM APFO (pH 8.5)	3.7 µM (L-SeMet) 3.1 µM (D-SeMet)	Determination of L-SeMet	[39]

Ala, alanine; Arg, arginine; Asp, aspartic acid; APFO, ammonium perfluorooctanoate; Car, carnitine; 18C6H4, (18-crown-6)-2,3,11,12-tetracarboxylic acid; Cit, citrulline; FLEC, (+)-1-(9-fluorenyl)ethyl chloroformate; FMOC, fluorenylmethyloxycarbonyl chloride; FITC, fluorescein isothiocyanate; Glu, glutamic acid; HP-β-CD, hydroxypropyl-β-CD; LEDIF, light-emitting diode-induced fluorescence; LIF, laser-induced fluorescence; NDA naphthalene-2,3-dicarboxaldehyde; PEO, ethylene oxide; Phe, phenylalanine; Pro, proline; S-γ-CD, sulfated- γ-CD; SDBS, sodium dodecylbenzene sulphonate; SDS, sodium dodecyl sulfate; SeMet, selenomethionine; STDC, sodium taurodeoxycholate; Succ-γ-CD, succinyl--cyclodextrin; Trp, tryptophan.

Some applications have also demonstrated the applicability of CEC to the enantioselective determination of chiral compounds in foodstuffs. Even though in some cases the methodologies developed were based on the use of detection systems such as laser-induced fluorescence (LIF), light-emitting diode-induced fluorescence (LEDIF), mass spectrometry (MS) or contactless conductivity detection (C_4D), UV continues to be the most widely used detection mode even taking into account that in some cases a derivatization step is needed to include a chromophore or fluorophore group enabling the detection. As Table **1** - **4** show, among the different derivatizing reagents, FMOC is the preferred choice. Since as it has been demonstrated in the previous section, sample preparation is of paramount importance in the chiral analysis of food samples due to the complexity of the matrices, all steps related to sample preparation including the use of off-line sample treatments or in-capillary pre-concentration techniques are also included in these tables.

Chiral Analysis of Amino Acids

All protein amino acids (except glycine) and most non-protein amino acids exhibit optical activity. Although the L-enantiomer is the natural form, the D-enantiomer can be found in foods as a consequence of a possible racemization of L-amino acids which may take place during food processing (heat treatments, fermentation, *etc.*), storage or microbiological processes [55, 56]. Also, D-forms can be found in foods due to the fraudulent addition of racemic mixtures in supplemented foodstuffs [57, 58] (regulatory agencies prohibit the use of D-enantiomers in the elaboration of foods and dietary supplements [59]). In this way, a high and valuable information, mainly related to food processing, storage or adulterations, among others, can be obtained from the enantiomeric analysis of protein and non-protein amino acids in foods.

During the period of time covered in this chapter, CD-MEKC with both UV and fluorescence detection (using LIF or LEDIF) was the preferred choice to perform the chiral determination of protein amino acids in different beverages such as rice wine, fruit juice, soymilk or beer [28, 32 - 34] (**see** Table **1**). In this line, a methodology based on the combined use of β-CD and hydroxypropyl-β-CD (HP-β-CD) as chiral selectors, SDS as surfactant, and isopropanol and D-fructose as BGE additives was developed to carry out the enantiomeric determination of different protein amino acids previously derivatized with FMOC [32]. The application of this method to the chiral analysis of glutamic and aspartic acids enabled to study the fermentation process of rice wine. The results obtained in this work showed that the percentages of the D-enantiomer of both amino acids determined in the samples were not significantly correlated with the rice wine age. The combination of CD-MEKC with LIF detection was used by Tezcan *et al.* to investigate the adulteration of pomegranate juice with apple juice [33]. First, the authors developed the analytical method by evaluating the use of FITC as derivatization reagent, β-CD as chiral selector and sodium dodecylbenzene sulphonate as surfactant (which enhances the fluorescence intensity of FITC-amino acids). With this aim, four pairs of amino acids (DL-arginine, DL-proline, DL- alanine, and DL-aspartic acid) were used as model compounds. Subsequently, they employed the methodology to obtain the amino acidic profile of pomegranate and apple juices. The comparison of these profiles demonstrated that apple juice contained a high amount of L-asparagine compared to pomegranate juice which made possible to propose this enantiomer as marker of pomegranate juice adulteration with apple juice. On the other hand, two interesting methodologies were reported by Hsieh *et al.* to carry out the sensitive determination of different protein amino acids in soymilk and beer samples [28, 34]. In both methods, the preconcentration of the analytes and their chiral separation were achieved employing a discontinuous CD-MEKC system. First, the sensitive determination of aspartic acid enantiomers (previously derivatized with naphthalene-2,3- dicarboxaldehyde (NDA)) in beers and soymilk was accomplished using a solution containing poly(ethylene oxide) (PEO, which acts as concentrating media), SDS (acting as pseudostationary phase) and HP-β-CD (which acts as chiral selector) in the buffer vial and a solution containing SDS and HP-β-CD into the capillary [34]. Under these conditions, the difference in viscosity between the sample zone and PEO gave rise to a preconcentration by stacking whereas the interaction between the amino acid enantiomers and SDS led to a preconcentration by sweeping. In this way, a 100 and 110-fold enhancement in the sensitivity of D- and L-enantiomers, respectively, was reached (LODs of 2.4×10^{-10} M for the D-enantiomer and 2.5×10^{-10} M for the L-enantiomer). Later, the authors modified slightly the components of the buffer vial solution (solution of PEO, sodium taurodeoxycholate (STDC), β- CD, and isopropanol) and the capillary solution (solution of STDC, β-CD, and isopropanol) to carry out the sensitive chiral analysis of tryptophan, phenylalanine and glutamic acid (derivatized with FMOC) in different beer samples [28]. In this case, the use of the discontinuous system provided over 100 to 1000-fold improvement in sensitivity which enabled to reach LODs in the nM level.

The chiral analysis of the non-protein amino acid citrulline derivatized with FMOC in food supplements has been achieved in the last ten years by using two different CE modes. On the one hand, the enantiomeric separation was accomplished by a CEC-UV methodology based on the use of cellulose tris(3-chloro-4- methylphenylcarbamate) as stationary phase [35]. Once demonstrated the suitability of the developed method through the evaluation of its analytical characteristics, it was successfully applied to the determination of citrulline enantiomers in food supplements. This approach enabled to detect values up to 0.15% of D-citrulline. On the other hand, the enantiomeric determination of this non-protein amino acid in food supplements was carried out by an EKC-UV methodology based on the use of sulfated-γ-CD as chiral selector [36]. In this case, it was possible to detect values of D-citrulline of 0.1% showing the high sensitivity of this method. The results obtained in this work demonstrated that the amount of L-citrulline decreased with respect to the labelled amount as a function of the storage time; however, considering that the D-enantiomer was not detected in any of the samples analyzed, the decrease of L-citrulline could not be attributed to a racemization process. Fig. (**2**) shows the electropherograms obtained for different food supplements non-spiked and spiked with D-citrulline by CEC and EKC.

Within the period of time covered by this chapter, two different works demonstrated the possibilities of EKC with MS detection for the chiral analysis of protein and non-protein amino acids. On the one hand, Lee *et al.* used (18-crow--6)-2,3,11,12- tetracarboxylic acid ($18C_6H_4$) as chiral selector to perform the simultaneous determination of 17 pairs of underivatized proteinogenic amino acids in different vinegar samples by EKC-MS [37]. Since the presence of nonvolatile chiral selectors hinders the EKC-MS hyphenation, the authors employed the partial filling technique to avoid the contamination of the MS ionization chamber. This approach was successfully applied to the determination of the amounts of L-amino acids and trace D-amino acids in vinegars samples which can provide information on the degree of fermentation during the manufacturing process. On the other hand, an EEKC-MS^2 methodology based on the use of succinyl-γ-CD as chiral selector was employed for the chiral analysis of carnitine enantiomers previously derivatized with FMOC [38]. In this case, the contamination of the MS source was avoided using a low amount of chiral selector (0.02%) in the BGE. The developed methodology allowed to reach a high sensitivity (LODs of 10 ng/mL) and was applied to the quantitative determination of L- and D-carnitine in twenty-two food supplements (capsules, tablets, drinks and biscuits). The results obtained demonstrated that one of the samples analyzed had been supplemented with racemic carnitine, which is not allowed by the legislation, whereas the others contained percentages of D-carnitine up to 6%.

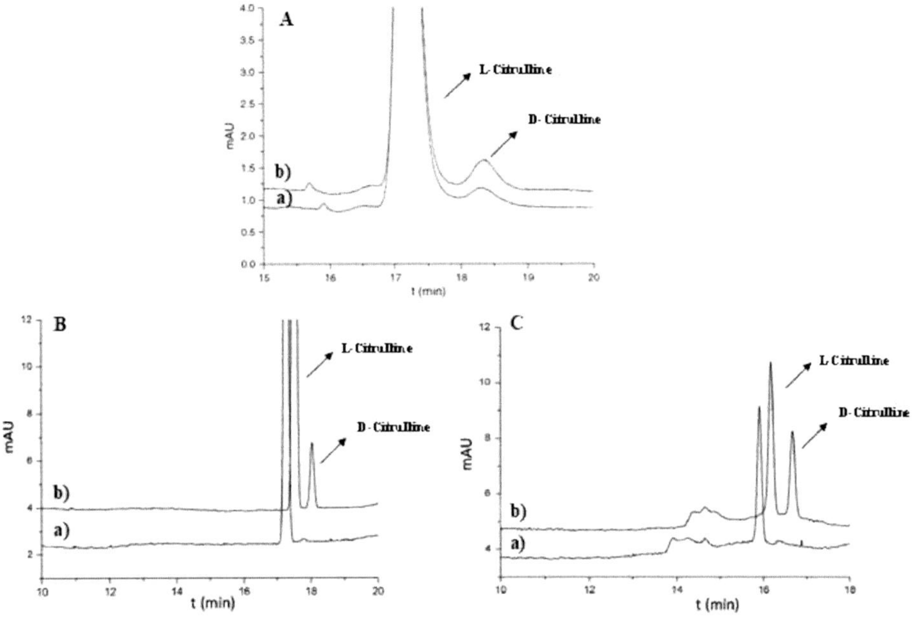

Fig. (2). Enantiomeric separation of citrulline enantiomers in three different food supplements by (A) CEC and (B and C) EKC. A and B correspond to food supplements of recent acquisition, whereas C correspond to food supplements submitted to a long storage time. Electropherograms corresponding to (a) non-spiked food supplement and (b) food supplement spiked with D-citrulline. Experimental conditions. CEC: capillary, 100 µm ID with 24 cm packed and 32.5 cm total length; BGE, 0.5 M ammonium formate/water/acetonitrile (1/19/80, v/v/v); voltage, -10 kV with a pressure of 12 bar in both vials; temperature, 35°C; injection, 10 bar x.

0.2 min of sample followed by a plug of mobile phase at 10 bar x 0.2 min, UV detection, 210 nm. EKC: uncoated fused-silica capillary, 48.5 cm (40 cm to the detector window) x 50 µm; BGE, 10 mM sulfated γ-CD in 100 mM formate buffer (pH 3.0); voltage, -20 kV; temperature, 25°C; injection, 50 mbar for 15 s; UV detection, 210 nm. Adapted and reprinted with permission from [35, 36].

All the methodologies described above were based on the use of the direct mode to obtain the chiral separation of amino acids. However, it is also possible to use an indirect approach in which a covalent interaction is established between each enantiomer and an enantiopure chiral derivatization reagent. This interaction gives rise to diastereoisomers that can be separated under achiral conditions because they have different physicochemical properties. This idea was applied to achieve the separation of selenomethionine enantiomers by MEKC-UV [39]. This non-protein amino acid was derivatized with (+)-1-(9-fluorenyl)ethyl chloroformate (FLEC) and the diastereomers obtained were subsequently separated using ammonium perfluorooctanoate (APFO) as pseudostationary phase. This strategy enabled to obtain the chiral separation of DL-selenomethionine in less than 6 min with LODs in the μM level. Then, it was successfully applied to the quality control of food supplements through the determination of L-selenomethionine (the D enantiomer was not detected in the samples analyzed).

Chiral Analysis of Organic Acids

The enantiomers of organic acids can provide a different taste or flavour to foods and beverages. Since the L-enantiomer is the natural form, the presence of L- and D-enantiomers in food matrices may indicate that they have been added as racemic additives which it is not always allowed by legal regulations [60]. In fact, one of the main applications of the chiral determination of organic acids in beverages is focused on the fight against fruit juice adulteration. In this line, different approaches based on the use of LECE or LECEC have been described in the literature during the period of time covered in this chapter (see Table **2**). Kodama *et al.* developed a LECE methodology with the purpose of separating DL-isocitric acid enantiomers since the ratio of citric acid to D-isocitric acid enables to differentiate authentic and adulterated fruit juices [40]. DL-Isocitric acid was enantiomerically separated with a sulfonated capillary using D-quinic acid as chiral selector, Ni(II) as central ion system and acetonitrile as BGE additive. By using this method, the contents of DL- isocitric and citric acids were successfully determined in three brands of orange juice, a pineapple juice, a grapefruit juice, an apple juice, and a grape juice. The values obtained met the requirements established on the fruit juice industry Code of Practice. Later, the same authors employed a similar LECE-UV method but in this case with a dual central metal ion system (Cu (II) and Sc (III)) to enantioseparate DL-malic acid, DL-tartaric acid, and DL-isocitric acid since the content of these α-hydroxy acids can also be used to distinguish authentic and adulterated fruit juices [41]. This methodology was successfully applied to the analysis of a broad variety of fruit juices. The results obtained showed that tartaric acid was not detected in fruit juices other than grape juices, the ratios of citric acid to D-isocitric acid met the requirements according to the Code of Practice, and D- malic acid was detected only in a grape juice sample suggesting that racemic malic acid was added to the juice. The potential of LECEC to the chiral determination of organic acids was demonstrated by Aydogan *et al.* who employed an open tubular CEC column, Cu (II) as central ion, and L-histidine as chiral ligand to achieve the enantiomeric determination of malic acid in apple juice solutions with different malic acid content [42].

Table 2. Representative applications published during the period 2010-2020 dealing with the chiral analysis of organic acids.

Analytes	Food Sample	Sample Preparation	Mode/Detection	Experimental Conditions	LOD	Application	Ref.
Isocitric acid (citric acid)	Fruit juices	Dilution with water (2-fold), centrifugation, and filtration	LECE/UV (λ= 200 nm)	20 mM $NiSO_4$ + 80 mM D-quinic acid + 30% acetonitrile in 20 mM acetic acid (pH 5.0)	-	Determination of the ratio of citric acid/D-isocitric acid to distinguish authentic and adulterated fruit juices	[40]
Malic, tartaric and isocitric acids (citric acid)	Fruit juices	*Apple, grape, pineapple juice:* dilution with water (8-fold), centrifugation and filtration *Orange and mixed fruit juice*: centrifugation, solid-phase extraction with C18 cartridge	LECE/UV (λ= 250 nm)	10 mM $CuSO_4$ + 1.8 mM $ScCl_3$ + 100 mM D-quinic acid in 20 mM acetate (pH 5.0)	3 mg/L (DL-malic and DL-isocitric acid) 2 mg/L DL-tartaric)	Determination of α-hydroxy acids enantiomers for detecting fruit juice adulteration	[41]
Malic acid	Apple juice	Centrifugation and dilution with water	LECEC/UV (λ= 214 nm)	Acetonitrile/5 mM $CuSO_4$ + 20 mM ammonium sulfate (60:40 v/v) (pH 3.0)	-	Determination of malic acid enantiomers	[42]
Tartaric and malic acid	Wine and wine precipitates	*Wine:* dilution with water solution (10 fold) and filtration. *Wine precipitates:* dilution in water and ultrasonication	LECE/UV (λ= 254 nm)	100 mM D-quinic acid + 10 mM Cu (II) + 0.5 mM Al (III) + 0.5 mM CTA-OH in 20 mM acetate	1.5 mg/L (D-tartaric acid) 3 mg/L (D-malic acid)	Chiral determination of tartaric and malic acids	[43]

(Table 2) cont.....

Tartaric acid	Wine and grapes	*Wine:* filtration and dilution with water (100 fold) *Grapes*: Squeeze, filtration, and dilution with water (200 fold)	LECE/C^4D	7 mM $CuCl_2$ + 14 mM trans-4-hydroxy-L-proline in 100 mM ε-aminocaproic acid (pH 5.0)	19 µM (D-enantiomer) 21 µM (L-enantiomer)	Enantiomeric separation of tartaric acid for quality control and safety	[44]
Lactic acid (α-hydroxybutyric acid, 2-hydroxycaproic acid, 2-hydroxyoctanoic acid, 2-hydroxydecanoic acid)	Milk and yogurt	Dilution in acetonitrile, centrifugation, solvent evaporation, redissolution in BGE and filtration	EKC-C^4D	5 mM vancomycin + 4.4 mM maleic acid + 0.03 mM CTAB in 10 mM Tris (pH 7.35)	2.8 µM (L- Lactic) 2.4 µM (D-Lactic)	Determination of lactic acid enantiomers to investigate possible microbial contamination	[45]

CTAB, cetyltrimethylammonium bromide; CTA-OH, hexadecyltrimethylammonium hydroxide.

The development of chiral methodologies capable to provide the enantioselective determination of organic acids is of high importance in winemaking where the application of these methods enables to acquire relevant data related to wine quality control. For example, DL-malic acid is a legal food additive which can be used in the acidification of wines. Conversely, the use of DL-tartaric acid is prohibited [42]. This fact makes very interesting the chiral determination of both short chain organic acids in wines. In this line, a LECE methodology was successfully applied to the simultaneous enantioseparation of tartaric and malic acids in wines and wines precipitates [43]. This method was based on the use of a bare fused capillary with EOF reversal *via* hexadecyltrimethylammonium hydroxide (CTA-OH), Cu (II) and Al (III) as dual central metal ion system, and D-quinic acid as chiral selector. Different amounts of D-malic acid were found in several of the wines analyzed, whereas D-tartaric acid was only found in two of the samples which appeared to be treated during manufacturing with racemic tartaric acid. Fig. (**3**) depicts the electropherograms obtained when the developed methodology was employed to analyze a wine sample containing D-malic acid and a wine sample containing D- tartaric acid. LECE was also coupled with C^4D for the quality and safety control of wines and grapes through the determination of tartaric acid enantiomers [44].

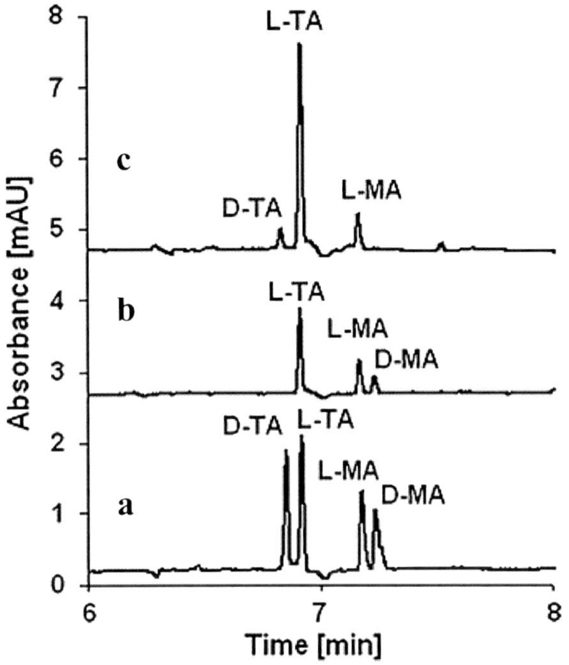

Fig. (3). Electropherograms of **(a)** calibration solution of D-tartaric, L-tartaric, L-malic and D-malic acids (50 mg/L each), **(b)** wine containing D-malic acid, and **(c)** wine containing D-tartaric acid. LECE Conditions: uncoated fused-silica capillary, 75 cm total length (65 effective length) x 50 μm; BGE, 100 mM D-quinic acid.

+ 10 mM Cu (II) + 0.5 mM Al (III) + 0.5 mM CTA-OH in 20 mM acetate buffer; voltage: -25 kV; temperature, 20°; injection: 30 mbar, 5 s; UV detection, 250 nm. Reprinted with permission from [44].

In this case, the EOF was suppressed using a polyacrylamide-coated capillary and the separation was accomplished using a Cu (II) as central metal ion together with L-hydroxyproline as chiral selector. Thus, LODs about 20 μM were achieved. Under these conditions, D-tartaric acid was not found in any of the samples analyzed, which demonstrated that a short-term bacterial contamination did not occur.

In the period of time considered in this book chapter, C^4D was also employed as detection system coupled to EKC to develop a methodology for the chiral determination of different α-hydroxy acids (namely lactic acid, α-hydroxybutyric acid, 2-hydroxycaproic acid, 2-hydroxyoctanoic acid and 2-hydroxydecanoic acid) and two protein amino acids (aspartic and glutamic acids) [45]. The developed procedure, based on the use of vancomycin as chiral selector, was applied to the enantioselective analysis of lactic acid in yogurt and milk. Since the presence of D- lactate in dairy products may indicate a microbial contamination, the present methodology may be considered an interesting tool to guarantee their quality and safety.

Chiral Analysis of Catechins

Catechins belong to flavonoids family which are one of the largest natural occurring phenolic compounds in food and beverages. As it can be seen in Table **3**, the use of CD-MEKC methodologies with UV detection for the chiral separation of these compounds in green tea samples enabled to obtain information related to the effects of storage [46], the geographical origin [47] or tea leaves treatment (fermentation, thermal treatment, *etc.*) [48]. For instance, a method based on the use of SDS as surfactant and HP-β-CD as chiral selector was proposed by Mirasoli *et al.* to evaluate quality changes in commercial green tea leaves during a long-term storage [46]. By using this methodology, it was possible to analyze different catechins along with the methylxanthines theobromine and caffeine as well as to selectively estimate (+)-gallocatechin in the presence of (---gallocatechin, the non-native enantiomer whose presence in tea samples can be associated to the epimerisation of (-)-epigallocatechin occurring during manufacturing processes or uncorrected storage conditions of fresh products. Interestingly, data obtained showed that epimerization did not significantly occur during aging since the enantiomeric ratio (+)-gallocatechin *versus* (---gallocatechin just slightly increased from "not aged" to "aged" samples. The combination of the results obtained by the application of the developed MEKC methodology and those obtained by an electronic nose analysis allowed the characterization of green tea samples taking into account that the catechins content is associated to the biological value of the product and the aroma is a decisive parameter for its acceptance. Later, other authors used the same CD-MEKC system for the quantitation of catechins and methylxanthines in ninety-two green tea samples from Japan and China [47]. This methodology presented an adequate selectivity, short analysis time (8 min) and provided the enantioselective separation of some of the catechins ((+)-catechin, (–)-catechin and (–)-epicatechin). Data analysis by chemometric techniques demonstrated the possibility of using catechins and methylxanthines content as phytomarkers of the geographical origin of green teas. In another work, the separation of six major green tea catechins including the enantioseparation of (±)-catechin and DL-theanine was also achieved by CD-MEKC-UV [48]. In this case, the poor detectability of theanine made necessary to carry out a derivatization step with o-phthaldialdehyde in the presence of N-acetyl-L-cystein. Subsequently, the separation was accomplished by adding SDS and (2,6-di-*O*-methyl)-β-CD to the running buffer. This methodology enabled the quantitation of a percentage of

0.5% of D-theanine with respect to the L- enantiomer. (Fig. **4**) shows the CD-MEKC electropherograms obtained in the analysis of three different green tea samples. From the results obtained in this work the authors suggested that the presence of (–)-catechin may be considered as a marker of thermal degradation, while the presence of D-theanine could rather be a marker of microbial or enzymatic processes.

Table 3. Representative applications published during the period 2010-2020 dealing with the chiral analysis of catechins and sweeteners.

Analytes	Food Sample	Sample Preparation	Mode/detection	Experimental Conditions	LOD	Application	Ref.
(±)-Gallocatechin (along with other catechins and methylxanthine)	Green tea	Pulverization of tea leaves, powderization, dissolution in water, filtration and dilution in water (1:1 v/v)	CD-MEKC/UV (λ= 200 nm)	25 mM HP-- -CD + 90 mM SDS in 25 mM borate-phosphate (pH 2.5)	0.4 µg/mL (+)-GC and (-)-GC 0.1 µg/mL (for all the other compounds)	Quality control of green tea samples subjected to long-term storage	[46]
(±)-Catechin (along with other catechins and methylxanthines)	Green tea	Grinding, dissolution in water, filtration, and dilution in water (1:2 v/v)	CD-MEKC/UV (λ= 200 nm)	25 mM HP-- -CD + 90 mM SDS in 25 mM borate-phosphate (pH 2.5)	-	Analysis of the principal catechins and methylxanthines in green tea	[47]
(±)-Catechin, DL-theanine (along with other catechins)	Green tea	Grinding, dissolution in water, filtration, and derivatization with OPA/NAC	CD-MEKC/UV (λ= 200 nm)	28 mM DM-- CD + 65 mM SDS in 25 mM borate-phosphate (pH 2.5)	0.1-02 µg/mL	Chiral analysis of theanine and catechin in the characterization of green tea samples	[48]
Neotame	Mango juice, cola soft drink, orange soft drink	Direct dissolution of neotame in each beverage	EKC/UV (λ= 200 nm)	30 mM TM-- -CD in 50 mM phosphate (pH 5.5)	0.018 mM (LL-neotame) 0.082 mM (DD-neotame)	Enantiomeric separation of neotame in spiked beverages	[49]

(Table 3) cont.....

Aldopentoses: arabinose, lyxose, ribose, xylose	Red seaweed	Freezing-drying, powderization, hydrolysis with trifluoroacetic acid and DBD-Apy derivatization	MEKC/LIF	50 mM SDS in 50 mM phenylboronic acid (pH 9.0)	0.3 μM	Analysis of DL-galactose in hydrolysates of dried red seaweed	[50]
Aldohexoses: allose, galactose, glucose, gulose, idose, mannose, talose			(use of a 588 nm band pass filter)				
			Indirect mode				

DBD-Apy, (S)-(+)-4-(N,N-dimethylaminosulfonyl)-7-(3-aminopyrrolidin-1-yl)-2,1,3-be nzoxadiazole; DM-β-CD, 2,6-di-O-methyl-β-CD; HP-β-CD, hydroxypropyl-β-CD; GC, gallocatechin; LIF, laser-induced fluorescence; OPA/NAC, o-phthaldialdehyde/ N-acetyl-L-cysteine; SDS, sodium dodecyl sulfate; TM-β-CD, trimethyl-β-CD.

Chiral Analysis of Sweeteners

The increasing prevalence of diseases related to sugar consumption has increased the use of sweeteners in foodstuffs [61]. As sweeteners can contain chiral centers in their structure, the development of chiral methodologies for their enantioselective determination is required.

In the time covered by this chapter, two different works demonstrated the possibilities of CE for the chiral determination of these food additives. Neotame is an artificial non-nutritive sweetener which is formed by four diastereomers (its structure contains two chiral centers). Its sweetness is attributed to the LL-diastereomer while DD-neotame is not sweet. As Table **3** shows, the chiral separation of neotame diastereomers (LL and DD) was performed by EKC using (2, 3, 6-tri-*O*-methyl)-β-CD as chiral selector and UV as detection mode [49].

Employing the developed methodology, it was possible to carry out a molecular docking study to elucidate the separation mechanism (which demonstrated the relevant role of electrostatic and hydrophobic interactions in the stabilization of the neotame-cyclodextrin inclusion complexes), and to achieve the determination of neotame in different spiked cola and orange soft-drinks and mango juice. Another research work was focused on the chiral separation of different DL-aldoses (arabinose, lyxose, ribose, xylose, allose, galactose, glucose, gulose, idose, mannose and talose) by an indirect MEKC-LIF approach [50]. Sugars were labeled with the chiral derivatization reagent (S)-(+)4-(N-N-dimethylaminosulfonyl)-7-(3- aminopyrrolidin-1-yl)-2,1,3-benzoxadiazole (DBD-Apy) and their separation was performed using phenylboronate buffer containing SDS as BGE [50]. Under these conditions, a good resolution of all the D/L isomers was reached with a high sensitivity (LOD around 0.3 μM).

The potential of this methodology to the quality control of edible seaweed was demonstrated by its successful application to the chiral analysis of DL-galactose in hydrolysates of dried red seaweed.

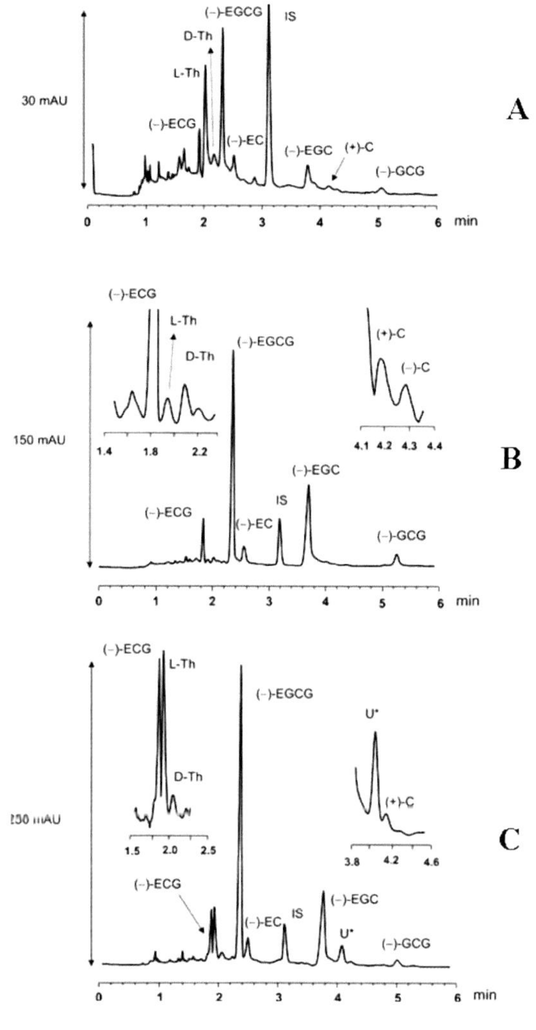

Fig. (4). CD-MEKC electropherograms of **(A)** Oolong Formosa (60% fermentation) tea sample; **(B)** Oolong (15% fermentation) tea sample. In the inset, the peak of (−)-ECG is out of the response scale; **(C)** Bi Luo Chun (green tea as such) sample. CD-MEKC Conditions: uncoated fused-silica capillary, 48.5 cm total length (8.5 effective length) x 50 µm; BGE, 65 mM SDS + 28 mM 2,6-di-O-methyl-β-CD in 25 mM phosphate-borate buffer (pH 2.5); voltage: 30 kV; temperature, 30°; injection: 50 mbar, 5 s; UV detection, 220 nm. Peak identification: C, (-)-catechin; EC, epicatechin; ECG, Epicatechin gallate; EGC, epigallocatechin; EGCG, epigallocatechin gallate; GCG, gallocatechin gallate; IS, internal standard; Th, theanine; U*, unknown. Reprinted with permission from [48].

Chiral Analysis of Other Compounds Present in Foods

The enantiomeric determination by CE of other chiral compounds (different from those included in the previous paragraphs) which can also be present in food matrices has also been reported in the last years as it can be seen in Table **4**.

Table 4. Representative applications published during the period 2010-2020 dealing with the chiral analysis of other compounds.

Analytes	Food Sample	Sample Preparation	Mode/detection	Experimental Conditions	LOD	Application	Ref.
DMAA	Food supplements	*Capsules or powder*: Dissolution in water, sonication and filtration.	EKC/UV (λ= 214 nm)	1.1% (w/v) S-α-CD + 0.2% (w/v) S-- -CD in 5 mM phosphate/Tris (pH 3.0) and 10 mM BTEAC	7.82-9.24 µg/mL	Determination of DMAA content and stereoisomeric composition to verify their potential natural origin	[51]
Lipoic acid	Dietary supplements	*Tablets or Capsules*: Dissolution in 70% methanol, ultrasonication, and centrifugation.	EKC/UV (λ= 200 nm)	8 mM TM-β-CD in 100 mM phosphate (pH 7.0)	1.5 mg/L	Determination of RS lipoic acid to control the use of racemic mixture or pure enantiomers	[52]
HETEs	Marine red algae	Freezing-drying, powderization, dissolution in water, incubation, filtration, liquid-liquid extraction with ethyl acetate, evaporation to dryness, and redissolution in methanol	CD-MEKC/UV (λ= 235 nm)	30 mM HP-γ-CD + 75 mM SDS in 30 mM phosphate-15 mM borate (pH 9.0)	0.95-0.99 µg/mL	Determination of the enantiomeric composition of the HETEs	[53]
Metalaxyl (tebuconazole, diniconazole, uniconazole, hexaconazole, benalaxyl, myclobutayl, futriafol)	Tap water	Solid-phase extraction with C18 cartridge, evaporation to dryness, and redissolution in acetonitrile/water (80/20, v/v)	CEC/UV (λ= 210 nm)	*Stationary phase*: cellulose tris(4-chloro-3-methylphenylcarbamate *BGE*: acetonitrile/water/formate (90/9/1, v/v/v) (pH 2.5)	1.4 mg/L (S-metalaxyl) 1.6 mg/L (R-metalaxyl)	Enantiomeric separation of metalaxyl in spiked tap water	[54]

BTEAC, benzyltriethylammonium chloride; CEC, capillary electrochromatography; DMAA, 1,3-dimethylamylamine; HETEs, hydroxyeicosatetraenoic acids; HP-γ-CD, hydroxypropyl-γ-CD; S-α-CD, sulfated-α-CD; S-β-CD, sulfated-β-CD; SDS, sodium dodecyl sulfate; TM-β-CD, trimethyl-β-CD.

Pribylka *et al.* developed an EKC methodology with indirect UV detection to achieve the separation of the four diastereoisomers of 1,3-dimethylamylamine (DMAA) with the aim of evaluating the quality of food supplements through the origin of this compound [51]. DMAA is a simple aliphatic primary amine that occurs naturally in plant species *Pelargonium* and *Geranium* and has a similar structure of amphetamine and analogous effects on the human organism. In this work, benzyltriethylammonium chloride was used as chromophoric additive and a dual system based on the combination of two negatively charged CDs (sulphated α-CD and sulphated β-CD) was employed as chiral selector. The diastereoisomeric ratios obtained in the analysis of DMAA in food supplements were identical to synthetic DMAA suggesting that this component probably was not of natural origin. A comparison between the results obtained in this work with those previously obtained using GC as separation technique [62], showed that the developed EKC method was faster and more economical for the evaluation of the stereoisomeric ratio and the quantification of the DMAA stereoisomers in dietary supplements.

Lipoic acid is an organosulfur compound with antioxidant properties whose R-enantiomer naturally occurs, while synthetic lipoic acid is racemic. While both racemic and R-lipoic acid are contained in food supplements, the potency of (S)-lipoic acid is not exactly known [52]. To achieve the enantiomeric separation of lipoic acid in dietary supplements, Kodama *et al.* proposed an EKC methodology in which a capillary coated with sulfonated groups and a BGE containing (2, 3, 6-tri-*O*-methyl)-β-CD as chiral selector were employed [52]. Most of the food supplements analyzed in this work contained the racemic mixture of lipoic acid, only three of them contained just the R-enantiomer as claimed in the label.

Hydroxyeicosatetraenoic acids (HETEs) are endogenous eicosanoids generated from arachidonic acid in mammalian systems which have important physiological and pathological functions [53]. A comprehensive analysis of natural and synthetic HETEs implied to carry out the separation of both regioisomers and enantiomers. The combination of hydroxypropyl-γ-CD with SDS in a CD-MEKC approach gave rise to the simultaneous enantioseparation of 8-, 11-, 12-, and 15-HETEs within 35 min [53]. Then, to demonstrate the potential of this methodology, it was applied to the chiral analysis of 8-, and 12-HETEs in *Gracilaria vermiculophylla* and *Gracilaria arcuate* (two species of red algae).

Many pesticides used in agriculture are chiral compounds. In most cases, one of the enantiomers has the pesticide activity while the other can present different activity toward the target organism [63]. For this reason, the development of enantioselective methodologies enabling the evaluation of the real toxicity of pesticide enantiomers in food samples and the study of their enantioselective degradation is needed. In this sense, Pérez-Fernández *et al.* optimized a CEC methodology based on the use of a polysaccharide-based chiral stationary phase (namely, cellulose tris(4-chloro-3-methylphenylcarbamate) and acetonitrile/water /formate (90/9/1, v/v/v) at pH 2.5 as mobile phase to perform the enantioseparation of sixteen different pesticides (including herbicides, insecticides, and fungicides) [54]. The chiral separation obtained by the developed CEC methodology and a nano-LC method under similar experimental conditions demonstrated advantages in the use of the CEC approach. Then, with the purpose of demonstrating the reliability of the developed methodology, it was applied to the separation of metalaxyl and its enantiomeric impurity in tap water samples spiked with the commercial product (see Fig. (**5**)).

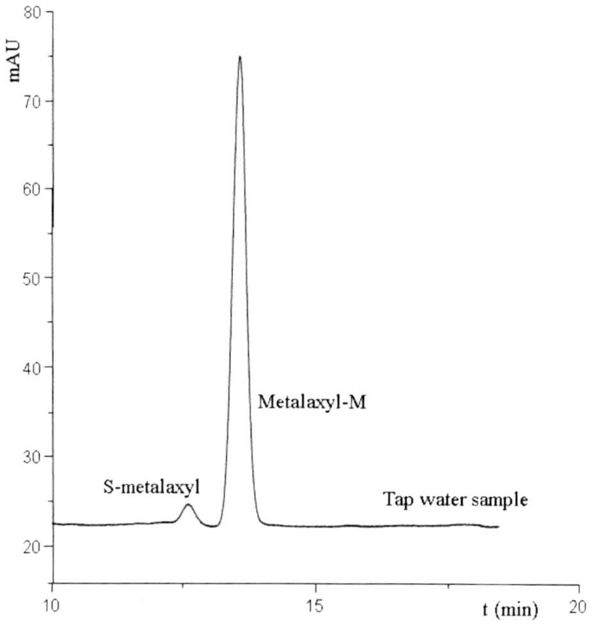

Fig. (5). CEC electrochromatogram corresponding to the chiral analysis of metalaxyl in tap water spiked with a commercial product containing only metalaxyl-M (at a concentration of approximately 75 mg/L according to its label). CEC conditions: capillary, 100 µm x 24 cm packed with Sepapak-4 and 32.5 cm total length; BGE, acetonitrile/water/formate pH 2.5 (90/9/1, v/v/v); voltage, -10 kV with 12 bar pressure in both inlet and outlet vials; temperature, 25°C; injection, 10 bar for 0.2 min (followed by a plug of mobile phase at 10 bar x 0.2 min); UV detection, 210 nm. Adapted and reprinted with permission from [54].

CONCLUDING REMARKS

This chapter describes the effectiveness of Capillary Electrophoresis in the chiral analysis of foodstuff and beverages including the most relevant applications reported from 2010 to present. Different CE modes using diverse detection systems have proven to enable the determination of a wide variety of chiral analytes such as amino acids, organic acids, catechins or sweeteners, among others, in food matrices including food supplements, fruit juices, tea, microalgae, wine, grapes, beer, milk, soymilk, yogurt, vinegars and tap water. In addition, the use of a broad range of chiral selectors together with the low consumption of solvents, reagents and samples, and short analysis times, make CE a powerful tool for chiral analysis.

Amino acids have been the most analyzed molecules followed by organic acids, probably due to their important role in several processes related to food elaboration. Their analysis can provide relevant information about fermentation processes, adulterations, food authenticity, or the effects of processing and storage, among others. MEKC has been the most used CE mode in the analysis of amino acids, mainly with UV detection although the use of LIF has also been reported, followed by EKC either with MS or UV detectors. In the analysis of catechins, similar trend was followed as MEKC-UV was the standard tool. The analysis of sweeteners, including neotame and sugars, was achieved by both EKC-UV and MEKC-LIF. Interestingly, the most common scenario for the analysis of organic acids has been LECE with UV detection.

Considering all the articles included in this chapter, CDs (as a single chiral selector or used in a dual system) were the most common chiral selectors employed due to their great potential for the stereoselective separation of different chiral compounds, as well as for their low cost, availability, and diversity. Particularly, CD-MEKC was the most frequent approach, using neutral CDs such as β-CD, HP- β-CD or 2,6-di-O-methyl-β-CD together with SDS or SDBS as surfactants. The good results observed with these strategies can be explained taking into account the interactions between the chiral analyte and the CD with the micellar phase, which often provide a successful enantioseparation. UV was the dominant detection mode due to its simplicity while LIF and MS were also employed to a lesser extent. These two detectors enabled to obtain low LODs (at nanomolar or ng/mL levels) allowing to expect an increase in their use in the future directed to the development of sensitive stereoselective analytical methodologies. It is important to keep in mind that in the particular case of CE-MS, it would be necessary to develop new compatible chiral selectors or new strategies making possible to carry out a chiral analysis in a compatible way.

It is clear that chirality is a very important topic in the food analysis field and, as this book chapter has shown, CE is a powerful analytical tool offering many possibilities to advance in the knowledge on this field.

CONSENT FOR PUBLICATION

Not applicable.

CONFLICT OF INTEREST

The author declares no conflict of interest, financial or otherwise.

ACKNOWLEDGEMENTS

Authors thank the Comunidad of Madrid and European funding from FSE and FEDER programs (project S2018/BAA-4393, AVANSECAL-II-CM). S.B.B and M.C.P. thank the Spanish Ministry of Economy and Competitiveness for their predoctoral (BES-2017-082458) and "Ramón y Cajal" (RYC-2013-12688) research contracts, respectively.

REFERENCES

[1] Rocco, A.; Aturki, Z.; Fanali, S. Chiral separations in food analysis. *TrAC. Trends Analyt. Chem.,* **2013**, *52*, 206-225. [http://dx.doi.org/10.1016/j.trac.2013.05.022]

[2] D'Orazio, G.; Fanali, C.; Asensio-Ramos, M.; Fanali, S. Chiral separations in food analysis. *TrAC. Trends Analyt. Chem.,* **2017**, *96*, 151-171. [http://dx.doi.org/10.1016/j.trac.2017.05.013]

[3] Alvarez-Rivera, G.; Bueno, M.; Ballesteros-Vivas, D.; Cifuentes, A. Chiral analysis in food science. *TrAC. Trends Analyt. Chem.,* **2020**, *123*, 115761. [http://dx.doi.org/10.1016/j.trac.2019.115761]

[4] Bernardo-Bermejo, S.; Sánchez-López, E.; Castro-Puyana, M.; Marina, M.L. Chiral capillary electrophoresis. *TrAC. Trends Analyt. Chem.,* **2020**, *124*, 115807. [http://dx.doi.org/10.1016/j.trac.2020.115807]

[5] Sánchez-López, E.; Castro-Puyana, M.; Marina, M.L. , **2019**.

[6] Simó, C.; García-Cañas, V.; Cifuentes, A.; Chiral, C.E-M.S. Chiral CE-MS. *Electrophoresis,* **2010**, *31*(9), 1442-1456. [PMID: 20376813]

[7] Wuethrich, A.; Haddad, P.R.; Quirino, J.P. Chiral capillary electromigration techniques-mass spectrometry-hope and promise. *Electrophoresis,* **2014**, *35*(1), 2-11. [http://dx.doi.org/10.1002/elps.201300377] [PMID: 24265218]

[8] Castro-Puyana, M.; Marina, M.L. Chiral capillary electrophoresis-mass spectrometry.*Chiral Separations-Methods and Protocols. Series Methods in Molecular Biology,* 3rd ed; Scriba, G.K.E., Ed.; Springer Science, **2019**, pp. 391-405. [http://dx.doi.org/10.1007/978-1-4939-9438-0_23]

[9] Herrero, M.; García-Cañas, V.; Simó, C.; Cifuentes, A. Recent advances in the application of capillary electromigration methods for food analysis and Foodomics. *Electrophoresis,* **2010**, *31*(1), 205-228. [http://dx.doi.org/10.1002/elps.200900365] [PMID: 19967713]

[10] Castro-Puyana, M.; García-Cañas, V.; Simó, C.; Cifuentes, A. Recent advances in the application of capillary electromigration methods for food analysis and Foodomics. *Electrophoresis,* **2012**, *33*(1), 147-167. [http://dx.doi.org/10.1002/elps.201100385] [PMID: 22147337]

[11] García-Cañas, V.; Simó, C.; Castro-Puyana, M.; Cifuentes, A. Recent advances in the application of capillary electromigration methods for food analysis and Foodomics. *Electrophoresis,* **2014**, *35*(1), 147-169.
[http://dx.doi.org/10.1002/elps.201300315] [PMID: 24114758]

[12] Acunha, T.; Ibáñez, C.; García-Cañas, V.; Simó, C.; Cifuentes, A. Recent advances in the application of capillary electromigration methods for food analysis and Foodomics. *Electrophoresis,* **2016**, *37*(1), 111-141.
[http://dx.doi.org/10.1002/elps.201500291] [PMID: 26256797]

[13] Álvarez, G.; Montero, L.; Llorens, L.; Castro-Puyana, M.; Cifuentes, A. Recent advances in the application of capillary electromigration methods for food analysis and Foodomics. *Electrophoresis,* **2018**, *39*(1), 136-159.
[http://dx.doi.org/10.1002/elps.201700321] [PMID: 28975648]

[14] Pérez-Míguez, R.; Marina, M.L.; Castro-Puyana, M. Capillary electrophoresis determination of non-protein amino acids as quality markers in foods. *J. Chromatogr. A,* **2016**, *1428*, 97-114.
[http://dx.doi.org/10.1016/j.chroma.2015.07.078] [PMID: 26233255]

[15] Pérez-Míguez, R.; Salido-Fortuna, S.; Castro-Puyana, M.; Marina, M.L. Advances in the determination of nonprotein amino acids in foods and biological samples by capillary electrophoresis. *Crit. Rev. Anal. Chem.,* **2019**, *49*(5), 459-475.
[http://dx.doi.org/10.1080/10408347.2018.1546113] [PMID: 30963784]

[16] Herrero, M.; Simó, C.; García-Cañas, V.; Fanali, S.; Cifuentes, A. Chiral capillary electrophoresis in food analysis. *Electrophoresis,* **2010**, *31*(13), 2106-2114.
[http://dx.doi.org/10.1002/elps.200900770] [PMID: 20533328]

[17] Yu, R.B.; Quirino, J.P. Chiral selectors in capillary electrophoresis: Trends during 2017–2018. *Molecules,* **2019**, *24*(6), 1135.
[http://dx.doi.org/10.3390/molecules24061135]

[18] Escuder-Gilabert, L.; Martín-Biosca, Y.; Medina-Hernández, M.J.; Sagrado, S. Cyclodextrins in capillary electrophoresis: recent developments and new trends. *J. Chromatogr. A,* **2014**, *1357*, 2-23.
[http://dx.doi.org/10.1016/j.chroma.2014.05.074] [PMID: 24947884]

[19] Otsuka, K.; Terabe, S. Chiral micellar electrokinetic chromatography.*Chiral Separations: Methods and Protocols*; Gübitz, G.; Schmid, M.G., Eds.; Humana Press Inc: Totowa, USA, **2004**, pp. 355-364.

[20] Mao, Z.; Chen, Z. Advances in capillary electro-chromatography. *J. Pharm. Anal.,* **2019**, *9*(4), 227-237.
[http://dx.doi.org/10.1016/j.jpha.2019.05.002] [PMID: 31452960]

[21] Yu, R.B.; Quirino, J.P. Chiral liquid chromatography and capillary electrochromatography: Trends from 2017 to 2018. *TrAC. Trends Analyt. Chem.,* **2019**, *118*, 779-792.
[http://dx.doi.org/10.1016/j.trac.2019.07.011]

[22] Ahmed, M.A.; Felisilda, B.M.B.; Quirino, J.P. Recent advancements in open-tubular liquid chromatography and capillary electrochromatography during 2014-2018. *Anal. Chim. Acta,* **2019**, *1088*, 20-34.
[http://dx.doi.org/10.1016/j.aca.2019.08.016] [PMID: 31623713]

[23] Sánchez-Hernández, L., Domínguez-Vega, E.; Montealegre, C.; Castro-Puyana, M.; Marina, M.L.; Crego, A.L. Potential of vancomycin for the enantiomeric resolution of FMOC-amino acids by capillary electrophoresis-ion-trap-mass spectrometry. *Electrophoresis,* **2014**, *35*(9), 1244-1250.
[http://dx.doi.org/10.1002/elps.201300489] [PMID: 24338646]

[24] Mikus, P.; Marakova, K. Chiral Capillary electrophoresis with on-line sample preparation. *Curr. Pharm. Anal.,* **2010**, *6*(2), 76-100.
[http://dx.doi.org/10.2174/157341210791202663]

[25] Quirino, J.P. On-line electrophoretic, electrochromatographic, and chromatographic sample

concentration in CE-MS., **2016**.
[http://dx.doi.org/10.1002/9783527693801.ch5]

[26] Breadmore, M.C.; Grochocki, W.; Kalsoom, U.; Alves, M.N.; Phung, S.C.; Rokh, M.T.; Cabot, J.M.; Ghiasvand, A.; Li, F.; Shallan, A.I.; Keyon, A.S.A.; Alhusban, A.A.; See, H.H.; Wuethrich, A.; Dawod, M.; Quirino, J.P. Recent advances in enhancing the sensitivity of electrophoresis and electrochromatography in capillaries and microchips (2016-2018). *Electrophoresis,* **2019**, *40*(1), 17-39.
[http://dx.doi.org/10.1002/elps.201800384] [PMID: 30362581]

[27] Zhang, L.H.; Wu, X.Z. Capillary electrophoresis with in-capillary solid-phase extraction sample cleanup. *Anal. Chem.*, **2007**, *79*(6), 2562-2569.
[http://dx.doi.org/10.1021/ac0621591] [PMID: 17298033]

[28] Lin, E.P.; Lin, K.C.; Chang, C.W.; Hsieh, M.M. On-line sample preconcentration by sweeping and poly(ethylene oxide)-mediated stacking for simultaneous analysis of nine pairs of amino acid enantiomers in capillary electrophoresis. *Talanta,* **2013**, *114*, 297-303.
[http://dx.doi.org/10.1016/j.talanta.2013.05.039] [PMID: 23953474]

[29] Glatz, Z. On-capillary derivatisation as an approach to enhancing sensitivity in capillary electrophoresis. *Electrophoresis,* **2015**, *36*(5), 744-763.
[http://dx.doi.org/10.1002/elps.201400449] [PMID: 25330133]

[30] Fradi, I.; Farcas, E.; Saïd, A.B.; Yans, M.L.; Lamalle, C.; Somsen, G.W.; Prior, A.; de Jong, G.J.; Kallel, M.; Crommen, J.; Servais, A.C.; Fillet, M. In-capillary derivatization with (-)-1-(9-fluorenyl)ethyl chloroformate as chiral labeling agent for the electrophoretic separation of amino acids. *J. Chromatogr. A,* **2014**, *1363*, 338-347.
[http://dx.doi.org/10.1016/j.chroma.2014.07.022] [PMID: 25082527]

[31] Martínez-Girón, A.B.; García-Ruiz, C.; Crego, A.L.; Marina, M.L. Development of an in-capillary derivatization method by CE for the determination of chiral amino acids in dietary supplements and wines. *Electrophoresis,* **2009**, *30*(4), 696-704.
[http://dx.doi.org/10.1002/elps.200800481] [PMID: 19156767]

[32] Miao, Y.N.; Liu, Q.; Wang, W.; Liu, L.; Wang, L. Enantioseparation of amino acids by micellar capillary electrophoresis using binary chiral selectors and determination of D-glutamic acid and D-aspartic acid in rice wine. *J. Liq. Chromatogr. Relat. Technol.,* **2017**, *40*(15), 783-789.
[http://dx.doi.org/10.1080/10826076.2017.1364263]

[33] Tezcan, F.; Uzaşçı, S.; Üyar, G.; Öztekin, N.; Erim, F.B. Determination of amino acids in pomegranate juices and fingerprint for adulteration with apple juices. *Food Chem.,* **2013**, *141*(2), 1187-1191.
[http://dx.doi.org/10.1016/j.foodchem.2013.04.017] [PMID: 23790902]

[34] Lin, K.C.; Hsieh, M.M.; Chang, C.W.; Lin, E.P.; Wu, T.H. Stacking and separation of aspartic acid enantiomers under discontinuous system by capillary electrophoresis with light-emitting diode-induced fluorescence detection. *Talanta,* **2010**, *82*(5), 1912-1918.
[http://dx.doi.org/10.1016/j.talanta.2010.08.009] [PMID: 20875595]

[35] Domínguez-Vega, E.; Crego, A.L.; Lomsadze, K.; Chankvetadze, B.; Marina, M.L. Enantiomeric separation of FMOC-amino acids by nano-LC and CEC using a new chiral stationary phase, cellulose tris(3-chloro-4-methylphenylcarbamate). *Electrophoresis,* **2011**, *32*(19), 2700-2707.
[http://dx.doi.org/10.1002/elps.201000701] [PMID: 21983819]

[36] Pérez-Míguez, R.; Marina, M.L.; Castro-Puyana, M. Enantiomeric separation of non-protein amino acids by electrokinetic chromatography. *J. Chromatogr. A,* **2016**, *1467*, 409-416.
[http://dx.doi.org/10.1016/j.chroma.2016.06.058] [PMID: 27372417]

[37] Lee, S.; Kim, S.J.; Bang, E.; Na, Y.C. Chiral separation of intact amino acids by capillary electrophoresis-mass spectrometry employing a partial filling technique with a crown ether carboxylic acid. *J. Chromatogr. A,* **2019**, *1586*, 128-138.
[http://dx.doi.org/10.1016/j.chroma.2018.12.001] [PMID: 30558847]

[38] Sánchez-Hernández, L.; Castro-Puyana, M.; García-Ruiz, C.; Crego, A.L.; Marina, M.L. Determination of L- and D-carnitine in dietary food supplements using capillary electrophoresis–tandem mass spectrometry. *Food Chem.*, **2010**, *120*(3), 921-928. [http://dx.doi.org/10.1016/j.foodchem.2009.11.004]

[39] Pérez-Míguez, R.; Marina, M.L.; Castro-Puyana, M. A micellar electrokinetic chromatography approach using diastereomeric derivatization and a volatile surfactant for the enantioselective separation of selenomethionine. *Electrophoresis,* **2019**, *40*(15), 1951-1958. [PMID: 31111508]

[40] Kodama, S.; Taga, A.; Yamamoto, A.; Ito, Y.; Honda, Y.; Suzuki, K.; Yamashita, T.; Kemmei, T.; Aizawa, S. Enantioseparation of DL-isocitric acid by a chiral ligand exchange CE with Ni(II)-D-quinic acid system. *Electrophoresis,* **2010**, *31*(21), 3586-3591. [http://dx.doi.org/10.1002/elps.201000320] [PMID: 20925051]

[41] Kodama, S.; Aizawa, S.; Taga, A.; Yamamoto, A.; Honda, Y.; Suzuki, K.; Kemmei, T.; Hayakawa, K. Determination of α-hydroxy acids and their enantiomers in fruit juices by ligand exchange CE with a dual central metal ion system. *Electrophoresis,* **2013**, *34*(9-10), 1327-1333. [http://dx.doi.org/10.1002/elps.201200645] [PMID: 23423790]

[42] Aydoğan, C.; Karakoç, V.; Denizli, A. Chiral ligand-exchange separation and determination of malic acid enantiomers in apple juice by open-tubular capillary electrochromatography. *Food Chem.,* **2015**, *187*, 130-134. [http://dx.doi.org/10.1016/j.foodchem.2015.04.062] [PMID: 25977007]

[43] Kamencev, M.; Komarova, N.; Morozova, O. Enantioseparation of tartaric and malic acids in wines by Ligand Exchange Capillary Electrophoresis using uncoated fused silica capillary. *Chromatographia,* **2016**, *79*(13-14), 927-931. [http://dx.doi.org/10.1007/s10337-016-3099-8]

[44] Knob, R.; Petr, J.; Sevčík, J.; Maier, V. Enantioseparation of tartaric acid by ligand-exchange capillary electrophoresis using contactless conductivity detection. *J. Sep. Sci.,* **2013**, *36*(20), 3426-3431. [http://dx.doi.org/10.1002/jssc.201300507] [PMID: 23946246]

[45] Pormsila, W.; Gong, X.Y.; Hauser, P.C. Determination of the enantiomers of alpha-hydroxy- and alpha-amino acids in capillary electrophoresis with contactless conductivity detection. *Electrophoresis,* **2010**, *31*(12), 2044-2048. [http://dx.doi.org/10.1002/elps.201000127] [PMID: 20496346]

[46] Mirasoli, M.; Gotti, R.; Di Fusco, M.; Leoni, A.; Colliva, C.; Roda, A. Electronic nose and chiral-capillary electrophoresis in evaluation of the quality changes in commercial green tea leaves during a long-term storage. *Talanta,* **2014**, *129*, 32-38. [http://dx.doi.org/10.1016/j.talanta.2014.04.044] [PMID: 25127562]

[47] Pasquini, B.; Orlandini, S.; Goodarzi, M.; Caprini, C.; Gotti, R.; Furlanetto, S. Chiral cyclodextrin-modified micellar electrokinetic chromatography and chemometric techniques for green tea samples origin discrimination. *Talanta,* **2016**, *150*, 7-13. [http://dx.doi.org/10.1016/j.talanta.2015.12.003] [PMID: 26838375]

[48] Fiori, J.; Pasquini, B.; Caprini, C.; Orlandini, S.; Furlanetto, S.; Gotti, R. Chiral analysis of theanine and catechin in characterization of green tea by cyclodextrin-modified micellar electrokinetic chromatography and high performance liquid chromatography. *J. Chromatogr. A,* **2018**, *1562*, 115-122. [http://dx.doi.org/10.1016/j.chroma.2018.05.063] [PMID: 29859684]

[49] Bathinapatla, A.; Kanchi, S.; Singh, P.; Sabela, M.I.; Bisetty, K. Determination of neotame by high-performance capillary electrophoresis using β-cyclodextrin as a chiral selector. *Anal. Lett.,* **2014**, *47*(17), 2795-2812. [http://dx.doi.org/10.1080/00032719.2014.924008]

[50] Yamamoto, S.; Tamata, Y.; Sejima, K.; Kinoshita, M.; Suzuki, S. Chiral separation of D/L-aldoses by

micellar electrokinetic chromatography using a chiral derivatization reagent and a phenylboronic acid complex. *Anal. Bioanal. Chem.*, **2015**, *407*(20), 6201-6206.
[http://dx.doi.org/10.1007/s00216-015-8802-3] [PMID: 26044740]

[51] Přibylka, A.; Švidrnoch, M.; Ševčík, J.; Maier, V. Enantiomeric separation of 1,3-dimethylamylamine by capillary electrophoresis with indirect UV detection using a dual-selector system. *Electrophoresis,* **2015**, *36*(23), 2866-2873.
[http://dx.doi.org/10.1002/elps.201500182] [PMID: 26249848]

[52] Kodama, S.; Taga, A.; Aizawa, S.I.; Kemmei, T.; Honda, Y.; Suzuki, K.; Yamamoto, A. Direct enantioseparation of lipoic acid in dietary supplements by capillary electrophoresis using trimethyl--cyclodextrin as a chiral selector. *Electrophoresis,* **2012**, *33*(15), 2441-2445.
[http://dx.doi.org/10.1002/elps.201100531] [PMID: 22887166]

[53] Kodama, S.; Nakajima, S.; Ozaki, H.; Takemoto, R.; Itabashi, Y.; Kuksis, A. Enantioseparation of hydroxyeicosatetraenoic acids by hydroxypropyl-γ-cyclodextrin-modified micellar electrokinetic chromatography. *Electrophoresis,* **2016**, *37*(23-24), 3196-3205.
[http://dx.doi.org/10.1002/elps.201600213] [PMID: 27649837]

[54] Pérez-Fernández, V.; Domínguez-Vega, E.; Chankvetadze, B.; Crego, A.L.; García, M.A.; Marina, M.L. Evaluation of new cellulose-based chiral stationary phases Sepapak-2 and Sepapak-4 for the enantiomeric separation of pesticides by nano liquid chromatography and capillary electrochromatography. *J. Chromatogr. A,* **2012**, *1234*, 22-31.
[http://dx.doi.org/10.1016/j.chroma.2012.01.035] [PMID: 22321947]

[55] Friedman, M. Origin, microbiology, nutrition, and pharmacology of D-amino acids. *Chem. Biodivers.,* **2010**, *7*(6), 1491-1530.
[http://dx.doi.org/10.1002/cbdv.200900225] [PMID: 20564567]

[56] Friedman, M.; Levin, C.E. Nutritional and medicinal aspects of D-amino acids. *Amino Acids,* **2012**, *42*(5), 1553-1582.
[http://dx.doi.org/10.1007/s00726-011-0915-1] [PMID: 21519915]

[57] Pérez-Míguez, R.; Marina, M.L.; Castro-Puyana, M. Capillary electrophoresis determination of non-protein amino acids as quality markers in foods. *J. Chromatogr. A,* **2016**, *1428*, 97-114.
[http://dx.doi.org/10.1016/j.chroma.2015.07.078] [PMID: 26233255]

[58] Giuffrida, A.; Maccarrone, G.; Cucinotta, V.; Orlandini, S.; Contino, A. Recent advances in chiral separation of amino acids using capillary electromigration techniques. *J. Chromatogr. A,* **2014**, *1363*, 41-50.
[http://dx.doi.org/10.1016/j.chroma.2014.08.041] [PMID: 25171947]

[59] EC on substances that may be added for specific nutritional purposes in foods for particular nutritional uses. *Official Journal of the European Communities*, **2001**. *L,* **2001/15**, *52*, 19-25.

[60] Pérez-Míguez, R.; Castro-Puyana, M.; Marina, M.L. , **2015**.

[61] Carocho, M.; Morales, P.; Ferreira, I.C.F.R. Sweeteners as food additives in the XXI century: A review of what is known, and what is to come. *Food Chem. Toxicol.,* **2017**, *107*(Pt A), 302-317.
[http://dx.doi.org/10.1016/j.fct.2017.06.046] [PMID: 28689062]

[62] Zhang, Y.; Woods, R.M.; Breitbach, Z.S.; Armstrong, D.W. 1,3-Dimethylamylamine (DMAA) in supplements and geranium products: natural or synthetic? *Drug Test. Anal.,* **2012**, *4*(12), 986-990.
[http://dx.doi.org/10.1002/dta.1368] [PMID: 22786761]

[63] Carrao, D.B.; Perovani, I.S.; de Albuquerque, N.C.P.; de Oliveira, A.R.M. Enantioseparation of pesticides: A critical review. *TRAC. Trends Analyt. Chem.,* **2020**, *122*, 115719.
[http://dx.doi.org/10.1016/j.trac.2019.115719]

CHAPTER 10

Food Analysis by Microchip Electrophoresis

Tania Sierra[1], Silvia Dortez[1], Agustín G. Crevillén[2,*] and Alberto Escarpa[1,3,*]

[1] *Department of Analytical Chemistry, Physical Chemistry and Chemical Engineering, University of Alcalá, Alcala de Henares, Madrid E-28871, Spain*

[2] *Department of Analytical Sciences, Faculty of Sciences, Universidad Nacional de Educación a Distancia (UNED), Madrid E-28040, Spain*

[3] *Chemical Research Institute "Andrés M. del Río" (IQAR), University of Alcalá, Alcalá de Henares, Madrid E-28805, Spain*

Abstract: The most significant advances in food analysis using microchip electrophoresis (ME) technology will be discussed in this book chapter, covering the works published from 2009 to 2019, which will be organized in three sections based on the detection mode employed: electrochemical and fluorescence detection, and bioanalytical-based approaches. The most innovative methodologies, relevant applications, and latest advances in instrumentation to achieve a truly portable lab-o--a-chip will be discussed. Commercial instruments will also be briefly mentioned as a demonstration of the maturity of ME technology.

Keywords: Biosensors, Bioassays, Electrochemical detection, Fluorescence detection, Food analysis, Lab-on-a-chip, Microfluidic, Microchip electrophoresis.

INTRODUCTION

In recent years, there is an increasing concern for health and the environment, including food quality and safety [1-3]. In fact, food is currently considered not only a source of nutrients but also an affordable way to prevent future illnesses [4].

Food analysis encompases a large variety of food-related molecules with different chemical properties, including amino acids, phenols, carbohydrates, DNAs, toxins, pesticides, residues, food additives, and small organic and inorganic compounds. In many cases, food contaminants and residues are at very low concentrations in very complex matrices.

* **Corresponding author Agustín G. Crevillen and Alberto Escarpa:** Department of Analytical Sciences, Faculty of Sciences, Universidad Nacional de Educación a Distancia (UNED), Madrid E-28040, Spain and Department of Analytical Chemistry, Physical Chemistry and Chemical Engineering, University of Alcalá, Alcalá de Henares, Madrid E-28871, Spain; E-mail: agustingcrevillen@ccia.uned.es, alberto.escarpa@uah.es

María Castro-Puyana, Miguel Herrero and María Luisa Marina (Eds.)
Capillary Electrophoresis in Food Analysis

This is why the development, validation and application of new or alternative analytical methods are highly relevant to characterize the identity of genuine foods, determine analytes of utmost interest such as allergens, toxins and foodborne pathogens, and detect frauds and adulterations [5, 6].

Microchip Electrophoresis (ME) is not only the miniaturized format of capillary electrophoresis but also a technology that provides additional functionalities such as parallelization (higher throughput), integration of multiple analytical steps and portability, and also several advantages such as shorter analysis times, lower sample and reagent consumption, and lower waste generation (green chemistry). In consequence, ME has become a good alternative to more conventional methods, which are time-consuming and expensive [7, 8]. Moreover, ME is a mature technology with a wide range of possibilities for lab-on-a-chip (LOC) applications [9]. In this sense, LOC technology aims to perform all functions carried out in laboratories, but on a much smaller scale (chip) and with new characteristics and capabilities [10]. One of the pillars of LOC systems is microfluidics, which is defined as the science and technology that process or manipulate small amounts of fluid (from 10^{-9} to 10^{-18} L) using channels measuring from tens to hundreds of micrometers [11]. Microfluidics offers new capabilities in controlling the concentration of molecules in space and time due to the omnipresence of laminar flow. The use of this technology in food analysis gave rise to a new concept, *Food Microfluidics* [9].

From the early times of the appearance of the ME technology, our research group has published several reviews dealing with the use of ME in food analysis [12-14]. In these reviews, the evolution in this field can be followed up; from the pioneers' works, which were focused on the exploration of fast separations of relevant compounds in simple food samples using home-made microchip systems (proof of concept) [12], to the use of the first commercial instruments based on microchip technology for food analysis [14].

In this chapter, we intend to review the most significant advances in food analysis by ME over the last decade (2009-2019). Innovative methodologies, relevant applications, and advances in instrumentation to achieve a truly portable lab-on-a-chip will be highlighted. In addition, commercial instruments and components will be cited throughout the manuscript for helping beginners to know companies involved in this field. As Professor George Whitesides stated about LOC technology, "*The science is working its way into technology, and preparing to become products*" [10] thus it is fundamental to point out commercial ME products achieved to date.

This chapter is structured in three sections. First, the main detection principles used in ME, electrochemical and fluorescence detection, will be discussed. Second, bioanalytical-based approaches will be described and third the main conclusions and ME future prospects for food analysis will be highlighted.

DETECTION MODES

In the context of food analysis by ME, both electrochemical (amperometry and conductometry) and laser induced fluorescence (LIF), are the most used detection modes. The use of mass spectrometry is residual in food ME. As it will be discussed in the next sections, LIF detection provides the lowest limits of detection, but it is often not miniaturized [15, 16]. In this sense, electrochemical detection is ideal in ME because it can be easily miniaturized without affecting the analytical performance [17, 18].

Electrochemical Detection

Electrochemical detection is an excellent approach for ME due to its easy miniaturization without losing analytical performance (unlike optical methods), good sensitivity and moderate selectivity. Also, the derivatization step is not widely needed [17-19].

In this section, amperometric and conductivity detection will be critically discussed, showing their advantages and limitations in some selected food analysis applications. Within each section, we will discuss the literature from two perspectives: works reporting methodological improvements (such as the use of nanomaterials) and/or relevant applications, and other research works proposing advanced technical innovations, such as fully integrated equipment.

Amperometric Detection

Among all electrochemical techniques, amperometric detection (AD) is the most used mode in ME. The basis of this technique is that a stable voltage is applied to a working electrode relative to a reference electrode, and the current generated from the oxidation or reduction reaction of electroactive species is measured on the working electrode surface. To increase this surface and improve the analytical performance, many of the works discussed in this section use mainly carbon and metallic-based nanomaterials [1, 20].

Marques Petroni *et al.* proposed a simple methodology to fabricate screen-printed carbon electrodes by manual deposition of home-made conductive carbon ink over patterned acrylic substrate [21]. The electrode was aligned at the end of the separation channel of a PDMS microchip with a cross-shaped pattern. Polymer microchips such as PDMS can be fabricated using mass-replication technologies, so that their fabrication cost is lower than glass microchips [22]. Indeed, polymer materials provide low-cost disposable chips for single-use applications (*in situ* analysis) and allow for the selection of the most suitable material for each application [23]. This ME-AD system was used for the determination of nitrite and ascorbic acid. Satisfactory limits of detection (LODs) were obtained in μM range. Separation of nitrite and ascorbic acid was successfully performed in 50 s with good resolution. The effectiveness of this method was tested by determining nitrite in drinking water samples with recoveries higher than 96%.

Godoy-Caballero *et al.* determined important phenolic compounds (tyrosol, hydroxytyrosol and oleuropeinglucoside) in extra virgin olive oil samples using ME-AD [24]. A 100 μm gold wire was used as a working electrode in an end-channel configuration. The best conditions for the detection and separation were achieved in 0.1 M sulphuric acid medium in 5% methanol, and tetraborate buffer pH 9.50 in 10% methanol, respectively. The three analytes were separated in only 100 s with LODs in μM range. Also, they used a sample stacking injection to improve sensitivity.

Escarpa´s group reported several works about the use of ME-AD for food analysis. In the first one, they developed a new method for the fast separation and detection of the fingerprint markers of *Vanilla planifolia* extracts (vanillin (VAN), *p*- hydroxybenzaldehyde (PHB), vanillic acid (VANA), ethyl vanillin (EVA) and *p*- hydroxybenzoic acid (PHBA)) in 300 s with good reproducibility (RSD < 10%), with the aim to assess possible frauds or adulteration [25]. The strategy used on board of a simple cross microchip layout in connection with the standards separation, is depicted in Fig. (**1**). The suitability of this method was demonstrated by the confirmation of natural or non□natural origin of *Vanilla* in several food samples containing this flavor. In another work, they studied the analytical potential of ME-AD for the determination of antioxidants, which are well recognized for their anticancer properties [26]. They introduced two different approaches, depending on the acid-base properties of the antioxidants: class-selective electrochemical index determination (CSEID) and individual antioxidant determination (IAD). In the first approach, the main antioxidant classes (flavonoids and phenolic acids) were separated in less than 100 s while in the second, an impressive separation of nine antioxidants was also offered in less than 260 s with adequate LODs. A good precision was obtained (RSDs < 9%) in both approaches. A set of representative samples was analyzed including apple and pear skins and pulps, red and white wine, and green tea.

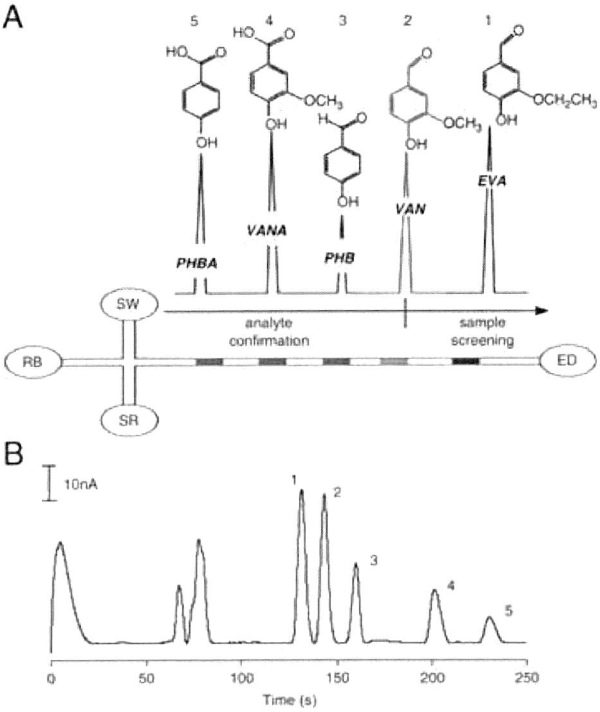

Fig. (1). (A) Schematic of microfluidic chip. It consisted of a glass plate (88 mm x 6 mm) with a four-way injection simple cross, a 74 mm long separation channel, and side arms measuring 5 mm in length. The dimensions of the channels were 20 µm depth and 50 µm width. RB, running buffer; SR, sample reservoir; SW, sample waste; ED, electrochemical detector. **(B)** Separation of *Vanilla* markers in optimum conditions: EVA (peak 1), VAN (peak 2), PHB (peak 3), VANA (peak 4), and PHBA (peak 5). Adapted with permission from [25].

It is well known that the use of nanomaterials in electrochemical detectors for ME can improve its analytical performance. This is due to the impressive properties of these materials such as high active surface, high conductivity, high passivation resistance, and electrocatalytic properties, which yield an improvement of ME-AD methods in terms of sensitivity, selectivity, and reproducibility [17, 18, 27]. For instance, Escarpa's group reported the coupling between ME and copper nanowires (CuNWs) for the selective analysis of monosaccharides in food samples [28]. Glucose and fructose were separated and detected in less than 250 s using a strong basic medium (20 mM NaOH + 10 mM H_3BO_3, pH 12). This basic media was required to ensure the electrocatalytic detection of these analytes on copper-based electrodes. A representative group of nine honey samples were analyzed by this method and the results were compared with those previously obtained by HPLC- refractive index, obtaining errors < 10%. Using the same ME-AD system, the authors reported a method for the determination of inulin, which is a polysaccharide with a diverse range of food and pharmaceutical applications [29]. They separated inulin from fructose and, interestingly, the sensitivity of inuline using these CuNWs was 40 times higher in comparison with its monomer fructose. LODs were below µM. The method was applied to analyze dietary supplements, showing excellent quantitative recoveries (97-103%).

The same group developed a new carbon nanotube-based detector for ME-AD [30]. These detectors were fabricated by press transferring of carbon nanotubes (CNTs) filtered dispersion on nonconductive PMMA substrates. In this approach, the electronic transfer was carried out directly on the target nanomaterial wherein CNTs is the only transducer in the electrochemical sensing. In this way, single-walled carbon nanotube press-transferred electrodes (SW-PTEs) exhibited an impressive antifouling performance (RSD < 9%, n=15) and an improvement of sensitivity. In the first application [31], an electrochemical isoflavone index determination was proposed for evaluation of antioxidant activity and bioavailability monitorization. Biological activity and metabolic fate of dietary soy isoflavones vary depending on their chemical form (aglycone or glycoside). So, a fast differentiation of both total aglycones (TA) and total glycosides (TG) should be very helpful. They were able to separate TA and TG in less than 250 s with RSD < 9% and with LODs in the µM range. The method was applied to analyze soy extracts and dietary supplements (E_r < 12%). In another work [32], they were able to separate and detect melatonin (MT) and its precursors, tryptophan (Trp) and serotonin (5-HT), in less than 150 s. SW-PTEs exhibited an impressive antifouling performance with RSD values of 2 and 4% for migration times and peak heights (n = 12). The LODs ranged from 1 to 5 µM. Melatonin capsules were analyzed by this method and excellent quantitative and reproducible recoveries were obtained with values of 96.2 ± 1.8%, 101.3 ± 0.2%, and 95.6 ± 1.2% for 5-HT, MT, and Trp, respectively.

Another interesting and simplified approach for developing transducers exclusively based on nanomaterials was developed by the same group [33]. In this case, the fabrication relies on the straightforward filtration of a homogeneous dispersion of the carbon nanomaterial on Teflon filter, which becomes the scaffold for the nanomaterial assembling. These new electrodes were named carbon nanomaterial scaffold films (CNSF) and, as in the aforementioned works, the carbon nanomaterial acts as the only conductive material. They employed graphene, SWCNTs and MWCNTs as nanomaterials. MWCNT-SFs showed better analytical behaviour to detect tyrosol (TY), oleuropein (OL) and hydroxytyrosol (HTY). These three components, present in olive oil, were separated and quantitatively determined in less than 240 s. Surface fouling was also carefully evaluated with excellent values (RSDs < 7%, n = 10) unveiling the antifouling properties of the transducer.

Dossi *et al.* developed a ME with a mixed-valence ruthenium oxide/hexacyanoruthenate polymeric film electrochemically deposited onto glassy carbon electrode for the detection of biogenic amines and their amino acid precursors [34]. In this case, they tested the separation and detection of ethanolamine, tryptamine and tryptophane in less than 500 s with good LODs (from 23 to 34 μM). The method was applied for the analysis of some dark beers.

Formaldehyde (FA) and acetaldehyde (AA) are indicators of food quality deterioration; however, they lack a chromophore for sensitive ultraviolet detection (UV) so their determination is not easy. In this sense, Zhang *et al.* proposed the use of 2-thiobarbituric acid (TBA) as the ideal derivatization reagent to facilitate the electrochemical determination of FA and AA [35]. They developed a novel miniaturized capillary electrophoresis with amperometric detection (mini-CE-AD) method for the separation of FA and AA in 13 min. The LODs of two aldehydes were 13.2 fg for FA and 23.2 fg for AA, respectively. The proposed method was successfully applied to determine FA and AA in wine.

Ding *et al.* developed a ME-AD method for the detection of five aminoglycoside antibiotics (AGs), namely spectinomycin, streptomycin, amikacin, paromomycin, and neomycin, in only 180 s [36]. One of the key-features of this method is the use of two auxiliary channels for post separation addition of alkaline solution, increasing the pH for AD of AGs without disturbing their separation in acidic running buffer. The working electrode was fabricated by electrodeposition of an alloy of Cu-Sn-Cr. RSDs in peak current and migration time were found to be less than 7.9 and 2.0%, respectively, and the LODs ranged from 2.1 to 4.6 μM. The suitability of the proposed method was evaluated by the determination of five selected AGs in milk samples obtaining recoveries from 50 to 77%. Since these recoveries were not satisfactory, authors hypothesized that this was due to the high Ca^{2+} concentration and fat present in milk sample.

The main goal of LOC systems is the miniaturization and integration of all steps or tasks carried out in a lab into a single device. We found a complete portable system that integrated the electrochemical detector and the high voltage power (Micrux HVStat, Spain). In this system, it is not necessary to have any external cable for the interconnection of both parts, making the use of the complete system easier [37]. Using this portable ME system, Escarpa´s team developed a method for the simultaneous measurement of total acids and flavonoids in less than 100 s [38]. The microchip integrated three Pt electrodes for AD of phenolic compounds (Fig. **2**). They demonstrated that ME-ED could become a good alternative to the traditional Folin-Ciocalteu antioxidant capacity assay. The method was applied to the analysis of complex antioxidant extracts.

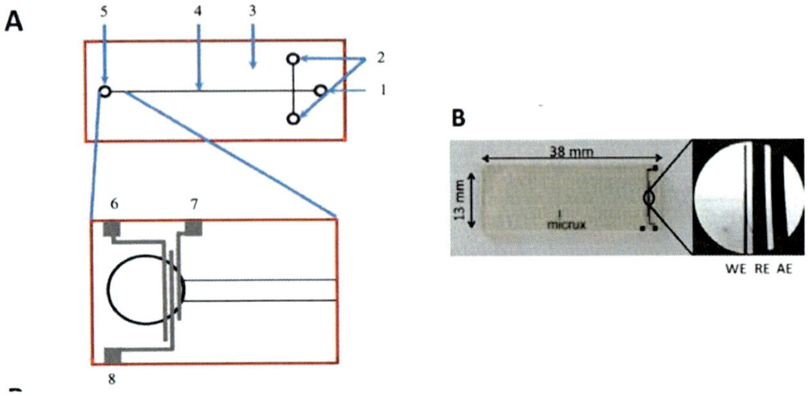

Fig. (2). **(A)** Scheme of simple-cross ME with integrated electrochemical detection; 1: running buffer reservoir, 2: sample reservoirs, 3: microchip substrate, 4: separation channel, 5: amperometric detection cell reservoir, 6: auxiliary electrode (AE), 7: working electrode (WE), 8: reference electrode (RE), AE = WE = RE= platinum (100 μm). **(B)** Photo of microchip with electrochemical detection integrated (from MicruX Technologies). Adapted with permission from [38].

Throughout this section, we have referred to ME-AD for the determination of small molecules with great interest in food quality (identification of frauds and degradation products), food safety (residual antibiotics) and health benefits (antioxidants). However, the sample treatment is performed off-chip, disclosing no advances in the fully integration of all analytical steps. On the other hand, AD provides LODs in μM range and it is not limited to electroactive species on carbon electrodes but, thanks to the use of modified electrodes or derivatizing agents, compounds such as carbohydrates, antibiotics and amine- or aldehyde derivatives can be successfully detected. Finally, it is interesting to point out the use of a truly portable ME-AD device with integrated amperometric detection in food analysis.

Table **1** summarizes the most relevant information about food analysis by ME with amperometric detection.

Table 1. Food analysis by ME with amperometric detection.

Analytes	Sample	Microchip Layout (and Features)	Electrophoretic Mode	Electrode/ Detection Mode	Analytical Features	Remarks	Refs.
Nitrite, ascorbic acid	Drinking water	Single-T PDMS microchip (L_{sep}=50 mm)	MZE (5 mM phosphate, 200 μM CTAB, pH=6.85)	Carbon/ Amperometry (E=+1.0 V)	LODs=8.2–12.7 μM Recoveries 96% t_{sep}=50 s	Very short separation time	[21]

(Table 1) cont.....

Analytes	Sample	Microchip Layout (and Features)	Electrophoretic Mode	Electrode/ Detection Mode	Analytical Features	Remarks	Refs.
Tyrosol, hydroxytyrosol, oleuropein glucoside	Extra virgin olive oil	Glass microchip (L_{sep}=40 mm)	MZE (15 mM sodium tetraborate, 10% methanol, pH=9.5)	Gold/ Amperometry (E=+0.90 V)	LODs=9–17 μM Recoveries 106–113% t_{sep}=100 s	Sample stacking injection	[24]
Vanillin, ethyl vanillin and related ahdehydes and acids	Natural Vanilla and vanilla-based foodstuffs	Simple-cross glass microchip (L_{sep}=74 mm)	MZE (10 mM borate, pH=9.5)	Carbon/ Amperometry (E=+1.0 V)	LOQs=0.06–0.3 ppm Recoveries 100% t_{sep}=300 s	Fast detection of frauds	[25]
Flavonoids, phenolic acids	Fruits, wines, tea	Simple-cross glass microchip (L_{sep}=74 mm)	MZE (10 mM borate, pH=9.0)	Carbon/ Amperometry (E=+1.0 V)	LOD=7.6 μM Recoveries 90.7–108% t_{sep}=260 s	Sequential total (<60s) and individual antioxidants determination (up nine <260s)	[26]
Glucose, fructose	Honey	Double-T glass microchip (L_{sep}=85 mm)	MZE (20 mM NaOH, 10 mM H_3BO_3, pH=12.0)	CuNWs / Amperometry (E=+0.70 V)	LODs=4–23 μM t_{sep}=250 s	Selective carbohydrate detection on copper nanowires	[28]
Inulin, free fructose	Dietary supplement	Double-T glass microchip (L_{sep}=85 mm)	MZE (20 mM NaOH, 10 mM H_3BO_3, pH=12.0)	CuNWs / Amperometry (E=+0.70 V)	LODs=3 μM Recoveries 97–103% t_{sep}=300 s	Selective carbohydrate detection on copper nanowires	[29]
Total aglycones, total glycosides	Soy extracts, dietary supplement	Glass microchip (L_{sep}=74 mm)	MZE (25 mM borate, pH=9.0)	SWCNT-PTEs/ Amperometry (E=+1.0 V)	LODs=20–30 μM t_{sep}=250 s	SWCNT press-transferred electrodes as exclusive transducers	[31]
Melatonin, tryptophan, serotonin	Melatonin extracts	Glass microchip (L_{sep}=75 mm)	MZE (20 mM di-sodium tetraborate, pH=9.2)	SWCNT-PTEs/ Amperometry (E=+0.70 V)	LODs=1–5 μM Recoveries 96–102% t_{sep}=150 s	SWCNT press-transferred electrodes as exclusive transducers	[32]
Tyrosol, oleupein, hydroxytyrosol	Olive oil	Glass microchip (L_{sep}=74 mm)	MZE (10 mM borax, 5% MeOH, pH=9.5)	MWCNT-SFs/ Amperometry (E=+0.70 V)	LOD_{TY}=10 μM t_{sep}=240 s	MWCNT scaffold films as exclusive transducers	[33]
Ethanolamine, tryptamine, tryptophane	Malt dark beers	Glass microchip (L_{sep}=80 mm)	MZE (20 mM phosphate, pH=2.5)	Ruthenium/ Amperometry (E=+1.05 V)	LODs=23–34 μM t_{sep}=500 s	Direct detection of amines and amino acids	[34]
Formaldehyde, acetaldehyde	Wines	Silica microchip (L_{sep}=195 mm)	MZE (80 mM H_3BO_3–$Na_2B_4O_7$, pH=9.0)	Carbon/ Amperometry (E=+0.95 V)	LODs=13.2–23.2 fg Recoveries 96–106% t_{sep}=13 min	Derivatization with TBA	[35]

(Table 1) cont.....

Analytes	Sample	Microchip Layout (and Features)	Electrophoretic Mode	Electrode/ Detection Mode	Analytical Features	Remarks	Refs.
Aminoglycoside antibiotics	Milk	Double-T PDMS microchip (L_{sep}=66 mm)	MZE (6 mM sodium acetate, 0.6 mM CTAB, pH=5.0)	Cu-Sn-Cr/ Amperometry (E=+0.65 V)	LODs=2.1–4.6 µM Recoveries 50–77% t_{sep}=180 s	Antibiotics detection	[36]
Phenolic acids, flavonoids	Natural extracts	SU-8/Pyrex single-channel microchip (L_{sep}=25 mm)	MZE (25 mM MES, pH=5.0)	Pt/ Amperometry (E=+1.0 V)	t_{sep}=100 s	Portable ME-AD system	[38]

CuNWs, Cu nanowires; **LOD**, the limit of detection; **LOQ**, limit of quantification; **MWCNT-SFs**, multi-walled carbon nanotubes scaffold films; **MZE**, microchip zone electrophoresis; **SWCNT-PTEs**, single-walled carbon nanotubes press transferring; **TBA**, 2-thiobarbituric acid.

Conductivity Detection

In recent years, conductivity detection has seen a renaissance in CE-ME systems. The main reason for this is that conductivity detection can be accomplished by contactless mode in which the electrodes are not in contact with the solution [39, 40]. This mode has several advantages over the contact mode, including the absence of problems associated with the electrode-solution contact (*e.g.* passivation, bubble formation, etc), effective isolation from high separation voltages, a simplified construction and alignment of the detector, and the use of narrow capillaries and microchannels [41]. Also, capacitively coupled contactless conductivity detector (C^4D) offers many advantages compared to the amperometric detector: simple electronic circuitry, low price, simple arrangement of the electrodes, decoupling from the high voltage used for separation and the universality of C^4D sensor, which detects all charged species [42, 43]. However, it provides lower sensitivity than amperometric detection. There are several papers related to this detection mode in the literature.

Rezende *et al.* [44] reported the use of microchip zone electrophoresis (MZE) with C^4D to investigate the authenticity of seized whiskey samples, based on the presence of inorganic anions like Cl^-, F^-, SO_4^{2-} and NO_2^-. Whiskey can be adulterated by simple dilution of authentic whiskey in water or ethanol, or addition of a mixture containing alcohol, water and caramel. When this happens, the anionic species detected in tampered samples have higher analytical response than those recorded for original whiskies. Electrophoresis experiments were performed on commercial glass microchips containing integrated electrodes for C^4D measurements. The best separations were achieved within 70 s using a running buffer composed of lactic acid and histidine (pH = 5.9) with a good reproducibility (RSDs < 7%). The limits of detection (S/N = 3) achieved for Cl^- and F^- were 6 and 16 µM, respectively.

Jozanović *et al.* published two works describing the use of ME-C⁴D for the determination of two imidazole dipeptides, carnosine (Car) and anserine (Ans), which are contained in the muscles, brains and olfactory bulbs of vertebrates in high concentrations [45,46]. The ratio of carnosine/anserine has been proposed as an aid to identify the species or origin of meat and can be a useful tool in food fraud prevention process. In the first paper, a portable homemade ME device with C^4D detection was constructed [45]. They obtained LODs ranged between 0.10 and 0.16 μM. They tested this new ME device in poultry meat and found that there was no significant difference between the results obtained by their homemade ME/C^4D and those obtained by a commercial ME/C^4D device. In the second work, an improved ME portable device was developed (Fig. **3**) [46]. Compared with their previous work, the new device has longer electrodes with a smaller detection gap between them and therefore exhibited higher sensitivity (LODs were 0.0086 and 0.0115 mg L^{-1} for Car and Ans). This method was also applied to analyze several meat samples, obtained recoveries of 96-105% for carnosine and 96-102% for anserine.

Thredgold *et al.* developed a ME-C^4D method for the direct detection of histamine in fish samples in less than 50 s [47]. Histamine was separated from other amines, such as 2-phenylethylamine and tyramine with good reproducibility (RSDs < 2%) and a LOD of 0.43 mg/L. This method was successfully applied to the determination of (spiked) histamine in yellowfin tuna flesh samples with recoveries from 88.8 to 112.5%.

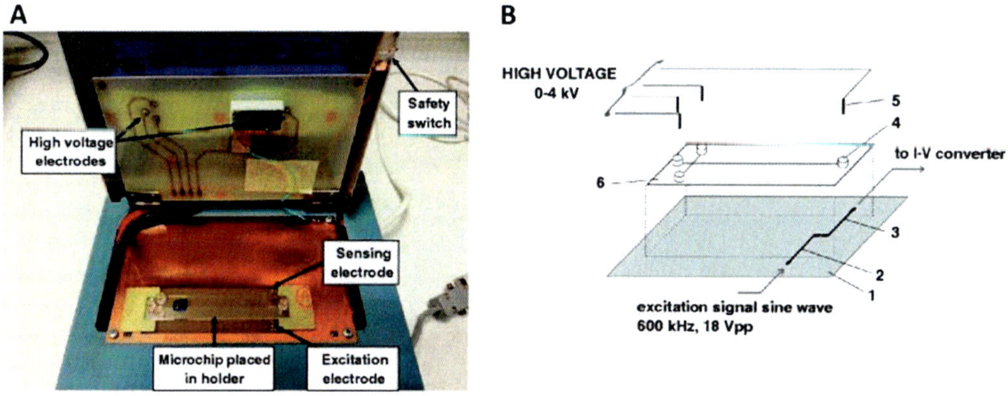

Fig. (3). Compact portable lab-on-a-chip instrument based on ME-C^4D. **(A)** Home-made portable ME instrument. **(B)** Schematic of the high-voltage electrodes and holder of a microchip showing: (1) custom made printed circuit board, (2) excitation electrode, (3) sensing electrode, (4) reservoirs of microchip, (5) high– voltage electrodes, (6) microchip. Adapted with permission from [46].

A portable LOC instrument based on ME with dual top–bottom capacitively contactless conductivity detection (dC^4D) in replaceable cell cartridge was developed by Ansari *et al.* [48]. The device has two compartments: a main body that is housing all the electrical and electronic components, which is shielded to prevent signal interferences from internal or external sources and a fluidic electrophoretic compartment, which includes the detection part. The microfluidic chip is placed on the chip stage in this compartment. They applied this equipment to the detection of cations and anions in several samples. For instance, ten cations and anions were separated and detected in fruit juice in less than 90 s with LOD between 12-45 μg/L. The RSD values for peak height and peak area were lower than 5% showing an acceptable reproducibility.

Mark *et al.* developed a new method for the efficient detection of volatile aliphatic amines, related with biodegradation processes [49]. First, the five aliphatic amines (methylamine, dimethylamine, trimethylamine, diethylamine and trimethylamine) were extracted *via* ultrasound-assisted head space SDME (off-chip) and then they were separated and detected by MZE-C^4D using a commercial equipment (ChipGenie edition E, Microfluidic Chipshop, Germany) in less than 40 s. A PMMA microchip (8.7 cm long separation channel) with integrated thin-film gold electrodes was used for C^4D. This method was successfully applied to the analysis of complex matrices, such as seafood samples, obtaining LODs ranging from 41 to 400 ng/mL.

Finally, Zhai *et al.* developed a ME system, which integrated the sample pretreatment, for auramine O determination (industrial dye that can lead to teratogenesis and cancer) [50]. In this way, sample extraction, injection, separation and detection were on-chip sequential and automatically performed. The microfluidic system consisted of a 0.17 mm-thick glass cover and a PDMS substrate, which due to its elasticity can help embedded SPE monolithic column into the glass/PDMS hybrid. SPE extraction was based on molecularly imprinted polymers (MIPs), which recognized and captured selectively auramine O. The SPE column yielded an on-line enrichment factor of about 12 and the LOD was 2.5μg mL^{-1}. The method accuracy was tested in spiked shrimp samples obtaining recoveries between 90.4–92.6%.

Contact conductivity detection (CD) has also been used in ME for food analysis. Henry's group developed a method for the determination of perchlorate in drinking water using ME-CD [51]. Although ion chromatography coupled with mass spectrometry (IC-MS) is the most commonly used technique in perchlorate detection, ME-CD present a number of advantages, such as reduced analysis times relative to IC-MS, inherent portability, high selectivity, and minimal sample pretreatment, which make this method a good alternative for the determination of perchlorate in food samples. To resolve perchlorate from other abundant anions present in samples, they used different zwitterionic surfactants (N-hexadecyl-N,N-dimethyl-3-ammonio-1-propane sulfonate (HDAPS) and N-tetradecyl-N,N-dimethyl-3-ammonio-1-propane sulfonate (TDAPS)). At concentrations above the critical micelle concentration (CMC), micelles selectively interact with perchlorate reducing its apparent mobility. TDAPS yielded the best analytical performance. The detection limit was 6 ppb and the analysis time was between 15 to 30 times shorter than current IC techniques. This method was successfully applied to drinking water samples.

In summary, ME with conductivity detection provide excellent separation and detection of small inorganic and organic ions present in food. There are even two works that assessed dipeptides (carnosine and anserine). Identification of frauds (whiskey and meat), detection of toxins (auramien O) and monitorization of biodegradation processes are the most relevant applications. Interestingly, there are several groups that developed portable ME-C^4D equipment and also there is one work that used a commercial instrument (ChipGenie edition E, Microfluidic ChipShop, Germany). Like ME-AD, sample treatment is performed off-chip but there is a promising work in which SPE is integrated on a chip.

The most important data involving food analysis by ME with conductivity detection is collected in Table **2**.

Table 2. Food analysis by ME with conductivity detection.

Analytes	Sample	Microchip Layout (and Features)	Electrophoretic Mode	Detection Mode	Analytical Features	Remarks	Refs.
Cl^-, F^-	Whiskeys	Glass microchip (L_{sep}=33 mm)	MZE (lactic acid and histidine, pH=5.9)	C^4D	LODs=6–16 µM t_{sep}=70 s	Fast detection of frauds	[44]
Carnosine, anserine	Poultry meat	TOPAS microchip (L_{sep}=95 mm)	MZE (200 mM acetic acid, 10 mM 2-hydroxyisobutyric acid, 0.3 mM potassium acetate, pH=2.7)	C^4D	LODs=0.10–0.16 µM Recoveries 96–106% t_{sep}=60 s	Home-made portable ME device for *in situ* analysis	[45]
Carnosine, anserine	Meat	PMMA microchip (L_{sep}=220 mm)	MZE (7.5 mM 2- hydroxy-2-methylpropanoic acid, 0.3 M acetic acid, 0.2 mM rubidium acetate, pH=2.6)	C^4D	LOD_{CAR}=0.0086 mg/L LOD_{ANS}=0.0115 mg/L Recoveries 97–105% t_{sep}=65 s	Method validation using HPLC	[46]
Histamine	Fish flesh	PDMS microchip (L_{sep}=37 mm)	MZE (50 mM 2-[4-(2-hydroxyethyl)piperaz-n-1-yl]ethanesulfonic acid, 5 mM histidine, pH=6.03)	C^4D	LOD=0.43 mg/L Recoveries 89–113% t_{sep}=50 s	Analysis of complex samples under very short separation time	[47]

(Table 2) cont.....

Analytes	Sample	Microchip Layout (and Features)	Electrophoretic Mode	Detection Mode	Analytical Features	Remarks	Refs.
Cations and anions	Fruit juice	Polycarbonate microchip (L_{sep}=90 mm)	MZE (10 mM His, 7 mM Glu, pH=5.53)	dC⁴D	LODs=12–45 µg/L t_{sep}=90 s	Separation (up 10 cations and anions) into a portable ME instrument	[48]
Aliphatic amines	Seafood	ChipGenie edition E (L_{sep}=87 mm)	MZE (20 mM MES, 80 mM His, pH=6.6)	C⁴D	LODs=41–400 ng/mL t_{sep}=40 s	Analysis of complex samples	[49]
Auramine O	Shrimp	PDMS/glass microchip (L_{sep}=50 mm)	MZE (5.0 mM lactic acid, 15% methanol)	C⁴D	LOD=2.5 µg mL^{-1} Recoveries 90.4–92.6% t_{sep}=60 s	Microchip integrating sample treatment, injection, separation and detection	[50]
Perchlorate	Drinking water	PDMS microchip (L_{sep}=50 mm)	MEKC (10 mM nicotinic acid, 1.0 mM TDAPS, pH=3.6)	CD	LOD=6 ppb Recoveries 99% t_{sep}=60 s	Fulfilled USEPA regulatory requirements	[51]

ANS, anserine; **CAR**, carnosine; **C⁴D**, capacitively coupled contactless conductivity detection; **CD**, contact conductivity detection; **LOD**, limit of detection; **MEKC**, micellar electrokinetic chromatography; **MZE**, microchip zone electrophoresis; **TOPAS**, thermoplastic olefin polymer of amorphous structure.

Laser Induced Fluorescence Detection

Ever since the inception of ME, LIF detection has been the most employed detection mode on account of its high sensitivity [14, 52, 53]. In the context of optical detection, UV detection is simpler than LIF and it is almost an universal detection mode; however, the sensitivity limitation of UV detection in CE (due to the short optical path) is even worsened when we switch to ME [54, 55]. LIF detection presents nevertheless a small drawback that is the need for a derivatization step, because the majority of analytes do not present intrinsic fluorescence. Also, often it is not a miniaturized approach.

Many on-line preconcentration methods have been developed and implemented in ME, including field-amplified sample stacking (FASS), large-volume sample stacking (LVSS), transient isotachophoresis (t-ITP), isoelectric focusing, transient trapping, sweeping, and reversed-field stacking [56]. Interestingly, some of them have been used for food analysis in ME as will be discussed below.

In this section, selected applications of ME-LIF in food analysis will be shown, in which not only small analytes were detected but also proteins.

For instance, a rapid ME method with deep UV-LIF detection was developed for the determination of biologically active compounds, such as dopamine, serotonin, tryptophan, tyrosine and salsolinol, in bananas in less than 1 min [57]. In addition, levodopa was also found in bananas that ripened during 10 days. These compounds exhibit an intrinsic fluorescence when excited in the deep UV spectral range, which enables a label-free detection. In order to confirm peak assignment and identify the unknown peak, they coupled ME with mass spectrometry (MS) using a self-made chip-interface. This MS-chip contained a nanoelectrospray emitter. For exact alignment of the emitter in front of the MS orifice, the microfluidic chip was mounted on a three-axis translation stage equipped with two CCD cameras.

Wang et al. developed a rapid and sensitive ME-LIF method for the determination of sulfonamide residues (antibacterial drug) in milk and chicken samples [58]. A microchip made of cyclic olefin copolymer was used. This copolymer is a good substrate due to its excellent optical transparency and low autofluorescence, besides its low cost. Sulfonamides (sulfamethazine, sulfamethoxazole, sulfaquinoxaline and sulfanilamide) were derivatized with fluorescamine. One of the most important advantages of the proposed method is its shorter analysis time (60 s), faster than CE and HPLC. LODs ranged from 0.2 to 2.3 µg/L.

Wei *et al.* detected glyphosate (GLYP) and glufosinate (GLUF) by MZE-LIF [59]. These two molecules are widely used as herbicides due to their excellent performance and low toxicity. However, there is a need for control over residues in agricultural products. Analytes were labeled with fluorescein isothiocyanate (FITC) in a short period of time (20 min). The electroosmotic flow (EOF) was suppressed by using hydroxypropyl cellulose (HPC). Due to the negative charge of FITC, GLYP and GLUF migrated anodically ahead of components of sample matrices that may interfere with the analysis, such as amino acids, small aminosugars, and proteins. LODs were 0.05 and 0.02 µg/L for GLYP and GLUF, respectively. The method was applied to the analysis of food samples (soybean and broccoli) without any preconcentration with good values of recovery (84-103%).

Yang *et al.* also used FITC as derivatizing reagent for domoic acid determination by ME-LIF method [60]. Domoic acid is the main constituent of amnesic shellfish poisoning. Under the optimized conditions (5 mM sodium tetraborate buffer at pH 9.2), domoic acid was completely separated of FITC in 60 s and the detection limit was 2.8×10^{-10} M. The method was applied to the analysis of shellfish tissue.

A novel ME-LIF method was developed for the separation and detection of trace of five β-casomorphins (β-CMs) [61]. They are a group of exogenous opioid peptides with physiological activity, which might be the cause of clinical pathology in the human body. β-CMs were off-chip labeled with FITC. The glass microchip consisted of a double-T injector with a serpentine separation channel (18 cm length). This long channel allowed for the baseline separation of all β-CMs, but the separation time was long (30 min). Detection limits between 18.7 and 75.1 nM were achieved for the five β-CMs, and the method was applied to the analysis of cheese samples with good recoveries between 89-109%.

Wang´s group presented three interesting papers in which the analytes were derivatized with sulfoindocyaninesuccinimidyl ester (Cy5) for LIF detection. In the first one, ME-LIF was used for the detection of free amino acids present in milk, including lysine (Lys), methionine (Met), phenylalanine (Phe), threonine (Thr), valine (Val), tryptophan (Trp) and leucine (Leu) [62]. The method provided LODs in μM range and good reproducibility (RSD < 6%). In the second, they combined field-amplified stacking and reversed-field stacking as preconcentration strategy for efficient and sensitive analysis of lysine (Lys), taurine (Tau) and niacinamide (NA) by MZE with LIF detection [63]. The use of reversed-field stacking lengthened the effective separation distance and provided longer time for FASS by pushing back the sample plug, so that the enrichment effect of FASS was increased without losing peak resolution. By this methodology, these compounds were preconcentrated around 200 times and the sensitivity was increased by 1–2 orders of magnitude respect to previous ME-LIF methods, achieving LODs below nM range. This method was successfully applied to eight functional drinks with recoveries between 97 and 108%. Later, they used the same methodology for the determination of acrylamide in food products [64], using a commercial ME system (p-800 model MCE system, Shanghai Spectrum Ltd. Co., Shanghai). Cy5 was also used as derivatizing agent and it was separated from the acrylamide in less than 150 s using 100 mM borate solution at pH 9.3 as running buffer (MZE). A sensitivity enhancement factor of 432-fold was obtained for acrylamide and the limit of detection was 1 ng/mL. The method was tested in potato chips and French fries with recoveries between 95 and 110%.

Food proteins were also assessed by ME-LIF. Diez-Masa's team reported two works about the determination of dairy proteins [65,66]. In the first, the main whey proteins (α-lactalbumin, β-lactoglobulin, and BSA) were separated and detected by SDS-microchip gel electrophoresis with LIF detection using SU-8 microchips in less than 300 s [65]. SU-8 is a kind of epoxy resin that is extensively used in microelectronics industry, but scarcely explored for electrophoresis separation. EOF was suppressed by using EOTrol® LN (a commercial polymer for EOF control) and dextran (M_w= 425–575 KDa) was used as sieving polymer. Proteins were off-chip labeled with Chromeo P540. LODs were in tens of nM range and these proteins were detected in milks (see Fig. **4**). In the second work, an on- chip method based on anionic transient ITP coupled to zone electrophoresis (t-ITP- MZE) was developed for the preconcentration and separation of the proteins α- lactalbumin and β-lactoglobulin [66]. Proteins were off-chip labeled with the fluorogenic reagent Chromeo P503 and detected by LIF. The use of N-(2- hydroxyethyl)piperazine-N′-(2-ethanesulfonic acid) (HEPES) as terminating anion was crucial for the success of t-ITP step. In optimized conditions, α-lactalbumin and β-lactoglobulin were separated in less than 4 min and preconcentration indexes around 10 were obtained in comparison with MZE-LIF without t-ITP. LODs were 55 nM and 380 nM, for α-lactalbumin and β-lactoglobulin, respectively.

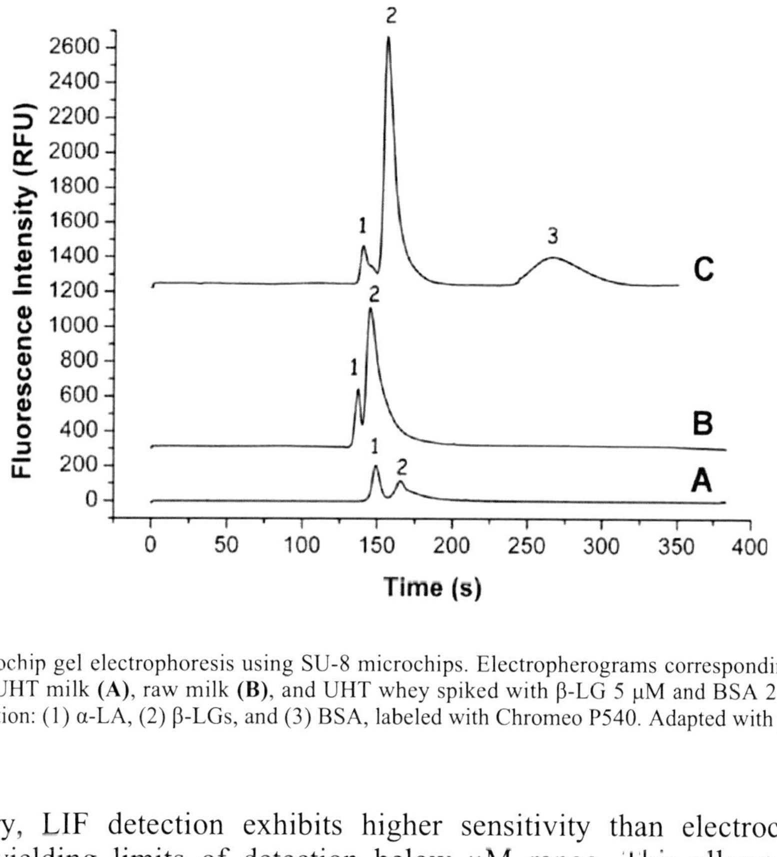

Fig. (4). Microchip gel electrophoresis using SU-8 microchips. Electropherograms corresponding to whey samples from UHT milk **(A)**, raw milk **(B)**, and UHT whey spiked with β-LG 5 μM and BSA 2.5 μM **(C)**. Peak identification: (1) α-LA, (2) β-LGs, and (3) BSA, labeled with Chromeo P540. Adapted with permission from [65].

In summary, LIF detection exhibits higher sensitivity than electrochemical detection, yielding limits of detection below μM range. This allows for the development of sensitive methods for the determination of dangerous residues in food such as β-casomorphins, sulfonamides and domoic acid. Different kind of organic compounds have also been determined by ME-LIF from small amino acids to proteins. In addition, while a commercial ME-LIF (SP-800 MCE system, Shanghai Spectrum, China) was used, portable ME-LIF systems were not reported yet.

Table **3** lists the most relevant information about ME-LIF in food analysis.

Table 3. Food analysis by ME with fluorescence detection.

Analytes	Sample	Microchip Layout (and Features)	Electrophoretic Mode	Detection Mode	Analytical Features	Remarks	Refs.
Dopamine, serotonin, tryptophan, tyrosine, salsolinol, levodopa	Bananas	Silica/glass microchip (L_{sep}=90 mm)	MZE (20 mM ammonium acetate, pH=9.2)	Laser deep UV fluorescence (λ_{exc}=266 nm) and MS	t_{sep}=60 s	ME-MS coupling	[57]
Sulfonamides	Milk, chicken	COC microchip (L_{sep}=50 mm)	MZE (5 mM boric acid, 1% (w/v) polyvinyl alcohol (PVA), pH=7.5)	LIF (λ_{exc}=405 nm)	LODs=0.2–2.3 µg/L Recoveries 83–101% t_{sep}=60 s	Derivatization with fluorescamine	[58]
Glyphosate, glufosinate	Water, soybean, broccoli	Simple cross microchip (L_{sep}=50 mm)	MZE (10 mM borax, 2.0% (m/v) HPC, pH=9.0)	LIF (λ_{exc}=473 nm)	LOD_{GLYP}=0.05 µg/L LOD_{GLUF}=0.02 µg/L Recoveries 84–103% t_{sep}=40 s	Derivatization with FITC	[59]
Domoic acid (separated FITC and DA-FITC)	Shellfish tissue	Simple cross glass microchip (L_{sep}=49 mm)	MZE (5 mM sodium tetraborate, pH=9.2)	LIF (λ_{exc}=473 nm)	LOD=0.28 nM t_{sep}=60 s	Derivatization with FITC	[60]
β-casomorphins	Cheese	Double-T glass microchip (L_{sep}=180 mm, serpentin)	MZE (15 mM boric acid–sodium borate, pH=10.5)	LIF (λ_{exc}=473 nm)	LODs=18.7–75.1 nM Recoveries 89–109% t_{sep}=30 min	Complex β-casomorphins separation on a serpentine channel	[61]
Aminoacids (Lys, Met, Phe, Thr, Val, Trp and Leu)	Milk	Simple cross glass microchip (L_{sep}=45 mm)	MZE (100 mM sodium borate, pH=10.0)	LIF (λ_{exc}=635 nm)	LODs=1.0–6.0 µmol/L Recoveries 95–104% t_{sep}=150 s	Derivatization with Cy5	[62]
Lys, Tau, niacinamide	Drink samples	Simple cross glass microchip (L_{sep}=45 mm)	MZE (10 mM borate, pH=9.88)	LIF (λ_{exc}=635 nm)	LODs=0.20–0.50 nM Recoveries 97–108% t_{sep}=170 s	High on chip pre-concentration factor (200 folds) by FASS and reverse field stacking	[63]

(Table 3) cont.....

Analytes	Sample	Microchip Layout (and Features)	Electrophoretic Mode	Detection Mode	Analytical Features	Remarks	Refs.
Acrylamide	Potato chips, French fries	SP-800 MCE system (L_{sep}=45 mm)	MZE (100 mM borate, pH=9.3)	LIF (λ_{exc}=635 nm)	LOD=1 ng/mL Recoveries 95–110% t_{sep}=150 s	High on chip pre-concentration factor (432 folds) by FASS	[64]
α-lactalbumin, β-lactoglobulin, BSA	Milk	Simple cross SU-8 microchip (L_{sep}=35 mm)	MGE (5 mM borate, 3.5 mM SDS, 10% EOTrol, pH=8.5)	LIF (λ_{exc}=532 nm)	LODs=17–70 nM Recoveries 89–109% t_{sep}=300 s	Detection of proteins in real samples Derivatization with Chromeo P540	[65]
α-lactalbumin, β-lactoglobulin	NR	Simple cross glass microchip (L_{sep}=80 mm)	t-ITP-MZE (100 mM borate, 5% (v/v) EOTrol, pH=8.0)	LIF (λ_{exc}=473 nm)	$LOD_{\alpha\text{-}LAC}$=55 nM $LOD_{\beta\text{-}LAC}$=380 nM t_{sep}=240 s	On chip pre-concentration factor (10 folds) by ITP	[66]

α-LAC, α-lactalbumin; **β-LAC**, β-lactoglobulin; **Cy5**, sulfoindocyaninesuccinimidyl ester; **COC**, cyclic olefin copolymer; **FASS**, field-amplified sample stacking; **FITC**, fluorescein isothiocyanate; **GLUF**, glufosinate; **GLYP**, glyphosate; **HPC**, hydroxypropyl cellulose; **ITP**, isotachophoresis; **Leu**, leucine; **LIF**, laser induced fluorescence; **LOD**, limit of detection; **Lys**, lysine; **Met**, methionine; **MGE**, microchip gel electrophoresis; **MZE**, microchip zone electrophoresis; **MS**, mass spectrometry; **NR**, not reported; **Phe**, phenylalanine; **t-ITP**, transient isotachophoresis; **Tau**, taurine; **Thr**, threonine; **Trp**, tryptophan; **Val**, valine.

Bioassays Using Microchip Electrophoresis

This section gathers works in which a bioassay is combined with a ME separation. This includes those works in which bio-reagents such as enzymes, antibodies, aptamers and DNA for detection and/or separation (affinity capillary electrophoresis) are used and/or employ biological methodologies such as PCR or DNA sequencing.

Most articles addressed in this section employed LED-induced fluorescence detection (LED-IF), due to the existence of robust commercial equipment that used this excitation source. LEDs are cheaper than lasers and show enough light power for fluorescence excitation.

However, there are two relevant works in which fluorescence detection was not used [67, 68]. In the first, Escarpa´s group proposed a ME-AD approach for the determination of aminoacids (AAs) involved in *Vibrio cholerae* as application in food safety [67]. This bacterium is able to specifically generate D amino acids (D-AA), which are not naturally present. Authors developed a new method for the racemic separation and enantiomeric detection of D-Methionine (D-Met) and D-Leucine (D- Leu). In this case, they used D-amino acid oxidase (DAAO) that oxidatively deaminates D-AAs to the corresponding alpha-keto acid, generating one mol of H_2O_2 which is the final product determined by AD, using a graphene-based detector. When only L-amino acids (L-Met and L-Leu) were injected, no peaks were detected, as the enzyme does not react with L-enantiomers. Using a simple cross glass microchip and electrokinetic flow, authors were able to smartly manipulate D-AAs and the class- enzyme DAAO inside the microchannels (avoiding the covalent immobilization of the enzyme into the wall channels or on the electrode surface) and to control both the D-AA separation and the specific reaction between DAAO and each D-AA (see Fig. **5**). Therefore, this is a paradigmatic example of *food microfluidics*, which demonstrates the potency of electrokinetics to perform analytical reactions integrated into a ME.

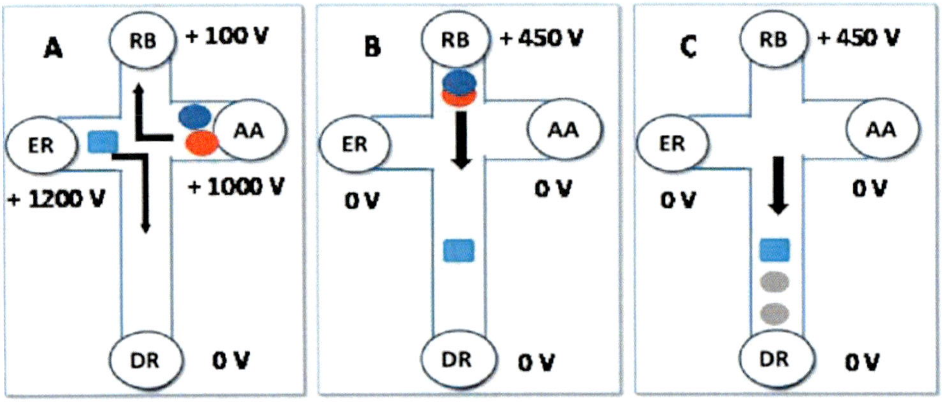

Fig. (5). Electrokinetic-electrochemical microfluidic strategy: **(A)** Enzyme and DAAs electrokinetic co-injection **(B)** Enzyme and D-AAs electro-focusing **(C)** D-AAs separation and in-channel enzymatic reaction and hydrogen peroxide detection. Adapted with permission from [67].

In a second work, Li *et al.* developed a novel method based on G-quadruplex/hemin DNAzyme-based ME chemiluminescence (CL) for biotin assessment [68]. This method is based on three principles: (i) the specific binding between biotin and streptavidin, (ii) the catalytic CL characteristics of biotin-labeled G- quadruplex/hemin DNAzyme (Bio-G4) to the oxidation-reduction reaction of hydrogen peroxide with luminol and (iii) the on-line separation function of ME. SA- Bio-G4 complex was formed when a limited amount of streptavidine (SA) bind with Bio-G4. However, due to free biotin is a small molecule whereas Bio-G4 is a larger molecule, free biotin can competitively bind with SA to release Bio-G4, so both SA- Bio-G4 complex and free Bio-G4 will be present in the reaction system, and the concentration of biotin is proportional to the amount of free Bio-G4. Both can be separated by MZE, due to the different charge-to-mass ratios. In addition, SA-Bio- G4 complex and free Bio-G4 can catalyze the oxidation-reduction reaction of H_2O_2 with luminol, producing a chemiluminescence signal. The LOD of this method was 6.4 nM and it was positively used to detect biotin in flour with recoveries between 94 and 103%.

(A) Enzyme and D-AAs electro-focusing **(C)** D-AAs separation and in-channel enzymatic reaction and hydrogen peroxide detection. Adapted with permission from [67].

In many cases, aptamers can be a good alternative for detecting different types of analytes in food samples, due to their high selectivity and specificity [69]. Aptamers are short and single-stranded DNA or RNA oligonucleotide molecules, which have a binding affinity towards specific targets, include small molecules, proteins, and even whole cells or bacteria. They are considered as nucleic acid versions of antibodies. However, unlike antibodies, aptamers can be easily labeled and synthesized, and are stable at a variety of temperatures as well as under a wide range of pH and salt concentrations. For these reasons, aptamers have been widely used as engineered molecular probes in different works and recently, some sensitive and selective sensors based on aptamer probes have been developed [69, 70].

Gan's group reported three interesting papers about the detection of antibiotics by ME using aptamers [71–73]. In all these works, a commercial microchip-based equipment was used (MultiNA 202 System, Shimadzu, Kyoto, Japan), which was equipped with a blue LED (470 nm; 20 mA). The apparatus can be equipped with double-T microchips with different lengths depending on the need. In the first work, they developed a high throughput aptasensor based on a ME platform for automatic

detection of chloramphenicol (CAP), which is a broad-spectrum antibiotic strictly prohibited to be used in foodstuffs [71]. The methodology was as follows: (i) CAP was mixed with its selective aptamer, forming aptamer-CAP adduct (Apt-CAP); (ii) then the complementary single strand of the aptamer (C-DNA) was added into the mixture, hybridizing with the excess of non-reactive aptamer (obtaining dsDNA);

(iii) finally, this mixture (C-DNA, dsDNA and Apt-CAP products) were on-line labeled with SYBR Gold and separated by microchip sieving electrophoresis (MSE) (Fig. **6**). Based on the ratio from signal of dsDNA and aptamer-CAP ($I_{dsDNA}/I_{Apt\text{-}CAP}$), the quantification of CAP was achieved. The whole detection time was limited within 3 min with a LOD of 0.003 ng/mL. The proposed method proved to be suitable for the analysis of milk and fish with good recoveries (91–108% and 86–114%, respectively).

In the second work [72], a novel ME-based aptasensor with functionalized magnetic beads for multiplex detection of antibiotics, kanamycin (KANA) and chloramphenicol (CAP), was developed using polymerase chain reaction (PCR) as signal amplification. In the presence of a target, the corresponding capture DNA (Cap-DNA) specifically binds to it and this induces the unwinding of double-strand structures, leading to the release of assistant DNA (A-DNA) into the supernatant after magnetic separation. Then the released A-DNA was co-amplified by PCR with an internal standard strand (I-DNA). After 20 cycles of PCR, the ratio (IA-DNA/II- DNA) between the above PCR products was proportional to the concentrations of the target. MSE was used for separating and detecting DNA products. To confirm the utility of the method, KANA and CAP were detected in milk and fish samples with recoveries between 85-107% and 88-104%, respectively.

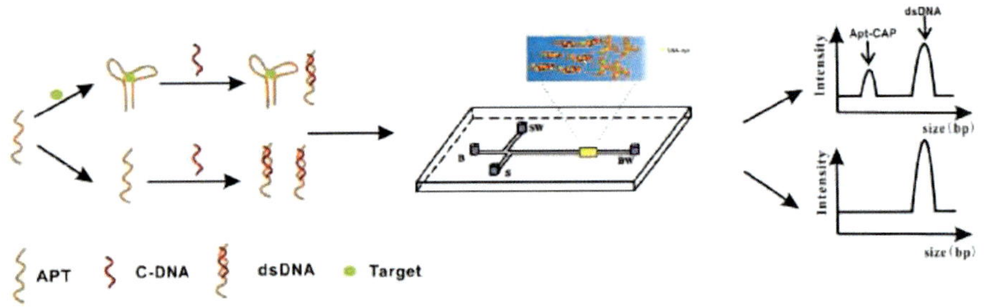

Fig. (6). Scheme of the aptasensor for determination of chloramphenicol in food samples. Adapted with permission from [71].

In the third work, authors developed a multiplex aptasensor based on ME for one-step and simultaneous detection of the previous antibiotics: chloramphenicol (CAP) and kanamycin (Kana) [73]. In this case, a gold stir-bar with encoded aptamer probes (selective to CAP and Kana) was used. It enabled signal conversion and facilitated phase separation for eliminating the matrix interference. Then the signal was amplified by a non-enzymatic multi-arm junctions recycling (MAJR) strategy, using 7 hairpin DNA probes. DNA three- and four-arm branched junctions were separated by MSE. The assay was tested using several food samples, such as milk and fish samples, obtaining recoveries between 90.2-108.3% for CAP; 98.0-112.0% for Kana.

Aflatoxin B1 (AFB1) and ochratoxin A (OTA) are mycotoxins that can be presented in foodstuffs and animal feeds, and they pose carcinogenic risks to human and animal health. In order to prevent AFB1 and OTA contamination, many countries have established strict regulations to limit these mycotoxins. Xiao *et al.* developed an aptamer-based ME-LIF method for fast determination of AFB1 and OTA in less than 3 min [74]. First, aptamers, which are specific to these two mycotoxins, were hybridized with their aptamer complementary oligonucleotides (dsDNA). Then aptamer solution was incubated with mycotoxin-containing sample for 30 min, yielding aptamer-mycotoxin complex, aptamer complementary single strand and dsDNA (competitive assay). These compounds were separated by MSE using hydroxyethyl cellulose (HEC) as sieving agent. Aptamers were on-line labeled with SYBR gold. The method exhibited extremely low LODs (0.026 ng/mL for AFB1 and 0.021 ng/mL for OTA) and good reproducibility with RSDs < 8%. It was tested in different food samples (rice, corn, peanut, barley and oil) with good recoveries (91- 123%).

Rapid and sensitive detection of foodborne pathogenic bacteria are also highly important for food safety and public health. Wang's group reported two interesting works about the detection of foodborne bacteria using specific aptamers in combination with ME-LED-IF (MCE-202 MultiNA, Shimadzu, Japan) [75,76]. Based on the differences in charge between bacteria-aptamer complexes and free aptamers, they were separated by ME. In the first case [75], *Salmonella entericaserovar Typhimurium* (*S. Typhimurium*) was detected, which is one of the most common and widespread foodborne pathogen and a major cause of illnesses and deaths every year. The separation of the bacteria-aptamer complex and free aptamers was achieved within 135 s with a limit of detection of 3.37×10^2 CFU mL^{-1}. The separation mechanism was complicated because the bacteria-aptamer complex migrated faster than free aptamer in contradiction with a sieving separation mechanism (DNA-100 kit was used that contains 3% hydroxyethylcellulose). Authors stated that bacteria supplied multiple negative charges to aptamer, so that its electrophoretic mobility was boosted. This method was applied to the detection of *S. Typhimurium* in fresh milk samples and a recovery rate of 95.8% was obtained. In the second work [76], another pathogenic bacteria was detected, *Escherichia coli* (*E. Coli*). Following the same procedure as in the previous work, bacteria-aptamer complexes and free aptamers were separated in less than 135 s with 7×10^2 CFU mL^{-1} as limit of detection. In this case, the bacteria-aptamer complex migrated slower than free aptamer according to a sieving separation mechanism. Analyzing spiked drinking water and milk samples, obtaining recovery rates of 94.7% and 92.8%, respectively, tested the method accuracy.

Chen *et al.* developed a new DNA-based method for the specific extraction of Hg^{2+} and Ag^{+} coupled to ME separation and LED induced fluorescence (LED-IF) detection [77]. In this case, the method consisted of three parts: (i) extraction, (ii) signal amplification and (iii) DNA separation. The extraction step utilized stir bars, which were functionalized with designed encoded hairpin DNA probes(H_0

with thymine-thymine (T-T) and H_0' with cytosine-cytosine (C-C) mismatches in stems).Hg^{2+} and Ag^{+} can specifically interact with T-T and C-C mismatches in DNA duplexes to form stable C-Ag^{+}-C and T-Hg^{2+}-T complexes, respectively. The signal amplification was based in DNA probes. Firstly, exonuclease III (Exo III) was used to release the complexed metal ions from the aptamer probes in the bars. Then the released targets participate in target recycling to amplify the signal. However, it exhibits limited activity. For that reason, catalytic hairpin assembly (CHA) is also used as another DNA amplification strategy for increasing sensitivity of aptasensors. Finally, produced duplexes can be monitored by ME-LED-IF in less than 135 s. The fluorescence intensity of CHA products (I_P) was increased and that of hairpin DNAs (I_R) was decreased with the increase of target concentrations. The LOD were 0.038 nM and 0.054 nM for Ag^{+} and Hg^{2+}, respectively. Several complex matrices such as milk, water and fish were analyzed obtaining recoveries ranging from 90 to 115%.

Polymerase chain reaction (PCR) based approaches in combination with separation techniques is a promising strategy for rapid detection of many different foodborne bacteria [78, 79]. Wang's group reported three interesting papers about PCR combined with ME. In the first work [80], three pairs of primers were specially designed for the amplification of target genes from *Escherichia coli* (*E. coli*), *Staphylococcus aureus* (*S. aureus*) and *Salmonella entericaserovar Typhimurium* (*S. Typhimurium*). The limits of detection of the three bacteria were concluded to be as low as 45 CFU mL^{-1} for *E. coli*, 62 CFU mL^{-1} for *S. aureus* and 42 CFU mL^{-1} for *S. Typhimurium*. DNA was separated and detected by MSE (MCE-202 MultiNA, Shimadzu). The method was tested in samples of pork, beef and chicken with good results. In the second work, the same strategy was used to detect simultaneously and sensitively *S. aureus*, *Proteus mirabilis*, and *Enterobacter sakazakiimilk* samples [81]. The limits of detection of the aim bacteria were calculated as 53 CFU mL^{-1} for *Enterobacter sakazakii*, 32 CFU mL^{-1} for *Proteus mirabilis* and 28 CFU mL^{-1} for *S. aureus*, respectively. Finally, a novel universal primer-duplex PCR process (UP- DPCR), was designed to simultaneously detect two kinds of bacteria [82]. The probe consists of an aptamer for specifically capturing bacterial cells and eventually releasing complementary DNAs (C1 and C2). The two released DNA strands (C1 and C2) can be simultaneously amplified by a pair of universal primers, because of the identical sequences designed at both ends of the two DNA strands. The UP-DPCR products of C1 and C2 can be separated and detected by ME-LIF, and the heights of the two peaks are correlated with the concentrations of the corresponding bacteria. The method was employed to determinate *Salmonella enterica serovar Typhimurium* (*S. Typhimurium*) and *Pseudomonas aeruginosa* (*P. aeruginosa*) yielding LODs of 15 CFU mL^{-1} for *S. Typhimurium* and 5 CFU mL^{-1} for *P. aeruginosa*.

This approach was successfully applied to the analysis of defatted milk. Multiplex PCR products of four foodborne pathogenic bacteria were detected by ME- LIF, whose separation conditions were optimized using a genetic algorithm-support vector regression (GA-SVR) [83]. To carry out the measurements, four pairs of oligonucleotide primers were designed to exclusively amplify the targeted gene of *Vibrio parahemolyticus*, *Salmonella*, *E. coli* O157:H7 and *Shigella*. After DNA extraction, the extract was used for quadruplex PCR amplification. The amplification products of the four pathogens were separated in less than 8 min by MSE using 2.5 % (w/v) hydroxypropilmethylcellulose as sieving agent. LODs were 1.2×10^2 CFU mL^{-1} for *Vibrio parahemolyticus*, 2.9×10^2 CFU mL^{-1} for *Salmonella*, 8.7×10^1 CFU mL^{-1} for *E. coli* O157:H7 and 5.2×10^1 CFU mL^{-1} for *Shigella*. This method was applied to the analysis of several food samples inoculated with two, three or four pathogens (beef, chicken, shrimp, fish, vegetables, noodle and rice).

Nan *et al.* developed a multi-channel ME system with a LIF detector for the simultaneous detection of rice mutants in genetically modified rice (GM) [84]. The authors designed a glass microchip with three uniform separation channels (effective length 50 mm) and 10 reservoirs, and it was applied for the high-throughput screening of amplified PCR products. EOF of glass microchip was suppressed by using a coating gel containing 1.0% (w/v) poly(vinylpyrrolidone) (PVP, Mr 1,000,000), and DNA fragments were separated in less than 4 min by a sieving gel containing 0.7% (w/v) polyethylene oxide (PEO, Mr 8,000,000). Furthermore, the multi channel ME method was more than 15 times faster than traditional slab gel electrophoresis, which is the most commonly used for the detection of amplified DNA fragments related to GM crops and GM foods.

Therefore, it can be said that the implementation of bioassays in ME provides multiple advantages such as low reagent and waste generation, high throughput, automation and speed. In addition, the existence of commercial ME-LED-IF equipment for RNA/DNA analysis has clearly increased the number of reports about food analysis by ME.

Finally, the most relevant information from bioassays using ME is depicted in Table **4**.

Table 4. Bioassays on ME for food analysis.

Analytes	Sample	Microchip Layout (and Features)	Electrophoretic Mode	Detection Mode	Bioassay Approach	Analytical Features	Remarks	Refs.
D-Met, D-Leu	NR	Glass microchip (L_{sep}=74 mm)	MZE (20 mM borate, pH=10.0)	Graphene-based electrode / Amperometry (E=+0.70 V)	D-amino acid oxidase class enzyme	Integrated electrokinetics: class enzyme and AAs co-injection, AAs separation, on channel enzymatic reaction and hydrogen peroxide detection on GO t_{sep}=800 s	Enantiomer analysis of D-AAs involved in Vibrio Cholera	[67]
Biotin	Flour	Double T PDMS/glass microchip (L_{sep}=40 mm)	MZE (15 mM borate, 5 mM SDS, 1.1 mM luminol, pH=9.3)	Chemiluminescence (Bio-G4 / luminol /H_2O_2 system)	Competitive immunoassay	LOD=6.4 nM Recoveries 94–103% t_{sep}=120 s	Very short separation time	[68]
CAP	Milk, fish	MCE-202 MultiNA (L_{sep}=23 mm)	MSE (DNA-500 kit)	LED-IF (λ_{exc}=470 nm)	Aptamer-based	LOD=0.003 ng/mL Recoveries 91–114% t_{sep}=180 s	Accurate $I_{dsDNA}/I_{Apt\text{-}CAP}$ ratio CAP quantification	[71]
CAP, KANA	Milk, fish	MCE-202 MultiNA (L_{sep}=23 mm)	MSE (DNA-500 Kit)	LED-IF (λ_{exc}=470 nm)	Aptamer-based magnetic beads	LOD_{CAP}=0.0025 nM LOD_{KANA}=0.006 nM Recoveries 90–112% t_{sep}=60 s	Co-amplification by PCR	[72]
CAP, KANA	Milk, fish	MCE-202 MultiNA (L_{sep}=23 mm)	MSE	LED-IF (λ_{exc}=470 nm)	Aptamer-based gold stir-bar PCR	LOD_{CAP}=0.52 pg/mL LOD_{KANA}=0.41 pg/mL Recoveries 85–107%	Sensitivity enhancement by MAJR	[73]
Aflatoxin B1, Ochratoxin A	Rice, corn, peanut, barley, different oils	Glass microchip (L_{sep}=23 mm)	MSE (10.00 mM Tris–HCl, 10 mM EDTA, 1% hydroxyethyl cellulose, pH=8.2)	LIF (λ_{exc}=488 nm)	Aptamer-based (competitive assay)	LOD_{AFB1} =0.026ng/mL LOD_{OTA}=0.021 ng/mL Recoveries 91–123% t_{sep}=180 s	Analysis of complex samples with good selectivity	[74]
S. Typhimurium	Fresh milk	Quartz microchip, MCE-202 MultiNA (L_{sep}=23 mm)	MZE (DNA-1000 Kit)	LED-IF (λ_{exc}=470 nm)	Aptamer-based	LOD=3.4×102 CFU/mL Recovery 95.8% tsep=135 s	Bacteria-aptamer complex migrated faster than free aptamer	[75]
E. coli	Drinking water, milk	MCE-202 MultiNA (L_{sep}=23 mm)	MSE (DNA-500 Kit)	LED-IF (λ_{exc}=470 nm)	Aptamer-based	LOD=7×102 CFU/mL Recoveries > 93% tsep=135 s	Very short separation time	[76]

(Table 4) cont.....

Analytes	Sample	Microchip Layout (and Features)	Electrophoretic Mode	Detection Mode	Bioassay Approach	Analytical Features	Remarks	Refs.
Hg^{2+}, Ag^{+}	Water, milk, fish	MCE-202 MultiNA (L_{sep}=23 mm)	MSE (DNA-500 Kit)	LED-IF (λ_{exc}=470 nm)	DNA probes-based	LODHg2+=0.054 nM LODAg+=0.038 nM Recoveries 90–115% tsep=135 s	Analysis of complex samples	[77]
E. coli, S. aureus, S. Typhimurim	Pork, beef, chicken	MCE-202 MultiNA (L_{sep}=23 mm)	MSE (DNA-1000 Kit)	LED-IF (λ_{exc}=470 nm)	PCR	LODs=42–62 CFU mL^{-1} t_{sep}=135 s	Analysis of complex samples	[80]
S. aureus, Proteus mirabilis, Enterobacter sakazakii	Milk	MCE-202 MultiNA (L_{sep}=23 mm)	MSE (DNA-500 Kit)	LED-IF (λ_{exc}=470 nm)	PCR	LODs=28–53 CFU mL^{-1} t_{sep}=135 s	Very short separation time	[81]
S. Typhimurium, P. Aeruginosa	Milk	MCE-202 MultiNA (L_{sep}=23 mm)	MSE (DNA-1000 Kit)	LED-IF (λ_{exc}=470 nm)	PCR	LODs=5–15 CFU mL^{-1} Recoveries 90–112% t_{sep}=135 s	Novel universal primer-duplex PCR amplification process	[82]
Vibrio parahemolyticu, Salmonella, E. coli, O157:H7, *Shigella*	Beef, chicken, shrimp, fish, vegetables, noodle, rice	PMMA microchip (L_{sep}=40 mm)	MSE (148 mM TBE, 2.5% hydroxypropyl methylcellulose (w/v), pH=8.5)	LIF (λ_{exc}=543.5 nm)	PCR	LODs=5.2×10^{1}-2.9×10^{2} CFU mL^{-1} t_{sep}=8 min	Genetic algorithm-support vector regression-based analysis	[83]
DNA fragments	Genetically modified (GM) rice	Glass multi-channel microchip (L_{sep}=60 mm)	MGE (sieving gel containing 0.7% (w/v) polyethylene oxide)	LIF (λ_{exc}=532 nM)	PCR	t_{sep}=240 s	Multi-channel ME Analysis times 15 folds faster than traditional slab gel electrophoresis	[84]

AAs, aminoacids; **AFB1**, aflatoxin B1; **CAP**, chloramphenicol; **D-Met**, D-methionine; **D-Leu**, D-leucine; **KANA**, kanamycin; **LED-IF,** induced fluorescence detection; **LIF**, laser induced fluorescence; **LOD**, limit of detection; **MAJR,** non-enzymatic multi-arm junctions recycling, **MGE**; microchip gel electrophoresis; **MSE**, microchip sieving electrophoresis; **MZE**, microchip zone electrophoresis; **OTA**, ochratoxin A; **PCR**, polymerase chain reaction; **NR**, not reported.

CONCLUDING REMARKS

Throughout this chapter, promising applications of ME in food analysis have been described, such as fast detection of frauds and adulterations, food quality assessment, and foodborne pathogens and toxins detection as examples of the important field of food safety. Fortunately, the appearance on the scene of the first commercial instruments in the last decade has boosted the knowledge about this topic, helping reduce the gap between research and market. At this point, we can easily distinguish between those detection modes that can be easily miniaturized (electrochemical) and those that cannot (fluorescence). Portable commercial ME equipment are based on electrochemical detection (amperometry and C4D); however, bench top commercial ME equipment are based on fluorescence detection. This detection mode has exhibited the lowest LODs, allowing for the development of extremely sensitive methods for pathogen and toxins determination. In fact, commercial ME-LED-LIF systems have been devised and developed for RNA/DNA analysis and they are proposed as a promising alternative to slab gel electrophoresis, which shows lower throughput, reproducibility, sensitivity and automation.

For ME fabrication, glass and PDMS are the most commonly used materials. The former is a very well established substrate with excellent properties such as optical transparency, chemical inertness and high EOF. However, glass microchips fabrication is a complex procedure, and this results in high cost yet. For that reason, polymeric substrates (such as PDMS, PMMA…) have become very popular in applications in ME, due to its compatibility with mass production techniques, which translates into lower costs. This is especially interesting in portable ME systems because microchip can be considered as a disposable component (single use microchip), avoiding the cleaning step for reusing it and thus pneumatic pumping systems are not needed. In fact, all portable ME systems involved in food analysis have employed polymer microchips.

LOC systems integrating all analytical steps in food analysis are still in their infancy. However, there are significant advances such as on-line derivatization, electrophoresis-based preconcentration strategies, and even the integration of SPE on a chip.

The existence of several commercial ME instruments with different detection modes can be an incentive not only for researchers in the food field but also for the food industry to replace their classical procedures with this advanced technology. Indeed, this chapter also outlines the potential to benefit greatly from food analysis using fluidic microsystems, as one of still underexplored areas in the field.

CONSENT FOR PUBLICATION

Not applicable.

CONFLICT OF INTEREST

The author declares no conflict of interest, financial or otherwise.

ACKNOWLEDGEMENTS

This work has been financially supported by the Spanish Ministry of Economy, Industry and Competitiveness (CTQ2017-86441-C2-1-R) and TRANSNANOAVANSENS program (S2018/NMT-4349) from the Community of Madrid. T.S. acknowledges the FPI fellowship from the University of Alcala. S.D. acknowledges the FPI fellowship from Spanish Ministry of Economy and Competitiveness.

REFERENCES

[1] Manikandan, V.S.; Adhikari, B.; Chen, A. Nanomaterial based electrochemical sensors for the safety and quality control of food and beverages. *Analyst (Lond.),* **2018**, *143*(19), 4537-4554. [http://dx.doi.org/10.1039/ C8AN00497H] [PMID: 30113611]

[2] Bunney, J.; Williamson, S.; Atkin, D.; Jeanneret, M.; Cozzolino, D.; Chapman, J.; Power, A.; Chandra, S. The Use of Electrochemical Biosensors in Food Analysis. *Curr. Res. Nutr. Food Sci.,* **2017**, *5*, 183-195. [http://dx.doi.org/10.12944/CRNFSJ.5.3.02]

[3] Zhao, X.; Li, M.; Liu, Y. Microfluidic-Based Approaches for Foodborne Pathogen Detection. *Microorganisms,* **2019**, *7*(10), 1-19. [http://dx.doi.org/10.3390/microorganisms7100381] [PMID: 31547520]

[4] García-Cañas, V.; Simó, C.; Herrero, M.; Ibáñez, E.; Cifuentes, A. Present and future challenges in food analysis: foodomics. *Anal. Chem.,* **2012**, *84*(23), 10150-10159. [http://dx.doi.org/10.1021/ac301680q] [PMID: 22958185]

[5] Li, S.; Kiehne, J.; Sinoway, L.I.; Cameron, C.E.; Huang, T.J. Microfluidic opportunities in the field of nutrition. *Lab Chip,* **2013**, *13*(20), 3993-4003. [http://dx.doi.org/10.1039/c3lc90090h] [PMID: 24056522]

[6] Weng, X.; Neethirajan, S. Ensuring food safety: Quality monitoring using microfluidics. *Trends Food Sci. Technol.,* **2017**, *65*, 10-22. [http://dx.doi.org/10.1016/j.tifs.2017.04.015]

[7] Atalay, Y.T.; Vermeir, S.; Witters, D.; Vergauwe, N.; Verbruggen, B.; Verboven, P.; Nicolaï, B.M.; Lammertyn, J. Microfluidic analytical systems for food analysis. *Trends Food Sci. Technol.,* **2011**, *22*, 386-404. [http://dx.doi.org/10.1016/j.tifs.2011.05.001]

[8] Ou, X.; Chen, P.; Huang, X.; Li, S.; Liu, B.F. Microfluidic chip electrophoresis for biochemical analysis. *J. Sep. Sci.,* **2020**, *43*(1), 258-270. [http://dx.doi.org/10.1002/jssc.201900758] [PMID: 31654552]

[9] Escarpa, A. Lights and shadows on food microfluidics. *Lab Chip,* **2014**, *14*(17), 3213-3224. [http://dx.doi.org/10.1039/C4LC00172A] [PMID: 24874074]

[10] Whitesides, G. The lab finally comes to the chip! *Lab Chip,* **2014**, *14*(17), 3125-3126. [http://dx.doi.org/10.1039/C4LC90072C] [PMID: 25046688]

[11] Whitesides, G.M. The origins and the future of microfluidics. *Nature,* **2006**, *442*(7101), 368-373. [http://dx.doi.org/10.1038/nature05058] [PMID: 16871203]

[12] Escarpa, A.; González, M.C.; Crevillén, A.G.; Blasco, A.J. CE microchips: an opened gate to food analysis. *Electrophoresis,* **2007**, *28*(6), 1002-1011. [http://dx.doi.org/10.1002/elps.200600412] [PMID: 17370302]

[13] Escarpa, A.; González, M.C.; López Gil, M.A.; Crevillén, A.G.; Hervás, M.; García, M. Microchips for CE: breakthroughs in real-world food analysis. *Electrophoresis,* **2008**, *29*(24), 4852-4861. [http://dx.doi.org/10.1002/elps.200800346] [PMID: 19130567]

[14] Martín, A.; Vilela, D.; Escarpa, A. Food analysis on microchip electrophoresis: an updated review. *Electrophoresis,* **2012**, *33*(15), 2212-2227. [http://dx.doi.org/10.1002/elps.201200049] [PMID: 22887146]

[15] Kuswandi, B.; Nuriman, ; Huskens, J.; Verboom, W. Optical sensing systems for microfluidic devices: a review. *Anal. Chim. Acta,* **2007**, *601*(2), 141-155.
[http://dx.doi.org/10.1016/j.aca.2007.08.046] [PMID: 17920386]

[16] Lara, F.J.; Airado-Rodríguez, D.; Moreno-González, D.; Huertas-Pérez, J.F.; García-Campaña, A.M. Applications of capillary electrophoresis with chemiluminescence detection in clinical, environmental and food analysis. A review. *Anal. Chim. Acta,* **2016**, *913*, 22-40.
[http://dx.doi.org/10.1016/j.aca.2016.01.046] [PMID: 26944987]

[17] Sierra, T.; Crevillen, A.G.; Escarpa, A. Electrochemical detection based on nanomaterials in CE and microfluidic systems. *Electrophoresis,* **2019**, *40*(1), 113-123.
[http://dx.doi.org/10.1002/elps.201800281] [PMID: 30132928]

[18] García-Carmona, L.; Martín, A.; Sierra, T.; González, M.C.; Escarpa, A. Electrochemical detectors based on carbon and metallic nanostructures in capillary and microchip electrophoresis. *Electrophoresis,* **2017**, *38*(1), 80-94.
[http://dx.doi.org/10.1002/elps.201600232] [PMID: 27412688]

[19] Nuchtavorn, N.; Suntornsuk, W.; Lunte, S.M.; Suntornsuk, L. Recent applications of microchip electrophoresis to biomedical analysis. *J. Pharm. Biomed. Anal.,* **2015**, *113*, 72-96.
[http://dx.doi.org/10.1016/j.jpba.2015.03.002] [PMID: 25840947]

[20] Sharma, S.; Jaiswal, S.; Duffy, B.; Jaiswal, A.K. Nanostructured Materials for Food Applications: Spectroscopy, Microscopy and Physical Properties. *Bioengineering (Basel),* **2019**, *6*(1), 1-17.
[http://dx.doi.org/10.3390/bioengineering6010026] [PMID: 30893761]

[21] Petroni, J.M.; Lucca, B.G.; Ferreira, V.S. Simple approach for the fabrication of screen-printed carbon-based electrode for amperometric detection on microchip electrophoresis. *Anal. Chim. Acta,* **2017**, *954*, 88-96.
[http://dx.doi.org/10.1016/j.aca.2016.12.027] [PMID: 28081818]

[22] Geschke, O.; Klank, H.; Telleman, P. Microsystem Engineering of Lab-on-a-chip Devices. Wiley-VCH Verlag GmbH & Co. KGaA: Weinheim, **2004**.

[23] Dawod, M.; Arvin, N.E.; Kennedy, R.T. Recent advances in protein analysis by capillary and microchip electrophoresis. *Analyst (Lond.),* **2017**, *142*(11), 1847-1866.
[http://dx.doi.org/10.1039/C7AN00198C] [PMID: 28470231]

[24] Godoy-caballero, M.P.; Acedo-Valenzuela, M.I.; Galeano-Díaz, T.; Costa-García, A. Fernández-Abedul, M. T., Microchip electrophoresis with amperometric detection for a novel determination of phenolic compounds in olive oil. *Analyst (Lond.),* **2012**, *137*, 5153-5160.
[http://dx.doi.org/10.1039/c2an35844a]

[25] Ávila, M.; Zougagh, M.; Escarpa, A.; Ríos, A. Fast single run of vanilla fingerprint markers on microfluidic-electrochemistry chip for confirmation of common frauds. *Electrophoresis,* **2009**, *30*(19), 3413-3418.
[http://dx.doi.org/10.1002/elps.200900238] [PMID: 19739138]

[26] Kovachev, N.; Canals, A.; Escarpa, A. Fast and selective microfluidic chips for electrochemical antioxidant sensing in complex samples. *Anal. Chem.,* **2010**, *82*(7), 2925-2931.
[http://dx.doi.org/10.1021/ac9029218] [PMID: 20210324]

[27] Pumera, M.; Escarpa, A. Nanomaterials as electrochemical detectors in microfluidics and CE: fundamentals, designs, and applications. *Electrophoresis,* **2009**, *30*(19), 3315-3323.
[http://dx.doi.org/10.1002/elps.200900008] [PMID: 19728305]

[28] García, M.; Escarpa, A. Microchip electrophoresis-copper nanowires for fast and reliable determination of monossacharides in honey samples. *Electrophoresis,* **2014**, *35*(2-3), 425-432.
[http://dx.doi.org/10.1002/elps.201300458] [PMID: 24115078]

[29] García, M.; García-Carmona, L.; Escarpa, A. Microfluidic system for enzymeless electrochemical determination of inulin using catalytically active metal nanowires. *Mikrochim. Acta,* **2015**, *182*, 745-

752.
[http://dx.doi.org/10.1007/s00604-014-1384-5]

[30] Vilela, D.; Garoz, J.; Colina, Á.; González, M.C.; Escarpa, A. Carbon nanotubes press-transferred on PMMA substrates as exclusive transducers for electrochemical microfluidic sensing. *Anal. Chem.*, **2012**, *84*(24), 10838-10844.
[http://dx.doi.org/10.1021/ac303049x] [PMID: 23170766]

[31] Vilela, D.; Martín, A.; González, M.C.; Escarpa, A. Fast and reliable class-selective isoflavone index determination on carbon nanotube press-transferred electrodes using microfluidic chips. *Analyst (Lond.)*, **2014**, *139*(10), 2342-2347.
[http://dx.doi.org/10.1039/C4AN00025K] [PMID: 24585070]

[32] Gomez, F.J.V.; Martín, A.; Silva, M.F.; Escarpa, A. Microchip electrophoresis-single wall carbon nanotube press-transferred electrodes for fast and reliable electrochemical sensing of melatonin and its precursors. *Electrophoresis*, **2015**, *36*(16), 1880-1885.
[http://dx.doi.org/10.1002/elps.201400580] [PMID: 25735903]

[33] Martín, A.; Vázquez, L.; Escarpa, A. Carbon nanomaterial scaffold films with conductivity at micro and sub-micron levels. *J. Mater. Chem. A Mater. Energy Sustain.*, **2016**, *4*, 13142-13147.
[http://dx.doi.org/10.1039/C6TA04870F]

[34] Dossi, N.; Toniolo, R.; Pizzariello, A.; Susmel, S.; Bontempelli, G. A modified electrode for the electrochemical detection of biogenic amines and their amino acid precursors separated by microchip capillary electrophoresis. *Electrophoresis*, **2011**, *32*(8), 906-912.
[http://dx.doi.org/10.1002/elps.201000690] [PMID: 21437919]

[35] Zhang, D.; Zhang, J.; Li, M.; Li, W.; Aimaiti, G.; Tuersun, G.; Ye, J.; Chu, Q. A novel miniaturised electrophoretic method for determining formaldehyde and acetaldehyde in food using 2-thiobarbituric acid derivatisation. *Food Chem.*, **2011**, *129*, 206-212.
[http://dx.doi.org/10.1016/j.foodchem.2011.04.025]

[36] Ding, Y.; Bai, L.; Suo, X.; Meng, X. Post separation adjustment of pH to enable the analysis of aminoglycoside antibiotics by microchip electrophoresis with amperometric detection. *Electrophoresis*, **2012**, *33*(21), 3245-3253.
[http://dx.doi.org/10.1002/elps.201200309] [PMID: 23065669]

[37] Fernández-la-Villa, A.; Sánchez-Barragán, D.; Pozo-Ayuso, D.F.; Castaño-Álvarez, M. Smart portable electrophoresis instrument based on multipurpose microfluidic chips with electrochemical detection. *Electrophoresis*, **2012**, *33*(17), 2733-2742.
[http://dx.doi.org/10.1002/elps.201200236] [PMID: 22965719]

[38] Castañeda, R.; Vilela, D.; González, M.C.; Mendoza, S.; Escarpa, A. SU-8/Pyrex microchip electrophoresis with integrated electrochemical detection for class-selective electrochemical index determination of phenolic compounds in complex samples. *Electrophoresis*, **2013**, *34*(14), 2129-2135.
[http://dx.doi.org/10.1002/elps.201300060] [PMID: 23595251]

[39] Brito-neto, J.G.A.; Fracassi da Silva, J.A.; Blanes, L.; Lucio do Lago, C. Understanding Capacitively Coupled Contactless Conductivity Detection in Capillary and Microchip Electrophoresis. Part 1. Fundamentals. *Electroanalysis*, **2005**, *17*, 1198-1206.
[http://dx.doi.org/10.1002/elan.200503237]

[40] Zemann, A.J. Capacitively coupled contactless conductivity detection in capillary electrophoresis. *Electrophoresis*, **2003**, *24*(12-13), 2125-2137.
[http://dx.doi.org/10.1002/elps.200305476] [PMID: 12858385]

[41] Pumera, M. Contactless conductivity detection for microfluidics: designs and applications. *Talanta*, **2007**, *74*(3), 358-364.
[http://dx.doi.org/10.1016/j.talanta.2007.05.058] [PMID: 18371649]

[42] Kubán, P.; Hauser, P.C. Ten years of axial capacitively coupled contactless conductivity detection for CZE--a review. *Electrophoresis*, **2009**, *30*(1), 176-188.

[http://dx.doi.org/10.1002/elps.200800478] [PMID: 19101932]

[43] Kubán, P.; Hauser, P.C. A review of the recent achievements in capacitively coupled contactless conductivity detection. *Anal. Chim. Acta,* **2008**, *607*(1), 15-29.
[http://dx.doi.org/10.1016/j.aca.2007.11.045] [PMID: 18155405]

[44] Rezende, K.C.A.; Moreira, R.C.; Logrado, L.P.L.; Talhavini, M.; Coltro, W.K.T. Authenticity screening of seized whiskey samples using electrophoresis microchips coupled with contactless conductivity detection. *Electrophoresis,* **2016**, *37*(21), 2891-2895.
[http://dx.doi.org/10.1002/elps.201600277] [PMID: 27542656]

[45] Jozanović, M.; Hajduković, M.; Galović, O.; Kralik, G.; Kralik, Z.; Sakač, N.; Medvidović-Kosanović, M.; Sak-Bosnar, M. Determination of anti-oxidative histidine dipeptides in poultry by microchip capillary electrophoresis with contactless conductivity detection. *Food Chem.,* **2017**, *221*, 1658-1665.
[http://dx.doi.org/10.1016/j.foodchem.2016.10.122] [PMID: 27979143]

[46] Jozanović, M.; Sakač, N.; Sak-Bosnar, M.; Carrilho, E. A simple and reliable new microchip electrophoresis method for fast measurements of imidazole dipeptides in meat from different animal species. *Anal. Bioanal. Chem.,* **2018**, *410*(18), 4359-4369.
[http://dx.doi.org/10.1007/s00216-018-1087-6] [PMID: 29707755]

[47] Thredgold, L.D.; Ellis, A.V.; Lenehan, C.E. Direct detection of histamine in fish flesh using microchip electrophoresis with capacitively coupled contactless conductivity detection. *Anal. Methods,* **2015**, *7*, 1802-1808.
[http://dx.doi.org/10.1039/C4AY02866J]

[48] Ansari, K.; Ying, J.Y.S.; Hauser, P.C.; de Rooij, N.F.; Rodriguez, I. A portable lab-on-a-chip instrument based on MCE with dual top-bottom capacitive coupled contactless conductivity detector in replaceable cell cartridge. *Electrophoresis,* **2013**, *34*(9-10), 1390-1399.
[http://dx.doi.org/10.1002/elps.201200592] [PMID: 23420647]

[49] Mark, J.J.P.; Kumar, A.; Demattio, H.; Hoffmann, W.; Malik, A.; Matysik, F.M. Combination of Headspace Single-Drop Microextraction, Microchip Electrophoresis and Contactless Conductivity Detection for the Determination of Aliphatic Amines in the Biodegradation Process of Seafood Samples. *Electroanalysis,* **2011**, *23*, 161-168.
[http://dx.doi.org/10.1002/elan.201000445]

[50] Zhai, H.; Li, J.; Chen, Z.; Su, Z.; Liu, Z.; Yu, X. A glass/PDMS electrophoresis microchip embedded with molecular imprinting SPE monolith for contactless conductivity detection. *Microchem. J.,* **2014**, *114*, 223-228.
[http://dx.doi.org/10.1016/j.microc.2014.01.006]

[51] Gertsch, J.C.; Noblitt, S.D.; Cropek, D.M.; Henry, C.S. Rapid Analysis of Perchlorate in Drinking Water at Electrophoresis. *Anal. Chem.,* **2010**, *82*, 3426-3429.
[http://dx.doi.org/10.1021/ac9029086] [PMID: 20334437]

[52] Verpoorte, E. Chip vision-optics for microchips. *Lab Chip,* **2003**, *3*(3), 42N-52N.
[PMID: 15100763]

[53] Ban, E.; Song, E.J. Recent developments and applications of capillary electrophoresis with laser-induced fluorescence detection in biological samples. *J. Chromatogr. B Analyt. Technol. Biomed. Life Sci.,* **2013**, *929*, 180-186.
[http://dx.doi.org/10.1016/j.jchromb.2013.04.028] [PMID: 23685428]

[54] Xu, X.; Li, L.; Weber, S.G. Electrochemical and optical detectors for capillary and chip separations. *Trends Analyt. Chem.,* **2007**, *26*(1), 68-79.
[http://dx.doi.org/10.1016/j.trac.2006.11.015] [PMID: 18167508]

[55] Götz, S.; Karst, U. Recent developments in optical detection methods for microchip separations. *Anal. Bioanal. Chem.,* **2007**, *387*(1), 183-192.
[http://dx.doi.org/10.1007/s00216-006-0820-8] [PMID: 17031620]

[56] Breadmore, M.C.; Grochocki, W.; Kalsoom, U.; Alves, M.N.; Phung, S.C.; Rokh, M.T.; Cabot, J.M.; Ghiasvand, A.; Li, F.; Shallan, A.I.; Keyon, A.S.A.; Alhusban, A.A.; See, H.H.; Wuethrich, A.; Dawod, M.; Quirino, J.P. Recent advances in enhancing the sensitivity of electrophoresis and electrochromatography in capillaries and microchips (2016-2018). *Electrophoresis,* **2019**, *40*(1), 17-39.
[http://dx.doi.org/10.1002/elps.201800384] [PMID: 30362581]

[57] Ohla, S.; Schulze, P.; Fritzsche, S.; Belder, D. Chip electrophoresis of active banana ingredients with label-free detection utilizing deep UV native fluorescence and mass spectrometry. *Anal. Bioanal. Chem.,* **2011**, *399*(5), 1853-1857.
[http://dx.doi.org/10.1007/s00216-010-4557-z] [PMID: 21181134]

[58] Wang, L.; Wu, J.; Wang, Q.; He, C.; Zhou, L.; Wang, J.; Pu, Q. Rapid and sensitive determination of sulfonamide residues in milk and chicken muscle by microfluidic chip electrophoresis. *J. Agric. Food Chem.,* **2012**, *60*(7), 1613-1618.
[http://dx.doi.org/10.1021/jf2036577] [PMID: 22277081]

[59] Wei, X.; Gao, X.; Zhao, L.; Peng, X.; Zhou, L.; Wang, J.; Pu, Q. Fast and interference-free determination of glyphosate and glufosinate residues through electrophoresis in disposable microfluidic chips. *J. Chromatogr. A,* **2013**, *1281*, 148-154.
[http://dx.doi.org/10.1016/j.chroma.2013.01.039] [PMID: 23398994]

[60] Cheng, Y.; Guo, C.; Zhao, B.; Yang, L. Fast analysis of domoic acid using microchip electrophoresis with laser-induced fluorescence detection. *J. Sep. Sci.,* **2017**, *40*(7), 1583-1588.
[http://dx.doi.org/10.1002/jssc.201600982] [PMID: 28160435]

[61] Wang, Z.; Wang, W.; Wang, W.; Xu, L.; Chen, G.; Fu, F. Separation and determination of β-casomorphins by using glass microfluidic chip electrophoresis together with laser-induced fluorescence detection. *J. Sep. Sci.,* **2011**, *34*(2), 196-201.
[http://dx.doi.org/10.1002/jssc.201000634] [PMID: 21246725]

[62] Wu, M.; Cheng, S.; Li, G.; Wang, Z.; Wang, Q.; He, P.; Fang, Y. Rapid Determination of Free Amino Acids in Milk by Microchip Electrophoresis Coupled with Laser-induced Fluorescence Detection. *Chin. J. Chem.,* **2012**, *30*, 2517-2521.
[http://dx.doi.org/10.1002/cjoc.201200446]

[63] Wu, M.; Gao, F.; Zhang, Y.; Wang, Q.; Li, H. Sensitive analysis of amino acids and vitamin B3 in functional drinks *via* field-amplified stacking with reversed-field stacking in microchip electrophoresis. *Talanta,* **2015**, *131*, 624-631.
[http://dx.doi.org/10.1016/j.talanta.2014.08.051] [PMID: 25281150]

[64] Wu, M.; Chen, W.; Wang, G.; He, P.; Wang, Q. Analysis of acrylamide in food products by microchip electrophoresis with on-line multiple-preconcentration techniques. *Food Chem.,* **2016**, *209*, 154-161.
[http://dx.doi.org/10.1016/j.foodchem.2016.04.065] [PMID: 27173547]

[65] Del Mar Barrios-Romero, M.; Crevillén, A.G.; Diez-Masa, J.C. Development of an SDS-gel electrophoresis method on SU-8 microchips for protein separation with LIF detection: Application to the analysis of whey proteins. *J. Sep. Sci.,* **2013**, *36*(15), 2530-2537.
[http://dx.doi.org/10.1002/jssc.201300275] [PMID: 23720160]

[66] Crevillén, A.G.; Frutos, M.; Diez-Masa, J.C. On-chip single column transient isotachophoresis with free zone electrophoresis for preconcentration and separation of α-lactalbumin and β-lactoglobulin. *Microchem. J.,* **2017**, *133*, 600-606.
[http://dx.doi.org/10.1016/j.microc.2017.04.040]

[67] Batalla, P.; Martín, A.; López, M.Á.; González, M.C.; Escarpa, A. Enzyme-based microfluidic chip coupled to graphene electrodes for the detection of D-amino acid enantiomer-biomarkers. *Anal. Chem.,* **2015**, *87*(10), 5074-5078.
[http://dx.doi.org/10.1021/acs.analchem.5b00979] [PMID: 25870911]

[68] Li, J.; Li, S.; Zhao, J.; Huang, Y.; Zhao, S. A G-quadruplex/hemin DNAzyme-based microchip

electrophoresis chemiluminescence assay for highly sensitive detection of biotin in flour. *Electrophoresis,* **2019**, *40*(16-17), 2157-2164.
[http://dx.doi.org/10.1002/elps.201800535] [PMID: 31025386]

[69] Li, F.; Yu, Z.; Han, X.; Lai, R.Y. Electrochemical aptamer-based sensors for food and water analysis: A review. *Anal. Chim. Acta,* **2019**, *1051*, 1-23.
[http://dx.doi.org/10.1016/j.aca.2018.10.058] [PMID: 30661605]

[70] Feng, C.; Dai, S.; Wang, L. Optical aptasensors for quantitative detection of small biomolecules: a review. *Biosens. Bioelectron.,* **2014**, *59*, 64-74.
[http://dx.doi.org/10.1016/j.bios.2014.03.014] [PMID: 24690563]

[71] Zhou, L.; Gan, N.; Zhou, Y.; Li, T.; Cao, Y.; Chen, Y. A label-free and universal platform for antibiotics detection based on microchip electrophoresis using aptamer probes. *Talanta,* **2017**, *167*, 544-549.
[http://dx.doi.org/10.1016/j.talanta.2017.02.061] [PMID: 28340759]

[72] Zhou, L.; Gan, N.; Hu, F.; Li, T.; Cao, Y.; Wu, D. Microchip electrophoresis array-based aptasensor for multiplex antibiotic detection using functionalized magnetic beads and polymerase chain reaction amplification. *Sens. Actuators B Chem.,* **2018**, *263*, 568-574.
[http://dx.doi.org/10.1016/j.snb.2018.02.136]

[73] Zhang, K.; Gan, N.; Shen, Z.; Cao, J.; Hu, F.; Li, T. Microchip electrophoresis based aptasensor for multiplexed detection of antibiotics in foods *via* a stir-bar assisted multi-arm junctions recycling for signal amplification. *Biosens. Bioelectron.,* **2019**, *130*, 139-146.
[http://dx.doi.org/10.1016/j.bios.2019.01.044] [PMID: 30735947]

[74] Xiao, M.W.; Bai, X.L.; Liu, Y.M.; Yang, L.; Liao, X. Simultaneous determination of trace Aflatoxin B_1 and Ochratoxin A by aptamer-based microchip capillary electrophoresis in food samples. *J. Chromatogr. A,* **2018**, *1569*, 222-228.
[http://dx.doi.org/10.1016/j.chroma.2018.07.051] [PMID: 30037541]

[75] Zhang, Y.; Luo, F.; Zhang, Y.; Zhu, L.; Li, Y.; Zhao, S.; He, P.; Wang, Q. A sensitive assay based on specific aptamer binding for the detection of Salmonella enterica serovar Typhimurium in milk samples by microchip capillary electrophoresis. *J. Chromatogr. A,* **2018**, *1534*, 188-194.
[http://dx.doi.org/10.1016/j.chroma.2017.12.054] [PMID: 29289340]

[76] Zhang, Y.; Zhu, L.; He, P.; Zi, F.; Hu, X.; Wang, Q. Sensitive assay of Escherichia coli in food samples by microchip capillary electrophoresis based on specific aptamer binding strategy. *Talanta,* **2019**, *197*, 284-290.
[http://dx.doi.org/10.1016/j.talanta.2019.01.040] [PMID: 30771937]

[77] Chen, X.; Hong, F.; Zhang, W.; Wu, D.; Li, T.; Hu, F.; Gan, N.; Lin, J.; Wang, Q. Microchip electrophoresis based multiplexed assay for silver and mercury ions simultaneous detection in complex samples using a stirring bar modified with encoded hairpin probes for specific extraction. *J. Chromatogr. A,* **2019**, *1589*, 173-181.
[http://dx.doi.org/10.1016/j.chroma.2019.01.004] [PMID: 30635170]

[78] Salihah, N.T.; Hossain, M.M.; Lubis, H.; Ahmed, M.U. Trends and advances in food analysis by real-time polymerase chain reaction. *J. Food Sci. Technol.,* **2016**, *53*(5), 2196-2209.
[http://dx.doi.org/10.1007/s13197-016-2205-0] [PMID: 27407185]

[79] Alarcón, B.; García-Cañas, V.; Cifuentes, A.; González, R.; Aznar, R. Simultaneous and sensitive detection of three foodborne pathogens by multiplex PCR, capillary gel electrophoresis, and laser-induced fluorescence. *J. Agric. Food Chem.,* **2004**, *52*(23), 7180-7186.
[http://dx.doi.org/10.1021/jf049038b] [PMID: 15537335]

[80] Zhang, Y.; Zhu, L.; Zhang, Y.; He, P.; Wang, Q. Simultaneous detection of three foodborne pathogenic bacteria in food samples by microchip capillary electrophoresis in combination with polymerase chain reaction. *J. Chromatogr. A,* **2018**, *1555*, 100-105.
[http://dx.doi.org/10.1016/j.chroma.2018.04.058] [PMID: 29724645]

[81] Zhu, L.; Zhang, Y.; He, P.; Zhang, Y.; Wang, Q. A multiplex PCR amplification strategy coupled with microchip electrophoresis for simultaneous and sensitive detection of three foodborne bacteria. *J. Chromatogr. B Analyt. Technol. Biomed. Life Sci.,* **2018**, *1093-1094*, 141-146. [http://dx.doi.org/10.1016/j.jchromb.2018.06.057] [PMID: 30015312]

[82] Luo, F.; Li, Z.; Dai, G.; Lu, Y.; He, P.; Wang, Q. Simultaneous detection of different bacteria by microchip electrophoresis combined with universal primer-duplex polymerase chain reaction. *J. Chromatogr. A,* **2020**, *1615*, 460734. [http://dx.doi.org/10.1016/j.chroma.2019.460734] [PMID: 31791592]

[83] Li, Y.; Li, Y.; Zheng, B.; Qu, L.; Li, C. Determination of foodborne pathogenic bacteria by multiplex PCR-microchip capillary electrophoresis with genetic algorithm-support vector regression optimization. *Anal. Chim. Acta,* **2009**, *643*(1-2), 100-107. [http://dx.doi.org/10.1016/j.aca.2009.04.006] [PMID: 19446070]

[84] Nan, H.; Lee, S.W.; Kang, S.H. Fast screening of rice knockout mutants by multi-channel microchip electrophoresis. *Talanta,* **2012**, *97*, 249-255. [http://dx.doi.org/10.1016/j.talanta.2012.04.026] [PMID: 22841075]

CHAPTER 11

Application of Capillary Electrophoresis to Food Authentication

František Kvasnička[1,*]

[1] *Department of food preservation, University of Chemistry and Technology, Technicka 3, 16628 Prague 6, Czech Republic*

Abstract: This chapter deals with the issue of food authenticity and applications of capillary electrophoresis in this field. An overview of food adulteration, including the frequency of adulteration of selected commodities and a list of methods used to prove authenticity or detect food adulteration, is presented. An overview of applications of capillary electrophoresis for food authentication is supplemented by specific cases described in more detail.

Keywords: Adulteration, Capillary electrophoresis, Falsification, Food authentication.

INTRODUCTION

Cases of food adulteration can be encountered from the time when they began to be manufactured for sale. Mentions on counterfeiting and penalties for dishonest producers and traders are found in a number of historical texts. In Biblical times, meat inspection rules and fitness criteria were specified. They can be found in the Book of Leviticus. Galen, an ancient physician who developed the principle of humours, linking body type with health and personality, warned against the adulteration of herbs and spices [1]. The Industrial Revolution brought the great age of food adulteration to its height. During this period, untainted, safe and healthy food was a difficult commodity to buy. The major problem was that adulterated food was not easy to detect. Science was based in universities and the practice was never applied to the food industry. Friedrich Engels (German-born political theorist) in his study "The Condition of the Working Class in England" stated: "*...workers were being exploited by the middle classes who were mixing sugar with ground rive, chicory with coffee, and earth with cocoa, then selling the adulterated products as pure and at full price...*" *[2]*.

* **Corresponding author František Kvasnička:**, Department of food preservation, University of Chemistry and Technology, Technicka 3, 16628 Prague 6, Czech Republic; E-mail: Frantisek.Kvasnicka@vscht.cz

María Castro-Puyana, Miguel Herrero and María Luisa Marina (Eds.)
Capillary Electrophoresis in Food Analysis

Methods have also been developed for years for counterfeit detection. In fact, already in 1820, the German chemist Frederick Accum published a treatise on food adulteration and culinary poisoning where he exposed fraudulent practice and toxic contamination in food and drink, providing methods for their detection. The treatise, popularly known as "There is Death in the Pot" aroused great attention and has spread throughout Europe [3].

In principle, counterfeiting procedures do not fundamentally change; the examples described in historical sources are similar. The main categories of counterfeiting are listed in Table **1**, while it is clear that the individual categories are not fundamentally defined and overlap [4 - 8].

Table 1. Examples of food adulteration issues.

Way of Adulteration	Examples
Abuse of a well-known brand	- fake sale of the product under a more expensive brand; - use of packaging, labels, names resembling a well-known brand, *etc.*;
Setting food cheaper component	- non-compliance with the requirements for the content of the so-called glaze in frozen meats; - the use of blood proteins instead of muscle in meat products; - additions of fibre to meat products; - undeclared additions of soya, cereal, pea and other vegetable proteins to meat products; - additions of cow's milk to buffalo in the production of true mozzarella; - dilution of milk with water; - partial or complete replacement of Basmati rice by cheaper varieties; - additions of common wheat flour (*Triticum aestivum*) to pasta presented as pasta made from semolina (made from wheat flour *Triticum durum*); - dilution of olive oil with other vegetable oils; - additions of starch hydrolysates or sugar syrups to honey; - additions of water, sugar, acids and dyes to fruit juices, nectars, beverages; - replacing more expensive fruits with cheaper ones (apple puree instead of strawberries, apple juice instead of more expensive juices, *etc.*); - reducing the cocoa butter content of chocolate by adding other oils; - addition of roasted skins to ground coffee or cocoa; - dilution of wine with water; - additions of water, sugar, acids and dyes to fruit juices, nectars, beverages

(Table 1) cont.....

Way of Adulteration	Examples
Presence of undeclared components	- undeclared use of other types of meat in meat products; - unauthorized or undeclared use of mechanically separated meat in meat products; - undeclared use of offal in meat products; - undeclared addition of substitutes such as maltodextrins in coffee; - addition of meat species in processed vegetarian foods; - undeclared use of genetically modified food
Food exchange for another cheaper	- issuing other vegetable oils as olive; - issuing sea trout for salmon; - issuing cheaper varieties as more expensive
Failure to comply with the declared technological procedure	- dispensing thawed meat and fish fresh; - declaring toasted bread as fresh - undeclared use of gamma radiation in production - dispensing the reconstituted juice from the concentrate with freshly pressed juice; - use of synthetic alcohol for the production of spirits; - issuing ordinary oil for cold-pressed oil (virgin olive oil); - incorrect indication of the age of the distillate
Incorrect indication of geographical origin or method of production	- farming of fish produced on farms; - labelling of usual production as bio (organic); - issuing imported wines for domestic - incorrect declaration of floral or geographical origin of honey
Declared content of the component higher than the actual content	- placing a higher number of eggs in pasta - declaration of the proportion of muscle proteins in meat products distorted by the addition of plant or blood proteins - the use of misleading labels and markings (images, names, graphic symbols inadequate to food composition)

Trends in commodity counterfeiting are being evaluated according to the frequency of publications in databases; they do not change in the long time span they develop according to current cases. Counterfeiting is motivated by economic profit, so expensive and luxurious foods (spirits, wine, spices) are most often counterfeited or, conversely, foods that are sold in large volumes (meat and dairy products, fats and oils, fruit juice). This also corresponds to the trends in the interest of the individual commodities expressed as the number of citations of articles with a given focus in various databases. Table **2** summarizes the trends of interest in individual commodities.

Table 2. Trends in food commodity adulteration by number of citations within last four decades (from 1980- present) [9].

Commodity	Number of Publications	(%)
Meat and meat products	1652	11
Milk and milk products	1996	14
Fruit, vegetables and beverages(juice)	2937	20
Oils and fats	4025	28
Honey	949	7
Coffee and tea	711	5
Spices	403	3
Wine and spirits	1778	12
Dietary/food supplements	588	4

Unlike the situation at the beginning of the 19th century, when food adulteration directly endangered consumers' health, most of the cases of adulteration described above are based on customer complaints because consumers do not get a product of the expected properties for their money. Nevertheless, there have been several cases of the harmful effects of food adulteration on human health in recent years. So far, perhaps the biggest case in terms of its consequences was the Chinese affair in 2007-2008 on the reduction of milk content in infant formula and the masking of lower milk protein content by the addition of melamine. The exposure led to kidney damage and kidney failure; according to data at the end of 2008, almost three hundred thousand children were affected and six died. An often-recurring example of adulteration is the addition of methanol to spirits or the replacement of ethanol with methanol. Every year, these cases of counterfeiting claim dozens of lives and many consumers have lifelong health consequences.

While adulteration practices remained simple, their detection by relatively straightforward analytical methods was possible. As new and advanced technologies have become widely available, fraudulent practices have also become more sophisticated. This situation provides analytical chemists with the most testing challenges. The need is not only for new analytical techniques to establish food authenticity, but also for a constant reappraisal of these techniques in order to keep pace with changing industry practices and the increasingly devious adulteration methods of unscrupulous producers. Bearing in mind the diversity of the authenticity criteria, it is obvious that a wide range of methods is required. These techniques must help pick up minor differences associated with the botanical origin or otherwise, with different manufacturing processes, and distinguish adulteration with materials chemically similar to genuine components, or detecting adulterants present at trace levels.

Counterfeit detection procedures include two approaches that are in principle intertwined and are based on the fact that the authentic product contains characteristic substances in a certain amount corresponding to the raw materials and technology used. One approach is to verify the hypothesis of a particular method of counterfeiting, which is confirmed by the analysis of specific markers related to the presumed process of counterfeiting. Examples are determination of undeclared addition of mechanically separated meat based on the presence of bone fragments and calcium content in the product, or detection of sorbitol in juice, which is typical for apples and not present in sufficient quantities in other fruits, *etc*. The second approach involves multi-ingredient analysis usually of a larger set of food samples; the obtained results are analyzed by the method of multidimensional statistical analysis (*e.g*., principal component analysis, cluster analysis) [7, 10].

Within the last decade, new approaches to detect food counterfeiting, are "-omics" techniques [11]. Proteomics, metabolomics and genomics tools represent a promising alternative to classical food analysis methods because of their sensitivity, high-throughput, multiplexing, robustness and discriminating power, suggesting that they will play an important role in food authenticity and detection of food adulteration. Table **3** provides examples of analytical techniques used for food authentication. In 2009 "*Foodomics*" was introduced by Cifuentes [12] as a novel discipline that uses omics technologies to assess relevant aspects related to food and nutrition such as safety, quality, production, processing and new ingredient design.

Table 3. Analytical techniques for food authentication.

Technique	Field of application
Pattern recognition (IR, NIR, FTIR, NMR)	- rapid, non-destructive spectroscopic methods for discrimination between authentic and non-authentic samples (orange juice, cereal grains, vegetable oil, *etc*.)
Structural analysis (MS, HR-NMR, IR, IA, optical and electron microscopy)	– methods for studying of physical and chemical properties (quality changes in fish products, evaluation of the size, shape and colour characteristics of the kernels from a cereal batch)
Stable isotope analysis (SIRA, SNIF-NMR, IRMS)	– pattern of naturally occurring stable isotopes of carbon, hydrogen, nitrogen and oxygen (botanical and geographical origin of both raw material and products, *i.e*., fruit, wine, coffee)
Trace element analysis (ICP-AES, AA, FES, NAA)	– indication of geographical origin and detection of specific contaminants (elemental analysis of potato samples by ICP-AES)

(Table 3) cont.....

Technique	Field of application
Biochemical techniques (DNA fingerprinting, PCR, ELISA)	- identification of plant species, cell lines and animal breeds, species identification of raw and processed meat, identification of genetically-modified organisms in food - combined use of molecular techniques (PCR) and capillary electrophoresis for protein fingerprinting technique for plant and animal species
Separation techniques (CE, GC, HPLC, TLC, CE-MS, GC-MS, HPLC-MS)	– characterization in terms of chemical composition – detection of specific adulterant or authenticity markers (3-methylhistidine for lean meat, ratio of citric and isocitric acids in citrus juice, D-malic acid in apple concentrate, *etc.*)
Sensory analysis	- identification of an authentic variety of fruit (mango, grape) - detection of counterfeit brand spirits - authentication of oils (olive, rapeseed)
Electronic sensors (E-tongue, E-nose, E-eye)	- botanical origin (vegetable oils, honey) - detection of adulterated meat (pork in beef, pork in mutton) - detection of adulterated oils (rapeseed oil in extra virgin olive oil)

Metabolomics also includes two possible approaches, chemometrics, in which data such as spectra, records of intensities of individual ions, molecular peaks, *etc.*, are processed by statistical methods without identification of individual substances. The purpose is to search for samples of remote properties or to create clusters of samples of similar properties. The second metabolomics approach, called quantitative metabolomics, aims to identify and quantify as many analytes as possible in a sample. Another term used in connection with the evaluation of food authenticity and the detection of adulteration is "fingerprinting", *i.e.*, essentially quantitative metabolomics, where the profile of the sample components is usually obtained by one analytical procedure.

The analysis of complex data sets, generated by a diverse range of analytical techniques, requires the use of multivariate statistical methods to enable the food scientist to analyze the interplay of the different variables. The computer-assisted multivariate pattern recognition software package has become an important tool to fight food adulteration. The interpretation of the obtained data is perhaps one of the most crucial steps of the analytical chain.

APPLICATION OF CAPILLARY ELECTROPHORESIS IN FOOD AUTHENTICATION

Capillary Electrophoresis (CE) is an electrokinetic separation technique that separate compounds based upon differences in electrophoretic mobility, phase partitioning, pI, molecular size or a combination of one or several of these properties. The separation of analytes takes place in narrow-bore capillary in free solution. CE have been used in several modes (*i.e.* Capillary Zone Electrophoresis (CZE), Capillary Isoelectric Focusing (CIEF), Capillary Isotachophoresis (CITP), on-line coupling of CITP and CZE, Capillary Gel Electrophoresis (CGE) and Capillary Electrochromatography (CEC)) to analyze and characterize a wide variety of analytes from simple inorganic ions, small organic molecules, peptides, proteins, nucleic acids to virus, microbes and particles. Chip-based separation is another promising field of capillary electrophoresis application in food authentication [13 - 15]. Microchip Electrophoresis (ME) is an arrangement of electrophoretic separation on a microfluidic chip [16, 17]. This approach allows CE methods to be translated to a smaller scale, requiring smaller sample volume, less reagents and most importantly, shorter analysis time. The fundamental factors governing the electrophoretic separation have been described and reader can find them elsewhere [18 - 20].

Food authentication needs highly selective and sensitive analytical methods that are also simple and cost-effective. CE can meet these requirements by offering high- resolution separations at a minimal cost in terms of sample size, reagents used and operator time. CE is used for these purposes, as evidenced by the number of articles in professional journals. Fig. (**1**) shows the increase in the number of publications on this issue over the last four decades. The number of publications on food authenticity or falsification terms has grown from 391 entries in the period 1980- 1989 up to 7650 items from 2010-present [21]. The number of publications dealing with CE in food analysis has grown from 35 (1980-1989) up to 3356 items (2010- present) and application of capillary electrophoresis to food authenticity or detection of food falsification has grown from 0 (1980-1989) up to 179 items in the period 2010-present.

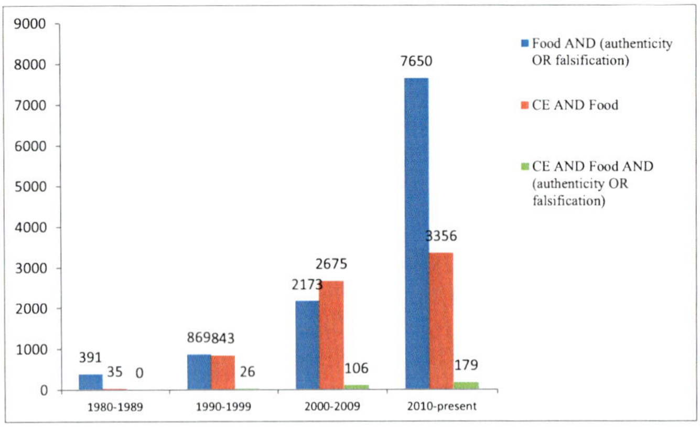

Fig. (1). Number of publications containing the term "capillary electrophoresis" and "food authenticity/authentication/falsification/adulteration" [21].

This chapter provides an overview of the CE application in food authentication. The reviewed articles are categorized according to commodities. Examples of application of CE in food authentication are listed in Table **4**.

Table 4. Applications of CE in food authenticity/detection of adulteration.

Meat, fish and meat products					
CE mode	**CE conditions**	**Detection**	**Analytes**	**Fields of application**	**Ref.**
CGE	fused-silica capillary (380x0.075 mm i.d.)	LIF	PCR products (DNA fragments)	Detection of soy in canned tuna	[22]
	20 mM Tris-HCl + 10 mM phosphate buffer + 2 mM EDTA + 2.5 mM YOPRO + 3% HEC (pH 7.3)	λ_{Ex} 488 nm			
	40°C, 12 kV	λ_{Em} 520 nm			
ME	Microfluidic electrophoresis instrument Agilent 2100 Bioanalyzer, Protein 80 Kit	-	Fish muscle proteins	Seafood species identification	[23]
MEKC	fused-silica capillary (570x0.05 mm i.d.)	UV	4-hydroxyproline	Collagene in meat products	[24]
	50 mM sodium tetraborate buffer + 100 mM SDS (pH 9.3)	214 nm			
	33°C, 12 kV				
CZE	fused-silica capillary (570x0.075 mm i.d.)	UV	Sarcoplasmic proteins pattern	Differentiation of fish species	[44]
	75-100 mM sodium phosphate buffer (pH 2.4)	214 nm			
	25°C, 10-20 kV				
CE	capillary (87.05 mm i.d.) coated with ethylpyrrolidine methacrylate-*N*,*N*-dimethylacrylamide	MS	Lysozyme	Meat species differentiation	[51]
	75 mM ammonium acetate/acetic acid (pH 5.5)	range = 800-2200 m/z			
	25°C, 25 kV,	target mass 1500 m/z			
MEKC	fused-silica capillary (380x0.075 mm i.d.)	UV	Haemoglobin, myoglobin, actin, myosin	Detection of MRM	[48]
	ProSort SDS-protein analysis buffer	214 nm			
	25°C, 12 kV				
CITP	LE: 5 mM ammonium hydroxide + 10 mM MES	Conductometric	Protein-bound 3-methyhistidine	Lean meat content	[49]
	TE: 10 mM 6-aminocaproic acid + 5 mM acetic acid				
CZE	fused-silica capillary (500-870x0.05-0,075 mm i.d)	UV	Isoflavones (daidzein, genistein)	Soy and lupin proteins detection in meat products	[50]
	200 mM boric acid (pH 8.6)	260 and 270 nm			
	25°C, 25-30 kV				

(Table 4) cont.....

Meat, fish and meat products					
CE mode	**CE conditions**	**Detection**	**Analytes**	**Fields of application**	**Ref.**
CZE	fused-silica capillary (170x0.025 mm i.d.)	UV	Cow milk proteins (serum -albumin, alpha and beta globulin, alpha and beta-caseins)	Detection of adulteration of buffalo milk with cow milk	[25]
	150 mM borate buffer (pH 9.9)	210 nm			
	7.8 kV, UV detection at 210 nm				
CZE	fused-silica capillary (170x0.025 mm i.d.)	UV	Cow milk proteins (albumin, globulin, beta-casein)	Detection of ewe milk in cow milk	[26]
	150 mM borate buffer (pH 9.9)	210 nm			
	7.8 kV				
CZE	methyl deactivated capillary (440-610x0.05 mm i.d.) 50-120 mM sodium borate (pH 9.2)	UV	Lactoglobulins	Differentiation of milk species	[52]
	25°C, 4-6 kV	200 nm			
CZE	fused-silica capillary (335x0.05 mm i.d.)	UV	Lactoglobulins	Differentiation of milk species	[53]
	50 mM iminodiacetic acid + 0.1% HPMC + 0.1-10% Tween 20 + 6 M urea (pH 3,1)	214 nm			
	25°C, 10-15 kV				
CZE	Hydrophilic coated capillary (C-Elect1, 500x0.05 mm i.d.)	UV	Caseins	Differentiation of milk species	[55]
	320 mM citric acid + 20 mM sodium citrate + 0.05% HPMC + 6 M urea (pH 8.6)	214 nm			
	25 kV				
CZE	fused-silica capillary (570x0.05 mm i.d.)	UV	CMP	Adulteration with whey rennet addition	[58]
	20 mM sodium citrate (pH 3.5)	200 nm			
	40°C, + 30 kV				
CZE	fused-silica capillary (300x0.05 mm i.d.)	UV	Furosine	Quality of dairy products	[62]
	100 M sodium phosphate + 1.2 mM -containing the additive hexadecyltrimethylammonium bromide (pH 7.0)	280 nm			
	30°C, - 10 kV				
MEKC	fused-silica capillary (485x0.05 mm i.d.)	UV	HMF	Heat load of milk	[64]
	50 mM phosphate buffer (pH 7.5) + 100 mM SDS	280 nm			
	25°C, 20 kV				
CZE	fused-silica capillary (600x0.05 mm i.d.)	UV	Polyphenoic compounds (proanthocyanidins)	Authentication of cranberry and grape based products	[27]
	35 mM sodium tetraborate (pH 9.4) + 5% (v/v) methanol	280 nm			
	25°C, 25 kV				
CZE	fused-silica capillary (480x0.05 mm i.d.)	indirect UV	Monosaccharides (glucose, fructose, sorbitol, sucrose,	Differentiation of juices and blends	[28]
	10 mM 2,6-pyridine dicarboxylic acid + 0.5 mM CTAB (pH 12.1)	575 nm			
	15 - 20°C, 25 kV,				

(Table 4) cont.....

Meat, fish and meat products					
CE mode	**CE conditions**	**Detection**	**Analytes**	**Fields of application**	**Ref.**
CZE	fused silica capillary (600 x 0.075 mm i.d)	indirect UV	Ca, K, Na, Mg	Authentication of orange juice	[70]
	5 mM UVCat-I, 8 mM HIBA (pH 4.4)	214 nm			
	20 kV				
CITP	LE: 7.5 mM sulphuric acid + 7 mM 18-crown-6 ether	Conductometric	NH_4, Mg, Na, Ca, K	Authentication of citrus juice	[68]
	TE: 10 mM BTP + 10 mM caproic acid				
CZE	capillary coated with polyacrylamide (570x0.05 mm i.d.)	UV	Citric, isocitric, malic and tartaric acids	Authentication of orange juice	[29]
	200 mM sodium phosphate buffer (pH 7.5)	200 nm			
	-14 kV				
CITP	LE: 10 mM HCl + beta-alanine+0.05% HPMC (pH 2.9)/ TE: 10 mM citric acid	Conductometric	Anionic profile of orange juice	Authentication of orange juice	[72]
	LE: 10 mM HCl + histidine+0.05% HPMC (pH 6.0)/ TE: 10 mM caproic acid				
	LE: 6 mM HCl + 12 mM histidine+2 mM $CaCl_2$+0.05% HPMC (pH 6.0)/ TE: 10 mM citric acid				
CITP	LE: 6 mM HCl + 3.8 mM BTP + 2 mM $CaCl_2$ + 0.05% HPMC	Conductometric	Isocitric, citric and malic acids	Authentication of citrus juice	[73]
	TE: 5 mM MES + 1 mM BTP				
CITP-CZE	LE: 10 mM HCl + beta-alanine + 5 mM beta-cyclodextrin + 0.05% HPMC, pH 3	UV	Fumaric acid	Authentication of apple juice (addition of synthetic malic acid)	[74]
	TE: 10 mM citric acid	254 nm			
	BGE: 20 mM citric acid + beta-alanine + 5 mM beta-cyclodextrin + 0.1% HPMC (pH 3.3)				
CZE	fused-silica capillary (700x0.005-0.075 mm i.d.)	UV-VIS	Polyphenols, flavonoids, phenolic amines, amino acids, vitamin C	Authentication of citrus juice (detection of pulpwash addition)	[30]
	50 mM sodium phosphate (pH 6.8) and/or 50 mM borax-boric (pH 7.6 – 9.2) and/or 50 mM borax-NaOH (pH 10.0)	200-500 nm			
MEKC	fused-silica capillary (450x0.05 mm i.d.)	LIF	7 DL-amino acids	Origin of orange juice, heat treatment	[78]
	100 mM sodium tetraborate +30 mM SDS) + 20 mM b-CD (pH 9.4)	λ_{Ex} 488 nm			
	15°C, 20 kV	λ_{Em} 520 nm			
CZE	fused-silica capillary (1040x0.05 mm i.d.)	indirect UV	Carbohydrates and organic acids	Juice composition	[80]
	20 mM 2,6-pyridinedicarboxylic acid + 0.5 mM cetyltrimethylammonium hydroxide (pH 12.1)	350 nm			
	20°C, - 25 kV				
Wine					

(Table 4) cont.....

CE mode	CE conditions	Detection	Analytes	Fields of application	Ref.
Meat, fish and meat products					
CZE	fused-silica capillary (370x0.075 mm i.d.)	UV	Polyphenols	Canary Islands wine differentiation	[97]
	25 mM sodium tetraborate (pH 9.5)	350 nm			
	25°C, 20 kV				
CITP-CZE	LE: 10 mM HCl + Tris + 0.2% HEC (pH 7.2)	UV	Gallic acid, caffeic acid, vanilic acid, quercetin, rutin, catechins, myricetin	Wine quality	[100]
	TE: 50 mM boric acid + barium hydroxide (pH 8.2)	254 nm			
	BGE: 25 mM TAPS or MOPS+50 mM Tris + 15-40 mM boric acid + 5 mM b-CD + 20% methanol + 0.2% HEC (pH 8.5-8.7)				
Chip-CZE	Glass microchip CE, 74x0.05x0.02 mm separation channel	Amperometric	Chlorogenic acid, vanillic acid, gentisic acid, and ferulic acid	Wine quality	[31]
	15 mM sodium borate (pH 9.5)				
	2 kV				
MEKC	fused-silica capillary (370x0.05 mm i.d.)	UV	Cis/trans-resveratrol	Wine quality	[98]
	25 mM sodium phosphate + 25 mM sodium borate + 75 mM SDS (pH 9.0)	310 nm			
	25 °C				
CZE	fused-silica capillary (700x0.075 mm i.d.)	indirect UV	Ca, Mn, K, Na, Li	Spanish red wine differentiation	[101]
	5 mM Cia-Pack UV-Cat1 buffer + 6.5 mM alfa-hydroxyisobutyric acid + 2 mM 18-crow--6 ether	214 nm			
	22 kV				
Cereals and cereal products					
CZE	fused-silica capillary (900x0.05 mm i.d. or 500x0.02 mm i.d.)	Conductivity	Glucose, fructose, xylose mannose, arabinose, sucrose, and galactose	Adulteration of processed coffee with cereals and coffee husk	[112]
	80 mM NaOH + 0.5 mM CTAB + 30% methanol				
	- 25 kV,				
CZE	fused-silica capillary (30-60x0.05-0.075 mm i.d.)	UV	Wheat gliadins	Differentiation of cultivar	[85]
	30-60 mM boric acid (pH 8-9) + 0 - 40% acetonitrile	200 nm			
	+ 0.1 - 1% SDS				
	20-50°C, 10 kV				
CZE	fused-silica capillary (27x0.02 mm i.d.)	UV	Wheat gliadins, wheat reduced glutenins	Differentiation of cultivar	[32]
	100 mM phosphate (pH 2.5) + 0.05% HPMC + 0-20% methanol/acetonitril/isopropanol/ethyleneglycol	200 nm			
	15 kV				

(Table 4) cont.....

CE mode	CE conditions	Detection	Analytes	Fields of application	Ref.
Meat, fish and meat products					
CZE	fused-silica capillary (20x0.025-0.05 mm i.d.)	UV	Cereal storage proteins (wheat, rye, oats, barley, rice)	Cultivar differentiation	[95]
	50 mM iminodiacetic acid + 20% acetonitrile + 0.05% HPMC	200 nm			
	45°C, 30 kV				
MEKC	fused-silica capillary (27x0.05-0.075 mm), 100-800 mM Tris-borate (pH 8.5) + 0.1% SDS + 10-20% dextran	UV	Wheat glutenins	Cultivar differentiation	[90]
	400 mM Tris-CHES (pH 8.5) + 0.1% SDS + 4% poly(ethyleneoxid)	214 nm			
	200 mM Tris-HEPES (pH 7.0) + 0.1% SDS+2% linear polyacrylamide,				
	8 kV				
MEKC	fused-silica capillary (27x0.055 mm i.d.)	UV	*Glu*-1 HMW-glutenin subunits of wheat	Breeding, bread-making quality of wheat cultivars	[33]
	ProSort SDS-Protein analysis kit (Applied Biosystems Division of Perkin-Elmer)	214 nm			
	30°C, 10 mA				
CZE (combined with	fused-silica capillary (270x0.05 mm i.d.), 100 mM phosphate (pH 2.5) + 20% acetonitrile + 0.05% HPMC, 45°C, 12.5 kV	UV	Gliadins and glutenins	Differentiation between closely related lines of wheat cultivars	[34]
RP-HPLC)		200 nm			
CZE (combined with PCR)	fused-silica capillary (30 cm × 0.075 mm i.d.)	LIF	DNA fragments from *Glu-1* genes high-molecular-mass (HMW) glutenin subunits encoded by the Glu-1 genes	Selection of useful wheat genotypes with good bread-making quality	[96]
	89 mM Tris base + 89 mM boric acid + 2mM EDTA + 1.0% HEC (pH 8.5)				
	- 10 kV				
Dietary supplements					
CITP	Column coupling 90x0.3 mm (preseparation capillary) coupled with 90x0.3 mm (analytical capillary)	Conductivity and UV (254 nm)	Chondroitin sulphate A and C	Dietary supplements (support of joint health)	[35]
	LE: 5 mM HCl + 10 mM glycylglycine + 0.05 % HEC				
	TE: 10 mM citric acid, anionic mode, driving current: 200 mA (preseparation capillary) and 50 mA decreased during detection to 20 mA (analytical capillary				
CITP	Column coupling 90x0.3 mm (preseparation capillary) coupled with 90x0.3 mm (analytical capillary)	Conductivity and UV (254 nm)	Glucosamine	Dietary supplements (support of joint health)	[35]
	LE: 10 mM NH_4OH + 20 mM acetic acid				
	TE: 10 mM citric acid, cationic mode, driving current: 250 mA (preseparation capillary) and 50 mA decreased during detection to 25 mA (analytical capillary)				

(Table 4) cont.....

Meat, fish and meat products					
CE mode	**CE conditions**	**Detection**	**Analytes**	**Fields of application**	**Ref.**
CITP-CZE	CITP: Column coupling 90x0.3 mm (preseparation capillary) coupled with 90x0.3 mm (analytical capillary); LE: 20 mM NH_4OH + 40 mM acetic acid; TE: 10 mM acetic acid, cationic mode, driving current: 250 mA (preseparation capillary) and 50 mA decreased during detection to 25 mA (analytical capillary), conductivity detection	Conductivity and UV (254 nm)	Carnitine	Dietary supplements (weight loss)	[36]
	CZE after FMOC derivatization of carnitine: fused-silica capillary (360x0.05 mm i.d.), 15 mM TRIS + 25 mM phosphate, 25°C, 15 kV,				
CZE	fused-silica capillary (370x0.05 mm i.d.)	UV	Theanine	Tea based products	[37]
	50 mM sodium borate (pH 8.)	200, 254 and 280 nm			
	30 °C, 10 mA				
Food additives, spices, miscellaneous					
CITP-CZE	LE: 10 mM ammonium hydroxyde+20 mM acetic acid	UV	Egg-white lysozyme	Differentiation between egg-white and whole egg	[105]
	TE: 5 mM 6-aminocaproic acid+5 mM acetic acid	254 nm			
	BGE: 20 mM 6-aminocaproic acid + 5 mM acetic acid + 0.1% HPMC				
MEKC	fused-silica (430x0.075 mm i.d.)	UV	Benzoic and sorbic acid	Detection of preservatives	[38]
	20 mM sodium phosphate + 50 mM SDS	230 nm			
	27°C, 25 kV				
MEKC	fused-silica (520x0.075 mm i.d.)	UV	BHA, BHT, gallate esters, ascorbic acid, and erythorbic acid	Detection of antioxidants	[39]
	20 mM sodium borate + 50 mM sodium cholate + 15 mM SDS + 10% methanol	214 and 254 nm			
	25°C, 18 kV				
MEKC	fused-silica (520x0.075 mm i.d.)	UV	Saccharine, acesulfame, aspartame, alitame, benzoic acid, and sorbic acid	Detection of sweeteners and preservatives	[40]
	10 mM sodium borate + 50 mM sodium deoxycholate + 10 mM potassium dihydrogenphosphate	230 nm			
	25°C, 20 kV				
CZE	fused-silica (600x0.05 mm i.d.)	UV	Erythrosine, fast green, light green, sunset yellow amaranth, tartrazine, SF yellow, new coccine	Detection of colorants	[41]
	20 mM sodium tetraborate (pH 7.5), 0-20 mM b-CD	230 nm			

(Table 4) cont.....

Meat, fish and meat products					
CE mode	**CE conditions**	**Detection**	**Analytes**	**Fields of application**	**Ref.**
MEKC	fused-silica (400x0.15 mm i.d.)	UV	Aspartame, saccharin, acesulfame-K, caffeine, sorbic acid, benzoic acid, brillant blue, quinoline yellow, green S, sunset yellow, ponceau 4R, black PN, carmoisine	Detection of sweeteners, preservatives, and colors	[42]
	20 mM sodium hydrogencarbonate +6 2 mM SDS (pH 9.2)	190-600 nm			
	25 °C, 20 kV				

Meat and Fish Products

Authenticity and adulteration issues in fish and meat products can be divided into the following four categories [43]:

- Origin: including factors such as geographic origin, sex, farmed or wild.
- Ingredients replacement: meat or fish protein (species, cheaper cuts, blood, offal), fat, other proteins (milk, egg, *etc*).
- Treatment: such as fresh or frozen/thawed, irradiation.
- Other subjects related to the addition of water or flavorings, the cooking method.

DNA analysis and protein analysis are the two main groups of techniques that could be used for the identification of meat or fish species used in processed products. By using a CZE is possible to distinguish among eight flatfish species in 35 min [44]. Likewise, linear discriminant analysis was applied to study the SDS-CE protein profile of water and saline extract of meat, being possible to carry out the identification and differentiation of meat species, including turkey, beef, and pork.

A common fraud involving Atlantic salmon (Salmo salar), which is an expensive species, is to use rainbow trout (Oncorhynchus mykiss) to substitute it. A CZE procedure was demonstrated to be able to identify raw species based on the analysis of muscle sarcoplasmic (water-soluble) proteins. Besides, for the identification of species in canned products, muscle proteins were solubilized with urea and sodium dodecyl sulfate (SDS) [46].

Fluorescent Polymerase Chain Reaction-restriction Fragment Length Polymorphism (PCR-RFLP) was reported for the identification of meat species [47]. Using CE with fluorescence detection, the species-specific DNA fingerprints of pork, goat, and beef were generated by restriction enzyme digestion following a fluorescence-labeling PCR amplification. The identification of the different species in meat mixtures was possible as well as in cooked and autoclaved samples with similar analytical accuracy. The semiquantitative detection levels were lower than 1% for binary mixtures of goat, beef, and pork.

Mechanically recovered meat (MRM) is prepared from meat generated after the hand-deboning (application of high pressures to bones to which residual meat is still attached). This can lead to MRM containing foreign material. As a result, the protein composition in MRM and hand-deboned meat (HDM) can be different. A CGE method was applied to differentiate both products [48]. Using the hemoglobin peak area as well as its ratio to other peaks, this procedure was tested to MRM HDM mixtures of chicken meat. The results obtained enabled the differentiation of mixtures containing 7.5% MRM from those free of MRM.

Since one of the best quality descriptors of meat products is their lean meat content, meat products can be classified into several quality classes (and prices) which are understandable for the consumer. 3-Methylhistidine ([S]---methylimidazole-4- alanine, or 3-MeHis) is an amino acid typically present in meat proteins. Protein- bound 3-MeHis levels in meats vary within a narrow range being practically constant among species. Moreover, 3-MeHis residues are not found in other protein-rich foodstuffs, such as eggs, soy, or milk. Therefore, levels of protein-bound 3-MeHis in meat products can be employed as a marker of lean meat content. A capillary isotachophoresis-capillary zone electrophoresis (CITP-CZE) methodology was developed for the determination of 3-MeHis in meat products and successfully applied to the determination of lean meat content [49].

A practice to increase products' shelf-life, the binding of fat by emulsification and gelation, and water retention consists of the addition of a certain amount of legume protein (lupin, pea, or soy). Typically, a percentage between 1-6% by weight of the final product is used and thus the addition of these lower value plant proteins to meat products can be very profitable. To detect this addition, a CZE methodology to analyzed specific isoflavones in meat products has been reported [50]. Additions up to 0.1% of soy protein could be detected in meat products, even in sausages heated to a high temperature or containing hydrolyzed soy proteins. On the other hand, a CE-MS strategy was used to detect minced meat adulterated with egg-white through the analysis of egg lysozyme [51]. Even after sterilization at 120 °C for 30 min, the existing lysozyme could be detected.

Milk and Dairy Products

Extension of a product with cheaper ingredients (*i.e.*, economic fraud) usually does not pose a health risk for consumers. However, in the particular case of milk and milk products, this statement may not be valid. Allergy to specific milk proteins means that sensitive consumers could suffer severe consequences if a type of milk is counterfeited with another different one. Authenticity topics related to milk and dairy products include:

- Addition of ingredients of dairy or non-dairy origin, including milk of different species, watering of milk, non-dairy proteins, animal or vegetable fats.
- Non-compliance with certain technological processes such as heat treatment or cheese ripening.
- Non-compliance with legal requirements in terms of composition or geographical origin.

  In this regard, to carry out the adulteration detection of milk and dairy products, the following components may be useful:
- Phospholipids, fatty acids, and sterols for the detection of non-dairy fats or buttermilk added to milk.
- Whey proteins, caseins, and glycomacropeptide for the differentiation of species, whey, and non-dairy proteins additions, and heat load.
- Furosine, lysinoalanine, and HMF presence to detect excessive heat load.

A common fraud is the addition of cheaper cow milk to ewe milk. To quantify the amount of ewe milk in ovine/bovine milk mixtures Trimboli *et al.* [26] developed a CE method. By using this methodology the recognition of ewe milk adulteration using cow milk was possible. To that aim, skimmed milk proteins were separated in alkaline buffer. Under optimized CE analytical conditions, the protein profiles of ovine and bovine milk were different, with a specific ewe peak showing a reproducible migration zone in ovine/bovine mixtures. As low as 5% of cow milk was detectable by this method.

CE is one of the most applied techniques for the separation of milk proteins in dairy products authentication. CE patterns of both whey proteins and casein can be employed for authentication. Whey proteins (lactoglobulins) corresponding to different animal species were used as authenticity markers after separation by CZE [52]. The use of sodium borate at pH 9.2 as background electrolyte enabled to fully resolve all the peaks of interest in methyl-silanized capillaries. The minimum amount detectable of cows' milk was 2% and 4% in milk mixtures and cheeses respectively. Further improvement gained advantage of a CE method for the determination of cow's milk in non-bovine cheese [53]. Proper separation of lactoglobulins in acidic, isoelectric buffers was achieved. This novel procedure allowed to determine 1% of cow milk in goat and ewe cheeses. Different CZE patterns of the casein fraction of bovine, ovine and caprine milk enabled to carry out the quantitative analysis of milk of each species in binary and ternary mixtures by multivariate regression analysis (prediction error lower than 3%) [54 - 56].

Whey rennet is a low value by-product of cheese making produced in high volumes. Thus, this is a potential material to be fraudulently added to whey to dairy products. Caseinomacropeptide (CMP) is produced from the cleavage of kappa-casein during renneting of milk, and can be considered as a marker of the presence of rennet whey. This compound was analyzed by a CZE method based on the use of citrate buffer and UV detection at 214 nm. Under the best conditions, it was possible to achieve detection limits as low as 0.4% of rennet whey solids [57 - 59].

Another different approach, based on the detection of fatty acids, for the detection of whey in milk has been presented by Mendes *et al.* [60]. C14:0, C16:0, C18:0, C18:1, C18:2 and C18:3 fatty acids were quantified by CZE in different adulterated milk samples. The correlation coefficient obtained between the values of whey added and the experimental values determined using this approach was 0.973. This methodology was useful to evaluate the adulteration of milk with whey, which contributes to the quality control of milk in the dairy industry.

A potential issue, interfering with the methods developed for rennet detection, is the existence of high proteolitic activity of psychrotropic bacteria during long cold storage of milk that may produce peptides similar to CMP [63]. The acidic hydrolysis of Amadori compounds (lactulose lysine, fructose lysine, and maltulose lysine,) may generate furosine, an amino acid derivative. This compound can be used as a marker of heat-load of milk, for determining whether milk powder has been added to fluid milk, or as a marker of reconstituted milk. In this sense, CE is a feasible tool to detect furosine in dairy products. A CE procedure based on the use of a 0.1 M phosphate buffer containing the additive hexadecyltrimethylammonium bromide (pH 7.0), allowed the determination of furosine in less than 10 min at ppm levels (LOQ was 0.5 mg/l) [61, 62]. Moreover, lactulose is formed in milk during heating by isomerisation of lactose. Thus, the content of lactulose was proposed as an analytical index to distinguish UHT (Ultra High Temperature Treated) from sterilized milk [63]. HMF is another heat-load marker of milk that may potentially be analyzed by MEKC (using a phosphate buffer containing SDS at pH 7.5) showing similar analytical capabilities as HPLC [64].

Fruit-based Products

Fruit-based products can be divided into two groups: nectars, purées, and fruit juices, and, and jams, jellies, marmalades, and conserves. The main authenticity issue of these products is the replacement of authentic material with cheaper alternatives. In juices is typical the addition of water, sugar, or fruit-derived extenders (pulpwash, cheaper fruit). In the case of jams and fruit preserves, the main difficulty is to carry out the accurate determination of the fruit content. For instance, the bulking of an expensive fruit (strawberry) with a relatively cheap one (apple pulp).

Regarding fruit-based products, authenticity issues include:

- the incorrect declaration of the fruit employed
- misrepresentation of the country of origin
- undeclared country of origin
- the addition of sugar is not declared
- the addition of other fruit types is not declared
- the addition of pulpwash and/or peel extract, organic acids, or undeclared compounds (colouring or artificial flavours)

Bearing this in mind, fruit-based products manufacturers need the appropriate methodologies to ensure that the type and the quantity of their supplied raw material is as declared. In this line, guidelines for assessing fruit-based product authenticity have been established. Even though different analytical methods are already available, new methodologies that enable lower cost and more rapid analysis are required. On this point, CE can be considered a potential alternative.

The composition of a fruit juice is defined by a set of physicochemical parameters which are typical of each fruit type. However, the chemical composition can considerably vary because it is derived from a natural product. Here, factors such as fruit variety, climatic conditions, geographic origin, or the technological processes used for juice production, can have a high influence. To evaluate the quality and authenticity of fruit juice, the AIJN (Association of the Industry of Juices and Nectars from Fruits and Vegetables) established A Code of Practice that includes guidelines and acceptable ranges and/or values for each parameter as quality and authenticity requirements [65]. CE enables the analysis of a high variety of fruit product components simultaneously including inorganic cations (minerals), organic acids, amino acids, phenolic compounds, *etc*. To decide about the authenticity of a fruit- based product a combination of different parameters should be employed. Some of these parameters are shown in Table **5**.

Table 5. Parameters employed for the authenticity evaluation of fruit-based products.

Parameter	Analyte	Applicability
Minerals	K, Na, Ca, Mg	pulpwash, water addition
Organic acids	citric, D-isocitric, L-malic, tartaric, fumaric	addition of another fruit type
Amino acids	Glutamic acid, pyroglutamic acid, aspartic acid, proline	fruit type
Phenolic compounds	Anthocyanidins, flavonols, flavones, flavanones,	fruit types and cultivars differentiation, admixtures of different fruit types
Sugar profile	Glucose, fructose, sucrose, sorbitol	fruit type added sugar
Oligosaccharides	di-, tri-, and tetrasaccharides	addition of invert syrup (beet, cane, and high fructose corn syrup)

As it can be seen in the table, the mineral content along with other parameters can be indicative of the geographical origin, variety, or processing technology of citrus juices. For instance, high content of Ca and an increase in the flavanone glycosides amount could be an indicator of pulpwash addition. Also, a low mineral content can be due to a water dilution. The separation of alkaline and alkali metal cations and ammonium was possible by CE using indirect UV detection [66, 67]. Also, the use of a CITP methodology with conductometric detection enabled to achieve the simultaneous separation of these compounds. Because potassium and ammonium effective mobilities are almost equal in aqueous electrolytes, their separation can be achieved using leading electrolyte at pH values higher than 8 when ammonium reduced its effective mobility. Another option is the use of non-ionic additives (crown ether [68] or polyethylene glycol [69]) in the electrolyte. The rapid analysis of Ca, Na, K, and Mg in orange juice in just 5 min with CE and indirect UV was described by Weston *et al.* [70]. The use of CE for the analysis of anions and cations in food and beverages, including fruit-based products, was reviewed by Sádecká and Polonský [71].

Organic acids are present in different fruit types in varying quantities which provide a unique organic acid profile to each fruit. These profiles can be used to determine whether another type of fruit has been added. For instance, the presence of tartaric acid in pineapple juice may indicate the addition of grape juice. Organic acid profiles can also be employed to determine juice authenticity. The content of isocitric acid and the ratio isocitric acid/citric acid are relevant to ascertain the quality and authenticity of fruit products. The content of citric and isocitric acids in citrus juice was determined by using both a CZE methodology [17] and a CITP method [72, 73]. A good agreement between the level of isocitric acid determined by CITP and enzymic method was obtained [73]. In well-prepared (authentic and not decayed) apple juice, the levels of fumaric acid do not exceed 3 mg/L, so that a higher value indicates microbial spoilage of juices or the processing of decayed fruits. The addition of synthetic malic acid is another source of fumaric acid is since it contains fumaric acid as a minor contaminant. Using a CITP/CZE procedure, it was possible to determine fumarate in apple juice with a limit of detection 0.1 mg/L [74]. Additionaly, the simultaneous separation of organic acids and minerals was also possible. In this line, Xiong and Li [75] proposed a CZE methodology for the analysis of organic acids and alkali metals in fruit juices. The separation of twelve anions and four cations in 6 min with detection limits from 0.08 to 5 mg/L was achieved by this methodology.

Free amino acids also establish a characteristic fruit pattern. For example, berry fruit does not contain proline (or a little amount) so that a high content of this amino acid can be considered as a marker of the addition of grape or cheaper pear juice (which are rich in proline). Cancalon *et al.* determined for the first time by CE three different amino acids (tyrosine, phenylalanine, and tryptophan) as well as biogenic amines, organic acids, polyphenols, and flavonoids in citrus juices, using borate buffer and UV detection [76]. The comparison of amino acid profile in orange juices by CE was performed by Klampfl *et al.* using acidic phosphate buffer containing alkyl sulfonic acid, and UV detection [77]. Also, MEKC mode with laser-induced fluorescence (LIF) detection was employed to perform the chiral analysis of different amino acids in orange juices [78]. More than 90% of the amino acid content found in orange juices is due to the presence of DL-Pro, DL-Arg, DL-Ser, DL-Asn, DL-Glu, DL-Ala, and DL-Asp. To carry out the detection of these seven amino acids they were derivatized with fluorescein isothiocyanate (FITC). The D enantiomers of Arg, Ala, Glu, and Asp were detected in fresh orange juices as natural constituents, in several orange concentrates from different geographical origins, and in pasteurized juices. This MEKC-LIF methodology was also employed to study the content of D-amino acids when orange juices were submitted to different temperature treatments. The results obtained showed that an increase in the thermal treatment gave rise to a higher racemization rate of D-Asp and D-Ala.

In most fruits, it is possible to find quantities up to 1000 mg/kg of phenolic compounds. As previously described for organic acids and amino acids, each plant also has a characteristic phenolic profile that can be employed to differentiate between fruit types and cultivars. For instance, hesperidin and narirutin are found in sweet oranges but they do not contain neohesperidine nor naringin. The presence of this last compound in orange juice may indicate the addition of grapefruit (where it is a bitter component). In this line, a CE methodology was developed to perform the analysis of different citrus products and the detection of grapefruit juice addition into orange juices [50]. The determination of phenolic compounds in pulpwash (a lower- quality juice product), showed that some of them were in much larger amounts than in pure juice. Among these compounds tryptophan, narirutin, hesperidin, synephrine, feruloyl, didymin, and sinapyl glucose are included, as well as at least four unidentified molecules, many having coumarin related spectra. Based on these differences, the development of a CE methodology for the quantitative determination of pulpwash added to orange juice was carried out.

The main carbohydrates found in fruit, are sucrose, glucose, and fructose. Their contents vary according to fruit type. The comparison of the sugar profile of a sample with the profile of a determined fruit type will provide information on a crude adulteration. Sorbitol, which is present in some fruits such as apple and grape, but not (or only in trace amount) in strawberries or other red fruits, is another sugar that can be used as an authenticity marker. The use CE methods for sugars analysis is limited since they are uncharged compounds and do not contain chromophores or fluorophores groups in their structures. The first drawback can be solved using indirect UV detection whereas the second one can be overcome by employing different strategies, such as complexation with anions or cations; ionization by increasing pH, or chemical derivatization [79]. To show an example, Soga and Serwe reported a CE methodology with indirect UV detection for the determination of carbohydrates in orange juice and pickled apricots [80]. To negatively charge every carbohydrate, a highly alkaline pH buffer was employed. Under the best experimental conditions, a total of 28 carbohydrates, including sugar alcohols, amino sugars, monosaccharides, and disaccharides, were separated in 25 min. Another simple CE strategy developed by Cabálková *et al.*, based on the use of phosphate/naphthol blue- black at pH 12.5 as buffer and indirect UV detection, was employed for the simultaneous analysis of glucose, sucrose, and fructose in juices [81]. The sugar contents determined in commercial juices were in agreement with those obtained by LC.

Authentication of olive oil by use of different modes of CE is summarized in a well- arranged review article published by Monasterio *et al.* [82]. Phenolic compounds, fatty acids, tocopherols, chlorophylls, betaines, proteins, and amino acids were used as markers of olive oils.

Cereals and Cereals Products

Cereal authentication is fundamentally related to variety since the properties of different grain varieties differ significantly. Thus, authenticity issues in this field are:

- the contamination of grains with cereals of similar appearance

- the contamination of grains with the wrong species
- the contamination of a specified wheat type with a different type
- the detection of cereals genetically modified.

Usually, the methodologies employed to identify different varieties of cereals are focused on specific compounds, such as protein and DNA [5]. CE methodologies can be used to characterize the protein composition of specific cereal grains or a flour sample in terms of the prolamines and/or glutenins profiles. The use of CE as analytical platform enables a more precise wheat differentiation than that obtained by classical methods (PAGE, SDS-PAGE). Thus, CE could be a tool for varietal identification and prediction of quality [83, 84]. CZE and MEKC with SDS are the two CE modes that have been applied to carry out cereal proteins analysis. The first one is analogous to PAGE and the second to SDS-PAGE.

To carry out the separation of cereal proteins by CZE, it is possible to employ different buffer systems. Bietz and Schmalzried demonstrated for the first time the potential of CZE to separate cereal proteins. Based on wheat gliadins patterns they could differentiate three wheat cultivars [85]. To do that, the authors employed an uncoated fused-silica capillary and an alkaline borate buffer plus 1% SDS and 20% acetonitrile. Under these conditions, up to twenty peaks were resolved within 25 min. A low pH buffer (namely aluminium lactate buffer (pH 2.3)) was employed by Werner *et al.* to separate wheat gliadins [86]. In this case, the analysis was performed using a coated capillary with a positively charged polymer which gives rise to a positive charge surface that generates a reversed electroosmotic flow and avoids the adsorption of positively charged proteins in the capillary wall. In this research line, a relevant work was performed by Bietz and Schmalzried in which it was demonstrated that a low pH phosphate buffer could be applied to the differentiation of wheat cultivars with high resolution [85]. The use of isoelectric buffers has substantially improved the separation of cereal proteins in terms of high-resolution separations in short analysis times. The first application of these novel buffers was presented by Capelli *et al.* who employed them for the separation of wheat gliadins [87]. In their work, the authors used 40 mM aspartic acid plus 7 M urea as buffer to carry out the differentiation of durum and bread wheat cultivars in a time below 15 min with good reproducibility. Low viscosity HEC (hydroxyethylcellulose) was employed as a dynamic buffer additive to prevent protein adsorption onto the capillary wall. The use of either beta-alanine/phosphate or glycine/phosphate buffers instead of the commonly used sodium phosphate buffers enabled to achieve faster separations (~ 10 min) with little loss in resolution [88]. 100 mM phosphate/glycine plus 20% acetonitrile and 0.05% HPMC (hydroxypropylmethylcellulose) was employed by

Lookhart *et al.* as separation buffer in a CZE methodology to differentiate barley cultivars [89].

The first use of an SDS-CE method for the analysis of wheat proteins using a commercial linear polyacrylamide polymer/buffer system (ProSort) was reported by Werner *et al.* [86]. Also, a systematic study in which different polymers (dextran, polyethyleneoxid, linear polyacrylamide) were employed to coat the capillaries was reported [90]. With this system, a very high resolution of wheat high-molecular- weight glutenin subunits (HMM GS) was possible. The patterns of these subunits were related with the bread-making quality of wheat [86, 91].

CE methodologies also enable to perform the differentiation of maize genotypes [92 - 94], sorghum genotypes [94], rice cultivars [95], wheat cultivars [92, 95], oats cultivars [95], and barley cultivars [89, 95]. The separation of storage proteins of oats, rice, barley and wheat, was performed using a buffer based on an isoelectric system with an extremely high resolution in less than 3 min. This methodology could be used to perform the high-throughput analysis (180-400 samples/24 h) of cereal proteins.

Finally, Salmanowicz and Moczulski combined CZE with Polymerase Chain Reaction (PCR) [96]. Taking into account that the quantity and composition of HMW glutenin subunits encoded by the *Glu-1* genes are closely related to the bread-making properties of wheat, these authors optimized a PCR method to carry out the identification of bread wheat genotypes carrying HMW glutenin allele composition of *Glu-1* complex loci by CE-LIF. The application of this CE methodology to obtain a rapid and accurate identification of molecular markers of *Glu-1* genes may be considered a potent alternative to standard procedure for the early selection of wheat genotypes with good bread-making quality.

Miscellaneous

Different analytical methods were developed in order to assess wine authenticity. They enabled the determination of different minerals (Ca, Mg, Cu, Fe, *etc*), organic acids (citric acid, malic acid, tartaric acid, *etc.*), and phenolic compounds (catechins, flavonoids, *etc.*). The determination of polyphenols in wines from different denominations of origin in Tenerife was achieved by CZE [97]. Characteristic polyphenols fingerprints are useful for wine differentiation. Phenolics are relevant compounds in the case of grape juices and wines since they contribute to different properties such as flavour, colour, oxidation, and other reactions. Analysis of phenolic compounds such quercetin, epicatechin, rutin, genistic and gallic acids, and cis and/or trans-resveratrol in wine by CE was reported [98 - 100]. Nuñez *et al.* [101] described the possibility of differentiating red wines from Ribeira Sacra Spanish using CZE to determine different metal ions content (Na, Ca, Mg, K, Mn, and Li).

MEKC showed its potential for separating a group of flavonoids such as naringenin, quercetin, and apigenin, among others, and CE was shown to have great potencial for the analysis of flavonoids in honey, and in the study of its floral and geographical origin [102]. Bronze *et al.* showed the ability of CE to achieve the analysis of brandy and wood extracts [103, 104]. The method enabled the separation of polyphenolic compounds using 100 mM borate buffer with 5% ethanol at pH 9.2, which is an interesting approach to study the authenticity and quality of brandies.

The analysis of lysozyme was achieved by a coupled capillary isotachophoresis-CZE methodology that was employed to detect eggs in different foods such as egg- melange or egg-liqueur, and pasta [105 - 107].

Regarding the determination of food additives such as colourings, sweeteners, antioxidants, or preservatives among others, numerous CE methods were developed and reviewed [108].

CE has shown to be a powerful analytical technique in food authentication mainly in specific areas such as peptide or protein analysis. Nevertheless, it should be taken into account that the use of protein patterns to carry out the identification of plant or animal species presents a higher difficulty once food has been processed and morphological characteristics are not visible. As a result, in the case of highly processed food products where proteins become denatured, CE presents limitations. In these cases, DNA-based methods are considered the most appropriate for species identification as DNA remains detectable even after a heavy food processing. Combining PCR with CE is a very promising approach to be applied to food authentication as reported in different works [47, 96, 109 - 111]. As an example, a methodology based on the extraction and amplification of a specific DNA fragment by PCR and its subsequent analysis by CGE-LIF enabled the detection of up to 0.01% genetically modified maize in a flour sample [111].

CONCLUDING REMARKS

The development of methodologies for assessing food authenticity or food adulteration is endless and a constant battle with the counterpart of dishonest producers. The choice of an appropriate analytical method depends on the case of food authentication. A broad range of analytical techniques including separation (chromatography and electrophoresis), spectroscopic (IR, Raman, NMR), spectrometric (MS, DART-MS, MALDI-TOF-MS), molecular biological (PCR, DNA sequencing), and electronic sensors (E-eye, E-nose, E-tongue) are used for this purpose. In this wide range of techniques, CE still has a place due to its high resolution, speed, and low running costs.

CONSENT FOR PUBLICATION

Not applicable.

CONFLICT OF INTEREST

The author declares no conflict of interest, financial or otherwise.

ACKNOWLEDGEMENTS

The author is grateful for the financial support provided from specific university research (MŠMT No 21-SVV/2020).

REFERENCES

[1] Gray, C. *History of food adulteration,* http://www.readymealsinfo.com/articles/foodadult.htm accessed in December 2004.

[2] Food adulteration http://reference.allrefer.com/encyclopedia/F/foodadul.html accessed in October 2004

[3] Accum, F. *A Treatise on Adulterations of Food, and Culinary Poisons: Exhibiting the Fraudulent Sophistications of Bread, Beer, Wine, Spiritous Liquors, Tea, Coffee, Cream, Confectionery, Vinegar, Mustard, Pepper, Cheese, Olive Oil, Pickles, and Other Articles Employed in Domestic Economy and Methods of detecting them. AB'M Small*; Philadelpia, **1820**.

[4] Ashurst, P.R.; Denis, M.J., Eds. *Analytical method of food authentication*; Blackie Academic Professional: London , **1998**.

[5] Lees, M., Ed. *Food authenticity Issues and Methodologies*; Eurofins Scientific, Nantes, France: Nantes, France , **1998**.

[6] Lees, M., Ed. Food authenticity and traceability. WP, Cambridge, UK, 2003

[7] Čížková, H.; Ševčík, R.; Rajchl, A.; Pivoňka, J.; Voldřich, M. Trends in Food Authenticity and Detection of Food Adulteration. *Chem. Listy,* **2012**, *106*, 903-910.

[8] Kvasnička, F. Capillary electrophoresis in food authenticity. *J. Sep. Sci.,* **2005**, *28*(9-10), 813-825. [http://dx.doi.org/10.1002/jssc.200500054] [PMID: 16013809]

[9] The Web of Science, Institute of Scientific Information, http://www.isinet.com/isi/products/citation/wos/index.html

[10] Medina, S.; Perestrelo, R.; Silva, P.; Pereira, J.A.M.; Câmara, J.S. Current trends and recent advances on food authenticity technologies and chemometric approaches. *Trends Food Sci. Technol.,* **2019**, *85*, 163-176.
[http://dx.doi.org/10.1016/j.tifs.2019.01.017]

[11] Böhme, K.; Calo-Mata, P.; Barros-Velázquez, J.; Ortea, I. Recent applications of omics-based technologies to main topics in food authentication. *TrAc. Trends Analyt. Chem.,* **2019**, *110*, 221-232. [http://dx.doi.org/10.1016/j.trac.2018.11.005]

[12] Cifuentes, A. Food analysis and foodomics. *J. Chromatogr. A,* **2009**, *1216*(43), 7109-7109. [http://dx.doi.org/10.1016/j.chroma.2009.09.018] [PMID: 19765718]

[13] Bodor, R.; Zuborová, M.; Olvecká, E.; Madajová, V.; Masár, M.; Kaniansky, D.; Stanislawski, B. Isotachophoresis and isotachophoresis-zone electrophoresis of food additives on a chip with column-coupling separation channels. *J. Sep. Sci.,* **2001**, *24*, 802-809. [http://dx.doi.org/10.1002/1615-9314(20010901)24:9<802::AID-JSSC802>3.0.CO;2-1]

[14] Masár, M.; Zuborová, M.; Kaniansky, D.; Stanislawski, B. Determination of oxalate in beer by zone electrophoresis on a chip with conductivity detection. *J. Sep. Sci.,* **2003**, *26*, 647-652. [http://dx.doi.org/10.1002/jssc.200301375]

[15] Obeid, P.J.; Christopoulos, T.K.; Ioannou, P.C. Rapid analysis of genetically modified organisms by in-house developed capillary electrophoresis chip and laser-induced fluorescence system. *Electrophoresis,* **2004**, *25*(6), 922-930. [http://dx.doi.org/10.1002/elps.200305772] [PMID: 15004856]

[16] Martín, A.; Vilela, D.; Escarpa, A. Food analysis on microchip electrophoresis: an updated review. *Electrophoresis,* **2012**, *33*(15), 2212-2227. [http://dx.doi.org/10.1002/elps.201200049] [PMID: 22887146]

[17] Manz, A.; Graber, N.; Widmer, H.M. Miniaturized total chemical analysis systems: A novel concept for chemical sensing. *Sens. Actuators B Chem.,* **1990**, *1*, 244-248. [http://dx.doi.org/10.1016/0925-4005(90)80209-I]

[18] Everaerts, F.M.; Beckers, J.L.; Verheggen, T.P.E.M. *Isotachophoresis: Theory, Instrumentation and Applications*; Elsevier: Amsterdam, **1976**.

[19] Foret, F.; Křivánková, L.; Boček, P. *Capillary zone electrophoresis*; VCH, Weinheim , **1993**.

[20] Weinberger, R. *Practical Capillary Electrophoresis*; Academic Press: New York , **2000**.

[21] The Web of Science, Institute of Scientific Information, http://www.isinet.com/isi/products/citation/wos/index.html

[22] Rodriguez-Ramirez, R.; Vallejo-Cordoba, B.; Mazorra-Manzano, M.A.; Gonzales-Cordova, A.F. Soy detection in canned tuna by PCR and capillary electrophoresis. *Anal. Methods,* **2015**, *7*, 530-537. [http://dx.doi.org/10.1039/C4AY01423E]

[23] Walker, C.C.; Lassitter, C.L.; Lynn, S.N.; Ford, C.B.; Rademacher, K.R.; Ruple, A.D.; Bell, J.W. Rapid Seafood Species Identification Using Chip-Based Capillary Electrophoresis and Protein Pattern Matching. *J. AOAC Int.,* **2017**, *100*(5), 1500-1510. [http://dx.doi.org/10.5740/jaoacint.17-0178] [PMID: 28721842]

[24] Mazorra-Manzano, M.A.; Torres-Llanez, M.J.; Gonzales-Cordova, A.F.; Vallejo-Cordoba, B. A Capillary Electrophoresis Method for the Determination of Hydroxyproline as a Collagen Content Index in Meat Products. *Food Anal. Methods,* **2012**, *5*, 464-470. [http://dx.doi.org/10.1007/s12161-011-9271-8]

[25] Trimboli, F.; Costanzo, N.; Lopreiato, V.; Ceniti, C.; Morittu, V.M.; Spina, A.; Britti, D. Detection of buffalo milk adulteration with cow milk by capillary electrophoresis analysis. *J. Dairy Sci.,* **2019**, *102*(7), 5962-5970. [http://dx.doi.org/10.3168/jds.2018-16194] [PMID: 31056332]

[26] Trimboli, F.; Morittu, V.M.; Cicino, C.; Palmieri, C.; Britti, D. Rapid capillary electrophoresis approach for the quantification of ewe milk adulteration with cow milk. *J. Chromatogr. A,* **2017**, *1519*, 131-136. [http://dx.doi.org/10.1016/j.chroma.2017.08.075] [PMID: 28888682]

[27] Navarro, M.; Núñez, O.; Saurina, J.; Hernández-Cassou, S.; Puignou, L. Characterization of fruit products by capillary zone electrophoresis and liquid chromatography using the compositional profiles of polyphenols: application to authentication of natural extracts. *J. Agric. Food Chem.,* **2014**, *62*(5), 1038-1046. [http://dx.doi.org/10.1021/jf404776d] [PMID: 24432703]

[28] Navarro-Pascual-Ahuir, M.; Lerma-García, M.J.; Simó-Alfonso, E.F.; Herrero-Martínez, J.M. Rapid differentiation of commercial juices and blends by using sugar profiles obtained by capillary zone electrophoresis with indirect UV detection. *J. Agric. Food Chem.*, **2015**, *63*(10), 2639-2646. [http://dx.doi.org/10.1021/acs.jafc.5b00122] [PMID: 25719749]

[29] Saavedra, L.; Rupérez, F.J.; Barbas, C. Capillary electrophoresis for evaluating orange juice authenticity: a study on Spanish oranges. *J. Agric. Food Chem.*, **2001**, *49*(1), 9-13. [http://dx.doi.org/10.1021/jf0004762] [PMID: 11170552]

[30] Cancalon, P.F.; Bryan, C.R. Use of capillary electrophoresis for monitoring citrus juice composition. *J. Chromatogr. A*, **1993**, *652*(2), 555-561. [http://dx.doi.org/10.1016/0021-9673(93)83278-Z] [PMID: 8287142]

[31] Scampicchio, M.; Wang, J.; Mannino, S.; Chatrathi, M.P. Microchip capillary electrophoresis with amperometric detection for rapid separation and detection of phenolic acids. *J. Chromatogr. A*, **2004**, *1049*(1-2), 189-194. [http://dx.doi.org/10.1016/S0021-9673(04)01275-0] [PMID: 15499932]

[32] Lookhart, G.L.; Bean, S.R. Improvements in cereal protein separations by capillary electrophoresis: Resolution and reproducibility. *Cereal Chem.*, **1996**, *73*, 81-87.

[33] Sutton, K.H.; Bietz, J.A. Variation among high molecular weight subunits of glutenin detected by capillary electrophoresis. *J. Cereal Sci.*, **1997**, *25*, 9-16. [http://dx.doi.org/10.1006/jcrs.1996.9999]

[34] Bean, S.R.; Lookhart, G.L. Separation of wheat proteins by two-dimensional reversed-phase high-performance liquid chromatography plus free zone capillary electrophoresis. *Cereal Chem.*, **1997**, *74*, 758-780. [http://dx.doi.org/10.1094/CCHEM.1997.74.6.758]

[35] Václavíková, E.; Kvasnička, F. Isotachophoretic Determination of Glucosamine and Chondroitin Sulphate in Dietary Supplements. *Czech J. Food Sci.*, **2013**, *13*, 55-65. [http://dx.doi.org/10.17221/107/2012-CJFS]

[36] Prokorátová, V.; Kvasnička, F.; Sevcík, R.; Voldřich, M. Capillary electrophoresis determination of carnitine in food supplements. *J. Chromatogr. A*, **2005**, *1081*(1), 60-64. [http://dx.doi.org/10.1016/j.chroma.2005.03.033] [PMID: 16013599]

[37] Čížková, H.; Voldřich, M.; Mlejnecká, J.; Kvasnička, F. Authenticity evaluation of tea-based products. *Czech J. Food Sci.*, **2008**, *26*, 259-267. [http://dx.doi.org/10.17221/10/2008-CJFS]

[38] Pant, I.; Trenerry, V.C. The determination of sorbic acid and benzoic acid in a variety of beverages and foods by micellar electrokinetic capillary chromatography. *Food Chem.*, **1995**, *53*, 219-226. [http://dx.doi.org/10.1016/0308-8146(95)90793-7]

[39] Boyce, M.C.; Spickett, E.E. Separation of food grade antioxidants (synthetic and natural) using mixed micellar electrokinetic capillary chromatography. *J. Agric. Food Chem.*, **1999**, *47*(5), 1970-1975. [http://dx.doi.org/10.1021/jf980959u] [PMID: 10552480]

[40] Thompson, C.O.; Trenerry, V.C.; Kemmery, B. Determination of cyclamate in low joule foods by capillary zone electrophoresis with indirect ultraviolet detection. *J. Chromatogr. A*, **1995**, *694*, 507-514. [http://dx.doi.org/10.1016/0021-9673(94)01224-3] [PMID: 7704194]

[41] Razee, S.; Tamura, A.; Masujima, T. Improvement in the determination of food additive dyestuffs by capillary electrophoresis using β-cyclodextrin. *J. Chromatogr. A*, **1995**, *715*, 179-188. [http://dx.doi.org/10.1016/0021-9673(95)00586-C]

[42] Frazier, R.A.; Inns, E.L.; Dossi, N.; Ames, J.M.; Nursten, H.E. Development of a capillary electrophoresis method for the simultaneous analysis of artificial sweeteners, preservatives and colours in soft drinks. *J. Chromatogr. A*, **2000**, *876*(1-2), 213-220.

[http://dx.doi.org/10.1016/S0021-9673(00)00184-9] [PMID: 10823516]

[43] Hargin, K.D. Authenticity issues in meat and meat products. *Meat Sci.,* **1996**, *43S1*, 277-289. [http://dx.doi.org/10.1016/0309-1740(96)00072-1] [PMID: 22060658]

[44] Gallardo, J.M.; Sotelo, G.; Pineiro, C.; Pérez-Martin, R.I. Use of Capillary Zone Electrophoresis for Fish Species Identification. Differentiation of Flatfish Species. *J. Agric. Food Chem.,* **1995**, *43*, 1238-1244.
[http://dx.doi.org/10.1021/jf00053a022]

[45] Vallejo-Córdoba, B.; Cota-Rivas, M. Meat species identification by linear discriminant analysis of capillary electrophoresis protein profiles. *J. Capillary Electrophor.,* **1998**, *5*(5-6), 171-175.
[PMID: 10812424]

[46] Acuña, G.; Ortiz-Riaño, E.; Vinagre, J.; García, L.; Kettlun, A.M.; Puente, J.; Collados, L.; Valenzuela, M.A. Application of capillary electrophoresis for the identification of Atlantic salmon and rainbow trout under raw and heat treatment. *J. Capill. Electrophor. Microchip Technol.,* **2008**, *10*(5-6), 93-99.
[PMID: 18982909]

[47] Sun, Y.L.; Lin, C.S. Establishment and application of a fluorescent polymerase chain reaction-restriction fragment length polymorphism (PCR-RFLP) method for identifying porcine, caprine, and bovine meats. *J. Agric. Food Chem.,* **2003**, *51*(7), 1771-1776.
[http://dx.doi.org/10.1021/jf020860u] [PMID: 12643628]

[48] Day, L.; Brown, H. Detection of mechanically recovered chicken meat using capillary gel electrophoresis. *Meat Sci.,* **2001**, *58*(1), 31-37.
[http://dx.doi.org/10.1016/S0309-1740(00)00127-3] [PMID: 22061916]

[49] Kvasnička, F. Isotachophoretic determination of 3-methylhistidine. *J. Chromatogr. A*, **1999**, *838*, 191-195. [[5]0] Mellenthin, O.; Galensa, R., Analysis of polyphenols using capillary zone electrophoresis and HPLC: Detection of soy, lupin, and pea protein in meat products. *J. Agric. Food Chem.,* **1999**, *47*, 594-602.

[51] Simo, C.; Gonzales, N.; Roman, J.S.; Barbas, C.; Cifuentes, A. Capillary electrophoresis-mass spectrometry of basic proteins using a new physically adsorbed polymer coating. Some applications in food analysis. *Electrophoresis,***2004**, *25*, 2056-2064. [[5]2] Cartoni, G.; Coccioli, F.; Jasionowska, R.; Masci, M., Determination of cows' milk in goats' milk and cheese by capillary electrophoresis of the whey protein fractions. *J. Chromatogr. A*, **1999**, *846*, 135-144.

[53] Herrero-Martínez, J.M.; Simó-Alfonso, E.F.; Ramis-Ramos, G.; Gelfi, C.; Righetti, P.G. Determination of cow's milk in non-bovine and mixed cheeses by capillary electrophoresis of whey proteins in acidic isoelectric buffers. *J. Chromatogr. A*, **2000**, *878*(2), 261-271.
[http://dx.doi.org/10.1016/S0021-9673(00)00299-5] [PMID: 10866071]

[54] Lee, S.C.; Chen, M.C.; Lin, C.W. Detection of cows' milk in goats' milk by capillary zone electrophoresis. *Aust. J. Dairy Technol.,* **2001**, *56*, 24-27.

[55] Molina, E.; Martin-Alvarez, P.J.; Ramos, M. Analysis of cows', ewes' and goats' milk mixtures by capillary electrophoresis: quantification by multivariate regression analysis. *Int. Dairy J.,* **1999**, *9*, 99-105.
[http://dx.doi.org/10.1016/S0958-6946(99)00028-X]

[56] Molina, E.; De Frutos, M.; Ramos, M. Capillary electrophoresis characterization of the casein fraction of cheeses made from cows', ewes' and goats' milks. *J. Dairy Res.,* **2000**, *67*(2), 209-216.
[http://dx.doi.org/10.1017/S0022029900004167] [PMID: 10840674]

[57] van Riel, J.; Olieman, C. Determination of caseinomacropeptide with capillary zone electrophoresis and its application to the detection and estimation of rennet whey solids in milk and buttermilk powder. *Electrophoresis,* **1995**, *16*(4), 529-533.
[http://dx.doi.org/10.1002/elps.1150160187] [PMID: 7588522]

[58] Cherkaoui, S.; Doumenc, N.; Neeser, J.R.; Veuthey, J.L. Development of a capillary zone electrophoresis method for caseinoglycomacropeptide determination. *J. Chromatogr. A*, **1997**, *790*, 195-205. [59] Recio, I.; Risc, M. R.; Fandino, R. L.; Olano, A.; Ramos, M., Detection of rennet whey solids in UHT milk by capillary electrophoresis. *Int. Dairy J.*, **2000**, *10*, 333-338.

[60] de Oliveira Mendes, T.; Porto, B.L.S.; Bell, M.J.V.; Perrone, I.T.; de Oliveira, M.A.L. Capillary zone electrophoresis for fatty acids with chemometrics for the determination of milk adulteration by whey addition. *Food Chem.*, **2016**, *213*, 647-653. [http://dx.doi.org/10.1016/j.foodchem.2016.07.035] [PMID: 27451230]

[61] De Block, J.; Merchiers, M.; Mortier, L.; Braekman, A.; Ooghe, W.; van Renterghem, E. Monitoring nutritional quality of milk powders: capillary electrophoresis of the whey protein fraction compared with other methods. *Int. Dairy J.*, **2003**, *13*, 87-94. [http://dx.doi.org/10.1016/S0958-6946(02)00152-8]

[62] Vallejo-Cordoba, B.; Mazorra-Manzano, M.A.; González-Córdova, A.F. New capillary electrophoresis method for the determination of furosine in dairy products. *J. Agric. Food Chem.*, **2004**, *52*(19), 5787-5790. [http://dx.doi.org/10.1021/jf049850e] [PMID: 15366821]

[63] Guingamp, M.F.; Hubert, G.; Midon, P.; Nicolas, M.; Linden, G. Screening procedure for evaluating heat load in commercial milks. *Lait,* **1999**, *79*, 457-463. [http://dx.doi.org/10.1051/lait:1999438]

[64] Morales, F.J.; Jiménez-Pérez, S. Hydroxymethylfurfural determination in infant milk-based formulas by micellar electrokinetic capillary chromatography. *Food Chem.*, **2001**, *72*, 525-531. [http://dx.doi.org/10.1016/S0308-8146(00)00284-3]

[65] Code of Practice for Evaluation of Fruit and Vegetable Juices, Association of the Industries of Juice and Nectars from Fruits and Vegetables (AIJN) of the EU, Belgium, **1994** [66] Šimuničová, E.; Kaniansky, D.; Lokšíkova, K., Separation of alkali and alkaline earth metal and ammonium cations by capillary zone electrophoresis with indirect UV absorbance detection. *J. Chromatogr. A,* **1994**, *665*, 203-208.

[67] Yang, Q.; Smeyers-Verbeke, J.; Wu, W.; Khots, M.S.; Massart, D.L. Simultaneous separation of ammonium and alkali, alkaline earth and transition metal ions in aqueous-organic media by capillary ion analysis. *J. Chromatogr. A,* **1994**, *688*, 339-348. [http://dx.doi.org/10.1016/0021-9673(94)00800-0]

[68] Kvasnička, F. Application of isotachophoresis in food analysis. *Electrophoresis,* **2000**, *21*(14), 2780-2787. [http://dx.doi.org/10.1002/1522-2683(20000801)21:14<2780::AID-ELPS2780>3.0.CO;2-W] [PMID: 11001284]

[69] Kaniansky, D.; Zelenský, I.; Valášková, I.; Marák, J.; Zelenská, V. Isotachophoretic separation of alkali and alkaline earth metal cations in water—polyethylene glycol mixtures. *J. Chromatogr. A,* **1990**, *502*, 143-151. [http://dx.doi.org/10.1016/S0021-9673(01)89571-6]

[70] Weston, A.; Brown, P. R.; Heckenberg, A, L.; Jandik, P.; Jones, W. R., Effect of electrolyte composition on the separation of inorganic metal cations by capillary ion electrophoresis. *J. Chromatogr. A,* **1992**, *602*, 249-256. [http://dx.doi.org/10.1016/0021-9673(92)80088-C]

[71] Sádecká, J.; Polonský, J. Determination of inorganic ions in food and beverages by capillary electrophoresis. *J. Chromatogr. A,* **1999**, *834*(1-2), 401-417. [http://dx.doi.org/10.1016/S0021-9673(98)00959-5] [PMID: 10189696]

[72] Ježek, J.; Suhaj, M. Application of capillary isotachophoresis for fruit juice authentication. *J. Chromatogr. A,* **2001**, *916*(1-2), 185-189. [http://dx.doi.org/10.1016/S0021-9673(00)01078-5] [PMID: 11382290]

[73] Kvasnička, F.; Voldřich, M.; Pyš, P.; Vinš, I. Determination of isocitric acid in citrus juice - A comparison of HPLC, enzyme set and capillary isotachophoresis methods. *J. Food Compos. Anal.*, **2002**, *15*, 685-691. [http://dx.doi.org/10.1006/jfca.2002.1101]

[74] Kvasnička, F.; Voldřich, M. Determination of fumaric acid in apple juice by on-line coupled capillary isotachophoresis-capillary zone electrophoresis with UV detection. *J. Chromatogr. A*, **2000**, *891*(1), 175-181. [http://dx.doi.org/10.1016/S0021-9673(00)00513-6] [PMID: 10999636]

[75] Xiong, X.; Li, S.F.I. Selection and optimization of background electrolytes for simultaneous detection of small cations and organic acids by capillary electrophoresis with indirect photometry. *J. Chromatogr. A*, **1998**, *822*, 125-136. [http://dx.doi.org/10.1016/S0021-9673(98)00561-5]

[76] Cancalon, P.F. Analytical monitoring of citrus juices by using capillary electrophoresis. *J. AOAC Int.*, **1999**, *82*(1), 95-106. [http://dx.doi.org/10.1093/jaoac/82.1.95] [PMID: 10028677]

[77] Klampfl, C.W.; Buchberger, W.; Turner, M.; Fritéz, J.S. Determination of underivatized amino acids in beverage samples by capillary electrophoresis. *J. Chromatogr. A*, **1998**, *804*, 349-355. [http://dx.doi.org/10.1016/S0021-9673(98)00047-8]

[78] Simó, C.; Barbas, C.; Cifuentes, A. Sensitive micellar electrokinetic chromatography-laser-induced fluorescence method to analyze chiral amino acids in orange juices. *J. Agric. Food Chem.*, **2002**, *50*(19), 5288-5293. [http://dx.doi.org/10.1021/jf0203848] [PMID: 12207463]

[79] Honda, S. Separation of neutral carbohydrates by capillary electrophoresis. *J. Chromatogr. A*, **1996**, *720*(1-2), 337-351. [http://dx.doi.org/10.1016/0021-9673(95)00022-4] [PMID: 8601200]

[80] Soga, T.; Serwe, M. Determination of carbohydrates in food samples by capillary electrophoresis with indirect UV detection. *Food Chem.*, **2000**, *69*, 339-344. [http://dx.doi.org/10.1016/S0308-8146(00)00044-3]

[81] Cabálková, J.; Žídková, J.; Přibyla, L.; Chmelík, J. Determination of carbohydrates in juices by high-performance liquid capillary electrophoresis, chromatography, and matrix-assisted laser desorption/ionization time of flight-mass spectrometry. *Electrophoresis*, **2004**, *25*, 487-493. [http://dx.doi.org/10.1002/elps.200305672] [PMID: 14760642]

[82] Monasterio, R.P.; Fernández, Mde.L.; Silva, M.F. Olive oil by capillary electrophoresis: characterization and genuineness. *J. Agric. Food Chem.*, **2013**, *61*(19), 4477-4496. [http://dx.doi.org/10.1021/jf400864q] [PMID: 23594110]

[83] Bean, S.R.; Bietz, J.A.; Lookhart, G.L. High-performance capillary electrophoresis of cereal proteins. *J. Chromatogr. A*, **1998**, *814*(1-2), 25-41. [http://dx.doi.org/10.1016/S0021-9673(98)00437-3] [PMID: 9718685]

[84] Bean, S.R.; Lookhart, G.L. Recent developments in high-performance capillary electrophoresis of cereal proteins. *Electrophoresis*, **2001**, *22*(8), 1503-1509. [http://dx.doi.org/10.1002/1522-2683(200105)22:8<1503::AID-ELPS1503>3.0.CO;2-P] [PMID: 11386662]

[85] Bietz, J.A.; Schmalzried, E. Capillary electrophoresis of wheat gliadin: Initial studies and application to varietal identification. *Lebensm. Wiss. Technol.*, **1995**, *28*, 174-184. [http://dx.doi.org/10.1016/S0023-6438(95)91346-7]

[86] Werner, W.E.; Wiktorowicz, J.E.; Kasarda, D.D. Wheat Varietal Identification by Capillary Electrophoresis of Gliadins and High-Molecular-Weight Glutenin Subunits. *Cereal Chem.*, **1994**, *71*, 397-402.

[87] Capelli, L.; Forlani, F.; Perini, F.; Guerrieri, N.; Cerletti, P.; Righetti, P.G. Wheat cultivar discrimination by capillary electrophoresis of gliadins in isoelectric buffers. *Electrophoresis,* **1998**, *19*(2), 311-318.
[http://dx.doi.org/10.1002/elps.1150190228] [PMID: 9548297]

[88] Bean, S.R.; Lookhart, G.L. Faster capillary electrophoresis separation of wheat proteins through modifications to buffer composition and sample handling. *Electrophoresis,* **1998**, *19*(18), 3190-3198.
[http://dx.doi.org/10.1002/elps.1150191823] [PMID: 9932814]

[89] Lookhart, G.L.; Bean, S.R.; Jones, B.L. Separation and characterization of barley (Hordeum vulgare L.) hordeins by free zone capillary electrophoresis. *Electrophoresis,* **1999**, *20*(7), 1605-1612.
[http://dx.doi.org/10.1002/(SICI)1522-2683(19990601)20:7<1605::AID-ELPS1605>3.0.CO;2-J] [PMID: 10424486]

[90] Bean, S.R.; Lookhart, G.L. Sodium dodecyl sulfate capillary electrophoresis of wheat proteins. 1. Uncoated capillaries. *J. Agric. Food Chem.,* **1999**, *47*(10), 4246-4255.
[http://dx.doi.org/10.1021/jf990413n] [PMID: 10552797]

[91] Weegels, P.L.; Hamer, R.J.; Schofield, J.D. RP-HPLC and Capillary Electrophoresis of Subunits from Glutenin Isolated by SDS and Osborne Fractionation. *J. Cereal Sci.,* **1995**, *22*, 211-224.
[http://dx.doi.org/10.1006/jcrs.1995.0058]

[92] Righetti, P.G.; Olivieri, E.; Viotti, A. Identification of maize lines *via* capillary electrophoresis of zeins in isoelectric, acidic buffers. *Electrophoresis,* **1998**, *19*(10), 1738-1741.
[http://dx.doi.org/10.1002/elps.1150191035] [PMID: 9719554]

[93] Olivieri, E.; Viotti, A.; Lauria, M.; Simò-Alfonso, E.; Righetti, P.G. Variety identification in maize lines *via* capillary electrophoresis of zeins in isoelectric acidic buffers. *Electrophoresis,* **1999**, *20*(7), 1595-1604.
[http://dx.doi.org/10.1002/(SICI)1522-2683(19990601)20:7<1595::AID-ELPS1595>3.0.CO;2-7] [PMID: 10424485]

[94] Bean, S.R.; Lookhart, G.L.; Bietz, J.A. Acetonitrile as a buffer additive for free zone capillary electrophoresis separation and characterization of maize (Zeamays L.) and sorghum (Sorghum bicolor L. Moench) storage proteins. *J. Agric. Food Chem.,* **2000**, *48*(2), 318-327.
[http://dx.doi.org/10.1021/jf990786o] [PMID: 10691635]

[95] Bean, S.R.; Lookhart, G.L. Ultrafast capillary electrophoretic analysis of cereal storage proteins and its applications to protein characterization and cultivar differentiation. *J. Agric. Food Chem.,* **2000**, *48*(2), 344-353.
[http://dx.doi.org/10.1021/jf990962t] [PMID: 10691639]

[96] Salmanowicz, B.P.; Moczulski, M. Multiplex polymerase chain reaction analysis of Glu-1 high-molecular-mass glutenin genes from wheat by capillary electrophoresis with laser-induced fluorescence detection. *J. Chromatogr. A,* **2004**, *1032*(1-2), 313-318.
[http://dx.doi.org/10.1016/j.chroma.2003.12.031] [PMID: 15065810]

[97] Pazourek, J.; González, G.; Revilla, A.L.; Havel, J. Separation of polyphenols in Canary Islands wine by capillary zone electrophoresis without preconcentration. *J. Chromatogr. A,* **2000**, *874*(1), 111-119.
[http://dx.doi.org/10.1016/S0021-9673(99)01348-5] [PMID: 10768505]

[98] Gu, X.; Chu, Q.; O'Dwyer, M.; Zeece, M. Analysis of resveratrol in wine by capillary electrophoresis. *J. Chromatogr. A,***2000**, *881*, 471-478. [99] Brandoliny, V.; Maietti, A.; Pedeshi, P.; Durini, E.; Vertuani, S.; Manfredini, S., Capillary electrophoresis determination, synthesis, and stability of resveratrol and related 3-O-beta-D-glucopyranosides. *J. Agric. Food Chem.,* **2002**, *50*, 7407-7411.

[100] Hamoudová, R.; Urbánek, M.; Pospísilová, M.; Polásek, M. Assay of phenolic compounds in red wine by on-line combination of capillary isotachophoresis with capillary zone electrophoresis. *J. Chromatogr. A,* **2004**, *1032*(1-2), 281-287.
[http://dx.doi.org/10.1016/j.chroma.2004.01.014] [PMID: 15065806]

[101] Núñez, M.; Pena, R.M.; Herrero, C.; Garcia-Martin, S. Analysis of some metals in wine by means of capillary electrophoresis. Application to the differentiation of Ribeira Sacra Spanish red wines. *Analusis,* **2000**, *28*, 432-437. [http://dx.doi.org/10.1051/analusis:2000129]

[102] Delgado, C.; Tomas-Barberan, F.A.; Talou, T.; Gaset, A. Capillary electrophoresis as an alternative to HPLC for determination of honey flavonoids. *Chromatographia,* **1994**, *38*, 71-78. [http://dx.doi.org/10.1007/BF02275729]

[103] Bronze, M.R.; Boas, L.F.V.; Belchior, A.P. Analysis of old brandy and oak extracts by capillary electrophoresis. *J. Chromatogr. A,* **1997**, *768*, 143-152. [http://dx.doi.org/10.1016/S0021-9673(96)01047-3]

[104] Bronze, M.R.; Boas, L.F.V. Characterisation of brandies and wood extracts by capillary electrophoresis. *Analusis,* **1998**, *26*, 40-47. [http://dx.doi.org/10.1051/analusis:1998109]

[105] Kvasnička, F. Determination of egg white lysozyme by on-line coupled capillary isotachophoresis with capillary zone electrophoresis. *Electrophoresis,* **2003**, *24*(5), 860-864. [http://dx.doi.org/10.1002/elps.200390108] [PMID: 12627448]

[106] Čížková, H.; Voldřich, M.; Prokorátová, V.; Kvasnička, F. Determination of egg yolk content in egg liqueurs. *Czech J. Food Sci.,* **2004**, *22*, 9-15. [http://dx.doi.org/10.17221/3401-CJFS]

[107] Čížková, H.; Prokorátová, V.; Voldřich, M.; Kvasnička, F.; Soukupová, V. Determination of egg content in pasta. *Czech J. Food Sci.,* **2004**, *22*, 197-203. [http://dx.doi.org/10.17221/3424-CJFS]

[108] Boyce, M.C. Determination of additives in food by capillary electrophoresis. *Electrophoresis,* **2001**, *22*(8), 1447-1459. [http://dx.doi.org/10.1002/1522-2683(200105)22:8<1447::AID-ELPS1447>3.0.CO;2-#] [PMID: 11386656]

[109] Siles, B.A.; O'Neil, K.A.; Fox, M.A.; Anderson, D.E.; Kuntz, A.F.; Ranganath, S.C.; Morris, A.C. Genetic fingerprinting of grape plant (Vitis vinifera) using random amplified polymorphic DNA (RAPD) analysis and dynamic size-sieving capillary electrophoresis. *J. Agric. Food Chem.,* **2000**, *48*(12), 5903-5912. [http://dx.doi.org/10.1021/jf000734v] [PMID: 11312766]

[110] Garcia-Cañas, V.; Gonzalez, R.; Cifuentes, A. The combined use of molecular techniques and capillary electrophoresis in food analysis. *TrAC. Trends Analyt. Chem.,* **2004**, *23*, 637-643. [http://dx.doi.org/10.1016/j.trac.2004.07.005]

[111] García-Cañas, V.; González, R.; Cifuentes, A. Ultrasensitive detection of genetically modified maize DNA by capillary gel electrophoresis with laser-induced fluorescence using different fluorescent intercalating dyes. *J. Agric. Food Chem.,* **2002**, *50*(16), 4497-4502. [http://dx.doi.org/10.1021/jf025585q] [PMID: 12137467]

[112] Nogueira, T.; do Lago, C.L. Detection of adulterations in processed coffee with cereals and coffee husks using capillary zone electrophoresis. *J. Sep. Sci.,* **2009**, *32*(20), 3507-3511. [http://dx.doi.org/10.1002/jssc.200900357] [PMID: 19777453]

CHAPTER 12

Chemical Food Safety Applications of Capillary Electrophoresis Methodologies

Maykel Hernández-Mesa[1,2,*], **David Moreno-González**[3], **Francisco J. Lara**[2], **Gaud Dervilly**[1] and **Ana M. García-Campaña**[2]

[1] *LABERCA, Oniris, INRAE, F-44307 Nantes, France*

[2] *Department of Analytical Chemistry, Faculty of Sciences, University of Granada, Av. Fuentenueva s/n, E-18071 Granada, Spain*

[3] *ISAS-Leibniz-Institut für analytische Wissenschaften, Bunsen-Kirchhoff-Str. 11, 44139 Dortmund, Germany*

Abstract: Chemical hazards may be present in food due to their intended use during food production (*i.e.*, residues), they may be intentionally added to food products to confer specific attributes (*i.e.*, food additives), or they may simply occur at any of the stages of the food supply chain (*i.e.*, contaminants). Since these chemical hazards represent a health risk to consumers, legislation has been developed to establish the maximum concentration levels of these substances in food, and to define control measures to monitor their presence in food products. In general, liquid chromatography (LC) and gas chromatography (GC) are used as analytical techniques in laboratories that are responsible for carrying out routine food safety analyses. During the last decades, capillary electrophoresis (CE) has been extensively investigated as an alternative (or complementary) separation tool to chromatographic techniques, and today, it is already a consolidated technique that can be implemented in routine food safety laboratories. This chapter presents the state of the art of CE in the field of chemical food safety and gives an overview of relevant applications in this area.

Keywords: Capillary electrophoresis, Chemical hazards, Contaminants, Food additives, Food safety, Residues.

INTRODUCTION

The intensification of agricultural and livestock activities related to an increasing population rate and a higher demand for food, as well as the current globalization of the food chain, poses numerous challenges for food safety. In this context, bacteria, viruses, parasites and chemical hazards can occur at any point in the food

* **Corresponding autor Maykel Hernández-Mesa:** LABERCA, Oniris, INRAE, F-44307 Nantes, France and Department of Analytical Chemistry, Faculty of Sciences, University of Granada, Av. Fuentenueva s/n, E-18071 Granada, Spain; E-mail: maykel.hernandez-mesa@oniris-nantes.fr / maykelhm@ugr.es

María Castro-Puyana, Miguel Herrero and María Luisa Marina (Eds.)
Capillary Electrophoresis in Food Analysis

chain, from primary production to delivery to the final consumer, endangering public health. It is the mission of policy-makers to promote health and ensure food safety by establishing food systems and infrastructures to respond to and manage food risks, such as the Rapid Alert System for Food and Feed (RASFF) implemented by the European Union (EU) to avoid food safety risks [1]. Consequently, food safety authorities can take actions and prevent similar risks. As part of food safety management, competent authorities have also implemented control plans to assess the prevalence of hazards in feed, food and animals [2]. In this sense, selective and analytical methods are required to successfully execute these control plans.

With respect to chemical hazards, and according to the Codex Alimentarius Commission, they can be classified into residues (*i.e.* substances intentionally used during food production for agricultural or livestock benefits), contaminants (*i.e.* substances that are not intentionally added to food or feed for food-producing animals, but are however present in such food or feed as a consequence of food-related activities such as processing, transport, storage, *etc.* or environmental contamination), and additives (*i.e.* substances intentionally added to confer food some specific organoleptic properties) [3]. This general classification may be even simpler, and chemical hazards can be differentiated according to whether their presence in food is due to intentional use (*i.e.* residues and additives) or not (*i.e.* contaminants). In order to standardize at an international level, and without prejudice to national regulations, the Joint Food and Agriculture Organization (FAO)/World Health Organization (WHO) Meeting on Pesticide Residues (JMPR) has established allowable daily intakes (ADIs, chronic values) and acute reference doses (ARfDs) of pesticide residues in food. In addition, the Joint FAO/WHO Expert Committee on Food Additives (JECFA) has set tolerable daily intakes (TDIs), ADIs and other guidance values for food additives, contaminants, such as naturally occurring toxins, and residues of veterinary drugs in food [4].

Nowadays, there is a wide range of regulations and guidelines aimed to control chemical hazards in food to ensure that they do not exceed concentrations levels considered 'safe' or levels related to a significant health risk. From the perspective of European regulations and guidelines on food safety, and as examples, the performance of analytical methods intended for the analysis of veterinary drugs, pesticides, and mycotoxins in food must meet the criteria established in Commission Decision 2002/657/EC [5], SANTE/12682/2019 [6] and SANTE/12089/2016 [7], respectively. These regulations and guidelines establish that mass spectrometry (MS) generally coupled with Liquid Chromatography (LC) or Gas Chromatography (GC) is the analytical technique of choice to control chemical hazards in food. However, GC-MS or LC-MS analysis of some chemical hazards may be limited by their volatility and polarity properties.

Capillary Electrophoresis (CE) is an alternative (or complementary) separation technique to LC and GC which offers several advantages, such as the requirement of low sample volumes and low solvent consumption in accordance with the principles of Green Chemistry. CE has been applied primarily to the separation of charged substances (*i.e.* Capillary Zone Electrophoresis (CZE) mode), although the separation of neutral compounds can also be achieved by generally applying the Micellar Electrokinetic Chromatography (MEKC) mode [8]. CE is recommended for the separation of polar substances that are difficult to be separated by chromatographic techniques, especially considering that Hydrophilic Interaction Chromatography (HILIC)-based methods are less robust than traditional Reverse Phase Liquid Chromatography (RPLC) methods implemented in food safety laboratories.

CE is routinely used in pharmaceutical applications [9] and in forensic laboratories, namely for the analysis of deoxyribonucleic acid (DNA) [10]. However, although the applicability of CE for food analysis has been extensively demonstrated [11 - 14], this technique has not yet been implemented in food safety laboratories for routine applications and still remains a research tool. In general, the implementation of CE has been limited by its lower concentration sensitivity and drawbacks associated to CE-MS hyphenation, especially considering that both are key aspects in the analysis of chemical hazards in food. Food safety applications involve the detection of substances at low concentration levels (within the ng kg^{-1} and μg kg^{-1} ranges), and as mentioned above, MS detection is necessary for the unambiguous determination of residues, contaminants and additives in food since they involve legal consequences. However, CE technology has rapidly evolved in recent years and numerous preconcentration strategies (on-line, in-line and off-line approaches) can be applied to achieve low limits of detection (LODs) [15, 16]. Efficient and robust interfaces for the hyphenation of CE with MS are now commercially available, contributing to the further development and implementation of CE-MS applications [17, 18].

In this context, this chapter shows the most outstanding current applications of CE in the field of chemical food safety, mainly focused on the analysis of residues (*i.e.* veterinary drugs and pesticides), contaminants (*i.e.* toxic metals, biogenic amines, natural toxins, food processing contaminants and contaminants related to food packaging) and food additives (*e.g.* preservatives, colorants and sweeteners). In addition, a particular section is dedicated to the CE analysis of nanoparticles in food, as this is an emerging concern in food safety and these materials can be either contaminants or food additives. In addition to the discussion on the most relevant CE separation conditions for the analysis of each type of chemical hazard, this chapter also presents the detection techniques and preconcentration approaches used, the performance characteristics of the CE methods, and the sample treatments applied before CE analysis.

CE ANALYSIS OF FOOD RESIDUES

Veterinary Drugs

The use of veterinary drugs in livestock practices is necessary to protect the health of animals, including the fight against parasites. However, this fact involves the risk of finding veterinary drug residues in food of animal origin which can have serious implications for human health, such as the transfer of antibiotic-resistant bacteria to man or allergic reactions in highly sensitive people. To ensure food safety, governments around the world and many organizations have established regulations and guidelines with lists of banned and authorized substances in veterinary medicine and setting maximum residue limits for these compounds in food [19 - 21]. In this context, a wide range of CE-based methods have been developed in recent years to control the presence of veterinary drugs in foods of animal origin, especially antibiotics (*i.e.* sulfonamides, quinolones, tetracyclines, β-lactams, aminoglycosides, macrolides and nitroimidazoles) [22 - 25]. The most relevant CE methods according to the family of veterinary drugs are presented below. Likewise, the most widely used preconcentration strategies in CE are also discussed, especially those relevant in the analysis of chemical hazards at low concentration levels. In this sense, CE-related preconcentration approaches are presented in a general way with a special focus on the determination of veterinary drugs, although these are applicable to other chemical hazards as shown in later sections.

In general, sulfonamides have being mainly separated by CZE using phosphate buffers around pH 7 or borate buffers at pH 9, and using ultraviolet (UV) detection at 270 nm. This is the case of the determination of up to fifteen sulfonamides using a quaternary ammonium chitosan derivative as dynamic coating [26]. CZE-MS using a sheath-flow interface has also been applied for the determination of sulfonamide residues in meat, using ammonium acetate as separation buffer and electrospray ionization (ESI) in positive mode, and combined with different sample treatments, such as Pressurized Liquid Extraction (PLE) [27] or extraction with acetonitrile (MeCN) and Solid Phase Extraction (SPE) cartridges for cleanup and preconcentration [28]. Other CE modes such as Capillary Electrochromatography (CEC) [29] and Microemulsion Electrokinetic Chromatography (MEEKC) [30] have also been applied to the analysis of sulfonamides in food. For example, a CEC-MS method has been reported for the determination of 9 sulfonamides in meat using step gradient elution [31], in which separation was stopped after 5 min and the initial mobile phase in the inlet vial containing 30% (v/v) MeCN was replaced by a new mobile phase containing 40% (v/v) MeCN. Poly(divinylbenzene-octyl methacrylate) monolith columns were prepared in the laboratory as they were not commercially available.

Quinolones, which are a relevant class of broad-spectrum antibiotics, are usually separated by CZE using phosphate buffers at pH between 8 and 9 when UV detection (around 270-280 nm) is applied, or 50 mM ammonium acetate at pH 9.1 for MS detection where they are separated as anions but detected as cations by ESI+ [32, 33]. Acidic buffers have also been used to determine ciprofloxacin and norfloxacin, in combination with reverse polarity and coated capillaries to reverse the electroosmotic flow (EOF). Cetyltrimethylammonium bromide (CTAB) [34] and poly(diallyldimethylammonium chloride) (PDDA) [35] have been used for this purpose with satisfactory results in milk samples. Since some quinolones show native fluorescence at low pH, CE has been proposed together with laser induced fluorescence (LIF) detection at 325 nm for their determination in bovine raw milk and pig kidney samples [36]. Also, 488 nm lasers can be used with a previous derivatization of quinolones with fluorescein isothiocyanate for the determination of theses antibiotics in milk, chicken, pork, and fish samples [37]. Other detectors such as capacitively coupled contactless conductivity detection (C^4D) [38, 39] or electrochemiluminescence (ECL) [40] have been used in combination with CE for the determination of quinolones.

Tetracyclines have been separated by CZE, generally at a pH above 9 using phosphate buffers [41 - 43], as well as by MEKC at lower pH [44]. When coupled with UV detection, the most common wavelength used is 360 nm [42, 43, 45]. In order to prevent interaction of the analytes with metal ions, ethylenediaminetetraacetic acid (EDTA) is usually added during the sample treatment [43, 46] and in the electrophoretic buffer [47]. Non-aqueous solvents can improve their native fluorescent emission and, for this reason, non-aqueous capillary electrophoresis (NACE) with LIF detection has been used to determine four tetracyclines in milk samples [48]. MS detection applying ESI+ has also been investigated by prior separation of tetracyclines as anions but MS detection as cations [46, 47].

CZE has also been proposed for the separation of β-lactam drugs, including penicillins, cephalosporins and carbapenems. Seven penicillins have been determined in milk using tris(hydroxymethyl)aminomethane buffer at pH 8.0 using UV detection [49], and in which separation resolution was improved by adding 20% (v/v) of ethanol to the background electrolyte (BGE). CZE-MS has also been applied to the analysis of nine penicillins in chicken meat using ammonium acetate buffer (pH 6) as BGE and ESI+ [50]. On the other hand, MEKC has been applied to the separation of three penicillins and three cephalosporins in milk and eggs, using UV detection at 200 nm [51]. Furthermore, different CEC approaches for the analysis of β-lactams in milk have been studied, including monolith columns made of poly(stearyl methacrylate–divinylbenzene vinylbenzyl trimethylammonium chloride) to determine eight penicillins with MS (ESI-) [52], and open tubular (OT)-CEC with o- cynaoethyl chitosan as stationary phase for the determination of four cephalosporins with UV detection at 254 nm [53].

Aminoglycosides are highly polar compounds; therefore, CE represents a powerful alternative to LC for their analysis. These substances are poorly retained in RPLC, and HILIC presents a major drawback related to the need to add salts to the mobile phase to improve resolution, which hinders the use of MS detectors. For example, CZE-tandem mass spectrometry (MS/MS) has been applied to the analysis of nine aminoglycosides in honey using an ion trap (IT) as mass analyzer and sheath-flow interface for CE-MS hyphenation [54]. In this method, the BGE consisted of ammonium formate buffer at pH 2.2 to avoid the analyte interactions with the silanol groups of capillary walls. Moreover, to improve the reproducibility of migration time, a pressure of 50 mbar was applied during the separation to counteract the absence of the EOF. Finally, LODs between 0.4 and 28.5 µg/kg were achieved, applying molecularly imprinted solid-phase extraction (MISPE) as selective sample treatment. Yu *et al.* have also recently proposed a CE-MS/MS method for the determination of four aminoglycosides in milk samples but using a sheathless interface [55]. LOQs of 0.67 µg/kg were obtained.

In the case of macrolides, MEKC analysis of five of these antibiotics has been explored using mixed micelles of CTAB and sodium cholate to achieve their separation, which could not have been resolved otherwise [56]. Another strategy to improve CE separation of macrolide compounds includes the use of ionic liquids (ILs) as additives in the BGE. Among them, long-chain alkylimidazolium-based ILs can act as pseudostationary phase in MEKC separations. They are capable of forming micelles and their critical micelle concentration (CMC) values are significantly lower than those of the classic cationic surfactants with a comparable alkyl chain length [57]. For example, 1-butyl-3-methylimidazolium tetrafluoroborate ($BMIMBF_4$) has been observed to improve the resolution but also the sensitivity of MEKC-ECL methods related to the determination of macrolides in egg samples [58] and milk [59].

Furthermore, 5-nitroimidazole antibiotics, which are not allowed as residues in foods by the EU [19], have been analyzed by MEKC-UV in eggs [60] and milk

[61]. CEC has also been proposed for the determination of eight 5-nitroimidazoles in milk (Fig. **1**) using packed laboratory-made columns [62]. The capillary columns were produced by carrying out a high pressure packing procedure using acetone as the driving solvent and uncapped C_{18} silica particles (5 µm particle size) as packing material. The frits in the column were produced by sintering the stationary phase, heating the packed material for 20 s with a nichrome ribbon connected to a power supply.

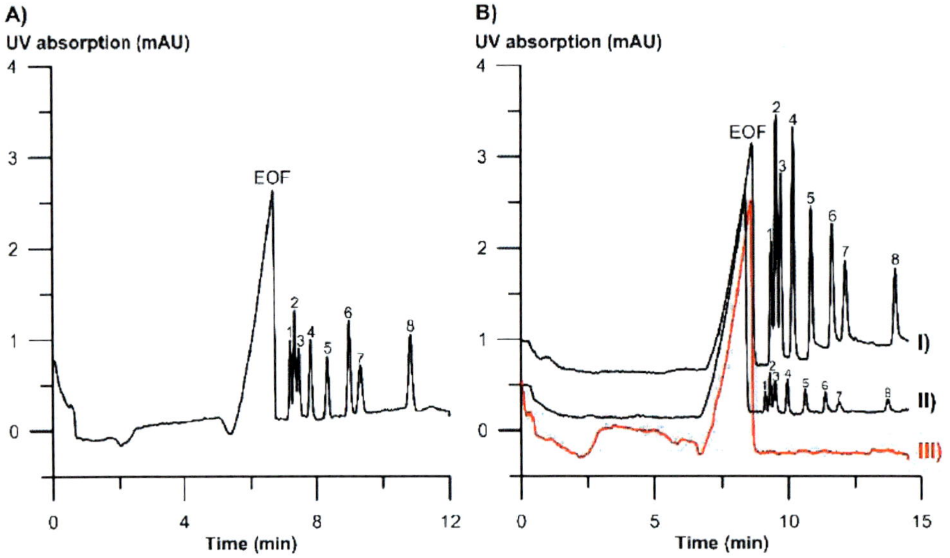

Fig. (1). Electrochromatograms related to the CEC-UV analysis of A) 5-nitroimidazole standards (1.2 mg/L of each compound, except carnidazole, 2.4 µg/L) and B) milk samples pretreated by a salt-assisted liquid-liquid extraction-SPE procedure and fortified at: I) 350 µg/L of each analyte (except carnidazole, 700 µg/L), II) 50 µg/L of each analyte (except carnidazole, 100.0 µg/L), and III) blank sample. Capillary dimensions: 40 cm x 50 mm inner diameter (i.d.). Separation conditions: 27 kV and 30°C. Mobile phase: 60:40 (v/v) MeCN:ammonium acetate buffer (2.5 mmol/L, pH 5). Sample injection: 120 s at 11.5 bar. The signals were monitored at 320 nm. Peaks identification: 1, ronidazole; 2, metronidazole; 3, TRZ, ternidazole; 4, secnidazole; 5, ornidazole; 6, dimetridazole; 7, carnidazole; 8, ipronidazole. Reprinted with permission from [62].

Of special interest are those CE-based methods intended for the simultaneous analysis of multiclass antibiotics. As an example, CZE-quadrupole-time-of-flight (Q- ToF)-MS has been evaluated for the determination of fifteen antibiotic residues (*i.e.* eight tetracyclines and seven quinolones) in milk samples, obtaining full scan and full MS/MS spectra for quantification/confirmation purposes in a single run. The method showed good linearity for all compounds studied and the relative standard deviations (RSDs) were below 13% in terms of peak area precision [63].

In addition to antibiotics, CE-based methods have been reported for the analysis of other types of veterinary drugs. β-adrenoceptor agonists (*i.e.* β-agonists) bind to β- receptors in cardiac and smooth muscle tissues, and are useful in treating cardiac decompensation or in the treatment of bronchial diseases. They also promote the reduction of body fat enhancing muscle growth in cattle, sheep and swine if used in high dosage, although this livestock practice is currently banned in EU. A very low MRL has been established for clenbuterol residues in meat products from different animals and milk samples [19**Error! Bookmark not defined.**], due to its risk to human health. Clenbuterol and salbutamol have been determined in meat by CZE- amperometric detection using borate buffer (pH 7.4) as BGE and a very short capillary (*i.e.* 8.5 cm) in less than one minute [64]. More volatile non-aqueous BGE should be used instead of borate buffer when applying MS detection [65]. In this sense, ionization suppression has been observed at the ESI interface when analyzing pork samples. In the mentioned method, the simultaneous combination of hydrodynamic and electrokinetic injection allowed LODs of 0.3 μg/L for clenbuterol, salbutamol and terbutaline.

Hormones can also be found as residues in animal-derived foods because they are widely used in intensive animal production, as veterinary drugs to increase the productivity of animals. In general, the EU has banned the use of substances having a hormonal action for growth promotion in farm animals. With regard to CE methods for the determination of hormones in food, MEKC analysis of testosterone and progesterone have been carried out successfully using borate buffer as running buffer and sodium dodecyl sulfate (SDS) as pseudostationary phase, and achieving satisfactory results for their analysis in milk samples [66, 67]. MEKC-MS has also been investigated for the analysis of hormones, using ammonium perfluorooctanoate (APFO) as both BGE and surfactant, and avoiding the use of SDS. In this sense, SDS is a non-volatile surfactant that produces suppression of the analyte signal in MS as well as contamination of the mass spectrometer. This approach has been used successfully in the determination of eleven compounds with estrogenic activity in milk and yogurt, which were previously treated by Dispersive Liquid-Liquid Microextraction (DLLME). LODs were obtained in the low μg/L range [68].

CE methods have also been reported for the analysis of anthelmintics and coccidiostats in food. Anthelmintics, mainly benzimidazoles and avermectins, are important veterinary drugs used in case of parasitic worm infections. Coccidiostacts are applied against *Coccidia* parasites. CZE-MS using an IT mass spectrometer has been applied in the determination of the coccidiostat amprolium in eggs, using ammonium acetate buffer (pH 4.5) as BGE [69]. It was necessary to add 40% (v/v) methanol (MeOH) to the BGE to separate amprolium from an interference identified as thiamine, and achieving a LOQ of 75 μg/kg. CZE-

MS/MS, using an IT mass spectrometer, has also been applied in determination of twelve benzimidazoles in chicken and pork meat [70]. LODs ranged from 1 to 5 µg/kg and the RSDs in terms of precision were less than 16%. Similar RSDs were achieved when a single quadrupole analyzer was used to determine benzimidazoles in eggs, but higher LODs were obtained (*i.e.* between 3 and 51 µg/L) [71]. Since single quadrupole mass spectrometers do not allow MS/MS experiments, this method does not allow the confirmation of residues in food. Fragmentation experiments must be carried out to achieve this purpose.

One of the main disadvantages of CE methodologies, especially when using UV detection, is the low detection sensitivity achieved due to low sample volume analyzed. This fact is critical in food safety applications where low concentration levels of chemical hazards must be detected. To overcome this inherent limitation of CE, analyte preconcentration to improve detection sensitivity can be achieved by applying sample treatments with high concentration factors (*i.e.* off-line preconcentration) or preconcentration approaches specifically related to CE-based methods (*i.e.* in-line and on-line preconcentration). Within this framework, the detailed discussion of sample treatments in food safety analysis is beyond the scope of this chapter. However, it should be noted that an ideal sample treatment for CE should preconcentrate the analytes and provide clean extracts, compatible with the CE method [8]. Despite progress in other methodologies, SPE remains one of the most efficient sample treatment techniques in food analysis, and specifically, in CE.

In addition to the off-line preconcentration strategies related to sample treatments for the analysis of residues of veterinary drugs in food, it is necessary to consider on-line methodologies to improve sensitivity in CE methods, such as the use of capillaries with extended optical path length (when using UV detection) or on-line preconcentration strategies that increase the amount of analyte injected. These last approaches, also called stacking techniques, are based on either the manipulation of differences in the electrophoretic mobility of analytes at the boundary of two buffers with different conductivity or the partitioning of analytes into a stationary or pseudostationary phase, and can be classified according to whether involve electrokinetic or hydrodynamic injection [15, 72, 73].

Stacking techniques based on hydrodynamic injection are more robust and allow repetitive injections of the same sample vial. Field-Amplified Sample Stacking (FASS) (also called normal stacking mode) is the simpler option. It is based on the differences in conductivity between the sample solution and the BGE. Up to 2% of the capillary volume is filled with the sample in a normal hydrodynamic injection, while 5% of capillary volume can be filled applying a FASS approach, as shown for the multiresidue determination of antibiotics in milk [74]. As a consequence, LODs ranged between 12 and 29 μg/L, which involved an improvement of an order of magnitude compared with regular hydrodynamic injection. Sensitivity improvements of 5 and 10 times have also been observed for the analysis of aminoglycosides in honey due to the application of FASS [54]. Large-Volume Sample Stacking (LVSS) is also based on hydrodynamic injection where almost the entire capillary is initially filled with the sample. Subsequently, the sample matrix is removed from the capillary before electrophoretic separation occurs, usually by polarity switching and monitoring the electric current, while the analytes are stacked in the interface between sample and BGE zone. As the matrix is expelled, the capillary is filled with separation buffer from the outlet and the electric current increases. This mode cannot be applied with MS detectors as these CE-MS setups do not have an outlet vial. Another limitation is that the separation buffer must be basic enough to generate a strong EOF able to refill the capillary. To avoid expelling the analytes from the capillary together with the sample matrix, this process is usually stopped when the current reach 90% of the steady current. This approach can increase sensitivity about 100-fold as it has been reported for the determination of fluoroquinolones and amphenicols in bovine milk samples [75]. LVSS with polarity switching, also called reverse electrode polarity stacking mode, has been widely applied for the determination of β-lactams in milk samples [49, 51, 76]. LODs below 30 μg/kg and preconcentration factors up to 600 have been obtained [51].

Stacking techniques based on electrokinetic injection provide greater sensitivity enhancements than those based on hydrodynamic injection but are less robust. Furthermore, analytes with lower electrophoretic mobilities are injected to a lesser extent with electrokinetic injection, whereas hydrodynamic injection does not make a difference between analytes, so all are equally injected. Field-Amplified Sample Injection (FASI), also known as Field-Amplified Sample Stacking with Electromigration Injection (FASS-EMI), is one of the most typical stacking techniques. In FASI, the sample is diluted in a solvent of lower conductivity than that of the BGE and is electrokinetically injected. Consequently, the electric field is higher in the sample zone and the charged analytes move rapidly until they reach the interface between the sample and the BGE zone, where they are slowed down and stacked or preconcentrated. As an example, FASI has been applied to the CE-UV analysis of five quinolones [77]. The combination of FASI preconcentration and the use of extended path length capillaries led to LODs between 0.4 and 1.3 μg/L. These LODs were from 460 to 1500 times lower than those obtained by a conventional CE method and lower than maximum residue limits for quinolones in milk. Similar LODs were obtained for the determination of fifteen sulfonamides in meat and honey [26]. LODs obtained by hydrodynamic injection were between 100 and 250 μg/L, while values between 0.1 and 0.5 μg/L were obtained with FASI (*i.e.*, between 200-1000 times lower).

Some other on-line FASI-based preconcentration approaches have also been evaluated in applications for residue analysis, such as anion selective exhaustive injection (ASEI) and cation selective exhaustive injection (CSEI). In these approaches, the electrokinetic injection is carried out for several minutes and there is no EOF during the separation because the BGE pH is very low. The absence of EOF allows selective injection of anions or cations depending on the polarity of the voltage applied for electrokinetic injection. For example, ASEI has been used for the CEC analysis of eight penicillins in milk samples, resulting in LODs between 0.05-0.2 μg/L [52]. A voltage of -10 kV was applied for sample injection for at least 90 s. On the other hand, CSEI is a very similar strategy to ASEI but in this case analytes are cations, so they are injected using a positive voltage. Furthermore, both ASEI and CSEI techniques can be applied to some analytes such as sulfonamides. In this case, it is possible to apply successive anion- and cation-selective injection (SACSI). Basically, two sample vials are required for analysis. The first vial contains the sample in a basic solvent and is injected by ASEI. The second contains the sample in an acidic solvent and CSEI is performed for the injection. Finally, a negative separation voltage is applied. This strategy has been followed by Lin *et al.* for the determination of nine sulfonamides in porcine liver, obtaining LODs between 0.9- 4.2 μg/L [30].

Finally, preconcentration techniques based on chromatographic principles, called in- line preconcentration approaches, are a useful alternative to monitor veterinary drug residues in food since sample clean-up, preconcentration and separation of analytes is carried out automatically in the CE system. In general, these approaches consist of a SPE performed in the separation capillary, in which small SPE columns (*i.e.* analyte concentrator or preconcentrators) are inserted near the capillary inlet. Analyte concentrators or preconcentrators are homemade with typical sorbents used in off- line SPE such as mixed-mode cation exchange (MCX) [32] or even molecularly imprinted polymers (MIPs) [33], both used in the analysis of quinolones in meat and milk, respectively. The MIP material provided greater selectivity but, when used as an in-line preconcentrator, sample loading had to be done at lower flows than for MCX to ensure that the specific interactions between the sorbent and analytes occurred. As a result, only 22 μL of sample was injected during a 15-min injection in contrast to 135 μL that were injected when using the MCX preconcentrator. LODs ranged from 40 to 140 ng/kg for the MCX preconcentrator, which was 10 times lower than those obtained using MIPs (1.0-1.4 μg/kg). Fig. (**2**) shows the electropherograms corresponding to the in-line MISPE–CE–MS/MS analysis of the eight quinolones in milk samples. It should be noted that the main difficulty of using analyte concentrators is to make the SPE procedure compatible with the optimal conditions for the electrophoretic separation.

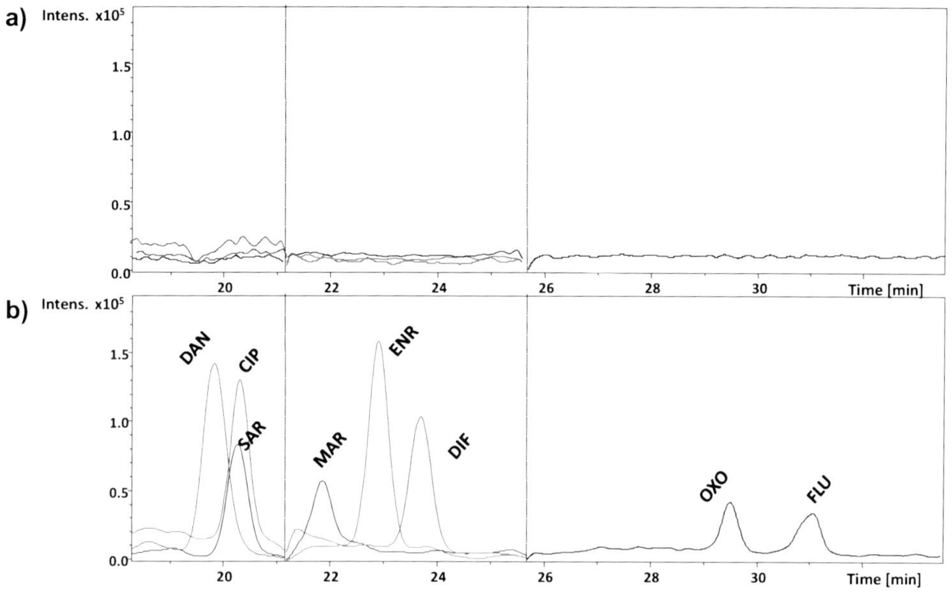

Fig. (2). Electropherograms resulted from the in-line MISPE–CE–MS/MS analysis of quinolones in milk samples: a) milk blank sample analyzed after analysis of a milk sample spiked at 100 µg/kg of each quinolone without observing carry over; b) bovine milk sample spiked at 10 µg/kg of each quinolone. Peaks identification: DAN (danofloxacin), SAR (sarafloxacin), DIF (difloxacin), ENR (enrofloxacin), CIP (ciprofloxacin), FLU (flumequine), MAR (marbofloxacin) and OXO (oxolinic acid). Reprinted with permission from [33].

Pesticides

The term pesticide encompasses insecticides, fungicides, herbicides, disinfectants, and any substance or mixture of substances intended for preventing, destroying, or controlling any pest or regulating plant growth [78]. It is well known that the use of pesticides provides great benefits because they are an inexpensive way to control pests. However, several studies have shown that the improper use of pesticides can cause adverse effects on human health such as anemia, cardiovascular and intestinal problems, kidney and liver damage, and endocrine disruption, among others [79, 80]. Aware of this problem, a large number of regulatory agencies around the world have set limits for pesticide residues in different food and feed products [81]. As an example, the EU has established maximum residue levels (MRLs) for pesticides in different food commodities under Regulation (EC) 396/2005 [82]. Within this framework, the development of sensitive and reliable methods that can detect these residues at very low concentration levels (*i.e.* µg/kg or µg/L) with an adequate selectivity is recommended.

Pesticides are classified in different families according to their chemical structure, being the most important organochlorines, organophosphates, carbamates, pyrethroids, sulfonylureas and triazines. In general, most of these compounds contain functional groups that can be ionized in a wide range of pH, so their separation is generally achieved by CZE methods. MEKC is typically applied to the separation of neutral pesticides such as carbamates [83]. CE with UV detection is the most common option for the determination of pesticides in food because UV detection presents several advantages, such as good linear dynamic range, robustness and great stability under variable separation conditions [84]. For example, Aparicio-Muriana *et al.* developed a MEKC-UV method for the determination of fipronil and fipronil-sulfone in eggs [85]. MEKC mode using APFO as micellar medium provided the best peak resolution among the different separation modes evaluated (*i.e.* CZE, NACE and MEKC with different micellar media). Baseline separation of both analytes was accomplished using a BGE consisting of 50 mmol/L APFO (pH 9.0) and 10% (v/v) of MeOH as organic modifier. In order to isolate these pesticides from the sample matrix, a simple sample treatment based on salting-out assisted liquid-liquid extraction (SALLE) was proposed using MeCN as extraction solvent and ammonium sulfate as salt. LOQs were 300 µg/kg for fipronil and 500 µg/kg for fipronil-sulfone, achieving good results in terms of precision and trueness. Although the concentration sensitivity achieved may be sufficient for some real cases, it would be necessary to improve the sensitivity for regulatory purposes. This fact is the main drawback of CE-UV methods, which is why the application of on-line and off-line sample preconcentration, the use of more sensitive detectors or a combination thereof have been proposed to overcome this drawback.

Off-line preconcentration approaches are the most widely used options to improve concentration sensitivity. In order to select a suitable sample treatment for CE, several points must be taken into account, such as good extraction efficiency, high analyte preconcentration, a sufficient selectivity and a final extract totally compatible with CE injection and separation. In this sense, DLLME is an excellent option to meet these requirements. As an example, a DLLME method was developed for the subsequent determination by NACE of imazalil, prochloraz and thiabendazole in fruits and juice samples [86]. The separation buffer consisted of a mixture of MeOH/MeCN (35:65, v/v) containing 30 mmol/L ammonium chloride and 0.5% (v/v) phosphoric acid. DLLME provided a preconcentration factor of 25. The main characteristics of the DLLME procedure include: trichloromethane as extraction solvent, tetrahydrofuran (THF) as dispersive solvent, pH of the sample solution at 6.0, amount of NaCl in the sample of 5% (w/v) and 10 s for the extraction time. Recoveries ranged from 72 to 102%, with RSDs less than 6.4%. LODs were below 0.8 µg/kg in all cases.

On-line preconcentration approaches may be related to the action of a pseudostationary phase or to the variation in the electrophoretic mobility of analytes at the boundary of two buffers with a distinct conductivity [15, 73, 87]. Sweeping can be considered as the most effective on-line preconcentration technique in MEKC mode. It is based on the interaction between the BGE, which contains micelles, and the analytes in a sample solvent similar to BGE but free of micelles. The focusing or accumulation effect occurs when the micelles penetrate the sample zone, 'sweep' the analytes and preconcentrate them in a narrow band at the boundary of the sample zone and the BGE. Sweeping can lead to preconcentration factors ranging from 10 to 5000 depending on the nature of the neutral analytes [88]. To date, the analysis of pesticide residues by CE applying sweeping as preconcentration approach has been reported in several articles [89 - 91]. For example, Fang *et al.* have applied the sweeping strategy to the analysis of triazine herbicide residues in cereals and vegetables [92]. First, the running buffer was optimized to obtain the separation of the eight triazines. It consisted of 25 mmol/L borate, 15 mmol/L phosphate, 40 mmol/L SDS and 3% (v/v) of 1-propanol (pH 6.5). The sample solution, consisting of 20% (v/v) MeOH, 25 mmol/L borate and 15 mmol/L phosphate (pH 6.5), was hydrodynamically injected into the capillary at 0.5 psi for 240 s. Compared to a conventional hydrodynamic sample injection in MEKC (*i.e.* 0.5 psi for 3 s), and as a consequence of the sweeping effect, sensitivity enhancement factors (SEFs) ranging from 496 to 603 were achieved. Applying this method, a LOQ of 0.1 μg/kg was obtained for all pesticides without any auxiliary enrichment method except sweeping. Finally, the concentration sensitivity achieved by this sweeping-MEKC method was compared to that obtained by a high performance liquid chromatography (HPLC)- UV method. The results revealed that the HPLC-UV method was unable to detect the analytes at concentration levels identical to the LOQ achieved by the proposed CE- based method.

Reversed electrode polarity stacking mode (REPS) is another on-line preconcentration technique aimed at improving concentration sensitivity. A REPS- MEKC-UV method has been developed for the simultaneous determination of carbamate residues in fruit juices [93]. The separation was accomplished in 8 min using 10 mmol/L borate buffer (pH 9.5) containing 60 mmol/L SDS as BGE. To perform REPS, the sample was hydrodynamically introduced into the capillary at 0.5 psi for 30 s (ca. 250 nL, 14% capillary length), using 10 mmol/L phosphate buffer (pH 7) as low-conductivity sample solvent. At this point, the sample matrix was removed by the action of the EOF when a negative voltage of -15 kV was applied for 30 s, while the stacked zone gradually moved back toward the inlet. Finally, the polarity was switched to +15 kV to carry out the separation. SEFs between 4 and 13 were obtained. The applicability of the method was tested on different juices using dilute-and-shoot approach as sample preparation. The LOD was less than 0.1 mg/L in all cases, being acceptable with respect to the MRLs for these insecticide residues in raw fruit samples.

In addition, various stacking modes can be combined to improve sensitivity as much as possible. As an example, two-step stacking modes, namely sweeping and micelle to solvent stacking (MSS) [94], have been combined for CE-UV analysis of quaternary ammonium herbicides (*i.e.*, paraquat and diquat) in beer [95]. Briefly, a plug of micelle solution consisting of 10 mmol/L SDS in 140 mmol/L sodium phosphate buffer (pH 2) was injected for 150 s at 50 mbar. Then, the basic conductive sample was injected at 50 mbar for 120 s. At this point, the focusing of analytes occurred between the micelle zone and the micelle-free sample zone. Finally, a solution rich in organic solvent (30%, v/v, MeCN) was introduced hydrodynamically at 50 mbar for 10 s in the capillary before applying the separation voltage. As a result, the focusing of analytes occurred a second time by reversing the effective electrophoretic mobility of the analytes through the use of an organic solvent and micelles. The application of this two-step stacking methodology allowed obtaining SEFs of 21 and 24 for paraquat and diquat, respectively. To isolate the analytes from the matrix, simultaneous electrophoretic concentration and separation was carried out as a simple sample treatment. A LOQ of 10 µg/L was achieved for both herbicides, also obtaining excellent results in terms of repeatability (RSDs < 3.9%). Thus, it is clear that CE-UV in combination with on-line sample preconcentration is a promising alternative in the food analysis field.

More sensitive CE-based methods can be developed using alternative detection systems to UV detectors. In this sense, ECL detection is a powerful analytical tool that combines the inherent sensitivity and wide linear range of chemiluminescence and the simplicity of electrochemistry. In ECL detection, the species generated at electrodes undergo high-energy electron transfer reactions. These reactions produce excited states that emit light [96]. However, the use of ECL to monitor pesticide residues in food is not frequent due to limited variety of CE-ECL commercial instruments and the strict control of the conditions that is necessary to carry out the ECL reaction. ECL has been used to detect three phenylurea herbicides in yam samples [97]. Analytes were electrokinetically injected into the capillary for 10 s at 10 kV and subsequently separated using 25 mmol/L phosphate buffer (pH 8.0) as BGE. In this work, the ECL reagent, namely tris(2,2-bipyridyl) ruthenium(II) or ($Ru(bpy)_3^{2+}$), was oxidized with a Pt disk (*i.e.* working electrode) using a detection potential of 1.13 V. Matrix solid phase dispersion (MSPD) was proposed as a sample treatment, achieving satisfactory recoveries which ranged from 90.0 to 99.2%. The LODs of the proposed method were equal or less than 0.008 µg/kg in all cases. The sensitive achieved by ECL reaction was comparable or even less than that obtained by HPLC-MS and GC-MS methods. This system (*i.e.* $Ru(bpy)_3^{2+}$–ECL) has also been explored for the analysis of glyphosate and aminomethylphosphonic acid in soybean

[98] and of pretilachlor in rice [99]. This type of detection shows great potential for monitoring pesticide residues in food due to its high concentration sensitivity. However, ECL detection is not suitable for implementation in routine analyses of pesticides because it lacks the specificity required for unambiguous identification of compounds, according to the most restrictive legislations [6].

MS is a selective detector that provides structural information about the molecules (*i.e.* the mass charge ratio, *m/z*), as well as great concentration sensitivity [100]. Some CE-MS methods have been developed for the analysis of various groups of pesticides in different matrices [83]. As an example, nine pesticides including thiabendazole, aminocarb, imazalil, atrazine, metazachlor, metoxuron, carbofuran, metosulam, and imazapyr were analyzed by CZE-MS/MS in corn using a triple-quadrupole (QqQ) mass spectrometer as analyzer [101]. Separation was accomplished in less than 6 min using 0.1 mol/L formic acid (pH 2.4) as BGE. In addition, a polyvinyl alcohol (PVA)- coated capillary was used to improve the reproducibility of migration times and reduce peak tailing. CE-MS hyphenation was performed by means of an ESI source, which was operated in positive mode. To close the electrical circuit, a sheath-liquid was used and was prepared by 5-fold dilution of the BGE with MeOH/water (1:1, v/v). The high sensitivity and specificity of MS detectors generally allows the use of generic sample preparation procedures such as Quick, Easy, Cheap, Effective, Rugged & Safe (QuEChERS) extraction. Thus, 15 g of corn sample was extracted with 15.0 mL of MeCN containing 1% (v/v) of acetic acid. Then, 6 g of magnesium sulfate and 1.5 g of sodium acetate were added and shaken for 1 min. Finally, a 1 mL aliquot of the supernatant was diluted with BGE (1:1, v/v). The proposed method allows the determination of the target pesticides using a very low injection volume (ca. 31 nL) but achieving low LOQs which ranged from 0.10 to 0.93 μg/kg. The trueness of the method was also satisfactory, achieving recovery values that ranged from 82 to 114% and RSDs not greater than 7.8%.

In general, CE-MS has been limited to the analysis of charged compounds by using CZE mode. The analysis of neutral pesticides by MEKC-MS presents several drawbacks such as suppression of the analyte signal and contamination of the MS related to non-volatile surfactants (*e.g.* CTAB or SDS). To overcome this latter limitation, the use of perfluorocarbon-based volatile surfactants such as APFO has become an alternative to coupling MEKC systems with MS. The compatibility of APFO as the pseudostationary phase in MEKC with MS has been successfully tested for the determination of carbamate residues in juices samples

[102]. The separation of 25 carbamates was carried out in approximately 35 min using a 100 mM APFO solution (pH 9.0) as running buffer (Fig. **3**). To improve the sensitivity of the method, a sweeping-based approach was applied in which sample solution consisted of 75 mmol/L APFO (pH 9.0). Vortex-assisted surfactant-enhanced emulsification liquid–liquid microextraction was proposed as sample treatment. Thus, 5 g of juice was placed in a 15-mL tube followed by injection of a mixture of 100 mmol/L APFO (pH 9.0), acting as an emulsifier (530 µL), and chloroform as extraction solvent (1300 µL). This mixture was subsequently shaken and centrifuged. Finally, the organic phase, containing the analytes, was collected and evaporated to dryness. The final residue was re-dissolved with the sample solvent. LOQs were less than 5 µg/kg, and below the MRLs established in raw fruits.

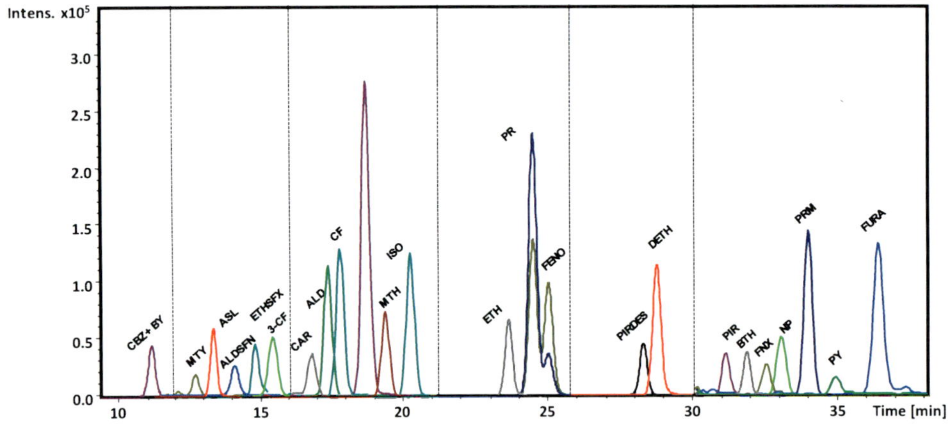

Fig. (3). Electropherogram of the analysis carried out by MEKC-MS/MS of carbamate residues in a spiked banana juice sample (10 mg/kg for each carbamate). Conditions: BGE, 100 mM APFO (pH 9.0); separation voltage, 23 kV; temperature of capillary, 24°C; sample injection, 50 mbar for 30 s; sample solvent, 75 mM APFO (pH 9.0); sheath-liquid, isopropanol/formic acid (99.9:0.1 v/v/v). Peak identification: carbendazim (CBZ), benomyl (BY), methomyl (MTY), asulam (ASL), aldicarb-sulfone (ALDSFN), ethiofencarb-sulfoxide (ETHSFX), carbofuran-3-hydroxy (3-CF), carbaryl (CAR), aldicarb (ALD), carbofuran (CF), propoxur (PX), methiocarb (MTH), isoprocarb (ISO), ethiofencarb (ETH), promecarb (PR), fenobucarb (FEN), pirimicarb- desmethyl (PIRDES), diethofencarb (DETH), pirimicarb (PIR), benthiocarb (BTH), fenoxycarb (FNX), napropamid (NP), propamocarb (PRM), pyraclostrobin (PY) and furathiocarb (FURA). Reprinted with permission from [102].

It should be noted that the analytical performances of these CE-MS methods mentioned above were comparable to those obtained by LC-MS or GC-MS. However, CE-MS is considered more cost-effective for routine analysis due to the relatively less expensive capillaries and small amounts of organic solvents used.

CE ANALYSIS OF FOOD CONTAMINANTS

Toxic Metals

Toxic metals can be present in food due to contamination during industrial processes or through environmental sources related to primary activities (*i.e.* agriculture and fishing). The risk of these substances is that they tend to bioaccumulate in biological tissues and, once found in certain concentrations, their related toxicity can cause serious health problems or even lead to death. Specifically, there is great concern about the presence of arsenic, tin, cadmium, lead and mercury in food because they are considered the most harmful metals. CE-UV has been proposed for the control of various metals including toxic ones (*i.e.* Ni^{2+}, Co^{2+}, Cu^{2+}, Cd^{2+} and Hg^{2+}), in food- related products (*i.e.* crayfish and beebread) [103]. Since some metals do not show UV-absorption properties, their complexion with complexing agents such as imidazole, 1,10-phenanthroline, 18-crown ether-6 or L-cysteine is required to apply UV-detection. As shown by Song *et al.*, metal complexion can be easily carried out on-line, as it is sufficient to inject the complexing agent just before the sample plug [103]. Briefly, the complexing agent (*i.e.* 1,10-phenanthroline) was injected at 50 mbar for 3 s into a capillary preconditioned with BGE. The sample was hydrodynamically injected immediately afterwards at 50 mbar for 60 s, and separation was subsequently carried out at 16 kV and 25°C using a 200 mmol/L sodium acetate solution containing 50% (v/v) MeOH (pH 5.5) as BGE. In addition, sample solution consisted of 7.2 mmol/L SDS and 180 mmol/L sodium acetate (pH 5.5), so detection sensitivity was improved due to the performance of a MSS approach. This preconcentration strategy provided a SEF of 45, which resulted in LODs between 2.66 – 27.9 µg/L for all analytes.

It is recommended that analytical methods intended to monitor toxic metals in food consider metal speciation because their toxicity depends on the concentration of specific species [104]. Various CE-based methods have been developed for mercury speciation [105]. For example, CZE-UV has recently been applied to the determination of different mercury species (*i.e.* methylmercury, ethylmercury, phenylmercury) in fish [106, 107]. However, CE together with inductively coupled plasma-mass spectrometry (ICP-MS) is preferred for metal speciation due to the selectivity and sensitivity of this detection technique [108]. In this sense, CE-ICP- MS has been shown to be even more sensitive than CE-ESI-ToF-MS, which is also a highly selective technique and is increasingly used in contaminant analysis, for the determination of different arsenic species (*i.e.* arsenobetaine, As(III), As(V), and dimethylarsinic acid) in fish samples [109].

CZE-ICP-MS has also been investigated for mercury speciation and the further determination of Hg^{2+}, methylmercury, and ethylmercury in fish samples [110]. Microwave-assisted extraction with HCl was proposed as sample treatment, and methylmercury was finally found in fish muscle samples at concentrations between 1.117–1.221 μg/g Hg of dry sample weight. Furthermore, CZE-ICP-MS has been evaluated for lead speciation and subsequent analysis of Pb^{2+}, trimethyl lead and triethyl lead in clam and oyster samples [111]. The running buffer consisted of 70 mmol/L H_3BO_3, 17.5 mmol/L $Na_2B_4O_7$, 0.4 mmol/L CTAB, 2% (v/v) MeOH solution (pH 8.85), and the voltage polarity was reversed at this method (*i.e.* -12.5 kV as optimum separation voltage). CTAB was added to the BGE at a concentration (*i.e.* 0.4 mmol/L) below its critical micelle concentration (CMC) (*i.e.* 0.9-1.0 mmol/L) to prevent adsorption of trimethyl lead and triethyl lead on the capillary surface, and reverse the EOF (from cathodic to anodic EOF). CTAB was found to improve separation between both analytes as well. Under these separation conditions, Pb^{+2} was not detected because it always migrated towards the cathode even at high EOF. Therefore, the complexion with EDTA was required to modify its migration nature. Finally, the application of this method led to the detection of Pb^{2+} at levels of 0.38 and 1.68 μg/g in clam and oyster samples, respectively.

One of the biggest challenges of metal speciation is not only related to the separation of metal species *per se*, but rather ensuring that there are no changes in species distribution during the execution of the analytical workflow [108]. In this sense, the extraction method must quantitatively extract each metal species from samples without altering the individual species. Microwave-assisted extraction is generally performed for extraction of metal species present in food prior to analysis by CE- ICP-MS, as already mentioned for extraction of mercury species [110], and also applied to the extraction of lead species [111]. This extraction method, using MeOH/water (1:1, v/v) as extraction solution, has also been used for the determination of arsenic species in clam and shrimp samples [112]. In this study, monomethylarsonic acid, dimethylarsinic acid, As(III) and As(V) were analyzed by CZE-ICP-MS, and the BGE consisted of 20 mmol/L NaH_2PO_4-5.00 mmol/L $Na_2B_4O_7$ (pH 6.50). Finally, As(III) and dimethylarsinic acid were found in clam samples while As(III) and monomethylarsonic acid were present in shrimp samples.

Biogenic Amines

Biogenic amines are nitrogenous organic bases that result from the decarboxylation of amino acids and are found mainly in fermented food and beverages (*e.g.* cheese, wine, and beer) as well as in fish, seafood, meat, fruit and vegetables. The presence of biogenic amines in food is an indicator of their freshness and spoilage because they are associated with unwanted microbial activity. High levels of these contaminants, which occur during food production and storage, can cause several physiological and toxicological effects including headaches, migraines, disturbances in blood pressure and adverse effects in cardiovascular, gastrointestinal and respiratory systems, among others. Although there is a wide variety of biogenic amines that may pose a risk to human health (*e.g*. putrescine, cadaverine, phenylethylamine, spermine, spermidine, tyramine, tryptamine, histamine, *etc*.), there is still a lack of information on toxic doses for each biogenic amine. As a consequence, current food safety legislation has only established thresholds for the presence of histamine in food, mainly in fishery products [113].

CE has been widely applied to the determination of biogenic amines in food and a large number of CE-based methods involving different detection techniques have been reported [114]. Since UV detection is generally integrated into commercial CE instruments, it is typically the first detection option considered in CE applications as is also seen in the analysis of biogenic amines in food. In general, biogenic amines have low-absorption properties, so they must undergo derivatization for UV detection in order to achieve satisfactory detection sensitivity [115]. Since derivatization can be time-consuming, the development of alternative strategies to avoid this analytical step is pursued. The analysis of thirteen aliphatic, aromatic and heterocyclic biogenic amines in beer samples has been carried out by CZE-UV without derivatization in less than nine minutes [116]. The addition of α-cyclodextrins (CDs) to the BGE was required to improve selectivity and achieve the separation of some pairs of compound such as octopamine and dopamine, which are structural isomers. Furthermore, the simultaneous detection (wavelength (λ) = 200 nm) of biogenic amines with and without UV absorption was achieved due to the addition of imidazole to the BGE, which acted as a UV absorbing probe (*i.e.* indirect UV detection approach). Biogenic amines with lower UV absorptivity than imidazole were detected as negative peaks (*i.e.* spermidine, putrescine, histamine, cadaverine), whereas the other analytes resulted in positive electrophoretic peaks (*i.e.* phenethylamine, tyramine, tryptamine, 5-hydroxytryptamine, octopamine, dopamine, norepinephrine, epinephrine, carnosine). In general, LODs less than 3.67 µmol/L were achieved for all compounds.

As an alternative to traditional CE-UV methods, CE coupled with LIF [117, 118], ECL [119], conductivity [120, 121] or MS [112, 123] detection has also been evaluated for the analysis of biogenic amines. Although high sensitivity can be achieved when using LIF detection, biogenic amines do not show native fluorescence. Therefore, as with UV detection, they must be derivatized to be detected by LIF. The list of compounds that can be used for amine labeling is large, including naphthalene-2,3-dicarboxaldehyde (NDA), 3-(4-fluorobenzoy-)-2- quinolinecarboxaldehyde (FBQCA), 3-(2-furoyl)-quinoline-2-carboxaldehyde (FQCA), *o*-phthalaldehyde (OPA), 4-fluor-7-nitro-1,2,3-benzaxadiazole (NBD-F). FBQCA is not fluorescent *per se*, but reacts with amines to form highly fluorescent isoindole derivatives in the presence of cyanide. This labeling approach has been proposed for MEKC-LIF analysis of seven biogenic amines in fish samples as well as in soy sauce and red wine [117]. Although derivatization was shown to be a solution for the analysis of biogenic amines by CE-LIF, it was also observed that this method cannot be applied to the analysis of polyamines such as cadaverine. Their derivatization can lead to various derivatized molecules, leading to double or more electrophoretic peaks. Therefore, CE-UV and CE-LIF methods involving derivatization can be limited to the analysis of biogenic amines having only one amine group where this issue is not a drawback. For example, a microchip capillary electrophoresis (MCE) with LIF detection has been developed for rapid determination (less than 65 s) of gizzerosine in fish meals [118]. Gizzerosine was fluorescently labeled with fluorescein isothiocyanate in an overnight reaction and, as expected, only one electrophoretic peak was identified with the derivatized gizzerosine compound. A LOD of 25 μg/kg was reached.

Since optical detectors have several limitations for the determination of biogenic amines (*i.e.* need of derivatization, low sensitivity of indirect UV detection methods), CE with conductivity detection is a feasible alternative for the analysis of these compounds. C^4D is typically used because the absence of physical contact between the electrodes and test solution prevents contamination and aging, which is related to adsorption of solution components or interfacial electrochemical processes [121]. For example, CZE-C^4D has been applied to the determination of eight biogenic amines (*i.e.* spermine, spermidine, histamine, putrescine, cadaverine, tryptamine, tyramine and β-phenylethylamine) in alcoholic beverages (*i.e.* brandy and Chinese hard liquors) [124], and to the analysis of five biogenic amines (*i.e.* cadaverine, histamine, spermidine, putrescine and tyramine) in fermented dairy products (*i.e.* cheese, yoghurt and kefir) [121]. An innovative CZE-C^4D method has recently been developed for the simultaneous analysis of cations and anions in beers, and including ten biogenic amines, eleven inorganic ions and sixteen amino acids [125]. Normally cations and anions are not analyzed simultaneously by CE because they tend to move towards opposite electrodes and, in addition, the detector is placed near or after the outlet electrode. Exceptions occur when the cathode is the outlet electrode and the electroosmotic mobility prevails over the electrophoretic mobility of anions. They are detected after cations and neutral compounds although, in these cases, the analysis time can be extremely long. In the mentioned work, Guo *et al.* have proposed a bi- end injection approach in which detection (*i.e.* C^4D) takes places 'at the middle' of the capillary to overcome this limitation. In short, the cations were injected into the anode by siphoning (*i.e.* 15 s at a drop height of 10 cm) followed by electrokinetic injection of anions into the cathode (*i.e.* 5 s at −1.2 kV). Separation was finally carried out in less than 24 min under a voltage of +18 kV, using 2.5 mol/L acetic acid as BGE. The distance between the anode and the detector was 50 cm, which was almost enough for the baseline separation of all cations (n = 34). A shorter capillary length (*i.e.*, 20 cm) was required for the separation of anions (n = 3). Finally, spermidine, putrescine and ethanolamine were detected at levels of 1.0-6.1 mg/L in the analyzed beers.

In general, coated capillaries are used in CE-MS methods intended for the analysis of biogenic amines [122, 126], although CE-MS methods using uncoated capillaries have also been reported [123, 127]. The coating prevents the adsorption of analytes on the capillary surface of silica, avoiding peak tailing and improving the reproducibility of retention times as well as the separation efficiency. In contrast, the EOF is suppressed, so the separation of compounds only depends on their electrophoretic mobility. For example, a commercial PVA-coated capillary has been investigated for the analysis of nine biogenic amines (*i.e.* spermine, spermidine, putrescine, cadaverine, histamine, 2-phenylethylamine, tryptamine, tyramine and urocanic acid), and 1,7-diaminoheptane used as internal standard, in beer and wine samples [122]. BGE consisted of 0.5 mol/L acetic acid (pH 2.5), and the separation time was less than 10 min. A QqQ mass spectrometer was used for detection and operated in positive multiple reaction monitoring (MRM). Therefore, two fragmentation ions were identified for each biogenic amine, providing their unambiguous identification. LODs ranged from 1 to 2 µg/L. Commercial PVA- coated capillaries cannot be used under certain experimental conditions; therefore, alternative laboratory-prepared coatings have also been investigated. Meixner *et al.* used a hydrophilic polymer, namely *N*-acryloylamido ethoxyethanol, as capillary coating, which showed high stability against abrasive solvents as a strong base and a strong acid (*i.e.* 1 mol/L NaOH or HCl, respectively) [126]. High reproducibility, in terms of migration time (RSDs < 3%) and peak areas (RSDs 6.4–17.3%), was observed for the analysis of polyamines (*i.e.* spermine, spermidine and putrescine) in fish eggs.

Natural Toxins

The WHO defines natural toxins as toxic compounds that are naturally produced by living organisms which are not harmful to the organisms themselves but can be toxic to other creatures, including humans, when eaten [128]. The JECFA has carried out formal health risk assessments of these contaminants [129]. Considering their origin, it is possible to establish several classes of natural toxins [130], differentiating between bio-contaminants, produced by microorganisms, which affect food and food products, including mycotoxins (produced by fungi), bacterial toxins (produced by bacteria, being mainly proteins) or phycotoxins (produced by algae); or other which are inherent components of plants and animals such as plant toxins or phytotoxins, produced by edible plant species, and zootoxins, produced by animals, such as snakes, scorpions, *etc*. The Codex Alimentarius Commission has established international standards, codes of practice and maximum levels of several chemical hazards in food and feed based on JECFA assessments [131], and specifically related to the category of natural toxins (*i.e.* mycotoxins, phycotoxins and phytotoxins) [129]. The implementation of international regulations for the control of these contaminants in food requires the establishment of efficient and sensitive analytical methods [132]. In general, LC-MS is the most widely used analytical technique for this purpose where sample preparation is a crucial step in achieving selective isolation and/or preconcentration of toxins from complex matrices and an adequate clean-up of the extract before instrumental analysis [133]. However, some CE-based methods have been successfully developed, mainly applied to the monitoring of mycotoxins and phycotoxins.

In food analysis, the most relevant class of natural toxins, which present a major food safety issue, are mycotoxins. They are toxic secondary metabolites produced by fungal species that easily colonize crops and contaminate them in the field or after harvest, causing important problems. They typically affect cereals, dried fruits, nuts and spices, derived products such as beer and wine, leading to huge agricultural and industrial losses [134, 135]. Mycotoxins can be also present in the human food chain through meat or other animal products such as eggs, milk and cheese, as a result of livestock eating contaminated feed, presenting adverse effects on animals due to their cytotoxic, genotoxic, immunotoxic, carcinogenic, or teratogenic properties [136, 137]. The mycotoxins currently considered most significant are aflatoxins, ochratoxin A, trichothecenes (HT-2, T-2 toxin and deoxynivalenol), zearalenone, fumonisins B1 and B2, patulin, citrinin and ergot alkaloids. In this sense, the European Commission has established maximum permitted contents of these compounds in different food commodities [138]. Subsequent amendments, scientific opinions on risks and recommendations have also been reported [139], along with sampling and analysis methods for their official control [140].

CE-based methods can been found to monitor patulin in fruit juices, since its occurrence as a natural apple contaminant is a worldwide problem. MEKC with UV detection is as an efficient alternative to LC to determine this mycotoxin, achieving its separation from 5-hydroxymethylfurfural (HMF), which is its main analytical interference and being both indicators of quality in apple juice [141]. BGE consisted of 35 mmol/L sodium tetraborate buffer (pH 9.0), 65 mmol/L SDS and 5% (v/v) MeCN. Liquid-liquid extraction was applied as sample treatment, using acetyl acetate as extraction solvent and achieving a LOD of 0.7 µg/L [142]. DLLME has also been proposed as an easy, cheap and environmentally friendly sample treatment for determination of patulin in juice samples and prior to analysis by MEKC-UV [143]. For the DLLME procedure, a mixture of 2-propanol and chloroform was used as dispersive and extraction solvent, respectively. Recoveries greater than 75% and a SEF of 10 were achieved. Additionally, CZE-UV has also been applied to the analysis of patulin in apple juice samples, using 25 mmol/L of sodium tetraborate buffer with 10% (v/v) MeCN (pH 10) as BGE. However, worse results were obtained in terms of sensitivity and trueness compared to MEKC-based methods [144].

MEKC-UV has also been applied to the determination of citrinin in food supplements based on red yeast rice, produced by the fermentation of rice substrates with *Monascus purpureus* [145]. These dietary supplements reduce blood-lipid levels in humans due to the presence of monacolin K, which is identical to lovastatin, the statin drug with the same effect. The proposed method was able to determine citrinin and the bioactive compound lovastatin in order to control the quality of the product. It is due to the controversy over red yeast rice, which started when its effectiveness was related to the fact that it contains this naturally occurring form of the statin drug, discovered as the major active ingredient. After extraction in MeOH/water (80:20, v/v), drying and reconstitution in MeOH, both compounds were analyzed in 2 min. BGE consisted of 20 mmol/L phosphate buffer (pH 7.0) and 30 mmol/L SDS. Significant differences were shown between the content of lovastatin and its acid form in the dietary supplements studied, suggesting that analytical controls are needed. CZE-UV has also been applied to evaluate the efficiency of immunoaffinity columns (IAC) manufacture to selectively extract citrinin from red yeast rice and monascus color [146].

CZE-UV has been used for the analysis of ergot alkaloids in cereal samples, applying cloud point extraction (CPE) as sample treatment and which involves extraction in acid conditions and preconcentration in a low volume of a non-ionic surfactant [147]. These compounds were determined together as total content (*i.e.* by a single identified peak and expressed as ergotamine), and separately (*i.e.* ergotamine and ergonovine) by adding β-CD to the BGE at pH 4 to achieve electrophoretic separation. The method showed LODs of 2.6 and 2.2 µg/kg for ergotamine and ergonovine, respectively, and recoveries greater than 92% in samples of flour of different origin, grain samples and infant cereal-based products.

Electrochemical detection (EC) has been applied for the screening and quantification of zearalenone and its metabolites (*i.e.* α- and β-zearalenol) in maize flour samples, since these compounds have a macrocyclic structure with a polyphenolic side susceptible to electrochemical oxidation. Supercritical fluid extraction was applied as sample treatment [148]. The screening method was carried out by CZE using borate (pH 9.2) as running buffer and monitoring the amperometric signal with a carbon paste electrode. The positive samples were then subjected to CZE analysis for quantification in which the BGE consisted of borate buffer modified with 30% (v/v) MeOH at pH 9.7. A similar CZE method was applied for the determination of zearalenone in feed, flour, grain, and cereal-based foods, using sodium tetraborate buffer (pH 9.0) with 15% (v/v) MeCN as BGE [149]. Samples were initially extracted with MeOH/1% (v/v) aqueous NaCl (60:40), defatted with n-hexane and dichloromethane, dried and redissolved in MeOH/water (40:60). Subsequently, they were subjected to a clean-up procedure based on SPE with C_{18} cartridges and resulting in a LOD of 8.25 μg/L.

LIF can also be combined with CE as sensitive detection for the control of mycotoxins that present native fluorescence such as certain aflatoxins, ochratoxins, zearalenone and related congeners. CDs has been investigated in CE-LIF methods as enhancers of native fluorescence of aflatoxins B1 and G1, zearalenone and the fluorescent pyrene derivative of T-2 toxin (T2-Pyr) [150]. The effect of heptakis (2,6- di-O-methyl)-β-cyclodextrin (DIMEB) was especially relevant for the enhancement of T2-Pyr fluorescence. Finally, the T-2 toxin was detected in maize samples using IAC as sample treatment and CE-LIF as analytical technique. DIMEB was added to the BGE to improve fluorescence and a He/Cd laser at 325 nm was used for detection. A LOD of 50 μg/kg was obtained. MEKC-LIF has also been evaluated for the simultaneous determination of aflatoxins B1, B2, G1 and G2 in rice samples [151]. In this method, BGE consisted of 20 mmol/L borate buffer (pH 8.5), 30 mmol/L SDS and 7% (v/v) MeCN, and sweeping was applied as on-line sample preconcentration procedure (Fig. **4**). Sample extraction was carried out with MeOH/water (70:30) and, subsequently, IAC was applied for sample clean-up. LODs were in the range of 0.04-0.52 μg/kg, with recoveries greater than 93.0%.

In recent years, microfluidic chips are emerging as on-site and lab-independent systems for mycotoxin detection, reducing analysis time and reagent consumption. Among them, the flow-through capillary electromigration microchip represents a typical format in microfluidic immunosensors in which analytes are injected into the inlet port and then flow perpendicularly through the microchannel by CE. Optical detection is typically used in this type of system, mainly LIF, as well as EC detection. Some applications of this methodology include the analysis of zearalenone in infant foods, ochratoxin A in coffee extracts or citrinin in rice [152]. Recently, the use of aptamers, as an upgraded version of antibodies for molecular recognition, has emerged as an interesting approach for mycotoxin determination. An aptamer-based microchip CE system has been used in conjunction with LIF detection for rapid determination of aflatoxin B1 and ochratoxin A in food, reaching LODs of 0.026 and 0.021 µg/L for each mycotoxin, respectively [153]. Analysis was accomplished in a single run in less than 3 min, although the aptamer binding procedure required 30 min. Satisfactory recoveries were obtained for cereals and vegetal oils. The specificity study indicated that other major mycotoxins would not cross-react with these two aptamers, demonstrating the good selectivity of the proposed method and its great potential in food safety screening.

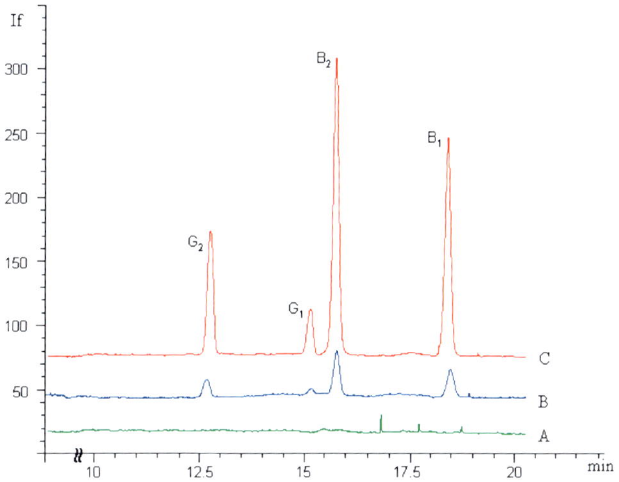

Fig. (4). Electropherogram obtained from the CE-LIF analysis of rice sample extracts: a) blank sample; b) extract spiked at 5 µg/L of aflatoxins B1 and G1 and 1.5 µg/L of aflatoxins B2 and G2; (c) extract spiked at 25 µg/L of aflatoxins B1 and G1 and 7.5 µg/L of aflatoxins B2 and G2. Reprinted with permission from [151].

Phycotoxins are produced by algae, such as dinoflagellates, other flagellated phytoplankton, and cyanobacteria that inhabit marine, brackish, or freshwater environments. Due to their expansion and adverse effects in humans, the European Food Safety Authority (EFSA) has recommended different permitted limits of these toxins in shellfish [154, 155] and regulations related to the performance of official controls have been established [156]. There are several phycotoxin groups depending on its chemical structure and toxic effects on humans, including: hydrophilic or polar toxins, such as domoic acid which causes amnesic shellfish poisoning (ASP), tetrodotoxins (TTX) and saxitoxins, causing paralytic shellfish poisoning (PSP); and lipophilic toxins including brevetoxins causing neurologic shellfish poisoning (NSP), okadaic acid, dinophysistoxins, and pectenotoxins that cause diarrheic shellfish poisoning (DSP), azaspiracids, responsible of azaspiracid shellfish poisoning (AZP) and ciguatoxins which belong to a family of toxins that causes ciguatera fish poisoning (CFP) and are linked to a wide variety of gastrointestinal, neurological, or cardiovascular symptoms [157]. The most important marine phycotoxins include saxitoxins, domoic acid, brevetoxins, and ciguatoxins. Among the most important cyanobacterial phycotoxins (cyanotoxins) are microcystins, nodularin, cylindrospermopsin, Lyngbya toxins, anatoxins, saxitoxins, yessotoxins and β-N-methylamino-L-alanine (BMAA). Exposure to these toxins in freshwater environments is more frequent through direct consumption of contaminated water.

In relation to the application of CE to the analysis of cyanotoxins, different modes and detections (*i.e.* CZE-UV, MEKC-UV and CZE-MS) have been evaluated for the determination of yessotoxin and its analog, 45-hydroxyyessotoxin, in mussel samples and algae cultures [158]. In this study, extraction was carried out with MeOH/water, followed by washing with hexane and sample clean-up by SPE with C_{18} cartridges. MEKC offered the best selectivity, allowing the resolution of both analogs, but CZE- MS was preferred for confirmation purposes. Sensitivity was significantly improved by applying on-line preconcentration approaches. LVSS was a more efficient approach than FASS for analyte enrichment but it was not compatible with CE-MS analysis. FASS-CE-ESI-MS has also been applied to the determination of yessotoxins and pectenotoxin (PTX-6) using ammonium acetate (pH 8.5) as running buffer and a sheath-liquid consisting of isopropanol/water (80:20, v/v) containing 0.1 mmol/L ammonium acetate [159]. LODs of 0.01 and 0.13 mg/kg were achieved for yessotoxin and PTX-6, respectively, with recoveries greater than 97.7% for their determination in mussel samples. No significant matrix effects were observed for the analysis of contaminated seafood samples.

CE is an ideal alternative for the analysis of polar phycotoxins, which are more challenging to analyze by MS-based methods than lipophilic toxins. The separation of polar toxins by RPLC requires either the use of strong ion-pairing agents that are not compatible with or may limit the sensitivity of ESI-MS, or the use of HILIC, which involves long separation time and desalting SPE procedures. CE-MS/MS has been applied for the monitoring of BMAA which is a cianobacterial neurotoxin [160]. The separation was carried out in 5 mol/L formic acid in water/MeCN (9:1, v/v) as BGE, and FASS preconcentration was applied. Previously, samples were submitted to cation-exchange clean-up to decrease their ionic strength. This method was characterized by a LOD of 16 μg/kg and provided several advantages such as excellent resolution of amino acid isomers and low matrix effect compared to a LC- MS/MS method. Finally, BMAA was detected in cycad, mussel and lobster samples, with recoveries between 70-100%. Following this strategy, the same research group has proposed a CE-MS/MS method for the sensitive multiclass analysis of paralytic shellfish toxins, tetrodotoxins, and domoic acid in seafood [161]. The method exhibited a high degree of selectivity between toxin isomers and matrix interferences. LODs in mussel tissue were 5.2 μg/kg for tetrodotoxins, 0.160 mg/kg for domoic acid, and between 1.8 μg/kg and 0.120 mg/kg for paralytic shellfish toxins.

Different CE modes and detectors such as UV, C^4D, fluorescence detection (FLD) and MS have been evaluated for the analysis of twelve PSP toxins in mussel samples [162]. CZE-UV and CZE-C^4D provided better resolution, selectivity and separation efficiency compared to CZE-MS and MEKC-FLD; however, greater sensitivity was achieved using CZE-C^4D and MEKC-FLD, with LODs between 140-715 and 60.9- 104 μg/L, respectively, and which are lower than the limits allowed in shellfish samples. Although CZE was not suitable for the separation of neutral C1 and C2 toxins, CZE-C^4D allowed the determination of ten of the charged toxins within a few minutes. The MEKC-FLD method was suitable for rapid screening of all twelve toxins in contaminated mussel samples, which were pretreated with 1% (v/v) acetic acid solution and subjected to SPE with C_{18} cartridges. Although the MEKC-FLD method provided lower sensitivity than the HPLC-FLD method used in this study, it allowed the direct quantification of saxitoxin, decarbamoylneosaxitoxin, and C1, C2 toxins. In a subsequent study, sensitivity was improved by applying a counter-flow transient isotachophoresis (tITP)-CZE method with UV and C^4D detection for the monitoring of ten PSP toxins in mussel samples, reducing matrix effects [163].

In addition, CE-ICP-MS has been used for determination of saxitoxin where the toxin was subjected to metal chelate labeling with diethylenetriamine-N,N,N',N",N"- pentaacetic acid as chelating reagent and Eu^{3+} as labeling element [164]. The analyte was previously extracted with dimethyl sulfoxide from dried mussel samples. The proposed method allowed the determination of traces of saxitoxin with a LOD of 0.38 fmol, without significant interferences, high sensitivity and robustness and low cost. Furthermore, the combination of CE-EC detection with the ligand specificity of immunoassay (IA) has demonstrated to be a useful alternative for the determination of these toxins. For example, brevetoxin-B has been determined in shellfish with this technique based on the competitive immunoreaction between the horseradish peroxidase (HRP)-labeled brevetoxin-B (Ag*) and the free brevetoxin-B (Ag) with a limited amount of antibrevetoxin-B antibody (Ab), and the immunological species were separated by CE [165]. The enzymatic product, 2-aminophenoxazine-3-one, produced from the oxidation of H_2O_2-o-aminophenol with H_2O_2 catalyzed by HRP, was detected amperometrically at a Pt electrode at the outlet of the reaction capillary. The method was finally applied to mussels, clams, cockles, scallops, and oysters, achieving a LOD of 0.1 μg/L and recoveries of around 95%. CE-IA-EC has also been used for the monitoring of saxitoxin, brevetoxin, domoic acid, and okadaic acid in shellfish samples by using gold nanoparticles (AuNPs), which modifies the mobility of analytes and consequently improves separation resolution [166], as well as for the determination of ciguatoxin CTX1B in fish samples [167].

Food Processing Contaminants

Various types of food contaminants are generated during food processing, especially during thermal processes. Heating treatments can lead to the formation of substances such as acrylamide, chloropropanols, chloroesters, polycyclic aromatic hydrocarbons (PAHs), HMF, furanes, acrolein and nitrosamines which pose a risk to human health due to their carcinogenic and mutagenic properties [168]. Therefore, analytical methods are required to control their presence in food and monitor their formation during food production [169]. CE analysis of a wide range of these contaminants has been proposed in the last years.

HMF represents one of the most common by-products generated in over-processed foods and is generally found in bakery products, malt, coffee, and vinegar [168]. Like other furfural compounds, it is a characteristic product of the Maillard reaction at an advanced stage. This non-enzymatic reaction occurs in foods containing amino acids and reducing sugars, and is responsible for browning the food during the heating treatment. The presence of HMF is generally used as an indicator of the possible loss of food quality due to excessive heat treatment or poor storage conditions [170]. HMF is a neutral molecule within a wide range of pH (pka 12.8), so MEKC mode using SDS as micellar medium is typically applied for CE analysis. MEKC-UV approaches have been developed for the determination of HMF in honey [171, 172] and milk [173, 174].

For example, Güray *et al.* have developed a MEKC-UV method for the determination of HMF in honey samples, in which the BGE consisted of 10 mmol/L tris buffer, 110 mmol/L SDS and 10% (v/v) MeOH (pH 9.1) [175]. HMF was found in fresh and processed (*i.e.* heated at 65°C, 85°C and 105°C for 5 h and 20 h) multifloral and honeydew honeys, and the results were compared with those obtained by HPLC-UV. HMF concentration ranged from 6.3 mg/kg in fresh honeydew honey (*i.e.* without thermal treatment) to 1.4 g/kg in floral honey processed at 105°C for 20 h. The results obtained by MEKC-UV and HPLC-UV were not statistically different, which shows that CE can be a real alternative to LC in applications related to food safety, such as the analysis of HMF. The validity of this hypothesis has been demonstrated by Teixidó *et al.* who determined the concentration of HMF in a wide range of food products (*i.e.*

cereals, toasts, honey, orange juice, apple juice, jam, coffee and chocolate) by MEKC-UV and LC-MS/MS, and similar results were achieved by both analytical techniques [176].

Furthermore, Bignardi *et al.* have also compared CE and HPLC techniques for the analysis of HMF in cereal-based baby foods, coffee, soft beverages and vinegars, and results were not significantly different either [177]. In this last study, CE was coupled with an IT mass spectrometer, and selected MS/MS ion monitoring (SMIM) was applied as acquisition mode. Unlike other CE methods intended for the determination of HMF, the separation was performed under CZE conditions and the migration time of HMF coincided with the EOF. BGE consisted of 50 mmol/L formic acid (pH 3), and despite the low pH, ensured that the EOF was not completely suppressed. On the other hand, this low pH should guarantee that most of the matrix molecules were charged so they did not move with the EOF and separated from the analyte of interest. In any case, HMF was distinguished from co-migrating compounds by the mass of its protonated ion (*m/z* 127) and its fragmentation ions (*m/z* 81 and 109). Finally, the method was fully validated in terms of LOD (30 µg/L), linearity (0.1 – 20 mg/L), and intra-day (RSDs < 6% for peak area) and inter-day (RSDs < 16% for peak area) precision. It should be noted that SDS, which is the most widely used surfactant in MEKC methods, is not volatile and therefore its use is not compatible with MS.

Like HMF, acrylamide is also a product of Maillard reaction and forms naturally when carbohydrate-rich foods are fried, baked, or roasted at high temperatures. Its presence in food is a health concern because it can induce carcinogenesis [168] and, as a consequence, the EU has launched several recommendations for monitoring acrylamide in food [178], including reference levels in different food products [179]. Acrylamide is an extremely weak base, so CZE is not the first choice for the analysis of this near-neutral molecule. However, derivatization has been proposed to modify the nature of acrylamide by introducing an ionizable group, so CZE can be applied for analysis. Acrylamine has been determined by CZE-UV in French fries, breakfast cereals, and biscuits after SPE, using Strata--C and Isolute ENV+ cartridges in tandem and further derivatization with 2-mercaptobenzoic acid [180]. This approach has also been applied to the analysis by CZE-MS of acrylamide in potato crisps, biscuits, crisp bread, breakfast cereals and coffee [181]. Derivatization with cysteine has also been proposed for the analysis of acrylamide by CZE-C^4D, using 10 mmol/L triethylamine as BGE, in freshly fried potato products and packaged potato chips previously subjected to QuEChERS extraction [182]. C^4D cannot be directly applied to the detection of acrylamide due to its near-neutral nature, so in this last study, derivatization played an important role in both CE analysis and subsequent detection by C^4D.

To avoid derivatization, MEEKC [183], MEKC [184], and NACE [185] have been investigated as alternatives to CZE for the analysis of acrylamide in food. Analyte focusing by micelle collapse (AFMC) has also been proposed for the determination of acrylamide in bread samples as an alternative to derivatization but still applying CZE as separation mode [186]. In AFMC-CZE, analyte injection solution contains micelles, usually SDS at a concentration above the CMC, and has a greater conductivity than BGE. Micelles are responsible for transporting non-ionized analytes. Since BGE does not contain micelles, the dilution of the SDS of the sample occurs at the boundary between the sample and the BGE (micellar dilution zone). As a consequence, SDS concentration falls below CMS, causing the collapse of the micelles and the release of the analytes. The continued transport and release produce the accumulation of the analytes in the micellar dilution zone, which in turn leads to their concentration [187]. An AFMC-CZ--UV method has been developed for the analysis of acrylamide using 12.5 mmol/L phosphate as BGE (pH 8.5) [186]. Instead of using SDS as the micellar medium, 1-butyl-3-methylimidazolium bromide (BMIMBr), which is an ionic liquid, was employed for this purpose. Finally, sample solvent consisted of 3.0 mol/L BMIMBr and samples were hydrodynamically injected at 25.0 mbar for 25.0 s, reaching a LOD for acrylamide of 0.71 µg/kg. The applicability of this method was demonstrated by determining acrylamide (2.99 µg/kg) in non-spiked white bread.

On the other hand, following the new trends in miniaturization of analytical systems, MCE-LIF has been applied to the analysis of acrylamide in potato chips and French fries [188]. Before analysis, the samples were subjected to solid-liquid extraction (SLE) with ultrapure water and defatted with hexane. After pH adjustment (pH 8.5), the samples were derivatized with sulfoindocyanine succinimidyl ester because acrylamide is not naturally fluorescent. Regarding the CE method, the separation was performed in 100 mM borate solution (pH 9.3). In addition, an on-line concentration approach was proposed to improve sensitivity and consisted of a combination of FASS and reversed-field stacking. A SEF of 432 was achieved by applying the concentration strategy, and a LOD of 1 µg/L was reached for acrylamide standards. The application of the proposed method to real samples showed acrylamide concentrations levels of 1.453 and 2.236 mg/kg in potato chips and French fries, respectively.

PAHs encompass a large class of more than 100 different organic compounds characterized by two or more fused benzene rings. Most of these substances are mutagenic and carcinogenic, or potentially so. They can occur in foodstuffs subjected to smoking treatment or intense thermal processing (> 120°C) (*e.g.* in meat) [168]. These contaminants can also be present in food not only as a result of thermal processes, but also from environmental contamination; therefore, PAHs can be found in vegetables due to their deposition in water and soil [189]. The United States (US) Environmental Protection Agency (US-EPA) and EFSA have established lists of PAHs that must be monitored in food and environmental samples because of their risk to human health. The combination of both lists includes up to 26 PAHs [190]. Due to the hydrophobic nature of these compounds, NACE, MEKC and CEC methods have been successfully applied to the determination of PAHs, but have been limited to analysis of environmental matrices and food samples have been little studied [190]. In this sense, most of the contributions are related to the analysis of PAHs in edible oils [191 - 193].

Despite the hydrophobic and neutral nature of PAHs, CZE has been proposed as electrophoretic separation mode using LIF detection. MeOH was added to the BGE to improve the solubility of PAHs. Finally, the separation of 19 PAHs was achieved in a BGE consisting of 10 mmol/L sodium borate buffer (pH 9.2), 600 mmol/L urea, 10 mmol/L sulfobutylether-β-CD, 20 mmol/L methyl-β-CD in 90:10 (v/v) water/MeOH mixture [193]. Under these conditions, baseline separation for all compounds was achieved in less than 18 minutes by applying a separation voltage of 16 kV and a capillary temperature of 25°C. The combination of CDs in the BGE provided high selectivity to the separation method, even higher than the selectivity obtained in MEKC using CDs or not. In this method, selectivity was extremely related to the concentration of the dual CD system because the complex formation between PAHs and each type of CD is different.

Contaminants Related to Food Packaging

There is a growing concern about the health risks associated with exposure to substances found in food packaging that may migrate to food, so methods must be developed to control their presence in these products [194]. Within this framework, compounds such as bisphenol A or phthalates have attracted the attention of risk assessors and risk managers due to their endocrine disrupting characteristics, although other food contact materials such as oligomers must also be investigated. The number of CE-based methods reported for the determination of this type of substances in food is very limited, although this analytical technique has been recommended for the separation of oligomers related to polar polymers [195]. Some CE methods have been proposed for the analysis of bisphenol A and phthalates in food, but always with UV detection [196 - 199].

CZE-UV has been applied using a BGE with a basic pH for the determination of bisphenol A ($pK_a \sim 9.6$) in milk [196] and in beverages [197, 200]. In this sense, MISPE has been evaluated as sample treatment, either in its off-line version [196] as its in-line version [197]. In-line MISPE involves a higher level of automation of the analytical process, reducing sample manipulation and improving method reproducibility. The main disadvantage of in-line MISPE is that capillaries containing the extraction phase for sample treatment are not commercial and must be prepared in the laboratory. However, this fact should not hinder the application of in-line SPE approaches in general, because the preparation of concentrators can be done in easy and reproducible manner. For example, Zhang *et al.* have evaluated an in-line MISPE concentrator, which was prepared directly in the separation capillary, for the detection of bisphenol A in several liquid matrices including beverages, urine and water samples [197]. The in-column MISPE concentrator was prepared by light- emitting diode (LED) induced polymerization, which only took 20 min. The polymerization solution consisted of bisphenol A as template molecule (0.05 mmol), methacrylic acid as monomer (0.2 mmol), and ethylene dimethacrylate cross-linking agent (1 mmol) in MeCN (3 mL). After polymerization, the capillary was rinsed with MeOH for 10 min to remove the template molecule. This step is particularly important when fabricating MISPE materials, not only to make available the cavities that will retain the target analytes during sample extraction, but also to avoid cross-contaminations. The proposed in-line-MISPE-CE-UV method provided high sensitivity (0.8 μg/L), as well as high recoveries (*i.e.* 87.2-108.3%) and repeatability (RSDs < 5.6%) in real samples.

Phthalates are neutral compounds, so various CE modes have been evaluated as an alternative to CZE. MEKC is the CE mode traditionally applied to the analysis of neutral compounds, and MEKC-UV has been successfully applied to the determination of dimethyl phthalate (DMP), diethylphthalate (DEP) and dibutyl phthalate (DBP) in olive oil [198]. Two MEKC systems using different surfactants were evaluated. In the first approach, the BGE consisted of 5 mmol/L ammonium tetraborate, 100 mmol/L SDS, 0.25% (w/v) hydro-xypropylmethyl cellulose (HPMC) and 5% (v/v) MeOH (pH 9). HPMC was added to the BGE as additive to improve peak symmetry and, as a collateral result, the viscosity was increased which favored the interactions between the micelles and the analytes, leading to longer migration times. In the second approach, the BGE consisted of 5 mmol/L sodium borate and 50 mmol/L sodium taurodeoxycholate, which is an aromatic surfactant, adjusted to pH 2.8. In comparison, the first approach provided a shorter analysis time (*i.e.* ~ 18 min *vs.* ~ 50 min), as well as higher peak efficiency.

The main drawback related to MEKC for the separation of hydrophobic analytes such as phthalates is that they are retained by the micelles strongly and in the same way, so the separation efficiency can be limited. Organic modifiers such as methanol, 1- butanol, MeCN and isopropanol can be added to the BGE when performing MEKC to improve separation efficiency, but may not be sufficient to achieve that goal. Furthermore, micelles can be destroyed in the presence of high concentration of organic solvents. Therefore, the development of alternative pseudostationary phases to surfactant-formed micellar phases has been pursued to overcome this MEKC limitation. With these principles in mind, MEEKC has been investigated as an alternative to MEKC to enhance separation efficiency, and a MEEKC-UV method has been applied to the analysis of DBP and di-(--ethylhexyl) phthalate (DEHP) in soft drinks [199]. In MEKKC, BGE is a microemulsion in which analytes are partitioned between the aqueous phase of the microemulsion and the oil droplets (*i.e.* the pseudostationary phase). In the mentioned work, BGE consisted of 3.46% (w/w) SDS, 0.5% (w/w) n-octane, 7% (w/w) 1-butanol, 0.25% (w/w) Pluronic® F-127 and 78.69% (w/w) 30 mmol/L NaH_2PO_4 (pH 2.5). Pluronic®, which is a material with polymeric and surfactant properties, was used as additive to stabilize the microemulsion and improve the selectivity. Separation was achieved in less than 15 min (including progesterone as internal standard) and LODs lower than 50 µg/L were obtained for both analytes. Finally, the application of this method to the analysis of six soft drinks resulted in the determination of 453.67 µg/L of DEHP in one of them.

On the other hand, Ni *et al.* have proposed the analysis by electrokinetic chromatography (EKC) of 15 phthalates in food packing materials using polymeric aggregates as pseudostationary phase [201]. BGE consisted of 1% (w/v) poly(methyl methacrylate-co-methacrylic acid), which is an amphiphilic copolymer, 40% (v/v) MeOH, 2% (v/v) 1-butanol, 5 mmol/L borax buffer solution (pH 9.2). As in traditional MEKC separations, organic solvents were added to improve selectivity. Compared to traditional MEKC, precision was not affected and RSDs less than 1.0% and 3.3% were observed for migration times and peak areas, respectively (n = 7; each phthalate was at 10 mg/L).

Other Contaminants

Other contaminants such as perfluoroalkyl substances (PFASs), dioxins, polychlorinated biphenyls (PCB), *etc*., can also be found in food products as result of environmental contamination, cleaning processes involved in the food chain, occurring during food transport, *etc* [202]. However, CE methods have hardly been investigated to monitor these substances in food products and are mainly limited to the analysis of these contaminants in environmental matrices [203, 204], so these approaches are not covered in this chapter.

APPLICATIONS RELATED TO FOOD ADDITIVES

Food additives encompass a wide range of compounds that can be found in processed food products to enhance their properties (*i.e.* attractiveness, texture, deliciousness, *etc.*) and extend their shelf life. As an example, the EU currently authorizes the use of approximately 330 food additives, covering antioxidants, dyes, emulsifiers, stabilizers, gelling agents, thickeners, preservatives and sweeteners [205]. A high consumption of these compounds can cause adverse health effects, so their concentration in food must be controlled to ensure food safety. CE, mainly coupled with UV detection, has been widely applied to the analysis of food additives [206], with special attention to preservatives, colorants and sweeteners [11 - 13].

Preservatives

Preservatives, generally organic acids and their salts, are added to foods to prevent the growth of microorganisms and extend the freshness of the product. Traditionally, CE methods for food safety are optimized using standards of the target analytes. However, Gonçalves *et al.* has investigated a novel approach based on computer simulations to optimize a CZE-UV method for the rapid analysis of citrate, phosphate and sulfite in seafood [207]. The robustness of the method was studied to validate the optimization strategy, which was only compromised by pH variations in the BGE above 4.7 and below 4.4. Separation of the three analytes and the internal standard (*i.e.* malonate) was finally accomplished in less than 3.5 min using 20 mmol/L benzenesulfonic acid and 45 mmol/L aminocaproic acid (pH 4.5) as BGE. LODs were less than 4.8 mg/kg for all preservatives, and of the fifteen samples of fish and shrimp analyzed, four of them were non-compliant with the Brazilian legislation in terms of phosphate and sulfite concentration.

In general, the UV absorbance of organic acids is poor, which limits the detection sensitivity in CE-UV methods. In fact, indirect UV detection is typically used when applying CE-UV to the determination of preservatives in food [207, 208]. As previously shown for biogenic amines and acrylamide, preservatives possessing amine groups can also be derivatized to be determined by UV detection. In-capillary derivatization with *o*-phthalaldehyde in the presence of 3-mercaptopropionic acid has been proposed for glutamate analysis [209]. The following requirements must be met by CE-UV methods that involve in-capillary derivatization: 1) the derivatization agent and its co-agent are added to the BGE, so they must be soluble in it; 2) the derivatization agent and its co-agent must show low volatility; 3) the derivatization reaction must occur at high rate; 4) the analyte to be derivatized must show a high electrophoretic mobility, so that the reaction takes place outside the sample zone, avoiding negative matrix effects; and 5) the analyte and its derivative should ideally have similar electrophoretic mobility to avoid band broadening. Finally, the aforementioned method allowed the simultaneous determination of derivatized glutamate and underivatized benzoic acid and sorbic acid in soy, fish, oyster, and sweet and sour chili sauces [209]. As an alternative to UV detection, CE-C^{4}D has been applied to the simultaneous determination of the preservatives propionate and sorbate in bread [210], and MEKC coupled to amperometric detection has been proposed for the detection of four electroactive preservatives (*i.e.* methylparaben, ethylparaben, propylparaben and butylparaben) and two non-electroactive preservatives (*i.e.* potassium sorbate and sodium lactate) in beverages, as well as in soy and oyster sauces [211].

On-line preconcentration has been widely studied to enhance the sensitivity of CE- based methods and, in the particular case of CE-UV methods for the analysis of preservatives, dynamic pH junction has been investigated as preconcentration approach [212, 213]. It involves sample injection in a solvent with a different than the BGE pH. The mobility of the analytes must be greater in the sample zone than in the BGE, so they focus on the boundary between both zones when they reach the BGE zone. Zhang *et al.* proposed the electrokinetic injection of benzoic acid and sorbic acid in 20 mmol/L phosphate buffer (pH 2.5) for 100 s at 10 kV, and subsequent separation in 50 mmol/L borate solution (pH 9.0) at 25 kV [212]. Analytes were neutral at low pH, but negatively charged at basic pH. Therefore, they initially migrated according to the EOF and, once they reached the BGE, they were negatively charged, thus reducing their migration velocity to the cathode (*i.e.* capillary outlet). The high sensitivity achieved with this dynamic pH junction-CZE- UV method (LODs < 30 μg/L) allowed the analysis of benzoic acid and sorbic acid in diluted and filtered beverages, jams and vinegar samples, without requiring any additional sample treatment. A dynamic pH junction strategy can also be applied in MEKC separations, while a combined preconcentration effect (*i.e.* sweeping and dynamic pH junction) can be achieved. This approach was studied by Hsu *et al.* for the analysis of benzoic acid and sorbic acid in beverages, juice, wine and soy sauce samples [213]. Briefly, the conditions of the method were as follows: 1) the capillary was filled with 15 mmol/L tetraborate buffer and 40 mmol/L SDS (pH 9.2), 2) samples prepared in 2.5 mmol/L phosphate (pH 3.0) solution were hydrodynamically injected, 3) separation took place in 15 mmol/L tetraborate buffer (pH 9.2) containing 40 mmol/L SDS and 0.1% (w/v) poly(ethylene oxide) solution. Although CZE was shown to provide the separation of both analytes [212], higher peak efficiency was observed in MEKC. In addition, in comparison to dynamic pH junction-CZ- -UV, dynamic pH junction-sweeping-MEKC-UV resulted in SEFs 10-fold greater.

Despite CZE, either using acid or basic buffers, and MEKC using SDS as pseudostationary phase are widely applied as CE modes in the analysis of preservatives, methods involving other CE modes such as CEC can be found in the literature. CEC is not typically used in food safety applications because CEC capillaries are generally required to be prepared in the laboratory due to the high cost of commercial capillaries. However, the preparation procedures of CEC capillaries do not necessarily have to be difficult and can be highly reproducible. As an example, the separation of benzoic acid and sorbic acid by OT-CEC has recently been reported [214]. The surface of capillary columns was coated in the laboratory with trimethylamine aminated polychloromethylstyrene nanolatex, and their performance was characterized by high reproducibility of the EOF after more than 600-700 runs in 6 weeks (RSD < 1.7%). The analysis of both preservatives in samples of soy sauce and vinegar was subsequently performed in less than 5 min, which was a shorter analysis time than that observed in similar applications using HPLC or GC. OT-CEC has also been proposed for the separation of parabens (*i.e.* methylparaben, ethylparaben and propylparaben), in which a zinc(II)-derived metal-organic cage (MOC) [Zn_2L] was used as capillary coating and, consequently, as stationary phase [215].

Furthermore, the use of additives in the BGE of traditional CE-UV methods has also been investigated to improve method selectivity. The simultaneous determination by CZE-UV of four parabens, which are esters of *p*-hydroxybenzoic acids, *m*-amino benzoic acid, *p*-hydroxybenzoic acid, benzoic acid and sorbic acid in beverages, has been carried out using 15.0 mmol/L borax and 25.0 mmol/L 1-butyl-2,3- dimethylimidazolium chloride ([BMMIm]Cl) (pH 10.0) as BGE [216]. In absence of [BMMIm]Cl, which is an ionic liquid, *m*-amino benzoic acid and sorbic acid co- migrated; therefore, ionic liquids were shown to improve separation selectivity. In fact, their use as BGE additives is recommended for the CE separation of chiral compounds. Finally, the eight preservatives were analyzed in less than 27 min, although seven analytes were separated in less than 15 min. High repeatability in terms of run-to-run RSD was observed for migration times (< 2.6%) and peak areas (< 3.3%).

Colorants

Natural or synthetic dyes used as food additives are added to give a more attractive appearance to food products and capture consumer attention. However, some colorants such as many azo dyes (*e.g.* Sudan dyes) are banned as food additives worldwide due to their toxicity, and their presence in food products must be controlled. Sudan dyes are weak acids ($pk_a \sim 11.5$) and neutral compounds over a wide pH range, so their CE separation is typically carried out under MEKC conditions. MEKC-UV has been applied to the determination of Sudan dyes (I, II, III and IV) in samples of chili powder [217]. The four analytes were separated in 20 min using 5 mmol/L borate (pH 9.3), 20 mmol/L SDS and 20% (v/v) MeCN as

BGE, and LODs ranged from 96 to 610 µg/L. As with many CE methods, the addition of organic solvents can greatly improve selectivity and peak resolution. In this case, Sudan dyes were not separated when MeCN concentration was below 15% (v/v). Moreover, the use of organic solvents in the BGE also increases the solubility of hydrophobic analytes, such as Sudan III and IV [218]. Moreno-González *et al.* have developed a MEKC-ESI-QqQ-MS method for the analysis of these four Sudan dyes in chili products, showing the importance for the CE separation of organic modifiers in the BGE (Fig. **5**) [218]. Initially, MeCN (40%, v/v) was added to the BGE to ensure the solubility of analytes in the separation buffer, but the compounds partially co-migrated. Therefore, different organic mixtures consisting of MeCN, MeOH and THF were evaluated as organic modifiers. Finally, BGE consisting of 60:20:10:10 (v/v/v/v) 150 mmol/L APFO (pH 9.0):MeCN:MeOH:THF provided baseline separation of all analytes. APFO was selected as surfactant for MEKC separation as an alternative to SDS, since this latter is not volatile and, consequently, is incompatible with MS detection. LOQs below 22 µg/kg were achieved for all Sudan dyes, complying with European recommendations that set an action limit of 500 µg/kg for Sudan dyes in foodstuff.

Although the use of volatile surfactants is recommended in MEKC-MS methods, non-volatile surfactants can be used for CE separation of analytes if the micelles do not reach the mass spectrometer. Partial filling-MEKC approach, in which only a fraction of the capillary length is filled with the pseudostationary phase, has long been applied to perform MEKC-MS analyses. For example, a partial filling-MEKC-IT- MS method has been proposed for the determination of Sudan dyes (I, II, III and IV) in chili powder samples [219]. Initially, and in a capillary previously rinsed with BGE (*i.e.* 40 mmol/L ammonium hydrogen carbonate (pH 8), 32.5% (v/v) MeCN), a plug of BGE containing 25 mmol/L SDS was hydrodynamically injected for 14.6 s at 10 psi, filling □25% (15 cm) of the capillary length. The sample was subsequently injected at 0.5 psi for 5 s, followed by MEKC separation at +30 kV. The inlet vial contained BGE without SDS during the separation, which was accomplished in about 10 min. In partial-filling MEKC, separation only occurs when the analytes pass through the surfactant zone and until they reach the micelle-free BGE zone. Therefore, the target compounds can be detected before the surfactant reaches the detector, but this strategy has some limitations. The separation efficiency is less than that achieved in tradition MEKC methods where separation takes place along the entire capillary length, as well as peak efficiency can also be compromised by a peak broadening mechanism that can occur in the micelle zone-buffer boundary [220].

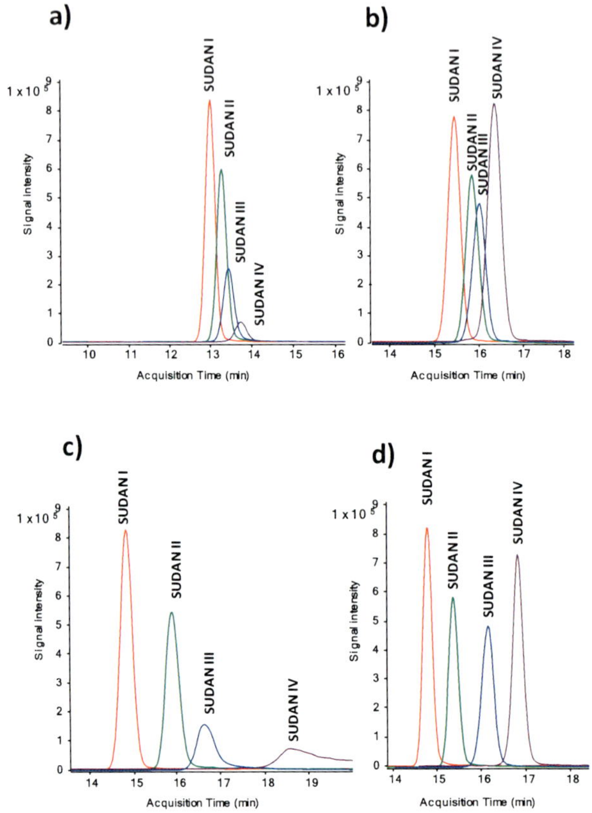

Fig. (5). Effect of the organic modifier in the BGE: a) 60:40 (v/v) 150 mmol/L APFO (pH 9.0):MeCN, b) 60:30:10 (v/v/v) 150 mmol/L APFO (pH 9.0):MeCN:THF, c) 60:30:10 (v/v/v) 150 mmol/L APFO (pH 9.0):MeCN:MeOH, and d) 60:20:10:10 (v/v/v/v) 150 mmol/L APFO (pH 9.0):MeCN:MeOH:THF. Figure edited with permission from [218].

In general, a large number of CE methods have been reported for the analysis of dyes in food, covering from more polar to nonpolar and high hydrophobic colorants such as Sudan dyes. CZE-C^{4}D has been applied to simultaneous determination of ponceau 4R, amaranth, tartrazine, sunset yellow and brilliant blue in commercial preserved fruit samples [221]. These compounds are charged at pH > 3 and are soluble is water, so they are susceptible to be separated in CZE mode. In the mentioned work, the BGE consisted of 25 mmol/L acetic acid and 2 mmol/L hydroxypropyl-β-CD (pH 3.5), and the use of CDs was essential to achieve the separation between ponceau 4R, amaranth and tartrazine that co-migrated in normal CZE conditions (*i.e.* BGE without additives). This work highlights the importance of the type of CD used as BGE additive, since no baseline separation between amaranth and ponceau R4 was observed when β-CD, which is the most commonly used CD, was added to the running buffer. CZE-UV using SDS below its CMC as a BGE additive (so no micelles were formed and CZE separation occurred rather than MEKC) [222] and EKC-UV using diamino moiety functionalized silica nanoparticles as pseudostationary phase [223] have also been investigated for the analysis of the tartrazine, sunset yellow, brilliant blue in food products. In both cases, β-CD was also used as a BGE additive, and a fourth but different colorant was included among the target analytes (*i.e.* red allure and carminic acid, respectively). However, the advantages of these two last methods compared to CZE separation [221] are not clear.

Sweeteners

Sweeteners impart a sweet flavor to food and are used as substitutes of sugar to reduce calorie intake. These compounds can have a natural origin (*e.g.* steviol glycosides, thaumatin, neohesperidine dihydrochalcone, *etc.*) or be synthetic substances (*e.g.* acesulfame K, aspartame, cyclamates, saccharin, sucralose, neotame, *etc.*), some of which can be harmful to human health. As a consequence, analytical methods are required to control the concentration of sweeteners in food, ensuring food safety and compliance with current legislations on food additives. For example, rapid analysis of aspartame, cyclamate, acesulfame-K and saccharin in sweeteners, soft drinks, and ice teas has been proposed by CZE using indirect UV detection [224]. The separation of the four analytes was achieved in approximately 1 min because short-end injection was carried out. In commercial CE-UV systems (from Agilent Technologies), capillaries provide a minimum effective separation length of 40 cm (total length 48.5 cm). However, in short-end injections, samples are injected into the capillary outlet (*i.e.* migration of analytes to the inlet vial), so the effective length is reduced to 8.5 cm. Since the capillary length for the separation is very limited in this CE approach, it is only applicable for separation of a few analytes and to samples without a significant matrix signal background.

Many sweeteners lack a chromophore or fluorophore group in their structure, so UV and LIF detection are not the most convenient for their analysis. In the case of applying CE coupled to one of these detection techniques, indirect UV or LIF detection is carried out, also avoiding the need of derivatization. Xiao *et al.* have proposed the analysis of xylitol, sorbitol, arabinitol and mannitol (as internal standard) by CZE-indirect LIF detection in xylitol mother liquor, xylitol orange juice, xylitol peanut milk and xylitol chewing gum [225]. Fluorescein (10^{-5} mol/L) was added to the BGE, consisting of 40 mmol/L sodium tetraborate and 5% (v/v) MeOH (pH 9.6), to carry out the indirect detection. The LODs ranged from 19.0

to 24.4 mg/L. As alternative to CE-indirect UV/LIF detection and in order to achieve greater detection sensitivity, CE-C^{4}D has been widely applied to the determination of sweeteners in food [226 - 228]. CZE-C^{4}D has been evaluated as analytical technique for the determination of cyclamate, acesulfame-K and saccharin in beverages, applying FASI as preconcentration technique [228]. Analytes were injected electrokinetically at -11 kV for 8 s after a plug of water introduced by gravity injection (*i.e.* FASI procedure), and subsequently separated in 20 mmol/L acetic acid (pH 3.3) as BGE at -12 kV. Compared to gravity sample injection, FASI provided a SEF of 560. The LODs ranged from 22 to 44 mmol/L.

Multiclass Methods

Since a food product may contain more than one type of food additive, CE applications related to the simultaneous determination of food additives with different properties are of special interest. In order to satisfy this purpose, Frazier *et al.* developed a MEKC-UV method for rapid analysis (□12 min) of thirteen food additives, including sweeteners, preservatives and colorants (*i.e.* caffeine, aspartame, sorbic acid, benzoic acid, saccharin, green S, acesulfame K, sunset yellow FCF, quinoline yellow, brilliant blue FCF, carmoisine, ponceau 4R, black PN) in soft drinks [229]. The separation was carried out at 20 kV and 25°C using 20 mmol/L carbonate and 62 mmol/L SDS (pH 9.5) as BGE, and the LOQ was set at 10 mg/L. In general, CE methods for the analysis of multiple food additives usually include the determination of preservatives together with artificial sweeteners [230], colorants [231], or acidulants [208]. Recently, an in-hous--made portable CE-C^{4}D instrument was developed and applied to the determination of different food additives in various food commodities in Vietnam [232], including oxalate and citrate in instant noodles, beer and tea drinks, acidulants in beverage, coffee, alcohol and vinegar samples, and artificial sweeteners in jelly and drink samples. The results were compared with those provided by a HPLC-photodiode array (PDA) method, demonstrating the reliability of the analytical data obtained by CE-C^{4}D.

Within this framework, the analysis of food additives has shown that CE can be applied as alternative to LC and is capable of being implemented in routine analysis as a cheaper technological option [232, 233]. In fact, CE has been shown to be better for typical, routine or screening analysis than HPLC, being a more environmental- friendly analytical tool. In this sense, HPLC can be limited to confirmatory analyses where more reliable quantifications and the minimization of false-negative results are required, and therefore, where the principles of green chemistry play a secondary role [231]. In order to demonstrate the applicability of CE methods in routine applications, a MEKC-UV has been recently developed for the determination of eleven food additives (*i.e.* caffeine, alitame, aspartame, neotame, dehydroacetic acid, sorbic acid, stevioside, quinine, benzoic acid, saccharin, acesulfame-K) in different food samples, and the accuracy of the method was demonstrated by a Food Analysis Performance Assessment Scheme (FAPAS®) proficiency test sample [230]. A FAPAS® proficiency test chocolate cake sample, containing caffeine and sorbic acid, was analyzed. The results in terms of *z*-score were of -1.0 and -1.1 for each analyte, respectively, and were comparable with the results obtained by other laboratories (n= 43-47) applying HPLC-based methods.

CE FOR THE ANALYSIS OF NANOPARTICLES IN FOOD

Nanoparticles do not constitute a family of chemical hazards *per se* since, for example, they can be used as food additives or can be contaminants from food packaging. However, the increasing use of nanoparticles in food products and food technology and the limited knowledge about the risk of these materials for human health, make nanoparticles an emerging concern in food safety, requiring their control in foods [234]. In this sense, CE is a useful analytical technique to characterize and determine nanoparticles in a wide range of matrices [235], but its applicability in food safety analysis is still in its early years.

Titanium dioxide nanoparticles (TiO_2NPs) may be present in sugary food products, as food-grade TiO_2 is a common food additive. Cytotoxic properties have been attributed to TiO_2NPs (< 100 nm), so screening methods are needed to monitor its presence in food and assess its impact on human health. Moreno *et al.* have proposed an analytical workflow to first detect by Raman spectroscopy the presence of TiO_2NPs and TiO_2 microparticles (>100 nm) in white powdered sugar used in industrial baking and then submit positive samples for CZE-UV analysis to distinguish both types of TiO_2NPs [236]. The separation was carried out at 25 kV and 20°C using 2.5 mmol/L Na_2HPO_4 (pH 9.5) as BGE. Under these CE conditions, nanoparticles of TiO_2 anatase (5 nm diameter) and TiO_2 rutile (60 nm diameter) were separated and distinguished from TiO_2 microparticles. Other CE applications for the determination of nanoparticles in food include a MEKC method for the analysis of silver nanoparticles (AgNPs) in orange juice and mussel tissue [237], and a CZE method for the separation of silica nanoparticles with different particle size (*i.e.* 20, 50, and 100 nm) and its subsequent analysis in sweeteners and salt samples [238].

CONCLUDING REMARKS

CE has shown to be an efficient technique for the analysis of chemical hazards in food, covering the determination of a wide range of compounds with different physicochemical properties in a wide variety of food matrices (*i.e.* from liquid matrices such as milk and juices and solid matrices such as meat, to products with low water content such as cereals or fatty products such as oils). Although the benefits it brings to the analytical workflow are well known (*i.e.* high peak efficiencies, need for low sample volumes, low solvent consumption and waste generation), its implementation in the field of food safety field has been limited due to the low detection sensitivity traditionally attributed to CE analysis. However, as shown in the analysis of various food residues, contaminants and additives, LODs at low levels of µg/L (or µg/kg) can be achieved by combining off-line and on-line preconcentration approaches, and with high repeatability. Off-line preconcentration is carried out through sample treatments that are also required in LC or GC applications related to food safety. In general, on-line preconcentration procedures as well as in-line preconcentration approaches are simple, robust and can be automated, making them capable of being implemented in routine applications.

There is still a lack of CE-MS methods intended to analyze chemical hazards in food, even though the use of MS in food safety applications is mandatory according to current legislation and guidelines. In this sense, CE can be coupled to any type of mass spectrometer, generally using sheath-flow and sheathless interfaces which are reliable and allow satisfactory detection sensitivity. Therefore, the development of more applications in this field using CE-MS is expected, since it is the only way to demonstrate that CE can really be an alternative, or at least complementary, to LC and GC in food safety. In this regard, particular attention should be paid when validating CE-based methods since, in general, the guidelines and legislation established for LC and GC methods are not fully followed. For example, decision limit ($CC\alpha$) and detection capability ($CC\beta$) are parameters that must be established in the methods for the analysis of veterinary residues [5], but they have not been typically reported in CE-based methods found in the literature. CE presents great potential to be the gold standard technique for the determination of polar or ionic compounds in routine analysis, but to reach acceptance and implementation in routine laboratories it still needs to be demonstrated that CE is able to comply with food safety legislation.

CONSENT FOR PUBLICATION

Not applicable.

CONFLICT OF INTEREST

The author declares no conflict of interest, financial or otherwise.

ACKNOWLEDGEMENTS

M.H.-M. has received funding from the European Union's Horizon 2020 research and innovation programme under the Marie Skłodowska-Curie grant agreement HAZARDOmics No 795946. D.M.-G has received funding from the European Union's Horizon 2020 research and innovation programme under the Marie Sklodowska-Curie grant agreement BIOplasma No 840743. A.M.G.-C. and F.J.L. acknowledge the financial support of the Spanish Ministry of Economy and Competitiveness (Project ref: RTI2018-097043-B-I00).

REFERENCES

[1] European Commission Implementing Regulation (EU) 2019/1715 of 30 September 2019 laying down rules for the functioning of the information management system for official controls and its system components ('the IMSOC Regulation'). *Off. J. Eur. Union,* **2019**, *L261*, 37-96.

[2] Regulation (EU) 2017/625 of the European Parliament and of the Council of 15 March 2017 on official controls and other official activities performed to ensure the application of food and feed law, rules on animal health and welfare, plant health and plant protection products. *Off. J. Eur. Union,* **2017**, *L95*, 1-142.

[3] Codex Secretariat (FAO/WHO) In: *Codex Alimentarius - Joint FAO/WHO Food Standards Programme,* 26th ed; FAO: Rome, **2018**.

[4] International Programme on Chemical Safety (IPCS): Inter-Organization Programme for the Sound Management of Chemicals (IOMC) In: *WHO Human Health Risk Assessment Toolkit: Chemical Hazards,* 1st ed; WHO: Geneva, **2010**.

[5] 2002/657/EC of 12 August 2002 implementing Council Directive 96/23/EC concerning the performance of analytical methods and the interpretation of results. *Off. J. Eur. Union,* **2002**, *L221*, 8-36.

[6] European Commission's Directorate-General for Health and Food Safety (DG SANTE) *Analytical quality control and method validation procedures for pesticide residues analysis in food and feed (SANTE/12682/2019)*, **2019**.

[7] European Commission's Directorate-General for Health and Food Safety (DG SANTE) *) Guidance document on identification of mycotoxins in food and feed (SANTE/12089/2016)*, **2016**.

[8] Moreno-González, D.; Lara, F.J.; del Olmo-Iruela, M.; García-Campaña, A.M. Trends in multiresidue analysis.*Encyclopedia of Analytical Chemistry: Applications, Theory and Instrumentation*; Valcárcel, M., Ed.; John Wiley & Sons: New Jersey, **2015**, pp. 1-39. [http://dx.doi.org/10.1002/9780470027318.a9520]

[9] Sastre Toraño, J.; Ramautar, R.; de Jong, G. Advances in capillary electrophoresis for the life sciences. *J. Chromatogr. B Analyt. Technol. Biomed. Life Sci.,* **2019**, *1118-1119*, 116-136. [http://dx.doi.org/10.1016/j.jchromb.2019.04.020] [PMID: 31035134]

[10] Hernández-Mesa, M.; Cruces-Blanco, C. Capillary electrophoresis in forensic analysis.*Capillary Electrophoresis in the Early Twenty-First Century*; Núñez, O.; Subirats, X., Eds.; Nova Science Publishers Inc.: New York, **2019**, pp. 59-146.

[11] García-Cañas, V.; Simó, C.; Castro-Puyana, M.; Cifuentes, A. Recent advances in the application of capillary electromigration methods for food analysis and Foodomics. *Electrophoresis,* **2014**, *35*(1), 147-169. [http://dx.doi.org/10.1002/elps.201300315] [PMID: 24114758]

[12] Acunha, T.; Ibáñez, C.; García-Cañas, V.; Simó, C.; Cifuentes, A. Recent advances in the application

of capillary electromigration methods for food analysis and Foodomics. *Electrophoresis,* **2016**, *37*(1), 111-141.
[http://dx.doi.org/10.1002/elps.201500291] [PMID: 26256797]

[13] Álvarez, G.; Montero, L.; Llorens, L.; Castro-Puyana, M.; Cifuentes, A. Recent advances in the application of capillary electromigration methods for food analysis and Foodomics. *Electrophoresis,* **2018**, *39*(1), 136-159.
[http://dx.doi.org/10.1002/elps.201700321] [PMID: 28975648]

[14] Hernández-Mesa, M.; Moreno-González, D.; Lara, F.J.; García-Campaña, A.M. Capillary electrophoresis | Food chemistry applications.*Encyclopedia of Analytical Science,* 3rd ed; Reedijk, J., Ed.; Elsevier: Amsterdam, **2019**, pp. 358-366.

[15] Kitagawa, F.; Otsuka, K. Recent applications of on-line sample preconcentration techniques in capillary electrophoresis. *J. Chromatogr. A,* **2014**, *1335*, 43-60.
[http://dx.doi.org/10.1016/j.chroma.2013.10.066] [PMID: 24210300]

[16] Breadmore, M.C.; Grochocki, W.; Kalsoom, U.; Alves, M.N.; Phung, S.C.; Rokh, M.T.; Cabot, J.M.; Ghiasvand, A.; Li, F.; Shallan, A.I.; Keyon, A.S.A.; Alhusban, A.A.; See, H.H.; Wuethrich, A.; Dawod, M.; Quirino, J.P. Recent advances in enhancing the sensitivity of electrophoresis and electrochromatography in capillaries and microchips (2016-2018). *Electrophoresis,* **2019**, *40*(1), 17-39.
[http://dx.doi.org/10.1002/elps.201800384] [PMID: 30362581]

[17] Zhang, W.; Hankemeier, T.; Ramautar, R. Next-generation capillary electrophoresis-mass spectrometry approaches in metabolomics. *Curr. Opin. Biotechnol.,* **2017**, *43*, 1-7.
[http://dx.doi.org/10.1016/j.copbio.2016.07.002] [PMID: 27455398]

[18] Stolz, A.; Jooß, K.; Höcker, O.; Römer, J.; Schlecht, J.; Neusüß, C. Recent advances in capillary electrophoresis-mass spectrometry: Instrumentation, methodology and applications. *Electrophoresis,* **2019**, *40*(1), 79-112.
[http://dx.doi.org/10.1002/elps.201800331] [PMID: 30260009]

[19] Commission Regulation (EU) No 37/2010 of 22 December 2009 on pharmacologically active substances and their classification regarding maximum residue limits in foodstuffs of animal origin. *Off. J. Eur. Union,* **2010**, *L15*, 1-72.

[20] The Code of Federal Regulations. Title 21, Food and Drugs, PART 556 https://www.accessdata.fda.gov/scripts/cdrh/cfdocs/cfcfr/CFRSearch.cfm?CFRPart=556

[21] FAO/WHO Codex Alimentarius Commission *Codex alimentarius: Maximum residue limits (MRLs) and risk management recommendations (RMRs) for residues of veterinary drugs in foods, CAC/MRL 2-2015,* **2015**.

[22] García-Campaña, A.M.; Gámiz-Gracia, L.; Lara, F.J.; del Olmo Iruela, M.; Cruces-Blanco, C. Applications of capillary electrophoresis to the determination of antibiotics in food and environmental samples. *Anal. Bioanal. Chem.,* **2009**, *395*(4), 967-986.
[http://dx.doi.org/10.1007/s00216-009-2867-9] [PMID: 19533105]

[23] Domínguez-Vega, E.; Montealegre, C.; Marina, M.L. Analysis of antibiotics by CE and their use as chiral selectors: An update. *Electrophoresis,* **2016**, *37*(1), 189-211.
[http://dx.doi.org/10.1002/elps.201500359] [PMID: 26471773]

[24] Greño, M.; Castro-Puyana, M.; García, M.A.; Marina, M.L. Analysis of antibiotics by CE and CEC and their use as chiral selectors: An update. *Electrophoresis,* **2018**, *39*(1), 235-259.
[http://dx.doi.org/10.1002/elps.201700306] [PMID: 28941242]

[25] Colombo, R.; Papetti, A. Advances in the analysis of veterinary drug residues in food matrices by capillary electrophoresis techniques. *Molecules,* **2019**, *24*(24), 4617.
[http://dx.doi.org/10.3390/molecules24244617] [PMID: 31861089]

[26] Ma, M.; Zhang, H.S.; Xiao, L.Y.; Xiao, L.; Wang, P.; Cui, H.R.; Wang, H. Quaternary ammonium

chitosan derivative dynamic coating for the separation of veterinary sulfonamide residues by CE with field-amplified sample injection. *Electrophoresis,* **2007**, *28*(22), 4091-4100. [http://dx.doi.org/10.1002/elps.200700317] [PMID: 17960537]

[27] Font, G.; Juan-García, A.; Picó, Y. Pressurized liquid extraction combined with capillary electrophoresis-mass spectrometry as an improved methodology for the determination of sulfonamide residues in meat. *J. Chromatogr. A,* **2007**, *1159*(1-2), 233-241. [http://dx.doi.org/10.1016/j.chroma.2007.03.062] [PMID: 17433345]

[28] Soto-Chinchilla, J.J.; García-Campaña, A.M.; Gámiz-Gracia, L. Analytical methods for multiresidue determination of sulfonamides and trimethoprim in meat and ground water samples by CE-MS and CE-MS/MS. *Electrophoresis,* **2007**, *28*(22), 4164-4172. [http://dx.doi.org/10.1002/elps.200600848] [PMID: 17960854]

[29] Lin, C.L.; Cheng, Y.J.; Huang, H.Y.; Lee, S. Analyses of sulfonamide antibiotics by CEC using poly(divinylbenzene-1-octadecene-vinylbenzyl trimethyl ammonium chloride) monolithic columns. *Electrophoresis,* **2009**, *30*(22), 3828-3837. [http://dx.doi.org/10.1002/elps.200900386] [PMID: 19885885]

[30] Lin, Y.T.; Liu, Y.W.; Cheng, Y.J.; Huang, H.Y. Analyses of sulfonamide antibiotics by a successive anion- and cation-selective injection coupled to microemulsion electrokinetic chromatography. *Electrophoresis,* **2010**, *31*(13), 2260-2266. [http://dx.doi.org/10.1002/elps.201000150] [PMID: 20593402]

[31] Cheng, Y.J.; Huang, S.H.; Singco, B.; Huang, H.Y. Analyses of sulfonamide antibiotics in meat samples by on-line concentration capillary electrochromatography-mass spectrometry. *J. Chromatogr. A,* **2011**, *1218*(42), 7640-7647. [http://dx.doi.org/10.1016/j.chroma.2011.06.027] [PMID: 21724191]

[32] Lara, F.J.; García-Campaña, A.M.; Alés-Barrero, F.; Bosque-Sendra, J.M. In-line solid-phase extraction preconcentration in capillary electrophoresis-tandem mass spectrometry for the multiresidue detection of quinolones in meat by pressurized liquid extraction. *Electrophoresis,* **2008**, *29*(10), 2117-2125. [http://dx.doi.org/10.1002/elps.200700666] [PMID: 18409156]

[33] Moreno-González, D.; Lara, F.J.; Gámiz-Gracia, L.; García-Campaña, A.M. Molecularly imprinted polymer as in-line concentrator in capillary electrophoresis coupled with mass spectrometry for the determination of quinolones in bovine milk samples. *J. Chromatogr. A,* **2014**, *1360*, 1-8. [http://dx.doi.org/10.1016/j.chroma.2014.07.049] [PMID: 25124225]

[34] Springer, V.; Jacksén, J.; Ek, P.; Lista, A.G.; Emmer, A. Determination of fluoroquinolones in bovine milk samples using a pipette-tip SPE step based on multiwalled carbon nanotubes prior to CE separation. *J. Sep. Sci.,* **2014**, *37*(1-2), 158-164. [http://dx.doi.org/10.1002/jssc.201300980] [PMID: 24227292]

[35] Springer, V.; Jacksén, J.; Ek, P.; Lista, A.G.; Emmer, Å. Capillary electrophoretic determination of fluoroquinolones in bovine milk followed by off-line MALDI-TOF-MS analysis. *Chromatographia,* **2015**, *78*, 285-290. [http://dx.doi.org/10.1007/s10337-014-2823-5]

[36] Lombardo-Agüí, M.; García-Campaña, A.M., Gámiz Gracia, L.; Cruces Blanco, C. Laser induced fluorescence coupled to capillary electrophoresis for the determination of fluoroquinolones in foods of animal origin using molecularly imprinted polymers. *J. Chromatogr. A,* **2010**, *1217*(15), 2237-2242. [http://dx.doi.org/10.1016/j.chroma.2010.02.016] [PMID: 20227081]

[37] Liu, C.; Feng, X.; Qian, H.; Fang, G.; Wang, S. Determination of norfloxacin in food by capillary electrophoresis immunoassay with laser-induced fluorescence detector. *Food Anal. Methods,* **2015**, *8*, 596-603. [http://dx.doi.org/10.1007/s12161-014-9936-1]

[38] Montes, R.H.O.; Marra, M.C.; Rodrigues, M.M.; Richter, E.M.; Muñoz, R.A.A. Fast determination of

ciprofloxacin by batch injection analysis with amperometric detection and capillary electrophoresis with capacitively coupled contactless conductivity detection. *Electroanalysis,* **2014**, *26*, 432-438. [http://dx.doi.org/10.1002/elan.201300474]

[39] Yang, Z.; Qin, W. Separation of fluoroquinolones in acidic buffer by capillary electrophoresis with contactless conductivity detection. *J. Chromatogr. A,* **2009**, *1216*(27), 5327-5332. [http://dx.doi.org/10.1016/j.chroma.2009.05.014] [PMID: 19481755]

[40] Zhou, X.; Xing, D.; Zhu, D.; Tang, Y.; Jia, L. Development and application of a capillary electrophoresis-electrochemiluminescent method for the analysis of enrofloxacin and its metabolite ciprofloxacin in milk. *Talanta,* **2008**, *75*(5), 1300-1306. [http://dx.doi.org/10.1016/j.talanta.2008.01.040] [PMID: 18585216]

[41] Mu, G.; Liu, H.; Xu, L.; Tian, L.; Luan, F. Matrix solid-phase dispersion extraction and capillary electrophoresis determination of tetracycline residues in milk. *Food Anal. Methods,* **2012**, *5*, 148-153. [http://dx.doi.org/10.1007/s12161-011-9225-1]

[42] Ibarra, I.S.; Rodríguez, J.A.; Miranda, J.M.; Vega, M.; Barrado, E. Magnetic solid phase extraction based on phenyl silica adsorbent for the determination of tetracyclines in milk samples by capillary electrophoresis. *J. Chromatogr. A,* **2011**, *1218*(16), 2196-2202. [http://dx.doi.org/10.1016/j.chroma.2011.02.046] [PMID: 21397241]

[43] Miranda, J.M.; Rodríguez, J.A.; Galán-Vidal, C.A. Simultaneous determination of tetracyclines in poultry muscle by capillary zone electrophoresis. *J. Chromatogr. A,* **2009**, *1216*(15), 3366-3371. [http://dx.doi.org/10.1016/j.chroma.2009.01.105] [PMID: 19233364]

[44] Lee, K.S.; Park, S.H.; Won, S.Y.; Shim, Y.B. Electrophoretic total analysis of trace tetracycline antibiotics in a microchip with amperometry. *Electrophoresis,* **2009**, *30*(18), 3219-3227. [http://dx.doi.org/10.1002/elps.200900117] [PMID: 19722202]

[45] Casado-Terrones, S.; Segura-Carretero, A.; Busi, S.; Dinelli, G.; Fernández-Gutiérrez, A. Determination of tetracycline residues in honey by CZE with ultraviolet absorbance detection. *Electrophoresis,* **2007**, *28*(16), 2882-2887. [http://dx.doi.org/10.1002/elps.200600614] [PMID: 17702063]

[46] Wang, S.; Yang, P.; Cheng, Y. Analysis of tetracycline residues in bovine milk by CE-MS with field-amplified sample stacking. *Electrophoresis,* **2007**, *28*(22), 4173-4179. [http://dx.doi.org/10.1002/elps.200700276] [PMID: 17957662]

[47] Juan-García, A.; Font, G.; Picó, Y. Simultaneous determination of different classes of antibiotics in fish and livestock by CE-MS. *Electrophoresis,* **2007**, *28*(22), 4180-4191. [http://dx.doi.org/10.1002/elps.200700383] [PMID: 17932875]

[48] Ma, T.Y.; Vickroy, T.W.; Shien, J.H.; Chou, C.C. Improved nonaqueous capillary electrophoresis for tetracyclines at subparts per billion level. *Electrophoresis,* **2012**, *33*(11), 1679-1682. [http://dx.doi.org/10.1002/elps.201100485] [PMID: 22887082]

[49] Bailón-Pérez, M.I.; García-Campaña, A.M.; Cruces-Blanco, C.; Del Olmo Iruela, M. Large-volume sample stacking for the analysis of seven β-lactam antibiotics in milk samples of different origins by CZE. *Electrophoresis,* **2007**, *28*(22), 4082-4090. [http://dx.doi.org/10.1002/elps.200700305] [PMID: 17960857]

[50] Bailón-Pérez, M.I.; García-Campaña, A.M.; del Olmo Iruela, M.; Cruces-Blanco, C.; Gámiz Gracia, L. Multiresidue determination of penicillins in environmental waters and chicken muscle samples by means of capillary electrophoresis-tandem mass spectrometry. *Electrophoresis,* **2009**, *30*(10), 1708-1717. [http://dx.doi.org/10.1002/elps.200800732] [PMID: 19391147]

[51] Shao, Y.X.; Chen, G.H.; Fang, R.; Zhang, L.; Yi, L.X.; Meng, H.L. Analysis of six β-lactam residues in milk and egg by micellar electrokinetic chromatography with large-volume sample stacking and polarity switching. *J. Agric. Food Chem.,* **2016**, *64*(17), 3456-3461. [http://dx.doi.org/10.1021/acs.jafc.6b00482] [PMID: 27088652]

[52] Liu, W.L.; Wu, C.Y.; Li, Y.T.; Huang, H.Y. Penicillin analyses by capillary electrochromatography-mass spectrometry with different charged poly(stearyl methacrylate-divinylbenzene) monoliths as stationary phases. *Talanta,* **2012**, *101*, 71-77.
[http://dx.doi.org/10.1016/j.talanta.2012.08.026] [PMID: 23158293]

[53] Lü, A.; Li, Q.; Lü, H.; Yu, J. Analysis of four cephalosporins in milk by capillary electrochromatography with o-cynaoethyl chitosan-coated open tubular column. *Sep. Sci. Technol.,* **2015**, *50*, 859-864.
[http://dx.doi.org/10.1080/01496395.2014.976874]

[54] Moreno-González, D.; Lara, F.J.; Jurgovská, N.; Gámiz-Gracia, L.; García-Campaña, A.M. Determination of aminoglycosides in honey by capillary electrophoresis tandem mass spectrometry and extraction with molecularly imprinted polymers. *Anal. Chim. Acta,* **2015**, *891*, 321-328.
[http://dx.doi.org/10.1016/j.aca.2015.08.003] [PMID: 26388393]

[55] Yu, Y.; Liu, Y.; Wang, W.; Jia, Y.; Zhao, G.; Zhang, X.; Chen, H.; Zhou, Y. Highly sensitive determination of aminoglycoside residues in food by sheathless CE-ESI-MS/MS. *Anal. Methods,* **2019**, *11*, 5064-5069.
[http://dx.doi.org/10.1039/C9AY01728C]

[56] Hong, Y. Q.; Guo, X.; Chen, G. H.; Zhou, J. W.; Zou, X. M.; Liao, X.; Hou, T. Determination of five macrolide antibiotic residues in milk by micellar electrokinetic capillary chromatography with field amplified sample stacking. *J. Food Saf,* **2017**, *e12382*, 1-7.

[57] Dong, B.; Zhao, X.; Zheng, L.; Zhang, J.; Li, N.; Inoue, T. Aggregation behavior of long-chain imidazolium ionic liquids in aqueous solution: micellization and characterization of micelle microenvironment. *Colloids Surf. A Physicochem. Eng. Asp.,* **2008**, *317*, 666-672.
[http://dx.doi.org/10.1016/j.colsurfa.2007.12.001]

[58] Liu, Y.M.; Yang, Y.; Li, J.; Peng, L.F.; Mei, L. A validated method for the sensitive determination of macrolide antibiotics by capillary electrophoresis with electrochemiluminescence detection and its applications. *Curr. Anal. Chem.,* **2011**, *7*, 325-332.
[http://dx.doi.org/10.2174/157341111797183029]

[59] Liu, Y.M.; Shi, Y.M.; Liu, Z.L.; Peng, L.F. Sensitive determination of tilmicosin, erythromycin ethylsuccinate and clindamycin by CE with electrochemiluminescence detection using azithromycin as internal standard and its applications. *J. Sep. Sci.,* **2010**, *33*(9), 1305-1311.
[http://dx.doi.org/10.1002/jssc.200900843] [PMID: 20235137]

[60] Airado-Rodríguez, D.; Hernández-Mesa, M.; García-Campaña, A.M.; Cruces-Blanco, C. Evaluation of the combination of micellar electrokinetic capillary chromatography with sweeping and cation selective exhaustive injection for the determination of 5-nitroimidazoles in egg samples. *Food Chem.,* **2016**, *213*, 215-222.
[http://dx.doi.org/10.1016/j.foodchem.2016.06.056] [PMID: 27451174]

[61] Hernández-Mesa, M.; García-Campaña, A.M.; Cruces-Blanco, C. Novel solid phase extraction method for the analysis of 5-nitroimidazoles and metabolites in milk samples by capillary electrophoresis. *Food Chem.,* **2014**, *145*, 161-167.
[http://dx.doi.org/10.1016/j.foodchem.2013.08.002] [PMID: 24128462]

[62] Hernández-Mesa, M.; Lara, F.J.; Cruces-Blanco, C.; García-Campaña, A.M. Determination of 5-nitroimidazole residues in milk by capillary electrochromatography with packed C18 silica beds. *Talanta,* **2015**, *144*, 542-550.
[http://dx.doi.org/10.1016/j.talanta.2015.06.049] [PMID: 26452860]

[63] Moreno-González, D.; Hamed, A.M.; Gilbert-López, B.; Gámiz-Gracia, L.; García-Campaña, A.M. Evaluation of a multiresidue capillary electrophoresis-quadrupole-time-of-flight mass spectrometry method for the determination of antibiotics in milk samples. *J. Chromatogr. A,* **2017**, *1510*, 100-107.
[http://dx.doi.org/10.1016/j.chroma.2017.06.055] [PMID: 28668369]

[64] Chu, Q.C.; Geng, C.H.; Zhou, H.; Ye, J.N. Fast determination of clenbuterol and salbutamol in feed

and meat products based on miniaturized capillary electrophoresis with amperometric detection. *Chin. J. Chem.*, **2007**, *25*, 1832-1835.
[http://dx.doi.org/10.1002/cjoc.200790338]

[65] Anurukvorakun, O.; Buchberger, W.; Himmelsbach, M.; Klampel, C.W.; Suntornsuk, L. A sensitive non-aqueous capillary electrophoresis-mass spectrometric method for multiresidue analyses of β-agonists in pork. *Biomed. Chromatogr.*, **2010**, *24*(6), 588-599.
[PMID: 19810005]

[66] Gañán, J.; Silva, M.; Morante-Zarcero, S.; Pérez-Quintanilla, D.; Sierra, I. Application of hybrid mesoporous silica for extraction of hormones in milk by matrix solid phase dispersion. *Mater. Lett.*, **2014**, *119*, 56-59.
[http://dx.doi.org/10.1016/j.matlet.2013.12.107]

[67] Gañán, J.; Morante-Zarcero, S.; Gallego-Picó, A.; Garcinuño, R.M.; Fernández-Hernando, P.; Sierra, I. Evaluation of a molecularly imprinted polymer for determination of steroids in goat milk by matrix solid phase dispersion. *Talanta,* **2014**, *126*, 157-162.
[http://dx.doi.org/10.1016/j.talanta.2014.03.041] [PMID: 24881547]

[68] D'Orazio, G.; Asensio-Ramos, M.; Hernández-Borges, J.; Rodríguez-Delgado, M.A.; Fanali, S. Evaluation of the combination of a dispersive liquid-liquid microextraction method with micellar electrokinetic chromatography coupled to mass spectrometry for the determination of estrogenic compounds in milk and yogurt. *Electrophoresis,* **2015**, *36*(4), 615-625.
[http://dx.doi.org/10.1002/elps.201400452] [PMID: 25394185]

[69] Martínez-Villalba, A.; Núñez, O.; Moyano, E.; Galceran, M.T. Field amplified sample injection-capillary zone electrophoresis for the analysis of amprolium in eggs. *Electrophoresis,* **2013**, *34*(6), 870-876.
[http://dx.doi.org/10.1002/elps.201200579] [PMID: 23334897]

[70] Tejada-Casado, C.; Moreno-González, D.; Lara, F.J.; García-Campaña, A.M.; Del Olmo-Iruela, M. Determination of benzimidazoles in meat samples by capillary zone electrophoresis tandem mass spectrometry following dispersive liquid-liquid microextraction. *J. Chromatogr. A,* **2017**, *1490*, 212-219.
[http://dx.doi.org/10.1016/j.chroma.2017.02.023] [PMID: 28222859]

[71] Domínguez-Álvarez, J.; Mateos-Vivas, M.; García-Gómez, D.; Rodríguez-Gonzalo, E.; Carabias-Martínez, R. Capillary electrophoresis coupled to mass spectrometry for the determination of anthelmintic benzimidazoles in eggs using a QuEChERS with preconcentration as sample treatment. *J. Chromatogr. A,* **2013**, *1278*, 166-174.
[http://dx.doi.org/10.1016/j.chroma.2012.12.064] [PMID: 23332305]

[72] Malá, Z.; Šlampová, A.; Křivánková, L.; Gebauer, P.; Boček, P. Contemporary sample stacking in analytical electrophoresis. *Electrophoresis,* **2015**, *36*(1), 15-35.
[http://dx.doi.org/10.1002/elps.201400313] [PMID: 25113855]

[73] Šlampová, A.; Malá, Z.; Gebauer, P. Recent progress of sample stacking in capillary electrophoresis (2016-2018). *Electrophoresis,* **2019**, *40*(1), 40-54.
[http://dx.doi.org/10.1002/elps.201800261] [PMID: 30073675]

[74] Vera-Candioti, L.; Olivieri, A.C.; Goicoechea, H.C. Development of a novel strategy for preconcentration of antibiotic residues in milk and their quantitation by capillary electrophoresis. *Talanta,* **2010**, *82*(1), 213-221.
[http://dx.doi.org/10.1016/j.talanta.2010.04.023] [PMID: 20685459]

[75] Springer, V.H.; Lista, A.G. Micellar nanotubes dispersed electrokinetic chromatography for the simultaneous determination of antibiotics in bovine milk. *Electrophoresis,* **2012**, *33*(13), 2049-2055.
[http://dx.doi.org/10.1002/elps.201100713] [PMID: 22806472]

[76] Giovannoli, C.; Anfossi, L.; Biagioli, F.; Passini, C.; Baggiani, C. Solid phase extraction of penicillins from milk by using sacrificial silica beads as a support for a molecular imprint. *Microchim. Acta,***2013**,

180, 1371-1377. [77] Deng, Y.; Gasilova, N.; Qiao, L.; Zhou, Y.; Zhang, X.; Girault, H. H., Highly sensitive detection of five typical fluoroquinolones in low-fat milk by field-enhanced sample injection-based CE in bubble cell capillary. *Electrophoresis,* **2014**, *35*, 3355-3362.

[78] FAO/WHO Joint Meeting on Pesticide Specifications (JMPS) In: *Manual on Development and Use of FAO and WHO Specifications for Pesticides,* 1st ed; World Health Organization: Geneva, **2016**.

[79] Gilden, R.C.; Huffling, K.; Sattler, B. Pesticides and health risks. *J. Obstet. Gynaecol. Neonat. Nur,* **2010**, *39*(1), 103-110.
[http://dx.doi.org/10.1111/j.1552-6909.2009.01092.x] [PMID: 20409108]

[80] Mostafalou, S.; Abdollahi, M. Pesticides and human chronic diseases: evidences, mechanisms, and perspectives. *Toxicol. Appl. Pharmacol.,* **2013**, *268*(2), 157-177.
[http://dx.doi.org/10.1016/j.taap.2013.01.025] [PMID: 23402800]

[81] New Zealand Government. *Pesticide maximum residue level legislation around the world.,* https://www.mpi.govt.nz/growing-and-harvesting/plant-products/pesticide-maximum-re-idue-levels-mrls-for-plant-based-foods/pesticide-maximum-residue-level-legislatio--around-the-world/

[82] 396/2005 of 23 February 2005 on maximum residue levels of pesticides in or on food and feed of plant and animal origin and amending council directive 91/414/EEC. *Off. J. Eur. Union,* **2005**, *L70*, 1-16.

[83] Bol'shakov, D.S.; Amelin, V.G. Determination of pesticides in environmental materials and food products by capillary electrophoresis. *J. Anal. Chem.*, **2016**, *71*, 965-1013. [⁸4] Araujo, L.; Rojas, C.; Cubillan, D.; Villa, N.; Mercado, J.; Prieto, A., Determination of trifloxystrobin, tebufenozide, and halofenozide in foods by micellar electrokinetic capillary chromatography. *Anal. Lett.,* **2010**, *43*, 2340-2348.

[85] Aparicio-Muriana, M.M.; Lhotská, I.; García-Campaña, A.M.; Lara, F.J. A first approach using micellar electrokinetic capillary chromatography for the determination of fipronil and fipronil-sulfone in eggs. *Electrophoresis,* **2020**, *41*(3-4), 202-208.
[http://dx.doi.org/10.1002/elps.201900291] [PMID: 31785119]

[86] Xu, L.; Luan, F.; Liu, H.; Gao, Y. Dispersive liquid-liquid microextraction combined with non-aqueous capillary electrophoresis for the determination of imazalil, prochloraz and thiabendazole in apples, cherry tomatoes and grape juice. *J. Sci. Food Agric.,* **2015**, *95*(4), 745-751.
[http://dx.doi.org/10.1002/jsfa.6834] [PMID: 25048793]

[87] Breadmore, M.C.; Tubaon, R.M.; Shallan, A.I.; Phung, S.C.; Abdul Keyon, A.S.; Gstoettenmayr, D.; Prapatpong, P.; Alhusban, A.A.; Ranjbar, L.; See, H.H.; Dawod, M.; Quirino, J.P. Recent advances in enhancing the sensitivity of electrophoresis and electrochromatography in capillaries and microchips (2012-2014). *Electrophoresis,* **2015**, *36*(1), 36-61.
[http://dx.doi.org/10.1002/elps.201400420] [PMID: 25330057]

[88] Silva, M. Micellar electrokinetic chromatography: a review of methodological and instrumental innovations focusing on practical aspects. *Electrophoresis,* **2013**, *34*(1), 141-158.
[http://dx.doi.org/10.1002/elps.201200349] [PMID: 23161127]

[89] Zhang, S.; Yang, X.; Yin, X.; Wang, C.; Wang, Z. Dispersive liquid-liquid microextraction combined with sweeping micellar electrokinetic chromatography for the determination of some neonicotinoid insecticides in cucumber samples. *Food Chem.,* **2012**, *133*(2), 544 550.
[http://dx.doi.org/10.1016/j.foodchem.2012.01.028] [PMID: 25683431]

[90] Moreno-González, D.; Gámiz-Gracia, L.; García-Campaña, A.M.; Bosque-Sendra, J.M. Use of dispersive liquid-liquid microextraction for the determination of carbamates in juice samples by sweeping-micellar electrokinetic chromatography. *Anal. Bioanal. Chem.,* **2011**, *400*(5), 1329-1338.
[http://dx.doi.org/10.1007/s00216-011-4682-3] [PMID: 21279629]

[91] Zhang, S.; Li, C.; Song, S.; Feng, T.; Wang, C.; Wang, Z. Application of dispersive liquid–liquid microextraction combined with sweeping micellar electrokinetic chromatography for trace analysis of six carbamate pesticides in apples. *Anal. Methods,* **2010**, *2*, 54-62.

[http://dx.doi.org/10.1039/B9AY00115H]

[92] Fang, R.; Chen, G-H.; Yi, L-X.; Shao, Y-X.; Zhang, L.; Cai, Q-H.; Xiao, J. Determination of eight triazine herbicide residues in cereal and vegetable by micellar electrokinetic capillary chromatography with on-line sweeping. *Food Chem.,* **2014**, *145*, 41-48. [http://dx.doi.org/10.1016/j.foodchem.2013.08.028] [PMID: 24128447]

[93] Santalad, A.; Srijaranai, S.; Burakham, R. Reversed electrode polarity stacking sample preconcentration combined with micellar electrokinetic chromatography for the analysis of carbamate insecticide residues in fruit juices. *Food Anal. Methods,* **2012**, *5*, 96-103. [http://dx.doi.org/10.1007/s12161-011-9219-z]

[94] Quirino, J.P. Two-step stacking in capillary zone electrophoresis featuring sweeping and micelle to solvent stacking: I. Organic cations. *J. Chromatogr. A,* **2010**, *1217*(49), 7776-7780. [http://dx.doi.org/10.1016/j.chroma.2010.10.020] [PMID: 21035122]

[95] Wuethrich, A.; Haddad, P.R.; Quirino, J.P. Simultaneous electrophoretic concentration and separation of herbicides in beer prior to stacking capillary electrophoresis UV and liquid chromatography-mass spectrometry. *Electrophoresis,* **2016**, *37*(9), 1122-1128. [http://dx.doi.org/10.1002/elps.201600050] [PMID: 26921124]

[96] Zhu, G.; Long, S.; Sun, H.; Luo, W.; Li, X.; Hao, Z. Determination of gibberellins in soybean using tertiary amine labeling and capillary electrophoresis coupled with electrochemiluminescence detection. *J. Chromatogr. B Analyt. Technol. Biomed. Life Sci.,* **2013**, *941*, 62-68. [http://dx.doi.org/10.1016/j.jchromb.2013.10.004] [PMID: 24184838]

[97] Hu, Y. Simultaneous determination of phenylurea herbicides in yam by capillary electrophoresis with electrochemiluminescence detection. *J. Chromatogr. B Analyt. Technol. Biomed. Life Sci.,* **2015**, *986-987*, 143-148. [http://dx.doi.org/10.1016/j.jchromb.2015.02.016] [PMID: 25732033]

[98] Chiu, H-Y.; Lin, Z-Y.; Tu, H-L.; Whang, C.W. Analysis of glyphosate and aminomethylphosphonic acid by capillary electrophoresis with electrochemiluminescence detection. *J. Chromatogr. A,* **2008**, *1177*(1), 195-198. [http://dx.doi.org/10.1016/j.chroma.2007.11.042] [PMID: 18061199]

[99] Cai, C.; Cheng, H.; Wang, Y. Determination of pretilachlor in soil and rice using matrix solid-phase dispersion extraction by capillary electrophoresis with field amplified sample injection and electrochemiluminescence detection. *Anal. Methods,* **2014**, *6*, 2767-2773. [http://dx.doi.org/10.1039/c3ay41950a]

[100] Masiá, A.; Suarez-Varela, M.M.; Llopis-Gonzalez, A.; Picó, Y. Determination of pesticides and veterinary drug residues in food by liquid chromatography-mass spectrometry: A review. *Anal. Chim. Acta,* **2016**, *936*, 40-61. [http://dx.doi.org/10.1016/j.aca.2016.07.023] [PMID: 27566339]

[101] Daniel, D.; do Lago, C.L. Determination of multiclass pesticides residues in corn by QuEChERS and capillary electrophoresis tandem mass spectrometry. *Food Anal. Methods,* **2019**, *12*, 1684-1692. [http://dx.doi.org/10.1007/s12161-019-01501-y]

[102] Moreno-González, D.; Huertas-Pérez, J.F.; García-Campaña, A.M.; Gámiz-Gracia, L. Vortex-assisted surfactant-enhanced emulsification liquid-liquid microextraction for the determination of carbamates in juices by micellar electrokinetic chromatography tandem mass spectrometry. *Talanta,* **2015**, *139*, 174-180. [http://dx.doi.org/10.1016/j.talanta.2015.02.057] [PMID: 25882424]

[103] Song, X.; Zhang, R.; Wang, Y.; Feng, M.; Zhang, H.; Wang, S.; Cao, J.; Xie, T. Simultaneous determination of five metal ions by on-line complexion combined with micelle to solvent stacking in capillary electrophoresis. *Talanta,* **2020**, *209*120578 [http://dx.doi.org/10.1016/j.talanta.2019.120578] [PMID: 31892076]

[104] Francesconi, K.A. Toxic metal species and food regulations--making a healthy choice. *Analyst*

(Lond.), **2007**, *132*(1), 17-20.
[http://dx.doi.org/10.1039/B610544K] [PMID: 17180173]

[105] Kubán, P.; Pelcová, P.; Margetínová, J.; Kubán, V. Mercury speciation by CE: an update. *Electrophoresis,* **2009**, *30*(1), 92-99.
[http://dx.doi.org/10.1002/elps.200800382] [PMID: 19072927]

[106] Yang, J.; Dong, X.; Zhen, X.; Chen, Y.; Wang, Y.; Wang, Q.; Hu, Y.; Xie, T.; Wang, S.; Cao, J. Metal organic framework assisted *in situ* complexation for miniaturized solid phase extraction of organic mercury in fish and *Dendrobium officinale. Talanta,* **2020**, *209*120598
[http://dx.doi.org/10.1016/j.talanta.2019.120598] [PMID: 31892039]

[107] Wang, S.; Song, X.; Hu, J.; Zhang, R.; Men, L.; Wei, M.; Xie, T.; Cao, J. Direct speciation analysis of organic mercury in fish and kelp by on-line complexation and stacking using capillary electrophoresis. *Food Chem.,* **2019**, *281*, 41-48.
[http://dx.doi.org/10.1016/j.foodchem.2018.12.083] [PMID: 30658763]

[108] Timerbaev, A.R. Element speciation analysis using capillary electrophoresis: twenty years of development and applications. *Chem. Rev.,* **2013**, *113*(1), 778-812.
[http://dx.doi.org/10.1021/cr300199v] [PMID: 23057472]

[109] Meermann, B.; Bartel, M.; Scheffer, A.; Trümpler, S.; Karst, U. Capillary electrophoresis with inductively coupled plasma-mass spectrometric and electrospray time of flight mass spectrometric detection for the determination of arsenic species in fish samples. *Electrophoresis,* **2008**, *29*(12), 2731-2737.
[http://dx.doi.org/10.1002/elps.200700902] [PMID: 18512677]

[110] Zhao, Y.; Zheng, J.; Fang, L.; Lin, Q.; Wu, Y.; Xue, Z.; Fu, F. Speciation analysis of mercury in natural water and fish samples by using capillary electrophoresis-inductively coupled plasma mass spectrometry. *Talanta,* **2012**, *89*, 280-285.
[http://dx.doi.org/10.1016/j.talanta.2011.12.029] [PMID: 22284493]

[111] Chen, Y.; Huang, L.; Wu, W.; Ruan, Y.; Wu, Z.; Xue, Z.; Fu, F. Speciation analysis of lead in marine animals by using capillary electrophoresis couple online with inductively coupled plasma mass spectrometry. *Electrophoresis,* **2014**, *35*(9), 1346-1352.
[http://dx.doi.org/10.1002/elps.201300410] [PMID: 24272447]

[112] Yang, G.; Xu, J.; Zheng, J.; Xu, X.; Wang, W.; Xu, L.; Chen, G.; Fu, F. Speciation analysis of arsenic in Mya arenaria Linnaeus and Shrimp with capillary electrophoresis-inductively coupled plasma mass spectrometry. *Talanta,* **2009**, *78*(2), 471-476.
[http://dx.doi.org/10.1016/j.talanta.2008.11.040] [PMID: 19203611]

[113] Ruiz-Capillas, C.; Herrero, A.M. Impact of biogenic amines on food quality and safety. *Foods,* **2019**, *8*(2), 62.
[http://dx.doi.org/10.3390/foods8020062] [PMID: 30744001]

[114] Papageorgiou, M.; Lambropoulou, D.; Morrison, C.; Kłodzińska, E.; Namieśnik, J.; Płotka-Wasylka, J. Literature update of analytical methods for biogenic amines. *TrAC Trend. Anal. Chem.,* **2018**, *98*, 128-142.

[115] Hernández-Cassou, S.; Saurina, J. Derivatization strategies for the determination of biogenic amines in wines by chromatographic and electrophoretic techniques. *J. Chromatogr. B Analyt. Technol. Biomed. Life Sci.,* **2011**, *879*(17-18), 1270-1281.
[http://dx.doi.org/10.1016/j.jchromb.2010.11.020] [PMID: 21185242]

[116] He, L.; Xu, Z.; Hirokawa, T.; Shen, L. Simultaneous determination of aliphatic, aromatic and heterocyclic biogenic amines without derivatization by capillary electrophoresis and application in beer analysis. *J. Chromatogr. A,* **2017**, *1482*, 109-114.
[http://dx.doi.org/10.1016/j.chroma.2016.12.067] [PMID: 28041619]

[117] Zhang, N.; Wang, H.; Zhang, Z-X.; Deng, Y-H.; Zhang, H-S. Sensitive determination of biogenic amines by capillary electrophoresis with a new fluorogenic reagent 3-(4-fluorobenzoyl-

-2-quinolinecarboxaldehyde. *Talanta,* **2008**, *76*(4), 791-797.
[http://dx.doi.org/10.1016/j.talanta.2008.04.027] [PMID: 18656660]

[118] Xiao, M-W.; Bai, X-L.; Xu, P-L.; Zhao, Y.; Yang, L.; Liu, Y-M.; Liao, X. Rapid determination of gizzerosine in fish meals using microchip capillary electrophoresis with laser-induced fluorescence detection. *Food Addit. Contam. Part A Chem. Anal. Control Expo. Risk Assess.,* **2017**, *34*(5), 760-765.
[http://dx.doi.org/10.1080/19440049.2017.1292053] [PMID: 28277175]

[119] An, D.; Chen, Z.; Zheng, J.; Chen, S.; Wang, L.; Huang, Z.; Weng, L. Determination of biogenic amines in oysters by capillary electrophoresis coupled with electrochemiluminescence. *Food Chem.,* **2015**, *168*, 1-6.
[http://dx.doi.org/10.1016/j.foodchem.2014.07.019] [PMID: 25172675]

Jastrzębska, A.; Piasta, A.; Szłyk, E. Simultaneous determination of selected biogenic amines in alcoholic beverage samples by isotachophoretic and chromatographic methods. *Food Addit. Contam. Part A Chem. Anal. Control Expo. Risk Assess.,* **2014**, *31*(1), 83-92.
[http://dx.doi.org/10.1080/19440049.2013.855326] [PMID: 24350674]

[121] Adımcılar, V.; Öztekin, N.; Bedia Erim, F. A direct and sensitive analysis method for biogenic amines in dairy products by capillary electrophoresis coupled with contactless conductivity detection. *Food Anal. Methods,* **2018**, *11*, 1374-1379.
[http://dx.doi.org/10.1007/s12161-017-1122-9]

[122] Daniel, D.; Dos Santos, V.B.; Vidal, D.T.R.; do Lago, C.L. Determination of biogenic amines in beer and wine by capillary electrophoresis-tandem mass spectrometry. *J. Chromatogr. A,* **2015**, *1416*, 121-128.
[http://dx.doi.org/10.1016/j.chroma.2015.08.065] [PMID: 26362807]

[123] Woźniakiewicz, M.; Woźniakiewicz, A.; Nowak, P.M.; Kłodzińska, E.; Namieśnik, J.; Płotka-Wasylka, J. CE-MS and GC-MS as "green" and complementary methods for the analysis of biogenic amines in wine. *Food Anal. Methods,* **2018**, *11*, 2614-2627.
[http://dx.doi.org/10.1007/s12161-018-1219-9]

[124] Li, W.; Pan, Y.; Liu, Y.; Zhang, X.; Ye, J.; Chu, Q. Simultaneous determination of eight typical biogenic amines by CZE with capacitively coupled contactless conductivity detection. *Chromatographia,* **2014**, *77*, 287-292.
[http://dx.doi.org/10.1007/s10337-013-2595-3]

[125] Guo, C.; Guo, Z.; Chen, Y. A bi-end injection capillary electrophoresis method for simultaneous determination of 37 cations and anions in beers. *Anal. Bioanal. Chem.,* **2019**, *411*(18), 4113-4121.
[http://dx.doi.org/10.1007/s00216-018-1507-7] [PMID: 30519960]

[126] Meixner, M.; Pattky, M.; Huhn, C. Novel approach for the synthesis of a neutral and covalently bound capillary coating for capillary electrophoresis-mass spectrometry made from highly polar and pH-persistent *N*-acryloylamido ethoxyethanol. *Anal. Bioanal. Chem.,* **2020**, *412*(3), 561-575.
[http://dx.doi.org/10.1007/s00216-019-02286-z] [PMID: 31872272]

[127] Simó, C.; Moreno-Arribas, M.V.; Cifuentes, A. Ion-trap *versus* time-of-flight mass spectrometry coupled to capillary electrophoresis to analyze biogenic amines in wine. *J. Chromatogr. A,* **2008**, *1195*(1-2), 150-156.
[http://dx.doi.org/10.1016/j.chroma.2008.05.004] [PMID: 18508068]

[128] World Health Organization *Natural Toxins in food (Factsheet),* **2018**.https://www.who.int/news-room/fact-sheets/detail/natural-toxins-in-food

[129] WHO Food Safety Programme In: *WHO global strategy for food safety: safer food for better health*; WHO: Geneva, **2002**.

[130] van Egmond, H.P. Natural toxins: risks, regulations and the analytical situation in Europe. *Anal. Bioanal. Chem.,* **2004**, *378*(5), 1152-1160.
[http://dx.doi.org/10.1007/s00216-003-2373-4] [PMID: 14673562]

[131] Food and Agriculture Organization of the United Nations (FAO); World Health Organization (WHO). In: *Codex Allimentarius International Food Standards. General standard for contaminants and toxins in food and feed. (Last amended 2019). CODEX STAN, 193-1995*; FAO: Rome, **2019**.

[132] Rodríguez, I.; González, J.M.; Botana, A.M.; Sainz, M.J.; Vieytes, M.R.; Alfonso, A.; Botana, L.M. Analysis of natural toxins by liquid chromatography, **2017**. [http://dx.doi.org/10.1016/B978-0-12-805392-8.00016-5]

[133] Casado, N.; Gañán, J.; Morante-Zarcero, S.; Sierra, I. New advanced materials and sorbent-based microextraction techniques as strategies in sample preparation to improve the determination of natural toxins in food samples. *Molecules*, **2020**, *25*(3), 702. [http://dx.doi.org/10.3390/molecules25030702] [PMID: 32041287]

[134] Pereira, V.L.; Fernandes, J.O.; Cunha, S.C. Mycotoxins in cereals and related foodstuffs: A review on occurrence and recent methods of analysis. *Trends Food Sci. Technol.*, **2014**, *36*, 96-136. [http://dx.doi.org/10.1016/j.tifs.2014.01.005]

[135] Boevre, M.; Mavungu, J.D.; Landshchoot, S.; Audenaert, K.; Eeckhout, M.; Maene, P. Natural occurrence of mycotoxins and their masked forms in food and feed products. *World Mycotoxin J.*, **2012**, *5*, 207-219. [http://dx.doi.org/10.3920/WMJ2012.1410]

[136] Stoev, S.D. Food safety and increasing hazard of mycotoxin occurrence in foods and feeds. *Crit. Rev. Food Sci. Nutr.*, **2013**, *53*(9), 887-901. [http://dx.doi.org/10.1080/10408398.2011.571800] [PMID: 23768181]

[137] Marin, S.; Ramos, A.J.; Cano-Sancho, G.; Sanchis, V. Mycotoxins: occurrence, toxicology, and exposure assessment. *Food Chem. Toxicol.*, **2013**, *60*, 218-237. [http://dx.doi.org/10.1016/j.fct.2013.07.047] [PMID: 23907020]

[138] 1881/2006 of 19 December 2006, setting maximum levels for certain contaminants in foodstuffs. *Off. J. Eur. Commun.*, **2006**, *L364*, 5-24.

[139] European Commission. **2020**. https://ec.europa.eu/food/safety/chemical_safety/contaminants/catalogue/mycotoxins_en

[140] 401/2006 of laying down the methods of sampling and analysis for the official control of the levels of mycotoxins in foodstuffs. *Off. J. Eur. Commun*, **2006**, *L70*, 12-34.

[141] Murillo-Arbizu, M.; González-Peñas, E.; Amézqueta, S. Comparison between capillary electrophoresis and high performance liquid chromatography for the study of the occurrence of patulin in apple juice intended for infants. *Food Chem. Toxicol.*, **2010**, *48*(8-9), 2429-2434. [http://dx.doi.org/10.1016/j.fct.2010.06.003] [PMID: 20566341]

[142] Murillo, M.; González-Peñas, E.; Amézqueta, S. Determination of patulin in commercial apple juice by micellar electrokinetic chromatography. *Food Chem. Toxicol.*, **2008**, *46*(1), 57-64. [http://dx.doi.org/10.1016/j.fct.2007.06.024] [PMID: 17707570]

[143] Víctor-Ortega, M.D.; Lara, F.J.; García-Campaña, A.M.; del Olmo-Iruela, M. Evaluation of dispersive liquid-liquid microextraction for the determination of patulin in apple juices using micellar electrokinetic capillary chromatography. *Food Control*, **2013**, *31*, 353-358. [http://dx.doi.org/10.1016/j.foodcont.2012.11.003]

[144] Güray, T.; Tuncel, M.; Uysal, U.D. A rapid determination of patulin using capillary zone electrophoresis and its application to analysis of apple juices. *J. Chromatogr. Sci.*, **2013**, *51*(4), 310-317. [http://dx.doi.org/10.1093/chromsci/bms142] [PMID: 22988003]

[145] Nigović, B.; Sertić, M.; Mornar, A. Simultaneous determination of lovastatin and citrinin in red yeast rice supplements by micellar electrokinetic capillary chromatography. *Food Chem.*, **2013**, *138*(1), 531-538. [http://dx.doi.org/10.1016/j.foodchem.2012.10.104] [PMID: 23265521]

[146] Zhu, D.; Zhang, H.; Bing, X. Preparation of an immunoaffinity column for the clean-up of fermented food samples contaminated with citrinin. *Food Addit. Contam. Part A Chem. Anal. Control Expo. Risk Assess.*, **2013**, *30*(2), 389-394. [http://dx.doi.org/10.1080/19440049.2012.743042] [PMID: 23157620]

[147] Felici, E.; Wang, C.C.; Fernández, L.P.; Gómez, M.R. Simultaneous separation of ergot alkaloids by capillary electrophoresis after cloud point extraction from cereal samples. *Electrophoresis,* **2015**, *36*(2), 341-347. [http://dx.doi.org/10.1002/elps.201400215] [PMID: 25257749]

[148] Arribas, A.S.; Bermejo, E.; Zapardiel, A.; Téllez, H.; Rodríguez-Flores, J.; Zougagh, M.; Ríos, A.; Chicharro, M. Screening and confirmatory methods for the analysis of macrocyclic lactone mycotoxins by CE with amperometric detection. *Electrophoresis,* **2009**, *30*(3), 499-506. [http://dx.doi.org/10.1002/elps.200800305] [PMID: 19156758]

[149] Güray, T.; Tuncel, M.; Uysal, U.D.; Oncu-Kaya, E.M. Determination of zearalenone by the capillary zone electrophoresis-UV detection and its application to poultry feed and cereals. *J. Liq. Chromatogr. Relat. Technol.,* **2013**, *36*, 1366-1378. [http://dx.doi.org/10.1080/10826076.2012.691437]

[150] Maragos, C.M.; Appell, M.; Lippolis, V.; Visconti, A.; Catucci, L.; Pascale, M. Use of cyclodextrins as modifiers of fluorescence in the detection of mycotoxins. *Food Addit. Contam. Part A Chem. Anal. Control Expo. Risk Assess.,* **2008**, *25*(2), 164-171. [http://dx.doi.org/10.1080/02652030701564555] [PMID: 18286406]

[151] Arroyo-Manzanares, N.; Gámiz-Gracia, L.; García-Campaña, A.M.; Soto-Chinchilla, J.J.; García-Ayuso, L.E. On-line preconcentration for the determination of aflatoxins in rice samples by micellar electrokinetic capillary chromatography with laser-induced fluorescence detection. *Electrophoresis,* **2010**, *31*(13), 2180-2185. [http://dx.doi.org/10.1002/elps.201000062] [PMID: 20593392]

[152] Li, P.; Zhang, Z.; Zhang, Q.; Zhang, N.; Zhang, W.; Ding, X.; Li, R. Current development of microfluidic immunosensing approaches for mycotoxin detection *via* capillary electromigration and lateral flow technology. *Electrophoresis,* **2012**, *33*(15), 2253-2265. [http://dx.doi.org/10.1002/elps.201200050] [PMID: 22887149]

[153] Xiao M-W.; Bai X-L.; Liu, Y-M.; Yang, L.; Liao, X., Simultaneous determination of trace aflatoxin B1 and ochratoxin A by aptamer-based microchip capillary electrophoresis in food samples. *J. Chromatogr. A,* **2018**, *2018*(1569), 222-228.

[154] Regulation (EC) No 853/2004 of the European Parliament and of the Council of 29 April 2004 laying down specific hygiene rules for food of animal origin. *Off. J. Eur. Union,* **2004**, *L139*, 55.

[155] Paredes, I.; Rietjens, I.M.C.M.; Vieites, J.M.; Cabado, A.G. Update of risk assessments of main marine biotoxins in the European Union. *Toxicon,* **2011**, *58*(4), 336-354. [http://dx.doi.org/10.1016/j.toxicon.2011.07.001] [PMID: 21777598]

[156] Commission Implementing Regulation (EU) 2019/627 of 15 March 2019 laying down uniform practical arrangements for the performance of official controls on products of animal origin intended for human consumption in accordance with Regulation (EU) 2017/625 of the European Parliament and of the Council and amending Commission Regulation (EC) No 2074/2005 as regards official controls. *Off. J. Eur. Union,* **2019**, *L131*, 51-100.

[157] Álvarez-Muñoz, D.; Llorca, M.; Blasco, J.; Barceló, D. Contaminants in the Marine environment.*Marine Ecotoxicology: Current Knowledge and Future Issues 2016*; Blasco, J.; Chapman, P.M.; Campana, O.; Hampel, M., Eds.; Academic Press: London, **2016**, pp. 1-34. [http://dx.doi.org/10.1016/B978-0-12-803371-5.00001-1]

[158] de la Iglesia, P.; Gago-Martínez, A.; Yasumoto, T. Advanced studies for the application of high-performance capillary electrophoresis for the analysis of yessotoxin and 45-hydroxyyessotoxin. *J. Chromatogr. A,* **2007**, *1156*(1-2), 160-166.

[http://dx.doi.org/10.1016/j.chroma.2006.12.084] [PMID: 17239891]

[159] de la Iglesia, P.; Gago-Martínez, A. Determination of yessotoxins and pectenotoxins in shellfish by capillary electrophoresis-electrospray ionization-mass spectrometry. *Food Addit. Contam. Part A Chem. Anal. Control Expo. Risk Assess.*, **2009**, *26*(2), 221-228. [http://dx.doi.org/10.1080/02652030802290522] [PMID: 19680893]

[160] Kerrin, E.S.; White, R.L.; Quilliam, M.A. Quantitative determination of the neurotoxin β--methylamino-L-alanine (BMAA) by capillary electrophoresis-tandem mass spectrometry. *Anal. Bioanal. Chem.*, **2017**, *409*(6), 1481-1491. [http://dx.doi.org/10.1007/s00216-016-0091-y] [PMID: 27909777]

[161] Beach, D.G.; Kerrin, E.S.; Thomas, K.; Quilliam, M.A.; McCarron, P. Capillary electrophoresis-tandem mass spectrometry for multiclass analysis of polar marine toxins. *Anal. Bioanal. Chem.*, **2018**, *410*(22), 5405-5420. [http://dx.doi.org/10.1007/s00216-018-1089-4] [PMID: 29767301]

[162] Keyon, A.S.; Guijt, R.M.; Gaspar, A.; Kazarian, A.A.; Nesterenko, P.N.; Bolch, C.J.; Breadmore, M.C. Capillary electrophoresis for the analysis of paralytic shellfish poisoning toxins in shellfish: comparison of detection methods. *Electrophoresis,* **2014**, *35*(10), 1496-1503. [http://dx.doi.org/10.1002/elps.201300353] [PMID: 24591173]

[163] Abdul Keyon, A.S.; Guijt, R.M.; Bolch, C.J.S.; Breadmore, M.C. Transient isotachophoresis-capillary zone electrophoresis with contactless conductivity and ultraviolet detection for the analysis of paralytic shellfish toxins in mussel samples. *J. Chromatogr. A,* **2014**, *1364*, 295-302. [http://dx.doi.org/10.1016/j.chroma.2014.08.074] [PMID: 25223612]

[164] He, Y.; Mo, F.; Chen, D.; Xu, L.; Wu, Y.; Fu, F. Capillary electrophoresis inductively coupled plasma mass spectrometry combined with metal tag for ultrasensitively determining trace saxitoxin in seafood. *Electrophoresis,* **2017**, *38*(3-4), 469-476. [http://dx.doi.org/10.1002/elps.201600411] [PMID: 27862045]

[165] Zhang, X.; Zhang, Z. Capillary electrophoresis-based immunoassay for the determination of brevetoxin-B in shellfish using electrochemical detection. *J. Chromatogr. Sci.*, **2013**, *51*(2), 107-111. [http://dx.doi.org/10.1093/chromsci/bms112] [PMID: 22781183]

[166] Zhang, Z.; Li, X.; Ge, A.; Zhang, F.; Sun, X.; Li, X. High selective and sensitive capillary electrophoresis-based electrochemical immunoassay enhanced by gold nanoparticles. *Biosens. Bioelectron.*, **2013**, *41*, 452-458. [http://dx.doi.org/10.1016/j.bios.2012.09.005] [PMID: 23062557]

[167] Zhang, Z.; Liu, Y.; Zhang, C.; Luan, W. Horseradish peroxidase and antibody labeled gold nanoparticle probe for amplified immunoassay of ciguatoxin in fish samples based on capillary electrophoresis with electrochemical detection. *Toxicon,* **2015**, *96*, 89-95. [http://dx.doi.org/10.1016/j.toxicon.2015.01.015] [PMID: 25637767]

[168] Koszucka, A.; Nowak, A. Thermal processing food-related toxicants: a review. *Crit. Rev. Food Sci. Nutr.*, **2019**, *59*(22), 3579-3596. [http://dx.doi.org/10.1080/10408398.2018.1500440] [PMID: 30311772]

[169] Wenzl, T.; Lachenmeier, D.W.; Gökmen, V. Analysis of heat-induced contaminants (acrylamide, chloropropanols and furan) in carbohydrate-rich food. *Anal. Bioanal. Chem.*, **2007**, *389*(1), 119-137. [http://dx.doi.org/10.1007/s00216-007-1459-9] [PMID: 17673989]

[170] Morales, F.J. Hydroxymethylfurfural (HMF) and related compounds.*Process-induced toxicants: occurrence, formation, mitigation, and health risks*; Stadler, R.H.; Lineback, D.R., Eds.; John Wiley & Sons, Inc.: Hoboken, NJ, USA, **2008**, pp. 135-174. [http://dx.doi.org/10.1002/9780470430101.ch2e]

[171] Rizelio, V.M.; Gonzaga, L.V.; Da Silva Campelo Borges, G.; Micke, G.A.; Costa, A.C.O. Development of a fast MECK method for determination of 5-HMF in honey samples. *Food Chem.*, **2012**, *133*, 1640-1645.

[http://dx.doi.org/10.1016/j.foodchem.2011.11.058]

[172] Biluca, F.C.; Della Betta, F.; de Oliveira, G.P.; Pereira, L.M.; Gonzaga, L.V.; Costa, A.C.O.; Fett, R. 5-HMF and carbohydrates content in stingless bee honey by CE before and after thermal treatment. *Food Chem.*, **2014**, *159*, 244-249. [http://dx.doi.org/10.1016/j.foodchem.2014.03.016] [PMID: 24767051]

[173] Morales, F.J.; Jiménez-Pérez, S. Hydroxymethylfurfural determination in infant milk-based formulas by micellar electrokinetic capillary chromatography. *Food Chem.*, **2001**, *72*, 525-531. [http://dx.doi.org/10.1016/S0308-8146(00)00284-3]

[174] Chen, Z.; Yan, X. Simultaneous determination of melamine and 5-hydroxymethylfurfural in milk by capillary electrophoresis with diode array detection. *J. Agric. Food Chem.*, **2009**, *57*(19), 8742-8747. [http://dx.doi.org/10.1021/jf9021916] [PMID: 19761188]

[175] Güray, T.; Tuncel, N.Y.; Tunçel, M.; Uysal, U.D. Validated micellar electrokinetic capillary chromatography (MECC) method for determination of 5-hydroxymethylfurfural in honey and comparison with HPLC. *Chem. Pap.*, **2019**, *73*, 2209-2220. [http://dx.doi.org/10.1007/s11696-019-00770-5]

[176] Teixidó, E.; Núñez, O.; Santos, F.J.; Galcerán, M.T. 5-Hydroxymethylfurfural content in foodstuffs determined by micellar electrokinetic chromatography. *Food Chem.*, **2011**, *126*(4), 1902-1908. [http://dx.doi.org/10.1016/j.foodchem.2010.12.016] [PMID: 25213975]

[177] Bignardi, C.; Cavazza, A.; Corradini, C. Selected product ion monitoring for quantification of 5-hydroxymethylfurfural in food products by capillary zone electrophoresis-tandem ion trap mass spectrometry. *Food Control*, **2014**, *46*, 41-48. [http://dx.doi.org/10.1016/j.foodcont.2014.04.049]

[178] (2010/307/EU) of 2 June 2010 on the monitoring of acrylamide levels in food. *Off. J. Eur. Union*, **2010**, *L137*, 4-10.

[179] (2013/647/EU) of 8 November June 2013 on investigations into the levels of acrylamide in food. *Off. J. Eur. Union*, **2013**, *L301*, 15-17.

[180] Bermudo, E.; Núñez, O.; Puignou, L.; Galcerán, M.T. Analysis of acrylamide in food samples by capillary zone electrophoresis. *J. Chromatogr. A*, **2006**, *1120*(1-2), 199-204. [http://dx.doi.org/10.1016/j.chroma.2005.10.074] [PMID: 16310205]

[181] Bermudo, E.; Núñez, O.; Moyano, E.; Puignou, L.; Galcerán, M.T. Field amplified sample injection-capillary electrophoresis-tandem mass spectrometry for the analysis of acrylamide in foodstuffs. *J. Chromatogr. A*, **2007**, *1159*(1-2), 225-232. [http://dx.doi.org/10.1016/j.chroma.2007.03.008] [PMID: 17383666]

[182] Yang, S.; Li, Y.; Li, F.; Yang, Z.; Quan, F.; Zhou, L.; Pu, Q. Thiol-ene click derivatization for the determination of acrylamide in potato products by capillary electrophoresis with capacitively coupled contactless conductivity detection. *J. Agric. Food Chem.*, **2019**, *67*(28), 8053-8060. [http://dx.doi.org/10.1021/acs.jafc.9b01525] [PMID: 31276393]

[183] Bermudo, E.; Ruiz-Calero, V.; Puignou, L.; Galcerán, M.T. Microemulsion electrokinetic chromatography for the analysis of acrylamide in food. *Electrophoresis*, **2004**, *25*(18-19), 3257-3262. [http://dx.doi.org/10.1002/elps.200406044] [PMID: 15472956]

[184] Zhou, X.; Fan, L-Y.; Zhang, W.; Cao, C-X. Separation and determination of acrylamide in potato chips by micellar electrokinetic capillary chromatography. *Talanta*, **2007**, *71*(4), 1541-1545. [http://dx.doi.org/10.1016/j.talanta.2006.07.037] [PMID: 19071489]

[185] Tezcan, F.; Erim, F.B. On-line stacking techniques for the nonaqueous capillary electrophoretic determination of acrylamide in processed food. *Anal. Chim. Acta*, **2008**, *617*(1-2), 196-199. [http://dx.doi.org/10.1016/j.aca.2008.01.008] [PMID: 18486658]

[186] Abd El-Hady, D.; Albishri, H.M. Simultaneous determination of acrylamide, asparagine and glucose in food using short chain methyl imidazolium ionic liquid based ultrasonic assisted extraction coupled

with analyte focusing by ionic liquid micelle collapse capillary electrophoresis. *Food Chem.*, **2015**, *188*, 551-558.
[http://dx.doi.org/10.1016/j.foodchem.2015.05.047] [PMID: 26041230]

[187] Quirino, J.P. Analyte focusing by micelle collapse in CZE: nanopreparation of neutrals. *Electrophoresis,* **2009**, *30*(5), 875-882.
[http://dx.doi.org/10.1002/elps.200801735] [PMID: 19260005]

[188] Wu, M.; Chen, W.; Wang, G.; He, P.; Wang, Q. Analysis of acrylamide in food products by microchip electrophoresis with on-line multiple-preconcentration techniques. *Food Chem.*, **2016**, *209*, 154-161.
[http://dx.doi.org/10.1016/j.foodchem.2016.04.065] [PMID: 27173547]

[189] Zelinkova, Z.; Wenzl, T. The occurrence of 16 EPA PAHs in food-A review. *Polycycl. Aromat. Compd.*, **2015**, *35*(2-4), 248-284.
[http://dx.doi.org/10.1080/10406638.2014.918550] [PMID: 26681897]

[190] Ferey, L.; Delaunay, N. Capillary and microchip electrophoretic analysis of polycyclic aromatic hydrocarbons. *Anal. Bioanal. Chem.*, **2015**, *407*(10), 2727-2747.
[http://dx.doi.org/10.1007/s00216-014-8390-7] [PMID: 25542576]

[191] Ferey, L.; Delaunay, N.; Rutledge, D.N.; Huertas, A.; Raoul, Y.; Gareil, P.; Vial, J. Use of response surface methodology to optimize the simultaneous separation of eight polycyclic aromatic hydrocarbons by capillary zone electrophoresis with laser-induced fluorescence detection. *J. Chromatogr. A,* **2013**, *1302*, 181-190.
[http://dx.doi.org/10.1016/j.chroma.2013.06.027] [PMID: 23831002]

[192] Ferey, L.; Delaunay, N.; Rutledge, D.N.; Cordella, C.B.Y.; This, H.; Huertas, A.; Raoul, Y.; Gareil, P. Optimizing separation conditions of 19 polycyclic aromatic hydrocarbons by cyclodextrin-modified capillary electrophoresis and applications to edible oils. *Talanta,* **2014**, *119*, 572-581.
[http://dx.doi.org/10.1016/j.talanta.2013.11.062] [PMID: 24401457]

[193] Ferey, L.; Delaunay, N.; Rutledge, D.N.; Huertas, A.; Raoul, Y.; Gareil, P.; Vial, J.; Rivals, I. An experimental design based strategy to optimize a capillary electrophoresis method for the separation of 19 polycyclic aromatic hydrocarbons. *Anal. Chim. Acta,* **2014**, *820*, 195-204.
[http://dx.doi.org/10.1016/j.aca.2014.02.040] [PMID: 24745754]

[194] Muncke, J.; Andersson, A-M.; Backhaus, T.; Boucher, J.M.; Carney Almroth, B.; Castillo Castillo, A.; Chevrier, J.; Demeneix, B.A.; Emmanuel, J.A.; Fini, J-B.; Gee, D.; Geueke, B.; Groh, K.; Heindel, J.J.; Houlihan, J.; Kassotis, C.D.; Kwiatkowski, C.F.; Lefferts, L.Y.; Maffini, M.V.; Martin, O.V.; Myers, J.P.; Nadal, A.; Nerin, C.; Pelch, K.E.; Fernández, S.R.; Sargis, R.M.; Soto, A.M.; Trasande, L.; Vandenberg, L.N.; Wagner, M.; Wu, C.; Zoeller, R.T.; Scheringer, M. Impacts of food contact chemicals on human health: a consensus statement. *Environ. Health,* **2020**, *19*(1), 25.
[http://dx.doi.org/10.1186/s12940-020-0572-5] [PMID: 32122363]

[195] Hoppe, M.; de Voogt, P.; Franz, R. Identification and quantification of oligomers as potential migrants in plastics food contact materials with a focus in polycondensates-A review. *Trends Food Sci. Technol.*, **2016**, *50*, 118-130.
[http://dx.doi.org/10.1016/j.tifs.2016.01.018]

[196] Alenazi, N.A.; Manthorpe, J.M.; Lai, E.P.C. Selective extraction of BPA in milk analysis by capillary electrophoresis using a chemically modified molecularly imprinted polymer. *Food Control,* **2015**, *50*, 778-783.
[http://dx.doi.org/10.1016/j.foodcont.2014.10.026]

[197] Zhang, X.; Zhu, D.; Huang, C.; Sun, Y. Lee, Y.-III., Sensitive detection of bisphenol A in complex samples by in-column molecularly imprinted solid-phase extraction coupled with capillary electrophoresis. *Microchem. J.,* **2015**, *121*, 1-5.
[http://dx.doi.org/10.1016/j.microc.2015.01.012]

[198] Morales-Cid, G.; Cárdenas, S.; Simonet, B.M.; Valcárcel, M. Comparison of aromatic and alkyl micelles for the electrokinetic determination of phthalates in virgin olive oil. *Electrophoresis,* **2009**,

30(4), 618-623.
[http://dx.doi.org/10.1002/elps.200800312] [PMID: 19180542]

[199] Hsieh, S-Y.; Wang, C-C.; Wu, S-M. Microemulsion electrokinetic chromatography for analysis of phthalates in soft drinks. *Food Chem.*, **2013**, *141*(4), 3486-3491.
[http://dx.doi.org/10.1016/j.foodchem.2013.06.023] [PMID: 23993511]

[200] Peng, X.; Wang, H.; Yang, B.; Zhan, X.; Wu, Y. Field-amplified sample injection-capillary electrophoresis for the determination of bisphenol A, α-naphthol and β-naphthol in drinks and lake water. *Chromatographia,* **2016**, *79*, 327-333.
[http://dx.doi.org/10.1007/s10337-016-3028-x]

[201] Ni, X.; Xing, X.; Cao, Y.; Cao, G. Determination of phthalates in food packing materials by electrokinetic chromatography with polymeric pseudostationary phase. *Food Chem.*, **2016**, *190*, 386-391.
[http://dx.doi.org/10.1016/j.foodchem.2015.05.134] [PMID: 26212987]

[202] Nerín, C.; Aznar, M.; Carrizo, D. Food contamination during food process. *Trends Food Sci. Technol.*, **2016**, *48*, 63-68.
[http://dx.doi.org/10.1016/j.tifs.2015.12.004]

[203] Hernández-Borges, J.; Rodríguez-Delgado, M.A.; García-Montelongo, F.J.; Cifuentes, A. Chiral analysis of pollutants and their metabolites by capillary electromigration methods. *Electrophoresis,* **2005**, *26*(20), 3799-3813.
[http://dx.doi.org/10.1002/elps.200500084] [PMID: 16217833]

[204] Trojanowicz, M.; Koc, M. Recent developments in methods for analysis of perfluorinated persistent pollutants. *Mikrochim. Acta,* **2013**, *180*, 957-971.
[http://dx.doi.org/10.1007/s00604-013-1046-z] [PMID: 23913984]

[205] Chazelas, E.; Deschasaux, M.; Srour, B.; Kesse-Guyot, E.; Julia, C.; Alles, B.; Druesne-Pecollo, N.; Galan, P.; Hercberg, S.; Latino-Martel, P.; Esseddik, Y.; Szabo, F.; Slamich, P.; Gigandet, S.; Touvier, M. Food additives: distribution and co-occurrence in 126,000 food products of the French market. *Sci. Rep.*, **2020**, *10*(1), 3980.
[http://dx.doi.org/10.1038/s41598-020-60948-w] [PMID: 32132606]

[206] Boyce, M.C. Determination of additives and organic contaminants in food by CE and CEC. *Electrophoresis,* **2007**, *28*(22), 4046-4062.
[http://dx.doi.org/10.1002/elps.200700280] [PMID: 17948272]

[207] Gonçalves, S.; Alves, V.R.; Pérez, S.O.; Ferreira, M.; Daguer, H.; de Oliveira, M.A.L.; Micke, G.A.; Vitali, L. Rapid method for the determination of citrate, phosphate and sulfite in seafood by capillary zone electrophoresis. *Food Chem.*, **2020**, *321*126705
[http://dx.doi.org/10.1016/j.foodchem.2020.126705] [PMID: 32276144]

[208] Yoshikawa, K.; Saito, S.; Sakuragawa, A. Simultaneous analysis of acidulants and preservatives in food samples by using capillary zone electrophoresis with indirect UV detection. *Food Chem.*, **2011**, *127*(3), 1385-1390.
[http://dx.doi.org/10.1016/j.foodchem.2011.01.126] [PMID: 25214142]

[209] Aung, H-P.; Pyell, U. In-capillary derivatization with *o*-phthalaldehyde in the presence of 3-mercaptopropionic acid for the simultaneous determination of monosodium glutamate, benzoic acid, and sorbic acid in food samples *via* capillary electrophoresis with ultraviolet detection. *J. Chromatogr. A,* **2016**, *1449*, 156-165.
[http://dx.doi.org/10.1016/j.chroma.2016.04.033] [PMID: 27156753]

[210] Guadalupe Coelho, A.; Fernanda Souza de Jesus, F.; de Assis Pallos, L.; Alberto Fracassi da Silva, J.; Pereira de Jesus, D. Capillary electrophoresis with capacitively coupled contactless conductivity detection for the determination of propionate and sorbate in bread. *J. Sep. Sci.*, **2018**, *41*(20), 3932-3937.
[http://dx.doi.org/10.1002/jssc.201800705] [PMID: 30137665]

[211] Wang, W.; Wang, Y.; Zhang, J.; Chu, Q.; Ye, J. Simultaneous determination of electroactive and non-electroactive food preservatives by novel capillary electrophoresis with amperometric detection. *Anal. Chim. Acta,* **2010**, *678*(1), 39-43.
[http://dx.doi.org/10.1016/j.aca.2010.08.018] [PMID: 20869502]

[212] Zhang, X.; Xu, S.; Sun, Y.; Wang, Y.; Wang, C. Simultaneous determination of benzoic acid and sorbic acid in food products by CE after on-line preconcentration by dynamic pH junction. *Chromatographia,* **2011**, *73*, 1217-1221.
[http://dx.doi.org/10.1007/s10337-011-2009-3]

[213] Hsu, S-H.; Hu, C-C.; Chiu, T-C. Online dynamic pH junction-sweeping for the determination of benzoic and sorbic acids in food products by capillary electrophoresis. *Anal. Bioanal. Chem.,* **2014**, *406*(2), 635-641.
[http://dx.doi.org/10.1007/s00216-013-7481-1] [PMID: 24247553]

[214] Tang, W.; Tian, X.; Zhang, Y.; Yin, D.; Zhao, W.; Zhang, W.; Zhang, S. Determination of benzoic acid and sorbic acid in seasoning by capillary electrophoresis with a new triethylamine aminated polychloromethyl styrene nanolatex coated capillary column. *Anal. Methods,* **2017**, *9*, 6280.
[http://dx.doi.org/10.1039/C7AY01930K]

[215] Li, Z.; Mao, Z.; Chen, Z. Polydopamine-assisted immobilization of a zinc(II)-derived metal-organic cage as a stationary phase for open-tubular capillary electrochromatography. *Mikrochim. Acta,* **2019**, *186*(7), 449.
[http://dx.doi.org/10.1007/s00604-019-3576-5] [PMID: 31197494]

[216] Sun, B.; Qi, L.; Wang, M. Determination of preservatives in soft drinks by capillary electrophoresis with ionic liquids as the electrolyte additives. *J. Sep. Sci.,* **2014**, *37*(16), 2248-2252.
[http://dx.doi.org/10.1002/jssc.201400444] [PMID: 24910409]

[217] Mejia, E.; Ding, Y.; Mora, M.F.; Garcia, C.D. Determination of banned sudan dyes in chili powder by capillary electrophoresis. *Food Chem.,* **2007**, *102*, 1027-1033.
[http://dx.doi.org/10.1016/j.foodchem.2006.06.038]

[218] Moreno-González, D.; Jáč, P.; Švec, F.; Nováková, L. Determination of Sudan dyes in chili products by micellar electrokinetic chromatography-MS/MS using a volatile surfactant. *Food Chem.,* **2020**, *310*125963
[http://dx.doi.org/10.1016/j.foodchem.2019.125963] [PMID: 31838374]

[219] Fukuji, T.S.; Castro-Puyana, M., Tavares, M F M.; Cifuentes, A. Sensitive and fast determination of Sudan dyes in chili powder by partial-filling micellar electrokinetic chromatography-tandem mass spectrometry. *Electrophoresis,* **2012**, *33*(4), 705-712.
[http://dx.doi.org/10.1002/elps.201100272] [PMID: 22451064]

[220] Muijselaar, P.G.; Otsuka, K.; Terabe, S. On-line coupling of partial-filling micellar electrokinetic chromatography with mass spectrometry. *J. Chromatogr. A,* **1998**, *802*, 3-15.
[http://dx.doi.org/10.1016/S0021-9673(97)01089-3]

[221] Yi, J.; Zeng, L.; Wu, Q.; Yang, L.; Xie, T. Sensitive simultaneous determination of synthetic food colorants in preserved fruit samples by capillary electrophoresis with contactless conductivity detection. *Food Anal. Methods,* **2018**, *11*, 1608-1618.
[http://dx.doi.org/10.1007/s12161-017-1141-6]

[222] Bordagaray, A.; Garcia-Arrona, R.; Vidal, M.; Ostra, M. Determination of food colorants in a wide variety of food matrices by microemulsion electrokinetic capillary chromatography. Considerations on the found concentrations and regulated consumption limits. *Food Chem.,* **2018**, *262*, 129-133.
[http://dx.doi.org/10.1016/j.foodchem.2018.04.086] [PMID: 29751899]

[223] Liu, F-J.; Liu, C-T.; Li, W.; Tang, A-N. Dispersive solid-phase microextraction and capillary electrophoresis separation of food colorants in beverages using diamino moiety functionalized silica nanoparticles as both extractant and pseudostationary phase. *Talanta,* **2015**, *132*, 366-372.
[http://dx.doi.org/10.1016/j.talanta.2014.09.014] [PMID: 25476319]

[224] Vistuba, J.P.; Dolzan, M.D.; Vitali, L.; de Oliveira, M.A.L.; Micke, G.A. Sub-minute method for simultaneous determination of aspartame, cyclamate, acesulfame-K and saccharin in food and pharmaceutical samples by capillary zone electrophoresis. *J. Chromatogr. A,* **2015**, *1396*, 148-152. [http://dx.doi.org/10.1016/j.chroma.2015.03.070] [PMID: 25895731]

[225] Xiao, Y.; Li, Y.; Ying, J.; Tian, Y.; Xiao, Y.; Mei, Z. Determination of alditols by capillary electrophoresis with indirect laser-induced fluorescence detection. *Food Chem.,* **2015**, *174*, 233-239. [http://dx.doi.org/10.1016/j.foodchem.2014.11.046] [PMID: 25529675]

[226] Bergamo, A.B.; Fracassi da Silva, J.A.; de Jesus, D.P. Simultaneous determination of aspartame, cyclamate, saccharin and acesulfame-K in soft drinks and tabletop sweetener formulations by capillary electrophoresis with capacitively coupled contactless conductivity detection. *Food Chem.,* **2011**, *124*, 1714-1717.
[http://dx.doi.org/10.1016/j.foodchem.2010.07.107]

[227] Stojkovic, M.; Mai, T.D.; Hauser, P.C. Determination of artificial sweeteners by capillary electrophoresis with contactless conductivity detection optimized by hydrodynamic pumping. *Anal. Chim. Acta,* **2013**, *787*, 254-259.
[http://dx.doi.org/10.1016/j.aca.2013.05.039] [PMID: 23830447]

[228] Yang, L.; Zhou, S.; Xiao, Y.; Tang, Y.; Xie, T. Sensitive simultaneous determination of three sulfanilamide artificial sweeters by capillary electrophoresis with on-line preconcentration and contactless conductivity detection. *Food Chem.,* **2015**, *188*, 446-451.
[http://dx.doi.org/10.1016/j.foodchem.2015.04.060] [PMID: 26041216]

[229] Frazier, R.A.; Inns, E.L.; Dossi, N.; Ames, J.M.; Nursten, H.E. Development of a capillary electrophoresis method for the simultaneous analysis of artificial sweeteners, preservatives and colours in soft drinks. *J. Chromatogr. A,* **2000**, *876*(1-2), 213-220.
[http://dx.doi.org/10.1016/S0021-9673(00)00184-9] [PMID: 10823516]

[230] Wu, Y.; Li, J.; Zhao, S.; Ding, X. Simultaneous determination of 11 food additives by micellar electrokinetic capillary chromatography. *Food Anal. Methods,* **2016**, *9*, 589-595.
[http://dx.doi.org/10.1007/s12161-015-0225-4]

[231] Nowak, P.M. Simultaneous quantification of food colorants and preservatives in sports drinks by the high performance liquid chromatography and capillary electrophoresis methods evaluated using the red-green-blue model. *J. Chromatogr. A,* **2020**, *1620*460976
[http://dx.doi.org/10.1016/j.chroma.2020.460976] [PMID: 32089292]

[232] Le, T.H.H.; Nguyen, T.Q.H.; Tran, C.S.; Vu, T.T.; Nguyen, T.L.; Cao, V.H.; Ta, T.T.; Pham, T.N.M.; Nguyen, T.A.H.; Mai, T.D. Screening determination of food additives using capillary electrophoresis coupled with contactless conductivity detection: A case study in Vietnam. *Food Control,* **2017**, *77*, 281-289.
[http://dx.doi.org/10.1016/j.foodcont.2017.02.020]

[233] Mai, T.D.; Le, M.D.; Sáiz, J.; Duong, H.A.; Koenka, I.J.; Pham, H.V.; Hauser, P.C. Triple-channel portable capillary electrophoresis instrument with individual background electrolytes for the concurrent separations of anionic and cationic species. *Anal. Chim. Acta,* **2016**, *911*, 121-128.
[http://dx.doi.org/10.1016/j.aca.2016.01.029] [PMID: 26893094]

[234] Anastasi, E.; Riviere, G.; Teste, B. Nanomaterials in food – Prioritisation & Assessment. *EFSA J.,* **2019**, *17*e170909

[235] Ban, E.; Yoo, Y.S.; Song, E.J. Analysis and applications of nanoparticles in capillary electrophoresis. *Talanta,* **2015**, *141*, 15-20.
[http://dx.doi.org/10.1016/j.talanta.2015.03.020] [PMID: 25966374]

[236] Moreno, V.; Zougagh, M.; Ríos, Á. Analytical nanometrological approach for screening and confirmation of titanium dioxide nano/micro-particles in sugary samples based on Raman spectroscopy - Capillary electrophoresis. *Anal. Chim. Acta,* **2019**, *1050*, 169-175.
[http://dx.doi.org/10.1016/j.aca.2018.10.067] [PMID: 30661586]

[237] Ruiz-Palomero, C.; Soriano, M.L.; Valcárcel, M. Sulfonated nanocellulose for the efficient dispersive micro solid-phase extraction and determination of silver nanoparticles in food products. *J. Chromatogr. A,* **2016**, *1428*, 352-358. [http://dx.doi.org/10.1016/j.chroma.2015.06.023] [PMID: 26116191]

[238] Adelantado, C.; Rodríguez-Fariñas, N.; Rodríguez Martín-Doimeadios, R.C.; Zougagh, M.; Ríos, Á. Analysis of silica nanoparticles by capillary electrophoresis coupled to an evaporative light scattering detector. *Anal. Chim. Acta,* **2016**, *923*, 82-88. [http://dx.doi.org/10.1016/j.aca.2016.03.055] [PMID: 27155305]

CHAPTER 13

The Role of Capillary Electrophoresis to Guarantee the Quality and Safety of Dietary Supplements

Enrica Donati[1] and **Zeineb Aturki**[1,*]

[1] *Istituto per I Sistemi Biologici, Consiglio Nazionale delle Ricerche, Rome, Italy*

Abstract: At present, dietary supplements are commercially available products, globally consumed as an addition to the usual diet. Considering that dietary supplements are a source of nutrients, they are widely utilized to improve human health and prevent various diseases; therefore they are expected to be safe. There is still no common definition regarding the role of supplements which cannot be considered functional foods nor drugs. Dietary ingredients in supplements are exempt from food additives or drugs regulations. For this reason, these supplements are marketed without any data on identity, including ingredient information, effectiveness, toxicology and safety. Therefore, efficacy and safety are necessary claims required to preserve consumer health. To face this imperative challenge, sensitive and selective analytical techniques capable of providing a full characterization of the supplements in terms of their components are needed. In the last decades, Capillary Electrophoresis (CE) has shown to be a powerful tool that offers solutions to almost any analytical issue arising in several application fields. Due to its simplicity of operation and versatility, it has become a complementary separation tool to other separation techniques such as gas and liquid chromatography in the analysis of dietary supplements. This chapter aims to give a comprehensive overview of the most important applications of CE for the analysis of dietary supplements in terms of their main key components.

Keywords: Amino acid, Antioxidant, Capillary electrophoresis, Contaminant, Dietary supplement, Vitamin.

INTRODUCTION

Dietary supplements are preparations taken in addition to the normal diet in order to assure a balanced nutrient intake for optimal health. Supplements are available on the market as tablets, capsules, powders, gels or liquids and include at least one of the following substances: vitamins, minerals, amino acids, fatty acids, proteins (enzymes), hormones or botanicals [1 - 3].

* **Corresponding author Zeineb Aturki:**, Istituto per I Sistemi Biologici, Consiglio Nazionale delle Ricerche, Rome, Italy; E-mail: zeineb.aturki@cnr.it

María Castro-Puyana, Miguel Herrero and María Luisa Marina (Eds.)
Capillary Electrophoresis in Food Analysis

In recent years, consumption of dietary supplements has greatly increased since their use has become a common habit to compensate for nutrient deficiency. Despite this, inadequate information relating to their safety, quality and benefits is provided [4].

The usage scenario of dietary supplements varies significantly over the world. Whereas in some countries supplements are considered as medicines for illnesses in others they are taken to enhance physical well-being. As a consequence, dietary supplements regulations vary considerably throughout the world [5, 6]. For instance, in the United States, dietary supplements are labelled as foods and include only the preparations orally taken under the Dietary Supplements Health Education Act of 1994 (DSHEA). As pre-market requirement, the U.S. Food and Drug Administration (FDA) usually demands data proving the harmlessness of these products to people. Therefore, food additives and ingredients in the supplements are classified as GRAS (generally recognized as safe) under the terms of use described on their labelling. However, for many ingredients there is no information about their safety, even though FDA requires from manufacturers to report side effects related to dietary supplements [3, 5, 7]. On the other hand, the European Union (EU) makes provision for two possible forms of dietary supplements regulation, depending both on products' nature and use. Namely, supplements are classified as "Food supplements" and follow the foodstuffs rules (Directive 2002/46/EC of the European Parliament) when they are sold to provide health benefits or to reduce the risks of chronic diseases (*e.g.*, cardiovascular diseases, osteoporosis, hypertension, overweight or obesity). In that case, centralized pre-market authorization is not needed. In the EU, the harmonized legislation regulates only vitamins and minerals ((EU), 2006; (EU), 2015b); for all other ingredients, the European Commission draws up a list of substances the use of which must be controlled because they are known or suspected to have an adverse effect on health (Annex III of Regulation (EC) No 1925/2006) ((EC), 2015a). All the non-harmonized ingredients are under national legislation and manufacturers must follow the rules established by the regulatory agencies for each country. On the contrary, botanical dietary supplements are classified as "Medicines" and are regulated under the EU medicinal law Directive 2004/24/EC (European Commission 2004) if therapeutic claims are made. Safety, effectiveness and premarket authorization of these products are regulated by the European Medicines Agency (EMA) [2, 6, 8].

The absence worldwide of a common legislation on dietary supplements causes lack of adequate information on the adverse effects of these products. Moreover, several reports on adulteration and contamination with unsafe substitutes, toxic metals, pharmaceuticals, pesticides, and pathogenic organisms are known. Lastly, supplements are often taken at the same time as other drugs and oftentimes there is no information on the safety of such combinations. Consequently, several people have experienced serious adverse reactions associated with some supplements or due to the interaction of supplements with prescription and over-the-counter drugs [4, 9].

In view of the foregoing, it is difficult to solve the challenge of the borderline between medicinal products and food supplements, owing to the lack of specific legislation so far. For these reasons, an effective collaboration between regulators is necessary since national decisions may have global consequences. Therefore, it is of utmost importance to evaluate efficient and sensitive analytical methods to guarantee quality, efficacy and safety of dietary supplements, taking into account legal requirements to maintain the associated claims [7, 8, 10, 11]. To realize these requirements, conventional chromatographic techniques including High Performance Liquid Chromatography (HPLC) and Gas Chromatography (GC) are the analytical methods commonly used. Nowadays, Capillary Electrophoresis (CE) is largely promoted as a valid and suitable alternative to conventional chromatographic techniques for the determination of compounds belonging to several application fields [12 - 15].

CE has emerged as a highly efficient, versatile and selective separation technique allowing fast and cost-effective analyses (simple fused-silica capillaries *vs.* LC columns), requiring only small amounts of sample and reagents [16, 17]. One of the main peculiarities of CE is the easy instrumentation needed. In addition, this analytical tool allows operating in a number of separation modes including among others, Capillary Zone Electrophoresis (CZE), Micellar Electrokinetic Capillary Chromatography (MEKC), Capillary Isotachophoresis (ITP) and Capillary Electrochromatography (CEC). This chapter aims at showing the capabilities of such methodologies in determining the dietary supplements' key components in order to assure their quality and safety.

CZE is the most widely used CE mode for the determination of the key ingredients of dietary supplements. The discrimination of the charged analytes is based on their different migration mobilities within fused-silica capillaries of narrow diameter (20 to 200 μm), filled with a background electrolyte (BGE), under the influence of a high electric field strength generated by applying a high voltage between the ends of the capillary. High separation efficiency and great resolution, rapid analyses and minimal consumption of reagents have made CZE a powerful tool and a valid alternative to liquid chromatography for the analysis of dietary supplements.

MEKC, on the other hand, expands the applicability of CZE to neutral molecules by combining the principles of electrophoresis and chromatography. MEKC allows the separation of both neutral and charged solutes, adding to the BGE surfactants at concentration levels higher than their critical micellar concentration (CMC). It also enhances the selectivity for charged analytes in complex mixtures [18]. More recently, Microemulsion Electrokinetic Chromatography (MEEKC) has been introduced as it offers the advantage to separate compounds simultaneously with a wide range of solubility (hydrophilic and lipophilic character). The principle on which this technique is based is similar to MEKC where micelles are replaced by dispersion of oil droplets in an aqueous buffer. Surfactants and cosurfactants are also needed in order to coat the droplets and reduce the surface tension between oil and water [19].

Capillary Isotachophoresis (cITP) is used for selective separation and concentration of ionic analytes. The sample is introduced in a discontinuous buffer system composed by two different buffer solutions containing the fast leading (LE) and the slow terminating electrolytes (TE), respectively. Under the influence of an applied voltage, the analytes, which have intermediate ionic mobility between those of the leading and terminating ions, are separated on the basis of their different migration and focused at the LE/TE interface. This electromigration technique is especially suitable to detect components at trace levels. Recently, cITP was coupled with CZE combining the enhancement of the analyte concentration and the reduction of matrix effects of cITP with the efficiency and resolution of CZE [20].

Capillary electrochromatography (CEC) is a hybrid electromigration technique combining the high separation efficiency of CE with the selectivity of HPLC. CEC is a powerful separation tool capable of analyzing both neutral and charged solutes. The separation is performed in capillary columns containing a stationary phase (SP). Compounds are moved into the column to the detector by a strong electroosmotic flow (EOF) generated by the application of a high voltage. The separation mechanism, based on both chromatographic and electrophoretic principles, allows to simultaneously analyze neutral and charged solutes, making CEC attractive for the separation of complex mixtures [21].

Despite the high separation efficiency, CE suffers from a lack of sensitivity and consequently, high limits of detection. This drawback is due to the low sample injection volumes (few nanoliters) and to the short optical path length of the most frequently employed detectors in CE methods (*i.e.,* UV-Visible detector). Among the different approaches evaluated to improve CE sensitivity in the last years, the most simple and suitable are based on the on-line pre-concentration strategies, including normal sample stacking, large-volume sample stacking (LVSS), field amplified sample injection (FASI), sweeping dynamic pH junction, among others. Online pre- concentration approach is carried out directly into the capillary, generating running buffer and sample solution zones with different compositions and introducing a large sample volume, thus obtaining an enrichment of the latter. The above reported methodologies are easy to perform as they do not require special equipment set-up configuration [22 - 24].

The on-line pre-concentration strategy is often combined with different sample preparation procedures (SPE, LLE, SPME, ect) for the simultaneous clean-up and analyte-enrichment. As reported previously, UV-Visible has been the most widely used optical detection method in CE, where UV-visible absorbance is directly proportional to the path length used. To improve the sensitivity of the method, in addition to the on-column pre-concentration of the analyte, the path length of the detector can be increased by employing bubble and z-shaped cells as extended path lengths which offer 3 to 5-fold signal-to-noise enhancements. On the other hand, the coupling of CE with other sensitive detectors such as laser-induced fluorescence (LIF), amperometric or mass spectrometric (MS) has been performed in the last years [16, 25]. LIF detection is characterized by high sensitivity and selectivity but it requires a derivatization procedure because the number of compounds with fluorescence properties is limited. In addition, the coupling of CE with LIF is quite a complex and expensive system due to the specific laser sources needed. Electrochemical detectors, especially the amperometric ones, gained popularity in the coupling with CE because their sensitivity and selectivity are higher than those obtained with UV detectors and they are less costly than LIF and MS. In particular, contactless conductivity detection (C^4D) has emerged as a low-cost detection system, suitable to detect compounds lacking chromophore or fluorophore groups [26]. The hyphenation of CE-MS is a powerful technique as it combines high-resolution separations of CE with high detection sensitivity and selectivity of MS. However, CE-MS requires specific interfaces compatible with some of the instrumental parameters of CE as the very low flow rates (nL/min) and the necessity to establish a closed electrical circuit in order to maintain high voltage across the separation capillary. Although different ionization interfaces have been proposed for online CE- MS coupling, electrospray ionization (ESI) is the most frequently used soft ionization technique that guarantees a high ionization efficiency for a wide range of compounds, minimizing the fragmentation process [27, 28].

APPLICATIONS OF CAPILLARY ELECTROPHORESIS IN DIETARY SUPPLEMENTS ANALYSIS

In this section, the role of capillary electrophoresis in dietary supplements analysis will be addressed. The potential of such methodologies for the determination of key components of dietary supplements, the quality control and safety of those products will be shown. Some relevant applications concerning the analysis of ingredients in supplements including amino acids, vitamins, selenium, phenetylamines, biomolecules and contaminants, will be described. Additional information including CE mode, experimental conditions, sample information are reported in Table **1**. The current Table is extracted from Ref [7]. and updated with studies published in 2018- 2019.

Table 1. Analysis of key components in dietary supplements by means of CE.

Analytes and Matrix	Analytical Technique and CE Conditions	LOD/LOQ	Ref
$_{D,L}$-homocysteine, glutathione, $_{L}$-cysteine, $_{N}$-acetyl-$_{L}$-cysteine *Matrix:* tablets	*CE-ED* Uncoated fused-silica capillary (50μm I.D. × 80cm Ltot). BGE: 0.05 M Tris/borate buffer, pH 8.4 Applied voltage 30kV Hydrodynamic injection 10 psi × 2s	LODs between 11-134 nM	[32]
$_{L}$-carnitine *Matrix:* Liquid and solid preparations	*ITP- ITP-conductivity detector* *CZE-UV (direct/indirect detection)* *λ=200-205-254nm* ITP method: PTFE pre-separation capillary (90mm×0.8mm I.D.) PTFE analytical capillary (90mm×0.3mm I.D.) Leading electrolyte 40mM acetic acid + 20mM NH_4OH. Terminating electrolyte 10mM acetic acid CZE – indirect UV detection: Uncoated fused-silica capillary (50μm I.D. × 255 mm Leff) BGE: 5mM TRIS+7mM H_3PO_4 + 0.5mM quinine CZE – direct UV detection: Uncoated fused-silica capillary (50 μm I.D. × 360 mm Ltot) BGE: 15mM TRIS + 25mM H_3PO_4 Experimental conditions described below are the same for all the methods described before: Applied voltage 15 kV Temperature 25°C. Hydrodynamic injection 25mbar × 5s	ITP-ITP LOD=0.5μg/mL LOQ=2.0μg/mL CZE (indirect UV) LOD=4.2μg/mL LOQ=14.0μg/mL CZE (direct UV) LOD=4.4μg/mL LOQ=14.7μg/mL	[34]

(Table 1) cont.....

Analytes and Matrix	Analytical Technique and CE Conditions	LOD/LOQ	Ref
$_{D,L}$-Carnitine *Matrix:* Infant Formulas (powdered milk)	*CE-ESI-MS² (Ion trap mass spectrometer)* Uncoated fused-silica capillary (50 µm I.D. × 83.5cm Ltot) BGE: 0.5M ammonium formate buffer pH 2.5 + 10mM Succ-γ-CD. Partial filling: 600 s x 50 mbar Applied voltage 25 kV Temperature 25°C Hydrodynamic injection 25mbar × 12s	LODs=16 ng/mL	[35]
$_{D,L}$-Carnitine *Matrix:* Tablets, capsules, drinks, biscuits	*CE-ESI-MS²* Uncoated fused-silica capillary (50 µm I.D. × 100 cm Ltot) BGE: 0.5 M ammonium formate buffer pH 2.5 + 0.2% (w/v) of succ-γ-CD Applied voltage 25 kV Temperature 25°C Hydrodynamic injection 25mbar × 12s	LOD 10 ng/mL LOQ 33 ng/mL	[36]
$_{D,L}$-Ornithine *Matrix:* Capsules, powder, tablets	*CZE-UV* *λ=240 nm* Uncoated fused-silica capillary (50 µm I.D. × 48.5cm Ltot) BGE: 100mM borate buffer pH 10.0 + 1mM γ-CD. Applied voltage 20 kV Temperature 25°C Hydrodynamic injection: 25mbar × 5s	LODs=1.6×10^{-7}M LOQs=5.4×10^{-7}M	[37]
$_{D,L}$-Arg, $_{D,L}$-Lys, $_{D,L}$-Orn *Matrix:* Capsules, powder	*CZE-UV* *λ=260 nm* Uncoated fused-silica capillary (50 µm I.D. × 72.5 cm Ltot). BGE: 50 mM phosphate buffer at pH 2.0 + HS-β-CD (5% m/v) and Ac-γ-CD (2% m/v). Applied voltage -25 kV Temperature 15°C	LODs between 6.4×10^{-6}M – 8.2×10^{-6}M LOQs between 2.1×10^{-5}M – 2.7×10^{-5}M	[38]
$_L$-Ala, $_L$-Arg, $_L$-Asp, $_L$-glutamic acid, glycine, $_L$-histidine, $_L$-methionine, $_L$-Ile, $_L$-Lys, $_L$-Leu, $_L$-Val $_L$-proline, $_L$-tyrosine, $_L$-Phenylalanine, $_L$-threonine *Matrix:* Tablets, capsules, drinks, raw materials of royal jelly	*CE-MS/MS* Uncoated fused-silica capillary (50 µm I.D × 100 cm Ltot). BGE: 1 M formic acid pH 1.8 Applied voltage 30 kV Temperature 20°C Hydrodynamic injection: 50mbar × 5s	LODs in the range 0.61-10.5µg/g	[39]

(Table 1) cont.....

Analytes and Matrix	Analytical Technique and CE Conditions	LOD/LOQ	Ref
10-hydroxy-2-decenoic acid, Lysine, Alanine, Proline, Glutamic acid, Aspartic acid *Matrix:* Fresh samples, colostrum, capsules, honey samples	*CE-C⁴D* Uncoated fused-silica capillaries (50 µm I.D × 60 cm Ltot). Channel 1: 10-HAD analysis BGE: 20 mM Tris/Ace pH 8.5 Applied voltage -17 kV Channel 2: Amino acids analysis BGE: 2 M Lactic acid Applied voltage -16 kV Experimental conditions described below are the same for both channels described before: Splitting injections: sample loop loading time 5s sample delivery to the capillary end 1s Hydrodynamic injection: 2s	LOD 0.039 mg/g and LOQ 0.13 mg/g for 10-HAD. LODs in the range 0.039-0.090 mg/g and LOQs in the range 0.13-0.30 mg/g for amino acids	[40]
2-aminoethanesulfonic acid, nicotinamide, thiamine nitrate, riboflavin sodium phosphate, piridoxine hydrochloride, caffeine *Matrix:* Vitamin-enriched drink	*MEKC-UV* *λ = 200-210 nm* Fused-silica capillaries (50 µm I.D. × 48.5cm Ltot) BGE: 0.2M borate buffer pH 8.0 + 150 mM SDS. Applied voltage: 12 kV Temperature: 37 °C Hydrodynamic injection: 3.4 kPa × 5s	LODs in the range 0.3-125µg/mL	[48]
CN-Cbl, OH-Cbl, Co(II) *Matrix:* Nutritive supplements, *Chlorella* samples	*MEKC-ICP-MS* Fused silica capillary (75 µm I.D. × 70 cm Ltot) BGE: 25mM Tris buffer pH 9.0 + 15 mM SDS Applied voltage: 20 kV Hydrodynamic injection: 8nL × 60s	LODs in the range 0.2-1.7 ng/mL	[49]
Vitamins B_1, B_2, B_3, B_5, B_6, B_8, B_9, B_{12}, C, PP *Matrix:* Energy drinks, tablets, powder, injectable solutions	*MEKC-UV* *λ = 214 nm* Fused silica capillary (50 µm I.D. × 60 cm Ltot) BGE: 20mM borate buffer pH 8.7 + 10.0% (v/v) ethanol, + 2.0% (w/v) SDS Applied voltage: 25 kV Temperature: 25 °C Hydrodynamic injection: 0.40 psi v 4 s	LODs between 0.20-4.98 mg/L LOQs between 0.62-15.09 mg/L	[50]

(Table 1) cont.....

Analytes and Matrix	Analytical Technique and CE Conditions	LOD/LOQ	Ref
Vitamins A, E, D_3, PP, B_1, B_2, B_3, B_5, B_6, H, M, B_{12}, C. *Matrix:* Tablets	*MEEKC-UV* *λ = 205 nm* Uncoated fused-silica capillary (75 µm I.D. × 66 cm Ltot) BGE: 20 mM borate buffer pH 8.7 + 1.2% (w/w) SDS + 21% (v/v) 1-butanol + 18% (v/v) ACN + 0.8% (w/w) n-hexane. Applied voltage: 25 kV Temperature: 25 °C Hydrodynamic injection: 0.4 psi × 4s	LOD between 0.2-12 µg/mL	[51]
EP, ψ-EP *Matrix:* Tablets, herbal preparations	*CE-LIF* *$λ_{ecc}$ = 488 nm* *$λ_{em}$ = 520 nm* Fused-silica capillary (75 µm I.D. × 47 cm Ltot) BGE: 60 mM borate pH 9.9 Applied voltage: 15 kV Temperature: 25 °C Hydrodynamic injection: 0.5 psi × 5s	LODs = 27 ng/mL	[52]
EP, ψ-EP *Matrix:* Capsules, pills	*MEEKC-LIF* *$λ_{ecc}$ = 488 nm* *$λ_{em}$ = 520 nm* Fused-silica capillary (75 µm I.D. × 47 cm Ltot) BGE: 20 mM borate pH 9.4 + microemulsion (23.3 mM SDS/180.85mM 1-butanol/16.4 mM n-heptane) + 8% ACN. Applied voltage: 20 kV Temperature: 25 °C Hydrodynamic injection: 0.5 psi × 5s	LOD = 5.3 ng/mL (EP) LOD =3.9 ng/mL (ψ-EP)	[53]
(-)-EP, (+)-ψ-EP, (+)-Me-ψ-EP, (+)-EP, (-)-ψ-EP, (+)-Me-EP, (---Me-EP, Nor-ψ-EP, (-)-NorEP, synephrine *Matrix:* Tablets, capsules, plant material	*CE-UV* *λ = 210 nm* Uncoated fused-silica capillary (50 µm I.D. × 110 cm Ltot) BGE: 55 mM phosphate buffer pH 8.6 + 10 mM sodium borate + 20 mM HP---CD + 7 mM β-CD + 30% MeOH Applied voltage: 30 kV Temperature: 37 °C	LODs between 1.0-10.0 µg/mL	[54]

(Table 1) cont.....

Analytes and Matrix	Analytical Technique and CE Conditions	LOD/LOQ	Ref
EP, ψ-EP, Me-EP, Nor-EP, Nor---EP, Me-ψ-EP *Matrix:* Plant material, solid oral dosage form, protein powder sample	*MEKC-ESI-MS* Uncoated fused-silica capillary (50 µm I.D. × 120 cm Ltot) BGE: 15 mM ammonium acetate, pH6.0 +35mM poly-L-SUCL + 30%(v/v) ACN Applied voltage: 30 kV Temperature: 20 °C Hydrodynamic injection: 5 mbar × 2s sheath liquid: 80:20 v/v methanol-water, 5 mM NH_4OAc at pH 8.5 delivered at 5 µL/min. Drying gas temperature: 250°C Capillary voltage: 3500 V	LODs in the range 0.0625-0.125 µg/mL LOQs in the range 0.25-0.5 µg/mL	[55]
Chondroitin sulphate, glucosamine *Matrix:* Tablets, capsules, liquid samples	*cITP-C⁴D/UV* *λ=254nm* Pre-separation FEP capillary (90 mm × 0.8 mm I.D.), analytical FEP capillary (90 mm × 0.3 mm I.D.). Cationic analysis of glucosamine, LE, 10mM NH4OH + 20mM acetic acid pH4.7, TE, 10mM citric acid. Anionic analysis of chondroit in sulphate, LE 5mM HCl + 10mM glycylglycine + 0.05% 2-hydroxyethylcellulose pH 3.2, TE, 10mM citric acid.	LOD 3mg/L and LOQ 9mg/L for CS. LOD 0.8mg/L and LOQ 2.4mg/L for GA.	[56]
Isoflavones: daidzin, genistin, daidzein, genistein, formononetin, biochanin A, glycitein *Matrix:* Soy drinks	*CZE–ESI-MS* *λ (UV) =214nm* Uncoated fused-silica capillary (50 µm I.D. × 87.5 cm Ltot and 21 cm to the UV detector). BGE: ammonium acetate buffer pH 11. Applied voltage 25 kV Temperature 25 °C Hydrodynamic injection 50mbar x 5s MS conditions: sheath liquid, isopropanol/water mixture (1:1, v/v) containing 0.05% (w/v) acetic acid capillary voltage 3500 V drying gas flow rate 6 L/min drying gas temperature 350 °C Nebulazing gas pressure 5 psi	LODs between 0.65-4.8µg/mL	[57]

(Table 1) cont.....

Analytes and Matrix	Analytical Technique and CE Conditions	LOD/LOQ	Ref
Palmitic, Stearic, Oleic, Linoleic, α-Linolenic fatty acids *Matrix:* Capsules	CZE-UV λ=224nm Uncoated fused-silica capillary (75 μm I.D. × 48.5 cm Ltot). BGE: 15 mM NaH_2PO_4/Na_2HPO_4 pH 6.8 + 4mM sodium dodecylbenzenesulfonate + 8.3 mM Brij 35 + 45% v/v ACN + 2.1% v/v octanol Applied voltage 19 kV Temperature 25 °C Hydrodynamic injection 12mbar x 4s	LODs in the range 0.01 – 0.02 mmol/L LOQs in the range 0.05 – 0.07 mmol/L	[58]
Eicosapentaenoic acid and Docosahexaenoic acid *Matrix:* Capsules	*CZE-UV* *λ=200nm* Uncoated fused-silica capillary (50 μm I.D. × 33.5 cm Ltot). BGE: 30 mM $Na_2B_4O_7$ + 12 mM Brij 35 + 33% v/v MeOH + 17% v/v ACN Applied voltage 27 kV Temperature 27 °C Hydrodynamic injection 25 mbar x 5s	LOD=0.0011 mmol/L and LOQ=0.0037 mmol/L for DHA LOD=0.0013 mmol/L and LOQ=0.0043 mmol/L for DHA	[59]
Metallic (Au, Pt, Pd) nanoparticles *Matrix*: Aqueous suspension	*CE-ICP-MS* Polymer coated fused-silica capillary (50 μm I.D. × 67 cm Ltot). BGE: 10 mM TRIS + 70 mM SDBS (dodecylbenzenesulfonate) pH 9.0 Make up solution: 1% HNO_3 and 10% MeOH in water. Temperature 23 °C Applied voltage 30 kV Hydrodynamic injection 30 mbar × 5s	LOD=6.2 ng/g	[60]
Ag NPs and Ag+ *Matrix*: Aqueous suspension	*CE-ICP-MS* Polymer coated fused silica capillary (100 μm I.D. × 60 cm Ltot). BGE: 30 mM Triton X-100 pH 9.5 + 10 mM cyclohexylaminoethane sulfonic acid. Make up solution: 1% HNO_3 and 10% 2-propanol in water. Applied voltage 25 kV Temperature 23 °C Hydrodynamic injection 30 mbar × 15s	LODs 0.05 μg/kg for Ag NPs and 0.03 μg/kg for Ag^+	[61]

(Table 1) cont.....

Analytes and Matrix	Analytical Technique and CE Conditions	LOD/LOQ	Ref
Trichlormethiazide, triamterene, spironolactone, phenolphthalein, phentermine, (±) -fenfluramine, mazindol, diazepam furosemide, bis (2-ethylhexyl) sulfosuccinate, sodium picosulfate, fluoxetine, 4-aminopyridine, sennoside A, sennoside B, N-didemethylsibutramine citrate, acetazolamide, N-nitrosofenfluramine hydrochlorothiazide, bisacodyl, sibutramine *Matrix:* Tablets, capsules, powders, biscuits	*CE-MS/MS* Uncoated fused-silica capillary (50 µm I.D. × 90 cm Ltot). CE conditions: BGE, 20mM ammonium formate buffer pH 8.0 in 20% v/v acetonitrile-water. Applied voltage 30 kV Temperature 40 °C Hydrodynamic injection 50mbar × 30s MS conditions: sheath liquid, mixture of 5 mM ammonium formate and 0.1% v/v formic acid in 50% v/v methanol-water capillary voltage (+ 4000 V) (- 3500 V) The nebulizing and drying gas conditions were 10 psi N_2 and10 L/min N_2 at 300 °C, respectively.	LODs in the range 1.2 - 750 µg/g	[62]
Furosemide, hydrochlorothiazide, chlorthalidone, amiloride, fluoxetine, paroxetine, phenolphthalein, amfepramone *Matrix:* Capsules, tablets, bulk	*CZE–C⁴D–UV* *λ=260nm* Uncoated fused-silica capillary (75 µm I.D. × 68 cm Ltot). BGE: 20 mM H_3PO_4 + 40 mM NaOH +30% MeOH (v/v) pH 9.2 Applied voltage 15 kV Temperature 25 °C Hydrodynamic injection	LODs (C^4D) between 1.54-3.30 mg/L LOQs (C^4D) between 5.14 -11.01 mg/L LODs (UV) between 4.80 -13.20 mg/L LOQs (UV) between 15.86 -43.58 mg/L	[63]

Amino Acids

CE, and in many instances CZE, is the technique of choice for the determination of amino acids in dietary supplements. Amino acids have a key role in the human metabolism and are a natural supply of nitrogen. Among the different types of amino acids, essential amino acids (EAAs), which cannot be synthesized by the human body, are noteworthy because they have the primary role in the regulation of muscle protein synthesis. Their ingestion provides a more efficient nutritional approach to enhance muscle protein synthesis compared with the ingestion of intact protein. The regular essential amino acid intake as part of an oral diet is effective in skeletal muscle maintenance and in restoring immunological functions [29]. Due to the involvement of EAAs in many physiological processes, their use as supplements to diet was adopted for the prevention and treatment of different diseases, increasingly in the last few years [30]. Capillary electromigration techniques have been widely employed for the determination of amino acids, in clinical studies, especially in metabolomic investigations. More recently, an increased number of CE methods have been optimized to analyse amino acids in food chemistry as markers of quality, safety, bioactivity and storage [31].

Following, the description of some applications concerning the determination of amino acids in dietary supplements. For those analyses, it is necessary to take into account that non-aromatic AAs are not UV absorbing compounds, and this is a limitation that can be overcome through a derivatization process or with the use of alternative detectors to optical ones.

One CE method coupled to an electrochemical detector (ED) was proposed by Inoue and Kirchoff [32] to analyze amino acids with thiol groups. A chemically modified electrode, consisting of pyrroloquinoline quinone (PQQ) biocatalyst entrapped into a conducting polypyrrole (PPy) matrix on a glassy carbon substrate, was developed. The CE method allowed to obtain cysteine, homocysteine, N-acetylcysteine, and glutathione separation. The PQQ/PPy electrode optimized showed high reproducibility and sensitivity. Limits of detection (LOD_S) for the four thiols were found in the range 11-134 nM. The CE method allowed the determination of cysteine in dietary supplements and the amount found resulted in good agreement with reported values.

$_L$-Carnitine, belonging to the family of non-protein amino acids, was deeply investigated in several CE studies. Carnitine is a crucial factor in fatty acid metabolism and in energy production processes. It is used in the therapy of primary and secondary deficiency and for monitoring the biochemical effect of carnitine therapy. In addition, $_L$-carnitine is an essential nutrient for infants because they are unable to synthesize it due to undeveloped biosynthesis pathway. The intake of carnitine in the form of supplements produces beneficial effects in patients with different metabolic diseases [33]. Bearing in mind that carnitine is a weak UV absorbing compound, Prokorátová *et al.* [34] developed three different CE approaches for the determination of $_L$-carnitine, compared each other and with an HPLC one. First, in the ITP-ITP method poly(tetrafluoroethylene) (PTFE) capillaries of different internal diameter were used as pre-separation and analytical columns coupled each other were used to achieve the separation. The detection of the AA was obtained with a conductivity detector. The second approach consisted of capillary zone electrophoresis (CZE) with indirect UV detection, achieved by adding quinine in the background electrolyte (BGE), and a CZE method with direct UV detection where a derivatizing step was introduced by using 9-fluorenyl-methylchlorformate (FMOC) as derivatizing reagent. In addition, an HPLC method based on a strong cation exchange (SCX) column and coupled with a conductivity detector was also validated. The best performance was obtained with the CZE-UV indirect method, considering the instrumentation, the possibility to perform automatic injection (ITP- ITP), the laborious sample preparation (derivatization step) and comparing the electrophoretic parameters of the three CE approaches. Furthermore, all three CE procedures showed comparable results with the HPLC method for the analysis of real samples. Nevertheless, it is noteworthy that CE analyses were faster, with lower solvents consumption and less expensive in comparison to HPLC method.

Carnitine is a chiral molecule in which the two enantiomers possess different pharmacological activities and health benefits. Specifically, L- carnitine is highly therapeutically effective and safe nutrient. D-Carnitine displays serious adverse and toxic effects. D-Isomer competes for uptake in small intestine and tubular re-absorption in kidneys for L-carnitine. Thus, the content of D-carnitine must be precisely determined and limited in nutritional formulations. Therefore, simple, rapid and accurate analytical methods for the discrimination of DL-carnitine are needed to assure the safety of food and dietary supplements.

CE has been recognized as the technique of choice for chiral separations due to its high efficiency, versatility and ease to change the chiral selector in the separation medium in order to obtain different enantioselectivities, and its cost effectiveness compared to HPLC. The following works have been selected and described to show the potential of CE for the separation of chiral molecules. Castro-Puyana *et al.* [35] proposed a chiral CE-MS^2 method for the enantiomeric separation of carnitine in infant formulas. As reported before, in newborns carnitine is an essential nutrient, and given that milk and soy-based infant formulas can often lack this AA, the addition of carnitine to infant samples is needed. Due to the toxic effect of $_D$-carnitine, the discrimination of the two enantiomers is paramount to assess the safety and quality of baby food products. The CE method was coupled to a mass spectrometer detector to obtain a highly sensitive method able to guarantee the certain identification and determination of both enantiomers in complex matrices such as infant formulas. Since MS detector is not capable to distinguish a couple of enantiomers due to the same mass-to-charge ratio, the chiral CE separation of carnitine is crucial. Several derivatized cyclodextrins (CDs) including [TM-β-CD, heptakis-(2,3,6-tri-*O*-methyl)- β-CD; DM-β-CD, heptakis-(2,6,-di-*O*-methyl)-β-CD; HP-β-CD, (2-hydroxy)propyl- β-CD; CM-- -CD, carboxymethyl-γ-CD; succinyl-γ-CD, Succ-γ-CD, ; γ-CD] were investigated. Among them succinyl-γ-CD exhibited the highest enantioresolution capability. Since the selected chiral selector was non-volatile, a partial filling approach was adopted to avoid contamination of the ionization source. A FMOC-derivative carnitine was prepared since the discrimination with CD resulted easier and it allowed to obtain higher sensitivity. Electrophoretic and spectrometric parameters were optimized and analyte's confirmation was achieved by performing MS^2 spectra in order to identify fragmentation patterns of carnitine, according to the European regulation 2002/657/EC. The developed method showed analysis time of about 20 min and LOD value of 16 ng/mL. The method was employed to quantify $_{L}$-carnitine in different infant formulas (IFs). $_{D}$-carnitine was found in amounts higher than 0.5% in 10 of 14 samples. These results indicated that although racemic mixture of carnitine was not used to supplement the target infant formulas, the amount of $_{D}$-carnitine was found to be higher than the limit reported in the European Pharmacopoeia.

A similar chiral CE-MS^2 method was proposed for the quantification of $_{L/D}$-carnitine in different dietary supplements, including drinks, biscuits, capsules and tablets [36]. In the current method, the enantiomeric separation of carnitine was achieved by using a minimal amount of CD directly introduced into the system, avoiding the partial filling approach, without a substantial reduction of the signal. Among the electrophoretic parameters investigated, the effect of the capillary length and two derivatized CDs, specifically acetyl-γ-CD (Ac-γ-CD) and Succ-- -CD were studied. The complete enatioresolution of carnitine was obtained in less than 30 min with 0.2% (w/v) of succ-γ-CD. No matrix interferences were detected for the optimized method. In addition, a good recovery was obtained. LOD and LOQ values for each enantiomer were found to be 10 and 33 ng/mL, respectively. The method sensitivity was higher than the one achieved in the previous work allowing the enantiomeric impurity detection up to 0.025%.

An amount of $_{D}$-carnitine higher than the LOQ was found in all the tested supplements. Specifically, one of them contained a content of $_{D}$-carnitine equal to 6% with respect to total carnitine. Racemic mixture was detected in another sample, proving the racemate addition, which is banned by EU regulations. This electromigration method resulted powerful to assess the safety of dietary supplements and consequently to protect consumers against health risks.

As mentioned before, amino acids need to be derivatized in order to improve UV detection and enantiomeric discrimination. In the study of Domínguez-Vega *et al.* [37], a chiral EKC method was developed for the enantiomeric separation of ornithine (Orn), another non-protein amino acid, added to dietary supplements in the L enantiomer because it facilitates a proper functioning of the immune system and liver regeneration. In this work the implementation of a fast derivatizing process was proposed. Specifically, the pre-capillary derivatization of the analyte with Fluorescein isothiocyanate (FITC), was performed by using an ultrasound probe resulting in reduced reaction time (from 16 h to 10 min). Racemic Orn was resolved investigating variables such as chiral selector type and concentration, temperature, and voltage. The enantiomeric separation of $_{D,L}$-Ornwas achieved with the neutral γ- CD within 10 min allowing also the enantioseparation of other three proteins AAs, Lysine (Lys), Arginine (Arg), and Aspartic acid (Asp), commonly included in food supplements. The LOD and LOQ values obtained for

Orn allowed the quantification of $_{L}$-Orn in 10 commercial samples and the amount found was in accordance with those stated on the label. In addition, the chiral EKC allowed to reveal amounts of up to 0.1% $_{D}$-Orn ensuring the absence of $_{D}$-Orn in all the samples examined.

In-capillary derivatization electrophoresis is an interesting alternative method to analyze amino acids. This approach presents some advantages over conventional pre- and post-capillary derivatization modes such as short derivatization time, fully automatization of the derivatization step, simple sample preparation, minimal use of both sample and derivatizing reagent. This approach was proposed by Martínez- Girón for the determination of Arg, Lys, and Orn in dietary supplements and wines [38]. The proposed methodology, consisted on using 6-aminoquinolyl-N- hydroxysuccinimidyl carbamate (AQC), as derivatizing agent. The three amino acids reported above were chosen owing to their presence in food supplements and their influence on the organoleptic properties of wines. The derivatization protocol was based on the tandem or sandwich modes in which the bands of aminoacid sample (AA) and AQC were introduced in the capillary and the order of the bands was studied. The best results were obtained when only a band of AQC was injected in tandem before or after the AA band, while the sandwich mode (AQC-AA-AQC) produced broad peaks and loss in resolution. Therefore, in the optimized derivatization procedure the AAs were introduced after the AQC reagent, mixed in the capillary, reacting to each other because of their different mobilities. This approach is called "zone-passing" derivatization technique.

The complete and simultaneous separation of the chiral amino acids was obtained within 16 min by using a mixture of Highly sulfated-β-CD (HS-β-CD) (5% m/v) and Ac-γ-CD (2% m/v) dissolved in the BGE. The developed method was validated in terms of linearity, sensitivity, precision and accuracy and applied to the analysis of food samples (dietary supplements and wines). D-enantiomers were not detected in the samples analyzed and the amounts found were in agreement with those indicated in the labels.

Fig. (**1**) shows: a) the approaches studied to achieve the in□capillary derivatization procedure; and b) the comparison between the electropherograms achieved by both methodologies (pre-capillary and in-capillary derivatization).

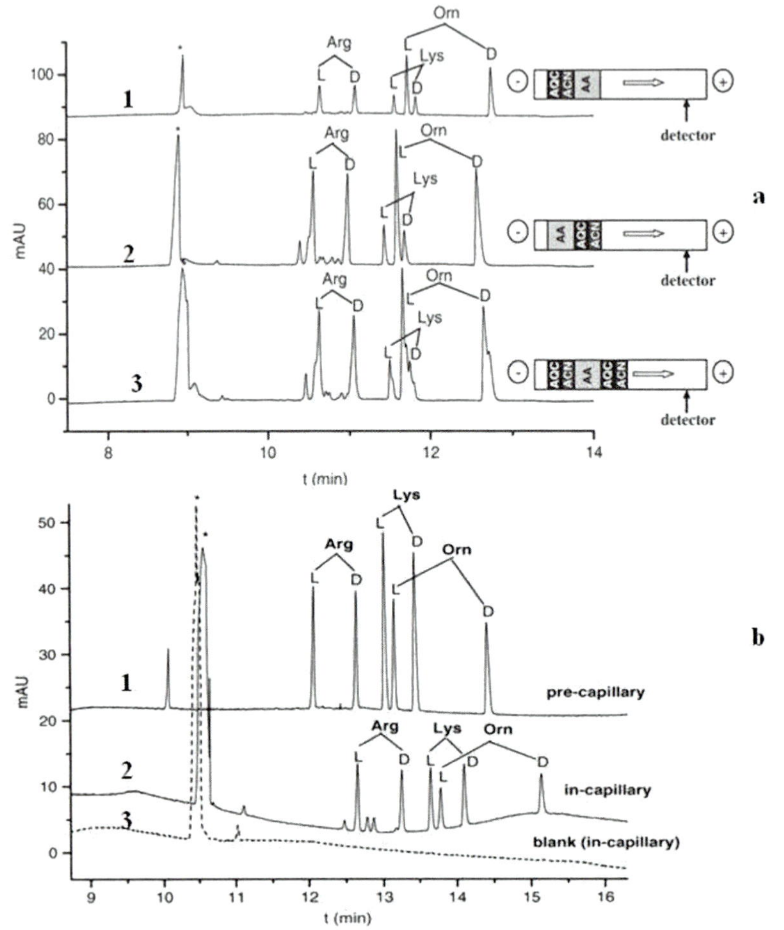

Fig. (1). (a) Approaches developed for the in□capillary derivatization procedures (A and B) tandem mode; (C) sandwich mode. Experimental conditions: BGE, 50 mM phosphate buffer at pH 2.0 containing 5% m/v HS□β□CD and 2% m/v Ac□γ□CD; capillary, 50 μm×72.5 cm uncoated fused□silica; injection, 50 mbar for 5 s of analyte (AA), 2 s of ACN, and 3 s of AQC reagent (plugs order indicated in the figure); AA concentration, 2 mM for dl □Arg and dl □Orn and 0.6 mM for dl □Lys; run voltage,−25 kV; temperature, 15°C; detection wavelength, 260 nm with a bandwidth of 5 nm. (caption as reported in Ref [38]. **(b)** Electropherograms achieved by the injections of racemic mixtures of Arg, Orn, and Lys with pre□ and in□capillary derivatization procedure. (A) Pre□capillary derivatization, 1:1:1: AA:borate:AQC. Injection; analytes derivatized in 200 mM borate at pH 8.8 (6 s) and BGE (5 s). (B) In□capillary derivatization of amino acids using a tandem mode with ACN (2 s)□AQC (3 s)□AA (6 s).

(A) Blank using the same in□capillary sequence as in (B) but with borate instead of AA in borate. Figure and caption adapted from Ref [38].

To avoid the derivatization procedure for the determination of amino acids, the coupling of CE with more sensitive detectors was carried out.

The analysis of free amino acids in dietary supplements containing royal jelly was performed optimizing a CE-MS/MS method [39] which provides high resolution, low cost and reduced preconditioning time by CE, high sensitivity and analytes identification by MS, without the derivatization step.

The present tendency in food science in order to identify the origin and shelf life of foodstuff is to refer to the profiling of free amino acids. In this work the composition of amino acids was studied to assess the quality and taste of royal jelly, that is usually evaluated by the content of trans-10-hydroxy-2-decenoic acid (10-HDA). This simple analytical method enabled the profiling of free amino acids and their composition gave useful information concerning the quality of the commercial RJ products. The separation of 16 amino acids was achieved within 20 min. In particular, Glycine (Gly) and Ala, owing to their low UV sensitivity, were determined by Selected ion monitoring (SIM) mode while due to their partial resolution the amino acids Ile and Leu were quantified in combination. The developed method showed LODs values in the range from 0.61 to 10.5 μg/g for each amino acid and recoveries in the range between 88.3-108.6% for tablets, liquid drinks, and raw materials. The electrophoretic method was used to analyse the amino acids profile of 17 commercial products containing royal jelly and honey products. This profile was found to be specific for each commercial sample, making it possible the differentiation among products.

A simple and inexpensive purpose-made dual channelled capillary electrophoresis coupled with capacitively contactless conductivity detection CE-C^4D method was recently proposed to determine in one single run 10-hydroxy-2-decenoic acid (10-HDA) and free amino acids, main constituents of RJ dietary supplements [40]. In the CE-C^4D system, a portable equipment let the simultaneous determination of the analytes of different charge due to the dual-channel setup. In addition, this approach allowed to optimize independently the experimental separation parameters for each class of compounds [41]. In this study, a BGE consisting of 20mM Tris adjusted to pH 8.5 with acetic acid was used for 10-HDA analysis, so that 10-HDA (pKa 6.5) was negatively charged and migrated towards the detector as opposed to the EOF. Conversely, free amino acids were separated in 2 M Lactic acid used as BGE in order to positively charge the analytes and concomitantly suppress or minimize the EOF. Method validation was performed and different RJ based dietary supplements were analysed. The detection limit for 10-HDA was 39 μg/g and those for free amino acids were in ranged of 39-90 μg/g. The results obtained were in good agreement with those achieved with the reference HPLC method. In particular, regarding 10-HDA only 2 out of 11 tested samples matched the amounts declared on the label. Concerning the amino acids composition, Pro, Lys and Glu were the most abundant in the samples studied.

-17 kV. Channel 2, Amino acids analysis, BGE, 2 M Lactic acid; applied voltage -16 kV. Reproduced with permission from Ref [40]. Fig. (**2B**), Scheme of the dual-channel CE apparatus (A) and view of the arrangement of the individual components on the supporting board (B). Reproduced with permission from Ref [41].

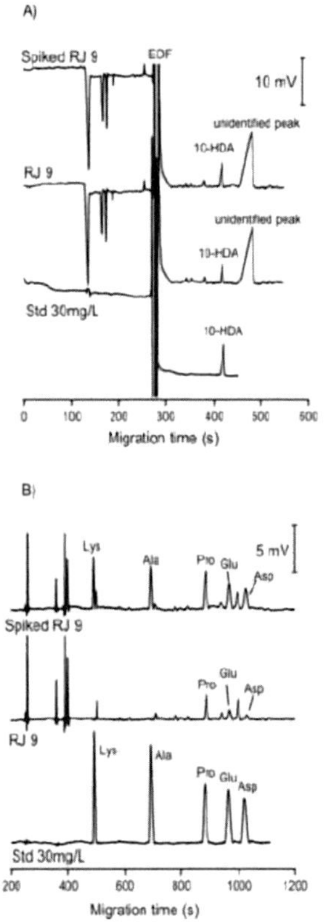

Fig. 2 A

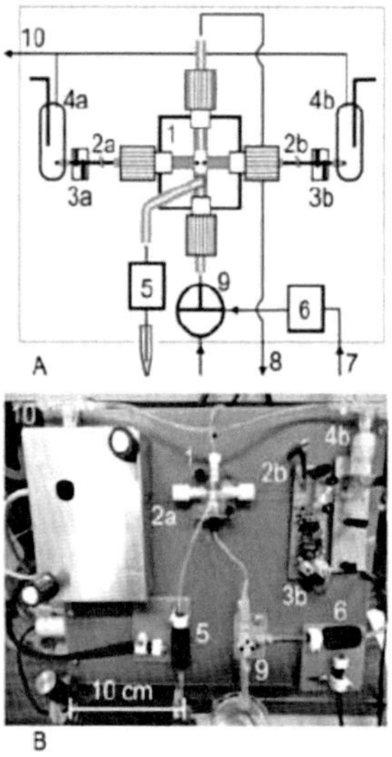

Fig. 2 B

Fig. (2). Fig. (**2A**), Electropherograms of a standard solution (each analyte at 30 mg/L) as well as a real sample with and without spiking. A) 10-HDA; B) Amino acids. Experimental conditions: Uncoated fused-silica capillaries (50 µm I.D × 60 cm Ltot), Channel 1, 10-HAD analysis, BGE: 20 mM Tris/Ace pH 8.5; applied voltage.

In Fig. (**2**), the analysis of a real sample of RJ in the optimum experimental conditions (2 A) and the dual-channel device (2B) are shown. Comparing this method with the CE-MS/MS one described above [39], the dual-channelled CE system showed several advantages such as, the simultaneous analysis of 10-HDA and free amino acids, faster analysis time and lower operation cost.

Dietary Supplements

Selenium Speciation

Electrophoretic methods have been proposed for the speciation of Selenium (Se) contained in dietary supplements. Se is a micronutrient essential for humans that, through the action of seleno-proteins with antioxidant and detoxification activities, plays a fundamental role in the human organism. The content of selenium is usually insufficient in the diet and is often balanced by selenium-enriched food supplements. Nowadays, Se dietary supplements or different Se-enriched foodstuff are available on the market including yeast, milk, broccoli, garlic, green tea and mushrooms. Usually, the concentration range of Se between deficiency and toxicity is narrow and depends on the Se chemical form. Despite this, many commercial formulations label only the total selenium content and not the speciation forms. In addition, the easy access to supplements, available in drugstores and supermarkets, makes their use uncontrolled and potentially dangerous [42]. Therefore, due to both beneficial and toxic effects on humans, analytical approaches for the speciation of Se are needed [43]. Selenized yeast is the most widely used Se-based dietary supplements and besides that its protection properties against several cancer diseases has been proven. Se speciation in yeast is crucial for evaluating selenized yeast quality and providing information about Se species with anti-cancer activity. For this purpose, inductively coupled plasma-mass spectrometry (ICP-MS) is the detector commonly used to detect specific elements such as Se.

An interesting application reporting the use of CZE coupled to ICP-MS was developed to analyze tablets of yeast-based Se nutritional supplements [45]. CE experiments were carried out in a poly(vinyl sulfonate) coated capillary in order to provide stable EOF regardless of the pH of the BGE. Thus, it is possible to carry out the analysis at acidic pH, aiming at separating as many Se compounds as possible, maintaining an adequate EOF towards the cathode. Hence a BGE consisting of 100 mmol/L formic acid pH 3.0, with 0.01% w/v poly(vinyl sulfonate) was selected to obtain the separation of 20 Se compounds in 13 min. An additional optimization of the method is still necessary owing to the co-migration of selenoethionine with SeMet and selenocystine co-migration with Se-methylselenocysteine (Se-MeSeCys). The proposed method allowed to achieve a

LOD value for Se of 15 µg/L. Bearing in mind the small sample volume injected in CE, this limit of detection can be accepted even though higher than other HPLC-ICP-MS methods. Therefore, the results obtained showed that CE-ICP-MS is a powerful technique to fingerprint selenized yeast preparations.

Another interesting application concerns the chiral speciation of Seleno amino acids, which is crucial because the two enantiomers may have different metabolic pathways and biological activity. The main Se species included in Se-enriched yeast is the peptidic SeMet, because of its lower toxicity compared to the inorganic forms. Consequently, an appropriate chiral method aimed at determining the two selenomethionine enantiomers in nutritional supplements is needed. Day *et al.* proposed a CE-ICP-MS method for the chiral speciation of $_{DL}$-selenomethionine in selenized yeast supplements [45] employing the Marfey's reagent (1-fluoro-2,4- dinitrophenyl-5-L-alanine amide) for the derivatization procedure. The two diastereoisomers were completely resolved in 14 min by using the reversed polarity mode. Comparing the data obtained with those achieved by the CE-UV system under identical conditions, it was proved that by using the ICP-MS detector the inorganic forms of Se can be determined simultaneously. In addition, LOD values for $_{D}$- and $_{L}$-SeMet of 250 and 50 ppb were achieved with CE-UV and CE-ICP-MS, respectively, with a general sensitivity increase of the CE-ICP-MS. Reproducibility between injections was about 2% for CE-ICP-MS. The optimized CE-ICP-MS method was employed for the enantioseparation of SeMet in selenized yeast digested with proteinase K (enzymatic hydrolysis), proving the presence $_{L}$-SeMet. ICP-MS detector enabled to reveal other Se species not observed with the UV detector, probably small peptides containing Se. A ligand-exchange micellar electrokinetic capillary chromatographic method was also utilized to get the chiral speciation of selenomethionine enantiomers in selenized yeast [46]. A new phenylalanine derivative [$_{L}$-N-(2-hydroxy-propyl)-phenylalanine] ($_{L}$HP-Phe) was synthesized and its chelate with Cu(II) (Cu(II)–($_{L}$-HP-Phe)$_2$) was employed as chiral selector. A two-step pre-concentration approach, made of an off-line solid phase extraction (SPE) followed by an online large volume sample stacking (LVSS) technique, was used to improve the sensitivity of the MECK-UV method in order to quantify the seleno amino acids in real samples. $_{D,L}$-SeMet was first pre-concentrated on a SPE based on the Cu(II) loaded mesoporous TiO_2adsorbent, eluted with ammonia and injected by hydrodynamic pressure for 1 min. With the aim to cut out the sample matrix, a reversed high voltage was applied on the capillary, whereas the anions were stacked at the interface between sample and BGE. The combined pre-concentration approach allowed an enrichment factor of 1400 and 1378 for the two enantiomers with corresponding LODs values of 0.44 and 0.60 ng/ mL for $_{L}$- and $_{D}$-SeMet, respectively. The proposed SPE–LE–LVSS–MEKC–UV method was employed to analyze a certified reference material of SELM-1 yeast and only the presence of L-SeMet was revealed even if in smaller quantities than the labelled amount of total SeMet. From the results it followed that D-SeMet could be present in a lesser amount than the LOD value. In addition, owing to the very low Se content in commercial yeast, no SeMet was found in this sample. In conclusion, the method proved to be highly sensitive, inexpensive, and applicable for the D,L-SeMet analysis in nutritional yeast samples.

Dietary Supplements

Vitamins

Nowadays, the health benefits of vitamin-based supplements are well recognized. Vitamins function predominantly as enzymatic cofactors and, other than vitamin D and vitamin K, they must be taken through the diet. Vitamins can be grouped into two categories: water-soluble and fat-soluble and, despite their health benefits, a high consumption of them could be dangerous, especially for fat-soluble vitamins that are not adequately metabolized by the human organism. For this reason, it is important that dietary supplements contain a proper amount of vitamins to avoid side effects [47]. Although HPLC remains the separation tool of choice for the determination of vitamins MEKC is a good alternative to LC for the analysis of both water and fat soluble vitamins, owing to the unique features of electromigration techniques. Within this context, in the first works performed by MEKC, SDS was employed as surfactant aimed at avoiding the comigration of noncharged vitamins with the EOF.

A MEKC-UV method was optimized to determine in one single run six active substances in a commercial vitamin-enriched drink, namely 2-aminoethanesulfonic acid, nicotinamide, pyridoxine hydrochloride, anhydrous caffeine, riboflavin sodium phosphate and thiamine nitrate. Analytes separation was achieved by using a BGE consisting of 150 mM SDS in borate buffer (pH 8.0). Moreover, sodium tetraborate solution, instead of water, was utilized to dissolve the sample in order to improve the analytes peak shape Validation of the method was carried out in terms of linearity, repeatability and sensitivity. LODs for all ingredients were found in the range of 0.3- 125 µg/mL. One drink containing vitamins was analysed with the method developed [48]. Particular attention should be paid to the speciation of cobalamin, the only vitamin that contains a transition metal as cobalt. For cobalt speciation in foodstuffs, several CE methods coupled with different detectors including UV detection, electrochemical detection, flame atomic absorption spectrometry (FAAS), fluorometry and inductively coupled plasma mass spectrometry (ICP-MS), have been developed.

A MEKC-ICP-MS based method for the determination of cobalamin in supplements and *Chlorella* samples was developed [49]. The Co-containing compounds studied included cyanocobalamin (CN-Cbl, vitamin B12), hydroxocobalamin (OH-Cbl), and Co(II). Analytes separation was performed in about 10 min, by using a Tris buffer (pH 9.0) containing 15 mM SDS, since it provides better resolution than phosphate and formate buffers. LOD values of 0.3, 0.2, and 1.7 ng/mL for CN-Cbl, OH-Cbl, and Co(II), respectively were achieved. All real samples were subjected to a simple microwave-assisted extraction using 5% v/v HNO_3 solution, prior analysis. The prevalent Co species found in supplements and *Chlorella* was CN-Cbl.

Cyanocobalamin concentration found in *Chlorella* was very close to that found in dietary supplements, proving that it can be considered as an alternative natural source of vitamin B_{12} Fig. (**3**) shows the MEKC-ICP-MS separation of the standard mixture of cobalamin and the determination of CN-Cbl in a nutrient supplement.

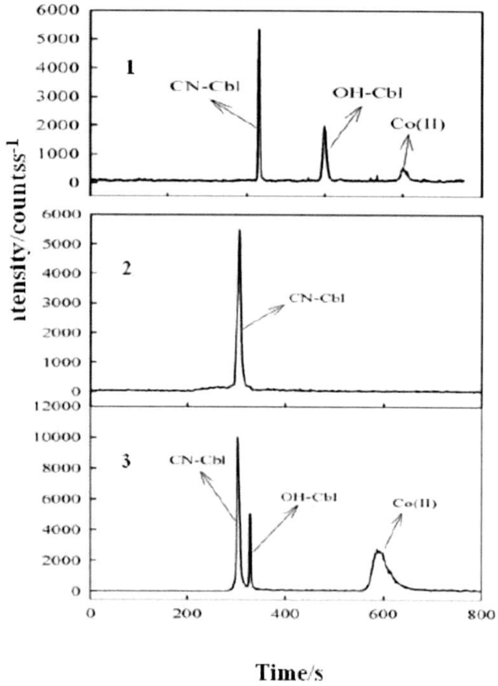

Fig. (3). 1) Electropherogram of the standard mixture composed by CN-Cbl, OH-Cbl, Co(II) achieved by MEKC-ICP-MS in the optimized conditions. 2) Electropherogram for the separation of cobalt species in the extract of a nutritive supplement; 3) Electropherogram of the nutritive supplement spiked with 2 µg/g each of cobalamins and 20 µg/g Co(II). Experimental conditions: BGE, 25 mM Tris, 15 mM SDS, pH 9.0; 20 kV; hydrodynamic injection (10 cm, 60 s). Reproduced with permission from Ref [49]

A fast and simple MEKC method, to simultaneously analyze ten water-soluble vitamins, namely vitamins B1, B2, B3, B5, B6, B8, B9, B12, C, and PP (nicotinamide), was developed by Da Silva *et al.* [50]. The concurrent separation of all the vitamins was obtained in about 20 min., with a BGE consisting of 10.0% (v/v) ethanol, 2.0% (w/v) SDS, 20 mmol/L sodium borate at pH 8.70. Bearing in mind that ethanol enhances the interactions between the analytes and micelles, it was added to the BGE aimed at improving selectivity and separation efficiency. The proposed method showed LODs in the range 0.20 and 4.98 mg/L and was employed to analyze different multivitamins preparations including energy drinks, tablets, injectable ampoules, commercial fruit pulp powders. A good recovery from all the matrices was achieved and the quantity of vitamins determined in the analyzed preparations was in accordance with those listed on the label and those determined by two conventional HPLC methods.

In addition to MEKC methods, microemulsion electrokinetic chromatography (MEEKC) tools have been recently introduced for the analysis of vitamins. MEEKC has emerged as an interesting alternative to MEKC owing to its features such as the concurrent separation of analytes with a large range of solubilities, or the separation of neutral and charged compounds. A MEEKC method aiming at simultaneously separating water- and fat soluble vitamins was proposed by Yin *et al.* [51]. A multicomponent solution consisting of 3 fat-soluble vitamins (retinol (vitamin A), tocopherol (vitamin E), cholecalciferol (vitamin D3)) and 10 water-soluble vitamins (PP, B1, B2, B3, B5, B6, H, M, B12, C) was resolved in the same run, within 30 min., employing a new microemulsion system composed of 1.2% (w/w) SDS, 21% (v/v) 1-butanol, 18% (v/v) ACN, 0.8% (w/w) n-hexane, 20 mM borax buffer (pH 8.7). The effect of experimental parameters such as the microemulsion composition in terms of the choice of organic solvent, buffer pH and nature of the co-surfactant on the selectivity and efficiency of the separation system was accurately studied. The validated method was utilized to quantify water and fat soluble vitamins in a commercial multivitamin tablet. The resulting data agreed with those on the label, except for vitamin C that may be degraded upon storage and vitamin M as a consequence of its low sensitivity.

Adrenergic Compounds

Safety of dietary supplements is another crucial item to assess. Many dietary supplements, consumed as weight-loss formulations, are prepared from botanicals (*e.g.* Ephedra, Citrus aurantium, Acacia) and synthetic biologically active phenethylamines (PEAs) (*e.g.* β-methylphenethylamines, methylsynephrine, α-ethyl- phenethylamine) or PEA drugs (*e.g.* clenbuterol, fenfluramine, sibutramine, lorcaserin). The presence of illicit pharmacologically active compounds, for instance ephedrine (EP) and norephedrine (Nor-EP), or the inclusion of adulterants as sibutramine and synephrine, is often detected in these herbal supplements. As reported in literature, CE is the technique of choice in the analysis of small molecules such drugs. Zhang *et al.* proposed a highly sensitive CE method coupled with LIF detector, aimed at determining ephedrine (EP) and pseudoephedrine (ψ-EP) in Chinese Ephedra herb and ephedrine-containing tablets [52]. Analytes derivatization with 4-chloro-7-nitrobenzo-2-oxa-1,3-diazol (NBD-Cl) as reagent was carried in order to use a LIF detector. Derivatization conditions were optimized by using 40 mM borate buffer pH 8.4 with a reaction time of 30 min at 50 °C. The analytes separation was achieved employing a borate buffer pH 9.9. The developed method allowed obtaining LODs values of about 27 ng/mL, improving the method sensitivity at least ten times with respect to the one reported in literature achieved with a UV detector. The optimized method was used to quantify ephedrine and pseudoephedrine in ephedra herb plants and the content found was in agreement with that on the label. Afterwards, a MEEKC method combined with LIF detection for the analysis of EP and ψ-EP was developed from the same group [53]. EP and ψ-EP were derivatized with NBD-Cl and the separation was obtained in 8 min. using a microemulsion solution. Method validation was performed assessing linearity, sensitivity, repeatability and reproducibility. The MEEKC-LIF method showed LOD values of 5.3 and 3.9 ng/mL for EP and ψ-EP, respectively. Hence, this method enabled to obtain sensitivity about 6 times higher than that achieved in the previous work. With respect to CZE method, the MEEKC-LIF one, owing to the microemulsion added to the BGE, improved the separation efficiency and selectivity as well as the fluorescence intensity of EP and ψ-EP. The determination of the two drugs in Chinese traditional medicines was performed with the optimized method.

A CE method was optimized for the enantioresolution of ephedrine alkaloids in dietary supplements [54]. Ephedrine, pseudoephedrine, N-methylephedrine, methylpseudoephedrine, norpseudoephedrine, together with synephrine were chirally resolved simultaneously. In this instance, enantioseparation of all analytes was performed by a dual CD system composed by HP-β-CD and β-CD as chiral selectors. Alkaloids separation was obtained within 23 min with LODs in the range 1.0-10.0 μg/mL. The developed method was employed to analyze ten dietary supplements including *E. sinica* together with other ingredients. The data obtained proved that the quantity of *Ephedra* herb provides information of total alkaloids content, conversely is not related to ephedrine amount. In addition, it was noted that (+)pseudoephedrine was the second most abundant alkaloid contained in all products; synephrine, a biologically active ephedrine-like compound found in *Citrus aurantium*, was detected in a commercial dietary supplement labelled as "*Ephedra free*", and in three other supplements along with ephedrine alkaloids.

The capillary electrophoretic method has proven to be a powerful tool to determine the possible illegal ephedrine alkaloids added to the supplements.

An interesting work concerns the development of a MEKC-ESI-MS method for the simultaneous separation of ten stereoisomers of ephedrine family and their quantification in *Ephedra* based dietary supplements [55]. Method optimization was carried out by employing polysodium N-undecenoxycarbonyl-L-leucinate (poly-L- SUCL) as polymeric surfactant allowing a large enantiomer window. Enantiomers and diastereoisomers of ephedrine, pseudoephedrine, N-methylephedrine, methylpseudoephedrine, norpseudoephedrine, norphedrine, were resolved in one single run within 30 min. Comparing the MEKC-UV and MEKC-MS methods with regard to precision, linearity and sensitivity, it was observed that the LODs obtained with MEKC-MS (0.125-0.0625 μg/mL) were 40-80 times lower than those achieved with MEKC-UV (5 μg/mL for each ephedrine alkaloid). Accordingly, the MEKC- MS method was then utilized to quantify ephedrine alkaloids in three different Standard Reference Materials (SRMs) including *Ephedra*, and a considerable amount of ephedrine alkaloids was determined. This study proved to be very effective both to quantify the EP alkaloids and to identify possible adulterations owing to the stereospecific data provided.

Biomolecules with Antioxidant Activity

Attractive works concerning the optimization of CE methods, aimed at determining bioactive molecules distributed in dietary supplements deserve to be mentioned. Among all biomolecules with healthy properties, plants bioactive compounds, as polyphenols, carotenoids, phytosterols and polyunsaturated ω-3 and ω-6 fatty acids, are the principal ingredients of different dietary supplements. These bioactive compounds, owing to their antioxidant, antimicrobial, immunostimulating, anti- inflammatory, antihypertensive and antitumor activities, are currently capturing the market attention.

Václavíková and Kvasnička have proposed a cITP method with conductivity detection in order to determine chondroitin sulfate (CS) and glucosamine (GA) in several commercial dietary supplements [56]. CS and GA, anionic and cationic glycosaminoglycans respectively, are present in dietary supplements for treating osteoarthrosis. The quality control of CS is of considerable importance since CS is characterized by a heterogenic structure and oftentimes impurities, due to the extraction process, can cause allergies or some other side effects. One issue of concern is the documented lack of quality in many of the marketed products. In the isotachophoretic method developed, capillaries consisting of fluorinated ethylene propylene (FEP) were used both for the pre-concentration step and for the analytical separation. The leading electrolyte, used for the separation of CS (negative ion), consisted of 5 mM HCl, 10 mM glycylglycine and 0.05% 2-hydroxyethylcellulose (pH 3.2), whereas the cationic analysis of GA was performed employing a leading electrolyte composed by 10 mM NH_4OH and 20 mM acetic acid (pH 4.7). In both instances, 10 mM citric acid was used as terminating electrolyte. Analytes were detected by conductivity detector achieving LOD of 0.8 mg/L for GA and 3 mg/L for CS. GA was analyzed in 35 commercial supplements and only in 21 samples the amount of GA found was more than 90% of that on the label. In addition, CS was also determined in 22 samples and, in this case, it was proved that CS was included in a quantity greater than 90% of the one stated on the label, in only 6 samples. It was assumed that these results were due to the low quality of chondroitin raw material. Moreover, on a set of 35 samples of marketed dietary supplements, it was shown that there was no accordance between the CE amount declared on the label and the one found in several commercial samples, which demonstrates the suitability of the developed cITP for the routine analysis of glucosamine and chondroitin sulfate.

Seven isoflavones in commercial soy drinks were determined by means of a CZE-ESI-MS method providing fast and highly sensitive results and demonstrating to be a valid alternative to conventional chromatographic tools [57]. The compounds examined included glucosides daidzin and genistin, and the aglycones daidzein, genistein, formononetin, biochanin A and glycitein. The complete separation of these compounds was achieved by utilizing a 15 mM ammonium acetate buffer, pH 11, in order to obtain the migration of analytes, as anions. MS detection was performed in ESI positive ion-mode, using an isopropanol/water mixture (1:1 v/v) containing 0.05% (w/v) acetic acid. To increase the isoflavones resolution and avoid drops in the current, a programmed nebulizing gas pressure (PNP) was employed. LODs were found to be 0.65 μg/L for glycitein and 4.8 μg/L for genistin and the developed method was used to determine the analytes in four commercial samples. A simple and efficient liquid-liquid extraction approach was employed for the sample preparation, optimizing the most important parameters such as solvent type, solvent ratio and the extraction time, achieving reproducible results. In each beverage, the content of isoflavones found was different; in particular the greatest quantities were always those related to the glycosides daidzin and genistin. On the other hand, the aglyconic forms, such as formononetin or glycitein, were always found in small amounts.

CZE was in recent years also employed for the analysis of unsaturated fatty acids, that have gained popularity as dietary supplements owing to antihypertensive, antihyperlipidemic, and antiaggregatory properties. Omega-3 and omega-6 polyunsaturated fatty acids (FA) are essential FA because our organism does not synthesize them, hence these FA should be included as part of a regular diet.

Amorim *et al.* [58] developed a CZE-UV method with indirect UV detection to quantify omega 3, 6 and 9 fatty acids in vegetable oil supplements. Specifically, samples of avocado, carthamus, evening primrose and linseed oils have been analyzed, in order to obtain a fast determination of palmitic, stearic, oleic, linoleic and α-linolenic FA. Although CE does not have the same peak capacity of chromatographic techniques, the CZE-UV validated method has some advantages in comparison with GC-FID, which is commonly used for FA determination. Concerning sample preparation, derivatization reactions are not necessary as in GC analysis but only a saponification reaction is performed, aiming at hydrolysing triacylglycerols. Thus, FA in the free form were separated as anions exploiting the counter electroosmotic flow. In addition, FA separation was achieved in a very short time, less than 9 min. This simple CE method has proven accurate and precise as the classic GC method and may be routinely employed for the determination of oil-based supplements.

In a further work the same research group proposed a fast CZE-UV method, with direct detection, with the aim of quantifying two omega-3 FA available in marine species, specifically eicosapentaenoic acid (EPA) and docosahexaenoic acid (DHA) in 10 supplements [59]. Direct UV detection was made possible by the presence of double bounds in the compounds chemical structure to obtain good selectivity by removing the interferences due to saturated FA which do not absorb in the UV range. As in the previous work, sample preparation consisted of only a saponification step and analysis time was very short (8 min). The method was validated in terms of linearity, sensitivity and precision, then applied to analyze 10 commercial samples of marine oil dietary supplements. Lastly, the quantitative results achieved with the CZE-UV method were compared with the conventional GC-FID ones and no significant differences were observed between them. On the whole, the validated method could be employed to quickly control the FA content in industrial production of omega-3 supplements to test its quality and declared content.

Contaminants

As stated in the Introduction, supplement safety is an ongoing challenge, thus contaminants and adulterants have to be absolutely detected to protect human health. In recent years, there has been a rapid development of nanotechnology in Food industry in which nanoparticles (NPs) or nanodelivery systems are used to increase the nutritive qualities of foods and dietary supplements. Consequently, it is of utmost importance to obtain details about the content and type of NPs, and nanodelivery systems in food. In addition, tools capable of determining physicochemical properties of nanomaterials are required. In the light of these considerations, eletrophoretic methods were developed to detect and characterize nanomaterials in foodstuffs. For the characterization of nanomaterials, ICP-MS is definitely the detector of choice since it allows the characterization of nanomaterials with good sensitivity and selectivity.

Hu *et al.* have suggested a CE- ICP-MS method for a rapid and high-resolution characterization of metallic (*e.g.*, gold, platinum, and palladium) NPs in commercial dietary supplements [60]. Several experimental parameters including surfactant type and concentration, pH of running buffer, and applied voltage, were evaluated to optimize the separation conditions. Among the surfactants investigated, the anionic surfactant sodium dodecyl benzenesulfonate (SDBS) added to the BGE provided an increase of the separation of NPs. A careful investigation showed that the type and concentration of the surfactant influenced the nanoparticles separation while applied voltage and pH values of the running buffers affected the elution. In the optimized conditions, the complete and efficient separation of gold NPs of 5, 15, 20, and 30 nm was achieved in less than 20 min. The accuracy and precision of this method was confirmed using 10 and 30 nm gold nanoparticle standard reference materials. The method was used to test two different dietary supplements containing the inclusion of nanometer sized gold particles. The electropherogram obtained revealed characteristic peaks corresponding to the gold NPs in the suspension with a size, calculated from calibration curves, of 17.0 and 6.0 nm for the first and second dietary supplements, respectively. The results were in good agreement with the ones obtained by transmission electron microscopy (TEM). In Fig. (**4**) the electrophoretic separation of the gold NPs and the TEM image are detailed. Furthermore, the feasibility of the CE/ICPMS method was confirmed since the optimized methodology allowed the characterization of a variety of dietary supplements claiming the inclusion of colloidal nanoparticles such as platinum (Pt), palladium (Pd), simply by tuning the CE conditions.

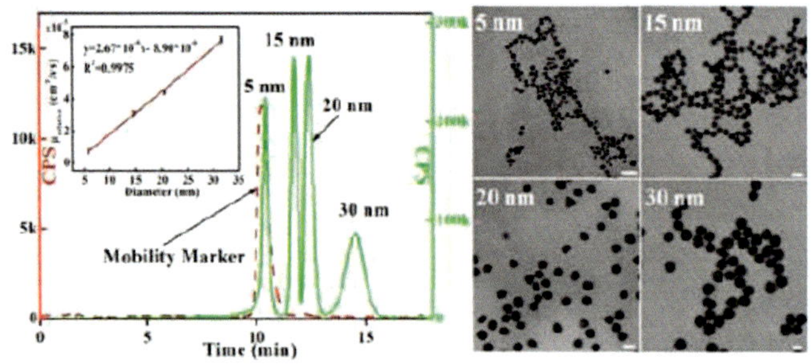

Fig. (4). Electropherogram of the separation of gold nanoparticles (5, 15, 20, and 30 nm). Separation conditions: SDBS, 70 mM; TRIS, 10 mM; pH 9.0; voltage, 30 kV. Inset: Relative electrophoretic mobility follows a liner relationship with the size of gold nanoparticles. Right: TEM images of gold nanoparticles. Scale bar = 20 nm). Reproduced with permission from Ref [60].

A MEKC-ICP-MS method was proposed for separation, size characterization, and speciation of gold (AuNPs) and silver (AgNPs) nanoparticles [61]. The optimized MEKC method allowed the resolution of polydispersed nanoparticle suspensions and the detection of ionic species and nanoparticles, simultaneously. Separation of a composite solution consisting of gold and silver NPs, different in size and with different surface chemistries (10 and 30 nm) was achieved in a polyimide-coated fused-silica capillary by means of a running buffer consisting of 60 mM SDS and 10 mM CAPS, pH 10. The addition of SDS to the BGE enabled the complete separation of the mixture since surfactant molecules attached to the surface of the NPs and determined their charge-to-size ratio. Therefore, small particles migrated before to large particles with the electrophoretic mobilities directly related to nanoparticle size. Stability of the EOF was checked with the ^{133}Cs signal, which verifies the constant transport of analyte from the CE capillary to the ICP-MS. Gold speciation was then performed on a nutritional supplement that contained AuNPs as well as ionic gold. The performance of the MEKC-ICP-MS method was shown finding only AuNPs in the supplement, as reported by the manufacturer.

There are actually numerous reports concerning the contamination of dietary supplements with pharmaceutical adulterants, banned substances, including anorexics, anxiolytics, antidepressants, diuretics and laxatives. Based on its unique characteristics such as high separation efficiency and rapid analysis, CE is a tool of election to analyse adulterants combinations. *Akamatsu and Mitsuhashi* optimized a simple CE-MS method to analyze in the same electrophoretic run, 20 pharmaceutical components such as cathartics and appetite suppressants in dietary supplements for weight loss [62]. The separation was achieved within 25 min on an uncoated fused- silica capillary using 20 mM ammonium formate in 20% v/v acetonitrile-water (pH 8.0) as BGE. Owing to the use of MS as detector, the optimized method allowed multidrug target screening and confirmation in differing radically in chemical structure. In addition, the method enabled reveal the examined analytes in dietary supplement, being the LODs for the target substances in the range 1.0–750 μg/g. The proposed method was utilized for the analysis of 12 commercial dietary supplements. In 10 of them pharmaceutical substances were not detected, instead phentermine and sibutramine, two hunger-suppression drugs, were detected in tablets in greater quantity than those permitted by the FDA organization.

A CE-C^4D method was developed for the quantification of pharmaceutical adulterants in dietary supplements by Müller *et al* [63]. The CE detection system used in this study consisted of C^4D and UV detectors, coupled in-line and worked independently. Commonly, C^4D detector is mainly hyphenated with UV detector because the latter provides information about changes in UV absorption whereas the C^4D detector gives information about changes in conductivity, thus enabling the detection of compounds with low UV absorption characteristics. The coupling of the two detectors prove to be valuable in analyzing complex matrices, when a single detector is not capable of revealing all the components of the matrix. In this study, the separation of furosemide, hydrochlorothiazide, chlorthalidone and amiloride (diuretics), fluoxetine and paroxetine (antidepressants), phenolphthalein (laxative) and amfepramone (anorexic) was obtained by using a BGE composed of 20 mmol L^{-1} phosphate buffer pH 9.2 containing 30% (v/v) of methanol. At first, separated ions were detected in the contactless conductivity cell and then the absorbing ions were revealed in a flow cell where the UV light beam was focused. Thus, the optimized method allowed the separation of all compounds in a single analysis, including fluoxetine and paroxetine which did not absorb UV light at 260 nm. The optimized method was validated and applied to the analysis of 113 dietary supplements samples, sold for the purpose of weight loss or metabolism acceleration, in order to identify their possible contamination with the target drugs. Taking into account all the samples studied, about 12% did not meet the label claims and turned out to be contaminated with diuretics, namely hydrochlorothiazide and furosemide.

CONCLUDING REMARKS

Although a significant part of the population regularly consumes dietary supplements, relatively little information exists on the safety, quality and the physiological effects of many available products. A nutrition survey must document the extent of supplement use, which will be a challenge given the large number of products now available. Because the nature of dietary supplements available in the marketplace is diverse, it is difficult to provide general recommendations regarding the intake.

By carefully examining the literature, the growing production of articles concerning the analysis of supplements is quite noticeable. Among the different separation techniques, CE is increasingly being employed due to its high separation efficiency, rapid separations, and low consumption of reagents as well as of solvents. This separation tool has shown to be a good complementary approach to LC and GC for the identification and quantification of the key components of dietary supplements. Nevertheless, CE is also characterized by some drawbacks including a limited sensitivity when dealing with conventional UV-detection systems and low sample loading. Many efforts have been made in this regard by employing off-line and on- line enrichment approaches which are more simple, easy to use and environment friendly.

In addition, the potential of the technique has been enhanced by the hyphenation with more sensitive and selective detectors including mass spectrometry, fluorescence and electrochemical detection.

In this chapter, an overview of the role of CZE in dietary supplements analysis has been presented by describing several applications. From the data collected in literature, CE is today and will continue to be a powerful complementary and alternative methodology to liquid chromatography and gas chromatography for quality and safety control of nutrient supplements, with an increasing number of articles dealing with this topic.

CONSENT FOR PUBLICATION

Not applicable.

CONFLICT OF INTEREST

The author declares no conflict of interest, financial or otherwise.

ACKNOWLEDGEMENTS

Declared none.

REFERENCES

[1] van der Voet, G.B.; Sarafanov, A.; Todorov, T.I.; Centeno, J.A.; Jonas, W.B.; Ives, J.A.; Mullick, F.G. Clinical and analytical toxicology of dietary supplements: a case study and a review of the literature. *Biol. Trace Elem. Res.*, **2008**, *125*(1), 1-12. [http://dx.doi.org/10.1007/s12011-008-8157-0] [PMID: 18709338]

[2] Coppens, P.; da Silva, M.F.; Pettman, S. European regulations on nutraceuticals, dietary supplements and functional foods: a framework based on safety. *Toxicology*, **2006**, *221*(1), 59-74. [http://dx.doi.org/10.1016/j.tox.2005.12.022] [PMID: 16469424]

[3] Oketch-Rabah, H.A.; Roe, A.L.; Muldoon-Jacobs, K.; Giancaspro, G.I. Challenges and Opportunities for Improving the Safety Assessment of Botanical Dietary Supplements: A United States Pharmacopeia Perspective. *Clin. Pharmacol. Ther.*, **2018**, *104*(3), 426-429. [http://dx.doi.org/10.1002/cpt.1113] [PMID: 30066384]

[4] Jordan, S.A.; Jack, S.A.; Pless, R.P. Safety Surveillance of Dietary Supplements: Importance, Challenges, and New Horizons. *Clin. Pharmacol. Ther.*, **2018**, *104*(3), 432-434. [http://dx.doi.org/10.1002/cpt.1101] [PMID: 30117148]

[5] Dwyer, J.T.; Coates, P.M.; Smith, M.J. Dietary supplements: regulatory challenges and research resources. *Nutrients*, **2018**, *10*(1), 41-65. [http://dx.doi.org/10.3390/nu10010041] [PMID: 29300341]

[6] Thakkar, S.; Anklam, E.; Xu, A.; Ulberth, F.; Li, J.; Li, B.; Hugas, M.; Sarma, N.; Crerar, S.; Swift, S.; Hakamatsuka, T.; Curtui, V.; Yan, W.; Geng, X.; Slikker, W.; Tong, W. Regulatory landscape of dietary supplements and herbal medicines from a global perspective. *Regul. Toxicol. Pharmacol.*, **2020**, *114*, 104647. [http://dx.doi.org/10.1016/j.yrtph.2020.104647] [PMID: 32305367]

[7] Rocco, A.; Donati, E.; Touloupakis, E.; Aturki, Z. Miniaturized separation techniques as analytical methods to ensure quality and safety of dietary supplements. *TrAC. Trends Analyt. Chem.*, **2018**, *103*, 156-183.
[http://dx.doi.org/10.1016/j.trac.2018.04.004]

[8] van Breemen, R.B. Development of safe and effective botanical dietary supplements. *J. Med. Chem.*, **2015**, *58*(21), 8360-8372.
[http://dx.doi.org/10.1021/acs.jmedchem.5b00417] [PMID: 26125082]

[9] Gershwin, M.E.; Borchers, A.T.; Keen, C.L.; Hendler, S.; Hagie, F.; Greenwood, M.R.C. Public safety and dietary supplementation. *Ann. N. Y. Acad. Sci.*, **2010**, *1190*(1), 104-117.
[http://dx.doi.org/10.1111/j.1749-6632.2009.05270.x] [PMID: 20388141]

[10] Rodrigues, D.; Rocha-Santos, T.A.P.; Freitas, A.C.; Gomes, A.M.P.; Duarte, A.C. Analytical strategies for characterization and validation of functional dairy foods, *TrAC. Trends Analyt. Chem.*, **2012**, *41*, 27-45.
[http://dx.doi.org/10.1016/j.trac.2012.08.009]

[11] Castro-Puyana, M.; Herrero, M. Metabolomics approaches based on mass spectrometry for food safety, quality and traceability, *TrAC. Trends Analyt. Chem.*, **2013**, *52*, 74-87.
[http://dx.doi.org/10.1016/j.trac.2013.05.016]

[12] Álvarez, G.; Montero, L.; Llorens, L.; Castro-Puyana, M.; Cifuentes, A. Recent advances in the application of capillary electromigration methods for food analysis and Foodomics. *Electrophoresis*, **2018**, *39*(1), 136-159.
[http://dx.doi.org/10.1002/elps.201700321] [PMID: 28975648]

[13] Zhu, Q.; Scriba, G.K.E. Analysis of small molecule drugs, excipients and counter ions in pharmaceuticals by capillary electromigration methods - recent developments. *J. Pharm. Biomed. Anal.*, **2018**, *147*, 425-438.
[http://dx.doi.org/10.1016/j.jpba.2017.06.063] [PMID: 28711220]

[14] Phillips, T.M. Recent advances in CE and microchip-CE in clinical applications: 2014 to mid-2017. *Electrophoresis*, **2018**, *39*(1), 126-135.
[http://dx.doi.org/10.1002/elps.201700283] [PMID: 28853177]

[15] Saz, J.M.; Marina, M.L. Recent advances on the use of cyclodextrins in the chiral analysis of drugs by capillary electrophoresis. *J. Chromatogr. A*, **2016**, *1467*, 79-94.
[http://dx.doi.org/10.1016/j.chroma.2016.08.029] [PMID: 27558357]

[16] Weinberger, R. *Practical Capillary Electrophoresis*, 2nd ed; Academic Press: San Diego, CA, USA, **2000**.

[17] Geiger, M.; Hogerton, A.L.; Bowser, M.T. Capillary electrophoresis. *Anal. Chem.*, **2012**, *84*(2), 577-596.
[http://dx.doi.org/10.1021/ac203205a] [PMID: 22148626]

[18] Ryan, R.; Altria, K.; McEvoy, E.; Donegan, S.; Power, J. A review of developments in the methodology and application of microemulsion electrokinetic chromatography. *Electrophoresis*, **2013**, *34*(1), 159-177.
[http://dx.doi.org/10.1002/elps.201200375] [PMID: 23161220]

[19] Ryan, R.; McEvoy, E.; Donegan, S.; Power, J.; Altria, K. Recent developments in the methodology and application of MEEKC. *Electrophoresis*, **2011**, *32*(1), 184-201.
[http://dx.doi.org/10.1002/elps.201000372] [PMID: 21171122]

[20] Malá, Z.; Gebauer, P.; Boček, P. Analytical capillary isotachophoresis after 50 years of development: Recent progress 2014-2016. *Electrophoresis*, **2017**, *38*(1), 9-19.
[http://dx.doi.org/10.1002/elps.201600289] [PMID: 27449988]

[21] Rocco, A.; D'Orazio, G.; Aturki, Z.; Fanali, S. *Capillary electrochromatography: a look at its features and potentiality in separation science. Liquid Chromatography: Fundamentals and Instrumentation*;

Elsevier: MA, USA, **2013**, pp. 469-487.

[22] Kitagawa, F.; Otsuka, K. Recent applications of on-line sample preconcentration techniques in capillary electrophoresis. *J. Chromatogr. A,* **2014**, *1335*, 43-60. [http://dx.doi.org/10.1016/j.chroma.2013.10.066] [PMID: 24210300]

[23] Breadmore, M.C.; Wuethrich, A.; Li, F.; Phung, S.C.; Kalsoom, U.; Cabot, J.M.; Tehranirokh, M.; Shallan, A.I.; Abdul Keyon, A.S.; See, H.H.; Dawod, M.; Quirino, J.P. Recent advances in enhancing the sensitivity of electrophoresis and electrochromatography in capillaries and microchips (2014-2016). *Electrophoresis,* **2017**, *38*(1), 33-59. [http://dx.doi.org/10.1002/elps.201600331] [PMID: 27678139]

[24] Šlampová, A.; Malá, Z.; Gebauer, P. Recent progress of sample stacking in capillary electrophoresis (2016-2018). *Electrophoresis,* **2019**, *40*(1), 40-54. [http://dx.doi.org/10.1002/elps.201800261] [PMID: 30073675]

[25] García, C.D.; Chumbimuni-Torres, K.Y.; Carrilho, E. *Capillary Electrophoresis and Microchip Capillary Electrophoresis: Principles, Applications, and Limitations*; John Wiley & Sons, Inc., **2013**. [http://dx.doi.org/10.1002/9781118530009]

[26] Elbashir, A.A.; Aboul-Enein, H.Y. Recent applications and developments of capacitively coupled contactless conductivity detection (CE-C4D) in capillary electrophoresis. *Biomed. Chromatogr.*, **2014**, *28*(11), 1502-1506. [http://dx.doi.org/10.1002/bmc.3230] [PMID: 24816678]

[27] Hommerson, P.; Khan, A.M.; de Jong, G.J.; Somsen, G.W. Ionization techniques in capillary electrophoresis-mass spectrometry: principles, design, and application. *Mass Spectrom. Rev.*, **2011**, *30*(6), 1096-1120. [http://dx.doi.org/10.1002/mas.20313] [PMID: 21462232]

[28] Stolz, A.; Jooß, K.; Höcker, O.; Römer, J.; Schlecht, J.; Neusüß, C. Recent advances in capillary electrophoresis-mass spectrometry: Instrumentation, methodology and applications. *Electrophoresis,* **2019**, *40*(1), 79-112. [http://dx.doi.org/10.1002/elps.201800331] [PMID: 30260009]

[29] Jonker, R.; Engelen, M.P.K.J.; Deutz, N.E.P. Role of specific dietary amino acids in clinical conditions. *Br. J. Nutr.*, **2012**, *108*(S2) Suppl. 2, S139-S148. [http://dx.doi.org/10.1017/S0007114512002358] [PMID: 23107525]

[30] Dioguardi, F.S. Clinical use of amino acids as dietary supplement: pros and cons. *J. Cachexia Sarcopenia Muscle,* **2011**, *2*(2), 75-80. [http://dx.doi.org/10.1007/s13539-011-0032-8] [PMID: 21766052]

[31] Poinsot, V.; Ong-Meang, V.; Ric, A.; Gavard, P.; Perquis, L.; Couderc, F. Recent advances in amino acid analysis by capillary electromigration methods: June 2015-May 2017. *Electrophoresis,* **2018**, *39*(1), 190-208. [http://dx.doi.org/10.1002/elps.201700270] [PMID: 28805963]

[32] Inoue, T.; Kirchhoff, J.R. Determination of thiols by capillary electrophoresis with amperometric detection at a coenzyme pyrroloquinoline quinone modified electrode. *Anal. Chem.*, **2002**, *74*(6), 1349-1354. [http://dx.doi.org/10.1021/ac0108515] [PMID: 11922303]

[33] Dąbrowska, M.; Starek, M. Analytical approaches to determination of carnitine in biological materials, foods and dietary supplements. *Food Chem.*, **2014**, *142*, 220-232. [http://dx.doi.org/10.1016/j.foodchem.2013.06.137] [PMID: 24001835]

[34] Prokorátová, V.; Kvasnička, F.; Sevcík, R.; Voldřich, M. Capillary electrophoresis determination of carnitine in food supplements. *J. Chromatogr. A,* **2005**, *1081*(1), 60-64. [http://dx.doi.org/10.1016/j.chroma.2005.03.033] [PMID: 16013599]

[35] Castro-Puyana, M.; García-Ruiz, C.; Crego, A.L.; Marina, M.L. Development of a CE-MS(2) method

for the enantiomeric separation of L/D-carnitine: application to the analysis of infant formulas. *Electrophoresis,* **2009**, *30*(2), 337-348.
[http://dx.doi.org/10.1002/elps.200800163] [PMID: 19123191]

[36] Sánchez-Hernández, L.; Castro-Puyana, M.; García-Ruiz, C.; Crego, A.L.; Marina, M.L. Determination of L- and D-carnitine in dietary food supplements using capillary electrophoresis-tandem mass spectrometry. *Food Chem.,* **2010**, *120*(3), 921-928.
[http://dx.doi.org/10.1016/j.foodchem.2009.11.004]

[37] Domínguez-Vega, E.; Martínez-Girón, A.B.; García-Ruiz, C.; Crego, A.L.; Marina, M.L. Fast derivatization of the non-protein amino acid ornithine with FITC using an ultrasound probe prior to enantiomeric determination in food supplements by EKC. *Electrophoresis,* **2009**, *30*(6), 1037-1045.
[http://dx.doi.org/10.1002/elps.200800358] [PMID: 19229843]

[38] Martínez-Girón, A.B.; García-Ruiz, C.; Crego, A.L.; Marina, M.L. Development of an in-capillary derivatization method by CE for the determination of chiral amino acids in dietary supplements and wines. *Electrophoresis,* **2009**, *30*(4), 696-704.
[http://dx.doi.org/10.1002/elps.200800481] [PMID: 19156767]

[39] Akamatsu, S.; Mitsuhashi, T. Development of a simple analytical method using capillary electrophoresis- tandem mass spectrometry for product identification and simultaneous determination of free amino acids in dietary supplements containing royal jelly. *J. Food Compos. Anal.,* **2013**, *30*(1), 3047-3051.
[http://dx.doi.org/10.1016/j.jfca.2013.01.006]

[40] Duong, H.A.; Vu, M.T.; Nguyena, T.D.; Manh Huy Nguyena, M.H.; Mai, T.D. Determination of 10-hydroxy-2-decenoic acid and free amino acids in royal jelly supplements with purpose-made capillary electrophoresis coupled with contactless conductivity detection. *J. Food Compos. Anal.,* **2020**, *87*, 103422.
[http://dx.doi.org/10.1016/j.jfca.2020.103422]

[41] Opekar, F.; Tůma, P. Dual-channel capillary electrophoresis for simultaneous determination of cations and anions. *J. Chromatogr. A,* **2016**, *1446*, 158-163.
[http://dx.doi.org/10.1016/j.chroma.2016.04.015] [PMID: 27083255]

[42] Constantinescu-Aruxandei, D.; Frîncu, R. M.; Capră, L.; Oancea, F. *Selenium analysis and speciation in dietary supplements based on next-generation Selenium ingredients. Nutrients,* **2018**. 10, 1466 (1 of 34).
[http://dx.doi.org/10.3390/nu10101466]

[43] Uden, P.C.; Boakye, H.T.; Kahakachchi, C.; Tyson, J.F. Selective detection and identification of Se containing compounds--review and recent developments. *J. Chromatogr. A,* **2004**, *1050*(1), 85-93.
[http://dx.doi.org/10.1016/j.chroma.2004.05.077] [PMID: 15503929]

[44] Bendahl, L.; Gammelgaard, B. Separation of selenium compounds by CE-ICP-MS in dynamically coated capillaries applied to selenized yeast samples. *J. Anal. At. Spectrom.,* **2004**, *19*(1), 143-148.
[http://dx.doi.org/10.1039/b307474a]

[45] Day, J.A.; Kannamkumarath, S.S.; Yanes, E.G.; Montes-Bayón, M.; Caruso, J.A. Chiral speciation of Marfey's derivatized DL-selenomethionine using capillary electrophoresis with UV and ICP-MS detection. *J. Anal. At. Spectrom.,* **2002**, *17*(1), 27-31.
[http://dx.doi.org/10.1039/B104679A]

[46] Duan, J.; He, M.; Hu, B. Chiral speciation and determination of selenomethionine enantiomers in selenized yeast by ligand-exchange micellar electrokinetic capillary chromatography after solid phase extraction. *J. Chromatogr. A,* **2012**, *1268*, 173-179.
[http://dx.doi.org/10.1016/j.chroma.2012.10.002] [PMID: 23141622]

[47] Ball, G.F.M. *Vitamins in Foods. Analysis, Bioavailability, and Stability*; CRC Press: Boca Raton, **2006**.

[48] Okamoto, H.; Nakajima, T.; Ito, Y. Simultaneous determination of ingredients in a vitamin-enriched

drink by micellar electrokinetic chromatography. *J. Pharm. Biomed. Anal.,* **2002**, *30*(3), 815-822. [http://dx.doi.org/10.1016/S0731-7085(02)00390-4] [PMID: 12367707]

[49] Chen, J.H.; Jiang, S.J. Determination of cobalamin in nutritive supplements and chlorella foods by capillary electrophoresis-inductively coupled plasma mass spectrometry. *J. Agric. Food Chem.*, **2008**, *56*(4), 1210-1215.
[http://dx.doi.org/10.1021/jf073213h] [PMID: 18247530]

[50] Da Silva, D.C.; Visentainer, J.V.; de Souza, N.E.; Oliveira, C.C. Micellar electrokinetic chromatography method for determination of the ten water-soluble vitamins in food supplements. *Food Anal. Methods,* **2013**, *6*(6), 1592-1606.
[http://dx.doi.org/10.1007/s12161-013-9576-x]

[51] Yin, C.; Cao, Y.; Ding, S.; Wang, Y. Rapid determination of water- and fat-soluble vitamins with microemulsion electrokinetic chromatography. *J. Chromatogr. A,* **2008**, *1193*(1-2), 172-177.
[http://dx.doi.org/10.1016/j.chroma.2008.04.016] [PMID: 18440539]

[52] Zhang, J.; Xie, J.; Chen, X.; Hua, Z. Sensitive determination of ephedrine and pseudoephedrine by capillary electrophoresis with laser-induced fluorescence detection. *Analyst (Lond.),* **2003**, *128*(4), 369-372.
[http://dx.doi.org/10.1039/b211862a] [PMID: 12741643]

[53] Zhang, J.; Xie, J.; Liu, J.; Tian, J.; Chen, X.; Hu, Z. Microemulsion electrokinetic chromatography with laser-induced fluorescence detection for sensitive determination of ephedrine and pseudoephedrine. *Electrophoresis,* **2004**, *25*(1), 74-79.
[http://dx.doi.org/10.1002/elps.200305643] [PMID: 14730571]

[54] Avula, B.; Khan, I.A. Separation and determination of ephedrine enantiomers and synephrine by high performance capillary electrophoresis in dietary supplements. *Chromatographia,* **2004**, *59*, 71-77.

[55] Hou, J.; Zheng, J.; Shamsi, S.A. Simultaneous chiral separation of ephedrine alkaloids by MEKC-ES--MS using polymeric surfactant II: application in dietary supplements. *Electrophoresis,* **2007**, *28*(9), 1426-1434.
[http://dx.doi.org/10.1002/elps.200600407] [PMID: 17465418]

[56] Václavíková, E.; Kvasnička, F. Isotachophoretic determination of glucosamine and chondroitin sulphate in dietary supplements. *Czech J. Food Sci.,* **2013**, *31*(1), 55-65.
[http://dx.doi.org/10.17221/107/2012-CJFS]

[57] Bustamante-Rangel, M.; Delgado-Zamarreño, M.M.; Carabias-Martínez, R.; Domínguez-Álvarez, J. Analysis of isoflavones in soy drink by capillary zone electrophoresis coupled with electrospray ionization mass spectrometry. *Anal. Chim. Acta,* **2012**, *709*, 113-119.
[http://dx.doi.org/10.1016/j.aca.2011.10.015] [PMID: 22122939]

[58] Amorim, T. L.; Duarte, L. M.; Chellini, P. R.; Leal de Oliveira, M. A. A validated capillary electrophoresis method for fatty acid determination in encapsulated vegetable oils supplements. *LWT - Food Sci Techn.,,* **2019**, 114-108380.

[59] Amorim, T.L.; Duarte, L.M.; de la Fuente, M.A.; de Oliveira, M.A.L.; Gómez-Cortés, P. Fast capillary electrophoresis method for determination of docosahexaenoic and eicosapentaenoic acids in marine oils omega-3 supplements. *J. Chromatogr. A,* **2020**, *1613*, 460641.
[http://dx.doi.org/10.1016/j.chroma.2019.460641] [PMID: 31677769]

[60] Qu, H.; Mudalige, T.K.; Linder, S.W. Capillary electrophoresis/inductively-coupled plasma-mass spectrometry: development and optimization of a high resolution analytical tool for the size-based characterization of nanomaterials in dietary supplements. *Anal. Chem.,* **2014**, *86*(23), 11620-11627.
[http://dx.doi.org/10.1021/ac5025655] [PMID: 25354835]

[61] Franze, B.; Engelhard, C. Fast separation, characterization, and speciation of gold and silver nanoparticles and their ionic counterparts with micellar electrokinetic chromatography coupled to ICP-MS. *Anal. Chem.,* **2014**, *86*(12), 5713-5720.
[http://dx.doi.org/10.1021/ac403998e] [PMID: 24854891]

[62] Akamatsu, S.; Mitsuhashi, T. Simultaneous determination of pharmaceutical components in dietary supplements for weight loss by capillary electrophoresis tandem mass spectrometry. *Drug Test. Anal.*, **2014**, *6*(5), 426-433.
[PMID: 23943335]

[63] Sabo Müller, L.; Tomazi Muratt, D.; Ramos Dal Molin, T.; Gonzalez Urquhart, C.; Viana, C.; Machado de Carvalho, L. Analysis of Pharmacologic Adulteration in Dietary Supplements by Capillary Zone Electrophoresis Using Simultaneous Contactless Conductivity and UV Detection. *Chromatographia,* **2018**, *81*(4), 689-698.
[http://dx.doi.org/10.1007/s10337-018-3496-2]

CHAPTER 14

An Overview of Food Metabolomics: CE-MS Based Targeted and Non-targeted Analysis

Tuba Reçber[1] and **Mustafa Çelebier**[1,*]

Hacettepe University, Faculty of Pharmacy, Department of Analytical Chemistry, Ankara, Turkey

Abstract: The safety and quality of food products, which is one of the most significant issues today, is a cause of increasing concern for consumers. For this reason, food policies are tightly determined by the governments with new conditions and regulations. Systematic monitoring of chemical pollutants such as pesticides, toxins, environmental and industrial contaminants, and residues in food products is critical in protecting public health. Capillary Electrophoresis with mass spectrometry (CE-MS) is frequently used in food analysis to ensure food safety and food quality. In this review, an updated overview of the targeted analysis of residues, contaminants, exogenous toxic ingredients, endogenous toxic ingredients, bioactive components, carbohydrates, amino acids, peptides and proteins in different food matrices through CE-MS is presented. In addition, the advantages of CE-MS based non-targeted analysis and its effectiveness in the field of food safety and quality are discussed in the light of recent studies. From a future perspective, the role of CE-MS based food metabolomics in food science is discussed together with recent developments on metabolomics applications.

Keywords: CE-MS, Food, Food safety, Food quality, Food metabolomics, Foodomics, Targeted analysis, Non-targeted analysis.

INTRODUCTION

Food safety and the need to control foods have become more important today than ever before. As a result of the incredible developments in the food industry and food trade today, new techniques are emerging every day in the production of foodstuffs, new product formulas are being designed and the product range is growing rapidly. Therefore, the importance of food safety services is obviously increasing.

Food products offered to consumers must be of high quality and produced in a safe manner with no harm to human health.

* **Corresponding author Mustafa Çelebier:** Hacettepe University, Faculty of Pharmacy, Department of Analytical Chemistry, Ankara, Turkey; E-mail: celebier@hacettepe.edu.tr

María Castro-Puyana, Miguel Herrero and María Luisa Marina (Eds.)
Capillary Electrophoresis in Food Analysis

This is why new legal arrangements are made in this regard, new audit mechanisms are established and proper standards are published. However, the implementation of these measures largely depends on the effectiveness of laboratory studies [1, 2].

In laboratories providing food analysis services, the points to be considered in sampling, preparation of samples for analysis and sampling for each food product to be analyzed are applied with great care. In addition, data obtained by analytical methods are evaluated and reported based on relevant standards. These laboratories are accredited through national and international accreditation organizations and they operate in accordance with the generally accepted test methods and test criteria valid globally. The purpose of all these analysis studies is ultimately to ensure food safety and quality assurance. Therefore, laboratories must keep up with developing technology, have experienced and educated staff and have up-to-date knowledge. Food analysis is the whole of detailed studies for the healthy consumption of food products. The content of this definition is very wide [3, 4]. Besides food safety, consumers have become more conscious about healthy food consumption especially since the 90's [5]. One of the biggest factors leading people to consume healthy foods may be the increasing number in cancer cases [6].

Cancer is one of the deadliest diseases at present, threatening today's people both socially and economically. As in the last century, it is one of the most important diseases that humanity will seek solution in the next century [7]. Cancer research continues to include studies to identify various molecular targets that are thought to be related to cancer prevention and treatment. However, the inability to reach the desired point with mono-therapeutic approaches with the use of a single drug leads researchers to apply combined therapy methods or to develop approaches that affect different metabolic pathways at the same time [8]. Another approach to fight cancer is to prevent it through small but effective lifestyle changes including a healthy diet and the use of food supplements [6].

Studies to establish a link between diet and the risk of developing cancer first started in the 80's and increased rapidly in the following years [9 - 12]. Cancer-related epidemiological data have been collected for the past 50 years, and the fact that some types of cancer are more common or less common in some cultures is evidenced [13 - 19]. For example, while lung, colon, prostate and breast cancer are common in western societies, they are less common in eastern societies [20 - 22]. Similarly, head and neck cancer are most common in India, while stomach cancer is most common in Japan [23]. Although it is possible to explain the data revealed by epidemiological research with genetic factors, when the post-migration cancer statistics of populations from the same ethnic group were examined, it was observed that in most cases similar results were obtained to those of the statistics of the geographic location they migrated from. Therefore, environmental factors and diet significantly increase the risk of developing cancer. For example, Ziegler *et al.* evaluated the risk of breast cancer for women migrating from China, Japan and Philippines to the United States [24]. The data obtained as a result of the study shows that; 4 to 7 times more cases of breast cancer detected in American women were almost identical to those of the Asian women settled in America as a result of migration. McMichael *et al.* evaluated the statistics of catching stomach, pancreas, colon and rectum cancer due to the change in the feeding styles of the communities that migrated from Europe to Australia. While the risk of continental European populations getting colon cancer is half of the Australian population, it was observed that after living in Australia for some time, this risk was equivalent to that of the Australian population [25]. It is possible to replicate similar examples [26 - 28]. Vecchia *et al.* show that some compounds in vegetables and fruits in the Mediterranean-style diet that are called 'micronutrients' reduce the risk of cancer [29]. In such a situation, a global evaluation must be concerned with evaluating the anticancer or cancer prevention properties of foods. Another issue discussed at this point includes the effect of consuming organic foods on increasing life quality and protecting against diseases [30]. Therefore, the content of foods caused by organic farming must be evaluated at a metabolome level to reveal the mechanisms of action on improving health.

In the light of this brief discussion above, the importance of food analysis could be discussed now in a better perspective than ever before since analysis of foods is a critical issue on both food safety and food quality which could be identified by determining the harmful and/or 'beneficial to health-life hacking' components in foods and food products.

Capillary Electrophoresis (CE) is an analytical technique, which theory was already known for years, but the fully-automated device was first introduced by Beckman Coulter (then Beckman Instruments) in 1989. Two different forces would change the migration of an analyte in CE. These are electroosmotic flow (EOF) and electrophoretic mobility. CE separates ions based on their electrophoretic mobility with the use of an applied voltage. The electrophoretic mobility depends on the charge of the molecule, the viscosity, and the atom's radius. Since neutral species do not have any charge, they have no electrophoretic mobilities and only ions move with the electric field. Ions with greater charge will move faster than the others. The other cause of ions separation in CE is their size. The smaller particle has less friction and an overall faster migration rate [31, 32]. This situation makes CE a powerful technique in comparison to High-Performance Liquid Chromatography (HPLC) where the polar-ionic compounds are hard to be separated and analyzed in reverse-phase mode and their separation is limited through Hydrophilic Interaction Liquid Chromatography (HILIC) [33, 34].

In this chapter, previous studies collecting the targeted and non-targeted CE-MS analysis of food samples are evaluated. Table **1** presents the CE-MS based targeted and non-targeted analyses of food components (residues, toxic exogenous compounds, endogenous compounds, bioactive compounds, proteins). Since 'the chemical components' inside the foods can be basically described as food metabolites, CE-MS based metabolite target analysis in food samples is discussed and recent developments on food metabolomics are evaluated. Thus, as a future perspective, how to transfer and why should the knowledge on clinical metabolomics to food analysis be transferred, is considered. (Fig. **1**) briefly illustrates the timeline for the simultaneous improvement of food products and food analysis techniques.

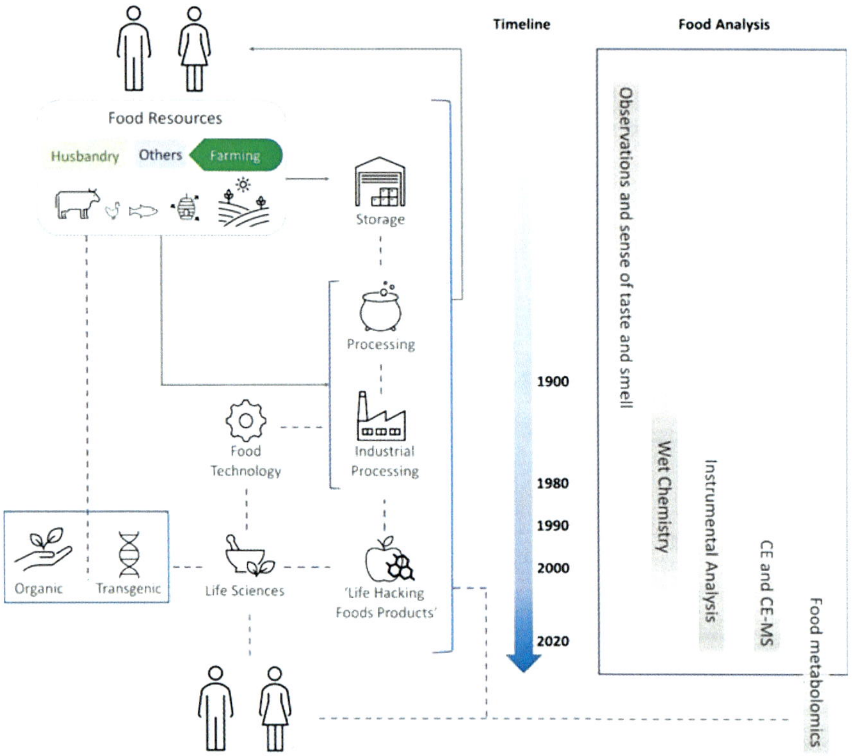

Fig. (1). A timeline presenting the simultaneous improvements on food technology and food analysis.

Table 1. CE-MS based targeted and non-targeted analysis of food samples.

Application	Food Matrices	Sample Preparation	CE Conditions	CE-MS Interface (Sheath liquid)	MS Conditions	LOD	REFERENCES
Paralytic poisoning toxins	Shellfish	SPE	CE capillary: Linear polyacrylamide-coated (100 cm x 50 μm I.D.) Buffer: 35 mM morpholine (pH 5)	Water:MeOH (50:50, v/v) with 0.2% FA, at 8 μL/min	Source: API (+) Analyzer: Triple quadrupole (SIM, SCAN, and MRM)	5–12 ng/mL	[35]
Peptides	Food dyes	n.i.	CE capillary: uncoated (100 cm x 75 μm I.D.) Buffer: 100 mM $(NH_4)_2CO_3$ (pH 9.2) Voltage: 17 kV	MeOH at 4 μL/min	Source: ESI (-) Analyzer: High-energy conversion dynode detector	n.i.	[36]
Analysis of Tetramine	Sea snails	LLE with diethyl ether	CE capillary: LPA-coated (90 cm x 50 μm ID) Buffer: 0.1 M FA Voltage: 5 kV	Water.MeOH (3:1, v/v) with 0.2% FA, at 7 μL/min	Source: API (+) Analyzer: Triple quadrupole (SIM and SRM)	4.4 μg/mL	[37]
Selenium speciation	Human milk	PP by centrifugation	CE capillary: uncoated (150 cm x 350 μm I.D.) Buffer: 10 mM sodium carbonate (pH 11.5) Voltage: 18 kV	10 mM sodium carbonate at 10 μL/h	Source: ICP Analyzer: Quadrupole	10 μg/L	[38]
Amino acid analysis	Soy sauce	Dilution With Water (1:100)	CE capillary: uncoated (50 cm x 100 μm I.D.) Buffer: 1 M FA Voltage: 30 kV	Water:MeOH (1:1, v/v) with 5 mM ammonium acetate at 10 μL/min	Source: ESI (+) Analyzer: n.i.	0.046-1.5 μg/mL	[39]
Determination of Carbohydrates	Wine	Dilution with Water (1:5)	CE capillary: uncoated (70 cm x 50 μm I.D.) Buffer: 300 mM diethylamine Voltage: 20 kV	Water:IPA (20:80, v/v) with 0.25% diethylamine at 4 μL/min	Source: ESI (-) Analyzer: Single quadrupole	0.5–3 mg/mL	[40]
Determination of Gellan gum	Fruit flavour drink	Dilution with Water, MeOH and digestion buffer (20 mM tris-HCl, pH 7.2)	CE capillary: uncoated (80 cm x 50 μm I.D.) Buffer: 15 mM ammonium formate (pH 8.0) Voltage: 30 kV	IPA:Water (1:1 v/v) with NH_3 solution 0.5% (v/v)	Source: ESI (-) Analyzer: Single quadrupole (SCAN and SIM)	81 μg/kg	[41]
Oligosacchari-des	Dextrin-15 starch hydrolysate from maize	Derivatization of carbohydrates with a fluorophore, 8-aminonaphthalen--1,3,6-trisulfonic acid (ANTS)	CE capillary: Polyacrylamide-coated (100 cm x 50 μm I.D.) Buffer: 160 mM acetic acid + 70 mM FA, (pH 2.0) Voltage: 30 kV	IPA 20.5% acetic acid:FA (1:1, v/v) (pH 3.5) at 5 μL/min	Source: ESI (-) Analyzer: Single quadrupole	n.i.	[42]
Amino acids	Infant foods	Dilution with Water (1:10)	CE capillary: uncoated (70 cm x 50 μm I.D.) Buffer: 0.3 M FA Voltage: 20 kV	Water:IPA (20:80, v/v) with 0.6% FA	Source: ESI (+) Analyzer: Single quadrupole	0.1–1.8 μg/mL	[43]
Analysis of Herbicides	Fruits and vegetables	SPE	CE capillary: uncoated (150 cm x 75 μm I.D.) Buffer: 12 mM ammonium formate (pH 3.5) 2% MeOH Voltage: 30 kV	Running buffer at 13 μL/min	Source: ESI (+) Analyzer: Single quadrupole (SCAN and SIM)	0.01–0.1 μg/mL	[44]
Glycoalkaloids	Potato tubers	Extraction with MeOH	CE capillary: uncoated (80 cm x 50 μm I.D.) Buffer: ACN-MeOH (90:10 v/v) with 50 mM ammonium acetate and 1.2 M acetic acid Voltage:30 kV	Water:MeOH (1:1, v/v) with 1% acetic acid at 2.5 μL/min	Source: ESI (+) Analyzer: Ion trap	10–50 μg/L	[45]
Glycoalkaloids	Potato tubers *(Transgenic potatoes)*	Extraction with MeOH	CE capillary: uncoated (80 cm x 50 μm I.D.) Buffer: ACN-MeOH (90:10 v/v) with 50 mM ammonium acetate and 1.2 M acetic acid Voltage:30 kV	Water:MeOH (1:1, v/v) with 1% acetic acid at 2.5 μL/min	Source: ESI (+) Analyzer: Ion trap (SIM, and MS/MS-SRM)	10–50 μg/L	[46]
Low molecular mass organic acids	Ale samples	Filtering	CE capillary: uncoated (103 cm x 50 μm I.D.) Buffer: 1,2,4-benzenetricarboxylic acid (TMA), 5 mM Tris, (pH 8.5) Voltage: 17 kV	0.25% v/v FA in IPA:Water (50/50, v/v) 4 μL/min	Source: ESI (-) Analyzer: Single quadrupole (SIM)	< 0.05–0.7 mg/L	[47]

(Table 1) cont.....

Application	Food Matrices	Sample Preparation	CE Conditions	CE-MS Interface (Sheath liquid)	MS Conditions	LOD	REFERENCES
Analysis of Herbicides	Wheat	Mixed with Water and acetone	CE capillary: uncoated (60 cm x 75 µm I.D.) Buffer: 1 mM ammonium acetate (pH 6.3) Voltage: 20 kV	Sheathless microelectrospray interface	Source: ESI (n.i.) Analyzer: Quadrupole Ion Trap	2.5 mM	[48]
Acidic pesticides	Fruits	SPME	CE capillary: uncoated (90 cm x 75 µm I.D.) Buffer: 32 mM ammonium formate (pH 3.1) Voltage: 25 kV	Running buffer 120% MeOH, at 14 µL /min	Source: ESI (-) Analyzer: Single quadrupole	0.1-3 µg/mL	[49]
Small carboxylic acids	Apple juice	Dilution with Water (1:200)	CE capillary: uncoated (50 cm x 80 µm I.D.) Buffer: 20 mM ammonium formate (pH 10) Voltage: 25 kV	5mM ammonium hydroxide:MeOH (pH 6.55) at 6 µL/min	Source: ESI (-) Analyzer: Ion trap	1.1-3.5 mg/L	[50]
Basic proteins	Chicken eggs, white wine, and minced meat from beef	Lyophilization	CE capillary: uncoated and EPyM-DMA-coated (87 cm x 50 µm I.D.) Buffer: 75 mM ammonium acetate (pH 5.5) Voltage: 25 kV	MeOH:0.05% acetic acid (50:50, v/v) at 4 µL /min	Source: ESI (+) Analyzer: Ion trap	1.3 µg/mL	[51]
Pesticides	Fruit juices	SPME	CE capillary: uncoated (87 cm x 50 µm I.D.) Buffer: 0.3 M ammonium acetate (pH 4) Voltage: 25 kV	IPA:Water (65:35, v/v) at 0.22 mL/h	Source: ESI (+) Analyzer: Ion trap	0.09-0.3 µg/mL	[52]
Polar antioxidants	Rosemary extracts	PLE using subcritical Water	CE capillary: uncoated (87 cm x 50 µm I.D.) Buffer: 40 mM ammonium acetate (pH 9) Voltage: 20 kV	IPA:Water (60:40, v/v) and 0.1% triethylamine at a flow rate of 0.24 mL/h	Source: ESI (-) Analyzer: Ion trap	n.i.	[53]
Peptides (Prediction of the amino acid composition)	Plant protein hydrolysa-te *(Rapeseed (Canola))*	Superdex™ Peptide HR 10/30	CE capillary: uncoated (100 cm x 50 µm I.D.) Buffer: 50 mM FA (pH 2.75) Voltage: 25 kV	100% MeOH with 0.2% FA at 6 µL /min	Source: ESI (+) Analyzer: Single quadrupole	n.i.	[54]
Anthocyanin	Wine and wine musts	SPE	CE capillary: uncoated (60 cm x 50 µm I.D.) Buffer: 1) acidic basic (chloroacetate ammonium, pH 2) 2) basic buffer (borate-ammonium, pH 9) Voltage: 20 kV	0.25% acetic acid in MeOH:Water mixture (80:20, v/v)	Source: ESI Analyzer: Ion trap	0.8-10 mg/L	[55]
Chiral amino acids	Orange juice	Filtration	CE capillary: uncoated (87 cm x 50 µm I.D.) Buffer: 0.1 M ammonium acetate (pH 6) with 5 mM β-CD Voltage: 15 kV	MeOH:Water (50:50, v/v) with 25% BGE without β -CD, at 3.5 µL/min	Source: ESI (+) Analyzer: Ion trap	0.16–3.1 µg/mL	[56]
Intact proteins	*Spirulina platensis* microalgae	PLE	CE capillary: uncoated (89 cm x 50 µm I.D.) Buffer: 40 mM ammonium hydrogen carbonate (pH 7.8) in Water + ACN:IPA (45:50:5 v/v/v) Voltage: 25 kV	Water:IPA (75:25, v/v) containing 0.5% v/v acetic acid at 6 µL/min	Source: ESI (+) Analyzer: Ion trap and TOF	n.i.	[57]
Intact proteins *(phycobilipro-teins)*	*Spirulina platensis* microalgae	PLE	CE capillary: uncoated (89 cm x 50 µm I.D.) Buffer: 40 mM ammonium hydrogen carbonate (pH 7.8) in Water + ACN: IPA (45:50:5 v/v/v). Voltage: 25 kV	Water:IPA (75:25, v/v) containing 0.5% v/v acetic acid at 6 µL/min	Source: ESI (+) Analyzer: Ion trap	n.i.	[58]
Intact glucosinolates	*Arabidopsis thaliana* seeds	Seeds were ground in liquid nitrogen	CE capillary: uncoated (88 cm x 50 µm I.D.) Buffer: 1 M FA Voltage: 25 kV	0.2% FA:IPA (1:1, v/v) at 4 µL/min	Source: ESI (-) Analyzer: TOF (SCAN, MS/MS)	n.i.	[59]
A bioactive Compound (γ-Glutamyl-S-ethenyl-cysteine)	*Vicia narbonensis L.* seeds	Extraction with ethanol-Water	CE capillary: uncoated (87 cm x 50 µm I.D.) Buffer: 20 mM ammonium hydrogen carbonate (pH 7) Voltage: 25 kV	MeOH:0.1% acetic acid (50:50, v/v) at 3 µL/min	Source: ESI (+) Analyzer: Ion trap	0.021 mg/mL	[60]

(Table 1) cont.....

Application	Food Matrices	Sample Preparation	CE Conditions	CE-MS Interface (Sheath liquid)	MS Conditions	LOD	REFERENCES
Pesticides	Fruits	SPE	CE capillary: uncoated (75 cm x 50 µm I.D.) Buffer: 0.3 M ammonium acetate (pH 4) with 10% MeOH Voltage: 30 kV	0.3 M ammonium acetate (pH 4) in 10% MeOH at 5 µL/min	Source: ESI (+)/(-) Analyzer: Quadrupole ion trap (MS/MS)	Using single Q: (0.05-0.5) µg/mL Using QIT: (0.001-0.015) µg/mL Using QIT: (0.005-0.2) µg/mL Using QIT: (0.009-1.4) µg/mL	[61]
Pesticides	Soy milk	SPE	CE capillary: uncoated (87 cm x 50 µm I.D.) Buffer: 24 mM FA and 16 mM $(NH_4)_2CO_3$ (pH of 6.4) Voltage: 25 kV	MeOH:Water (82.5:17.5, v/v) with 2% TEA at 0.35 mL/h	Source: ESI (-) Analyzer: Ion trap	(74-150) µg/L	[62]
Sulfonamide residues	Milk	•PP with metahanol •SPE	CE capillary: uncoated (70 cm x 75 µm I.D.) Buffer: 50 mM ammonium acetate (pH 8.5) Voltage: 20 kV	MeOH:1.5% FA (50:50, v/v) at 0.2-0.4 µL/min	Source: ESI-API Analyzer: Single quadrupole	0.6-1.1 µg/L	[63]
Phenolic compounds	Rosemary honey	SPE	CE capillary: uncoated (100 cm x 75 µm I.D.) Buffer: 100 mM ammonium acetate with 10% of IPA (pH 10) Voltage: 25 kV	IPA:Water (60:40, v/v) and 0.1% (v/v) of triethylamine at 3 µL/min	Source: ESI (-) Analyzer: Ion trap	1–25 mg/L	[64]
Quinolone residues	Chicken and fish	•Solvent extraction with dichloromethane •SPE	CE capillary: uncoated (75 cm x 75 µm I.D.) Buffer: 60 mM $(NH_4)_2CO_3$ (pH 9.2) Voltage: 20 kV	Running buffer at 10 µL/min	Source: ESI (+) Analyzer: Quadrupole ion trap (QITMS[3])	20 ng/g	[65]
Quinolone Antibiotics	Bovine raw milk	SPE	CE capillary: uncoated (96 cm x 50 µm I.D.) Buffer: 70 mM ammonium acetate (pH 9.1) Voltage: 25 kV	IPA:Water:FA (50:49:1, v/v/v) at 3 µL/min	Source: ESI (+) Analyzer: Ion trap (MS/MS, MRM)	4-6 ng/mL	[66]
Acrylamide	Food products *(potato crisps, crisp bread, biscuits, breakfast cereals and coffee)*	SPE (after derivatization with 2-mercaptobenzoic acid)	CE capillary: uncoated (80 cm x 50 µm I.D.) Buffer: 35 mM ammonium formate (pH 10) Voltage: 25 kV	IPA:Water (75:25, v/v) at 8 µL/min	Source: ESI (-) Analyzer: Ion trap (MS/MS and MS[3])	8 ng/g	[67]
Sulfonamides	Pork meat	PLE	CE capillary: uncoated (75 cm x 75 µm I.D.) Buffer: 50 mM ammonium acetate (pH 4.16) Voltage: 23 kV	MeOH:Water:FA with (49.5:49.5:1, v/v/v) at 10 µL/min	Source: ESI (+) Analyzer: Ion trap (MRM, MS[2])	1.56-12.5 µg/kg	[68]
Sulfonamides	Beef meat	SPE	CE capillary: uncoated (96 cm x 50 µm I.D.) Buffer: ammonium acetate 70 mM (pH 9.5) Voltage: 28 kV	IPA:Water:FA (50:49:1, v/v/v) at 3 µL/min	Source: ESI (+) Analyzer: Ion trap (MRM, MS/MS)	CE-MS: 5.5-65.4 µg/L CE-MS/MS: 5.1-80.2 µg/L	[69]
12 antibacterial residues	Fish and livestock *(bovine meat)*	•Solvent extraction with ACN •SPE	CE capillary: uncoated (75 cm x 75 µm I.D.) Buffer: 60 mM ammonium acetate (pH 8) with 10% of MeOH Voltage: 25 kV	MeOH:1% FA (50:50, v/v) at 10 µL/min	Source: ESI (+) Analyzer: Ion trap (MRM)	1-10 µg/kg	[70]
12 antibacterial residues	Fish and livestock *(bovine meat)*	•Solvent extraction with ACN •SPE	CE capillary: uncoated (75 cm x 75 µm I.D.) Buffer: 60 mM ammonium acetate (pH 8) with 10% of MeOH Voltage: 25 kV	MeOH:1% FA (50:50, v/v) at 10 µL/min	Source: ESI (+) Analyzer: Ion trap (MRM)	1-10 µg/kg	[70]
Tetracycline residues	Bovine milk	SPE	CE capillary: uncoated (60 cm x 50 µm I.D.) Buffer: 35 mM Tris, 1.1% FA, 5% MeOH, and 15% ACN Voltage: 28.5 kV	100% MeOH with 0.8% FA at 3 µL/min	Source: ESI (+) Analyzer: single quadrupole	7.14-14.9 ng/mL	[71]
Comparative metabolomic study	Soybean	Extraction with MeOH/Water (80/20, v/v)	CE capillary: uncoated (90 cm x 50 µm I.D.) Buffer: 50 mM ammonium hydrogencarbonate (pH 9) Voltage: 28 kV	Water:IPA (50/50, v/v) with 0.1% ammonium hydroxide at 0.24 mL/h	Source: ESI (-) Analyzer: TOF	n.i.	[72]

(Table 1) cont.....

Application	Food Matrices	Sample Preparation	CE Conditions	CE-MS Interface (Sheath liquid)	MS Conditions	LOD	REFERENCES
Biogenic amines	Wine	Dilution 1:1 with Water and filtering	CE capillary: uncoated (90 cm x 50 µm I.D.) for CE-IT-MS and (60 cm x 50 µm I.D.) for CE-TO--MS Buffer: 1M FA (pH 2) Voltage: 28 kV	IPA:Water (50:50, v/v) at (5 µL/min for CE-IT-MS) and (3 µL/min for CE-TOF-MS)	Source: ESI (+) Analyzer: Ion trap and TOF	CE-IT-MS: 0.08-2.81 µg/mL CE-TOF-MS: 0.011-0.071 µg/mL	[73]
Untargeted metabolome analysis	*Arabidopsis thaliana*	Extraction with MeOH:Chloroform:Water [3:1:1; v/v/v]	CE capillary: uncoated (100 cm x 50 µm I.D.) Buffer: 1 M FA Voltage: 30 kV	MeOH:Water containing 0.5 mM reserpine (50:50, v/v) at 10 µL/min	Source: ESI (+) Analyzer: TOF	n.i.	[74]
Comparative Metabolomics of Transgenic *versus* Conventional Maize	Maize	Extraction with MeOH/Water mixture (50:50)	CE capillary: uncoated (80 cm x 50 µm I.D.) Buffer: 5% FA (pH 1.9) Voltage: 20 kV	IPA:Water (50:50, v:v), at 0.24 mL/h	Source: ESI (+) Analyzer: TOF	n.i.	[75]
Anthocyanins	*Hibiscus sabdariffa L.* (Karkade)	SPE	CE capillary: uncoated (80 cm x 50 µm I.D.) Buffer: 200 mM boric acid (pH 9) Voltage: 25 kV	IPA:0.1% FA (60:40 v/v) at 0.24 mL/h	Source: ESI (+) Analyzer: Ion trap and TOF	n.i.	[76]
Determination of arsenic species	Fish	Extraction with acetone and Water/MeOH (1:1 v/v)	CE capillary: uncoated (50 cm x 75 µm I.D.) Buffer: 25 mM borate buffer (pH 9.2) with 10% v/v of MeOH Voltage: 30 kV	Water and MeOH (1:1, v/v) with 0.01% FA at 0.3 mL/h	Source: ESI (+)/(-) and ICP Analyzer: TOF Quadrupole	CE-ESI-TOF-MS: $1.0x10^{-7}$ $-1.0x10^{-6}$M CE-ICP-MS: $6.2x10^{-8}$ $-9.5x10^{-8}$M	[77]
Cobalamin *(Cobalt speciation)*	Nutritive Supplements and Chlorella Foods	MAE	CE capillary: uncoated (70 cm x 75 µm I.D.) Buffer: 25 mM Tris buffer (pH 9.0) with 15 mM SDS Voltage: 20 kV	CETAC CEI-100 at 5.4 µL/min	Source: ICP Analyzer: Quadrupole	0.2-1.7 ng/mL	[78]
Metabolomics of transgenic maize	Maize	PLE	CE capillary: uncoated (80 cm x 50 µm I.D.) Buffer: 5% FA (pH 1.90) Voltage: 20 kV	IPA:Water (50:50, v/v) at 0.24 ml/h	Source: ESI (+) Analyzer: TOF	n.i.	[79]
Determination of yessotoxins and pectenotoxins	Shellfish	SPE	CE capillary: uncoated (80 cm x 50 µm I.D.) Buffer: 10mM ammonium acetate (pH 8.5) Voltage: 20 kV	IPA:0.1 mM ammonium acetate (80:20, v/v)	Source: ESI (-) Analyzer: single quadrupole	0.02-0.25 µg/mL	[80]
Bioactive peptides	Functional food (hypoallergenic infant milk formulas)	SPE	CE capillary: uncoated (57 cm x 50 µm I.D.) Buffer: 1 M acetic acid Voltage: 15 kV	IPA:0.05% FA (60:40, v/v) at 3.3 µL/min	Source: ESI Analyzer: Ion trap and TOF	n.i.	[81]
Quinolizidine alkaloids	Lupinus species	Solvent extraction with dichloromethane and MeOH	CE capillary: uncoated (90 cm x 50 µm I.D.) Buffer: 100 mM ammonium acetate in MeOH/ACN/Water (70/20/10) with 1% acetic acid Voltage: 25 kV	MeOH:Water (1/1), with 0.1% acetic acid at 0.20 ml/h	Source: ESI (+) Analyzer: n.i.	6.5-30.5 µg/mL	[82]
Determination of carnitine enantiomers	Dietary food supplements (drink, biscuit, capsule and tablet)	Drinks: Dilution with Water (1:100) Biscuits, capsules, and tablets: Extraction with Water •Derivatization with FMOC (All samples)	CE capillary: uncoated (100 cm x 50 µm I.D.) Buffer: 0.5 M ammonium formate (pH 2.5) with 0.2% (m/v) succ-c-CD (4 succinyl groups/CD ring) Voltage: 25 kV	IPA:Water (50:50, v:v) with 0.1% FA at 3.3 µL/min	Source: ESI (+) Analyzer: Ion trap (MS/MS)	10 ng/mL	[83]
Covering Chemical Diversity of Genetically-Modified Tomatoes	Tomatoes	Extraction with MeOH:Chloroform:Water [3:1:1; v/v/v]	CE capillary: uncoated (100 cm x 50 µm I.D.) Buffer: 1 M FA Voltage: 30 kV	50:50 (v/v) MeOH-Water with 0.5 mM reserpine at 10 µL/min	Source: ESI (+) Analyzer: TOF	n.i.	[84]
Changes in the Charged metabolite and Sugar Profiles	Japanese sake (rice wine)	Dilution with Water	CE capillary: •uncoated (100 cm x 50 µm I.D.) for cationic metabolites •coated with a cationic polymer (110 cm x 50 µm I.D.) for anionic metabolites Buffer: • 1 mol/L formic acid for cationic metabolites •50 mM ammonium acetate (pH 8.5) for anionic metabolites Voltage: 30 kV	•MeOH:Water (1:1. v/v) and 0.1 µmol/L hexakis (2,2-difluoroethoxy) phosphazene (hexakis) at 10 µL/min for cationic metabolites • 5 mM ammonium acetate in 1:1 (v/v) MeOH:Water, containing 0.1 µmol/L hexakis for anionic metabolites	Source: •ESI (+) for cationic metabolites • ESI (-) for anionic metabolites Analyzer: TOF	n.i.	[85]

(Table 1) cont.....

Application	Food Matrices	Sample Preparation	CE Conditions	CE-MS Interface (Sheath liquid)	MS Conditions	LOD	REFERENCES
Metabolic Profiling	Cheonggukjang *(Fast-Fermented Bean Paste)*	Extraction with MeOH	CE capillary: uncoated (100 cm x 50 μm I.D.) Buffer: •1 M formic acid for cation analysis •20 mM ammonium formate (pH 10.0) for anion analysis Voltage: 30 kV	MeOH:Water (1:1 v/v) with 0.5 M Reserpine at 10 L/min	Source: •ESI (+) for cationic metabolites • ESI (-) for anionic metabolites Analyzer: TOF	n.i.	[86]
Determination of furosine	Flours, pasta, and milk	Hydrolyzation with 5 mL of 8 N HCl	CE capillary: uncoated (60 cm x 50 μm I.D.) Buffer: FA 50 mM (pH 2.7) Voltage: 25 kV	50% MeOH and 0.1% FA at 3 μL/min	Source: ESI (+) Analyzer: TOF (MS/MS)	0.07 mg/L	[87]
Forensic analysis of mesembrine alkaloids	Sceletium tortuosum seeds	Extraction with MeOH	CE capillary: uncoated (70 cm x 50 μm I.D.) Buffer: 75 mM ammonium acetate in a mixture of 9:1 ACN and glacial acetic acid Voltage: 30 kV	5% glacial acetic acid in IPA:Water (66:34) at 4 μL/min	Source: ESI (n.i.) Analyzer: Ion trap	n.i.	[88]
The effect of different cooking treatments	Flour-derived products	Hydrolyzation with 5 mL of 8 N HCl	CE capillary: uncoated (60 cm x 50 μm I.D.) Buffer: 50 mM FA (pH 2.7) Voltage: 25 kV	MeOH:0.1% FA (50:50, v/v) at 3 μL/min	Source: ESI (+) Analyzer: TOF (MS/MS)	0.07 mg/L	[89]
Identification of bioactive peptides	Hypoallergenic infant milk formulas	SPE	CE capillary: uncoated (57 cm x 50 μm I.D.) Buffer: 1 M acetic acid (pH 2.3) Voltage: 15 kV	IPA:0.05% FA (60:40, v/v) at 3.3 μL/min	Source: ESI (+) Analyzer: TOF	n.i.	[90]
Metabolic Profiling (targeted and non-targeted)	Avocado fruits (Persea americana)	Extraction with MeOH	CE capillary: uncoated (85 cm x 50 μm I.D.) Buffer: 40 mM ammonium acetate (pH 9.5) Voltage: 30 kV	IPA:Water (60:40, v/v) at 0.24 mL/h	Source: ESI (+) and (-) Analyzer: TOF (MRM and Full Scan)	MRM: 20.1-203 μg/mL Full scan: 0.22-1.90 μg/mL	[91]
Analysis of lysozyme	Cheese	Extraction with 1 M acetic acid	CE capillary: polyacrylamide-coated (80 cm x 50 μm I.D.) Buffer: 100 mM FA Voltage: 20 kV	MeOH:Water:FA (50:49.5:0.5, v/v/v)	Source: ESI (n.i.) Analyzer: Triple quad.	3.6 mg/kg	[92]
Determination of nucleotides	Infant formulas	Centrifugal ultrafiltration	CE capillary: uncoated (87.5 cm x 50 μm I.D.) Buffer: 30 mM ammonium formate (pH 9.6) Voltage: 30 kV	IPA:Water (50:50, v/v) at 10 μL/min	Source: ESI (-) Analyzer: Single quadrupole (SIM)	0.8-1.8 μg/g	[93]
Metabolic profiling	Breast milk	Extraction chloroform/MeOH (2:1) and Water	CE capillary: uncoated (60 cm x 50 μm I.D.) Buffer: 0.1 mol/L FA Voltage: 25 kV	MeOH:1.0 mM FA (70:30, v/v) at 6 μL/min	Source: ESI (+) Analyzer: TOF	n.i.	[94]
Anionic metabolite profiling	Orange juice and wine	Filtering	CE capillary: PTH-coated (50 μm I.D.) Buffer: 1 M FA Voltage: 20 kV	IPA:Water (50:50, v/v) at 0.24 mL/h	Source: ESI (-) Analyzer: TOF	0.1-16.4 μg/mL	[95]
Determination of nucleosides and nucleotide mono-, di- and tri-phosphates	Baby foods (fish, meat, fruits, and baby dairy desserts)	Centrifugal ultrafiltration	CE capillary: uncoated (87.5 cmx50 μm I.D.) Buffer: 76 mM Hexafluoro-2- propanol at (pH 10.2) Voltage: 30 kV	1.6% (v/v) of Hexafluoro-2-propanol in Water at 1 mL/min	Source: ESI (-) Analyzer: Single quadrupole (SIM)	14-53 ng/mL for Nucleo-sides. 7-2) ng/mL for the others	[96]
Detection of coffee adulteration	Soybean and corn	Acid hydrolysis with H2SO4	CE capillary: uncoated (90 cm x 20 μm I.D.) Buffer: 500 mM TEA (pH 12.3) Voltage: 25 kV	6 μL/min	Source: ESI (-) Analyzer: Triple-quadrupole (MRM)	0.001-0.003 mM	[97]
Identification of antihypertensive peptides *(bioactive compounds)*	Nutraceuticals (nutraceutical tablets)	Extraction with warm water	CE capillary: uncoated (72 cm x 75 μm I.D.) Buffer: 1 M acetic acid (pH 2.3) Voltage: n.i.	IPA:0.05% FA (60:40, v/v) at 3.3 μL/min	Source: ESI (+) Analyzer: TOF	n.i.	[98]
Metabolic fingerprinting	Green and roasted coffee	SLE with 25% (v/v) MeOH in Water	CE capillary: uncoated (100 cm x 50 μm I.D.) Buffer: 1 M FA (pH 1.8) Voltage: 30 kV	MeOH:1 M acetic acid (50:50, v/v) at 8 μL/min	Source: ESI (+) Analyzer: TOF (MS/MS)	n.i.	[99]
Pesticides Residues	Corn	QuEChERS extraction	CE capillary: poly(vinyl alcohol)-coated (58 cm x 50 μm I.D.) Buffer: 0.1 M FA (pH 2.4) Voltage: 28 kV	Running buffer with MeOH:Water (1:1, v/v) at 5 μL/min	Source: ESI (n.i.) Analyzer: Triple quadrupole (MRM)	0.03-0.28 μg/kg	[100]

(Table 1) cont.....

Application	Food Matrices	Sample Preparation	CE Conditions	CE-MS Interface (Sheath liquid)	MS Conditions	LOD	REFERENCES
Chiral separation of intact amino acids	Vinegar	Filterin and dilution with Water	CE capillary: uncoated (100 cm x 50 μm I.D.) Buffer: 1 M FA Voltage: 24 kV	MeOH:Water (1:1, v/v) at 10 μL/min	Source: ESI (+) Analyzer: QTOF	0.07-1.03 μg/mL	[101]
Determination of aminoglycoside residues	Milk	SPE	CE capillary: uncoated (90 cm x 30 μm I.D.) Buffer: 10% acetic acid Voltage: 25 kV	Sheathless	Source: NanoSpray III source Analyzer: Triple Quad (MRM, MS/MS)	0.2 μg/kg	[102]
Determination of organotin compounds	Edible oils (sesame oil, olive oil, peanut oil, soybean oil and corn oil)	•Low-temperature freezing extraction •SPE	CE capillary: uncoated (90 cm x 30 μm I.D.) Buffer: 10% acetic acid in MeOH/ACN (20:80, v/v) Voltage: 30 kV	Sheathless	Source: ESI (+) Analyzer: Triple TOF	0.016-0.05 ng/mL	[103]

LOD: Limit of Detection; Ref: References; n.i: not indicated; PP: Protein Precipitation; LLE: Liquid-Liquid Extraction; SPE: Solid Phase Extraction; SLE: Solid-Liquid Extraction; SPME: Solid Phase Micro Extraction; MAE: Microwave-Assisted Extraction; PLE: Pressurized Liquid Extraction; MeOH: Methanol; ACN: Acetonitrile; $(NH_4)_2CO_3$: Ammonium carbonate; FA: Formic Acid; IPA: Isopropyl alcohol.

CE-MS BASED TARGETED AND UNTARGETED ANALYSIS TO ENSURE FOOD SAFETY AND FOOD QUALITY

Targeted Analysis of Residues and Toxic Exogenous Compounds in Food Samples

Foods can face factors that disrupt their safety at every stage of the chain from production to consumption. Reliable foods are foods that do not bear physical, chemical and biological risks from their production to consumption. Unwanted chemicals in foods are especially pesticides and antibiotics, environmental pollutants, agrochemicals, some other drugs, hormones and plastics used in packaging. Each chemical substance can act as a toxic agent according to the dose taken. All these substances, called residues, are metabolites that accumulate in the structures of animals or plants consumed as food and used to produce food products [104, 105].

All chemical drugs used against living organisms that harm plants and animals are called pesticides. Pesticides are both naturally and synthetically produced chemicals that kill harmful weeds, diseases and pests in various cultivated plants that form the basis of the foods we consume and decrease their yield and quality. The term pesticide includes insecticide (insecticide), herbicide (weed killer), fungicide (mold killer), rodenticide (rodent killer), *etc* [106]. Some CE-MS based targeted approaches were published to achieve the analysis of pesticides in food samples to ensure food safety. Rodrigues *et al.* performed analysis of thiabendazole and procymidone in fruits and vegetables by CE-MS [44]. In their study, they used SPE to prepare the samples and an uncoated capillary was used for separation where positive ionization mode on ESI was used for determination. The sensitivity of the developed method was enough to analyze the approved quantity of herbicides to be used in daily applications. In another study performed by same group, SPME was used to prepare the samples of fruits to be analyzed and acidic pesticides in fruits were successfully analyzed through CE-MS [49]. Pyrimethanil, pyrifenox, cyprodinil, cyromazine, and pirimicarb are some other pesticides behind thiabendazole and procymidone. In a study published in 2004, CE-MS was successfully applied to quantify these compounds in fruit juices [52]. SPME was used for sample preparation and an uncoated capillary was used in positive ionization mode for analysis. Juan-Garcia *et al.,* developed a CE-MS methodology to quantify six pesticides in fruits [61]. These pesticides separated and analyzed through an uncoated capillary while working in both positive and negative ionization modes are dinoseb, pirimicarb, procymidone, pyrifenox, pyrimethanil, and thiabendazole. A SPE method was used for sample preparation.

In recent years, soybean products have been a highly emphasized product and become the economical source of proteins especially in developed countries. Soy milk and its products (soy milk, soy yoghurt, tofu, *etc.*) increased interest on soybean. Soy milk is more and more easily accessible type of milk consumed by people. Unlike cow's milk, soy milk does not contain lactose, and this makes soy milk a good alternative for people who are lactose intolerant. Fortified soy milk can be a good source of protein, iron, vitamin B, and calcium [107]. However, pesticides used in the growing stage of soybean can prevent its usage on babies. In such a situation the analysis of pesticides need to guarantee the undesired exogenous contents of soy milk products. A study published in 2005 describes a SPE-CE-ESI-MS procedure for the simultaneous determination of a group of triazolopyrimidine sulfoanilide herbicides at ppb levels in soy milk [62]. This study shows the capability of CE-MS to be used on analysis of exogenous compounds in milk products.

QuEChERS extraction is a relatively new extraction technique in comparison to SPE and SPME. It was primarily used to extract pesticide residues in food matrices [108]. Its compatibility with MS made it a preferred sample preparation technique to clean- up food matrices before analysis [109]. One of the recent CE-MS application combined with QuEChERS extraction was published by Daniel *et al.,* and they determined multiclass pesticides residues in corn [100].

Wheat ranks first among cultivated plants used in human nutrition in terms of cultivation and production in the world. This is because the wheat plant has wide adaptability. In addition, wheat grain is the basic food of approximately 50 countries due to its suitable feeding value, convenience in storage and processing. Wheat provides about 20% of the total calories supplied to the world population from plant- based foods [110]. This situation makes the quality control of wheat a vital issue besides the analysis of other food products. Since herbicides usage on wheat farming is common practice and it is under the control of legal authorities, analytical methods for the precise and accurate determination of herbicides in wheat is a focused subject by researchers and CE-MS based methodologies were presented in literature for this purpose [48].

Selenium is a trace element found in trace amounts in the human body, but it is long known that selenium deficiency causes different diseases and that this problem could be solved through a balanced diet. However, there are many studies on the toxic effects of a high intake of selenium. Selenium is found in biological tissues and fluids bond to proteins, especially in milk [111]. A capillary zone electrophoresis (CZE) method combined with ICP-MS is a promising powerful tool for metal speciation and separations of selenium species, both, organic and inorganic were achieved in human milk by Michalke *et al.* [38]. In this study, the precipitation of milk proteins using organic solvents facilitated the analysis of the sample matrix on CE.

Its use in wide industrial applications has increased the production of organotin compounds worldwide in the last 50 years, since tin contains many organometallic derivatives. This has led to a significant amount of organotin entering various ecosystems. Sn is not toxic in its inorganic form, but the toxicological pattern of organotin compounds is complex. Some organotins exhibit specific toxic effects to different organism, even at low exposures, depending on the characteristics of organic groups linked to the Sn cation. Therefore, specific determination of the individual organotin compounds is required [112]. He *et al.* presented an ultrasensitive determination of organotin compounds in plastic food packaging and edible oils which are sesame oil, olive oil, peanut oil, soybean oil and corn oil [103]. An uncoated capillary was used for separation and ESI source was utilized in positive ionization mode. Low temperature freezing extraction and SPE were used for sample preparation.

Meermann *et al.* presented a study to analyze arsenic species in fish samples through extraction with acetone and water/methanol (1:1 v/v) prior to analysis with CE-ICP- MS [77]. In their study they indicated that CE-ICP-MS provided high sensitivity and better reproducibility for quantitative measurements but CE-ESI-MS with a TOF mass analyzer proved to be valuable for species identification.

Acrylamide is a compound that occurs in foods produced or cooked at high temperatures and its amount gradually increases if heat is applied for a long time. Acrylamide can form with several different mechanisms as a result of the reactions of carbohydrates, proteins, amino acids, lipids and other possible components in the composition of heat-treated foods. Although its effects on humans are not fully determined, the existence of genotoxic and carcinogenic effects on experimental animals has been proven. Since it occurred through the cooking methodology, it could be referred as an exogenous toxic compound found in foods [113]. One of the first papers to demonstrate the analysis of acrylamide in food products (potato crisps, crisp bread, biscuits, breakfast cereals and coffee in the scope of the study) showed that after derivatization with 2-mercaptobenzoic acid, SPE was a suitable extraction technique when it was coupled with CE-MS/MS for the analysis of acrylamide in food samples [67]. Although the amount of formed acrylamide was relatively low in food products, the methodology was successful to perform sensitive analyses.

Nowadays, with the rapid increase in world population, antibiotics or similar substances have become widespread in order to deal with the animal protein gap, to rapidly remove diseases in animal husbandry and to prepare for production, to increase animal development or the yield of animal foods. The causes of drug residues in animals are preparing meat products without waiting for a certain period after the drug treatment applied to animals. In addition, the consumption of products (meat, milk, eggs) obtained from animals treated with drugs causes another risk factor [114].

Sulphonamides, due to their important antibacterial effects are widely used in veterinary practice and animal husbandry. The analysis of sulphonamides in pork meat and beef meat was successfully carried out through CE-MS [68, 69]. Pressurized liquid extraction (PLE) was used to prepare pork meat samples [68] while SPE was used to prepare the beef meat samples [69]. CE coupled with ESI source in positive ionization mode was capable of determining sulfonamide in these studies. Since sulphonamide residues could pass through the gastrointestinal system, it is obvious that they could be found in milk. Therefore, Santos *et al.* presented a CE-MS methodology to analyze sulfamethazine, sulfadiazine, sulfathiazole, sulfadimethoxine, sulfamethizole, and sulfamerazin in milk [63].

Quinolones are broad-spectrum antibiotics active against both Gram-positive and Gram-negative bacteria, including mycobacteria, and anaerobes [115] and are widely used for treating a variety of infections. In a study published by Juan□García *et al.*, the analysis of the residues of some quinolone antibiotics in chicken and fish was achieved through CE-MS [65]. CE coupled with quadrupole ion trap was used for this purpose where ESI was used on positive ionization mode. In another study presented by Lara *et al.*, SPE was used for the extraction of danofloxacin, sarafloxacin, ciprofloxacin, marbofloxacin, enrofloxacin, difloxacin, oxolinic acid, and flumequine, then CE-MS/MS was used for determination of these compounds in bovine raw milk [66].

Tetracyclines are broad-spectrum antibiotics that disrupt protein synthesis at the ribosomal level [116]. Tetracycline residues were analyzed in bovine milk through CE combined with single quadrupole MS when SPE was used for sample preparation [71].

A CE-MS procedure was developed for the confirmation and quantification of 12 antibacterial residues that belong to three different chemical classes: β-lactams, quinolones, and sulfonamides. Confirmation and quantitation were achieved at concentrations below tolerance. When tested on fish and bovine meat intended for human consumption, the method was able to identify and quantify the antibiotic present in the samples [70].

Targeted Analysis of Toxic Endogenous Compounds in Food Samples

Paralytic shellfish poisoning is related with the ingestion of mussels, clams, and oysters. The primary paralytic shellfish poisoning toxins are heterocyclic compounds that block nerve and muscle action potentials by binding to sodium channels at the same site as tetrodotoxin, thereby resulting in paralysis. Saxitoxin and gonyautoxin are the primary toxins associated with this situation and they are produced by marine microalgae dinoflagellates [117]. A SPE procedure prior to CE-MS analysis in API positive ionization mode achieved the sensitive determination of paralytic poisoning toxins [35]. In another study, shellfish was extracted through SPE and yessotoxins and pectenotoxins [118] were analyzed through CE-MS in negative ionization mode of ESI on a single quadrupole instrument [80].

Tetramine (tetramethylammonium ion) is a typical name for a chemical containing four amine groups. Tetramine is stable and its metabolism is slow [119]. It is an agent blocking autonomic ganglionic and it is found in several marine gastropod species. It is responsible for numerous incidents of human intoxication in Japan, Europe and Canada, due to the consumption of whelks (sea snails). In order to analyze this toxic, LLE with diethylether was performed on sea snails and a LPA-coated capillary used on CE-MS determination where the API was in a positive ionization mode [37].

Targeted Analysis to Investigate Food Quality

Since the beginning of 00's, the capability of CE-MS to analyze amino acids is a well-known issue [39]. The ionic structures of these compounds make their separation easy through CE. Since amino acids are vital components, their amount in infant foods is an indicator of the quality of these commercial products and need to be detected both on food technology and quality control processes. Klampfl *et al.* developed a CE-MS methodology to analyze free amino acids in infant food after dilution with water (1:10 v/v) [43]. An ESI source in positive ionization mode was used in a single quadrupole instrument. Yu *et al.* published a paper to determine aminoglycoside residues in milk by sheathless CE-ES--MS/MS [102].

In another study, chiral amino acids in orange juice were analyzed in an uncoated capillary. On-line coupling between CE and MS was formed through an electrospray-coaxial sheath flow interface. Enantiomer separation was performed by using β-cyclodextrin in the background electrolyte (BGE) [56]. Beside orange juice, chiral amino acids in vinegar were successfully determined through CE-MS/MS after filtering procedure and dilution with water [101].

Human milk has a higher concentration of nucleotides than bovine milk which is the source of most infant formulas. As the composition of human milk is considered the 'gold standard,' an increasing number of infant formulas are supplemented with nucleotides [120, 121]. In literature, determination of nucleotides in infant formulas was performed through CE coupled with single quadrupole MS [93, 96]. In these studies, ultrafiltration procedure to prepare the samples was successful to separate nucleotides from matrix components.

Determination of carbohydrates/oligosaccharides in food samples is essential to evaluate food quality as an energy source. Larsson *et al* performed a targeted approach to analyze carbohydrates in dextrin-15 starch hydrolysate from maize after derivatization with 8-aminonaphthalene-1, 3, 6-trisulfonic acid [42].

Gellan gum, a polysaccharide, is a food additive that was discovered in the 1970s. First used as a substitute for gelatin and agar agar, it's currently found in a variety of processed foods including jams, candy, meats, and fortified plant milks. Determination of gellan gum was indirectly obtained monitoring the peak of the corresponding tetrasaccharide fragment from lyase digestion *via* CE-MS in a positive ionization mode of ESI when single quadrupole was used. The developed methodology was applied to the analysis of fruit flavour drink [41]. A dilution process was fine to detect carcohydrates in wine when an ESI negative mode was performed on CE analysis [40].

Potatoes are inexpensive low-fat food sources providing energy (starch), high-quality protein, fiber, and vitamins. Potatoes also produce biologically active secondary metabolites which may have both, adverse and beneficial effects in the diet. One of these secondary metabolites is glycoalkaloids which are reported to be involved in host-plant resistance and to have a variety of adverse as well as beneficial effects on cells, animals, and humans, and this is why there is need for a better understanding of their roles both in plants and the diet [122]. Therefore, the analysis of these unique compounds is essential to evaluate the quality of potato products. NACE buffer which is well compatible with ESI-MS in terms of enhanced sample ionization was used in two different studies to analyze α-chaconine, α-solanine, α-tomatine and relative aglycones solanidine and tomatidine in potato tubers [45, 46].

Analysis of small acidic compounds, gluconic, isosaccharinic, oxalic, lactic, malonic, maleic, citric, tartaric, adipic, glutaric, and succinic acid, in ale samples after filtering were possible in a negative ionization mode of ESI in a single quadrupole instrument after separation of these compounds with CE [47]. Another application for the analysis of small acidic compounds, carboxylic acids, in fruit juice was achieved by Sawada *et al* [50].

L-Furosine is an analytical indicator of heat treatment damage or storage deterioration in milk, dried foods, pasta, tomato paste and eggs. Therefore, analysis of furosine in these products ensures food safety. In a study, a CE-MS/MS method was proposed for determination of furosine in flours, pasta, and milk [87]. Hydrolyzation with 5 mL of 8 M HCl achieved to detect furosine in positive ionization mode of ESI source. The effect of different cooking treatments needs to be addressed if food quality and safety are investigated. In order to investigate the effect of cooking treatment on foods, furosine analysis was performed on flour-derived products *via* CE-MS/MS [89].

The quality of coffee can be determined through the adulteration on coffee samples. In such a situation, CE-MS/MS could be an alternative method to determine monosaccharides concentration which could be used as a quality marker [97].

Targeted Analysis of Bioactive Compounds in Foods

Free radicals and antioxidants, which are among the most studied topics in recent years, are becoming more and more important. Under normal conditions, antioxidants and free radicals are in balance in a healthy metabolism. However, when this balance changes in favor of free radicals, a susceptibility towards disease caused by oxidative stress is observed. Especially an increase in exogenous free radical sources such as environmental pollution, alcohol and tobacco use, forest fires, X-rays and UV rays can cause damage to carbohydrates, fats, proteins and DNA in the human body, leading to oxidation. With the increase of free radicals, endogenous antioxidants may be inadequate, and this may require the removal of exogenous antioxidants. The Mediterranean diet can be assumed as a life-style supporting the body with lots of antioxidants which are beneficial particularly for cancer prevention [123 - 125]. The antioxidant activity of the essential oils of five spice plants including rosemary, widely used in the Mediterranean diet has been known for a long time [126]. The antioxidant polyphenolic compounds in rosemary is a subject that must be investigated since they have beneficial effects against cancer. Some studies showed the *in-vitro* effect of rosemary polyphenols against colon cancer [127 - 129]. Some other clinical trials showed the benefits of these polyphenols against cancer [124]. CE-MS/MS studies were performed to analyze the antioxidant compounds in rosemary extract [53] and rosemary honey [64], respectively.

Anthocyanins are a type of flavonoid, a class of compounds with antioxidant effects. Found naturally in a number of foods, anthocyanins are the pigments that give red, purple, and blue plants their rich coloring. Analysis of anthocyanins in wine and wine musts was achieved by Bednář *et al.* [55]. In another study, these compounds were analyzed in *Hibiscus sabdariffa* L. when ESI was operated in positive ionization mode and an uncoated capillary was used [76].

Chlorella is a kind of fresh algae in the green algae branch, 2-8 micrometers in size, spherical or elliptical. *Chlorella* is seen as a source of healthy protein and used as a dietary supplement [130]. Cobalamin (Vitamin B12) is an exogenous compound for humans but it could be indicated as an endogenous bioactive compound in foods. In a study published by Chen *et al.,* cobalamin was analyzed through CE combined with ICP-MS in nutritive supplements and chlorella foods [78]. This study shows the potential of using CE in combination with ICP-MS for a wide variety of samples and compounds.

γ-Glutamyl-S-ethenyl-cysteine, was first isolated and crystallised by Enneking *et al* [131]. γ-Glutamyl-S-ethenyl-cysteine is an antinutritional dipeptide and a CE-MS/MS methodology for its targeted analysis in *Vicia narbonensis* L. seeds were published by Arias *et al* [60].

Glucosinolates, defined for the first time in the early 17th century, are secondary plant metabolites that contain nitrogen and sulfur in their structure, which are responsible for the sharp taste and smell of *Brassica* vegetables such as cabbage, broccoli, cauliflower [132, 133]. Therefore, analysis of intact glucosinolates is an important topic to evaluate possible anticancer activities of foods. The mass accuracy and true isotopic pattern of the ESI-TOF-MS achieved the determination of the elemental composition of CE separated glucosinolates in Arabidopsis thaliana seeds [59].

Alkaloid-containing plants have been used by humans since ancient times for therapeutic and recreational purposes [134]. In literature, CE-MS/MS based analysis of alkaloids serve two different aims. The first one is the usage of these methodologies for forensic applications [88] and the other focus on the determination of alkaloids in plants and foods where alkaloids could be assigned in positive ESI mode at corresponding $[M+H]^+$ values [82].

Targeted Analyis of Intact Proteins in Foods

Although CE cannot find its rightful place in pharmaceutical analysis applications, it has been seen that CE has theoretical advantages in analyzing protein structures when the production of biotechnological drugs have been globally improved [135, 136]. The capability of CE to analyze intact proteins can be a useful tool to analyze bioactive enzymes in plants, vegetables and fruits. Simo and Herrero presented that allophycocyanin- α chain, allophycocyanin- β, c-phycocyanin- α, c-phycocyanin- β, and phycobiliproteins could be analyzed in *Spirulina platensis* microalgae through CE-MS [57, 58]. They indicated that, CE-IT-MS and CE-TOF-MS, allow the characterization of these proteins with coherent results, although CE-TOF-MS provides higher resolution and better mass accuracy than CE-IT-MS. PLE was used as extraction technique in these studies. In another CE-MS study, basic proteins (lysozymes, cytochrome c, ribonuclease A) were analyzed in chicken eggs, white wine, and minced meat from beef [51]. Such studies show the opportunities to analyze intact bioactive proteins through CE-MS especially on food products and food supplements. For instance, lysozyme, also known as muramidase or N- acetylmuramide glycanhydrolase, is an antimicrobial enzyme produced by animals that forms part of the innate immune system [137]. It is a commercially valuable enzyme for food industry and pharmaceutical applications, with potential use for cancer chemotherapy and antimicrobial approaches [138]. Analysis of lysozyme in cheese was successfully performed through CE-MS/MS [92].

Non-targeted Analysis and Metabolite Profiling Studies on Food Samples

One of the first metabolomics approach on food analysis was presented by García- Villalba *et al* [72]. In their study, a comparative metabolomics approach was performed through CE-ESI-TOF/MS for the untargeted analysis of genetically modified organisms. The control group in this study was conventional soybeans. In a negative ionization mode, tentative identification of more than 40 compounds including isoflavones, aminoacids, carboxylic acids, peptides and other analytes were achieved. Although, comparative analysis of these metabolites was performed, one of the most interesting results was the disappearing of 4-hydroxy-l-threonine in the transgenic soybean samples. Other untargeted metabolomics approaches performed on maize samples presented differences on food metabolome when transgenic seed were used on farming [75, 79]. CE-ES--TOF/MS was the instrument used on these studies since tentative identifications were performed through database search.

Extraction with MeOH:chloroform:Water [3:1:1; v/v/v] mixture is a well-known procedure to separate polar and non-polar metabolites [139, 140]. In a study published by Kusano *et al.,* CE-ESI-TOF/MS was used to investigate the chemical diversity of genetically-modified tomatoes after sampling with this unique extraction solvent [84].

Other comparative examinations *via* CE-MS based untargeted metabolomics were performed on various food samples. For instance, monitoring the changes during the fermentation process for Cheonggukjang (Fast-Fermented Bean Paste) was performed through an untargeted metabolomics approach [86]. ESI was operated on both positive and negative ionization modes and this provided to detect 7820 and 3057 detected variables in the positive and negative modes, respectively. In another study, monitoring the metabolic profile during lactation of breast milk was performed through CE-ESI-TOF/MS [94].

Non-targeted analysis of bioactive compounds in food samples has been a novel issue, since the registered food metabolites on databases became accessible. In a study published by Català□Clariana *et al.,* 24, 30, and 38 bioactive peptides were confirmed through CE-ESI-TOF/MS in three infant milk formulas, respectively [90]. Targeted (10 metabolites, including phenolic acids, flavonoids, a carbohydrate, an organic acid, a vitamin and a phytohormone) and non-targeted analysis of Avocado fruits was performed when a MS/MS for targeted approaches and Q-TOF MS for untargeted approaches were used [91]. Metabolite profiling experiments for orange juice and coffee was performed in negative ionization mode of CE-ESI-TOF/MS instrument [95].

CONCLUSION AND FUTURE PERSPECTIVES

Nutrition is one of the most basic human needs and one of the basic criteria for health. For a better life, healthy nutrition comes first and one of the most important principles of healthy nutrition is food safety which can be briefly described as the safety from food resources to food products. Food safety covers all stages from primary production (vegetative production, harvesting, care, irrigation, animal health, animal vaccination, control and transportation of feeds they consume, *etc.*) to the consumption of the food products.

The policies of the countries regarding food safety, their ability to provide food security and the goals of the people to reach safe food vary according to the development stage of those countries. Food safety risks and public health risks increase in parallel. In addition, strong economies in developed countries face higher levels of obesity although they have access to a more varied and healthy diet and this issue has become another face of food safety.

Although debates and precautions regarding food safety are as old as human history, approaches to regulate food safety are almost an issue of the past 50 years. With the arrival of the 80's, the developments in chromatographic techniques and the availability of commercial forms of mass spectrometry devices led to a new era in food analysis and the regulations were improved in parallel with improvements on advanced analytical techniques. In the 90's, the importance of food quality and safety began to be understood and their impact on human health started to be evaluated together.

In parallel with the developments in advanced analytical techniques, the developments in food technology have also increased, and in some cases, known analytical approaches have become insufficient in today's food technology. Especially with the increase in the variety of food products, the sample matrix has become more complicated. In some cases, pharmaceutical technology has been transferred into food technology to produce outstanding products. The exogenous substances added for the purpose of increasing food quality or preparing regular food products as dietary supplements have become part of the recent food technology and this situation caused another insufficiency on targeted analysis. For instance, drug active substances (sildenafil, *etc.*), life-hacking bioactive ingredients (carnitine, taurine, *etc.*), peptide supplements added intentionally or fraudulently in a fruit juice is a relatively new issue when it is compared with adding preservatives or minerals/vitamins into such products. In addition, consumption of these improved- super foods has become part of popular culture and their usage has been encouraged. Therefore, routine analysis of such products to ensure food safety is more important and harder than ever.

Another insufficiency of targeted analysis of food samples occurred when transgenic foods and organic farming faced us with new products. In such a situation, targeting some components in these products might not provide reliable results when a real comparison is necessary to understand the potential effects of transgenic seeds or organic farming on food quality in whole cases.

Foodomics is a new field of study that includes genomic, transcriptomic, proteomic and metabolomic techniques, and intends to investigate the effects of food on human health at molecular level by using advanced analytical techniques [141 - 147]. Metabolomics is the vast scale investigation of small particles, generally known as metabolites, inside cells, bio fluids, tissues or life forms. These little particles and their connections inside an organic framework are known as the metabolome. Metabolomic studies are divided into two main titles, namely 'metabolite profiling' and 'targeted metabolomic'. The metabolite profiling step is the study conducted to observe possible differences in the level of metabolome, between the groups defined on a determined variable [148].

With the data obtained from the metabolite profiling step, targeted metabolomic studies are carried out to quantify the selected metabolites to evaluate the differentiation in metabolome level [149].

Today, thanks to innovative technological developments, metabolomic profiling studies have become more comprehensive and stronger than in the past in terms of the number of metabolite peaks detected and the number of metabolites that can be identified [150]. As a result of new findings in recent years, the number of detected metabolites has increased significantly compared to a decade ago (to> 100,000 from <10,000) and Human metabolome Database (HMDB) [151] has been upgraded to version 4.0. In this version, lots of food metabolites are also included. Thus, it is possible to evaluate food quality and food safety by transferring metabolite profiling technologies performed in clinical applications to food science with the help of the improvements in databases [152 - 154].

The variety in the chemical structures of metabolites makes hard to profile a whole metabolome in a single injection even with an advanced analytical instrument. The advantages of CE over HPLC is its capability to analyze polar-ionic compounds in a sample and discard matrix effect in analysis, partly as a result of its separation mechanism for analytes. At this point, metabolomics approaches to ensure food safety and quality through CE-MS could provide valuable information on food researches and food quality control with the contribution of scientists working on CE- MS-based food analysis.

CONSENT FOR PUBLICATION

Not applicable.

CONFLICT OF INTEREST

The author declares no conflict of interest, financial or otherwise.

ACKNOWLEDGEMENTS

None declared.

REFERENCES

[1] Kotsanopoulos, K.V.; Arvanitoyannis, I.S. The role of auditing, food safety, and food quality standards in the food industry: A review. *Compr. Rev. Food Sci. Food Saf.*, **2017**, *16*(5), 760-775. [http://dx.doi.org/10.1111/1541-4337.12293] [PMID: 33371608]

[2] Chassy, B.M. Food safety risks and consumer health. *N. Biotechnol.*, **2010**, *27*(5), 534-544. [http://dx.doi.org/10.1016/j.nbt.2010.05.018] [PMID: 20621653]

[3] Nielsen, S.S. Introduction to food analysis.*Food analysis*; Springer, **2017**, pp. 3-16. [http://dx.doi.org/10.1007/978-3-319-45776-5_1]

[4] Van der Heijden, K. *International food safety handbook: Science, international regulation, and control*; Routledge, **2019**. [http://dx.doi.org/10.1201/9780203750346]

[5] Golian, J.; Nagyová, Ľ.; Andocsová, A.; Zajác, P.; Palkovič, J. J. P. *Food safety from consumer perspective: health safety. Potravinarstvo Slovak J. Food Sci.*, **2018**, *12*(1)

[6] Kumar, P. Role of Food and Nutrition in Cancer.*The Role of Functional Food Security in Global Health*; Elsevier, **2019**, pp. 193-203. [http://dx.doi.org/10.1016/B978-0-12-813148-0.00012-8]

[7] Siegel, R.L.; Miller, K.D.; Jemal, A. Cancer statistics, 2020. *CA Cancer J. Clin.*, **2020**, *70*(1), 7-30. [http://dx.doi.org/10.3322/caac.21590] [PMID: 31912902]

[8] Sauter, E.R. Cancer prevention and treatment using combination therapy with natural compounds. *Expert Rev. Clin. Pharmacol.*, **2020**, *13*(3), 265-285. [http://dx.doi.org/10.1080/17512433.2020.1738218] [PMID: 32154753]

[9] Parkin, D.M.; Coleman, M.P. Changes in diet and changes in cancer risk: observational studies. *IARC Sci. Publ.*, **1990**, *103*(103), 93-111. [PMID: 2279800]

[10] Cerrato, P.L. Low-fat diet, lower risk of colorectal cancer. *RN,* **1991**, *54*(9), 71-73. [PMID: 1653448]

[11] Dwyer, J.T. Diet and nutritional strategies for cancer risk reduction. Focus on the 21st century. *Cancer,* **1993**, *72*(3) Suppl., 1024-1031. [http://dx.doi.org/10.1002/1097-0142(19930801)72:3+<1024::AID-CNCR2820721314>3.0.CO;2-M] [PMID: 8334654]

[12] Hill, M.J. Diet, physical activity and cancer risk. *Public Health Nutr.*, **1999**, *2*(3A), 397-401. [http://dx.doi.org/10.1017/S1368980099000543] [PMID: 10610079]

[13] Kolonel, L.N.; Altshuler, D.; Henderson, B.E. The multiethnic cohort study: exploring genes, lifestyle and cancer risk. *Nat. Rev. Cancer,* **2004**, *4*(7), 519-527. [http://dx.doi.org/10.1038/nrc1389] [PMID: 15229477]

[14] Espey, D.; Paisano, R.; Cobb, N. Regional patterns and trends in cancer mortality among American Indians and Alaska Natives, 1990-2001. *Cancer,* **2005**, *103*(5), 1045-1053. [http://dx.doi.org/10.1002/cncr.20876] [PMID: 15685622]

[15] Parkin, D.M.; Pisani, P.; Ferlay, J. Global cancer statistics. *CA Cancer J. Clin.*, **1999**, *49*(1), 33-64, 1. [http://dx.doi.org/10.3322/canjclin.49.1.33] [PMID: 10200776]

[16] Jemal, A.; Siegel, R.; Ward, E.; Hao, Y.; Xu, J.; Murray, T.; Thun, M.J. Cancer statistics, 2008. *CA Cancer J. Clin.*, **2008**, *58*(2), 71-96. [http://dx.doi.org/10.3322/CA.2007.0010] [PMID: 18287387]

[17] Parkin, D.M. Global cancer statistics in the year 2000. *Lancet Oncol.*, **2001**, *2*(9), 533-543. [http://dx.doi.org/10.1016/S1470-2045(01)00486-7] [PMID: 11905707]

[18] Baykan, A.; Zorluoğlu, A.; Geçim, E.; Terzi, C. *Kolon ve Rektum Kanserleri*; Türk Kolon ve Rektum Cerrahisi Derneği: İstanbul, **2010**.

[19] Boran, M.; Hakan, G. Özofagus Kanserinde Epidemiyoloji. *Turkiye Klinikleri J. Surg. Med. Sci.*, **2007**, *3*(29), 1.

[20] Yoder, L.H. Lung cancer epidemiology. *Medsurg Nurs.*, **2006**, *15*(3), 171-174. [PMID: 16817299]

[21] Potter, J.D.; Slattery, M.L.; Bostick, R.M.; Gapstur, S.M. Colon cancer: a review of the epidemiology. *Epidemiol. Rev.*, **1993**, *15*(2), 499-545. [http://dx.doi.org/10.1093/oxfordjournals.epirev.a036132] [PMID: 8174669]

[22] Grönberg, H. Prostate cancer epidemiology. *Lancet,* **2003**, *361*(9360), 859-864.
[http://dx.doi.org/10.1016/S0140-6736(03)12713-4] [PMID: 12642065]

[23] Aggarwal, B.B.; Shishodia, S. Molecular targets of dietary agents for prevention and therapy of cancer. *Biochem. Pharmacol.,* **2006**, *71*(10), 1397-1421.
[http://dx.doi.org/10.1016/j.bcp.2006.02.009] [PMID: 16563357]

[24] Ziegler, R.G.; Hoover, R.N.; Pike, M.C.; Hildesheim, A.; Nomura, A.M.; West, D.W.; Wu-Williams, A.H.; Kolonel, L.N.; Horn-Ross, P.L.; Rosenthal, J.F.; Hyer, M.B. Migration patterns and breast cancer risk in Asian-American women. *J. Natl. Cancer Inst.,* **1993**, *85*(22), 1819-1827.
[http://dx.doi.org/10.1093/jnci/85.22.1819] [PMID: 8230262]

[25] McMichael, A.J.; McCall, M.G.; Hartshorne, J.M.; Woodings, T.L. Patterns of gastro-intestinal cancer in European migrants to Australia: the role of dietary change. *Int. J. Cancer,* **1980**, *25*(4), 431-437.
[http://dx.doi.org/10.1002/ijc.2910250402] [PMID: 7372370]

[26] Haenszel, W.; Kurihara, M. Studies of Japanese migrants. I. Mortality from cancer and other diseases among Japanese in the United States. *J. Natl. Cancer Inst.,* **1968**, *40*(1), 43-68.
[PMID: 5635018]

[27] Haenszel, W.; Kurihara, M.; Segi, M.; Lee, R.K. Stomach cancer among Japanese in Hawaii. *J. Natl. Cancer Inst.,* **1972**, *49*(4), 969-988.
[PMID: 4678140]

[28] Lubin, J.H.; Boice, J.D., Jr Lung cancer risk from residential radon: meta-analysis of eight epidemiologic studies. *J. Natl. Cancer Inst.,* **1997**, *89*(1), 49-57.
[http://dx.doi.org/10.1093/jnci/89.1.49] [PMID: 8978406]

[29] La Vecchia, C.; Bosetti, C. Diet and cancer risk in Mediterranean countries: open issues. *Public Health Nutr.,* **2006**, *9*(8A), 1077-1082.
[http://dx.doi.org/10.1017/S1368980007668475] [PMID: 17378944]

[30] Smith-Spangler, C.; Brandeau, M.L.; Hunter, G.E.; Bavinger, J.C.; Pearson, M.; Eschbach, P.J.; Sundaram, V.; Liu, H.; Schirmer, P.; Stave, C.; Olkin, I.; Bravata, D.M. Are organic foods safer or healthier than conventional alternatives?: a systematic review. *Ann. Intern. Med.,* **2012**, *157*(5), 348-366.
[http://dx.doi.org/10.7326/0003-4819-157-5-201209040-00007] [PMID: 22944875]

[31] Li, S.F.Y. *Capillary electrophoresis: principles, practice and applications*; Elsevier, **1992**.

[32] Grossman, P.D.; Colburn, J.C. *Capillary electrophoresis: Theory and practice*; Academic Press, **2012**.

[33] Kok, M.G.; Somsen, G.W.; de Jong, G.J. Comparison of capillary electrophoresis-mass spectrometry and hydrophilic interaction chromatography-mass spectrometry for anionic metabolic profiling of urine. *Talanta,* **2015**, *132*, 1-7.
[http://dx.doi.org/10.1016/j.talanta.2014.08.047] [PMID: 25476271]

[34] McCalley, D.V. Understanding and manipulating the separation in hydrophilic interaction liquid chromatography. *J. Chromatogr. A,* **2017**, *1523*, 49-71.
[http://dx.doi.org/10.1016/j.chroma.2017.06.026] [PMID: 28668366]

[35] Locke, S.; Thibault, P. Improvement in detection limits for the determination of paralytic shellfish poisoning toxins in shellfish tissues using capillary electrophoresis/electrospray mass spectrometry and discontinuous buffer systems. *Anal. Chem.,* **1994**, *66*(20), 3436-3446.
[http://dx.doi.org/10.1021/ac00092a022]

[36] Wheat, T.E.; Lilley, K.A.; Banks, J.F. Capillary electrophoresis with electrospray mass spectrometry detection for low-molecular-mass compounds. *J. Chromatogr. A,* **1997**, *781*(1-2), 99-105.
[http://dx.doi.org/10.1016/S0021-9673(97)00372-5] [PMID: 9368381]

[37] Zhao, J-Y.; Thibault, P.; Tazawa, T.; Quilliam, M.A. Analysis of tetramine in sea snails by capillary electrophoresis-tandem mass spectrometry. *J. Chromatogr. A,* **1997**, *781*(1-2), 555-564.

[http://dx.doi.org/10.1016/S0021-9673(97)00729-2]

[38] Michalke, B.; Schramel, P. Application of capillary zone electrophoresis-inductively coupled plasma mass spectrometry and capillary isoelectric focusing-inductively coupled plasma mass spectrometry for selenium speciation. *J. Chromatogr. A,* **1998**, *807*(1), 71-80. [http://dx.doi.org/10.1016/S0021-9673(97)01231-4] [PMID: 9646489]

[39] Soga, T.; Heiger, D.N. Amino acid analysis by capillary electrophoresis electrospray ionization mass spectrometry. *Anal. Chem.*, **2000**, *72*(6), 1236-1241. [http://dx.doi.org/10.1021/ac990976y] [PMID: 10740865]

[40] Klampfl, C.W.; Buchberger, W. Determination of carbohydrates by capillary electrophoresis with electrospray-mass spectrometric detection. *Electrophoresis,* **2001**, *22*(13), 2737-2742. [http://dx.doi.org/10.1002/1522-2683(200108)22:13<2737::AID-ELPS2737>3.0.CO;2-Z] [PMID: 11545400]

[41] Craston, D.; Farnell, P.; Francis, J.; Gabriac, S.; Matthews, W.; Saeed, M.; Sutherland, I. Determination of gellan gum by capillary electrophoresis and CE-MS. *Food Chem.,* **2001**, *73*(1), 103-110. [http://dx.doi.org/10.1016/S0308-8146(00)00320-4]

[42] Larsson, M.; Sundberg, R.; Folestad, S. On-line capillary electrophoresis with mass spectrometry detection for the analysis of carbohydrates after derivatization with 8-aminonaphthalene-1,3-6-trisulfonic acid. *J. Chromatogr. A,* **2001**, *934*(1-2), 75-85. [http://dx.doi.org/10.1016/S0021-9673(01)01274-2] [PMID: 11762766]

[43] Klampfl, C.W.; Ahrer, W. Determination of free amino acids in infant food by capillary zone electrophoresis with mass spectrometric detection. *Electrophoresis,* **2001**, *22*(8), 1579-1584. [http://dx.doi.org/10.1002/1522-2683(200105)22:8<1579::AID-ELPS1579>3.0.CO;2-C] [PMID: 11386673]

[44] Rodríguez, R.; Picó, Y.; Font, G.; Mañes, J. Analysis of thiabendazole and procymidone in fruits and vegetables by capillary electrophoresis-electrospray mass spectrometry. *J. Chromatogr. A,* **2002**, *949*(1-2), 359-366. [http://dx.doi.org/10.1016/S0021-9673(01)01395-4] [PMID: 11999753]

[45] Bianco, G.; Schmitt-Kopplin, P.; De Benedetto, G.; Kettrup, A.; Cataldi, T.R. Determination of glycoalkaloids and relative aglycones by nonaqueous capillary electrophoresis coupled with electrospray ionization-ion trap mass spectrometry. *Electrophoresis,* **2002**, *23*(17), 2904-2912. [http://dx.doi.org/10.1002/1522-2683(200209)23:17<2904::AID-ELPS2904>3.0.CO;2-1] [PMID: 12207298]

[46] Bianco, G.; Schmitt-Kopplin, P.; Crescenzi, A.; Comes, S.; Kettrup, A.; Cataldi, T.R. Evaluation of glycoalkaloids in tubers of genetically modified virus Y-resistant potato plants (var. Désirée) by non-aqueous capillary electrophoresis coupled with electrospray ionization mass spectrometry (NACE-ESI-MS). *Anal. Bioanal. Chem.*, **2003**, *375*(6), 799-804. [http://dx.doi.org/10.1007/s00216-003-1831-3] [PMID: 12664181]

[47] Hagberg, J. Analysis of low-molecular-mass organic acids using capillary zone electrophoresis-electrospray ionization mass spectrometry. *J. Chromatogr. A,* **2003**, *988*(1), 127-133. [http://dx.doi.org/10.1016/S0021-9673(02)02022-8] [PMID: 12647827]

[48] Goodwin, L.; Startin, J.R.; Keely, B.J.; Goodall, D.M. Analysis of glyphosate and glufosinate by capillary electrophoresis-mass spectrometry utilising a sheathless microelectrospray interface. *J. Chromatogr. A,* **2003**, *1004*(1-2), 107-119. [http://dx.doi.org/10.1016/S0021-9673(03)00572-7] [PMID: 12929967]

[49] Rodríguez, R.; Mañes, J.; Picó, Y. Off-line solid-phase microextraction and capillary electrophoresis mass spectrometry to determine acidic pesticides in fruits. *Anal. Chem.,* **2003**, *75*(3), 452-459. [http://dx.doi.org/10.1021/ac025884e] [PMID: 12585470]

[50] Sawada, H.; Nogami, C. Capillary electrophoresis-electrospray ionization mass spectrometry using

uncoated fused-silica capillaries and alkaline buffer solution for the analysis of small carboxylic acids. *Anal. Chim. Acta,* **2004**, *507*(2), 191-198. [http://dx.doi.org/10.1016/j.aca.2003.11.047]

[51] Simó, C.; Elvira, C.; González, N.; San Román, J.; Barbas, C.; Cifuentes, A. Capillary electrophoresis-mass spectrometry of basic proteins using a new physically adsorbed polymer coating. Some applications in food analysis. *Electrophoresis,* **2004**, *25*(13), 2056-2064. [http://dx.doi.org/10.1002/elps.200305790] [PMID: 15237406]

[52] Hernández-Borges, J.; Rodríguez-Delgado, M.Á.; García-Montelongo, F.J.; Cifuentes, A. Highly sensitive analysis of multiple pesticides in foods combining solid-phase microextraction, capillary electrophoresis-mass spectrometry, and chemometrics. *Electrophoresis,* **2004**, *25*(13), 2065-2076. [http://dx.doi.org/10.1002/elps.200405938] [PMID: 15237407]

[53] Herrero, M.; Arráez-Román, D.; Segura, A.; Kenndler, E.; Gius, B.; Raggid, M.A.; Ibáñez, E.; Cifuentes, A. Pressurized liquid extraction-capillary electrophoresis-mass spectrometry for the analysis of polar antioxidants in rosemary extracts. *J. Chromatogr. A,* **2005**, *1084*(1-2), 54-62. [http://dx.doi.org/10.1016/j.chroma.2004.10.003] [PMID: 16114236]

[54] Tessier, B.; Schweizer, M.; Fournier, F.; Framboisier, X.; Chevalot, I.; Vanderesse, R.; Harscoat, C.; Marc, I. Prediction of the amino acid composition of small peptides contained in a plant protein hydrolysate by LC–MS and CE–MS. *Food Res. Int.,* **2005**, *38*(5), 577-584. [http://dx.doi.org/10.1016/j.foodres.2004.11.011]

[55] Bednár, P.; Papousková, B.; Müller, L.; Barták, P.; Stávek, J.; Pavloušek, P.; Lemr, K. Utilization of capillary electrophoresis/mass spectrometry (CE/MSn) for the study of anthocyanin dyes. *J. Sep. Sci.,* **2005**, *28*(12), 1291-1299. [http://dx.doi.org/10.1002/jssc.200500071] [PMID: 16138681]

[56] Simó, C.; Rizzi, A.; Barbas, C.; Cifuentes, A. Chiral capillary electrophoresis-mass spectrometry of amino acids in foods. *Electrophoresis,* **2005**, *26*(7-8), 1432-1441. [http://dx.doi.org/10.1002/elps.200406199] [PMID: 15759299]

[57] Simó, C.; Herrero, M.; Neusüss, C.; Pelzing, M.; Kenndler, E.; Barbas, C.; Ibáñez, E.; Cifuentes, A. Characterization of proteins from Spirulina platensis microalga using capillary electrophoresis-ion trap-mass spectrometry and capillary electrophoresis-time of flight-mass spectrometry. *Electrophoresis,* **2005**, *26*(13), 2674-2683. [http://dx.doi.org/10.1002/elps.200500055] [PMID: 15929060]

[58] Herrero, M.; Simó, C.; Ibáñez, E.; Cifuentes, A. Capillary electrophoresis-mass spectrometry of Spirulina platensis proteins obtained by pressurized liquid extraction. *Electrophoresis,* **2005**, *26*(21), 4215-4224. [http://dx.doi.org/10.1002/elps.200500230] [PMID: 16200528]

[59] Bringmann, G.; Kajahn, I.; Neusüss, C.; Pelzing, M.; Laug, S.; Unger, M.; Holzgrabe, U. Analysis of the glucosinolate pattern of Arabidopsis thaliana seeds by capillary zone electrophoresis coupled to electrospray ionization-mass spectrometry. *Electrophoresis,* **2005**, *26*(7-8), 1513-1522. [http://dx.doi.org/10.1002/elps.200410255] [PMID: 15776481]

[60] Arias, M.; Simó, C.; Ortiz, L.T.; de los Mozos-Pascual, M.; Barbas, C.; Cifuentes, A. Detection and quantitation of a bioactive compound in Vicia narbonensis L. seeds by capillary electrophoresis-mass spectrometry: a comparative study with UV detection. *Electrophoresis,* **2005**, *26*(12), 2351-2359. [http://dx.doi.org/10.1002/elps.200410224] [PMID: 15966020]

[61] Juan-García, A.; Font, G.; Picó, Y. Quantitative analysis of six pesticides in fruits by capillary electrophoresis-electrospray-mass spectrometry. *Electrophoresis,* **2005**, *26*(7-8), 1550-1561. [http://dx.doi.org/10.1002/elps.200410104] [PMID: 15759305]

[62] Hernández-Borges, J.; Rodríguez-Delgado, M.Á.; García-Montelongo, F.J.; Cifuentes, A. Analysis of pesticides in soy milk combining solid-phase extraction and capillary electrophoresis-mass spectrometry. *J. Sep. Sci.,* **2005**, *28*(9-10), 948-956.

[http://dx.doi.org/10.1002/jssc.200500014] [PMID: 16013821]

[63] Santos, B.; Lista, A.; Simonet, B.M.; Ríos, A.; Valcárcel, M. Screening and analytical confirmation of sulfonamide residues in milk by capillary electrophoresis-mass spectrometry. *Electrophoresis,* **2005**, *26*(7-8), 1567-1575.
[http://dx.doi.org/10.1002/elps.200410267] [PMID: 15765476]

[64] Arráez-Román, D.; Gómez-Caravaca, A.M.; Gómez-Romero, M.; Segura-Carretero, A.; Fernández-Gutiérrez, A. Identification of phenolic compounds in rosemary honey using solid-phase extraction by capillary electrophoresis-electrospray ionization-mass spectrometry. *J. Pharm. Biomed. Anal.,* **2006**, *41*(5), 1648-1656.
[http://dx.doi.org/10.1016/j.jpba.2006.02.035] [PMID: 16574364]

[65] Juan-García, A.; Font, G.; Picó, Y. Determination of quinolone residues in chicken and fish by capillary electrophoresis-mass spectrometry. *Electrophoresis,* **2006**, *27*(11), 2240-2249.
[http://dx.doi.org/10.1002/elps.200500868] [PMID: 16736458]

[66] Lara, F.J.; García-Campaña, A.M.; Alés-Barrero, F.; Bosque-Sendra, J.M.; García-Ayuso, L.E. Multiresidue method for the determination of quinolone antibiotics in bovine raw milk by capillary electrophoresis-tandem mass spectrometry. *Anal. Chem.,* **2006**, *78*(22), 7665-7673.
[http://dx.doi.org/10.1021/ac061006v] [PMID: 17105157]

[67] Bermudo, E.; Núñez, O.; Moyano, E.; Puignou, L.; Galceran, M.T. Field amplified sample injection-capillary electrophoresis-tandem mass spectrometry for the analysis of acrylamide in foodstuffs. *J. Chromatogr. A,* **2007**, *1159*(1-2), 225-232.
[http://dx.doi.org/10.1016/j.chroma.2007.03.008] [PMID: 17383666]

[68] Font, G.; Juan-García, A.; Picó, Y. Pressurized liquid extraction combined with capillary electrophoresis-mass spectrometry as an improved methodology for the determination of sulfonamide residues in meat. *J. Chromatogr. A,* **2007**, *1159*(1-2), 233-241.
[http://dx.doi.org/10.1016/j.chroma.2007.03.062] [PMID: 17433345]

[69] Soto-Chinchilla, J.J.; García-Campaña, A.M.; Gámiz-Gracia, L. Analytical methods for multiresidue determination of sulfonamides and trimethoprim in meat and ground water samples by CE-MS and CE-MS/MS. *Electrophoresis,* **2007**, *28*(22), 4164-4172.
[http://dx.doi.org/10.1002/elps.200600848] [PMID: 17960854]

[70] Juan-García, A.; Font, G.; Picó, Y. Simultaneous determination of different classes of antibiotics in fish and livestock by CE-MS. *Electrophoresis,* **2007**, *28*(22), 4180-4191.
[http://dx.doi.org/10.1002/elps.200700383] [PMID: 17932875]

[71] Wang, S.; Yang, P.; Cheng, Y. Analysis of tetracycline residues in bovine milk by CE-MS with field-amplified sample stacking. *Electrophoresis,* **2007**, *28*(22), 4173-4179.
[http://dx.doi.org/10.1002/elps.200700276] [PMID: 17957662]

[72] García-Villalba, R.; León, C.; Dinelli, G.; Segura-Carretero, A.; Fernández-Gutiérrez, A.; Garcia-Cañas, V.; Cifuentes, A. Comparative metabolomic study of transgenic *versus* conventional soybean using capillary electrophoresis-time-of-flight mass spectrometry. *J. Chromatogr. A,* **2008**, *1195*(1-2), 164-173.
[http://dx.doi.org/10.1016/j.chroma.2008.05.018] [PMID: 18508066]

[73] Simó, C.; Moreno-Arribas, M.V.; Cifuentes, A. Ion-trap *versus* time-of-flight mass spectrometry coupled to capillary electrophoresis to analyze biogenic amines in wine. *J. Chromatogr. A,* **2008**, *1195*(1-2), 150-156.
[http://dx.doi.org/10.1016/j.chroma.2008.05.004] [PMID: 18508068]

[74] Watanabe, M.; Kusano, M.; Oikawa, A.; Fukushima, A.; Noji, M.; Saito, K. Physiological roles of the β-substituted alanine synthase gene family in Arabidopsis. *Plant Physiol.,* **2008**, *146*(1), 310-320.
[http://dx.doi.org/10.1104/pp.107.106831] [PMID: 18024555]

[75] Levandi, T.; Leon, C.; Kaljurand, M.; Garcia-Cañas, V.; Cifuentes, A. Capillary electrophoresis time-of-flight mass spectrometry for comparative metabolomics of transgenic *versus* conventional maize.

Anal. Chem., **2008**, *80*(16), 6329-6335.
[http://dx.doi.org/10.1021/ac8006329] [PMID: 18613645]

[76] Segura-Carretero, A.; Puertas-Mejía, M.A.; Cortacero-Ramírez, S.; Beltrán, R.; Alonso-Villaverde, C.; Joven, J.; Dinelli, G.; Fernández-Gutiérrez, A. Selective extraction, separation, and identification of anthocyanins from Hibiscus sabdariffa L. using solid phase extraction-capillary electrophoresis-mass spectrometry (time-of-flight /ion trap). *Electrophoresis,* **2008**, *29*(13), 2852-2861.
[http://dx.doi.org/10.1002/elps.200700819] [PMID: 18546170]

[77] Meermann, B.; Bartel, M.; Scheffer, A.; Trümpler, S.; Karst, U. Capillary electrophoresis with inductively coupled plasma-mass spectrometric and electrospray time of flight mass spectrometric detection for the determination of arsenic species in fish samples. *Electrophoresis,* **2008**, *29*(12), 2731-2737.
[http://dx.doi.org/10.1002/elps.200700902] [PMID: 18512677]

[78] Chen, J-H.; Jiang, S-J. Determination of cobalamin in nutritive supplements and chlorella foods by capillary electrophoresis-inductively coupled plasma mass spectrometry. *J. Agric. Food Chem.,* **2008**, *56*(4), 1210-1215.
[http://dx.doi.org/10.1021/jf073213h] [PMID: 18247530]

[79] Leon, C.; Rodriguez-Meizoso, I.; Lucio, M.; Garcia-Cañas, V.; Ibañez, E.; Schmitt-Kopplin, P.; Cifuentes, A. Metabolomics of transgenic maize combining Fourier transform-ion cyclotron resonance-mass spectrometry, capillary electrophoresis-mass spectrometry and pressurized liquid extraction. *J. Chromatogr. A,* **2009**, *1216*(43), 7314-7323.
[http://dx.doi.org/10.1016/j.chroma.2009.04.092] [PMID: 19477454]

[80] de la Iglesia, P.; Gago-Martínez, A. Determination of yessotoxins and pectenotoxins in shellfish by capillary electrophoresis-electrospray ionization-mass spectrometry. *Food Addit. Contam. Part A Chem. Anal. Control Expo. Risk Assess.,* **2009**, *26*(2), 221-228.
[http://dx.doi.org/10.1080/02652030802290522] [PMID: 19680893]

[81] Català-Clariana, S.; Benavente, F.; Giménez, E.; Barbosa, J.; Sanz-Nebot, V. Identification of bioactive peptides in hypoallergenic infant milk formulas by capillary electrophoresis-mass spectrometry. *Anal. Chim. Acta,* **2010**, *683*(1), 119-125.
[http://dx.doi.org/10.1016/j.aca.2010.10.002] [PMID: 21094390]

[82] Ganzera, M.; Krüger, A.; Wink, M. Determination of quinolizidine alkaloids in different Lupinus species by NACE using UV and MS detection. *J. Pharm. Biomed. Anal.,* **2010**, *53*(5), 1231-1235.
[http://dx.doi.org/10.1016/j.jpba.2010.05.030] [PMID: 20580181]

[83] Sánchez-Hernández, L.; Castro-Puyana, M.; García-Ruiz, C.; Crego, A.L.; Marina, M.L. Determination of L-and D-carnitine in dietary food supplements using capillary electrophoresis–tandem mass spectrometry. *Food Chem.,* **2010**, *120*(3), 921-928.
[http://dx.doi.org/10.1016/j.foodchem.2009.11.004]

[84] Kusano, M.; Redestig, H.; Hirai, T.; Oikawa, A.; Matsuda, F.; Fukushima, A.; Arita, M.; Watanabe, S.; Yano, M.; Hiwasa-Tanase, K.; Ezura, H.; Saito, K. Covering chemical diversity of genetically-modified tomatoes using metabolomics for objective substantial equivalence assessment. *PLoS One,* **2011**, *6*(2), e16989.
[http://dx.doi.org/10.1371/journal.pone.0016989] [PMID: 21359231]

[85] Sugimoto, M.; Kaneko, M.; Onuma, H.; Sakaguchi, Y.; Mori, M.; Abe, S.; Soga, T.; Tomita, M. Changes in the charged metabolite and sugar profiles of pasteurized and unpasteurized Japanese sake with storage. *J. Agric. Food Chem.,* **2012**, *60*(10), 2586-2593.
[http://dx.doi.org/10.1021/jf2048993] [PMID: 22352923]

[86] Kim, J.; Choi, J.N.; John, K.M.; Kusano, M.; Oikawa, A.; Saito, K.; Lee, C.H. GC-TOF-MS- and CE-TOF-MS-based metabolic profiling of cheonggukjang (fast-fermented bean paste) during fermentation and its correlation with metabolic pathways. *J. Agric. Food Chem.,* **2012**, *60*(38), 9746-9753.
[http://dx.doi.org/10.1021/jf302833y] [PMID: 22913417]

[87] Bignardi, C.; Cavazza, A.; Corradini, C. Determination of furosine in food products by capillary zone electrophoresis-tandem mass spectrometry. *Electrophoresis,* **2012**, *33*(15), 2382-2389. [http://dx.doi.org/10.1002/elps.201100582] [PMID: 22887159]

[88] Roscher, J.; Posch, T.N.; Pütz, M.; Huhn, C. Forensic analysis of mesembrine alkaloids in Sceletium tortuosum by nonaqueous capillary electrophoresis mass spectrometry. *Electrophoresis,* **2012**, *33*(11), 1567-1570. [http://dx.doi.org/10.1002/elps.201100683] [PMID: 22736358]

[89] Bignardi, C.; Cavazza, A.; Rinaldi, M.; Corradini, C.; Massini, R. Evaluation of thermal treatment markers in wheat flour-derived products cooked in conventional and in low-emissivity ovens. *Food Chem.,* **2013**, *140*(4), 748-754. [http://dx.doi.org/10.1016/j.foodchem.2012.12.006] [PMID: 23692762]

[90] Català-Clariana, S.; Benavente, F.; Giménez, E.; Barbosa, J.; Sanz-Nebot, V. Identification of bioactive peptides in hypoallergenic infant milk formulas by CE-TOF-MS assisted by semiempirical model of electromigration behavior. *Electrophoresis,* **2013**, *34*(13), 1886-1894. [http://dx.doi.org/10.1002/elps.201200547] [PMID: 23564639]

[91] Contreras-Gutiérrez, P.K.; Hurtado-Fernández, E.; Gómez-Romero, M.; Ignacio Hormaza, J.; Carrasco-Pancorbo, A.; Fernández-Gutiérrez, A. Determination of changes in the metabolic profile of avocado fruits (Persea americana) by two CE-MS approaches (targeted and non-targeted). *Electrophoresis,* **2013**, *34*(19), 2928-2942. [http://dx.doi.org/10.1002/elps.201200676] [PMID: 24228266]

[92] Kondeková, M.; Maier, V.; Ginterová, P.; Marák, J.; Sevčík, J. Analysis of lysozyme in cheese samples by on-line combination of capillary zone electrophoresis and mass spectrometry. *Food Chem.,* **2014**, *153*, 398-404. [http://dx.doi.org/10.1016/j.foodchem.2013.12.078] [PMID: 24491746]

[93] Rodríguez-Gonzalo, E.; Domínguez-Álvarez, J.; Mateos-Vivas, M.; García-Gómez, D.; Carabias-Martínez, R. A validated method for the determination of nucleotides in infant formulas by capillary electrophoresis coupled to mass spectrometry. *Electrophoresis,* **2014**, *35*(11), 1677-1684. [http://dx.doi.org/10.1002/elps.201300550] [PMID: 24615919]

[94] Andreas, N.J.; Hyde, M.J.; Gomez-Romero, M.; Lopez-Gonzalvez, M.A.; Villaseñor, A.; Wijeyesekera, A.; Barbas, C.; Modi, N.; Holmes, E.; Garcia-Perez, I. Multiplatform characterization of dynamic changes in breast milk during lactation. *Electrophoresis,* **2015**, *36*(18), 2269-2285. [http://dx.doi.org/10.1002/elps.201500011] [PMID: 25959062]

[95] Acunha, T.; Simó, C.; Ibáñez, C.; Gallardo, A.; Cifuentes, A. Anionic metabolite profiling by capillary electrophoresis-mass spectrometry using a noncovalent polymeric coating. Orange juice and wine as case studies. *J. Chromatogr. A,* **2016**, *1428*, 326-335. [http://dx.doi.org/10.1016/j.chroma.2015.08.001] [PMID: 26296988]

[96] Mateos-Vivas, M.; Domínguez-Álvarez, J.; Rodríguez-Gonzalo, E.; Carabias-Martínez, R. Capillary electrophoresis coupled to mass spectrometry employing hexafluoro-2-propanol for the determination of nucleosides and nucleotide mono-, di- and tri-phosphates in baby foods. *Food Chem.,* **2017**, *233*, 38-44. [http://dx.doi.org/10.1016/j.foodchem.2017.04.060] [PMID: 28530588]

[97] Daniel, D.; Lopes, F.S.; Santos, V.B.D.; do Lago, C.L. Detection of coffee adulteration with soybean and corn by capillary electrophoresis-tandem mass spectrometry. *Food Chem.,* **2018**, *243*, 305-310. [http://dx.doi.org/10.1016/j.foodchem.2017.09.140] [PMID: 29146342]

[98] Benavente, F.; Pero-Gascon, R.; Pont, L.; Jaumot, J.; Barbosa, J.; Sanz-Nebot, V. Identification of antihypertensive peptides in nutraceuticals by capillary electrophoresis-mass spectrometry. *J. Chromatogr. A,* **2018**, *1579*, 129-137. [http://dx.doi.org/10.1016/j.chroma.2018.10.018] [PMID: 30361036]

[99] Pérez-Míguez, R.; Sánchez-López, E.; Plaza, M.; Marina, M.L.; Castro-Puyana, M. Capillary

electrophoresis-mass spectrometry metabolic fingerprinting of green and roasted coffee. *J. Chromatogr. A,* **2019**, *1605*, 360353.
[http://dx.doi.org/10.1016/j.chroma.2019.07.007] [PMID: 31307792]

[100] Daniel, D.; do Lago, C.L. Determination of Multiclass Pesticides Residues in Corn by QuEChERS and Capillary Electrophoresis Tandem Mass Spectrometry. *Food Anal. Methods,* **2019**, *12*(7), 1684-1692.
[http://dx.doi.org/10.1007/s12161-019-01501-y]

[101] Lee, S.; Kim, S-J.; Bang, E.; Na, Y-C. Chiral separation of intact amino acids by capillary electrophoresis-mass spectrometry employing a partial filling technique with a crown ether carboxylic acid. *J. Chromatogr. A,* **2019**, *1586*, 128-138.
[http://dx.doi.org/10.1016/j.chroma.2018.12.001] [PMID: 30558847]

[102] Yu, Y.; Liu, Y.; Wang, W.; Jia, Y.; Zhao, G.; Zhang, X.; Chen, H.; Zhou, Y. Highly sensitive determination of aminoglycoside residues in food by sheathless CE-ESI-MS/MS. *Anal. Methods,* **2019**, *11*(39), 5064-5069.
[http://dx.doi.org/10.1039/C9AY01728C]

[103] He, H.; Tian, M.; Hu, L.; Yang, L. Ultrasensitive determination of organotin compounds in plastic food packaging and edible oils by sheathless capillary electrophoresis-electrospray ionization-mass spectrometry. *Analyst (Lond.),* **2020**, *145*(6), 2286-2296.
[http://dx.doi.org/10.1039/C9AN02331C] [PMID: 32003368]

[104] Thomson, B.; Poms, R.; Rose, M. Incidents and impacts of unwanted chemicals in food and feeds. *Qual. Assur. Saf. Crops Foods,* **2012**, *4*(2), 77-92.
[http://dx.doi.org/10.1111/j.1757-837X.2012.00129.x]

[105] Schrenk, D.; Cartus, A. *Chemical contaminants and residues in food*; Woodhead Publishing, **2017**.

[106] Winter, C.K. Pesticide residues in foods.*Chemical contaminants and residues in food*; Elsevier, **2017**, pp. 155-169.
[http://dx.doi.org/10.1016/B978-0-08-100674-0.00007-2]

[107] Woodside, J.V.; Brennan, S.; Cantwell, M. Are Soy-Milk Products viable Alternatives to Cow's Milk?*Beverage Impacts on Health and Nutrition*; Springer, **2016**, pp. 151-162.
[http://dx.doi.org/10.1007/978-3-319-23672-8_10]

[108] Wilkowska, A.; Biziuk, M. Determination of pesticide residues in food matrices using the QuEChERS methodology. *Food Chem.,* **2011**, *125*(3), 803-812.
[http://dx.doi.org/10.1016/j.foodchem.2010.09.094]

[109] Lehotay, S.J. QuEChERS sample preparation approach for mass spectrometric analysis of pesticide residues in foods.*Mass spectrometry in food safety*; Springer, **2011**, pp. 65-91.
[http://dx.doi.org/10.1007/978-1-61779-136-9_4]

[110] Nakka, S.; Jugulam, M.; Peterson, D.; Mohammad, A. Herbicide resistance: Development of wheat production systems and current status of resistant weeds in wheat cropping systems. *Crop J.,* **2019**, *7*(6), 750-760.
[http://dx.doi.org/10.1016/j.cj.2019.09.004]

[111] Brigelius-Flohé, R. Selenium in human health and disease: An overview.*Selenium*; Springer, **2018**, pp. 3-26.
[http://dx.doi.org/10.1007/978-3-319-95390-8_1]

[112] Hoch, M. Organotin compounds in the environment-an overview. *J. Appl. Geochem.,* **2001**, *16*(7-8), 719-743.
[http://dx.doi.org/10.1016/S0883-2927(00)00067-6]

[113] Semla, M.; Goc, Z.; Martiniaková, M.; Omelka, R.; Formicki, G. Acrylamide: a common food toxin related to physiological functions and health. *Physiol. Res.,* **2017**, *66*(2), 205-217.
[http://dx.doi.org/10.33549/physiolres.933381] [PMID: 27982682]

[114] Tang, K.L.; Caffrey, N.P.; Nóbrega, D.B.; Cork, S.C.; Ronksley, P.E.; Barkema, H.W.; Polachek, A.J.;

Ganshorn, H.; Sharma, N.; Kellner, J.D.; Ghali, W.A. Restricting the use of antibiotics in food-producing animals and its associations with antibiotic resistance in food-producing animals and human beings: a systematic review and meta-analysis. *Lancet Planet. Health,* **2017**, *1*(8), e316-e327. [http://dx.doi.org/10.1016/S2542-5196(17)30141-9] [PMID: 29387833]

[115] Pham, T.D.M.; Ziora, Z.M.; Blaskovich, M.A.T. Quinolone antibiotics. *MedChemComm,* **2019**, *10*(10), 1719-1739. [http://dx.doi.org/10.1039/C9MD00120D] [PMID: 31803393]

[116] Grossman, T.H. Tetracycline antibiotics and resistance. *Cold Spring Harb. Perspect. Med.,* **2016**, *6*(4), a025387. [http://dx.doi.org/10.1101/cshperspect.a025387] [PMID: 26989065]

[117] Fox, J.W. 114 - Venoms and Poisons from Marine Organisms. In Goldman's Cecil Medicine (Twenty Fourth Edition), Goldman, L.; Schafer, A. I., Eds. W.B. Saunders: Philadelphia, , **2012**; pp. 697-700.

[118] Alfonso, A.; Tobio, A.; Louzao, M.C. Yessotoxins and Pectenotoxins.*Seafood and Freshwater Toxins: Pharmacology, Physiology, and Detection*; CRC Press: Boca Ratón, **2014**, pp. 657-676. [http://dx.doi.org/10.1201/b16662-22]

[119] Zhang, Y.; Su, M.; Tian, D-P. Tetramine poisoning: A case report and review of the literature. *Forensic Sci. Int.,* **2011**, *204*(1-3), e24-e27. [http://dx.doi.org/10.1016/j.forsciint.2010.06.025] [PMID: 20678875]

[120] Yu, V.Y. The role of dietary nucleotides in neonatal and infant nutrition. *Singapore Med. J.,* **1998**, *39*(4), 145-150. [PMID: 9676143]

[121] Wang, L.; Liu, J.; Lv, H.; Zhang, X.; Shen, L. Effects of Nucleotides Supplementation of Infant Formulas on Plasma and Erythrocyte Fatty Acid Composition: A Meta-Analysis. *PLoS One,* **2015**, *10*(6), e0127758-e0127758. [http://dx.doi.org/10.1371/journal.pone.0127758] [PMID: 26102282]

[122] Friedman, M. Potato glycoalkaloids and metabolites: roles in the plant and in the diet. *J. Agric. Food Chem.,* **2006**, *54*(23), 8655-8681. [http://dx.doi.org/10.1021/jf061471t] [PMID: 17090106]

[123] Toobert, D.J.; Glasgow, R.E.; Strycker, L.A.; Barrera, M., Jr; Radcliffe, J.L.; Wander, R.C.; Bagdade, J.D. Biologic and quality-of-life outcomes from the Mediterranean Lifestyle Program: a randomized clinical trial. *Diabetes Care,* **2003**, *26*(8), 2288-2293. [http://dx.doi.org/10.2337/diacare.26.8.2288] [PMID: 12882850]

[124] Petiwala, S.M.; Puthenveetil, A.G.; Johnson, J.J. Polyphenols from the Mediterranean herb rosemary (Rosmarinus officinalis) for prostate cancer. *Front. Pharmacol.,* **2013**, *4*, 29. [http://dx.doi.org/10.3389/fphar.2013.00029] [PMID: 23531917]

[125] La Vecchia, C. Mediterranean diet and cancer. *Public Health Nutr.,* **2004**, *7*(7), 965-968. [http://dx.doi.org/10.1079/PHN2004562] [PMID: 15482626]

[126] Viuda□Martos, M.; Ruiz Navajas, Y.; Sánchez Zapata, E.; Fernández□López, J.; Pérez□Álvarez, J.A. Antioxidant activity of essential oils of five spice plants widely used in a Mediterranean diet. *Flavour Fragrance J.,* **2010**, *25*(1), 13-19. [http://dx.doi.org/10.1002/ffj.1951]

[127] Çelebier, M.; Ibáñez, C.; Simó, C.; Cifuentes, A. A Foodomics Approach: CE-MS for Comparative Metabolomics of Colon Cancer Cells Treated with Dietary Polyphenols.*Electrophoretic Separation of Proteins*; Springer, **2019**, pp. 303-313. [http://dx.doi.org/10.1007/978-1-4939-8793-1_26]

[128] Çelebier, M.Ç.; Simo, C.; Altınöz, S.; Cifuentes, A. Comparative proteomics to investigate the *in vitro* antiproliferative effect of dietary polyphenols against K562 leukemia cells/Gıda kaynaklı polifenollerin K562 lösemi hücreleri üzerine *in vitro* çoğalmayı önleyici etkisinin karşılaştırmalı

proteomik çalışmalarla incelenmesi. *Turkish J. Biochem.*, **2015**, *40*(2), 95-104.

[129] Ibáñez, C.; Simó, C.; García-Cañas, V.; Gómez-Martínez, A.; Ferragut, J.A.; Cifuentes, A. CE/LC-MS multiplatform for broad metabolomic analysis of dietary polyphenols effect on colon cancer cells proliferation. *Electrophoresis,* **2012**, *33*(15), 2328-2336. [http://dx.doi.org/10.1002/elps.201200143] [PMID: 22887153]

[130] Görs, M.; Schumann, R.; Hepperle, D.; Karsten, U. Quality analysis of commercial Chlorella products used as dietary supplement in human nutrition. *J. Appl. Phys.*, **2010**, *22*(3), 265-276.

[131] Enneking, D.; Delaere, I.M.; Tate, M.E. γ-Glutamyl-S-ethenylcysteine: a dipeptide from Vicia narbonensis. *Phytochemistry,* **1998**, *48*(4), 643-645. [http://dx.doi.org/10.1016/S0031-9422(98)00025-9]

[132] Dinkova-Kostova, A.T.; Kostov, R.V. Glucosinolates and isothiocyanates in health and disease. *Trends Mol. Med.*, **2012**, *18*(6), 337-347. [http://dx.doi.org/10.1016/j.molmed.2012.04.003] [PMID: 22578879]

[133] Hayes, J.D.; Kelleher, M.O.; Eggleston, I.M. The cancer chemopreventive actions of phytochemicals derived from glucosinolates. *Eur. J. Nutr.*, **2008**, *47*(2) Suppl. 2, 73-88. [http://dx.doi.org/10.1007/s00394-008-2009-8] [PMID: 18458837]

[134] Hesse, M. *Alkaloids: nature's curse or blessing?*; John Wiley & Sons, **2002**.

[135] Sekhon, B.S. An overview of capillary electrophoresis: pharmaceutical, biopharmaceutical and biotechnology applications. *J. Pharm. Ed. Res.*, **2011**, *2*(2), 2.

[136] Pioch, M.; Bunz, S.C.; Neusüss, C. Capillary electrophoresis/mass spectrometry relevant to pharmaceutical and biotechnological applications. *Electrophoresis,* **2012**, *33*(11), 1517-1530. [http://dx.doi.org/10.1002/elps.201200030] [PMID: 22736352]

[137] Kutter, E. Lysozyme.*Encyclopedia of Genetics*; Brenner, S.; Miller, J.H., Eds.; Academic Press: New York, **2001**, pp. 1131-1132. [http://dx.doi.org/10.1006/rwgn.2001.0782]

[138] Grumezescu, V.; Holban, A.M.; Barbu, I.; Popescu, R.C.; Oprea, A.E.; Lazar, V.; Grumezescu, A.M.; Chifiriuc, M.C. Nanoarchitectonics Used in Antiinfective Therapy.*Antibiotic Resistance*; Kon, K.; Rai, M., Eds.; Academic Press, **2016**, pp. 145-166. [http://dx.doi.org/10.1016/B978-0-12-803642-6.00007-1]

[139] Woldemariam, E.T. *Isolation of lipid classes from nutritional marine oils: improved extraction strategy and discrimination studies based on the positional distribution of omega-3 polyunsaturated fatty acids on triacylglycerol structures using liquid chromatography tandem mass spectrometry and principal component analysis*; The University of Bergen and National Institute of Nutrition and Seafood, **2014**.

[140] Sinclair, K.; Dudley, E. Metabolomics and Biomarker Discovery.*Advancements of Mass Spectrometry in Biomedical Research*; Springer, **2019**, pp. 613-633. [http://dx.doi.org/10.1007/978-3-030-15950-4_37]

[141] Ibáñez, C.; Simó, C.; García-Cañas, V.; Cifuentes, A.; Castro-Puyana, M. Metabolomics, peptidomics and proteomics applications of capillary electrophoresis-mass spectrometry in Foodomics: a review. *Anal. Chim. Acta,* **2013**, *802*, 1-13. [http://dx.doi.org/10.1016/j.aca.2013.07.042] [PMID: 24176500]

[142] Capozzi, F.; Bordoni, A. Foodomics: a new comprehensive approach to food and nutrition. *Genes Nutr.*, **2013**, *8*(1), 1-4. [http://dx.doi.org/10.1007/s12263-012-0310-x] [PMID: 22933238]

[143] Valdés, A.; Simó, C.; Ibáñez, C.; Rocamora-Reverte, L.; Ferragut, J.A.; García-Cañas, V.; Cifuentes, A. Effect of dietary polyphenols on K562 leukemia cells: a Foodomics approach. *Electrophoresis,* **2012**, *33*(15), 2314-2327. [http://dx.doi.org/10.1002/elps.201200133] [PMID: 22887152]

[144] Ibáñez, C.; Valdés, A.; García-Cañas, V.; Simó, C.; Celebier, M.; Rocamora-Reverte, L.; Gómez-Martínez, A.; Herrero, M.; Castro-Puyana, M.; Segura-Carretero, A.; Ibáñez, E.; Ferragut, J.A.; Cifuentes, A. Global Foodomics strategy to investigate the health benefits of dietary constituents. *J. Chromatogr. A*, **2012**, *1248*, 139-153. [http://dx.doi.org/10.1016/j.chroma.2012.06.008] [PMID: 22727325]

[145] García-Cañas, V.; Simó, C.; Herrero, M.; Ibáñez, E.; Cifuentes, A. Present and future challenges in food analysis: foodomics. *Anal. Chem.*, **2012**, *84*(23), 10150-10159. [http://dx.doi.org/10.1021/ac301680q] [PMID: 22958185]

[146] Celebier, M.; Ibáñez, C.; Simó, C.; Cifuentes, A. A Foodomics approach: CE-MS for comparative metabolomics of colon cancer cells treated with dietary polyphenols. *Methods Mol. Biol.*, **2012**, *869*, 185-195. [http://dx.doi.org/10.1007/978-1-61779-821-4_15] [PMID: 22585486]

[147] Cifuentes, A. Food analysis and foodomics. *J. Chromatogr. A*, **2009**, *1216*(43), 7109. [http://dx.doi.org/10.1016/j.chroma.2009.09.018] [PMID: 19765718]

[148] Patti, G.J.; Yanes, O.; Siuzdak, G. Innovation: Metabolomics: the apogee of the omics trilogy. *Nat. Rev. Mol. Cell Biol.*, **2012**, *13*(4), 263-269. [http://dx.doi.org/10.1038/nrm3314] [PMID: 22436749]

[149] Liu, X.; Locasale, J.W. Metabolomics: a primer. *Trends Biochem. Sci.*, **2017**, *42*(4), 274-284. [http://dx.doi.org/10.1016/j.tibs.2017.01.004] [PMID: 28196646]

[150] Fernie, A.R.; Trethewey, R.N.; Krotzky, A.J.; Willmitzer, L. Metabolite profiling: from diagnostics to systems biology. *Nat. Rev. Mol. Cell Biol.*, **2004**, *5*(9), 763-769. [http://dx.doi.org/10.1038/nrm1451] [PMID: 15340383]

[151] Wishart, D.S.; Feunang, Y.D.; Marcu, A.; Guo, A.C.; Liang, K.; Vázquez-Fresno, R.; Sajed, T.; Johnson, D.; Li, C.; Karu, N.; Sayeeda, Z.; Lo, E.; Assempour, N.; Berjanskii, M.; Singhal, S.; Arndt, D.; Liang, Y.; Badran, H.; Grant, J.; Serra-Cayuela, A.; Liu, Y.; Mandal, R.; Neveu, V.; Pon, A.; Knox, C.; Wilson, M.; Manach, C.; Scalbert, A. HMDB 4.0: the human metabolome database for 2018. *Nucleic Acids Res.*, **2018**, *46*(D1), D608-D617. [http://dx.doi.org/10.1093/nar/gkx1089] [PMID: 29140435]

[152] Castro-Puyana, M.; Pérez-Míguez, R.; Montero, L.; Herrero, M. Reprint of: Application of mass spectrometry-based metabolomics approaches for food safety, quality and traceability. *TrAC. Trends Analyt. Chem.*, **2017**, *96*, 62-78. [http://dx.doi.org/10.1016/j.trac.2017.08.007]

[153] López-Ruiz, R.; Romero-González, R.; Frenich, A.G. Metabolomics approaches for the determination of multiple contaminants in food. *Curr. Opin. Food Sci.*, **2019**, *28*, 49-57. [http://dx.doi.org/10.1016/j.cofs.2019.08.006]

[154] Pinu, F.R. Metabolomics: Applications to food safety and quality research.*Microbial Metabolomics*; Springer, **2016**, pp. 225-259. [http://dx.doi.org/10.1007/978-3-319-46326-1_8]

SUBJECT INDEX

María Castro-Puyana, Miguel Herrero and María Luisa Marina (Eds.)

B

C

D

E

F

G

H

K

L

M

N

O

P

R

S

T

U

V

W

Y

Made in United States
North Haven, CT
04 November 2026

21727485R00310